SPATIAL, LATTICE and TENSION STRUCTURES

Proceedings of the IASS-ASCE International Symposium 1994

Held in conjunction with the ASCE Structures Congress XII
April 24-28, 1994
Georgia World Congress Center Atlanta, Georgia, USA

jointly organized by
The International Association for Shell and Spatial Structures
and Committee on Special Structures, Structural Division,
American Society of Civil Engineers

Edited by John F. Abel, John W. Leonard, and Celina U. Penalba

Published by the
American Society of Civil Engineers
345 East 47th Street
New York, New York 10017-2398

ABSTRACT

This proceedings, *Spatial, Lattice and Tension Structures,* contains over 100 papers presented at the IASS-ASCE International Symposium 1994, held in conjunction with the ASCE Structures Congress XII, April 24-28, 1994, Atlanta, Georgia. These papers present issues related to both traditions and innovations in the conception, analysis, design, construction, aesthetics, and evaluation of lightweight and long-span structural systems. They are grouped in five main sections: 1) Spatial structures; 2) lattice structures; 3) tension structures; 4) shell structures; and 5) structures and architecture. Within each section, typical topics addressed by the sessions are case studies of new or existing structures, structural design, structural theory, experimental studies, structural analysis, structural dynamics, earthquake engineering, and computation.

Library of Congress Cataloging-in-Publication Data

IASS-ASCE International Symposium (1994: Atlanta, Ga.)
    Spatial, lattice and, tension structures: proceedings of the IASS- ASCE International Symposium 1994, held in con junction with the ASCE Structures Congress XII, April 24-29, 1994, Georgia World Congress, Atlanta, Georgia, USA/joint ly organized by the International Association for Shell and Spatial Structures and Committee on Special Structures, Structural Division, American Society of Civil Engineers; edited by John F. Abel, John W. Leonard, and Celina U. Penalba.
    p. cm.
    Includes index.
    ISBN 0-87262-953-8
    1. Space frame structures—Congresses. 2. Shells (Engineering)— Congresses. 3. Structural analysis (Engineering)—Congresses. I. Abel, John Fredrick, 1940- II. Leonard, John W. (John William) III. Penalba, Celina U. IV. Structures Congress '94 (1994: Atlanta, Ga.) V. International Association for Shell and Spatial Structures. VI. American Society of Civil Engineers. Committee on Special Structures. VII. Title.
TA660.S63l17 1994   94-7103
624.1'77—dc20   CIP

FOREWORD

One manifestation of humankind's desire for an enhanced quality of life is the conception, analysis, design, and construction of lightweight, long-span structural systems which offer functionality, economy, and beauty. There is a renewed interest in these structures consistent with evolving functional demands and the availability of new materials and technologies. The potential of lightweight, long-span structural systems for vital solutions can be appreciated, for example, in urban design, where the concept of macro-envelopes or megastructures can be used to protect sectors of cities from harsh environments. Furthermore, the feasibility of using these structural systems for marine and space habitats seems enhanced by today's new developments.

The joint efforts of the **International Association of Shell and Spatial Structures** (IASS) and the **Committee on Special Structures** of the American Society of Civil Engineers (ASCE) to hold the International Symposium "Spatial, Lattice and Tension Structures" deserves special recognition. The Symposium, held at the Georgia World Congress Center and in conjunction with the ASCE Structures Congress XII, is a forum for the exchange of knowledge and for the advancement of a wide variety of structural systems that reflect dynamic creativity and innovative technologies. Indeed, these factors are typified at the very site of the Symposium by the Georgia Dome, a state-of-the-art hypar tensegrity structure which is the world's largest single-span fabric roof and which is the subject of some papers in this Proceedings.

This Proceedings represents the contribution of leading world experts and provides an incentive for further developments. The organizers and editors wish to express appreciation to the authors for their contributions and to all those who participated in the organization of the Symposium for their diligent and fruitful efforts. Membership in the Organizing Committee, the International Advisory Committee, and the Scientific Committee was drawn from the IASS and the ASCE Committee on Special Structures. In addition to the editors, members of the Organizing Committee included Peter Glockner (University of Calgary, Edmonton), Barry J. Goodno (Chair of the Structures Congress XII Organizing Committee, Georgia Institute of Technology, Atlanta), Witold Gutkowski (Polish Academy of Sciences, Warsaw), Matthys P. Levy (Weidlinger Associates, New York), John K. Parsons (John K. Parsons & Assocs., Las Vegas), Alex C. Scordelis (University of California, Berkeley), Ronald E. Shaeffer (Florida A&M University, Tallahassee), and N. K. Srivastava (University of Moncton, Moncton, New Brunswick). The Chairman of the International Advisory Committee was Stefan J. Medwadowski (President of the IASS, San Francisco). Over 60 engineers, architects, and academics from all over the world served on the Scientific Committee. Finally, assistance in organizing specific sessions was received from ACI-ASCE Committee 334 on Concrete Shell Design and Construction through Stuart E. Swartz (Kansas State University, Manhatten), from the ASCE Task Committee on Tension Fabric Structures through Ronald E. Shaeffer (Florida A&M University, Tallahassee), from IASS Working Group 12 on Spatial Wood Structures through Siegfried M. Holzer (Virginia Polytechnic and State University, Blacksburg), and from the ASCE Committee on Tubular Structures through Jeffrey A. Packer (University of Toronto).

The abstract of each of the technical-session papers included in the Proceedings has been reviewed by at least two and as many as four members of the IASS Symposium

iii

'94 Scientific Committee, and abstracts accepted for conversion to full papers have received at least two positive peer reviews. The keynote and plenary-session papers are invited contributions. Each of the papers included in the Proceedings has been accepted for publication by the Proceedings Editors. All papers are eligible for discussion in the ASCE Journal of Structural Engineering. All papers are eligible for ASCE and IASS awards.

John F. Abel, Chair, Scientific Committee
(Cornell University, Ithaca, NY)
John W. Leonard, Vice Chair, Scientific Committee
(University of Connecticut, Storrs, CT)
Celina U. Penalba, Chair, Organizing Committee
(California Polytechnic University, San Luis Obispo, CA)

CONTENTS

Note: Sessions listed are not necessarily in order of presentation

SPATIAL STRUCTURES
Plenary Session: Spatial Structures

Session M3S3: Spatial Structures: Case Studies

Session T3S2: Spatial Structures: Wood Structures

Session W2S2: Spatial Structures: Design and Theory

Session R2S2: Spatial Structures: Experiments, Analysis and Computation

Session T1S2: Spatial Structures: Design with Hollow Structural Steel Sections

**Session R1S3: Spatial Structures: Structural Dynamics and Earthquake
Engineering I**

**Session R2S3: Spatial Structures: Structural Dynamics and Earthquake
Engineering II**

Session W2S3: Lightweight Structures: Some Recent Developments

LATTICE STRUCTURES

Plenary Session: Lattice Structures

Session R1S1: Lattice Structures: Case Studies

Session R2S1: Lattice Structures: Design and Theory

vii

Session T1S3: Lattice Structures: Analysis and Experiments I

Session T3S3: Lattice Structures: Analysis and Experiments II

TENSION STRUCTURES

Plenary Session: Tension Structures

Plenary Session: The Georgia Dome

ix

x

Session M1S2: Concrete Shells: Industrial Shells

Session S3S3: Shell Structures

STRUCTURE AND ARCHITECTURE

Plenary Session: Structure and Architecture

Plenary Session: The Tensile Architecture of the New Denver Airport

Session S3S1: Structure and Architecture I

xi

Session M1S1: Structure and Architecture II

Glass-covered Lightweight Spatial Structures

Jörg Schlaich[1] and Hans Schober[2]

Introduction

Glass roofs are attractive from an architectural as well as a climatical point of view. Having already been the symbol of the new architecture of the Industrial Revolution during the 18th and 19th century, they experienced a revival during the second half of this century through the work of pioneers like Walther Bauersfeld, Konrad Wachsmann, Buckminster Fuller, Max Mengeringhausen, Frei Otto and others.

Obviously the most favourable basis for a translucent roof is the double curved reticulated spatial structure with triangular mesh. Such structure, however, especially if directly glazed without intermediate glass frames, evokes three basic problems:

- The geometrical problem to cover a double curved, i.e. non-developable surface with triangles, having for manufactural simplicity as many as possible members and nodes of equal size. (This problem obviously got some relief through recent progress in CNC-manufacturing.)

- Glass panels are preferably produced in quadrangles, of course permitting a variation of their angles. Therefore only two out of three members of the triangle constituting the structure should support the glass.

- Especially for double glazing the quadrangular glass panels must either be produced double curved to fit the structure's surface, or the geometry of the structure must be chosen so that the four node points of each mesh are in one plane and may be glazed with plane glass panels. For single glazing, however, some warp of the glass panels is acceptable.

It would by far exceed the scope of one paper to discuss this whole issue and all possible solutions. Therefore, one solution, recently developed by the authors, shall be presented here and exemplified by three applications: A roof over an indoor swimming-pool in the shape of a regualr spherical segment, and a roof over a museum's courtyard, its irregular free shape resulting from the intersection of two unequal cylindrical surfaces, and finally a glazed dome covering an oval plan and a trapezoidal courtyard.

----

[1] Director,Institute for Structural Design, University of Stuttgart, Pfaffenwaldring 7, D-70569 Stuttgart and Schlaich Bergermann und Partner, Civil Engineering Consultants, Stuttgart.
[2] Schlaich Bergermann und Partner, Hohenzollernstr. 1, D-70178 Stuttgart

The Historical Background and the Geometrical Principle of a New Glazed Grid Dome

How can straight bars be arranged on a double curved, i.e. nondevelopable surface in a triangular grid with as many bars as possible, identical in length - or as few as possible, deviating in length, and what has to be observed with respect to the most favourable glazing of such structure? There can be no real or general solution to this problem - as proven by Mercator's (1512 - 1594) early attempts to accurately represent the earth's curved surface on a map. The issue can only be, to reach - as ingenious as possible - the optimum compromise between production engineering, loadbearing behaviour, material consumption, cost, aesthetics, balancing this between structure and glazing, with the progress in CAD and CAM obviously playing a decisive role. It would go far beyond the scope of this paper to document this highly interesting development, which started in 1811 with the first cast-iron grid dome, the Halle-au-Blé (now: Bourse du Commerce) by F. J. Bélanger (architect) and F. Brunet (engineer and entrepreneur). There are some good, summarizing papers (no comprehensive ones yet) thanks to O. Büttner/E. Hampe [5], Z. S. Makowski [6], H. Klimke [7], [8]. The purpose of the following is only to assist in understanding the further development of a glazed grid dome described here, and its first application:

Initially and up to the "grid shells" by Frei Otto, as in Montreal 1967 and Mannheim 1975 [9], geometrical efforts essentially focused on the axirespectively rotationally symmetrical spherical segment. The first ones, as the already mentioned Halle-au-Blé or that of the Galleria V. Emmanuele in Milan, were framed domes with bars placed only in the circular and meridional direction, i. e. with quadrangular mesh, relying on the frame-effect respectively the bending stiffness of the bars (Fig. 1).

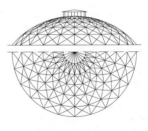

Fig. 1: Dome of the Bourse de Commerce, Paris 1811

Fig. 2: Schwedler's dome

The fact that these early glass domes appear so light, much lighter than if designed today, may possibly be attributed to a stiffening effect due to the glass panels. Unfortunately, this concentrical arrangement of members results in a material congestion at the zenith, exactly where the roof is expected to expose its utmost translucence. Unsatisfactory remedies are to continuously reduce the number of meridional bars there, or to provide a classical lantern. A decisive and successful development is due to Schwedler who in 1863 proposed to provide diagonals to connect the ring and meridional bars (Fig. 2). Of course, the lengths of the bars still vary layer by layer, but thanks to the rotational symmetry there is frequent repetition, but not without this congestion at the zenith. It was Buckminster Fuller, who in 1954 took a completely different approach with his geodesic domes, profiting from the work of Walther Bauersfeld, who, as early as 1922, started to approximate the sphere with an icosahedron. By projecting this polyeder, consisting of 20 equal triangles from the inside onto the surface of a sphere, it is subdivided into 20 spherical triangles. These principal triangles may be further subdivided in a suitable manner into a hexagonal and triangular grid, minimizing the number of bar lengths and shape of the nodes. Where the principal triangles meet in their corners, pentagons show up and the edges of the principal triangles in view expose these typical ondulating lines (Fig. 3). The geodesic domes are ideal for spheres which go beyond the semi-sphere and the US-pavillion for the Expo '67 in Montreal (Fig. 4) and the EPCOT-Center in Orlando, Florida are the best-known examples. In case of flat segments the ondulating lines are rather disturbing and there, even today, the simple dome with ring and meridional bars is preferred as was the case even for the huge Stockholm Globe, which is almost a semi-sphere with a diameter of 110.4 m [10].

Fig. 3: Geodesic dome. A structure that has evolved from icosahedron

Fig. 4: The US exposition building designed by R. B. Fuller for the EXPO '67 in Montreal

None of these solutions was able to offer a sound technical and geometric solution to the problem of shapes without rotational symmetry or even of free-form dome-shapes. Here the grid shells and the principle of square-meshed cable nets [12] present a possible solution, whereas a complete different approach has to be taken in the case of triangular-meshed cable nets [13]. The principle of the grid dome described here, is primarily based on the quadrangular net of the lattice dome, but the use of diagonal cables consequently turns it into a triangular net without leading to any technical disadvantages. Therefore, grid domes are ideal single-membrane shells which can be used for any type of dome as well as for the special case of the spherical segment.

Another feature of this grid dome is the fact that the quadrangular mesh, forming the bars, directly supports the glass, obviating the need for a glazing frame on a structure. This results not only in savings but also in an optimum translucence. At first the constantly changing angles of the square mesh seemed to be disadvantageous with respect to the manufacture of the glass panels; but since they can be exactly pre-fabricated according to computer analysis this is hardly an issue today, and does definitely not outweigh the advantage of the simple manufacture and erection of this structure.

Fig. 5: Glazed grid dome, Neckarsulm, featuring a dome radius of 16.5 m and a span of 25 m. The glazing uses spherically curved antisolar insulating glass.

The Structural Principle

The basic grid of the structure when developed into a plane is a square net consisting of flat bars (Fig. 6). This plane square net may be turned into almost any type of shape by changing the original 90 ° mesh angle - a fact commonly known from the wire mesh strainer. The quadrangles become rhomboids (Fig. 8, 9). This way any double curved structure suddenly becomes "developable", and accordingly simple is the assembly of the basic grid. It entirely consists of bars, identical in length, bolted together, pivoting at their intersections. Bars of different length occur only at the outer edge as dictated by the structural geometry. The mesh angles are determined by the intended structural shape.

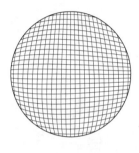

a) View

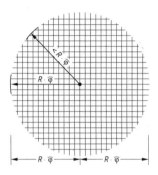

c) The bar grid, developed into a plane (= plane square net)

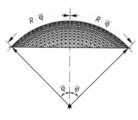

b) Elevation (with diagonal cables)

Fig. 6: The structure, when developed into a plane is a right-angled, square grid of bars of equal length.

However, this basic quadrangular mesh pattern does not yet have the favourable structural characteristics of a shell to withstand wind and snow loads. Hence, the square mesh is braced diagonally with thin cables to achieve the required triangles. Diagonal bars would all vary in

length, entailling the never ending task of cutting and fitting them. Therefore, instead of bars, cables are used, running beneath the bars from one
edge to the other, fixed by clamping plates at the joints. Therefore, there
is no need for measuring the constantly changing length of the diagonals. Later the cables are simply fitted, prestressed and clamped over
the entire length without any problems. The prestressed cables work
both ways, in tension and compression, and very efficiently support the
structure. The glazing is clamped directly onto the flat bars. Consequently the glass panels are rhomboids with constantly changing angles.

In view of corrosion protection the durability of this construction is ideal:
Solid (not hollow) galvanized and painted bars without any welding. The
small structural elements are ideally suited for galvanizing.

Fig. 8: The strainer, too,
is originally a plane
square net.

Fig. 7: Inside view of the
Neckarsulm grid dome

Fig. 9: The make-up of the
loadbearing structure:
A quadrangular grid of flat bars,
diagonally braced with thin,
prestressed cables.

Fig. 10: Inside view of the Neckarsulm grid dome

Fig. 11: Glazed grid dome Neckarsulm

First Example:
The Spherical Glass Covered Grid Dome at Neckarsulm

Dome: Radius 16.50 m
Max. span 25.00 m
Max. rise 5.75 m
Mesh width 1.0 m x 1.0 m
Mesh angle ß = 90° to 65°

Flat bars:
Flat bars 60 x 40 mm, St 37-2, moulded, hot galvanized, curved in the
radius of the dome (R = 16.50 m)
System length l = 1.00 m
All bars are equally long, only at the edge the length varies.
Joint straps 60 x 20 mm, St 37-2
All joint straps are identical.
All joints are screwed together with 2 M 12 HV, 100 % prestressed.
The structure weighs about 20 t.

Cables:
Spiral cables ⌀ 5 mm, $ß_N$ = 1570 N/mm², hot galvanized and plastic-
coated, prestressed.
They run from edge to edge as twin cables.
The twin cables facilitate the clamping, using only one screw at the inter-
section.

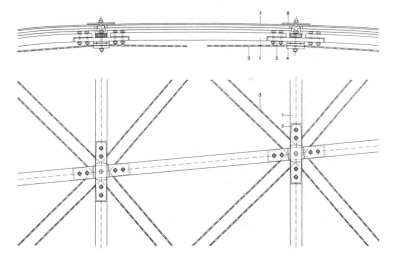

Fig. 12: Flat bars and node configuration

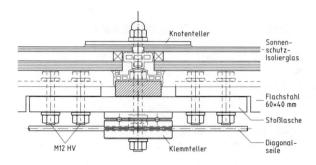

Fig. 13: Cross section of a joint

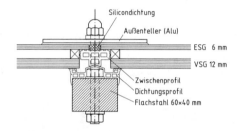

Fig. 14: Glazing detail. The flat bars directly carry the glazing, which is fixed only at the nodes using circular-shaped cover plates.

Fig. 15: A typical node

Nodes:
Node screw M 12, 10.9, 100 % prestressed.
3-piece clamping plate ⌀ 90 mm made of St 52-3, moulded.
All clamping plates are identical. For the connection only one screw was used at each intersection, allowing the node to pivot in the structural plane.

Glazing:
Spherically curved sun protection insulating glass (double glazing).
6 mm single-pane safety glass on the outside,
2 x 6 mm laminated safety glass on the inside,
12 mm air cushion.
A heating wire was inserted at the seam, ventilation is accomplished using ventilation screws.
Due to symmetry and by combining panels within pre-determined tolerable limits, only 32 different sizes had to be used for all of the 524 panes.
Corrosion protection:
All steel parts were hot galvanized and coated during manufacture. White top layer. The screws were hot galvanized and painted on site. Hot galvanized, plastic-coated ropes. The gaps in the clamping plate were filled with silicone.

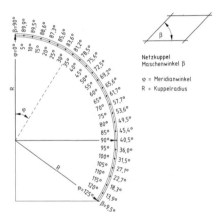

Fig. 16: Mesh angle ß along the 45° axes.

As mentioned before, the basic grid consists of identical flat bars. The bars' dead load causes only minimum changes in length, therefore, for the determination of the joints on the dome surface only a constant distance of 1.000 mm in net direction has to be kept in mind. For this purpose a computer program was developed, starting at the center of the dome and moving iteratively along the dome surface joint by joint. This results in the cartesian co-ordinates (respectively the spherical co-ordinates) of the joints and the mesh angle ß depending on the intended structural shape. Fig. 16 shows the result for the Neckarsulm dome. The principle axes and the 45° axes are symmetrical. The mesh is practically right-angled along the main axes, the greatest deviation from a square mesh occurs along the 45° axes. Fig. 16 shows the mesh angles ß increasing with the meridian angles φ in the case of the 45° axes.

The variation of the mesh angles is achieved by deviating the flat bars in the joints. These angles of deviation ε, made possible by a clearance in the two outer drill holes, are practically identical along the bar (Fig. 18). The flat bars do not twist, despite running across the sphere. Each bar is part of a principle circle.

It is a well-known fact that the diagonals of squares and rhomboids are in a right angle to each other. Consequently, the grooves in the clamping plates intersect also at right angles - all plates can be manufactured identically.

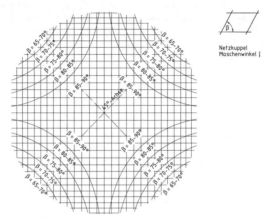

Netzkuppel
Maschenwinkel ƒ

Fig. 17: Distribution of mesh angles over the surface area of the dome.

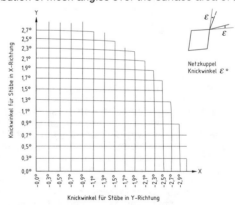

Netzkuppel
Knickwinkel ε °

Fig. 18: Angles of deviation ε of flat bars over the surface area of the dome.

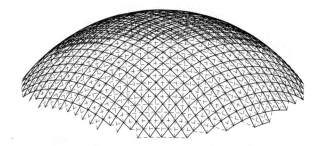

Fig. 19: Theoretical model.

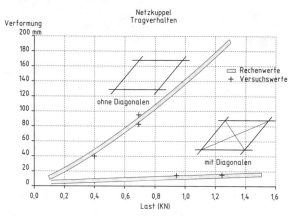

Fig. 20: Test results for a braced and unbraced dome.

The dome, though double curved, consists entirely of identical parts. Different bar lengths occur only at the edge, but they could be exactly calculated using the computer program described above. Since the geometry of the dome is exactly determined by the length of the edge bars, the erection was accordingly simple.

Starting at the edge the pre-cut starter bars and the remaining bars of constant length are installed and bolted step by step (Fig. 24). In order to obtain the exact shape, great precision has to be observed when cutting the bars and fitting the pre-drilled edge panel.

After installing the flat bars, the cables were placed beneath them, running from edge to edge and partially prestressed. The remaining prestress was achieved by tightening the clamps, initially placed at 5-mm-intervals to the flat bars.

Prefabrication of all parts and the use of CNC-equipment guaranteed the required precision. Consequently the size of all glass panels did not have to be measured on site, but could be pre-determined by computer and prefabricated, therefore, the glazing could start right after the construction. All panels fitted, none had to be exchanged.

The entire installation including the glazing was accomplished within 2 months without any welding endangering the corrosion protection.

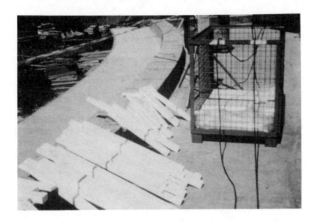

Fig. 21: Flat bars of equal length (1.0 m)

Fig. 22: Pivoted joint straps of equal length. The pivoting feature provides for variable mesh angles.

Fig. 24: Method of field
erection: Starting from the
periphery and progressing
toward the centre, the flat
bars are fitted into place
one by one.

Fig. 23: View of the dome in the field
erection phase. The desired geometry of the dome takes shape "by
itself".

Fig. 25: The surface of the dome forms a perfect sphere.

Second Example:
A Free Shape Glazed Grid Dome at Hamburg
(Hamburg City Historical Museum)

Courtyard approx. 900 m^2
Barrel-vaults:
Spans 14 m resp. 18 m
Rise 3.80m resp. 5.10 m
Mesh width 1.17 x 1.17 m
Flat bars:
Flat bars 60 x 40 mm, St. 52-3, moulded, straight, hot galvanized and coated. Total length: 2.400 m. Joint straps 60 x 20 mm, St. 52-3. All joints are screwed together with 2 M 12 10.9, 100 % prestressed.
Cables:
Spiral cables ⌀ 6 mm, β_N = 1570 N/mm^2, stainless steel cables. Twin cables running from edge to edge. Total length: 6000 m.
Nodes:
Node screw M 12, 10.9, 100 % prestressed, 3-piece clamping plate ⌀ 90 mm made of St. 52-3, moulded.
Glazing:
2 x 5 mm laminated safety glass (single glazing), sun protection glass, entire glazing surface approx. 1.000 m^2.

Total weight of the structure including the point-supported gutter approx. 50 t.

Fig. 26: Inside view

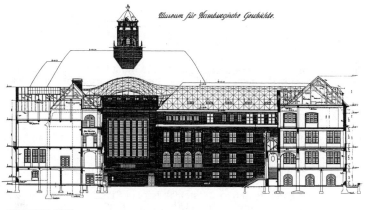

Schnitt und Ansicht   M 1 : 500

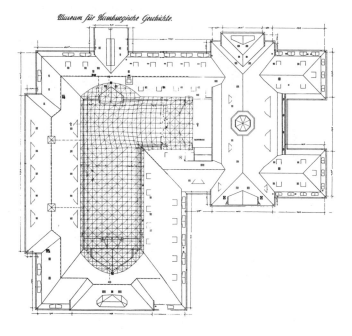

Fig. 27: Hamburg City Historical Museum. Plan, sectional view and view.

Between 1914 and 1923 the well-known architect Fritz Schumacher created this museum.

It is a historical monument and therefore protected. Even though the possibility of a glass-covered courtyard was considered from the beginning, the courtyard remained open until recently.

For the L-shaped courtyard, the lightest and most transparent structure had to be designed, which would burden the historical building as little as possible - in both senses of the word: it was not to alter the overall appearance, and must only transmit minimal additional loads to the historical building.

The net dome has two barrel-vaulted sections, with spans of 14 m resp. 18 m, with a smooth transition between them. There the geometry is the result of an optimization, transferring the majority of the roof loads via membrane compressive forces and avoiding bending stresses.

The structure consists of 60 x 40 mm flat bars of St. 52.3, galvanized and painted white - in other word, hardly any more than the minimal dimensions of supporting members for a glass cover. These bars form a quadrangular net with a uniform mesh of about 1.17 x 1.17 m. Cables, installed afterwards, prestressed and clamped down at the joints form the varying diagonals.

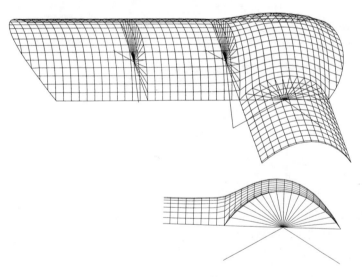

Fig. 28: Three-dimensional view of the grid shell.

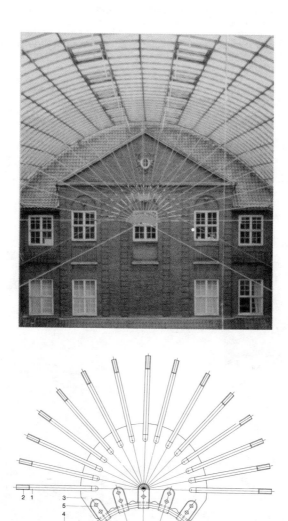

Fig. 29: View of a "hub".

These minimal cross-sections enable the grid shell structure to transfer deadload as well as snow and wind loads. Since extremely high snow loads on one side due to drifting or trapping of snow in the roof valley could not be ruled out, the "somewhat softer" areas of the barrel vault were additionally stiffened with spokes wheels consisting of cables radiating from a "hub".

The glazing, 2 x 5 mm laminated safety glass, was placed directly on the flat bars and secured with plates at the joints. A heating wire was inserted between the glass support and the steel bars to prevent condensation. A gutter heating system ensures trouble-free roof drainage.

A continuous rigid edge girder was installed along the rim, approximately 70 to 90 cm above the existing roof and supported at several points through the roof by the reinforced concrete ceilings resp. walls. Since the new glass roof structure is so light, no special underpinning of the historical building was necessary.

The roof was designed and built in just 6 months.

Fig. 30: View of the glazed grid dome

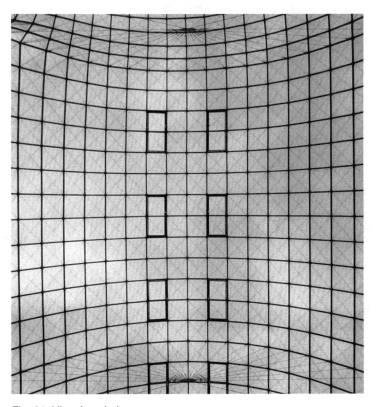

Fig. 31: View from below.

Fig. 32: Glazing in the spandril of the two barrel-vaults.

Fig. 33: Inside view.

Fig. 34: Inside view.

Third Example:
Glazed Grid Dome with Plane Glass Panels

To glaze freely shaped domes or domes with an amorphous plan, the panels either have to be curved according to the structure's surface or the geometry of the dome and the mesh has to be such, that each glazed mesh is in a plane. This is especially desirable when using double glazing insulating glass.

Simple shapes such as barrel-shaped shells may be covered with plane panels.

But even freely shaped domes can be built as grid domes using plane panels and bars of equal length (identical mesh width), if the chosen structure of the grid dome bars allows any curve, acting as generating line, to move along any other curve, acting as directrix (translational surface). Generating line and directrix form the glazed, quadrangular basic grid of the flat bars. There are some simple examples such as the elliptical paraboloid with an elliptical curve in plan, the rotation paraboloid with a circular curve in plan and the hyperbolical paraboloid.

In Leipzig the trapezoidal courtyard of a historical building had to be covered with a glass roof consisting of plane panels. The shape of the roof was achieved by moving a parabolic curve, open at the bottom, along another one, which was open at the top. All quadrangular mesh, except the rim spandrel consist of equally long bars (1.20 m) and are plane. The roof is presently under construction.

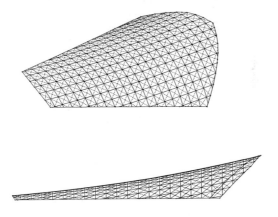

Fig. 35: Glass dome covering an oval plan. All bars are equally long (1.20 m). All glazed quadrangular meshes are in plane.

At present the design work proceeds for a sporting arena near Hamburg with a glass dome covering an oval plan. The dome is to be covered by plane insulating glass panels. The dome surface was produced by moving one parabolic curve as the generating line along a parabolical directrix. The result is a curve elliptical in plan.

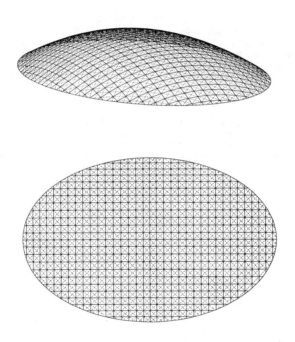

Fig. 36: Glass dome covering a trapezoidal plan. All bars are equally long (1.20 m). All glazed quadrangular meshes are in plane.

References

[1] An goldenen Ketten ... Hofüberdachung des Museums für Hamburgische Geschichte. Von Gerkan, Marg und Partner; Schlaich Bergermann und Partner. Deutsche Bauzeitung, Vol. 124, July 1990, pp. 32 - 40

[2] v. Gerkan, M., Marg, V., Schlaich, J.: Copertura vetrata della Corte del Museo di Storia della Citta. Amburgo, Domus No. 719, September 1990.

[3] Herr, M.: Ingenieurbau-Preis 1990. Beton- und Stahlbetonbau 86 (1991), Vol. 5, pp. 105 - 106, Stahlbau 60 (1991), Vol. 5, pp. 129 - 130.

[4] Museum für Hamburgische Geschichte: Glasdach über dem Innenhof. Glasforum 2.90, Vol. 40.

[5] Büttner, O., Hampe, E.: Bauwerk Tragwerk Tragstruktur. Klassifizierung - Tragqualität - Bauwerksbeispiele. Berlin: Ernst und Sohn, 1985.

[6] Makowski, Z. S.: History of the development of braced domes. IASS-Bulletin Vol. 30 (1989).

[7] Klimke, H.: Zum Stand der Entwicklung von Stabwerkskuppeln. Der Stahlbau 1983, Vol. 9, pp. 257 - 262

[8] Klimke, H.: Entwurfsoptimierung räumlicher Stabwerksstrukturen durch CAD-Einsatz. Bauingenieur 61 (1986), pp. 481 - 489.

[9] Otto, F.: Multihalle Mannheim, Vol. IL 13. Mitteilungen des Instituts für leichte Flächentragwerke (IL), University of Stuttgart, 1978.

[10] Klimke, H., Kemmer, W., Rennon, N.: Die Stabwerkskuppel der Stockholm Globe Arena. Stahlbau 58 (1989), Vol. 1, pp. 1 - 8.

[11] Otto, F.: Gitterschalen, Vol. IL 10. Mitteilungen des Instituts für leichte Flächentragwerke (IL), University of Stuttgart, 1975.

[12] Leonhardt, F., Schlaich, J.: Vorgespannte Seilnetzkonstruktionen - Das Olympiadach in München. Der Stahlbau (1972), Vol. 9, pp. 257 - 266, Vol. 10, pp. 298 - 301, Vol. 12, pp. 367 - 378. Der Stahlbau (1973), Vol. 2, pp. 51 - 58, Vol. 3, pp. 80 - 86, Vol. 4, pp. 107 - 115, Vol. 6, pp. 176 - 185.

[13] Schlaich, J., Mayr, G.: Naturzugkühlturm mit vorgespanntem Membrandachmantel. Der Bauingenieur 49 (1974), pp. 41 - 45.

A DESIGN FOR A FOOTBALL STADIUM ROOF ENABLING
GROWTH OF A NATURAL GRASS PLAYING SURFACE
SYDNEY FOOTBALL STADIUM - AUSTRALIA

G.S. Ramaswamy; D.S. Thomas; I. Norrie; L. Thorogood

ABSTRACT

This paper discusses the development of a design for a
stadium roof which will provide conditions allowing the
growth of a wearable natural grass surface suitable for
sports including Rugby, Soccer and American Football and
which uses mainly passive energy methods to maintain
spectator and player comfort with minimum energy cost.

A preliminary design for a spaceframe which is elliptic
in plan with a major axis of 185m and a minor axis of
148m supported at four points is described.

1.0 INTRODUCTION

As a result of difficulty in growing a wearable natural
grass playing surface at the Houston Astrodome in 1960,
indoor stadia for American football have used artificial
"grass" playing surfaces. Soccer and Rugby football
have continued in outdoor stadia, artificial surfaces
being considered too hazardous for players in those
Codes. The open stadia suffer from extreme conditions
for players and spectators and limited revenue
opportunity from multi-purpose usage as a result of
absence of protection from the weather.

It has been considered that an indoor grass surface
cannot be sufficiently tough to withstand vigorous play
and that even if horticultural concerns could be
resolved, there would still remain a problem of
unacceptably high cost of removing the heat associated
with the high level of light required for healthy grass
growth. As a result, the majority of international
stadia are open and suffer a revenue loss owing to bad
or uncertain weather conditions.

Chairman, Octatube Space Structures (India) Pty Ltd,

2.0 THE SYDNEY FOOTBALL STADIUM (Fig 1)

Opened in 1986 the Stadium has 42,000 seats of which 20,000 (approx) are covered. The Stadium is used for Rugby and Soccer with a minimum of additional events.

In 1991 after several events were severely rain-affected causing community and media criticism, the owner of the stadium, The Sydney Cricket and Football Ground Trust, commissioned an Australian group with a background in the design and construction of lightweight roofs, to report on the technical feasibility of extending the existing roof to provide cover over all of the seats.

The curator of the Stadium defined the extent of natural light required on the grass surface, assuming a change to shade tolerant grasses under development elsewhere. The curator became an integral member of the study group. It was specified that the extended roof must allow at least 70% light transmission.

The total amount of sunlight falling onto the grass before and after addition of a 70% translucent roof was established. It was noted that the amount of sunlight was already insufficient before the proposed roof extension and the playing surface is returfed at the end of each football season.

A cable supported design concept was proposed to minimise sunlight transmission loss.

A "fabric" cladding was sought. However reinforced fabrics provided no more than 30% light transmission after which unacceptably low tensile and tear strength are experienced, rendering the material unsuitable for use as a tensioned membrane. These fabrics also block most UVA wavelength required for strong leaf growth and colour. An unreinforced material with suitable light transmission properties and durability was identified.

A structural solution was developed comprising a 600 MPa steel compression ring mounted adjacent to the front edge of the existing roof, an elliptical inner cable tension ring comprising four 94mm diameter cables, fixed to the compression ring by radial cables, to which the cladding in aluminium framed panels would be attached. The roof extension would slope down to improve spectator protection and not unduly interfere with the existing architecture. (Fig 2)

In considering the group's report in July 1991, the owner observed that although the proposed design provided the specified protection to spectators from

direct rain (and sun), it left unresolved the covering
of the playing area necessary to increase event
opportunities and revenue. The owner noted that the
results were sufficiently encouraging to suggest that
with further study, total enclosure of the roof might be
possible.

3.0 TOTAL ROOF ENCLOSURE

In the first study, several operations problems at the
Stadium had been identified:

* Maintenance costs are high. These costs include
  annual replacement of the grass surface and
  irrigation and fertiliser losses. Because of fast
  release drainage to disperse peak rain and to avoid
  flooding during events, irrigation is applied at a
  rate more the 2.5 times Sydney rainfall to maintain
  soil moisture levels.
* Humid conditions on the infield cause physical
  stress to players owing to limited air circulation.
* Spectators in the covered grandstands became wet
  from wind driven rain and cold from wind through
  openings in the Stadium at upper seating levels.
* Extended use of the Stadium is limited owing to
  noise restrictions imposed by local Council, the
  Stadium being sited in a residential area.

Further analysis of the problem of the deterioration of
the grass each year showed the main factors to be:

* Early dormancy of the grass owing to the cold
  night-sky effect through the aperture of the roof.
* Softening of grass leaf because of high level
  irrigation.
* Increasing shadow effect and progressively reducing
  light during winter months, the time of highest use
  and wear.

Weather records for a 130 year period were reviewed;
main features noted were:

* The maximum temperature in a given 24 hour period
  is consistently $10-12^0C$ higher than the minimum.
* On the majority of days a minimum windspeed of
  11 km/h is experienced.

Sunlight exposure diagrams revealed that shadows from
the existing roof prevent sufficient light falling onto
the grass despite the extent of the roof having been cut
back during initial design (and thus compounding the
problem of inadequate coverage of spectators).

The data showed that before the proposed roof extension,
maximum sunlight exposure varies between 4 and 6 hours
over an elliptical area located centrally covering
approximately 25-30% of the playing area, reducing to 2
to 4 hours at the sidelines.

4.0 SUMMARY OF DESIGN REQUIREMENTS

* Allow sufficient light penetration in the specified
  range to ensure vigorous grass growth.
* Identify most suitable grass varieties and develop
  grass management procedures.
* Develop clear span structural framing independent
  of the existing roof, with minimum obstructions to
  daylight and with a minimum of columns to avoid
  loss of lines of sight from existing seating.
* Develop a ventilation system to provide spectator
  and player comfort with a minimum of operating
  costs.
* Use existing proven technologies in combination to
  achieve a new, low risk, design solution.

5.0 STRUCTURE

The structural arrangement for the proposed roof had to
accommodate constraints imposed by the existing roof
structure, as well as the requirement that there would
be no loss of sight lines from spectator seating through
the introduction of support columns.

After considering various design options, the Sydney
group selected a structural arrangement consisting of a
double layer tubular spaceframe comprising circular
hollow sections joined together with hollow spherical
connectors.

The spaceframe is an elliptical paraboloid dome with a
structural depth of 5 metres. Several topologies were
tried based on the need to minimise deflections. A
square on square 5m grid offset configuration set
diagonally was chosen. The spaceframe is supported at
only four points, at the four corners of the stadium at
the back of the top level of seating. (Fig 3)

Structural analysis was undertaken in Australia using
MicroSTRAN 3-D software, assuming the space frame to be
pin-jointed. The space frame behaves as part arch and
part in flexure. The main support to the structure are
bands running diagonally between the four corners acting
as arches. The remainder of the structure acts
flexurally. (A summary of the loads used in the
analysis is presented in Fig 4)

The design concept was referred to a spaceframe supplier in the Netherlands and India for confirmation and refinement.

6.0 CLADDING

The fabric cladding, contained in rectangular aluminium panels, will be fastened to a steel support layer on top of the spaceframe. Each panel will contain two layers of film separated by a constant supply of low pressure air forming a pillow.

The double layer pillows will provide insulation benefits, and the convex internal surfaces will offset the refocussing of sound which would otherwise be created by the global concave profile of the roof. Acoustics will be further improved by the damping effect of the air gap formed in the pillows and the selection of different film thicknesses for the internal and external film layers.

A secondary partial ceiling layer, using a fabric with specific acoustical and thermal properties, will be fixed beneath the spaceframe providing a means to control thermal and acoustical effects.

7.0 PASSIVE ENERGY DESIGN

A feature of the proposed enclosed stadium design is the extent to which passive energy design has been used for environmental control resulting in low energy costs.

Spectator and player comfort will be maintained under all weather conditions and an optimum environment for grass growth and management will be provided when the stadium is not in use for events.

The passive design features include provisions by which the extent and location of diffuse and direct solar radiation may be controlled for optimum levels of natural lighting and solar transmission, depending on climatic conditions and the use of the stadium at the time.

The design also takes advantage of the influence of the stadium roof geometry in relation to natural convection and wind induced air movement. The naturally ventilated air flow resulting from these influences will be automatically controlled to suit the complete range of ventilation requirements. These will be augmented under extreme weather conditions by a supplementary mechanical ventilation system and by the control of stratified air layers.

The estimated 900 tonnes of air in the stadium will be changed by entirely passive means, at rates varying from 25 minutes in winter with still external air, to 7.7 minutes in summer with 20km/h external wind. Supplementary fans with a capability of 7.8 minutes per air change are included for smoke exhaust purposes.

The passive energy design uses the massive thermal-energy storage of the stadium construction and the playing field area, with the volume of the stadium space to moderate atmospheric conditions during periods of extreme hot and cold climatic conditions (Fig 5).

Use is made of the diurnal temperature conditions to pre-cool the stadium prior to summer events, and conversely to maximise the storage of energy during winter months. This process is assisted by the spectrally selective transmission qualities of the film cladding in the near and far infrared spectrum.

8.0 PLAYING SURFACE

When a natural turf grass playing surface is exposed to the elements throughout the changing seasonal conditions, it is virtually impossible to achieve an ideal playing surface except for a limited period of the year. This is because of the competing influences of the soil base, irrigation, drainage, extremes of weather, wide variation in soil and grass temperatures from exposure and the natural seasonal behaviour of all grasses.

Notwithstanding these highly variable influences, the ideal grass will be required to have a preferred leaf size, suitable colour, good salt tolerance, excellent recuperation, resistance to disease, shade tolerance, high shoot density and very importantly, a high wear tolerance.

The proposed roof enclosure of the Sydney Football Stadium eliminates the extremes of weather influence and exposure, and enables for the first time an ideal environment to be provided indoors, essentially by passive means, to meet the specific and essential requirements of the grass.

Tests carried out over complete cycles of seasons under simulated conditions, show that the design features of the transparent roofing system can be engineered to successfully grow and maintain high quality natural turfgrass playing fields under virtually all climatic conditions with resultant increase in revenue and reductions in operating costs. Annual savings in reduced irrigation and fertiliser costs alone will amount to around $US350,000 at the Sydney Football Stadium.

9.0 PROJECT DELIVERY

The Sydney Cricket and Football Ground Trust has resolved to adopt the design subject to finance. Timing will be influenced by decisions now being made regarding usage of various sporting venues during and after the Sydney 2000 Olympics. On site construction time will be approximately nine months.

The structural, mechanical/thermal and grass technologies developed for the Sydney Football Stadium and capable of adaptation to suit specific conditions at alternative sites will be provided by Sydney based Stadium Technologies Pty Limited.

10.0 ACKNOWLEDGMENTS

DESIGN CONCEPT AND DEVELOPMENT:

Stadium Technologies Pty Limited
255-259 Pacific Highway, NORTH SYDNEY   NSW   2000, AUSTRALIA   Fax: 61-2-9297975

Project Management L. Thorogood
Mechanical   D.S. Thomas
Structural   I. Norrie
Hydraulics   R. Brell

in association with:

The Sydney Cricket & Football Ground Trust
N.E. Neate; P. Leroy

SPACEFRAME DEVELOPMENT

Octatube Space Structures (India) Pty Ltd, "Sans Souci"
E-119, 16th Cross Road, BESANT NAGAR, MADRAS   600 090
INDIA
Professor G.S. Ramaswamy

in association with:

Octatube Space Structures B.V., Rotterdamseweg 200
2628 AS Delft, The Netherlands
Dr M. Eekhout

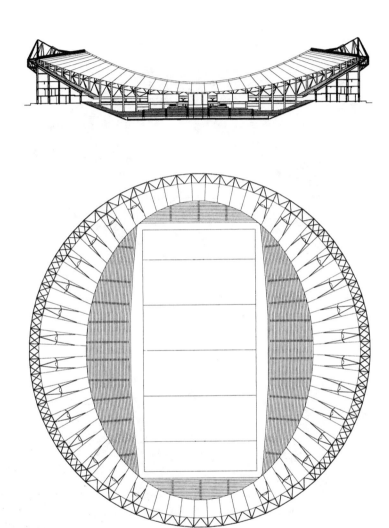

Fig 1   Sydney Football Stadium

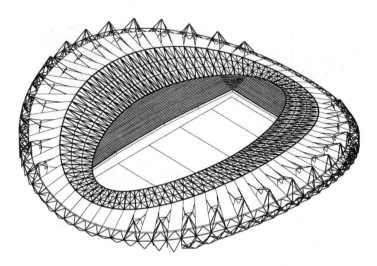

Fig 2   Cable Supported Concept

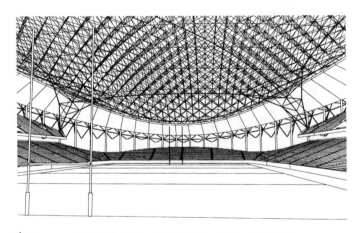

Fig 3   Spaceframe Supported Four Corners

DESIGN LOADS	Mass/m²	Tonne
(i) DEAD LOAD		
Space frame self weight	65kg/m²	1415
Nodes additional	3140 @ 35kg ea	110
Fabric Support Framing	8kg/m²	175
Fabric Membrane, A1, clamps	2kg/m²	43
Sunscreen mechanical layer	15kg/m²	325
Catwalks, lights, loudspeakers	5kg/m²	110
Hanging exhibits	16 @ 5t each	80
Hanging scoreboard centre	4 @ 25t each	100
Operable wall down centre	30kg/m²	150
Perimeter louvres	50kg/m run	25
Perimeter edge deflector	50kg/m run	25
Perimeter exhaust vents	28 @ 1t each	28
Corner columns	4 @ 74 tonne	296
TOTAL DEAD LOAD		2882
(ii) LIVE LOAD	25kg/m²	543
TOTAL DEAD & LIVE LOAD		3425
(iii) WIND LOAD	0.98(coeff) kPa	2345

Fig 4   Design Load Summary

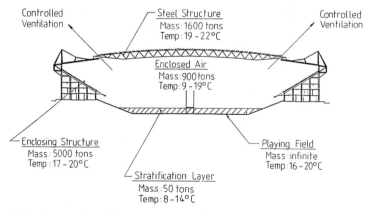

THERMAL STORAGE INFLUENCES

Controlled Ventilation

Steel Structure
Mass: 1600 tons
Temp: 19 - 22°C

Controlled Ventilation

Enclosed Air
Mass: 900 tons
Temp: 9 - 19°C

Enclosing Structure
Mass: 5000 tons
Temp: 17 - 20°C

Playing Field
Mass: infinite
Temp: 16 - 20°C

Stratification Layer
Mass: 50 tons
Temp: 8 - 14°C

Fig 5   Passive Energy Design

Phoenix Central Library
Reading Room Roof

Michael W. Ishler[1]
Member ASCE

Abstract

This paper discusses the design and analysis of the cable
truss roof over the fifth floor reading room of the new
Phoenix Central Library. Forty pairs of diagonally
oriented cable trusses located on a 32'-8" square grid
carry vertical load and provide lateral stability for the
precast columns. The top and bottom cable chords of each
truss are separated by a pair of vertical flying struts.
These struts subdivide the span and support parallel
lines of type steel purlins at 16'-4" on center. A deep
metal deck spans between the purlins. Although the roof
deck, insulation and weather membranes are of
conventional construction, they do not directly touch the
columns or side walls. The cable truss network elevates
the roof plane above the columns and allows a visual
separation between the roof and the side walls.

Introduction

The new Phoenix Central Library is located in the urban
core of the 10th largest U.S. city. The 280,000 sq. ft.
floor area is accommodated in a five level precast
concrete box. Restrooms, emergency stairs, freight
elevators, mechanical equipment, electrical equipment and
other services are located adjacent to the central floor
plate in the east and west steel framed saddlebags. The
Sonoran desert provides the context for the use of

---

[1]Associate, Ove Arup & Partners Calif.; during the design
Ishler Design & Engineering Associates; current
2314 Pearl Street
Santa Monica, California 90405

38

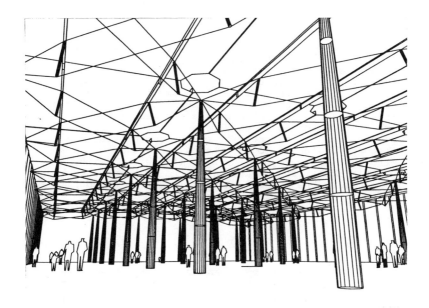

Figure 1. Interior Perspective of the Reading Room

concrete, copper and glass in this urban mesa. A
stainless steel canyon in the mesa highlights the
entrance. Precast concrete was selected for the primary
structural system. Eight foot two inch wide two story
wall panels support the floors, provide thermal inertia
to delay the heat of the desert sun and work as the
interior and exterior finish surfaces. The eight
foot two inch wide (cast from standard ten foot molds)
precast double-T floor system supports the heavy design
live load, provides a chase for the service systems and
is the finish ceiling. The vertical organization of the
library takes advantage of the first two floors for the
most common public services: best sellers, children's
collection, video, newspapers, magazines and government
publications. The third and fourth floors house the
special collections, closed stacks and library services.
The fifth floor has been set aside to house the city's
640,000 book fiction collection in an open public reading
room. Not restricted by the need to support heavy loads,
the reading room roof is free to rise in height with an
exposed lightweight structure. Its volumetric
proportions were selected in the tradition of the great
reading rooms of Henri Labrouste at the Bibliotèque
Nationale and the great hall/office at the Johnson Wax
complex by Frank Lloyd Wright.

Structural Systems

The reading room roof is supported by a network of cable
trusses. The cable trusses spring from the exterior
precast walls and interior candlestick columns. Each
cable truss supports two tube steel purlins spaced
parallel to the exterior walls. A skylight atrium is
carved out between the central purlins. Metal deck spans
between the purlins and supports the insulation, weather
membrane and roof ballast. The deck profile follows an
arc over the taller interior columns to provide the
necessary drainage slope. The galvanized deck has
perforated sides to provide a finish ceiling with
acoustic absorption capacity. At the east and west edges
of the roof, the metal deck cantilevers to an edge
skylight.

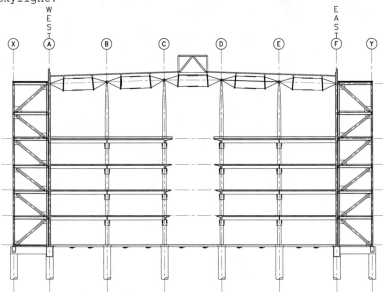

Figure 2. East-West Building Section

The cable trusses are continuous between perimeter
supports. They are supported by and anchored to the east
and west saddlebag walls. Anchorage to the building
frame is provided at grids 1,2,3,4,5,6,7 and 8. The
precast shear walls on grids A and F provide lateral
restraint in the north/south direction. Horizontal
trusses in the saddlebag roof are used to transfer the
east/west lateral loads to the steel braced frames on
grids 0.5,0.9,2,3,5,7 and 8.5.

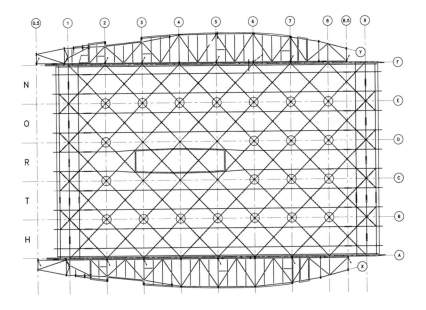

Figure 3. Roof Plan

Horizontal stability at the top of each column is
provided by four cable trusses. The diagonal arrangement
of the trusses restricts horizontal racking of the
columns and roof. Although the metal deck will provide
significant diaphragm shear stiffness in the plane of the
roof, this stiffness is not necessary for stability.

At the north and south edges of the roof, the tension
from the cable trusses is balanced by compression in the
tube steel purlins transferred to the column capital by
the starburst struts. The purlins provide a continuous
compression link between the north and south edges of the
roof. Buckling of the purlins is restrained by the cable
trusses and metal deck. The purlins also transfer
pressure loads from the north and south curtain walls in
compression to the interior of the roof where the they
are balanced by tension in the diagonal cable trusses
anchored to the saddlebags.

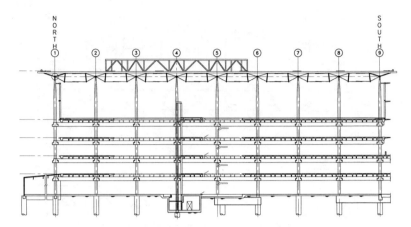

Figure 4. North-South Building Section

The southern edges of the saddlebags stop at grid 8.5.
Consequently they do not provide lateral stability to the
roof trusses at A9 of F9. Grid 9 has compression struts
joining the tops of the columns in the east-west
direction to balance the east-west component of the
diagonal cable trusses. Unbalanced pressure loads on the
concrete walls at the southern edge of the building are
transferred by these compression struts to the interior
of grid 9 where the diagonal tension lines can be used to
transfer the load back to the saddlebags.

Each diagonal cable truss consists of a lower and upper
cable chord separated by two compression struts. The
struts support the roof purlins. The roof purlins are
continuous in the north-south direction. They are one-
quarter of a bay offset from the column grid and are
spaced at intervals of one-half of the column grid. The
purlins are propped by the starburst compression struts
at the north and south cantilever roof edges. Two steel
tubes in the plane of the metal deck extend to the
precast walls in each corner to limit the vertical
deflection at the cantilever edges.

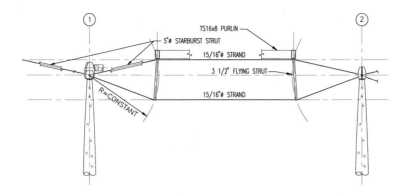

Figure 5. Cable Truss at Roof Edge

Load Carrying Behavior

Symmetric vertical loads are resisted by an increase in
the top or bottom cable tension. As the downward load in
a bay is increased, the prestress in the top cable will
be relaxed and the tension in the lower cable will
increase. The vertical reaction of the cables is
transferred to the column. Any change in the net
horizontal reaction is shared by the top and bottom
chords of the adjacent cable truss and carried back to
the saddlebag anchors. The relative depth of the top and
bottom chords above and below the column capital
centerline was selected to equalize the factored tension
in these cables under the extreme upward and downward
load combinations. This justifies using the same size
cable for both chords. The cable length from the column
capital to the bottom of the flying struts was set equal
to the top length to simplify fabrication. This forces
the struts slightly off vertical alignment allowing their
dance to follow the arc of the roof without the
distraction of apparent misalignment.

Non-uniform loading of a cable truss can be considered as
the super-position of symmetric and asymmetric loading.
Asymmetric loading causes asymmetric deformation of the
cable truss. Although the cable truss is quite stiff
under symmetric load it offers very little resistance to
asymmetric load. The center cables must move out of
horizontal alignment to provide a net vertical reaction.
Although the truss is stable under asymmetric load, cable
truss deflections could be excessive. Both the purlins

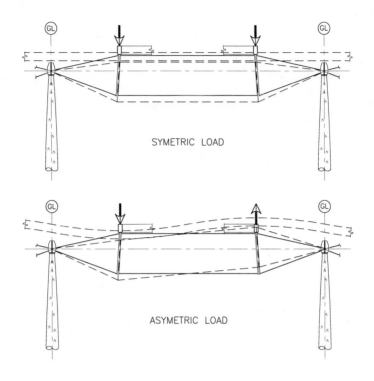

Figure 6. Symmetric and Asymmetric Load

and the metal deck are continuous over the cable trusses. Asymmetric loads force the purlins and deck to deform into a series of reverse curves. The flexural stiffness of the purlins and deck combine to provide the primary resistance to unacceptably large asymmetric deformations.

Dynamic behavior of the roof system was investigated. Phoenix is in UBC seismic zone 1. With the light mass of the structure and the considerable shear strength of the diagonally oriented cables, dynamic horizontal loads are not problematic. Vortex shedding under modest but steady winds could potentially excite the relatively flexible roof at the north and south cantilever edges. The stiffness and mass of the roof were adjusted to keep its first three modes of vibration beyond of the range where wind flutter could introduce a significant amplification to the static design pressures.

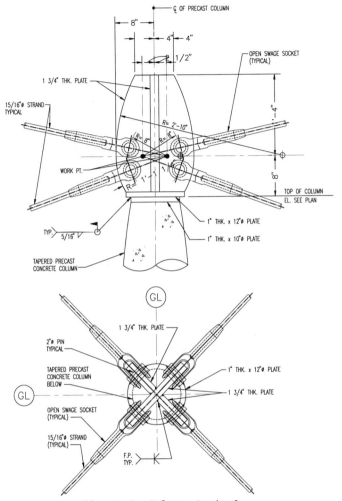

Figure 7. Column Capital

Construction Considerations

The columns, cables, flying struts and purlins are all prefabricated to specified tolerances. The roof installation presumes the attachment of these components to the site-constructed anchors. The cable trusses are continuous over the interior columns. There is no site

length adjustment capability in these pin connections.
Consequently, the equilibrium locations of the column
capitals are not established until the roof is fully
tensioned. The vertical alignment of the columns is
determined by the constructed base and capital positions.

The cables are continuous under and over the flying
struts. They are held in position by friction and a
cable collar clamp. The clamps can be field adjusted to
make up for small misalignments between the top of the
flying struts and the purlins.

Four NW-SE and four NE-SW diagonal cable truss lines
extend uninterrupted between grids A and F. These can be
tensioned prior to the placement of the purlins. The
other diagonal lines intersect either the north or south
curtain wall edges. The purlins must be in place to
balance the net horizontal inward reaction before these
diagonal cables can be tensioned. The metal deck can be
attached after all purlins are in place and all cable
trusses are fully tensioned. Splice locations in the
metal deck were selected to provide the greatest possible
continuity across purlins. Where the deck is spliced at
a purlin, it is lapped and fastened at the top and bottom
to provide flexural stiffness. The flexural continuity
of the deck gives it a stiffness comparable to that of
the purlins in resisting asymmetric deformations. The
installation of the rigid insulation, weather membrane
and ballast are comparable to that of a standard roof.

Summary

It is the composition of exposed structural cables
trusses to support metal deck above the tapered precast
candlestick columns that provides the intrigue for the
reading room roof of the New Phoenix Central Library.
Although each of the structural components of the reading
room roof has been used in numerous buildings, the design
for this public space incorporates them in a unique
combination.

Acknowledgements

The architects for the library are the joint venture
bruder/DWL of Phoenix. Thomas Hartman, computer
specialist for bruder/DWL provided figure 1. The
structural engineers are Ove Arup & Partners of
California. They provided figures 2,3,4,5 and 7. The
assistance of the architects and engineers in preparing
this paper is greatly appreciated. The design of the
structure was lead by the author while an associate with
Ove Arup & Partners.

Space Frame Solution for the Design of
Large Pyramid Structures

T.J. DeGanyar[1], M.D. Griffin[2], M.R. Patterson[3]

Abstract

This paper presents the engineering design methodology, structural analysis, system detailing, and installation procedures for a large span space frame structure currently under construction at California State University in Long Beach, California. The Physical Education Addition is designed as a general use athletic facility and basketball arena. On completion (spring, 1994) it will be the largest space frame pyramid in the world. The space frame solution proved to be a lighter weight and more economical alternative than an original design utilizing massive box trusses and conventional framing. This case study is significant in revealing circumstances in which a light weight space frame solution to a large span building problem was significantly more efficient and cost effective than a conventional one-way truss solution.

Introduction

Innovative design professionals have for decades recognized the attributes of space frame structures in large span applications. Yet their attempts to apply space frame solutions in specific building projects have frequently been frustrated due to economic factors. While many successful space frame solutions have been realized over the years, this building technology has yet to reach its potential in terms of range and frequency of applications.

----

[1] Vice President of Technology, Advanced Structures Inc., 1350 Abbot Kinney Blvd. 202, Venice, California, 90291

[2] Vice President of Eng., Advanced Structures Inc., 1350 Abbot Kinney Blvd. 202, Venice, California, 90291

[3] President, Advanced Structures Inc., 1350 Abbot Kinney Blvd. 202, Venice, California, 90291

47

A space frame solution will often be significantly less in weight of material than a conventional design in a given application, but a higher cost per unit weight of installed material for the space frame can offset the weight advantage and create a net economic disadvantage. This is especially true when other factors are not accounted for, as in the impact of the lighter structure of the space frame on the supporting foundation. These second order costs are seldom considered in the comparative cost analysis but can have a substantial favorable cost impact on the overall project incorporating the space frame design.

It is useful and necessary to recognize the applications in which space frame is currently competitive with conventional building practices. This paper is a case study of one such application.

The new Physical Education Addition for California State University, Long Beach, (CSULB), is a pyramid arena 105.23-M at it's base dimension with a height of 56.62-M. The structure has an opening centered on each of its base sides with a length of 25.5-M. (Fig. 1) The structure designed by the original project engineer utilized massive box trusses at the edges of the pyramid with conventional one-way truss framing. The design included a custom metal deck to enclose the structure.

Figure 1. Artists Rendering of the CSULB Physical
Education Addition

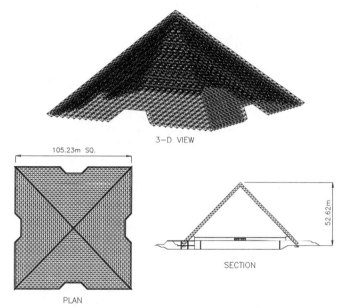

3-D VIEW

105.23m SQ.

52.62m

SECTION

PLAN

Figure 2. Space Frame Solution

A space frame solution to the enclosure was considered early in the design process and space frame vendors were consulted by the architect. However, it was decided instead to develop a conventional solution to the structure so as to not restrict the bidding. There was concern that a space frame structure might prove more expensive than a conventional truss approach. However, accommodation was provided to allow for the bidding of a space frame alternate, and a page of schematic details and a performance specification for the space frame were included in the architectural bid package.

Nine general contractors bid the project. They received bids from at least four companies specializing in proprietary space frame technology. Although some of the space frame costs received at bid time were substantially lower than the prices received from structural steel companies bidding the conventional structure, only one general contractor chose to base its bid on a less expensive space frame solution. This general contractor ultimately received the project award, with an overall bid approximately 20% less than the nearest competitor.

The structural framing system successfully

developed for the project is a multiple layer space grid
utilizing a prefabricated connection system (Fig. 2). The
connection detailing was developed for this project with
a primary design objective of utilizing the most common
materials and processes in its manufacture. The intent
here was to open the system's fabrication to the highly
competitive environment of steel fabrication job shops.
This objective was successfully realized resulting in per
pound fabrication costs competitive with conventional
structural steel. An overall economic advantage was thus
gained as the space frame alternative was lighter than
the conventional approach (by a factor of 2).

Design and Analysis

The space frame system is made of the assembly of
individual tubular steel strut members joined at their
ends through a plate node. The joining process is
accomplished by bolting through connection plates at
strut ends to mating plates on a node assembly (Fig. 3).
The assembled space frame behaves as a three dimensional
truss system with individual struts primarily carrying
axial loads. The frame is constructed from 3.64-M chords
and 3.17-M webs to form a 45 degree pentahedron pyramid.

The struts are fabricated from round and square
ASTM A-500 Grade B tubing with a minimum certified yield
strength of 290-MPa. The horizontal members on the outer
chord are square sections to accommodate the direct
attachment of metal decking to the outer surface of the
structure. The strut end plates are fabricated form 20-mm
ASTM A-36 steel plate. The weld between the end plates
and struts is a 6-mm factory applied fillet weld.

The nodes are fabricated via the welding of three
20-mm ASTM A-36 steel plates. The steel plates are
precision drilled prior to assembly with hole patterns to
match adjoining strut end configurations.

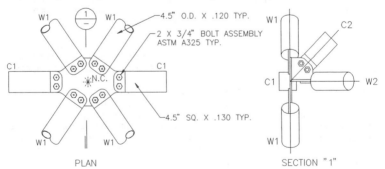

PLAN   SECTION "1"

Figure 3. Typical Connection Detail

The struts are field connected to the nodes via A325X bolts ranging in size from 20-mm to 40-mm. The connection includes at least two bolts at the end of each strut. The connections are designed as bearing connections, hence the bolts need to be snug-tight only.

The foundation system for the space frame consists of eight grade beams which are 36.3-M long, 1.82-M wide and .76-M deep (Fig. 4). The beams have a continuous 36.3-M long transverse shear key, .66-M deep and .3-M wide, and six 3.5-M longitudinal shear keys 1.42-M deep and .3-M wide. The foundation analysis combines the soils passive resistance with the frictional lateral resistance. The grade beams are designed as flexural members, and the soils is considered as elastic foundation with stiffness computed based on the modulus of the subgrade reaction and the soils contact area.

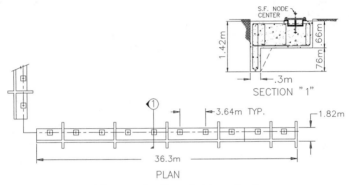

Figure 4. Foundation System

The space frame is analyzed utilizing SPACE2, a proprietary special purpose finite element computer program developed solely for the three dimensional analysis and design of space frames. The computer model is constructed by definition of nodal coordinates and connecting truss elements with the appropriate cross-sectional properties between these nodes. The supports are modeled as three dimensional linear spring elements with spring constants approximating the stiffness of the foundation. The model is automatically generated and the openings are cut out graphically. The space frame is constructed from 14,400 struts and 3,150 nodes, and is supported at 80 anchor locations.

Design Loads are applied at the nodes for the various load cases, and the resulting member axial forces are combined with the proper load combination factors.

The member sizing and component design is carried out automatically and the bill of materials and erection drawings are generated by the SPACE2 program. The entire space frame is assembled from 57 different strut types and 22 different node types.

The design loads are computed based on the 1988 edition of Uniform Building Code (UBC). The space frame weight is approximately 31.7 Kg/m^2 per plan area and an additional 58.6 Kg/m^2 of additional dead load has been applied for the roofing and mechanical equipment. A lateral seismic load of three times the code required load is used to design the connections. The majority of the exposed structure of the pyramid can be classified as a roof. Wind tunnel testing of similar structural profiles have indicated that UBC code over-estimates the total lateral force but provides an accurate value for the loading of the elements and the local areas of discontinuities. While the code contains the coefficients for wind directions perpendicular to vertical faces, a second case is considered as quartering direction by taking the same pressures for the perpendicular faces and applying them at a 45 degree angle. Due to the symmetrical nature of the structure only 12 different loading combinations needed to be considered.

The charts in Figure 5 present the member force and the stress distributions for the gravity, the lateral wind, and the seismic loads. Figure 6 depicts the graphical distribution of the member forces for gravity and wind load combinations. The vertical deflections and the lateral drift for all of the load combinations are well under conventional levels.

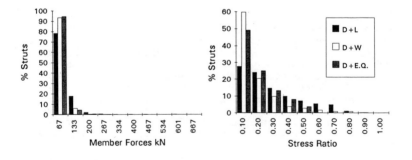

Figure 5. Force and Stress Distribution

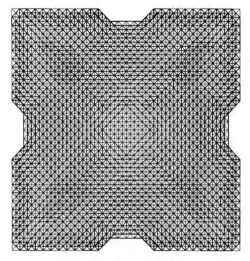

(a) Gravity Loading

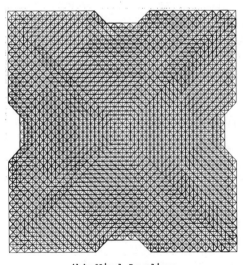

(b) Wind Loading

Figure 6. Gravity and Wind Load Force Distribution

In addition to the applied load cases an extreme case of settlement was also considered to examine the effect of possible soils liquefaction at the site. In this study two of the highly loaded footings (adjacent to the entrances) were removed and the new stress distribution was computed. The resulting peak stresses were shown to remain within the elastic levels.

Fabrication

Manufacture of this plate node space frame system utilizes a combination of conventional steel fabrication shop methods along with machine shop dimensional control practices. Dimensional control depends heavily on accurate fixtures to position the components during initial tack welding.

Overall dimensional tolerances for the pyramid are primarily determined by the chordal plane elements; the continuous face-plate of the node and the end to end length of the struts. Using typical tolerance controls for space frames of 0.5-mm per 1.8-M, the 3.6-M grid must maintain a total node-center to node-center fabrication tolerance of 1.0-mm. The node face-plate is made from one piece and hole location tolerances can be held to 0.10 to 0.20-mm, leaving the remaining tolerance of 0.80 to 0.90-mm for strut length.

The system bolts act in shear and bear on the surfaces of the holes. Thus, the finished hole diameters must be large enough to allow for mean fabrication and assembly tolerances, yet small enough to maintain accurate geometric configuration. A finished hole diameter 0.40-mm larger than the bolt accomplished this goal.

Tube cutting, attachment of hinges to tubes and node plate connections utilize conventional structural steel fabrication shop practices. Tube diameters of 114 to 168 mm allow for welding of the hinge elements on the inside of the tube. This makes for a visually simple, aesthetically appealing detail.

The strut and node assemblies are first tack welded into position in the fixture assembly. Tacked parts are removed from the fixture for final welding. While some minor distortion of the node plates does occur during this weld-up, the majority of movement is in the weak axis of the plate. This movement was analyzed and determined to be not critical for the dimensional control of the structure as a whole.

As mentioned earlier, square tubing is used where required in the outer chordal planes of the structure to facilitate the direct connection of a custom steel roof deck. This strategy saves the expense of a secondary support system as is often required in conventional one-way truss structures and ball node type space frames.

Struts are left open at the tube ends and hot dipped galvanized to provide superior corrosion protection inside and out. Node assemblies and hardware are also hot dipped galvanized. The natural galvanized coating was selected as the desired final finish by the architect and no color topcoat was applied.

Installation

The space frame is assembled on the ground into subsections of 3 to 6 metric tons. (Fig. 7) Erection of the structure is then accomplished by crane-lifting these subassemblies into position and fixing them to adjoining structure. Exposed, bolted bearing type connections allow for maximum ease of assembly and erection. The galvanized finish, as opposed to a paint finish, minimized material handling and touch-up considerations while providing superior corrosion protection.

A variety of secondary analysis was completed to support an optimized assembly and erection strategy. A level by level analysis of the 30 vertical levels of the frame determined anticipated construction deflections. Starting at base level 1, deflections peaked at level 10. (Fig. 8) By providing temporary bracing adjacent to the openings in the center of each side, peak deflections are minimizes and correct alignment of the structure maintained during the erection process. Once the structure height exceeds 12 levels, the temporary bracing is no longer needed.

Figure 7. Typical Space Frame Subassembly

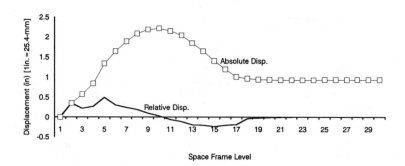

Figure 8. Sequential Displacement Analysis

Summary and Conclusions

The source of the cost disparity between space frame and conventional structural solutions is rooted in a fundamental difference of the two technologies; space frame designs represent a highly optimized, *information-based* building approach while conventional one-way truss designs typically represent a more *material-based* building practice. In today's marketplace materials are still relatively cheap and information is expensive.

It is important to recognize that the comparative costs of lightweight and conventional building technologies are very dynamic. Each passing day yields a world increasingly lean in material resources and rich in information. Environmental and energy forces are acting with constant pressure to force the cost of materials ever higher. Simultaneously, information processing technology combined with rapidly evolving CAD/CAM systems are driving the cost of information down. The consequences of this dynamic is that *process and material efficient* building systems are constantly becoming increasingly competitive with conventional *material intensive* building practices.

Acknowledgments

The space frame contract was awarded to ASI Advanced Structures Incorporated/Harmon Contract, W.S.A., Inc.; a joint venture. The project general contract was awarded to Nielsen Construction Company of San Diego, California. The project was designed by Hugh Gibbs & Donald Gibbs, Architects FAIA, of Long Beach, California. Special thanks to Mr. Arnold Garcia of ASI for preparation of graphic materials presented in this paper.

ROOF SPACE FRAME DESIGN OF THE
C. K. S. AIRPORT PASSENGER TERMINAL II

CHEN, YEN-PO[1] CHAO, HUNG-TAO[2]

Abstract

This article attempts to provide a description of the characteristics of the Second Terminal Airport. To comply with the functional requirement of the magnificent facade and spacious design of the building, high strength aluminum space frame with mirror effect coating is selected as the roof structural system, contributing to the prominent feature of the national gate. In this article, brief reports are furnished concerning the influential factors on the project, selection of the structural material, wind tunnel test, analysis, design and construction method of the space frame, expecting to share our experience and achievement with experts and scholars in the structural field. Any comments on this article will be appreciated.

1. Introduction

The Terminal Building is located in the west of the airport tower. The entire structure, including the

----

1. Chen, Yen-Po, Vice President, China Engineering Consultant, Inc., Taiwan, R.O.C

2. Chao, Hung-tao, Section Chief Architect Office, China Engineering Consultant, Inc., Taiwan, R.O.C.

carport, is roofed with large span space frame, forming a
continuous interior space measuring 198 M in length, 156
M in width and 34 M in height above ground. It is a
4-story building with one level basement, made of
steel reinforced concrete system with 18 M by 18 M
module. The construction is scheduled to be completed in
1996 (See Fig. 1).

2. Module and General Arrangement of spaceframe

The roof spaceframe of the Terminal Building is made of
strut members with two force members hinged together to
form a pyramid of 3.6 M x 3.6 M module with 2.5 M in
height (See Fig. 2) to comply with the design requirements
of the large space and shape of the building. These basic
pyramids are joined together to form a three-dimensional
indeterminate structural grid network with fair stiffness
and high degree of safety. The whole spaceframe structure
is composed of 26728 nos. of struts and 6441 nos of hubs,
supported on 50 nos of SRC columns. The total surface area
is about 39000㎡(See Fig. 3) .

3. Design Consideration

3.1 Geography and Climate

The Terminal Building is located in the coastal region
with high content of chlorine ion in the atmosphere.
Together with the high humidity in the air, it forms a
highly corrosive environment. Therefore the
requirements of corrosion resistant material are
strict and stringent.

3.2 Airport Operation and Maintenance

The Terminal building is the gate to our country and
each year tens of millions of passengers pass through
it. Therefore the frequency of maintenance work
required for the spaceframe has to be considered
during the design stage.

3.3 Selection of Shapes of Strut and Hub

Coping with the requirements of the shape of the
building, the Terminal Spaceframe is composed of
squared pyramid modules to form a plate grid network.
In selecting the shapes of the strut, the following

considerations were made.

(1) The strut can be easily fit and connected to the hub yet retain the required strength while conical transition can be used at both ends of the strut to reduce the connection area.

(2) It can be precambered as per design requirements.

(3) Plenty of sectional area for option.

(4) Aesthetically pleasing and able to match the design of the building.

In selecting the shape of the connecting hub the following points were considered.

(1) It can accommodate at least 12 nos of strut with the axial force of each strut passing through the center of the hub without introducing secondary stresses.

(2) Its strength must meet the design requirements.

(3) Struts can be bolt joined to the hub. This will make the assembly work simple and fast, thus reduces the time of assembly and installation.

(4) Aesthetically pleasing and able to match the design of the building.

3.4 Selection of Material

There are, in general, three kinds of materials suitable for spaceframe structures. They are stainless steel, structural steel and aluminum alloys. After comprehensive comparison among different materials by considering design requirements and characteristic of the Terminal spaceframe, the 6061-T6 extruded aluminum circular strut and solid spherical hub, with the strength of Fy=35 ksi, is selected as the material of the Terminal Spaceframe.

4. Analysis of Spaceframe

4.1 Geometrical Coordinates and numbering of members

The geometrical coordinates of each connection point

of the spaceframe are first established, and the struts are layed out and each given a number. The extent of the spaceframe is so huge and complicated that it will be very time-consuming to input all data into the computer by using softwares available in the market. In addition, errors are easily made and are difficult to be checked out. We have therefore developed a conversational preprocessing program which includes connection coordinates, numbering of strut members, cross section of members, boundary conditions, and properties of materials, etc. Data is input through a HP work station and transformed into COSMOS 6 analysis program input files.

4.2 Design Dead Load and Live Load

The roofing and cladding system above the spaceframe, including metal deck, glazing skylights, insulation (heat and sound), waterproofing system and supporting materials, weigh about 105 kg/m². This does not include the weight of spaceframe itself, air ducts, water pipes, drainage, gutters, and lighting fixtures, etc. whose weight is to be estimated according to the actual design layout. As for live load, Taiwan Building Regulation specifies that roof not designed for special purposes shall have a live load of 60 kg/m².

4.3 Design Wind Load

Due to the peculiar shape of the roof it is determined that besides computing the wind pressure conservatively using the specifications of (1) ANSI (2) "Design specifications for Wind Load in the Taiwan Region" by professor Tsai, Y.T. (3) "Building Technical Regulations" published by the Ministry of Interior, we have employed RWDI of Canada to perform a wind tunnel test with a 1:400 scale model of the Phase II Terminal Building. Results of the test were then compared to the calculated design wind load. Basic wind velocity data were obtained from the Taiwan Central Observatory which include the record of wind velocity of the Taiwan Region in the last 39 years and wind velocity of the CKS airport in the last 10 years. We have decided to use a 100-year return period, an average wind speed of 48.5 m/sec. and a gust velocity of 69.7 m/sec. to supplement the data obtained. Results of the test were analysed by statistical method and then compiled into wind Pressure Distribution Diagrams. These diagrams were then compared to the calculated wind loads. Finally they were summarized into design wind load diagrams which include 12 load cases considering all directions. The biggest wind suction force is 580 kg/ m² (See Fig. 4) and the biggest wind pressure is 260 kg/m². Generally speaking, the design wind load for the roof cladding is approximately 1.5 times that of the spaceframe structure.

4.4 Design Seismic Load

Since the Terminal Spaceframe is a large span, flexible and light weight structure, the calculated seismic load is much smaller in magnitude than the design wind load. However, there is a part of the spaceframe at the entrance of the carriageway on the departure level which has a cantilever of 18 M, and the effect of seismic force must not be neglected. Lateral seismic force for the spaceframe is H=0.135W, where W indicates the self weight of structure.

4.5 Thermal Load

The subtropical climate of Taiwan has noticeable difference in the weather of the four seasons. The spaceframe with an area of about 39000 m² has no expansion joint. This is to comply with the architectural requirements. The design thermal load is based on a reference temperature of 20 ℃ while the highest and lowest temperatures are 50 ℃ and 0 ℃ respectively. Because the whole Terminal Building is air-conditioned all year round, the temperature changes of the spaceframe are considered to be trapezoidal distributions.

4.6 Structural Analysis

Since the data are so huge in volume, besides using the conversational drawing program to establish input data file directly at the HP work station, some regular loadings such as live load, thermal load, etc. are subprogrammed to help developing input data after all the above basic data were checked and confirmed. They were transformed into input file of the structural analysis program COSMOS 6. Ordinarily buildings of beam and column construction can be easily designed by simple calculation (or sometimes even by experience) to obtain sections of members to a satisfactory degree. But the Terminal Spaceframe has as many as 26728 nos of struts arranged in a three – dimensional grid network system. Slight alteration of any strut section may cause load redistributions to the neighboring struts. Therefore, the structural analysis and the engineering must work alternately. Each strut is firstly given an assumed minimum section and the initial stress of all members is obtained through analysis. The design program is executed. Struts with stress in excess of their design allowables are recorded, and their cross section will be adjusted so that they are within limit. Then the analysis program is executed again, and the cross section adjusted again. This recurring process will continue until the stress of each and every strut is within the allowable stress limit. By this time, the structural analysis and the engineering work are finished at the same time. One point that must be reminded is that loadings used in the above recurring process are the combination of the basic loadings as per the specification requirements.

5. Design of Spaceframe

5.1 Strength and Properties of Material

Struts of the Terminal Spaceframe shall be 6061-T6 extruded aluminum circular pipes. All hubs except anchor hubs, are 6061-T6 extruded aluminum solid sphere. Anchor hubs shall be 6061-T6 extruded aluminum solid hemispheres. Connecting sleeves (collars) shall be 6061-T6 extruded aluminum. Yield tensile stress of 6061-T6 shall be Fty = 2460 kg/cm² (35 ksi), Elastic modulus E = 710,000 kg/cm² and coefficient of thermal expansion α = 2.3 x 10⁻⁶mm/mm/ ℃. Safety factors of all tensile and compressive members are between 1.65 and 1.95. Equations used for the design are listed in the "Specifications for Aluminum Structures by Aluminum Assocation, U.S.A." and will not be discussed any further here. 2014-T4 aluminum bolts or cadmium plated high strength steel bolts of A325 or A490 grade shall be used to join the struts to the hubs.

5.2 Load Combination

There are 93 cases of load combinations used for evaluating the max. stress of each strut and then

design their cross sections. These combinations are
established in accordance with the specification
requirements and have included all basic design
loadings such as dead and live loads, 12 cases of wind
loads, 4 cases of seismic loads, 2 cases of thermal
loads, etc.

5.3 Strut Size Summary

The Terminal Spaceframe will have 17 different sizes
of circular struts with diameters ranging from 7.5 cm
(3") to 25 cm (10"). About 99% of them are smaller
than 15 cm (6") in diameter. The remaining 1% are
larger struts located above the 50 supporting columns.
This is generally in line with the stress distribution
as computed by the structural analysis.

6. Fabrication and Installation

All fabrication works, including cutting and welding,
shall be done in the factory. All components are clearly
labeled with different codes for easy identification.
This will make the installation work easier and faster,
and thus effectively shorten the installation time.

During the preparation of shop drawings, it is important
that the manufacturer liaise closely with the designer to
solve any problem of detail design that may arise. This is
especially important for the design of the hubs which not
only have holes drilled for connection bolts, but also
must have holes prepared for hanging other facilities,
such as water pipes, electrical conduits, air ducts,
lighting fixtures, etc. After approval of the shop
drawings the manufacturer shall start their production. To
ensure the quality and performance of the component to
comply with design standards, the technical specifications
requires strength proof tests to be carried out besides
the usual random quality control test. The strength proof
test will verify that the allowable stress of the member
are in accordance with the design values.

After the material is delivered to the job site, the
contractor shall carry out installation works according to
the installation method previously prepared to suit site
working conditions. The operation of space frame elements,
pre-assembled in blocks, duly arranged in order on ground
and then lifted and joined partially through manual
approach, is considered as the best performance.

7. Conclusion

The phase II terminal building of the CKS International Airport will be "the Gate to the Republic of China". It is one of the most important construction projects of the country. Needless to say, the quality of the project must be kept at the highest standards and all design requirements must be strictly enforced. Although aluminum spaceframe structures are rather common in the rest of the world, it is the largest aluminum spaceframe to be built in Taiwan.

It is our pleasure to share this incredible experience with the people all over the world who are in the field of building constructions.

8. Appendix

Engineering Conversion

1 MPa. = 0.145 ksi
1 kPa. = 0.0102 kg/cm² = 102 kg/m²

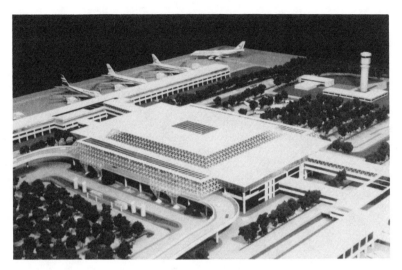

Figure 1. CKS International Airport Passenger Terminal

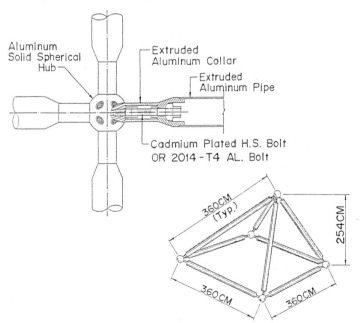

Figure 2. Space Frame Elements and Module

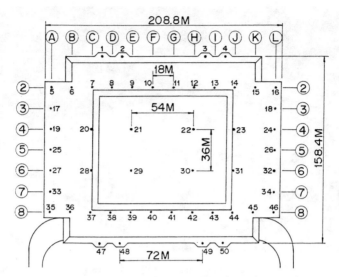

Figure 3. Layout of Space Frame Supporting Columns

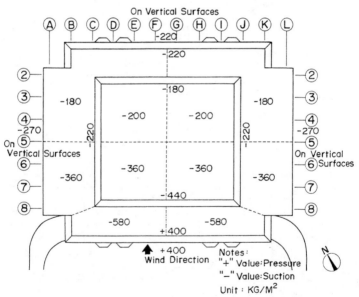

Figure 4. Wind Pressure Distribution Diagram

ANALYSIS OF GLULAM-FRP BEAMS
BY A LAYER-WISE FORMULATION

Julio F. Davalos[1], A.M. ASCE,   Youngchan Kim[2], S.M. ASCE,
and Somnath S. Sonti[3], S.M. ASCE

Abstract

For applications in large-span roof structures, the stiffness and strength of glued-laminated timber beams (Glulam) can be efficiently increased by reinforcing the members with fiber-reinforced plastic (FRP) composites. To analyze Glulam-FRP sandwich beams, the formulation of a laminated beam finite element with layer-wise constant shear (BLCS) is presented. The BLCS element can accurately predict the response of soft-core beams. The constant shear stress on each layer is approximated by a quadratic function in a post-processing operation. The accuracy of BLCS is validated using experimental results for twelve, 20-ft long, yellow-poplar Glulam-FRP beams tested in bending.

Introduction

Glued-laminated timber beams (Glulam) are used extensively for large-span bridges (Davalos et al. 1992) and roof structures, such as frames, arches, and reticulated domes (Holzer et al. 1990). The advantages of Glulam are economical production of tapered and curved members, excellent energy absorption characteristics, high chemical and corrosion resistance, and better fire resistance than steel. However, due to the relatively low bending stiffness and strength of Glulam, the construction of large structures usually requires Glulam members of large depths, which in turn

---

[1]Assistant Prof. of Civil Engineering, Constructed Facilities Center (CFC),
 West Virginia University (WVU), Morgantown, WV 26506-6101

[2]Graduate Research Assistant, Civil Engineering, CFC, WVU

[3]Research Engineer, Civil Engineering, CFC, WVU

requires bracing of the members to prevent lateral buckling.   To significantly
increase the stiffness and strength of Glulam, the members can be reinforced with
fiber-reinforced plastic (FRP) composites at top and bottom surfaces. Since the
stiffness of the FRP reinforcement can be four to ten times greater than that of
wood, Glulam-FRP laminates are sandwich beams with relatively soft cores, for
which the common beam theory assumption of plane cross sections may not apply.

Based on the Generalized Laminate Plate Theory (GLPT; see Reddy 1987), we
present in this paper the formulation and verification of a beam finite element with
layer-wise constant shear (BLCS), which is equivalent to a first order shear
deformation beam theory (Timoshenko beam) on each layer. While retaining the
simplicity of a beam theory, the BLCS element gives results as accurate as much
more complex three-dimensional elasticity analyses.   The layer-wise linear
representation of in-plane displacements permits accurate computation of normal and
shear stresses on each layer for laminated beams with dissimilar ply stiffnesses, such
as Glulam-FRP beams.   For the accurate computation of inter- and intra-laminar
shear stresses, the constant shear stress on each layer is approximated by a quadratic
function in a post-processing operation.   The displacements are formulated with
respect to an arbitrary reference axis, and therefore, BLCS is basically a one-
dimensional element suitable to model complex frame-type structures. The BLCS
model predicts the linear response of rectangular laminated beams with symmetric
or asymmetric laminates.   The model predictions are correlated with the
experimental response of small soft-core sandwich beams and twelve 20-ft long
yellow-poplar glulam beams reinforced with pultruded glass-fiber-
reinforced/vinylester plastic (FRP) composite.

Modeling of Laminated Beams

Elasticity solutions, although limited in scope and applications, have been developed
for the analysis of laminated beams (Rao and Ghosh 1979, Cheng et al. 1989). As
shear deformation is not negligible in composite materials, a number of laminated
beam models using first-order or higher-order shear deformation theories have been
presented (Yuan and Miller 1989, Kant and Manjunath 1989). In this study, a beam
theory with layer-wise linear representation of in-plane displacements is presented,
which is equivalent to a first-order shear deformation theory (Timoshenko beam) on
each layer.

Layer-Wise Formulation: The kinematic assumptions used in the present beam
finite element with layer-wise constant shear (BLCS) are transverse incompressibility
and layer-wise representation of in-plane displacements (see Reddy et al. 1989).
Using linear interpolation functions through the thickness, the displacements of a
point (x-z plane) in the laminated beam is expressed as

$$u_1(x, z) = u(x) + \sum_{j=1}^{n} U^j(x)\phi^j(z), \quad u_2(x, z) = w(x) \tag{1}$$

where, u and w are, respectively, the longitudinal and transverse displacements of a point on the reference axis of the laminate, and $U^j(x)$ are in-plane displacements. The transformed stress-strain relation of an orthotropic lamina, under the assumption of plane stress in the x-y plane and without the transverse normal stress component (Jones 1975), can be written as

$$\begin{Bmatrix} \sigma_x \\ \sigma_y \\ \sigma_{xy} \\ \sigma_{yz} \\ \sigma_{xz} \end{Bmatrix} = \begin{bmatrix} \overline{Q}_{11} & \overline{Q}_{12} & \overline{Q}_{16} & 0 & 0 \\ \overline{Q}_{12} & \overline{Q}_{22} & \overline{Q}_{26} & 0 & 0 \\ \overline{Q}_{16} & \overline{Q}_{26} & \overline{Q}_{66} & 0 & 0 \\ 0 & 0 & 0 & \overline{Q}_{44} & \overline{Q}_{45} \\ 0 & 0 & 0 & \overline{Q}_{45} & \overline{Q}_{55} \end{bmatrix} \begin{Bmatrix} e_x \\ e_y \\ \gamma_{xy} \\ \gamma_{yz} \\ \gamma_{xz} \end{Bmatrix} \tag{2}$$

where, $\overline{Q}_{ij}$ are the transformed reduced stiffnesses. The state of stress in each lamina is approximated as follows: $\sigma_y = \sigma_{yz} = 0$, and, for a cross-ply lamina, $\sigma_{xy} = 0$. Imposing these conditions in Eq. (2), we obtain

$$\sigma_x = E_x e_x, \quad \sigma_{xz} = G_{xz} \gamma_{xz} \tag{3}$$

where

$$E_x = \overline{Q}_{11} + \overline{Q}_{12} \frac{\overline{Q}_{16}\overline{Q}_{26} - \overline{Q}_{12}\overline{Q}_{66}}{\overline{Q}_{22}\overline{Q}_{66} - \overline{Q}_{26}\overline{Q}_{26}} + \overline{Q}_{16} \frac{\overline{Q}_{12}\overline{Q}_{26} - \overline{Q}_{16}\overline{Q}_{22}}{\overline{Q}_{22}\overline{Q}_{66} - \overline{Q}_{26}\overline{Q}_{26}}, \quad G_{xz} = -\frac{\overline{Q}_{45}^2}{\overline{Q}_{44}} + \overline{Q}_{55}$$

The constitutive equation of the laminate is derived by integrating stresses through the thickness and substituting Eq. (3) into the resultant forces. In the finite element formulation of BLCS, the strain-displacement relation of m-node element is defined as

$$\{e^o\} = [B_L]\{\Delta^o\}, \quad \{e^j\} = [\overline{B}_L]\{\Delta^j\}, \quad \{\Delta^o\}^T = \{u_1 w_1 \ldots u_m w_m\}, \quad \{\Delta^j\} = \{U^j\} \tag{4}$$

where, the compatibility matrices $[B_L]$ and $[\overline{B}_L]$ can be expressed in terms of the interpolation functions H_i as

$$[B_L] \atop (2x2m) = \begin{bmatrix} \dfrac{\partial H_1}{\partial x} & 0 & \dfrac{\partial H_2}{\partial x} & 0 & .... & \dfrac{\partial H_m}{\partial x} & 0 \\[2ex] 0 & \dfrac{\partial H_1}{\partial x} & 0 & \dfrac{\partial H_2}{\partial x} & .... & 0 & \dfrac{\partial H_m}{\partial x} \end{bmatrix}$$

$$[\bar{B}_L] \atop (2xm) = \begin{bmatrix} \dfrac{\partial H_1}{\partial x} & \dfrac{\partial H_2}{\partial x} & .... & \dfrac{\partial H_m}{\partial x} \\[2ex] H_1 & H_2 & .... & H_m \end{bmatrix}$$

Using Eq. (4) and the constitutive equation of the laminate, and applying the principle of virtual work to the equilibrium equation, we obtain the element model as follows:

$$\begin{bmatrix} [A_1] & [B_1] & [B_2] & .... & [B_N] \\ [B_1]^T & [D_{11}] & [D_{12}] & .... & [D_{1N}] \\ . & . & . & .... & . \\ [B_N]^T & [D_{N1}] & [D_{N2}] & .... & [D_{NN}] \end{bmatrix} \begin{Bmatrix} \{\Delta^0\} \\ \{\Delta^1\} \\ . \\ \{\Delta^N\} \end{Bmatrix} = \frac{1}{b} \begin{Bmatrix} \{F\} \\ \{F_x^1\} \\ . \\ \{F_x^N\} \end{Bmatrix} \tag{5}$$

where, b is the beam width, $\{F\}$ includes transverse (f_z) and axial (f_x) force vectors applied at the reference axis, $\{F_x^j\}$ contains axial (f_x^j) force vectors applied at the laminate interfaces, and the submatrices are defined as

$$[A_1] = \int_0^L [B_L]^T [A][B_L] dx$$

$$[B_i] = \int_0^L [B_L]^T [B^i][\bar{B}_L] dx$$

$$[D_{ij}] = \int_0^L [\bar{B}_L]^T [D^{ij}][\bar{B}_L] dx$$

The dimensions of $[A_1]$, $[B_i]$ and $[D_{ij}]$ are (2m x 2m), (2m x m), and (m x m) respectively. Therefore, the total degrees of freedom per element are $2m+mN$. Since a 2-node linear element can only model constant moment, a 3-node quadratic isoparametric element is used in this study. Also, to avoid "shear locking" a 2-point Gauss integration along the reference axis is adopted (see Cook et al. 1989).

Computation of parabolic shear stress: The constant shear stress $\sigma^{(i)}_{xz}$, computed from the constitutive equation after recovery of displacements in the ith layer, is interpolated using quadratic functions to obtain a parabolic distribution as follows:

$$\sigma_{xz}^{(i)}(\bar{z}) = \sum_{j=1}^{3} \phi_j(\bar{z}) \sigma_j^{(i)} \qquad (6)$$

where $\bar{z}$ is a non-dimensional local coordinate with origin at the bottom surface of the ith layer ; $\sigma_1^{(i)}$, $\sigma_2^{(i)}$, and $\sigma_3^{(i)}$ are, respectively, the shear stresses at the bottom, middle, and top of the ith layer; and ϕ_j are the second order Lagrange polynomials. As mentioned in Chaudhuri and Seide (1987) and Reddy et al. (1989), the required number of equations can be derived from the following four conditions : (1) N equations are used to equate the average shear stress with the constant shear stress from constitutive equation on each layer; (2) two equations are used to impose zero-shear condition at the surfaces of the beam; (3) N-1 equations are used to satisfy continuity of shear stresses at the interfaces, and (4) N-1 equations are used to define the slope discontinuities $\sigma_{xz,z}$ at each interface. Using Eq. (6), the condition (1) yields

$$\sigma_{xz}^{(i)} = \frac{1}{6}[\sigma_1^{(i)} + 4\sigma_2^{(i)} + \sigma_3^{(i)}] \qquad (7)$$

The shear-free conditions on the top and bottom surfaces of the beam lead to

$$\sigma_1^{(1)} = \sigma_3^{(N)} = 0 \qquad (8)$$

At the interfaces the continuity of shear stress satisfies the condition

$$\sigma_3^{(i)} - \sigma_1^{(i+1)} = 0 \qquad (9)$$

Finally, the slope discontinuity of the shear stress at an interface becomes

$$\sigma_{xz,z}^{(i+1)} - \sigma_{xz,z}^{(i)} = \frac{1}{h^i}[-\sigma_1^{(i)} + 4\sigma_2^{(i)} - 3\sigma_3^{(i)}] + \frac{1}{h^{i+1}}[-3\sigma_1^{(i+1)} + 4\sigma_2^{(i+1)} - \sigma_3^{(i+1)}]$$

From the equilibrium equation and neglecting body forces, the variation of the shear stress can be equated to the variation of the normal stress as

$$\sigma_{xz,z} = -\sigma_{xx,x} \qquad (11)$$

By arranging Eqs.(7) through (11) with respect to unknowns $\sigma_j^{(i)}$, 3N simultaneous equations can be obtained, whose solution gives the parabolic distribution of shear stress. For convenience, the stresses calculated at the Gauss points are extrapolated to the nodes using linear interpolation functions, as suggested by Cook et al. (1989). Further details of the element formulation is given by Kim et al. (1993). The accuracy of BLCS is evaluated by predicting the response of sandwich beams and large-scale Glulam beams reinforced with fiber-reinforced plastic (FRP) composite.

Numerical Examples

1. Sandwich beams: Using photoelasticity, Kemmochi and Uemura(1980) measured the stress distribution in sandwich beams, which were tested in bending under symmetric two-point loading (Fig. 1a). The laminate analyzed in this paper is "model 4", which has a very low core stiffness compared to face stiffness. The material properties of the face and core layers given in MPa are as follows: E_{face} = 2440, G_{face} = 875, E_{core} = 4.34, G_{core} = 0.53, and the beam width is 6.57 mm. Kemmochi and Uemura predicted the response of the test-beams by a multi-layer built-up theory. In BLCS, 16 elements with eight layers is used to compute the stresses at sections s-s (constant shear) and m-m (maximum moment). In Fig. 1b, the BLCS and experimental normal and shear stress distributions are compared at section s-s, and in Fig. 1c, the BLCS and multi-layer built-up theory normal stress predictions are compared to experimental measurements at section m-m. The BLCS predictions agree closely with the experimental results.

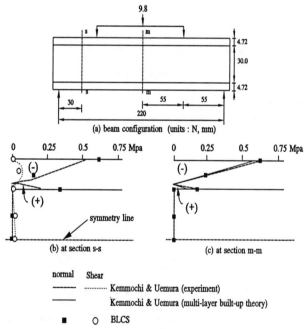

Figure 1.   Sandwich beam configuration and stress comparison

2. Glulam-FRP beams: The linear response of Glulam beams reinforced with FRP are presented here. The beams were tested under 2-point bending, and displacements (at midspan and loading points) and strains (at midspan) were recorded with a data acquisition system. Each 5.8-m beam consisted of ten wood layers and two FRP layers (Fig. 2). The average material properties of the wood and FRP layers given in GPa are: $E_{wood/face} = 13.5$, $E_{wood/core} = 11.3$, $G_{wood} = 0.732$, $E_{FRP} = 41.6$, and $G_{FRP} = 7.0$. The load-deflection curves showed almost linear response up to failure, and for the linear analysis the experimental data of up to one half of the failure load were used. As shown in Table 1, the BLCS predictions of displacement and strains are very close to the experimental values. By adding two layers of 4.77 mm thick FRP (3% by volume of the beam), the increase in stiffness was approximately 20%.

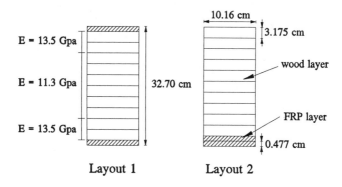

Figure 2. Beam section lay-up

Summary and Conclusion

A relatively simple laminated beam finite element with layer-wise constant shear (BLCS) is used for the analysis of Glulam-FRP beams, which, depending on the amount of FRP reinforcement, can be treated as soft-core sandwich beams. The comparisons presented in this paper show that BLCS can accurately predict the experimental response of large-scale Glulam-FRP beams. The laminates are analyzed with a 3-node BLCS element, which assumes transverse incompressibility and layer-wise linear distribution of in-plane displacements. BLCS is essentially a one-dimensional N-layer element with $2+N$ degrees of freedom per node. The layer-wise constant shear stresses obtained from constitutive relations are transformed into parabolic distributions, and for convenience, the stresses computed at the Gauss integration points are extrapolated to the nodes using interpolation functions. BLCS can be used for the analysis of soft-core laminated frame-type structures.

Table 1. Comparison of displacement, strains, and stiffness

Reinforcement Layout	Displacement Ratio (BLCS/Exp.)	Strain Ratio (BLCS/Exp.)		Stiffness Increase $(E_{exp.}/E_{wood\ alone})$
		Top	Bottom	
Bottom Reinforcement	0.99	0.99	1.11	1.16
	0.93	1.16	1.02	1.23
	0.99	-	1.10	1.15
	1.01	1.07	1.13	1.16
	0.97	1.10	1.11	1.20
Top & Bottom Reinforcement	1.02	0.85	1.10	1.14
	0.93	1.04	1.05	1.24
	0.99	1.05	1.07	1.17
	1.00	1.04	0.94	1.17
	0.88	1.11	1.13	1.33
	0.95	1.07	1.04	1.22

Acknowledgements

The experimental program of Glulam-FRP beams is partially supported by the USDA-FS-FPL grant FP-92-1845. The authors thank the participation of Russell C. Moody and Roland Hernandez of FPL.

References

Chaudhuri, R. A. and Seide, P. (1987). "An approximate semi-analytical method for prediction of interlaminar shear stresses in an arbitrarily laminated thick plate," **Computers and Structures**, 25(4), 627-636.

Chen, S., Wei, X., and Jiang, T. (1989). "Stress distribution and deformation of adhesive-bonded laminated composite beams," **Journal of Engineering Mechanics**, 115(6), 1150-1162.

Cook, R. D., Malkus, D. S., and Plesha, M. E. (1989). *Concepts and Application of Finite Element Analysis*, 3rd Ed., John Wiley & Sons, New York.

Davalos, J. F., Salim, H. A., and Munipalle, U. (1992). "Glulam-GFRP composite beams for stress-laminated T-system timber bridges," **1st International Conference on Advanced Composite Materials in Bridges and Structures**, CSCE-CGC, Sherbrooke (Quebec), Canada, 455-463.

Holzer, S. M., Davalos, J. F., and Huang, C. Y. (1991). "A review of finite element stability investigation of spatial wood structures," **Bulletin of the International Association of Shell and Spatial Structures**, 31(3), 161-171.

Jones, R. M. (1975). *Mechanics of Composite Materials*, Scripta Book Co., Washington, DC.

Kim, Y., Davalos, J. F., and Barbero, E. J. (1993). "A composite beam element with layer-wise plane sections," **Journal of Engineering Mechanics**, to appear.

Kemmochi, K. and Uemura, M. (1980). "Measurement of stress distribution in sandwich beams under four-point bending," **Experimental Mechanics**, 20, 80-86.

Kant, T. and Manjunath, B. S. (1989). "Refined theories for composite and sandwich beams with Co finite elements," **Computers and Structures**, 33(3), 755-764.

Rao, K. M. and Ghosh, B. G. (1979). "Exact analysis of unsymmetric laminated beam," **Journal of The Structural Division**, ASCE, 105(11), 2313-2325.

Reddy, J. N. (1987). "A generalization of two-dimensional theories of laminated composite plates," **Communication of Applied Numerical Method**, 3, 173-180.

Reddy, J. N., Barbero, E. J., and Teply, J. L. (1989). "A plate bending element based on a generalized laminate plate theory," **International Journal for Numerical Methods in Engineering**, 28, 2275-2292.

Yuan, F. G. and Miller, R. E. (1989). "A new finite element for laminated composite beams," **Computers and Structures**, 14, 125-150.

LRFD vs. ASD FOR WOOD STRUCTURES

David G. Pollock[1], M. ASCE
Thomas G. Williamson[2], F. ASCE

ABSTRACT

There is a growing preference within the structural engineering community for the adoption of reliability-based load and resistance factor design (LRFD) procedures in lieu of traditional allowable stress design (ASD) procedures. The LRFD format introduces new terminology and nomenclature, and provides a consistent methodology for quantitative assessment of structural safety and reliability. Comparisons of ASD and LRFD procedures for design of wood structures, based on provisions of the *National Design Specification® for Wood Construction* (NDS®) and the draft *LRFD Specification for Engineered Wood Construction*, indicate that required sizes of structural members and connections will change only slightly ($\pm$ approximately 10%) with the adoption of new LRFD procedures.

INTRODUCTION

The development of reliability-based design (RBD) procedures to replace traditional allowable stress design procedures (ASD) in the field of structural engineering has gained significant momentum in recent years. Within the United States, a load and resistance factor design (LRFD) format has been chosen for the presentation of RBD concepts in specifications and standards governing the design of bridges (1) and buildings (7) constructed of steel (4,5), reinforced concrete (2) and wood (6). RBD procedures have already been adopted in Canada in a similar

---

[1] Director of Engineering, American Forest & Paper Association, Washington, DC

[2] Executive Vice President, APA-American Wood Systems, Tacoma, WA

"limit states design" (LSD) format, and have been proposed for adoption in Europe and Pacific Rim nations.

The primary motivation for switching from ASD to LRFD procedures is the ability within LRFD to more explicitly quantify the reliability inherent in design procedures based on statistical analyses of material strength properties as related to maximum anticipated design loads. LRFD procedures facilitate the establishment of consistent levels of safety for all components within a structure, regardless of construction material.

STRUCTURAL SAFETY

A primary difference between ASD and LRFD procedures is the manner in which safety provisions are incorporated in design checking equations. In ASD procedures, the designer must ensure that published "allowable stresses" exceed the induced stresses from maximum loads expected to occur during the life of a structure. This can be represented conceptually by the following general equation:

$$R_{ASD} \geq \Sigma Q_i \qquad \text{[Eq.1]}$$

in which R_{ASD} =   strength or capacity of a structural member or connection based on allowable design values
ΣQ_i =   combined effect of maximum design loads (Q_i)

Structural safety is attained in ASD procedures by applying safety adjustments to material strength properties in the determination of published "allowable stresses." For example, safety adjustments considered in the development of allowable stresses for wood products are typically provided in the following two steps:

1. Determine a lower 5th percentile strength value based on statistical analysis of material strength test data (10,11), and then
2. Apply a reduction factor to account for additional safety considerations, and to adjust from a 5-10 minutes test duration basis to a load duration basis of 10 years (9,10).

Thus, while acceptable levels of safety are obtained using ASD procedures, the magnitudes of applied safety adjustments are typically unknown to the designer since they are implicit in the published allowable design values.

In LRFD procedures, overall safety is attained by applying load factors ($\gamma \geq 1.0$) to code-specified design loads and resistance factors ($\phi < 1.0$) to material strength properties, as indicated by the following general equation:

$$\phi R_{LRFD} \geq \Sigma \gamma_i Q_i \qquad \text{[Eq.2]}$$

in which R_{LRFD} = resistance of a structural member or connection

ϕ = resistance factor

$\Sigma\gamma_i Q_i$ = combined effect of factored design loads

γ_i = load factor associated with maximum design load Q_i

As with ASD design values, published LRFD resistances for wood products are also based on 5th percentile strength properties. However, additional safety is then provided through the explicit application of resistance factors and load factors in design checking equations. The magnitude of each resistance factor is based on assessments of variability for specific material strength properties (16). A reliability normalization factor (K_R) is also applied in the determination of reference resistance values to ensure consistent reliability across a wide range of structural wood products (12). The magnitude of each load factor is based on the probability of occurrence of maximum design loads during the anticipated life of the structure, and on the probability of simultaneous occurrence of multiple maximum loads in a combination of design loads. Thus, design load magnitudes, load factors and load combinations can be determined based on building occupancy, usage and geographic location of various building types, and are equally applicable to all construction materials (7).

LOAD DURATION EFFECTS

Although many structural materials exhibit viscoelastic behavior due to time under load, this effect is more pronounced in materials such as wood, glass and plastic. Explicit recognition of this time dependent behavior has historically been provided by testing wood products and assemblies to failure within 5-10 minutes (8), reducing the resulting strength properties to the strength associated with a load duration of 10 years (9), and then providing load duration factors (C_D) for designers to adjust to the higher material strengths associated with shorter duration design loads (3).

In LRFD procedures, published resistances are based directly on test data for load durations of 5-10 minutes, and time effect factors (λ) are provided for designers to adjust to the lower resistances associated with longer duration design loads. A comparison of ASD load duration factors and LRFD time effect factors is provided in Table 1.

In addition to the different load duration basis for published LRFD resistances and ASD design values, two further changes from traditional practice may be observed. First, historical ASD procedures specify separate load duration factors for floor live loads ($C_D = 1.0$), snow loads ($C_D = 1.15$) and roof live (construction) loads ($C_D = 1.25$) based on estimated cumulative load durations of 10 years, 2 months and 7 days, respectively. However, recent research indicates that durations of maximum design live loads for floors and roofs will not exceed 1-2 weeks during the anticipated life of a structure (14,18). As a result, a single time effect factor (λ = 0.8) has been specified for floor live loads, snow loads and roof live loads in LRFD procedures based on probabilistic analyses of load histories, load durations and load combinations, coupled with stochastic analyses of material properties, A similar change has been proposed for the next (1996) edition of the *National Design Specification®* *for Wood Construction* (NDS®), in order to provide a common load duration basis for ASD and LRFD procedures.

Secondly, the time effect factor (λ = 1.0) for wind and seismic loads in LRFD parallels the 5-10 minutes load duration factor (C_D = 1.6) currently specified in ASD procedures (3). Recent research indicates that durations of peak loading during earthquakes and extreme wind events are typically recorded in seconds, leading to cumulative durations of no more than a few minutes during the expected life of a structure (13). The adoption of a short term (5-10 minutes) load duration basis for wind and seismic loads is thus consistent with recorded data, and provides consistency with international codes and standards.

TABLE 1 - ASD Load Duration Factors and LRFD Time Effect Factors

Typical Design Loads	ASD Load Duration Factors, C_D	LRFD Time Effect Factors, λ
Permanent (Dead) Load	0.9	0.6
Storage Live Load	1.0	0.7
Occupancy Live Load	1.0	0.8
Snow Load	1.15	0.8
Roof Live Load	1.25	0.8
Wind Load	1.6	1.0
Seismic Load	1.6	1.0
Impact Load	2.0	1.25

PUBLISHED CAPACITIES

When comparing LRFD and ASD procedures, probably the most noticeable distinction is that published LRFD resistances are of significantly higher magnitude than published ASD capacities. This difference in magnitude is attributable to:

1. the load duration basis for published LRFD resistances (5-10 minutes duration) vs. that for published ASD design values (10 years duration), and
2. the fact that safety adjustments are explicitly applied by the designer in LRFD design checking equations, while safety adjustments have inherently been applied to published ASD design values.

Since designers apply appropriate time effect factors (λ), resistance factors (ϕ) and load factors (γ) in LRFD design checking equations, overall levels of safety are actually quite similar for wood structures designed using LRFD or ASD procedures, as indicated by the following illustration (15):

> For a typical structural wood bending member supporting a combination of dead load and snow load, a reduction factor of 2.1 is applied to the 5th percentile bending strength (R_{05}) to determine the ASD bending design value (9,10), and a load duration factor ($C_D = 1.15$) is utilized in the design checking equation (3). The LRFD moment resistance for the same member is also based on the 5th percentile bending strength (R_{05}), with a reliability normalization factor (K_R) correlated to variability in bending strength (12). For example, $K_R = 1.1$ for products with a coefficient of variation (CV) in bending strength of approximately 25%. In addition, a resistance factor ($\phi = 0.85$), time effect factor ($\lambda = 0.8$), and load factors for snow load ($\gamma = 1.6$) and dead load ($\gamma = 1.2$) are applied in LRFD design checking equations as indicated below (6):

ASD

LRFD

$C_D R_{ASD} \geq \Sigma Q_i$

$\lambda \phi R_{LRFD} \geq \Sigma \gamma_i Q_i$

$(1.15)(R_{05}/2.1) \geq D + S$

$(0.8)(0.85)(1.1)R_{05} \geq 1.2D + 1.6S$

$R_{05} \geq 1.83D + 1.83S$

$R_{05} \geq 1.60D + 2.14S$

for S = 3D:

$R_{05} \geq 1.83D + (1.83)(3D)$

$R_{05} \geq 1.60D + (2.14)(3D)$

$R_{05} \geq 7.3D$

$R_{05} \geq 8.0D$

Thus, for the given load combination and variability in bending strength, overall safety levels for ASD and LRFD procedures are within approximately 9%. For the same load combination and a product with CV = 18%, overall safety levels for ASD and LRFD procedures are within 1%. Similar analyses for other load combinations and variability in material strength indicate that overall safety levels for ASD and LRFD wood design procedures are typically within $\pm$ approximately 10%.

DESIGN IMPACTS

Basic behavioral equations for structural wood members and connections are identical in ASD and LRFD specifications (3,6). Thus, while designers must familiarize themselves with different nomenclatures and formats for design checking equations, the resulting designs from both ASD and LRFD procedures are very similar, as indicated in the following examples:

1. Determine the cross-sectional dimensions for a laterally braced 24F-V4 Douglas fir glued laminated timber roof beam spanning 6.1 m (20 ft), and supporting a dead load (w_D) of 1460 N/m (100 plf) and a roof live load (w_L) of 2920 N/m (200 plf).

$$F_b = 16,550 \text{ kN/m}^2 \text{ (2400 psi)}$$

$$R_{LRFD} = 42,060 \text{ kN/m}^2 \text{ (6100 psi)}$$

Determine the section modulus required by ASD procedures (S_{ASD}):

$$(C_D)(F_b)(S_{ASD}) = (w_D + w_L)(L)^2/8$$

$$(1.25)(16,550,000)(S_{ASD}) = (1460 + 2920)(6.1)^2/8$$

$$S_{ASD} = 985 \text{ cm}^3 \text{ (60.0 in}^3)$$

Use a 79mm x 305mm (3-1/8" x 12") 24F-V4 Douglas fir glued laminated timber beam, with S = 1230 cm^3 (75.0 in^3)

Determine the section modulus required by LRFD procedures (S_{LRFD}):

$$(\lambda)(\phi)(R_{LRFD})(S_{LRFD}) = (1.2w_D + 1.6w_L)(L)^2/8$$

$$(0.8)(0.85)(42,060,000)(S_{LRFD}) = [(1.2 \times 1460) + (1.6 \times 2920)](6.1)^2/8$$

$$S_{LRFD} = 1040 \text{ cm}^3 \text{ (63.6 in}^3)$$

Use a 79mm x 305mm (3-1/8" x 12") 24F-V4 Douglas fir glued laminated timber beam, with S = 1230 cm^3 (75.0 in^3)

Although the required section modulus is approximately 6% larger using LRFD procedures, the same size beam would be specified using ASD or LRFD procedures. Similar analyses show that a 79mm x 343mm (3-1/8" x 13-1/2") 24F-V4 Douglas fir glued laminated timber beam would be required to support a snow load of 4380 N/m (300 plf) and dead load of 1460 N/m (100 plf) over a 6.1 m (20 ft) span, using either ASD or LRFD procedures (15,19).

2. Determine the cross-sectional dimensions for a southern pine No.2 post supporting a residential deck with a dead load (P_D) of 5340 N (1200 lb), a live load (P_L) of 21,350 N (4800 lb) and a height of 2.44 m (8 ft).

Assume wet service conditions since the post will be in contact with soil. Start with design values for a 89mm x 89mm (4x4) cross section:

$$E * C_M = 9930 \text{ MN/m}^2 \text{ (1440 ksi)}$$

$$F_c * C_M = 9100 \text{ kN/m}^2 \text{ (1320 psi)}$$

$$R_{LRFD} * C_M = 21,840 \text{ kN/m}^2 \text{ (3170 psi)}$$

Adjusting for column lateral stability:

$$F_c' = 3520 \text{ kN/m}^2 \text{ (510 psi)}$$

$$R_{LRFD}' = 7040 \text{ kN/m}^2 \text{ (1020 psi)}$$

Determine the cross-sectional area required by ASD procedures (A_{ASD}):

$$(C_D)(F_c')(A_{ASD}) = P_D + P_L$$

$$(1.0)(3520)(A_{ASD}) = 5340 + 21,350$$

$$A_{ASD} = 75.9 \text{ cm}^2 \text{ (11.8 in}^2\text{)}$$

Use a 89mm x 89mm (4x4) southern pine No.2 post, with A = 79.0 cm^2 (12.25 in^2).

Determine the cross-sectional area required by LRFD procedures (A_{LRFD}):

$$(\lambda)(\phi)(R_{LRFD}')(A_{LRFD}) = 1.2 \, P_D + 1.6 \, P_L$$

$$(0.8)(0.9)(7,040,000)(A_{LRFD}) = (1.2 \text{x} 5340) + (1.6 \text{x} 21,350)$$

$$A_{LRFD} = 80.1 \text{ cm}^2 \text{ (12.4 in}^2\text{)}$$

A 89mm x 89mm (4x4) southern pine No.2 post, with A = 79.0 cm^2 (12.25 in^2) is not sufficient. Use a 89mm x 114mm (4x5) or larger southern pine No.2 post, with A $\geq$ 102 cm^2 (15.75 in^2).

The required cross-sectional area is approximately 5% larger using LRFD procedures, resulting in a slightly larger post to carry the specified design loads. However, for a slightly smaller design load or a slightly lower deck height, a 89mm x 89mm (4x4) post would be sufficient using either ASD or LRFD procedures.

Similar comparisons of ASD and LRFD procedures for connection design in wood structures yield the following general conclusions (17):

TABLE 2 - Comparison of Connection Capacities in LRFD vs. ASD

Connections Supporting	Conclusion
Dead Load plus Floor Live Load	LRFD is 9-15% less conservative than ASD
Dead Load plus Snow Load	LRFD is within ± 5% of ASD
Dead Load plus Roof Live Load	LRFD is 1-8% more conservative than ASD
Wind Uplift minus Dead Load	LRFD is within ± 5% of ASD
Wind Shear Load	LRFD is 4% less conservative than ASD

CONCLUSIONS

Rather than introducing radical changes to the traditional safety levels associated with ASD procedures, the adoption of LRFD procedures provides an opportunity to "smooth" reliability over the broad spectrum of structural components and assemblies which make up wood construction.   The fact that LRFD-based member and connection capacities are typically within ± 10% of member and connection capacities using traditional ASD procedures is an indication that historical safety levels are adequate, and will continue to be maintained using both design formats.   However, the consistent levels of reliability provided by LRFD procedures for all assemblies in a given structure permit the realization of some cost savings for systems which may have been slightly overdesigned historically, while requiring a higher level of safety (consistent with safety levels throughout the structure) for systems which may have been slightly underdesigned.   The LRFD methodology also provides a platform for quantitative improvements in design procedures as additional data is collected and analyzed regarding material strength properties and load histories for various types of structures.

REFERENCES

1.   American Association of State Highway and Transportation Officials (AASHTO). *Standard LRFD Specifications for Highway Bridges* (draft). Washington, DC.

2.   American Concrete Institute (ACI). 1989. ACI 318-89 - *Building Code Requirements for Reinforced Concrete*. Detroit, Michigan.

3.   American Forest & Paper Association (AF&PA). 1991. ANSI/NFoPA NDS-1991 - *National Design Specification® for Wood Construction.* Washington, DC.

4.   American Institute of Steel Construction (AISC). 1986. *Load and Resistance Factor Design Specification for Structural Steel Buildings.* Chicago, Illinois.

5.   American Iron and Steel Institute (AISI). 1991. *Load and Resistance Factor Design Specification for Cold-Formed Steel Structural Members.* Washington, DC.

6.   American Society of Civil Engineers (ASCE). *Load and Resistance Factor Design Specification for Engineered Wood Construction* (draft). New York, New York.

7.   American Society of Civil Engineers (ASCE). 1988. ANSI/ASCE 7-88 - *Minimum Design Loads for Buildings and Other Structures.* New York, New York.

8.   American Society for Testing and Materials (ASTM). 1984. ASTM D198 - *Standard Methods of Static Tests of Timbers in Structural Sizes.* Philadelphia, Pennsylvania.

9.   American Society for Testing and Materials (ASTM). 1992. ASTM D245 - *Standard Practice for Establishing Structural Grades and Related Allowable Properties for Visually Graded Lumber.* Philadelphia, Pennsylvania.

10.   American Society for Testing and Materials (ASTM). 1991. ASTM D1990 - *Standard Practice for Establishing Allowable Properties for Visually-Graded Dimension Lumber from In-Grade Tests of Full-Size Specimens.* Philadelphia, Pennsylvania.

11.   American Society for Testing and Materials (ASTM). 1988. ASTM D2555 - *Standard Test Methods for Establishing Clear Wood Strength Values.* Philadelphia, Pennsylvania.

12.   American Society for Testing and Materials (ASTM). ASTM D5457 - *Standard Specification for Computing the Reference Resistance of Wood-based Materials and Structural Connections for Load and Resistance Factor Design* (draft). Philadelphia, Pennsylvania.

13.   Caldwell, R.M., B.K. Douglas and D.G. Pollock. 1991. *Load Duration Factor for Wind and Earthquake.* Wood Design Focus, Vol.2, No.2.

Wood Products Information Center/Wood Forest Institute. Portland, Oregon.

14. Ellingwood, B.R. 1992. *Load Combinations and Time Effect Factors* (draft), submitted for publication in Journal of Structural Engineering. American Society of Civil Engineers. New York, New York.

15. Gromala, D.S. 1993. *LRFD in Wood Design.* Presentation at Structural Engineers Association of Northern California 1993 Spring Seminar. San Mateo, California.

16. Gromala, D.S., Sharp, D.J., Pollock, D.G., and Goodman, J.R. 1990. *Load and Resistance Factor Design for Wood: The New U.S. Wood Design Specification.* International Timber Engineering Conference. Tokyo, Japan.

17. Pollock, D.G. 1993. *Design of Connections in Wood Structures: LRFD vs. ASD.* Proceedings of Structures Congress '93. American Society of Civil Engineers. New York, New York.

18. Rosowsky, D.V. and B.R. Ellingwood. 1990. *Stochastic Damage Accumulation and Probabilistic Codified Design for Wood.* Johns Hopkins University Civil Engineering Report No. 1990-02-02. Baltimore, Maryland.

19. Williamson, T.G. 1993. *LRFD for Timber Structures.* Proceedings of ASCE Annual Convention & Exposition. American Society of Civil Engineers. New York, New York.

What Controls the Ultimate Load of a Glulam Dome?

S. M. Holzer*, Member ASCE, S. A. Kavi**, S. Tongtoe**,
and J. D. Dolan***, Member ASCE

Abstract

A glulam dome is analyzed with the finite element method to determine the governing failure mode and ultimate snow loads. The dome consists of a triangulated network of curved glulam beams, a decking supported by curved purlins, and a steel tension ring. Geometric and material nonlinearities are considered. The analysis centers on the effect of the beam-decking connectors on the dome behavior. The results include ultimate snow loads, buckling modes, and the status of the material prior to buckling.

Introduction

The paper is concerned with finite element analyses of a glued-laminated timber (glulam) dome to predict the governing failure mode and ultimate snow loads. It represents a continuation of experimental and numerical investigations of spatial wood structures (Davalos, 1989; Davalos et al., 1991; Holzer et al., 1990; Holzer et al., 1992).

The principal objective of this study is to test the hypothesis that the beam-decking connectors form the weakest link of the dome. Connection failures are likely to impair the integral action of the dome and lead to failure of the dome. The beam-decking connectors are represented by nonlinear springs which model the load-slip behavior up to failure. Geometric and material nonlinearities are considered in the dome model. The status of the connectors and the stability of the dome are monitored as the nonlinear equilibrium path is tracked with the Riks method (ABAQUS).

---

*Professor, **Graduate Student, Civil Engineering, ***Assistant Professor; Wood Science and Forest Products, Virginia Polytechnic Institute and State University, Blacksburg, VA 24061.

Modeling of Dome

The dome under study (Fig. 1) is a 3-way grid dome (Tsuboi, et al., 1984) whose design is based on the Crafts Pavilion dome in Raleigh, N.C. The dome has a span of 133 ft (40.5 m), a rise of 18 ft (5.5 m), and a radius of 133.3 ft (40.6 m). It consists of a triangulated network of curved glulam beams, a tongue-and-groove decking supported by curved purlins, and a steel tension ring. The glulam beams are joined by patented steel hubs, and the decking is nailed to the beams and purlins.

The geometry of the dome, which is characteristic of Triax (Neal, 1973) and Varax (Eshelby and Evans, 1988) domes, is obtained by projecting a plane network of equilateral triangles onto the spherical surface of the dome. The projection is defined by rays that originate at the center of the sphere. Thus, all members lie in great circle planes and have the same radius of curvature. The dome is cyclically symmetric and composed of six identical sectors (Figs. 1 and 2).

Programs. Modeling, analysis, and post-processing of the dome are performed with the commercial finite element programs I-DEAS and ABAQUS. The components of the model that are not available in the ABAQUS library, specifically the beam-decking connector element and the nonlinear material law for wood (Telang, 1992), are introduced through user-supplied subroutines. This is a powerful option of ABAQUS.

Framework. Each beam of the dome (Fig. 1) is modeled by two straight, three-dimensional, Bernoulli/Euler, beam finite elements. The purlins and the tension ring are represented by truss elements. It was found that the large curvature and slenderness of the beams make it unnecessary to use curved beam finite elements or to include shear deformation. Although the discretization error can be reduced by using a finer mesh, we selected the two-element mesh because at this stage our main interest is the qualitative behavior of the dome. Moreover in this study, the joints corresponding to the steel hubs are assumed to be rigid. Sensitivity analyses indicated that variations in the joint stiffness of a wide range do not have a significant effect on the ultimate load capacity of the dome (Holzer, et al., 1992). However, the effect of a pin joint, representing a Triax joint, has not been investigated.

Decking. The effect of the decking on the behavior of the dome is investigated in three ways: (1) it is neglected; (2) it is represented by truss bracings, which restrain beam elements at the nodes (Fig. 2); and (3) it is reflected in the beam-decking connector model, called connector element (Fig. 3).

The connector element includes 16 nonlinear springs, one is shown in Fig. 3, that provide lateral support for the beam element. Each spring is connected at one end through a rigid link, that positions the spring at the top of the beam, to the beam element and at the other end to a fixed support. The fixed support signifies that the membrane deflections of the decking are neglected relative to the lateral displacement and rotation of the beam element and the elongation of the spring. The elongation of each spring is determined through interpolation functions from the nodal displacements of the beam element. Hence, the springs do not add degrees of freedom to the assembly. This modeling technique is frequently used for connectors in timber structures (Dolan, 1989). The load-deflection curve of a spring, representing two 16 d nails, was obtained experimentally and is idealized for one nail in Fig. 3b. When the elongation of a spring reaches 1 in. (2.54 cm), the spring is disconnected.

Material properties. Although wood is an inhomogeneous, anisotropic, and highly variable material, specific properties of glulam beams meeting fabrication standards allow one to make simplifying assumptions: (1) The longitudinal fibers are approximately parallel to the axis of the beam. Hence, the normal stress-strain relation of the beam can be defined by a single parameter, a longitudinal modulus. (2) The orientations of the growth rings in the cross section may vary greatly from lamina to lamina. Thus, it may be assumed that in an average sense the cross section is isotropic, which permits one to select a single shear modulus. This assumption of transverse isotropy is supported by an investigation of southern pine glulam beams (Davalos et al., 1989). Moreover, it was concluded that the shear modulus can be obtained from torsion tests on the basis of Saint-Venant's isotropic torsion solution.

The material properties used for the southern pine glulam beams are defined in Fig. 4. Three normal stress-strain relations are used: linear, bilinear, and nonlinear (Conners, 1989). The bilinear relation is used to match the initial moduli E_t and E_c of Conners' nonlinear model; thus it provides a transition from linear ($E = E_t$) to nonlinear material behavior.

Loading. The dome is subjected to a pressure composed of independent load distributions (Davalos, 1989)

$$p = p_D + \lambda p_L$$

where p_D = 16 psf (766 Pa) is the dead load pressure; p_L = 20 psf (958 Pa) is the live load pressure resulting from snow load over the entire dome or over half of the dome; and λ is the proportionality factor of the live load, the load parameter.

The pressure loads are discretized by using shell elements and performing a linear analysis with I-DEAS. Since the loading is proportional, the shell elements are removed once the nodal forces are computed.

Analysis Results

A selection of analysis results is discussed in the context of Table 1. The critical pressure, corresponding to a bifurcation or limit point on the equilibrium path, is the ultimate pressure of the dome model.

Table 1. Critical pressure: $p_{cr} = p_D + \lambda_{cr}p_L$

Material law	Decking model	λ_{cr}	
		Full snow	Half snow
Linear	none	3.6	5.3
	truss bracing	9.3	7.2
	connector elements	8.5	7.6
Bilinear	none	4.4	6.5
	truss bracing	10.3	8.6
	connector elements	9.0	8.9
Nonlinear	none	4.4	5.9
	truss bracing	10.2	7.7
	connector elements	9.0	8.3

The response comparisons are confined to models with bilinear and nonlinear material, no decking and decking represented through connector elements, full snow and half snow.

Full snow. The connectors cause the critical load parameter, λ_{cr}, to increase by 105% (from 4.4 to 9.0) for bilinear and nonlinear material. Material nonlinearity is not a factor. In all cases, the critical points, the transitions from stable to unstable equilibrium, are bifurcation points. Just before the critical pressure is reached, half of the connectors of some beam finite elements are already disconnected. The resulting buckling mode is shown in Fig. 5.

Half snow. The connectors effect an increase in the critical load parameter of 41% from (5.9 to 8.3) for nonlinear material. The bilinear material law is no longer valid since the maximum compressive stresses in some regions of the dome exceed the ultimate compressive stress by a large margin. For the nonlinear material model, the maximum compressive stresses are equal to the ultimate compressive stress, while the maximum tensile stresses are near the proportional limit (Fig. 4). As the critical point, a limit point, is reached, two thirds of the connectors of some beam finite elements are disconnected. The buckling mode is shown in Fig. 6.

Conclusion

As expected, the decking contributes significantly to the ultimate load capacity of the dome model, and connector failures seem to trigger failure of the dome model. However, for nonuniform snow load, in this study snow over half of the dome, nonlinear material behavior is closely linked with the failure mode.

References

ABAQUS, General-Purpose Finite Element System, Hibbit, Karlsson and Sorensen, Inc., 1080 Main Street, Pawtucket, RI 02906.

Davalos, J. F., Geometrically Nonlinear Finite Element Analysis of a Glulam Timber Dome, PhD Dissertation, Virginia Polytechnic Institute and State University, Blacksburg, VA, July, 1989.

Davalos, J. F., Loferski, J. R., Holzer, S. M., and Yadama, V., "Transverse Isotropy Modeling of 3-d Glulam Timber Beams," Journal of Materials in Civil Engineering, ASCE, Vol. 3, No. 2, May 1991, pp. 125-139.

Dolan, J. D., The Dynamic Response of Timber Shear Walls, PhD Dissertation, The University of British Columbia, Vancouver, Canada, October 1989.

Eshelby, R. W. and Evans, R. J., "Design Procedures for Reticulated Timber Domes," Proceedings of the 1988 International Conference on Timber Engineering, Seattle, Washington, Vol. 1, pp. 283-287.

Holzer, S. M., Davalos, J. F., and Huang, C. Y., "A Review of Finite Element Stability Investigations of Spatial Wood Structures," Bulletin of the International Association for Shell and Spatial Structures, Vol. 31, No. 3, November 1990, pp. 161-171.

Holzer, S. M., Wu, C. H., and Tissaoui, J., "Finite Element Stability Analysis of a Glulam Dome," International Journal of Space Structures, Vol. 7, No. 4, 1992, pp. 353-361.

I-DEAS, Engineering Analysis, Model Solution, and Optimization, MacNeal-Schwendler, Structural Dynamics Research Corporation, 2000 Eastman Drive, Milford, Ohio, U.S.A.

Neal, D. W., The Triax Dome, Culbertson, Noren, and Neal, Consulting Engineers, 1410 S. W. Morrison St., Portland, Oregon 97205, March, 1973.

Telang, N. M., Stability Analysis of a Glulam Dome with Nonlinear Material Law, MS thesis, Virginia Polytechnic Institute and State University, Blacksburg, VA, January 1992.

Tsuboi, Y., et al., "Analysis, Design and Realization of Space Frames," A State-of-the-Art Report by the I.A.S.S. Working Group on Spatial Steel Structures, Bulletin of the International Association for Shell and Spatial Structures, Vol. XXV-1/2, April-August 1984.

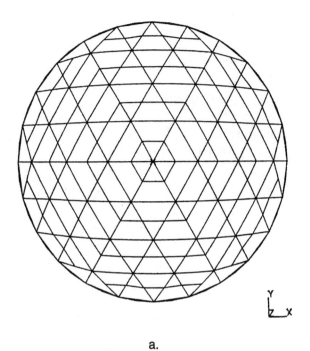

a.

b.

Fig 1. Geometry of dome: (a) plan view, (b) elevation

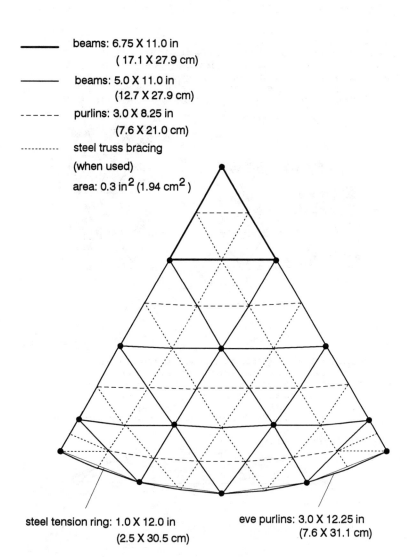

beams: 6.75 X 11.0 in
(17.1 X 27.9 cm)

beams: 5.0 X 11.0 in
(12.7 X 27.9 cm)

purlins: 3.0 X 8.25 in
(7.6 X 21.0 cm)

steel truss bracing
(when used)

area: 0.3 in^2 (1.94 cm^2)

steel tension ring: 1.0 X 12.0 in
(2.5 X 30.5 cm)

eve purlins: 3.0 X 12.25 in
(7.6 X 31.1 cm)

Fig. 2 Sector of dome

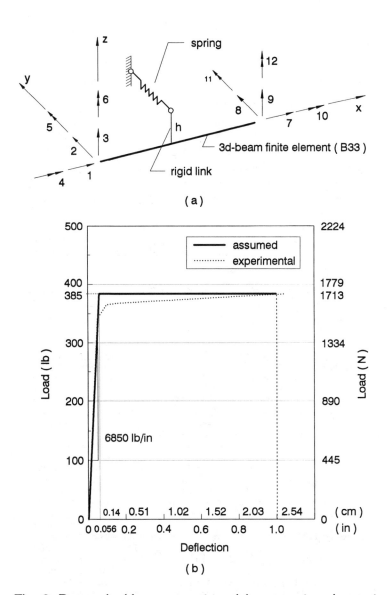

Fig. 3 Beam-decking connector : (a) connector element, (b) load-slip curve for one 16 d nail

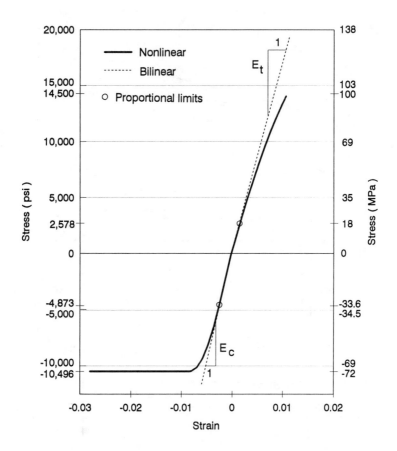

Fig. 4 Material properties : $G = 1.6 \times 10^5$ psi (1100 MPa), $E_t = 1.8 \times 10^6$ psi (12400 MPa), $E_c = 2.2 \times 10^6$ psi (15200 MPa), MC = 12 %, SG = 0.52

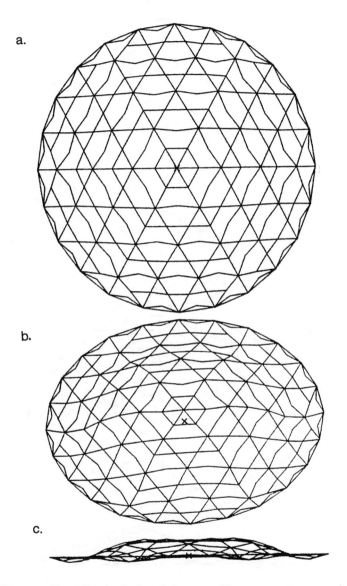

Fig. 5 Buckling mode of dome with connectors and full snow: (a) plan view, (b) inclined view, (c) elevation

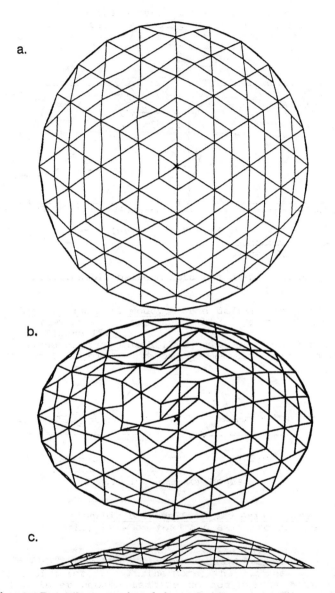

Fig. 6 Buckling mode of dome with connectors and
half snow: (a) plan view, (b) inclined view, (c) elevation

TIMBER STRUCTURES
by
Daniel F. Tully[1]

Abstract

This paper deals with three different curvilinear timber structures of moderate spans - 120 feet to 200 fee +/-. In each of the three cases cited, the roof deck and boundary members are designed to interact and be interdependent. All of the structures have been erected in the last 16 years.

All of the structures are composed of heavy timber laminated beams made from long leaf Southern yellow pine, laminated with phenolic resin glue (waterproof).

There are 5 illustrations:

1. Photograph of interior view of the University of Hartford Arena *(Figure 1)*.
2. Graphic Plan of the University of Hartford *(Figure 2)*.
3. Perspective of the University of Maine at Farmington *(Figure 3)*.
4. Graphic Plan of the University of Maine at Orono Describing the Sequence *(Figure 4)*.
5. Perspective of the University of Maine, Orono *(Figure 5)*.

University of Hartford - The Design Challenge:

The first of the three structures is the University of Hartford Arena and is fabricated of nominal 4" Southern Pine deck, double tongue and grooved, and of heavy timber laminated beams *(Figure 1)* . The design challenge at the University of Hartford was to design an arena with clear spans to allow seating of 3,000 to 6,5000 seats depending upon funding availability.

In addition to the normal requirements of architectural and structural design, the imposition of a "constantly flexible" scheme in the **design** stage of a project is a difficult one.

The design task required the use of a shape which was first determined by the size of the playing surface and last by the flexible seating requirement. The latter caused

----

[1] President, Daniel F. Tully Associates, Inc.; and Shell and Spatial Structures, Inc.; Architect, Structural Engineer and Builder, 99 Essex Street, Melrose, MA 02176

the "cruiciform" plan to be developed which has at its intersecting axes the required geometry of the basketball court and at its extending arms the seating for up to 6,000 (*See Figure 2*). The seating was finally limited during the design to 4,500. The horizontal roof thrusts were to be carried by moment resistant abutments which had the spring line of the parabolic arches 25' above the playing floor. However, the construction cost of these tall abutments was prohibitive, and steel ties at the spring lines were introduced to resolve these forces. The ties were placed on the perpendicular lines, not the diagonals, so that the space within the vault was preserved without any obstacles to the eye or to sports purposes.

The structure used is two parabolic cylinders intersecting perpendicular to one another forming in the process a groined vault. The span at the diagonal or at the intersection is 169.7 feet. The rise of the arch is 28 feet.

Figure 1

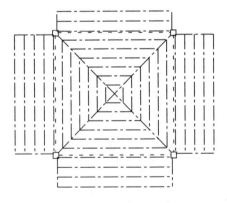

Figure 2

University of Maine at Farmington - The Design Challenge:

Real estate, sports dimensions and economy dictated the design solution to the design challenge at the University of Maine at Farmington *(Figure 3)*. The user wanted to have six intramural basketball courts, or six tennis courts. The intramural basketball courts dictate the width of each space, 60 foot bays for a total length of 300 feet. The tennis court is the limiting sport in determining the span of the structure, 120 feet. It was also stated that in the future when funds become available, the sixth bay would be used as a space within which to build a six lane swimming pool. Thus the building dimensions of 120 feet by 300 feet composed of shells with dimensions of 30.0' x 30.0' x 16.62'. The gross height of the building above the abutments (work points) is 33.23'.

The roof deck is composed of four layers of 1/2" plywood fabricated in rectalinear pieces such that the elements could be easily shipped. At the shell to shell boundary or refabrication joint is a double tongue and groove formed by offsetting the plywood layers while in fabrication. After assembly in the field, the refabrication joint is covered with a 1/2" x 12" plywood joint cover and nail glued to each side of the joint on the upper and lower surface of the shell.

The building is in the snow belt in Maine and the design live load from snow was taken at 65 pounds per square foot.

The roof geometry of Farmington is a hyperbolic paraboloid *(see Figure 3)*. Measured dead load deflections after the removal of the erection towers was 3/8" or less in all locations. Since the completion of the initial open structure, five of the courts have been coated with a 3/8" polyurethane sports surface for tennis and intramural basketball, and the swimming pool has been designed and is now in operation. Programs for the University students, faculty and staff are ongoing and, in addition, community programs for health and wellness are in place.

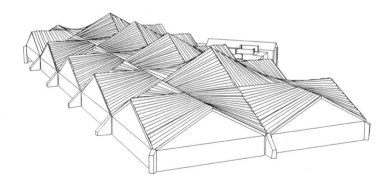

Figure 3

The University of Maine, Orono - The Design Challenge:

An existing arena designed and built by the writer in 1974 to seat 3,000 needed to be expanded to seat approximately 6,000 spectators. The existing structure was a 140'-0" x 280'-0" clear span wood hyperbolic paraboloid consisting of 35'-0" square modules with a module height of 21.62' *(see Figure 4 - Existing)*

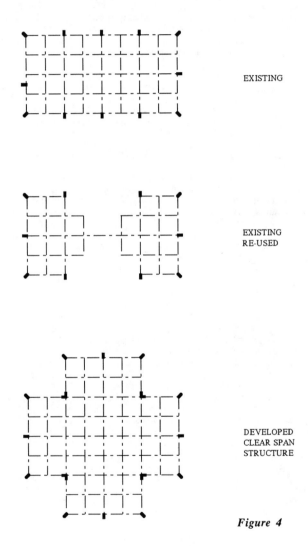

EXISTING

EXISTING
RE-USED

DEVELOPED
CLEAR SPAN
STRUCTURE

Figure 4

Because the structural support abutments were located around the perimeter at 70'-0" intervals, any expansion needed to occur at locations which would place the abutment within the gross clear span, a contradiction that required a design which totally removed the abutment and extended the sphere of influence of different boundaries. The requirement for the total absence of columns or obstacles is a function of the need for unobstructed vision of the spectators.

The evolved solution changed the simple rectangular plan to one which had identical internal spans in both axes, a plan which is photo-identical about both axes *(See Figure 4 - Developed Clear Span and Figure 5).*

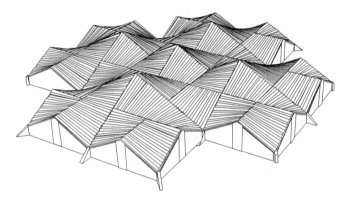

Figure 5

There are three new abutments on each side of the building on a line about 70 feet away and parallel to the original sidewalls. There are two modified abutments ("L" shaped) which supplant the existing sidewall abutments at the intersection of the new structure and the existing. Further, there is one abutment on each side in the middle of the longest sides which are removed. The removal of these two abutments provided the designer with the opportunity to design an obstruction free arena, adding most of the new seating required in each addition. To complement the new and old construction, lobbies and public service spaces were added.

Nevertheless, the elimination of two abutments required a new system of structural elements which provided the capacity. A large pair of beams were fabricated out of laminated timber which were placed parallel to the original sidewalls and placed such that the height of the intersection of these sloped beams was twice the module height (45.23'). Due to the high compression forces in these members, the laminated timber was supplemented with a 1" flitch plate of A-36 steel.

The structure was shored and all the intermediate elements (demonstrated by *Figure 4 Existing Re-Used)* were removed. Although several of the existing shells were re-used, most of the construction of the roof geometry was composed of four layers of 1/2" plywood fabricated and erected as in the Farmington project.

All plywood used was 1/2" structural 1 C/D exterior grade Douglas Fir plywood. Assembly in the plant was by waterproof glue (phenolic resin) machine spread and staples. Fabricated parts of the plywood assembly were randomly shear tested to a minimum strength of 100 psi.

The University of Maine, Orono hockey rink and basketball arena was completed in 1992 and has completed two active and successful winter sports seasons for the University of Maine "Bears" hockey team, who were the National Collegiate Champions for the 1992-1993 season *(see Figure 5)*.

Deployable Mesh Reflector

Z. You[1] and S. Pellegrino[2]

Abstract

This paper presents a new deployable reflector for space applications, based on a deployable edge frame that deploys and prestresses a "rigid" cable network. The edge frame consists of a three-dimensional ring pantograph, a set of passive cables that become taut when the pantograph is deployed and are then pre-tensioned to stiffen the pantograph , and one/two active cables that control the deployment of the pantograph and the prestressing of the passive cables. The cable network consists of a 3-way regular tessellation of hexagons — for the top part — and a 4-way semi-regular tessellation of hexagons and triangles — for the bottom part —connected by a series of cable ties. A global state of prestress can pretension all cables. Tests on a 3.5 m model of the reflector have demonstrated the feasibility of the new concept.

Introduction and Background

This paper presents a new deployable reflector for space applications. The reflector, shown in Fig. 1, consists of a bar-and-cable deployable frame of toroidal shape, which deploys and pretensions a cable network spanning the region within the frame. A reflective wire mesh (not shown in the figure) can be connected to the cable net, to form a reflective surface that approximates closely to a paraboloid. Structures of this type are required for high-gain telecommunication antennae on spacecraft: they are neatly folded on the side of the spacecraft during launch, because they would not fit within the envelope of the launcher, and are automatically deployed once the spacecraft has been placed into orbit.

General reviews of deployable reflectors have been compiled by Freeland (1982) and, more recently, by Roederer and Rahmat-Sahmii (1989). Broadly speaking, there are two types of reflectors: solid-surface reflectors, where the reflective surface is a continuous dish, and mesh reflectors, where the reflective surface consists of interlaced, electrically conductive thin wires. The second type is

---

[1] Research Student, Department of Engineering, University of Cambridge.
[2] Lecturer, Department of Engineering, University of Cambridge, Trumpington Street, Cambridge CB2 1PZ, U.K.

suitable only for transmission of low frequency waves. From a structural viewpoint, standard requirements for deployable reflectors include: low mass; compact packaged volume; simple and reliable deployment mechanisms; and sufficient stiffness to avoid dynamic coupling between the reflector and the spacecraft attitude control system. A final requirement, specific to reflector structures, is that the shape error of the reflective surface, i.e. its deviation from the best-fit paraboloid, should be within a specified limit, which is related to the shortest wave-length to be transmitted. Typically, surface accuracies in the range 1 - 10 mm are required.

Fig. 1. 3.2 m model of the new reflector.

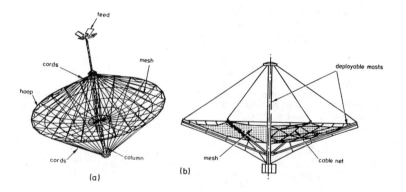

Fig. 2. (a) Hoop-Column and (b) Tension-Truss antennas.

Before introducing the new mesh reflector, two existing concepts are reviewed briefly. The Hoop-Column Antenna (HCA), shown in Fig. 2(a), consists of a telescopic column that deploys from a central hub, and of a deployable hoop supporting the edge of the mesh reflector through a system of cables. A 15 m diameter model has been made and tested (Harris Corporation 1986): the hoop is a tubular structure, with 24 segments connected by radial hinges. Some of the hinges are driven directly by an electric motor, while the others are coupled to the motors by connecting rods. The Tension-Truss Antenna (TTA), shown in Fig. 2(b), is based on a kinematically determinate cable net forming a coarse triangular network, pre-tensioned against a back-up structure by "soft" tie cords. The back up structure consists of 7 deployable masts, all originating from a rigid hub (Miura and Miyazaki, 1990). The reflective mesh is attached to the cable network.

The new reflector has some similarities with these two deployable reflectors. It has a deployable edge frame, like the HCA, and a kinematically determinate cable network, like the TTA, however both the edge frame and the cable network are of a new type. Furthermore, a central mast is not an inherent part of the new concept, and hence a wide range of options can be considered, to support the subreflector/feed.

Deployable Edge Frame

The structural concept for the edge frame is an extension of the cable-stiffened pantographic masts recently developed by Kwan, You and Pellegrino (1993). These masts consist of a three-dimensional *pantograph*, a set of *passive cables* that become taut when the pantograph is deployed and are then pre-tensioned to stiffen the mast, and one/two *active cables* that control the deployment of the mast and the prestressing of the passive cables.

Ring Pantograph A previous investigation of foldable ring structures (Pellegrino and You 1993) has identified three concepts for ring pantographs that allow strain-free folding/unfolding. The three-dimensional pantograph in the deployable reflector shown in Fig. 1 is based on *concept (b)* from that study, with 12 sides. It consists of 48 pantograph units lying on the side faces of 24 triangular prisms. A pantograph unit is an assembly of two rods joined by a pin, which allows free rotation of one rod relative to the other, about the axis of the pin. Any other relative motion of the rods is prevented. In plan view, Fig. 3, the prisms form a chain of interlaced isosceles triangles: 12 long-base triangles form the outer edge of the ring, while 12 short-base

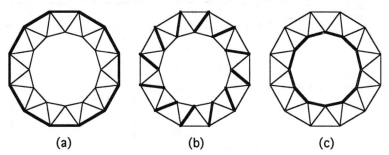

(a)   (b)   (c)

Fig. 3. Possible arrangements of passive cables; top view.

triangles form the inner edge. Of the 48 pantograph units, 24 have connecting pins in the middle; the 12 units on the inner face of the pantograph have rods of length 2ℓ, while the 12 units on the outer face have rods of length $2L$. The remaining 24 units, which connect the inner and outer face of the pantograph have connecting pins not in the middle. All of these units have rods of length $\ell + L$, and the connecting pin is located at distance ℓ from the inner end. Pellegrino and You (1993) have shown that, for 12-sided pantographs, strain-free folding is obtained if

$$\frac{L}{\ell} = 1.582 \tag{1}$$

Adjacent pantograph units are connected by pins at an upper and a lower joint; these pins allow free rotation between the rods and the joints. In practice, the axes of these pins cannot all meet at the same point, and hence the axes of rotation of the rods are a small distance away from the ideal joint. To avoid compromising the strain-free behaviour of the pantograph, outer and inner joint eccentricities must be in the same ratio as L/ℓ (Pellegrino and You 1993).

Layout of Passive and Active Cables Passive and active cables (Kwan, You and Pellegrino 1993) are used (i) to control deployment and (ii) to increase the stiffness of the ring pantograph in its fully deployed configuration. Their first function, deployment control, is based on the observation that during deployment of the pantograph the relative distances between many joints will vary. For example, two adjacent top joints get further apart during deployment, while a top joint and the corresponding bottom joint on the same face of the ring get closer during deployment. Of course, the relative distances of joints connected by rods remain unchanged. An active cable is connected to a motor-driven drum, and hence can vary its length as needed. Thus, winding in an active cable that links several joints which get closer during deployment will cause the pantograph to deploy. Passive cables link joints which get further apart during deployment. A passive cable connecting two joints has a fixed length, equal to the distance between these joints in the fully-deployed configuration. Hence, this cable will be slack during deployment, but will become taut when the pantograph is fully deployed, thus terminating deployment. Only one passive cable would be sufficient to terminate deployment, but the second function of active and passive cables, stiffening the pantograph, requires many more passive cables. The best strategy to transform a pantograph into a stiff structure (Kwan, You and Pellegrino 1993) is to consider the pantograph as a pin-jointed truss, thus neglecting the internal pin in each pantograph unit, and to rigidise this truss by means of passive cables. Usually, there are many different ways of doing this, but only those arrangements where all passive cables can be pretensioned by a small number of active cables are worth pursuing.

The truss model of the 12-sided pantograph of Fig. 1 has 48 pin joints and 96 pin-jointed bars. The total number of cables that have to be introduced to obtain a rigid structure can be estimated from the generalised Maxwell's rule:

$$3j - b = m - s \tag{2}$$

where j = no of joints, b = no of bars, m = no of mechanisms, and s = no of states of self-stress. Here, j = 48, m = 6 because the pantograph has 6 degrees of freedom as a rigid body in 3D space, and s = 1, at least, so that the passive cables may be pretensioned. Substituting these values into Eq. (2) yields $b = 139$ and, since the truss already has 96 bars, no fewer than 43 cables need to be introduced, say 42 passive cables and 1 active cable.

It is desirable that the passive cables should be arranged in a symmetric way, for uniform structural properties in all directions. Figure 3 shows in plan view three rotationally symmetric ways of introducing 12 passive cables into the pantograph. In Fig. 3(a) the cables form continuous outer hoops, at the top and/or at the bottom of the pantograph, while in Fig. 3(c) they form inner hoops. In Fig. 3(b) they are arranged in near-radial directions, and rotated in a clockwise sense, i.e. the arrangement shown in the figure, or in an anti-clockwise sense, and either on the top face or on the bottom face of the pantograph. Thus, there are 8 different ways of arranging a set of 12 passive cables, and hence a maximum of 96 passive cables. To find a suitable set of passive cables, we select 4 out of these 8 sets and take out 6 cables. Then, we look for an active cable that will impose a state of pretension onto all of the passive cables. This is an iterative process that requires computer software for the analysis of the static and kinematic properties of a pin-jointed assembly (Pellegrino 1993).

A good choice for the passive cables is to take two near-radial sets of cables, of the type shown in Fig. 3(b), e.g. the clockwise set on the top face of the pantograph and the anti-clockwise set on the bottom face; plus the two hoops shown in Fig. 3(a), but without 3 cables in each hoop. An analysis of the equilibrium matrix for an assembly which includes 96 pin-jointed bars, representing the rods of the pantograph, and 42 additional bars, representing the passive cables, shows that this assembly has $m = 7$ and $s = 1$, instead of the expected $m = 6$ and $s = 0$. The state of self-stress does not involve the cables, but only the 96 bars, and the accompanying internal mechanism is, actually, a high stiffness mode that requires the rods of the pantograph to bend and twist.

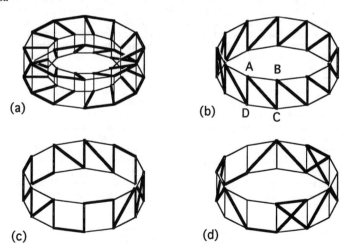

Fig. 4. Arrangement of passive and active cables.

No single active cable has been found able both to deploy the pantograph and to impose a state of pretension onto all of the passive cables. However, the passive cables can be pretensioned uniformly by a combination of two active cables, both running on the outer face of the ring, as discussed next.

Active cable 1 is shown in Fig. 4(b). It starts near joint A, runs parallel to rod AC, at joint C it goes over a pulley with axis perpendicular to a facet of the ring (ABCD), then at joint B it goes over a pulley with axis perpendicular to the next facet of the ring, and so on until it reaches joint A. A unit tension in this active cable is in equilibrium with uniform tensions of 0.6617 in the 24 near-radial passive cables and with zero tension in the remaining 18 passive cables.

Active cable 2 is shown in Fig. 4(c). Its route, rather similar to the route of active cable 1, includes the 6 gaps left in the two hoops by the passive cables. A unit tension in this active cable is in equilibrium with uniform unit tensions in the 18 hoop cables, and with unit uniform compressions of 0.6617 in the 24 near-radial cables.

A uniform state of pretension of all passive cables is set up by imposing tensions in the ratio 2.5513 : 1 in active cables 1 and 2, respectively.

This solution, involving a pair of active cables, is entirely satisfactory in terms of prestress control, but it is not ideal for deployment control because the lengths of the two active cables need to vary in a non-proportional fashion during deployment. Figure 5 is a plot of active cable length for different configurations of the pantograph. The *deployment angle* is defined as the angle between the two rods in a pantograph unit. This angle has the same value in all units, for any configuration of the pantograph. It is about 180° when the pantograph is fully folded, and about 90° when it is fully deployed. As the pantograph deploys, we move from left to right in the graph. During this process the length of active cable 1 decreases all the time, and hence deployment can be controlled by winding in this cable. However, the length of active cable 2 increases during the first phase of deployment and then decreases, which means that the direction of motion of the motor controlling this active cable has to be reversed at some point. Furthermore, because the ratio between the rate of variation of the lengths of these two cables is not a constant, the rotational speed of the second motor needs to be continuously adjusted during deployment.

Figure 4(d) shows an alternative route for the second active cable, which will be referred to as active cable 3. Like active cable 2, this cable spans the 6 gaps in the hoop passive cables, whose length increases during deployment. Active cable 3, though, does not connect any joints whose distance increases. Thus, as shown in Fig. 5, the length of this cable would increase throughout deployment, which makes it unsuitable to control deployment, but suitable to control *retraction*. In "equilibrium" terms, a unit tension in active cable 3 is in equilibrium with unit tension in the 18 hoop cables, like active cable 2, and with compressions of 1.1324 in the 24 near-radial cables, double the magnitude obtained with active cable 2. Hence, a larger force ratio between cables 1 and 2 would be required, for uniform pretension of all passive cables.

The main reason why this route was not chosen for the edge frame of the mesh reflector is because active cable 3 would require 3 pulleys at each joint, while active cable 2 requires only one pulley per joint.

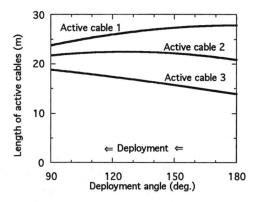

Fig. 5. Active cable length during deployment

Kinematically Determinate Cable Network

The deployable edge frame described in the previous section provides 24 "hard" support points for the cable network that supports the mesh surface. This section presents a kinematically determinate cable net (m = 0) whose state of prestress can be easily controlled (s ≅ 1). The top side of the cable network has to approximate to a paraboloidal, concave surface. The bottom side could be of any shape but, for simplicity, will be assumed to be paraboloidal as well. Thus, the cable network consists of two curved nets connected by a series of cable ties: the key problem is to identify a suitable structural form which is both "rigid" and has one state of self-stress, at least, involving tension in all cables. This problem has been tackled only once before, for the TTA (Miura and Miyazaki 1990), as discussed in the Introduction.

Miura's solution is a triangulated network where each joint is connected to other joints by 6 cables, hence forming a curved, 6-way grid. Because this network, by itself, does not admit any states of self-stress, an additional tie cable is included at each joint of the grid. The resulting assembly has many states of self-stress, which make it quite difficult to accurately control its prestress. This difficulty can be resolved by pursuing a global solution, rather than by adding extra cables to a rigid net.

In the following analysis each cable will be modelled as a pin-jointed bar, and hence the cable network will be modelled as a 3D pin-jointed truss. This approach is only acceptable, of course, if it can be assumed that all cables will remain taut throughout, which will be ensured by means of a global state of pretension. The general static and kinematic properties of such an assembly can be investigated by Maxwell's rule (Eq. 2). To achieve m ≅ 0 and s ≅ 0, we need to consider networks where the number of cables is approximately 3 times the number of joints, i.e. 6-way networks. Note that each cable contributes only 1/2 to this count, because it is linked to 2 joints. The only 6-way network forming a single rigid surface is a tessellation of triangles but, because we are interested in two curved nets joined by some tie cables, each net must have a connectivity lower than 6, on its own.

The chosen network is shown in Fig. 6. It is based on a 3-way regular tessellation of hexagons — forming the top net — and a 4-way semi-regular tessellation of hexagons and triangles — forming the bottom net. Each node of the top net is connected by 3 ties to the triangle directly below, in the bottom net, and hence each node of the bottom net is connected by ties to 2 nodes of the top net. Each

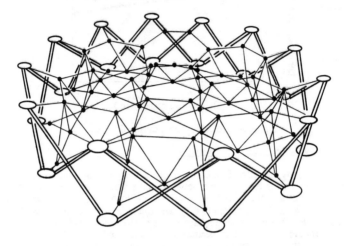

Fig. 6. Cable network, connected by inner edge of pantograph.

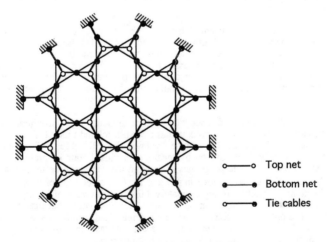

Fig. 7. Cable network; top view.

o———o Top net

●———● Bottom net

o———● Tie cables

edge joint of the network (24 in total: 12 top joints and 12 bottom joints) is connected to a rigid support, i.e. to the edge frame, by a tie cable. This assembly has 66 joints (respectively 24 and 42 in the upper and bottom nets), and 198 cables (respectively 30, and 72 in the top and bottom nets, 72 net-net ties, and 24 edge ties). Hence Eq. 2 gives $m - s = 0$, and this result is valid for any 3D network obtained by adding complete layers of hexagons and triangles at the edges of the tessellation of Fig. 7.

It turns out that the net shown in Figs 6, 7 has $m = s = 2$, regardless of the curvature of the two nets. The first state of self-stress is highly symmetric, it involves uniform tensions in the top and bottom edge ties (the ratio between these tensions depends on the curvature of the nets). All cable elements are in a state of tensions, with the highest tension in the edge ties and the lowest in some of the net-net ties. The second state of self-stress is less intuitive; it involves alternate tensions and compressions in pairs of edge ties, of much larger magnitude in the bottom ties than in the top ones, and large tensions and compressions in the tie members. Obviously, the fist state of self-stress is the desirable one, in order to pretension the network.

Regarding the 2 inextensional mechanisms associated with this solution, they are stabilised by the symmetric state of prestress described above. They could be eliminated by adding a single tangential restraint to the top and bottom nets, but this is not needed in practice.

Experiments

To verify the feasibility of the concept described in the previous sections, a physical model of the deployable reflector has been made, Fig. 1, based on a 12-sided pantograph with $\ell = 0.366$ m and L = 0.579 m. Its fully deployed dimensions, for a deployment angle of 90° are: outer diameter 3.164 m, inner diameter 2 m, and vertical distance between inner joints 0.518 m. Fully folded, the model has outer diameter of 0.6 m and height of 0.819 m, but a smaller outer radius could be obtained by redesigning the inner joints. The model has two active cables that follow the routes shown in Fig. 4(b, c), and whose length is controlled by 2 electric motors, each driving a slender drum whose axis is parallel to a rod of the pantograph. Figure 1 shows also a cable network consisting of 198 cables, connected by Al-alloy adjustable joints.

The model works very well, even though all experiments were without the cable net. Two types of tests have been completed. First, the lengths of the passive cables were set to their nominal values and the motor current cut-offs were set to 0.5 A and 0.15 A, in order to a achieve a ratio of about 2.5 between the tension in active cables 1 and 2. Then, having deployed the model, all cable tensions were measured with a tension meter. The measured tensions were 68 N in active cable 1, 27 N in active cable 2, and varying in the range 24 - 30 N in the passive cables. Note that, according to the section Deployable Edge Frame one would expect to find 26.6 N in active cable 2 and in all passive cables, for a tension of 68.0 N in active cable 1.

Finally, the model was folded and deployed 10 times, to measure how accurately it reproduces its shape. After each cycle, the 3D coordinates of 6 joints on the inner edge of the pantograph were measured with an Industrial Measurement System, and the distances between the joints were computed. The total errors from the mean distance are plotted in Fig. 8.

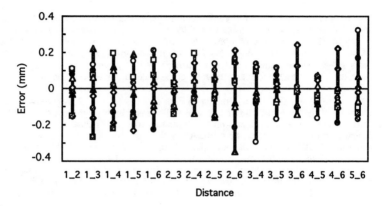

Fig. 8. Deployment test results.

Final remark

A new concept for deployable mesh reflectors has been proposed and its feasibility has been demonstrated by experiment. A key property of the new concept is that both the deployable edge frame and the cable network have only a small degree of redundancy, which makes it easy to set up the global state of prestress required for the successful implementation of the concept.

Acknowledgements

Z. You has been supported by the Cambridge Overseas Trust. Financial support from the SERC (grant GR/F57113) is gratefully acknowledged.

Appendix. References

Freeland, R. E. (1982). "Survey of deployable antenna concepts." In: E. B. Lightner, ed., *Large Space Antenna Systems Technology-1982, NASA-CP-2269* , pp 381-421.
Harris Corporation (1986). "Development of the 15 meter diameter Hoop Column Antenna." *NASA-CR-4038.*
Kwan, A. S. K., You, Z. and Pellegrino, S. (1993). "Active and passive cable elements in deployable masts." *Int. J. Space Structures* 8(1-2), 29-40.
Miura, K. and Miyazaki, Y. (1990), "Concept of the tension truss antenna." *AIAA Journal* 28(6), 1098-1104.
Pellegrino, S. (1993). "Structural computations with the Singular Value Decomposition of the equilibrium matrix." *Int. J. Solids Structures* 30(21), 3025-3035.
Pellegrino, S. and You, Z. (1993). "Deployable ring beams for space applications." In: G. A. R. Parke and C. M. Howard, eds, *Space Structures 4*, Thomas Telford, London, pp 783-792.
Roederer, A. G. and Rahmat-Samii, Y. (1989). "Unfurlable satellite antennas: a review." *Ann. Telecommun.* 44(9-10), 475-488.

AN APPLICATION OF INVERSE PROBLEM TECHNIQUES TO SPATIAL STRUCTURES

Pei Shan Chen[1], Mamoru Kawaguchi[2] and Masaru Abe[3]

Abstract

The present paper attempts to give a definition of inverse problems for spatial structures, and to develop a design method to find the optimum configurations and shapes for such structures by applying inverse problem techniques to them. As an example, the present paper proposes a method to find the optimum shape of a shell or a space frame that has a maximum buckling capacity.

Introduction

The conventional procedure of structural analysis has normally been to find the displacements and the internal forces of a structure with given informations such as topological characters (configuration), dimensions and shape or coordinates, and material character under given loading conditions. These informations are predetermined on the basis of experiences, guess work or conjecture. Occasionally, some of them are altered when they are found unsuitable for design as a result of analysis. By this method, however, the capabilities of the structure and its materials cannot be fully exploited, especially when we design large span spatial structures, for which we have only limited experiences. To circumvent the above drawbacks of conventional design methods, it may be suitable to apply the inverse problem techniques to the design of spatial structures .

––––––––––––––––––––––

[1]Ph.D.-student, M.Sc., Department of Architecture, Hosei University,
          3-7-2 Koganei 184, Tokyo Japan.
[2]Professor, Dr.Eng., Department of Architecture, Hosei University ;
[3]Lecturer, M.Eng., Department of Architecture, Hosei University .

With the development of computer science, the complex inverse problems can be solved quickly and more easily, and it seemes that the time will soon come when we can design the configuration and shape of a structure, especially a large-span spatial structure, by means of inverse porblem techniques.

1. Inverse Problems For Spatial Structure

1.1 Definition of Inverse Porblem

Many people have been trying to give the inverse problem a definition(ref. Sabatier,P.C., 1986), but there has not seem to be a unified definition for discrimination between direct and inverse problems. Before giving the definition of the inverse problems on spatial structures, we should analyze the general inverse problems mathematically. Generally, a problem can be expressed as a transformation from one space to another.

$$S \to R . \tag{1}$$

In the source space S, there is always a subspace Ω which is constituted by some invariable-sets with constant elements, conditions or some specific units. We call Ω the invariable space of the direct problem. So, the problem can be expressed as

$$\mathcal{F}(\Omega, \Upsilon) = R; \tag{2}$$

where $\Omega \cup \Upsilon = S$, and $\mathcal{F}$ is the tansformation, and we call such a problem a direct problem.

There are many problems in which the source subspace Ω is unknown and we have to find it by the informations of the Υ and its response R. Such a problem is called an inverse problem.

Definition : A problem is called an inverse problem if it is a problem to find its invariable space Ω of its direct problem.

1.2 The Inverse Problem For Spatial Structure

There are only few papers which give a definition of the inverse problem on spatial structures. Before defining it, the direct problem should be analyzed first. The spatial structure problems are always constituted by considering some main elements, which are

τ : Topological charactor of the structure ;

ζ : Shape, coordinates or major dimensions of the structure ;

ϵ : The material charactor of the structure ;

ρ : The prestresses (self-balancing internal forces) of the structure ;

γ : The working surroundings, temperature change

κ : The steady or dynamic (moving) loading system exerting on the structure ;

δ : The displacements, deformation of the structure ;

σ : The stresses, internal forces of the structure.

With the conventional design method, the invariable space of the problem is

$$\Omega = (\tau, \zeta, \epsilon, \rho). \tag{3}$$

The direct problem of the structural design is to find one or more elements of $\{\gamma, \kappa, \delta, \sigma\}$ corresponding to a given space Ω . For example, to find the deformation of a given structure subjected to a certern loading system is one of the direct problem, which is a transformation $T(\Omega, \kappa) = \delta$. To find the buckling-load of a space frame or a shell is a direct problem $T(\Omega) = \kappa$, etc. .

Contrarily, the inverse problem is a problem to find the elements of the invariable space Ω of the direct problem.

Definition : The inverse problem for spatial structure is a problem to find the elements of invariable space $\Omega = (\tau, \zeta, \epsilon, \rho)$ of its direct problem .

So, the inverse problems for spatial structures can be classified as :

1. **Topological problems :** To find the topological configuration of a structure;

2. **Shape-problems :** To find the shape, or some of the dimensions or nodal coordinates of the structure;

3. **Material-problems :** To find the best mixture of materials of some parts, some members of the structure or whole of the strucuture;

4. **Prestress-problems :** To find the optimum distribution of prestresses;

For the topological problems, the important thing is not only the mathematical analysis but also the structural design idea. These problems are challenged by many researchers, and some encouraging progress have been obtained (ref. Rozvany, G.I.N.,1989) (ref. Lin,J.H., Che,W.Y. and Yu,Y.S.,1982). Now, there are many important topics on this research, for example, optimum configuration of a space frame and multicriteria optimization with topological problem. As an example, the present paper proposes a design method for finding a space frame or a shell suitably strong against buckling.

Shape problems are these of shape-finding, shape optimization(ref.Chen,P.S., Abe,M. and Kawaguchi,M.,1993) and shape analysis of movable structures or folding-structures (ref. Kawaguchi, K., Hangai,Y. and Nabana,K.,1993). The prestress problems are always concerned with the shape-problems, because the distribution of the internal forces due to prestressing depends on the shape, and must be in equilibrium in the given shape before loading.

Now the computers have become popular and powerful enough to deal with inverse problems and we may be able to believe that there will be a revolution on the design of spatial structures in this direction. So, it is very important to develop the desgin method by application of the inverse problems.

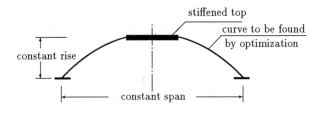

Fig.1 The section of a space frame of revolution stiffened at its top

2. Shape With Maximum Buckling-Load

2.1 Introduction of The Design Method

In design of a shell or a space frame one of the most important problems is the buckling problem, especially for large span structures. It is not rational or economical to assume a uniform stiffeness for the whole area or all of the members of the structure. We can stiffen some particular parts which are susceptible to buckling, or can be stiffened with ease. The other parts can be found by shape optimization for maximum buckling-load.

For example, a dome of revolution with a given rise can be stiffened at its top so as not to buckle (Fig.1), and the curvature of other areas can be found by optimization for maximum buckling load.

2.2 The Basic Functions

For the nonlinear analysis, by the finite difference method, finite element method or Galerkin's method, the relationship between the loads and the displacements is expressed in form, (ref. Thompson, J.M.T. and Hunt,G.W.,1973) (ref. Hangai,Y. and Kawamata, S., 1973)

$$f_r(D_1, D_2, \ldots, D_n, \Lambda) \equiv f_r(D_i, \Lambda) = 0, \quad (r = 1, 2, \ldots, n), \tag{4}$$

where $\{D_i\}$ are the displcement parameters, Λ the load parameter, and n the degrees of freedom. For changes in displacement parameter $\{d_i\}$ and in load λ, by Taylor's expansion, the following equation can be obtained

$$
\begin{aligned}
f_r(D_i^o + d_i, \Lambda^o + \lambda) &= f_r(D_i^o, \Lambda^o) + \left(d_i \frac{\partial}{\partial D_i} + \lambda \frac{\partial}{\partial \Lambda}\right) f_r^o \\
&+ \frac{1}{2!}\left(d_i \frac{\partial}{\partial D_i} + \lambda \frac{\partial}{\partial \Lambda}\right)^2 f_r^o + \cdots = 0 ; \\
&(r = 1, 2, \ldots, n) \quad .
\end{aligned}
\tag{5}
$$

The change in displacements and load (d_i, λ) can be indicaed as functions of a parameter t with conditions $d_i(0) = 0; \lambda(0) = 0$,

$$d_i(t) = d_i't + \frac{1}{2!}d_i''t^2 + \cdots, \tag{6}$$

$$\lambda(t) = \lambda't + \frac{1}{2!}\lambda''t^2 + \cdots. \tag{7}$$

Substituding (6) and (7) into (5), and considering that equation (5) should be tenable for arbitrary t, we obtain the perturbation equation

$$\sum_{i=1}^{n} k_{ri}d_i' = f_{r\lambda}\lambda', \tag{8}$$

$$\sum_{i=1}^{n} k_{ri}d_i'' + 2\left(\sum_{i=1}^{n}\sum_{j=1}^{n} k_{rij}d_i'd_j' + \sum_{i=1}^{n} k_{ri\lambda}d_i'\lambda' - f_{r\lambda\lambda}\lambda'^2\right) = f_{r\lambda}\lambda''. \tag{9}$$

where,

$$k_{ri} = \frac{\partial f_r}{\partial D_i}; \quad k_{rij} = \frac{1}{2}\frac{\partial^2 f_r}{\partial D_i \partial D_j}; \quad k_{ri\lambda} = \frac{\partial^2 f_r}{\partial D_i \partial \Lambda};$$

$$f_{r\lambda} = -\frac{\partial f_r}{\partial \Lambda}; \quad f_{r\lambda\lambda} = -\frac{1}{2}\frac{\partial^2 f_r}{\partial \Lambda^2}.$$

Equation (8) can be indicated in matrix form

$$\mathbf{K}\mathbf{d}' = \mathbf{f}\lambda', \tag{10}$$

where $\mathbf{K}$ is the tangential stiffness matrix, $\mathbf{d}$ is vector of change in displacements and $\mathbf{f}$ is the load model vector. By load incremental method, taking $t = \lambda$, $\lambda' = d\lambda/dt = 1$, $\lambda'' = 0$, omitting the high order, equation(10) takes the form

$$\mathbf{K}\mathbf{d} = \lambda\mathbf{f}, \tag{11}$$

which is the basic equation of load incremental method for finding the load deformation curve and the buckling analysis. Equation (9) is used to analyze the buckling-type.

2.3 Objective Function And Mathematical Model

For a space frame with a certain configuration and subjected to a certain load system, the buckling-load is sensitive to its erection shape, or to its nodal coordinates. So, it is very important to find its optimum shape (nodal coordinates) with maximum buckling-load. The objective function to be maximized is the load-parameter Λ .

By load incremental method, the iteration is done step by step with the change of load-parameter. In the iteration, the buckling point is reached if $\det \mathbf{K} = \mathbf{0}$. For an iteration reaching the buckling point from step 0 to step T, we have

$$\sum_{s=0}^{T} (\mathbf{K}^s \mathbf{d}^s - \lambda^s \mathbf{f}) = \mathbf{0} \ . \tag{12}$$

Take

$$\sum_{s=0}^{T} \lambda^s = \Lambda \tag{13}$$

as the objective function to be maximized, equation (12) can be rewritten as

$$\sum_{s=0}^{T} \mathbf{K}^s \mathbf{d}^s - \Lambda \mathbf{f} = \mathbf{0} \ . \tag{14}$$

Hence, the objective function Λ takes the form

$$\Lambda = \mathbf{g} \sum_{s=0}^{T} \mathbf{K}^s \mathbf{d}^s \quad , \tag{15}$$

where $\mathbf{g} = \frac{1}{n} \left(f_1^{-1}, \ f_2^{-1}, \ \ldots, \ f_n^{-1} \right)$. Then the problem can be expressed as

$$\begin{cases} \text{Maximize} & \Lambda = \mathbf{g} \sum_{s=0}^{T} \mathbf{K}^s \mathbf{d}^s \\ \text{Subject to constraints} & \\ & \Psi(\mathbf{X}) = \mathbf{0} \\ & \psi_i(\mathbf{X}) \leq 0 \ . \end{cases}$$

2.4 The Iteration

In general, the geometrical and mechanical constraints can be divided into active constraints (equalities) and inactive constraints (inequalities).

Some of the geometrical constraints may be given by the architectural requirements or conditions for some specific purposes. Inactive constraints should be adopted to meet the architectural requirements, and to control the locations of the joints not to be too high or too low, or to make it a dome-like one.

The mechanical constraints are always given according to the mechanical principles. If prestresses are introduced in the structure, however the shape be changed by the optimization, the internal forces introduced by prestressing must be in equilibrium without loads, which is the active constraints (ref. Chen,P.S., Kawaguchi,M.,1993).

The geommetrical constraint equations, linear or nonlinear, make a constraint space for the nodal cartesian-coordinates of the structure. The nodal cartesian-coordinates can be defined by a base of the constraint space, or in other words, the nodal cartesian-coordinates $\mathbf{X} = (x_1, x_2, \ldots, n)$ can be indicated by a generalized coordinate system $\mathbf{q} = (q_1, q_2, \ldots, q_m)$ $n \geq m$, which is the independent variable of the optimization.

$$
\left.
\begin{aligned}
x_1 &= \mathcal{X}_1 \left(q_1, q_2, \ldots, q_m \right) \\
x_2 &= \mathcal{X}_2 \left(q_1, q_2, \ldots, q_m \right) \\
&\cdots\cdots\cdots\cdots\cdots\cdots \\
x_n &= \mathcal{X}_n \left(q_1, q_2, \ldots, q_m \right)
\end{aligned}
\right\}
\quad ; \quad (n \geq m) \ .
\tag{16}
$$

Probably, transformation (16) have to be expressed in implicit function, but the Jacobian matrix of the transformation can be obtained

$$
\mathbf{J} =
\begin{pmatrix}
\partial x_1/\partial q_1 & \cdots & \partial x_n/\partial q_1 \\
\vdots & & \vdots \\
\partial x_1/\partial q_m & \cdots & \partial x_n/\partial q_m
\end{pmatrix} \ .
\tag{17}
$$

The incremental direction of the objective function can be obtained by its derivative with respect to the nodal coordinates.

$$\dot{\Lambda}_i = \dot{g}_i \sum_{s=0}^{T} \mathbf{K}^s \mathbf{d}^s + g \sum_{s=0}^{T} \dot{\mathbf{K}}_i^s \mathbf{d}^s \quad . \tag{18}$$

where $\dot{\Lambda}_i = \partial \Lambda / \partial x_i$. Generalized incremental direction respecting the generalized coordinates $\mathbf{q}$ takes form

$$\begin{pmatrix} \Lambda_1' \\ \Lambda_2' \\ \vdots \\ \Lambda_m' \end{pmatrix} = \mathbf{J} \begin{pmatrix} \dot{\Lambda}_1 \\ \dot{\Lambda}_2 \\ \vdots \\ \dot{\Lambda}_n \end{pmatrix} \quad , \tag{19}$$

where $\Lambda_i' = \partial \Lambda / \partial q_i$. The direction moving to the next step of the iteration takes the form

$$\forall i \quad \begin{cases} \Gamma_i = 0 \quad , \quad \text{if} \quad \Lambda_i' \le 0 \\ \Gamma_i = \Lambda_i' \quad , \quad \text{otherwise} \quad . \end{cases}$$

If $\mathbf{\Gamma} = \mathbf{0}$ then stop the calculation, or else change the coordinates, and continue the calculation,

$$\{q_i\}^{T+1} = \{q_i\}^{T} + \alpha \{\Gamma_i\}^{T} \quad ; \qquad (i = 1, 2, \ldots, m) \quad , \tag{20}$$

where α is the step-length which is defined by

$$0 \le \alpha \le \min \left\{ \Psi_j \left(\mathbf{q}^T \right) / \nabla^{\mathsf{T}} \Psi_j^T \Lambda_i'^T \mid \nabla^{\mathsf{T}} \Psi_j^T \Lambda_i'^T \ge 0 \right\}. \tag{21}$$

Conclusion

The present analysis is the first step of our research, and there are many important topics need deeper researches, for example, the multicriteria optimization for maximum backing-load and minimum volume with sensitivity analyses etc..

The method presented is very convenient, and is proposed as a useful aid in design and design automation.

References

Chen,P.S., Abe,M. and Kawaguchi,M., Shape of Tensegrity Frames With an Optimum Rigidity, Space Structure 4 (the proceedings of Furth International Conference on Space Structures held at the University of Surrey,UK, Sep. 5-10th, 1993),Vol.1 .

Chen,P.S., Kawaguchi,M., Minimum-Deformation-Shape of Prestressed Bar-Structures, Proceedings of Seiken-IASS Symposium on Nonlinear Analysis And Design For Shell And Spatial Structures held at the University of Tokyo Oct. 19-22th 1993.

Hangai,Y. and Kawamata,S., The Analysis of Geometrically Nonlinear And Stability Problems By Static Perturbation Method, Report of the Institute of Industrial Science, The University of Tokyo,Vol.22, No.5, January, 1973.

Kawaguchi,K., Hangai,Y. and Nabana,K., Numerical Analysis For Folding of Structures, Space Structure 4 (the proceedings of Furth International Conference on Space Structures held at the University of Surrey,UK,Sep. 5-10th, 1993),Vol.1 .

Lin,J.H., Che,W.Y. and Yu,Y.S., Structural Optimization on Geometrical Configuration And Element Sizing With Statical And Dynamical Constraints, Int. J. Comp. & Srtuct., Vol.15(5), 1982, pp507-515.

Rozvany,G.I.N., Structural Design via Optimality Criteria : the Prager Approach to Structural Optimization, Kluwer Academic Publishers, Boston, 1989.

Sabatier,P.C., Introduction And A Few Questions, Inverse Problems: An Interdisciplinary Study (the proceedings of a meeting held at Montpellier Dec.1st-5th 1986), pp1-5 .

Thompson,J.M.T. and Hunt,G.W., A General Theory of Elastic Stability, John Wiley, 1973.

Structural Failure and Quality Assurance
of Space Frames

Tien T. Lan[1] M. ASCE

Abstract

Space frame has been developed rapidly in China and used to wide extent to cover various types of buildings. The large scale of application will unavoidably cause quality problems. Damages or even collapse of space frames have occured. Typical cases of failure are described in the paper. The reason of these failures varies and flaws in design, fabrication and erection causing the failure are analysed. Measurements were taken to assure the quality of space frames. In China, a complete system of quality control has been established. Codes and Standards are powerful tools for quality assurance.

Introduction

Since the construction of the first space frame in 1964, it has been developed ralidly in China. Space frames were applied to cover various types of buildings, from short to long span. The widest field of application for space frame is in sports buildings, the maximum span has attained 110 m. It has also been applied to roof the single-storey industrial buildings. Many workshops with area over 10,000 sq. m. were covered by space frames supported on intermediate columns with large grids. Nowadays in China, due to the large demand, there are

---

[1] Research Fellow and Professor, Institute of Building Structures, Chinese Academy of Building Research, P.O. Box 752, Beijing 100013, China.

more than 50 manufacturers, large and small, specialised
in the production and erection of space frames.

Such large scale application will unavoidably cause
quality problems. Damages or even collapse of space
frames have occured. Since space frame is frequently
used to cover public buildings of high population density
such as for sporting or cultural events, any catastrophic
failure will introduce serious results and consequently
danger to people. Therefore, the quality problem of
space frame has attracted much concern from the public.
It is necessary to investigate the direct or indirect
causes of failures which are generally due the deficiencies in design, fabrication or erection.

Cases of Failure

1. Meeting Room for a Communication Building, Taiyuan

On the top floor of a communication building for
Gujiao Mining District in Taiyuan, Shanxi Province, a
meeting room was covered by a space frame, 13m x 18m
rectangular in plan. The space frame is of the square
on diagonal type and supported along perimeters. The
modules are 5 x 7 with a depth of 1m. Steel circular
tubes are used as members connected by welded spherical
nodes.

The space frame was assembled by welding on a
neighboring roof top and then erected by a cumulative
sliding method. After the construction of the overlaid
r.c. slabs, the insulation on the roof and suspended
ceiling on bottom chords was finished, the space frame
suddenly collapsed without any warning at the dawn of
June 7th, 1988. Before the accident, there were moderate and heavy rains in that area for several days.

Figure 1 shows the scene after the collapse, the
space frame, r.c. slabs and the ceilings tangled in a
heap of rubbles. The supporting walls were not damaged
and the supports on one shorter side still remained on
the top of the wall. Along other three sides, all the
embeded elements of the supports were pulled down. The
space frame, together with the roof slabs and ceilings,
fell on the floor. It was found that most web members
were seriously buckled in a S shape. Some circular tubes
as the web member were broken near the connection with
nodes, showing a distinct neck-down.

The failure is mainly due to the serious mistake
in design which was accomplished by an unqualified designer. The space frame was checked after the collapse,
it was found that 20 web members, i.e. nearly 1/3 of

Figure 1   Collapse of space frame
for a Communication Building

the total, were overstressed.   The maximum compressive
stress of certain member was 2.24 times the allowable
stress.   The buckling of the web members was the main
cause of collapse.   Besides, the poor quality of construction was another reason of failure.   Many welds
connecting the steel tubes and spherical nodes were
damaged due to unthorough welding.   Instead of single
bevel groove welding as required by the Specifications,
fillet welds were used.

2. Carpet Production Shop of Dongfeng Chemical Works,
   Beijing

     One of the main shops with an area of 54m x 90m
employed 18m x 18m square on larger square space frame
as basic unit to cover the roof.   The space frame, composed of circular tubes and bolted nodes, was assembled
on the ground.   Each unit was lifted up by crane and
rested on column capitals.   In order to hold the column
in its right position, four steel pipes were assigned
to each column capital as temporary struts (Figure 2).
The erection of the space frames was proceeded according
to the sequence as shown in Figure 3.   After the fifth
unit was settled on the column top, suddenly the column
capital No. 1 turned over first and then all five units
of space frame fell to the ground.   The arrows in Figure
3 show the direction of turn over of column capitals.
It was found that many top and bottom chords suffered
large deformation and some of the high tensile bolts
were broken by shearing.

Figure 2   Erection of Space Frame
for Carpet Production Shop

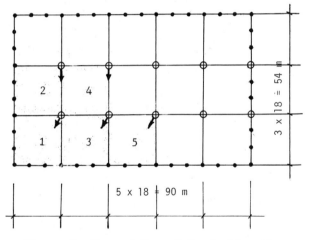

Figure 3   Plan of Carpet Production Shop

It was evident that   deficiency in erection caused
the collapse of space frames. In the erection of previous building, the construction was going so smoothly
that the temporary struts supporting the column capitals
were omitted in the later. In addition, the anchor bolts
connecting the capital to the column top were not welded
thoroughly. The column capitals were turned down easily
under eccentric loadings and the anchor bolts were
pulled out from the column. The collapsed site was
cleared immediately. After rigorous inspection of the

damaged space frame, only qualified members and nodes
were used for rehabilitation.

3. Theater in Wuqia County, Sinkiang Autonomous Region

The theater is composed of three parts, i.e. the
lobby, the audience hall and the stage. Figure 4 shows
the plan and section of the building. A 24m x 27m space
frame was used to cover the audience hall, the module
and depth being 3m and 2.7m respectively. The stage by
the size of 24m x 10.8m employed r.c. girders as the main
roof structure. One end of the girder is supported on
r.c. posts extending from a transverse girder across the
stage opening, while the other end is supported on the
brick gable wall. The lobby was constructed with rigid
frames which also serve as support for space frame.

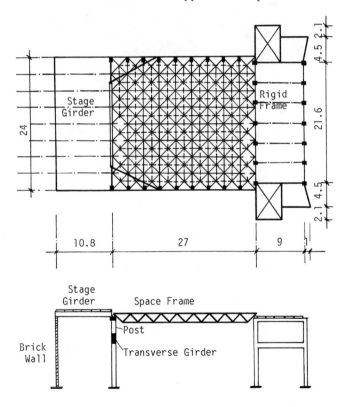

Figure 4 Plan and Section of the Theater

In August and September of 1985, strong earthquakes have been occured in Wuqia County. The magnitude of the shock was registered as high as 7.4 on Richter scale. The main supporting structure and the roof of the theater did not collapse, but the space frame suffered serious damage. During the earthquake, all r.c. slabs were laid on the roof, except the waterproofing and ceilings were not completed. About half of the design load was acting on the space frame. Bowing of member appeared on most upper chords in the first module near the stage. Some supports near the stage were also damaged resulting large pieces of concrete spalled and reinforcements exposed.(Figure 5) Similar damages also occured in the upper chords near the lobby, but to a lighter degree.

Figure 5   Bowing of Upper Chords due to Earthquake

Computer analysis based on the Response Spectrum Method was used to estimate the earthquake effect. The dynamic characteristic reveals that the space frame is quite sensitive both to the horizontal and vertical com- ponents of ground motion. The inertia effect of the supporting structures of space frame should not be ignored. In this case, the inertia effect due to the transverse girder across the stage opening as well as the stage girders is the main cause of failure. During the earthquake, a strong inertia force was transmitted to the upper chords of the space frame, causing a drastic increase of internal forces in members, especially near the stage opening. The consequence was the buckling of upper chords and damage of supports.

The structural layout of supporting both the stage girders and space frame on a single transverse girder

is most unfavorable with regard to earthquake resistance.
The stage roof, constructed with r.c. slabs, is possessed
of a large mass, yet its supporting girder has a very
small lateral stiffness.   Thus the upper chords of the
space truss were compelled to transmit the large inertia
forces.   The failure of members near the end by buckling
was inevitable.

4. Shenzhen International Exhibition Center

    The exhibition area includes 5 exhibition halls, of
which Hall No. 1, 3, 5 is 45m x 45m with an area of 2025
sq m and Hall No. 2, 4 is 22.5m x 28.5m with an area of
640 sq m.   These halls which use space frame with bolted
spherical nodes as roof, are linked closely to each
other at different elevations.   The space frame for Hall
No. 4 is composed of orthogonal square pyramids and supported on 4 columns.

    After more than three years of operation, the space
frame of Hall No. 4 suddenly collapsed in a mornig of
September 7th, 1992 during a rainstorm.   On the site,
the space frame fell down completely from roof.   On one
side, the roof elements scattered on the ground and the
rest part were still being supported on the column (Fig.
6).   Some of the longitudinal lower chords and web members were buckled.   Many high tensile bolts appeared to
have experienced tension failure and many sleeves were
yielded by bending.   There were no distinctive breaking
nor neck-down of tube member.

Figure 6   Collapse of space frame
for Exhibition Hall No. 4

The investigation into the collapse of space frame
revealed the defect in the drainage system of Hall No. 4.
It was found that the drainage gutter on the roof of
neighboring Hall No. 6 was often full during heavy rain-
ing and the rainwater overflowed down to Hall No. 4.
Owing to the fact that the discharge ability of gulleys
was insufficient, there have been several times that
rainwater detained on one side of the roof up to a considerable depth. In the morning of the accident, there was
rainstorm in Shenzhen area with a very intense rainfall.
The situation was worsened by the fact that the gulleys
were partly clogged. The overloading on the roof due to
rainwater ponding was the main cause of collapse. This
is justified by the reanalysis of the space frame considering the actual load. A trapezoidal distribution of
load was assumed. When the maximum depth of water
attains 45cm, the compressive force in those web members
near the support on the higher loading side will exceed
the critical buckling load. Meanwhile, the force in
tension members will also exceed the bearing capacity of
high tensile bolt. Thus the buckling of compression
members and breaking of the bolts lead to the total collapse of the structure.

5. Bowing and Breaking of Members of Space Frame

Several cases of bowing of members occured during or
after the erection of space frame. Figure 7 is a typical
example showing the damage of space frame for a rolling
stock plant in Qingdao. It also happened recently during
the erection of the space frames for the grandstnds of
Shenzhen Stadium. The reasons of bowing have been manifold. Some were due to unproper erection, such as the
compulsory fitting of members with large tolerance. Some
were due to design by using slim members or large difference of tube section intersecting at the node. When the
'Specifications for the Design and Construction of Space
Trusses' in China was revised, based on the experiences
in practice all the requirements for the allowable slenderness ratio of compression members and minimum section
of all members were corrected to a more strict side.

The breaking of members is also one of the few drawbacks in the construction of space frames. Figure 8 shows
the breaking between the cone and tube of a member. Such
breaking, caused by the defect in welding, usually occured when part or all loadings were imposed on the space
frame. For example, during the erection of two space
frames of Zhengzhou International Exhibition Center, 45m
x 45m each, three lower chords near the center suddenly
broke and the space frame sagged 5-6cm. The examination
of the broken section revealed very poor quality of welding. The welding did not meet the design requirement of

Figure 7   Bowing of Member

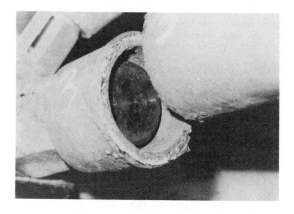

Figure 8   Breaking of Member

single-V groove weld with minimum thickness of 12mm. In reality, the thickness of the weld was less than 3mm with many cavities.

Quality Assurance

Due to the increased amount of space frame construction in China, a large amount of design institutes and factories were engaged in the design, fabrication and erection, of which many were newly established. The

production of space frame is getting more and more specialized and industrialized. Nevertheless, due to the large
difference in the technical level of these designers and
manufacturers, there have been occured accidents of different degrees of flaws. Fortunately there were no serious
injuries or casualties until now, yet the quality inspection and control is still an important task. A complete
system of quality assurance for space frame, starting
from the design stage and up to the check and acceptance
of the structure, has been established. Codes and Standards were issued which are mandatory for all designers
and manufacturers. The revelant technical documents are:

JGJ 75-91 Standard for Steel Space Frame Structures.

JGJ 7-91 Specifications for the Design and Construction of Space Trusses

JGJ 78-91 Standard for Quality Inspection and
Assessment of Space Trusses

The Standard JGJ 78-91 was compiled especially for
the purpose of quality control. The compilation of the
Standard was based on the former design and construction
specifications and exiting experience in China. It contains the following seven chapters and an appendix:

1. General principles
2. Welded hollow spherical nodes
3. Bolted spherical nodes
4. Welded plate joints
5. Members
6. Erection of space truss
7. Painting works

In each chapter, the corresponding technical requirements,
quality assurance, tolerance, method of examination and
assessment of quality grade are provided. The prescribed
indices, not only take into consideration of the needs of
quality control, but also fulfill the requirements of ins-
pecting judgement.

Quality Inspection Office has been established in
each city which undertakes the duty of quality assurance
of all construction project. The Office will inspect
every space frame that is being built and assess the quality grade which will be counted as a part of the general
grade of the project.

ADAPTABILITY OF TRANSLATION-GRID SHELLS

Dr.-Ing. J.G. Oliva Salinas
Member of the IASS

Abstract

The aim of this work is to show the adaptability of translation-grid shells (TGS) in order to cover a building with a predesigned plan. Although a TGS obeys to a geometrical ordinance, which might appear very rigid and with not too many possibilities of adaptation to free plans, the project for the cover for an Auditorium called "Sala del Pleno" at the Federal Electing Tribunal in Mexico City will be described. The plan of this building was modified three times. Every one of these three different adaptations will be described.

Introduction

Before we begin with the adaptation of the structure and in order to understand the structural behavior of the TGS, some properties of this system will be briefly defined*.

The translation surface is generated with two curves which are orthogonal between themselves. Both curves lies on two planes, which are normal to the horizontal plane.

The selection of catenaries by translation surfaces with positive gaussian curvature gives the form of a hanging net. The form finding procedure is made with the computer program GEOG. The building of a measurement model is not necessary.

GEOG delivers in a PC the whole geometry, coordinates, rode lengths, angles and diagonal lengths of the structure.
Nevertheless, the designed TGS with GEOG has only two possibilities of edges:
- TGS with curvilinear edge on a horizontal plane.
- TGS with straight edges.

Co-ordinator of the Research Center of Architecture and Townplanning
National Autonomous University of Mexico
Edificio Unidad de Posgrado, 2do. piso
Ciudad Universitaria
Mexico, D.F. CP 04510
Mexico
* See "Innovative Large Span Structures", IASS-CSCE
   International Congress 1992, Volume 2, Pages 633 and 634.

The Federal Electing Tribunal is an independent organism, which sanctions the federal elections. The building consists mainly in offices, an auditorium, squares, yards, gardens and parking.

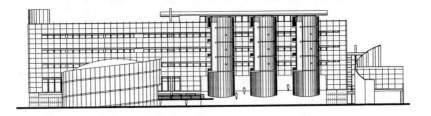

Federal Electing Tribunal
Fig. 1

Now we will discuss the Auditorium only, which is completely separated from the rest of the building.

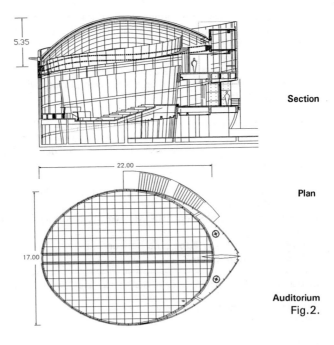

Section

Plan

Auditorium
Fig. 2.

1st. Adaptation:

This adaptation was the easiest one, because the task was only to cover an horizontal circular plan with a diameter of approximately 21.00 m and a height of apex of 4.00 m. With GEOG, it was determinated the form of a TGS in which the directrix and generatrix was a catenary with the same value of the parameter a = 14.402558. The dimensions of the TGS are: x = 21.00m; y = 21.00 m; z = 4.00 m; MW = 1.00 m.

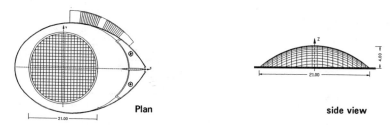

Plan   side view

Translation Grid Shell (1st. Attempt.)
Fig.3

The plan is not exactly a circle, but the edge differences can be considered in two different ways:
- Edge differences considered along the edge beam, on a horizontal plane, **Fig. 4 case A.**
- Edge differences considered by taking the axis of the edge beam and growing up until we cut the translation surface, such as it is shown in the next drawing, **Fig. 4 case B.**

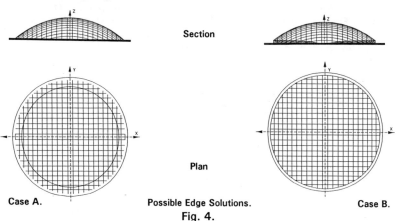

Section

Plan

Case A.   **Possible Edge Solutions.**   Case B.
Fig. 4.

2nd. Adaptation:

In this case the structure should cover the whole building. The plan was generated with two circumferences. The radius of the biggest circumference was equal to the diagonal length of the quadrant which involved the smaller circumference. The end plan was an ovoid.

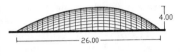

side view ·

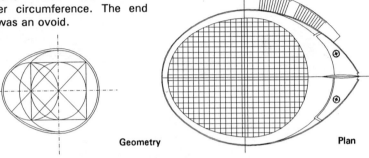

Geometry

Plan

Translation Grid Shell (2nd. Attempt.)
Fig. 5.

To cover the ovoid plan, it was necessary to generate a translation surface consisting of two different shells. The first shell obeys to the plan of a half circle. For this reason it was used exactly the same geometry of the shell designed by the first attempt. Both translation curves, generatrix and directrix obey to a catenary with the same value of the parameter a. **Fig. 6 case A.** For the second shell, the directrix remained the same as before but the generatrix changed its parameter a, in such a way that the edge of the translation surface could lie on the three points of the ovoid plan, such as it is shown in the **Fig. 6 case B.**

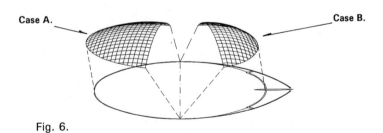

Case A.   Case B.

Fig. 6.

By this case two possible structural systems were evaluated. The first one consisted in metal rips with a mesh width of 1.00 m. **Fig. 7 A.** The second structural system would be resolved with prefabricated concrete plates. In this case by using the geometrical property of the translation surface consisting in the cooplanarity of the four nods. The mesh width was 4.00 x 4.00 m.**Fig. 7.B.**

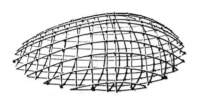

A. TGS with metal rips.

B. TGS with prefabricate concrete plates

Fig. 7.

The next drawing shows some of the cutting patterns (one quarter), which were previously calculated with GEOG and that correspond to the concrete plates.

Cutting Patterns
Fig. 8.

3rd. Adaptation:

The third and final designed changed to an almost elliptical plan, which was generated with circumferences. **Fig. 9.A.** The length axis had an unevenness of approximately 2.20 m. See **Fig. 2.** page 2 and **Fig. 9.B.**
With an analog process used by the second attempt, the translation surface was generated with two different shells, in which three different catenaries were applied. **Fig.** 9 - 11.

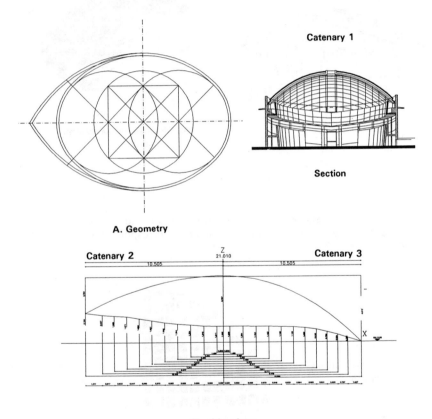

Catenary 1

Section

A. Geometry

Catenary 2

Catenary 3

B. side view

Fig. 9.

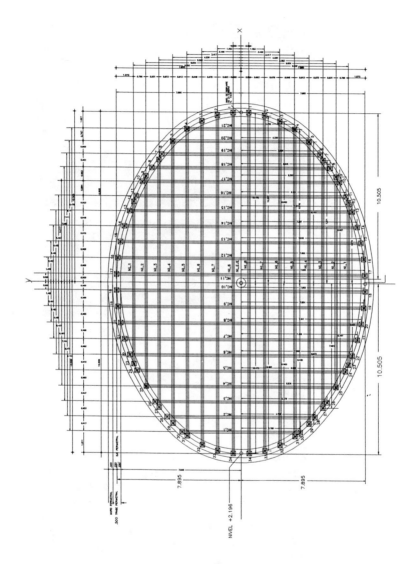

Plan
Translation-Grid Shell
Fig. 10.

1) Auditorium

2) Prefabricated
   timber plates

3) Translation-Grid
   Shell

Fig. 11.

References

Faber,C., 1965 "Candela und seine Schalen". Verlag Georg D.W. Callwey, München.

Nooshin, H., 1984, "Third International Conference on Space Structures". Elsevier Applied Science Publishers LTD, London and New York.

Oliva Salinas, J.G., 63/1982, "Über die konstruktion von Gitterschalen". Selbstverlag SFB 64, Universität Stutgart.

Otto, Frei und Mitarbeiter, 1975, "Gitterschalen", Sebstverlag Institut für leichte Flächentragwerke, Universität Stutgart.

Srivastava N.K., Sherbourne A.N., Roorda J., 1992, "Innovative Large Span Structures", IASS-CSCE International Congress 1992.

Thierauf, G., 1985, "Benutzerhandbuch für das Programmsystem B&B zur Berechnung und Bemessung allgemeiner Flächentragwerke, Universität-GH-Essen.

PRISMS AND ANTIPRISMS

Dr. Pieter Huybers[1]
and
Gerrit van der Ende[1]

Abstract

Prisms form a group of mathematical figures, that have found wide-spread application in many disciplines, but especially in architecture and in building structures. They have two identical parallel polygonal faces that are kept apart by a closed ring of squares or of triangles. The two polygons and the square or triangular faces of the mantle enclose a portion of space, that is competely surrounded by regular polygons. They have therefore very much in common with the Platonic and Archimedean - often called 'uniform' - polyhedra.

Both groups form endless rows as the parallel polygons can have any number of sides. They were first mentioned and shown in sketch by Kepler in the 16th century [Ref. 1]. The present paper deals in detail with these figures and their duals, as well as with similar solids, having polygrams (or star-shaped) parallel faces. Attention will be paid to practical applications in architecture or in engineering of some representants.

Geometrical properties

The geometrical properties of prisms and antiprisms can be determined in a similar way as is done for the uniform polyhedra. These properties are all laid down in the pyramidal caps that can be cut off from the corners of these solids at a distance of the unit edge length. The basis of such a 'vertex pyramid' is called: 'vertex figure'. This is identical for all vertices of any polyhedron and it is therefore characteristic, as it contains all geometric data that are specific for such a polyhedron.

---

[1] Delft University of Technology, Civil Engineering Department,
1 Stevinweg, Delft 2628 CN, The Netherlands

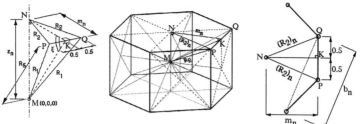

Fig. 1. *Principal data of prismatic polyhedra.*

Prisms

The prism has a triangular vertex figure with the sides: $\sqrt{2}$ - $\sqrt{2}$ - b_n, where b_n is the so-called 'lesser diagonal' which connects two alternate corners in one of the two parallel regular polygons.

$$b_n = 2 \cos\varphi_n \qquad \{1\}$$

$$\text{with } \varphi_n = \frac{180^\circ}{\pi} \qquad \{2\}$$

so that the circumscribed circle of a polygon with n sides is:

$$(R_2)_n = \frac{1}{2 \sin\varphi_n} \qquad \{3\}$$

The inscribed circle radius:

$$M_n = (R_2)_n \cos\varphi_n \qquad \{4\}$$

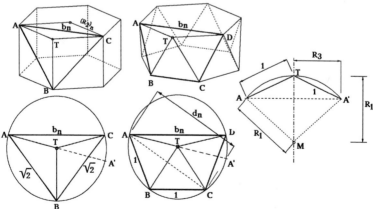

Fig. 2. *Vertex figures of n-sided prisms and antiprisms.*

The radius of the circumscribed circle of the vertex figure, that connects the other ends of the edges that meet in the vertex of a prism:

$$R_3 = \frac{a\ b\ c}{4\ \text{Area(ABC)}} = \frac{\sqrt{2}\ \sqrt{2}\ b_n}{4\ \sqrt{(\sqrt{2} + \cos\varphi_n)\ \cos\varphi_n\ \cos\varphi_n\ (\sqrt{2} + \cos\varphi_n)}}$$

$$= \frac{1}{\sqrt{(2 - \cos^2\varphi_n)}} \qquad \{5\}$$

The radius of the circumscribed sphere of the prism:

$$R_E = \sqrt{R_n^2 + 0.25} = \sqrt{\frac{1}{4\ \sin^2\varphi_n} + 0.25} = \frac{\sqrt{1 + \sin^2\varphi_n}}{2\ \sin\varphi_n} \qquad \{6\}$$

The radius of the 'inter-sphere' [Ref. 3]:

$$R_5 = \sqrt{(R_1^2 - 0.25)} \qquad \{7\}$$

The inter-sphere connects the mid-points of the edges. The radius of the 'in-sphere', which is the sphere that touches the reciprocal faces:

$$R_6 = \sqrt{(R_5^2 - 0.25\ R_3^2)} \qquad \{8\}$$

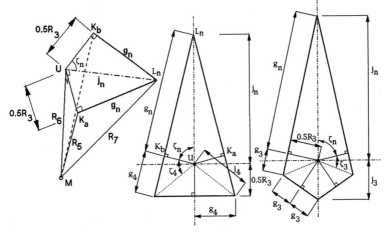

Fig. 3. *Faces of the reciprocals or duals.*

$$\zeta_n = \arcsin \frac{\cos\varphi_n}{R_3} \qquad \{9\}$$

$$g_n = 0.5\ R_3\ \tan\zeta_n \qquad \{10\}$$

$$j_n = \frac{g_n}{\sin\zeta_n} \qquad \{11\}$$

The distance of a vertex from the system centre:

$$(R_7)_n = \sqrt{(R_6^2 + j_n^2)} \qquad \{12\}$$

Antiprisms

The antiprism has a trapezoidal or quadrangular vertex figure with the sides: $1 - 1 - 1 - b_n$. This has a diagonal with the length:

$$d_n = \sqrt{1 + 2 \cos\varphi_n} \qquad \{13\}$$

The circumscribed circle of this vertex figure:

$$R_3 = \frac{1*1*d_n}{4 \, \text{Area(BCD)}} = \frac{d_n}{4 \sqrt{(1 + 0.5 \, d_n) \, 0.5 \, d_n*0.5 \, d_n \, (1 - 0.5 \, d_n)}}$$

$$= \frac{1}{2 \sqrt{1 - (0.5 \, d_n)^2}} = \frac{1}{\sqrt{3 - 2 \cos\varphi_n}} \qquad \{14\}$$

The radius of the circumscribed sphere:

$$R_F = \frac{1}{2 \sqrt{1 - R_3^2}} = \frac{1}{2 \sqrt{1 - \dfrac{1}{3 - \cos\varphi_n}}} = \frac{\sqrt{3 - 2 \cos\varphi_n}}{2 \sqrt{2 - 2 \cos\varphi_n}} \qquad \{15\}$$

As $1 - \cos 2\varphi_n = 2 \sin^2\varphi_n$:

$$R_F = \frac{\sqrt{3 - 2 \cos\varphi_n}}{4 \sin(\varphi_n/2)} \qquad [\text{Ref. 3}] \qquad \{16\}$$

By substitution of the cosine, the equation is obtained that can be found at Bruckner [Ref. 2]:

$$R_F = \frac{\sqrt{3 - 2 + 4 \sin^2(\varphi_n/2)}}{4 \sin(\varphi_n/2)} = 0.5 \sqrt{1 + (\frac{1}{2 \sin(\varphi_n/2)})^2} \qquad \{17\}$$

The values of R_5, R_6, ζ_n, g_n, j_n and $(R_7)_n$, as well as those for n=3 or n=4 in both cases, can similarly be calculated with the formulas $\{7\text{-}12\}$.
The dihedral angles in an antiprism are a summation of one part which is contributed by the triangle and another part by the n-gon, thus:

$$\xi_n = \text{arctg} \, \frac{z_n}{M_n} \quad (\text{Bruckner: } = \text{arctg} \, \frac{\sqrt{\sin(\varphi_n/2)}}{\cos\varphi_n} \sqrt{1 + \cos\varphi_n}) \qquad \{18\}$$

$$\xi_3 = \text{arctg} \, \frac{\sqrt{1 + 2 \cos\varphi_n}}{2 \sin(\varphi_n/2)} \qquad \{19\}$$

$$\xi_{total} = \xi_n + \xi_3 \qquad \{20\}$$

The dihedral angles in the reciprocal figure of an antiprism are all alike and equal to:

$$\vartheta = 2 \, \text{arctg} \, \frac{2 \, R_6}{R_3} \qquad \{21\}$$

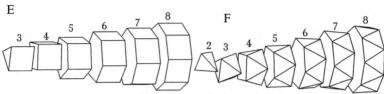

Fig. 4. *Rows of prisms (E) and antiprisms (F) with 2 to 8 sides. Note that E3 is identical to a cube, F2 to a tetrahedron and F3 to an octahedron.*

Star-shaped or polygrammatic versions

The two parallel polygonal faces can be substituted by regular stars or polygrams. This produces two new families: star-prisms and star-antiprisms. In the first group a mutual distance, equal to the unit edge length can be chosen, as in the normal prism. The resulting figure has a mantle, consisting of rectangles. The star-antiprisms have a somewhat unexpected appearance [Ref. 4]. At closer examination the forms with even numbers of sides seem to be composed out two antiprisms of half the number of sides, but with an edge length b_n {1}.

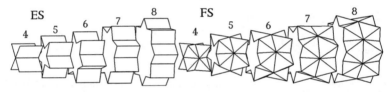

Fig. 5. *Polygrammatic prisms (ES) and antiprisms (FS). The square FS4 is identical to the Stella Octangula [Ref. 1].*

Form generation

Forms like these can be generated in a fully automatic way, requiring the input of only very few parameters. If the number of sides is known, all other geometric data can be computed, using the preceeding formulas, for any of the four families that were indicated before. The two parallel faces are first generated and placed at a distance y = + or - z_n from the Z-X-plane, with

$$z_n = \sqrt{(R_1^2 - R_{2n}^2)} \quad \text{(see Fig. 1)} \quad \{22\}$$

This is equal to 0.5 for prisms. The mantle is formed by rotational reproduction of a square or a triangle. For the formation of prisms a square is generated in the XY-plane, then translated over the distance $M_n = z_4$ along the Z-axis and finally placed n times under mutual angles of 2φ around the vertical Y-axis. The mantle of an antiprism is formed in two stages. One triangle is placed point-down at a distance z_3, under an angle of $+\alpha = 90° - \xi_{total}$ to the X-axis and n times rotationally reproduced under mutual angles of 2φ with the Y-axis. The second row is placed point-upwards under an angle $-\alpha$ and similarly reproduced with a starting angle of φ.

Reciprocal prisms can be generated similarly by rotational reproduction of the isosceles triangles under an angle of $+$ or $-\vartheta/2$. The quadrangular sides of the reciprocal antiprism have an inclination of $+$ or $-\alpha$ = arcos (j_n/R_{7n}). Compounds of the original figures and their reciprocals are made with the same routine and using the faces of the reciprocals, upon which small vertex pyramids are placed.

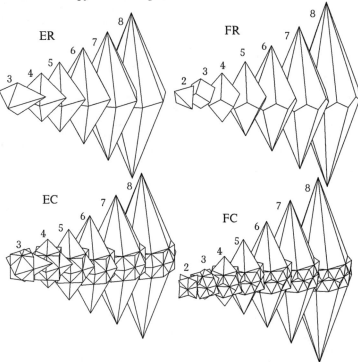

Fig. 6. *The reciprocal prisms (ER) and antiprisms (FR), and their compounds (EC and FC) with the original figures. Note that ER4 is an octahedron, FR2 a tetrahedron and FR3 a cube; FC2 is the Stella Octangula as FS4 (see Fig. 5).*

Prismatic forms

The most simple structural forms are the prismatical shapes. They fit usually well together and they allow the formation of close-packings in many varieties. The accompanying figure shows examples of such applications: matrices can be formed with regular or deformed prisms, parts can be linked up in rows to make cylinders or elongated prisms can serve as the elements of space frames.

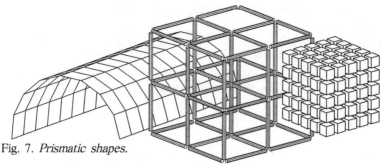

Fig. 7. *Prismatic shapes.*

Antiprismatic folding structures

If a number of antiprisms is put together according their polygonal faces, a geometry is obtained which is often used as the basis for structural applications - usually in a more or less adapted form. The outer mantle has the appearance of a cylindrical, concertina-like folded plane. It can be described by a combination of 3 angles: α, β and γ. The element in Fig. 8 represents 2 adjacent triangles, that in this case are not taken equilateral, so that a more general character is obtained.

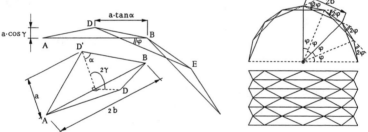

Fig. 8. *Geometric data of antiprismatic structures.*

α = half the top angle of the isosceles triangle ABC with height $\underline{a}$ and base length $\underline{2b}$.
γ = half the dihedral angle between the 2 triangles along the basis.
φ_n = half the angle under which this basis with the length 2b is seen from the cylinder axis.
The relation of these angles α, γ and φ_n [Ref. 5]:

$b = a \tan\alpha = a \cos\gamma \cot\varphi_n + a \cos\gamma / \sin\varphi_n$

$\tan\alpha = \cos\gamma (\cos\varphi_n + 1) / \sin\varphi_n$

$\tan\alpha = \cos\gamma \cot \varphi_n/2$ {23}

These three parameters define together with the base length (or scale factor 2b) the shape and the dimensions of a section in such a structure. This provides an interesting tool to describe any antiprismatic configuration. Two additional data must be given: the number of elements in transverse direction (p) and that in length direction (q).

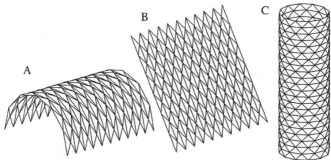

Fig. 9. *Antiprismatically folded forms (A:* α *= 65 and* γ *= 73.6, B:* α *= 65 and* γ *= 90, C:* α *= 45, and* γ *= 82.4).*

Manipulation of prismatic and antiprismatic forms

O.L. Tonon describes methods to modify the general shape of such anti-prismatically folded planes [Ref. 6]. In the present paper this is being worked for circular transformations, so that toroidal and spherical overall forms are found on the basis of polygonal or star-formed prisms and antiprisms. Parts of these can be combined for larger compounds.

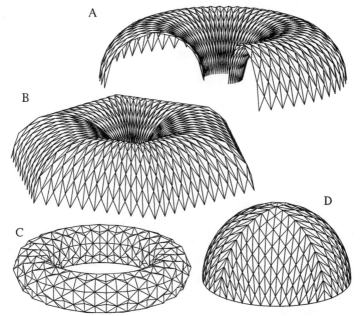

Fig. 10. *Modifications of antiprismatic forms (C = torus of FS8's).*

Sphere subdivisions on the basis of prisms or antiprisms

As these forms can be generated by the rotation of polygons in space, it is imaginable that any object can be rotated instead, using the same rotation angles. This offers the opportunity to start from polygons that are subdivided appropriately into smaller parts, preferably triangles. In this way suitable patterns can be depicted on the faces of such a figure. As a next step the whole pattern can be projected upon a sphere. This is done by converting the cartesian coordinates, that are found with the previously described rotation procedure, into polar coordinates and by substituting the direction vector for the radius of the circumsphere.

This is demonstrated in Fig. 11 for a hexagonal prism and an antiprism. Hexagonal pyramids with subdivided faces are in this case placed on their top and bottom. This method reminisces very much to that where the icosahedron is used as the starting point, which has a pentagonal basis. It had allready been pointed out by the first author [Ref.7] and by Nooshin et al. [Ref. 8], that any of the regular and semi-regular polyhedra are usefull in this respect. But the prisms and antiprisms were never mentioned for this purpose - presumably as their automatic generation is not familiar. It would be interesting to further investigate, what particular aspects arise from this approach.

Acknowledgements

The data for this paper and the visual output has been generated with the help of the computer programme CORELLI, being developed by the authors in GFAbasic both for ATARI and for IBM- environments. Both versions are at present meant for study purposes. The output can be obtained in alphanumeric form and in the form of vector files, that are compatible with most of the commercially available computation or presentation software.

References

1. Kepler, J., Harmonices Mundi, Liber II (1571-1630).
2. Bruckner, M., Vielecke und Vielflache, Theorie und Geschichte, Druck und Verlag von B.G.Teubner, Leipzig, 1900.
3. Cundy, H.M. and A.P. Rollett, Mathematical models, Oxford University Press, 1968.
4. Holden, A., Shapes, space and symmetry, Columbia University Press, New York, 1971.
5. Huybers, P., See-through structuring, a method of construction for large span plastics structures, Delft University Press, Delft, 1972.
6. Tonon, O. L., Geometry of the spatial folded forms, 4th Conf. on Space Structures, 5-10 Sept. 1993, Guildford, England, p. 2042-2052.
7. Huybers, P., The formation of polyhedra by the rotation of polygons. 4th Conf. on Space Structures, 5-10 Sept. 1993, Guildford, England, p. 1097-1108.
8. Nooshin, H. and D. Tzourmakliotou, 4th Conf. on Space Structures, 5-10 Sept. 1993, Guildford, England, p. 1085-1096.

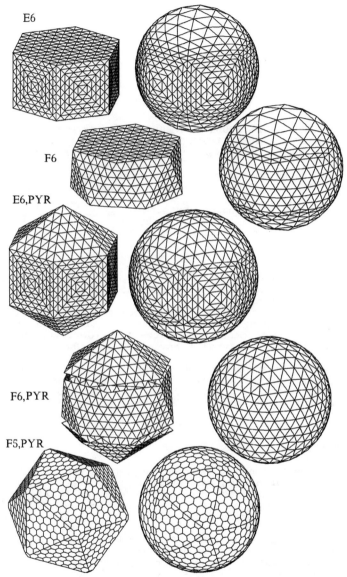

Fig. 11. *Prism and antiprism based triangular subdivisions of spheres. Irregularities in the pattern can be reduced by the addition of n-sided pyramids, that are equally subdivided (Note: P5-PYR is an icosahedron).*

EXPERIMENTAL STUDY ON A BUCKLING BEHAVIOR
OF A WOODEN SINGLE-LAYER SPACE FRAME
OF RETICULAR SHELL ON A HEXAGONAL PLAN

-- Influence due to the Bolted Joint Looseness
with Axial and Bending Resistance --

Kouichi Matsuno[1]
and
Shigeru Aoki[2]

Abstract

The authors have investigated about the structural
behavior of a wooden single-layer space frame structure
with the bolted joints. We have carried out a buckling
tests of a wooden single-layer space frame of reticular
shell on a hexagonal plan. The purpose of our experimental study is to clear the bolted joint activity which
affects to the structural behavior of the whole structure.
In this paper,we have described the buckling behavior of a wooden reticular shell on a hexagonal plan.
Especially, we have discussed the influence due to the
bolted joint looseness with axial and bending resistance.

Introduction

In general,the bolted joint system has been applied
to the joint of a long-span wooden structures. As the
wooden members are fastened with several bolts at each
joint and looseness exists between bolts and bolt's hole
, the state of connection is neither rigid nor pin
connection.
Up to the present, only few studies have so far
been made at the influence of a bolted joint looseness.
However, the authors have been studying the influence of
_ _

[1]Manager, JDC Corporation, Engineering Dept., 4-9-9
Akasaka Minato-ku, Tokyo 107, JAPAN
[2]Professor, Dr., Department of Architecture, Hosei
University, Koganei, Tokyo 107, JAPAN

a bolted joint looseness with axial and bending resistance. Furthermore, we have suggested that a bolted
joint looseness should be substituted axial and bending
rigidity in the analytical model.

On the other hand, although a large number of studies
have been made on a buckling behavior of a single-layer
space frame, little is known about a buckling behavior
of a wooden reticular shell with a bolted joint system.

In this paper, we have described the results of
buckling tests for a wooden single-layer space frame of
reticular shell with a bolted joint system on a hexagonal plan. Furthermore, in order to observe the stiffness of a bolted joint, the tensile and bending tests
were carried out. These results of joint tests were
described, too.

In this study, to make clear the influence of the
number of bolts, the joint of test specimens were consisted of one-bolt, two-bolts and four-bolts.

Purpose of Experimental Study

Tensile Tests of the Bolted Joint. In order to
observe the structural behavior under the axial force
and the axial rigidity of the bolted joints, the tensile
tests were carried out.

Bending Tests of the Bolted Joint. In order to
observe the structural behavior under the bending moment
and the bending rigidity of the bolted joint, the bending tests were carried out.

Buckling Tests of the Reticular Shell. In order to
observe the buckling behavior of the reticular shell
under the vertical load, the load carrying tests were
carried out.

Outline of Test Specimens

All wooden members of the specimens were consisted
of two lumbered rectangular cross-sectional Oregon pines
which were 30mm X 100mm in dimension.

Tensile and Bending Tests of the Bolted Joint. The
size of test specimens were decided from the inner
wooden members of the reticular shell specimens. They
are shown in Figure 1 with mark of specimens.

Buckling Tests of the Reticular Shells. A inserted
steel plate for a bolted joint was adopted. The thickness of a steel plate is 9mm. All the wooden members
were fastened with a high-strength bolt of 12mm diameter

at each joint.   All joints are positioned on the same
spherical surface.   The span of the reticular shell is
3600mm and the rise to span ratio is 0.10.   This specimens was consisted of triangular grids which had approximately 745mm length in each member.   At the twelve
supporting points which are on the periphery of the
specimens they are supported by pin in rotational and
roller in diagonal direction.   The shape of the specimen is shown in Figure 2 with mark of specimens.   Photo
1 and 2 show the side view and the overall view of the
reticular shell specimen, respectively.

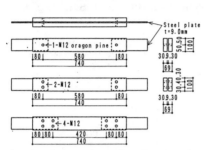

Specimen's Mark

Number of Bolt	Tensile Test	Bending Test
1-Bolt	J1B-T	*****
2-Bolts	J2B-T	J2B-M
4-Bolts	J4B-T	J4B-M

Fig.1 Size and Mark of Tensile and Bending
        Test Specimens

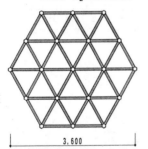

Specimen's Mark

Number of Bolt	Concentrated Loading	Distributed Loading
1-Bolt	RS1B-C	******
2-Bolts	RS2B-C	RS2B-D
4-Bolts	RS4B-C	******

Fig.2 Shape and Mark of the Reticular Shell Specimen

Photo 1. Side View of   Photo 2. Overall View of
         Reticular Shell   Reticular Shell

Photo 3 to 5 show one bolt joint system, two bolts joint system and four bolts joint system for the reticular shell specimens, respectively.

Photo 3. One Bolt Joint
System (RS1B)

Photo 4. Two Bolts Joint
System (RS2B)

Photo 5. Four Bolts Joint
System (RS4B)

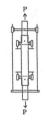

Fig.3
Method of Tensile Test

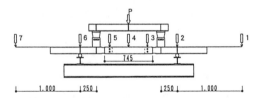

Fig.4 Method of Bending Test

Method of Load Carrying

Tensile Tests of the Bolted Joint. Loading was carried out by using a universal testing machine. The 980N increments of loading was applied. The way of loading are shown in Figure 3.

Bending Tests of the Bolted Joint. Loading was carried out by using a hydraulic jack. The 490N increments of loading was applied. The way of loading is shown in Figure 4.

Buckling Test of the Reticular Shell.   Loading was
carried out by using a hydraulic jack.   We had two
loading type, that is to say a concentrated loading at
the center of the reticular shell and a distributed
loading.   The distributed loading at each node was
executed by a tournament system in order to divide the
total load in proportion to the shared areas of the
node.   Photo 6 shows a concentrated loading system
which made of H-shaped steel.   Photo 7 shows a detail
of this loading system.

Photo 6. Overall View   Photo 7. Detail of System

< Concentrated Loading System >

Results of Tensile Tests of Bolted Joint

The relationship between loading and displacement
which give average performance in all specimens are
shown in Figure 5.   The displacement of the bolted
joint is the sum of a joint looseness and the joint
deformation by compressive strain inclined to the grain.
The experimental values are shown in Table 1.

N_0: Tensile Strength of a Wooden Member

Fig.5 Relationship between
      Loading and Displacement

Table 1.Experimental Values

	1-Bolt	2-Bolts	4-Bolts
Rigidity at a Slipping Region (kN/cm)	38.3	113.8	240.0
Rigidity at an Elastic Region (kN/cm)	146.2	290.4	477.8
Strength (kN)	32.4	61.8	112.8

Results of the Bending Tests of Bolted Joint

The relationship between bending moment and rotational angle which give average performance in all specimens are shown in Figure 6. The experimental values are shown in Table 2.

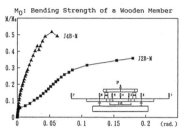

M_0: Bending Strength of a Wooden Member

Fig.6 Relationship between
Bending Moment and Rotational Angle

Table 2. Experimental Values

	1-Bolt	2-Bolts	4-Bolts
Rigidity at a Slipping Region (kN·cm/rad)	*****	773.0	5130.6
Rigidity at an Elastic Region (kN·cm/rad)	*****	2038.5	5130.6
Strength (kN m)	*****	1.5	1.8

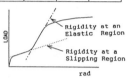

Results of the Buckling Test of the Reticular Shell
(The influence of the number of bolts at the joint)

In this chapter, we will discuss the results of a concentrated loading tests of the reticular shell.

The Relationship between Load and Displacement.
The relationship between load and displacement are shown in Figure 7 and 8. In these figures, the vertical displacements are divided by h(=9cm) which is rise of a hexagonal unit at the center of the reticular shell.

These figures prove clearly that there is a remarkable influence due to the number of bolts at the joints. The buckling load and the ultimate strength of the reticular shell grow larger with the increase of the number of bolts. The buckling behavior of the RS1B-C and the RS2B-C have a snap-through buckling.

In the case of the RS1B-C, a joint buckling occurs at the center of the reticular shell, and the vertical displacements of six joints neighboring the center joint grow larger. Namely, a overall buckling occurs at low loading region.

In the case of the RS2B-C, a joint buckling occurs at the center of the reticular shell, but the vertical displacements of six joints neighboring the center joint are smaller than the RS1B-C.

In the case of the RS4B-C, at the end of a wooden members which compose the center joint of the reticular shell, bending failure occurs after the load reached maximum value and the load decreased rapidly.

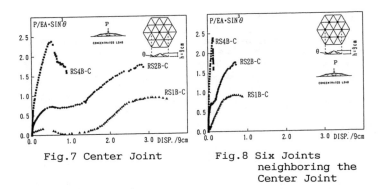

Fig.7 Center Joint   Fig.8 Six Joints
                            neighboring the
                            Center Joint

< Relationship between Load and Displacement >

   The Deformation Mode.   The deformation modes of
test specimens (RS1B-C,RS2B-C,RS4B-C) are shown in
Figure 9.   The buckling behavior mentioned as above is
revealed in this figure.   In particular, in the case of
RS1B-C, the vertical displacements increase rapidly
after the joint buckling occurs at the center of the reticular shell.

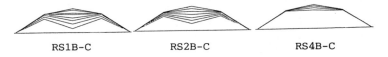

   RS1B-C   RS2B-C   RS4B-C

Fig.9 Deformation Modes

   The Relationship between Load and Axial Force.   The
nature of axial forces of three test specimens are shown
in Figure 10 to 12.   In these figures, the axial forces
of a wooden member divided by N_0 which is decided from a
member buckling load.   This member buckling load is
based on Euler's equation.   Moreover, this member
buckling load is approximately equal to the tensile
strength of the bolted joint which is about 61kN.   The
tensile strength is judged from the results of a tensile
tests which have been already described.
   In the case of the RS1B-C and the RS2B-C which have
shown a buckling behavior, the axial force of the radial
members and the hexagonal members are compressive and
tensile force, respectively.   After a joint buckling
occurred at the center of the reticular shell, the compressive forces of the radial members change to tensile
force.

On the other hand, the tensile forces of the hexagonal members change to compressive force.
   In the case of the RS4B-C which a joint buckling did
not occur, the axial forces of the radial members are
compressive force all along.   However, the axial forces
of the hexagonal members are small tensile forces.
After the loading reached maximum value, the small tensile forces change to compressive ones.

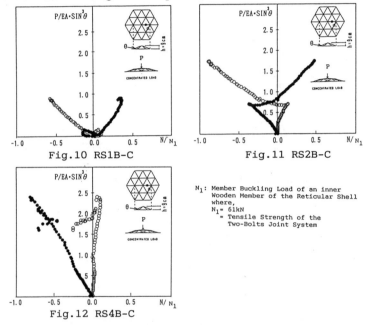

Fig.10 RS1B-C   Fig.11 RS2B-C

Fig.12 RS4B-C

N_1: Member Buckling Load of an inner
     Wooden Member of the Reticular Shell
     where,
     N_1= 61kN
     = Tensile Strength of the
       Two-Bolts Joint System

< Relationship between load and axial force >

Results of the Buckling Test of the Reticular Shell
(The influence of a different loading type)

   In this chapter, we will compare the results of a
concentrated loading test with a distributed loading
test.

   The Relationship bet..een Load and Displacement.
The relationship between load and displacement are shown
in Figure 13 and 14.
   These figures prove clearly that there is a remarka-
ble influence of a different loading type.
   In the case of the RS2B-D which is a test specimen

of distributed loading type, the buckling phenomenon did
not occur.   After the load reached maximum value, the
bolted joint of a periphery wooden member occurred tensile failure.

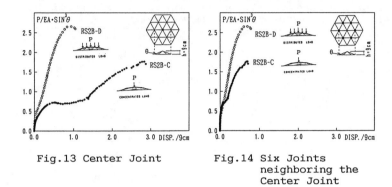

Fig.13 Center Joint   Fig.14 Six Joints
                                   neighboring the
                                   Center Joint

< Relationship between Load and Displacement >

The Deformation Mode.   The deformation modes of
test specimens which are the RS2B-C and the RS2B-D are
shown in Figure 15.
     In the case of the RS2B-D, the vertical displace-
ments of all joints are approximately equal to each
other.

RS2B-C   RS2B-D

Fig.15 Deformation Modes

The Relationship between Load and Axial Force.
The relationship between load and axial force are shown
in Figure 16.
     In the case of the RS2B-D, the axial forces of the
inner wooden members without peripheral members are
compressive force.   Furthermore, the compressive forces
of the inner wooden members do not change to tensile
force.   This nature of axial forces indicates that a
joint buckling did not occur.

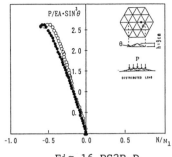

N$_1$: Member Buckling Load of an inner
Wooden Member of the Reticular Shell
where,
N$_1$= 61kN
= Tensile Strength of the
Two-Bolts Joint System

Fig.16 RS2B-D

< Relationship between Load and Axial Force >

Conclusion

1. There is a remarkable influence due to a number of bolts at the joint. The buckling load and the ultimate strength of the reticular shell grow larger with the increase of the number of bolts.

2. In the case of one-bolt joint system (RS1B-C), as the state of connection is pin connection, the overall buckling behavior is confirmed after the joint buckling occurred at the center joint of the reticular shell.

3. In the case of two-bolts joint system (RS2B-C), as the bolted joint has bending rigidity, the only joint buckling occurs at the center joint of the reticular shell.

4. In the case of a distributed loading (RS2B-D), the reticular shell with two-bolts joint system dose not have a buckling behavior.

References

1. Erki, M.A.: Modeling the Load-Slip Behavior of Timber Joints with Mechanical Fasteners, Can. J. Civ. Eng., Vol.18,1991,pp.607-616.
2. Hangai, Y.: Load Carrying Test of a Wooden Lattice Dome for EXPO'90 Osaka,Printed in "Innovative Large Span Structures" Vol.1,Proc. of IASS Symp.,Copenhagen 1991
3. Matsuno,K. : Experimental Study on a Wooden Single - Layer Space Frame of EP Shell on a Rectangular Plan, Proc. of IASS Symp., Istanbul,1993

On Some Problems of the Multicriteria Optimization
of the Spatial Grid Structures

Jan A.Karczewski[1] and Witold M.Paczkowski[2]

Abstract

One can not think contemporary structure designed without con-
ciously comply with optimization methods.creative activity of engineers have to be optimal or semi-optimal.It is necessary to be in
search of optimal solutions of:subsequent stages of investigations,
designing process,cost of the structure,technological process as well
as others values which have influence upon quolity,reliability, funcionality aesthetic as well as material and energy consumptions.
Authors discuss some problems of optimization of the space grid struc
tures The problems which ones among well known methods of optimization are more helpful in spatial grid structure designing: continuous or discrete,linear or nonlinear,single or multicriteria and one
or multilevel are considered.The paper is illustrated with results of
exemplary structural optimization of above mentioned structures realized in compliance with methods discussed.

Introduction

Optimal shaping of the structure until quite lately met with numerous of barriers.Now in view of the tempestuous developing of the
computational technics optimization of the real structures became
possible.Nevertheless the proper organize of the analysis is of great
importance.It can reduce significantly time needed to complete ne-
cessery calculations.It is natural in the structural designing to get
a structure under given code limitations which is the best according
to a chosen criteria.It means that the choice of a definitive variant
of the structure should be introduce with optimization process.Very
often a spatial grid strucure should satisfy the conflicting requirements coming from serviceability,technological and mechanical condi-
tions.There is a question how to formulate optimization problem and

---

[1]Professor,Warsaw University of Technology,Civil Engineering Faculty,
Armii Ludowej 16, 00-637 Warsaw,Poland, Fax: 0048-22-258899
[2]Senior Lecturer,Technical University of Szczecin, Poland

to choose the method of searching for solution of the problem to obtain satysfying result in real time of computing.

Discrete- or continuous optimization

Technolgical and computational limitations in discrete optimization are resulting with discrete character of the design variables whereas design variables in continuous optimization can take any values from the continuous sets Brandt (1984). The following design variables can be assumed as continuous: interlayer distance,rise or camber in the case of flat structure,prestressing forces,span of the structure and so on.Numerous of design values,from the theoretical point of view,could be assumed as continuous.However,in practice,for majority of them only discrete character is possible.It result with data describing:properties of material,technological conditions, catalogues of elements and so on.Discrete variables occure mostly in civil engineering design.Type of cross-section of the strut,catalogue

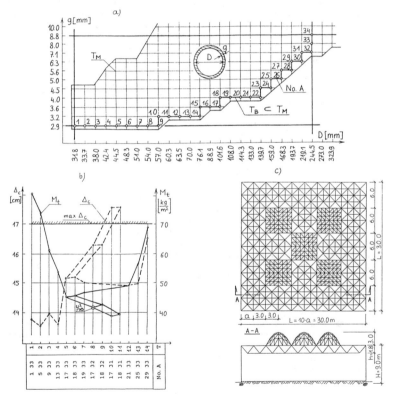

Fig.1

of cross-section of the strut,inter-nodal distance,grade of steel,
partition of stiffness zones in structure, support properties, stiffness of the nodes,structural and assembling system,numberof the
prestressed member are typical ones among discrete variables. The
spatial grid structure designing is connected mostly with operations
on the sets of discrete values.Rounding the continuous solution to
the nearest discrete solution can lead to the solution which is far
from optimal discrete solution.In such case discrete optimization is
recommended Gutkowski,Bauer and Iwanow (1993).

    As a typical example of discrete optimization let us consider a
catalogue of ring pipe cross-section with limited number of elements.
The choice of elements should result with minimum mass of the spatial
grid structure composed with struts of cross-section given in catalogue analyzed.The basic catalogue T_B consisting t=34 elements was chosen from full catalogue T_M issued by metalurgic factories-see Fig.1a.
Catalogue T_B fulfill the criteria of minimum of mass of the double
layer,orthogonal spatial grid structure span 30 x 30 m. and simulta-
neity of the critical as well as yielding force increasing for subsequent cross-sections (No.A).Then catalogue with two elements t=2,mi-
nimizing mass of the spatial grid structure with interlayer distance
h=1.8 m and span 30 x30 m was selected from basic catalogue T_B .The
graph of the functions representing mass of the structure for square
meter and deflection of the central node of the structure Δ_c are
shown at Fig.1b.In the case of catalogue T_B among 182 possible combinations of diameter D and wall thickness g, were chosen 34,while in
the case of catalogues T_O with t=2, two elements were chosen from
catalogue T_B consisting 34 elements.

Linear- or nonlinear optimization

    Linear objective function f(x) and linear constraints $g_k(x)$
with respect to design variables lead to the so called linear optimization or linear programming,Gass(1973).If at least one of the function f(x) or $g_k(x)$ is nonlinear than the optimization problem can be
solved by nonlinear programming,Wit(1986).In linear optimization,in
the limited admissible area X ,the solution x occures always at
the cutting point of the constraints or on one constraint line.In
nonlinear optimization the solution x can occure inside feasible
domain X or on the boundary.In optimum structural design it is difficult to avoid nonlinear functions.Usually such quantities like:cross
-section areas of strut,stiffness or mechanical properties of material lead to nonlinear objective functions like:mass,elastic energy,displacement,frequency of vibration,critical load or stress.Line-
arization such functions,in the case of real structures,is usually
difficult even impossible,in majority of considered problems,due to
algorithmic form of objective functions,Majid(1981).The example of
nonlinear optimization is presented at previous point.One can easy
notice that linearization of the function $M_t(T)$ and $\Delta_c(T)$ is difficult at least.Optimal geometrical shaping of the industrial hall
roofing by the double layers orthogonal spatial grid structure span
24 x24 m is presented by Karczewski and Paczkowski(1992),see Fig.2a.
The objective functions describing mass of the spatial grid structure
members - Fig.2b,mass of the hall elements made with steel - Fig.2c,

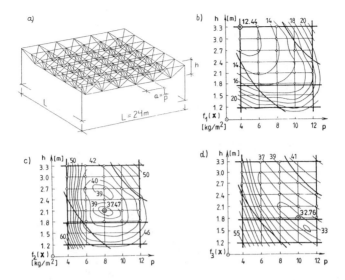

Fig.2

and laboriousness of the spatial grid structure realization - Fig.2d
were analyzed in the decisive space defined by interlayer distance h
and density of the net in particular layers p=L/a.The last one among
above mentioned functions is formulated as some function of geometry
of the structure.It is following example of nonlinearity in optimal
designing of the spatial grid structure.

Single- or multicriteria optimization

Multicriteria optimization,Jendo (1990), takes into
account the set of J-elements vector of the objective functions

$$y = f(x) = [f_1(x)....f_j(x)....f_J(x)]$$

contrary to single criteria or scalar optimization where only one
criterion is considered.The choice of the criterion of optimization
is essential to the result of solution.The most popular criteria used
in structural optimization are as follows,Brandt(1984):minimum mass
or cost,maximum of stiffness,minimum displacement at certain point of
structure,minimum laboriousness,maximum funcionality of the structure,maximum load bearing capacity in elastic as well as elastic-plastic phase. These criteria very often are in conflict,Paczkowski
(1988).If chosen variant of the structure should satisfy a lot of
equivalent criteria being in conflict one with others,only and only
the multicriteria optimization approach can give desirable results.In
vector optimization expected solution is seeked simultaneously in the
space of solution A and the objective space B-Fig.3.The estimation of
solutions which can not to be univocally improved in compliance with

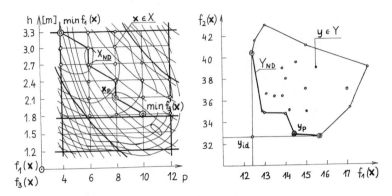

Fig.3

assumed criteria vector $f(x)$ are selected from the set of estimations
of solutions Y. They form the set of undominated estimations Y_{ND} related to the set of undominated solutions X_{ND}. Next,the so called
preferable solution x_p should be chosen taking into account an additional criterion or using so called global criteria like utility function,distance function or hierarchical method,Jendo (1990)The
results of analysis shown at Fig.2b and 2d well illustrate   the
conflict can occure in shaping of the spatial grid structure.Mass of
the struts and laboriousness function have reached their minimum on
the opposite restraints of the admissible area.The set of permissible
solutions and adequate set of estimations Y are shown at Fig.3.
Next,the undominated sets Y_{ND} and X_{ND} were formed.It enable to find
the preferable solution using superior criterion of the cost of
structure.

One- or multilevel optimization

Multilevel optimization can be applied to solve large scale optimization systems using decomposition and coordination method,Łubiń-
ski,Karczewski and Paczkowski(1985).Large scale system can be decomposed into a few subsystems Which can be optimized separately -Fig.4.
Coordination level allows to get the solution of decomposed system
which converges to the solution of original system. Many civil engineering structures can be decomposed into subsystems. For example the
hall -Fig.4,can be divided into:roofing covers, spatial grid structure,columns,walls and foundations.In optimization process it is necessary to choose the local design variables x_L concerning with particular subsystems and global design variables x_G common for at least
two or more subsystems.The particular subsystems are optimized
independently of each other with respect to local variables. Next,co-
ordination is performed according to global objective function with
respect to the coordination variables.

Decomposition of the optimization problem

Decomposition can be used with respect not only to analyzed

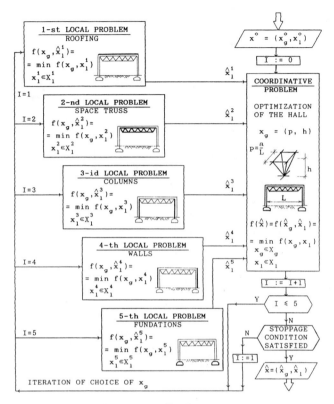

Fig.4

structure but also to method of optimization.Discrete formulation of the optimization problem in the case of hall roofed by spatial grid structure lead to the great amount of permissible solutions exceed sometimes 10^{30} variants,Karczewski and Paczkowski(1992).Decomposition of the structure-Fig.5, enable to reduce the number of the permissible solutions for particular local problems up to 10^4-10^{12}.Taking into account the effective optimization algorithm as well as actual computational possibilities solution of such problem is impossible. The problem can be simplyfied additionally in relation to monotonic properties of the sets of design variables and estimations of solutions.Decomposition of design variables vector $\mathbf{x}$ as well as vector of the objective functions $\mathbf{y}=\mathbf{f}(\mathbf{x})$ reduce number of permissible solutions to 10^2-10^4 variants.Such problem can be solved in real time between 2 and 20 hours.The points corresponding with minimal values of particular objective functions as well as two or three-criteria compromises belong to the set Y_{ND}. It result with monotonic character of objective functions vector. If consecutive j-th objective function, where $j \in \overline{1,J}$ will added it can result only with increasing amount of

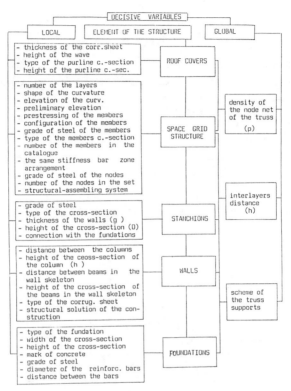

Fig. 5

elements belonging to the set $Y_{ND}(j-1)$ and dimension of the convex envelope of the set $Co[Y_{ND}(j-1)]$. The points $\mathbf{y}_{ND}$ belonging to $Y_{ND}(j-1)$ belong to the set $Y_{ND}(j)$ too. Then occure following relation

$$\bigwedge_{j<J} X_{ND}(j-1) \subset X_{ND}(j) \subset X_{ND}(J).$$

Addition of the n-th variable, where $n \in \overline{1,N}$, can only improve actual set $Y_{ND}(n-1)$. It result with monotonic property of the decisive variable. It does mean also that the sets $Y_{ND}(n)$ and $Y_{ND}(n-1)$ belong to conical relation of the order $\angle_\Lambda$ and exist such value $\mathbf{x}_{n(k)}$ of active decisive variable $\mathbf{x}(n)$ for which the common part of the sets $X_{ND}(n-1) \cap X_{ND}(n)$ is no empty.

$$\bigwedge_{n\leq N} [\bigvee_{x_{n(k)}} X_{ND}(n-1) \cap X_{ND}(n) \neq \emptyset] \wedge [(Y_{ND}(n), Y_{ND}(n-1) \in \angle_\Lambda].$$

The example of polyoptimal analysis with the vector of objective functions decomposition is shown by Karczewski Niczyj and Paczkowski

(1993).The variability of three objective functions describing: production cost of the exhibition hall,elastic energy accumulated in the spatial grid structure and horizontal displacement of the supporting node are analyzed.Design variables assumed in analysis describe:shape of the roof,dimension of the camber and number of the elements in catalogue of the struts - Fig.6.

Analysis of the objective functions vector or decisive variables vector with decomposition of the structure result usually in approximate solution - optimal or undominated.It is so called problem

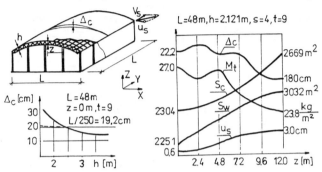

Fig.6

of the satisfaction.Here calculated solution improve values of the criteria assumed, for starting structure,by experienced designer taking no account optimal analysis.Influence of the spatial grid struc-

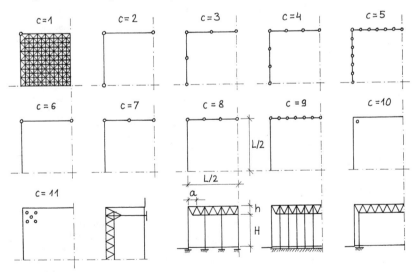

Fig.7

ture support in different manner on behaviour of chosen objective
functions can be consider as following example of decomposition.The
analyzed supports are shown at Fig.7.Fundamental elements of the
structure:spatial grid structure,stanchions,purlines,wall skeleton as
well as roof and wall panels were considered.The graphs of the objective functions representing:mass of thh spatial grid structure mem-
ber-M_b, mass of all others elements of the hall made with steel-M_c,
total steel consumptions per 1 m^2 of the ground plane-M_t and vertical
displacement of the spatial grid structure central node-Δ_c are shown
at Fig.8.Here the choice of subsequent values of design variables c,h
and a depend significantly on the M_t values.The function Δ_c,showing
stiffness of the structure rather assist, especially in the choice of
the preferable solution $x_p = [c,a,h]^T = [4, 2.1, 3.0]^T$

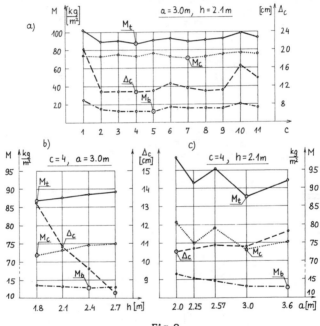

Fig.8

Final remarks

Shaping an industrial or exibition halls roofed by the spatial
grid structure lead to formulating and solving the problem of optimi-
zation.In majority of the cases they are discrete,multicriteria
problems with nonlinear objective functions and restraints.Through
application decomposition of structure,objective functions vector
and/or decisive variables vector one can obtain the solution in real
time of computing.Solutions obtained on this way are approximated in

majority.It denote that can exist variant better than indicated as optimal one.Optimization of the structure independently of assumed structural assumptions and dimension of the problem significantly increase laboriousness of the calculations.Therefore the question when optimization can lead to an economical effects become actual.

Structural optimization is particulary important in two cases,namely:for systems applied many times and for new large structure designed without previous engineering experience.In both above mentioned cases the spatial grid structure is often used as roof or slab.Now the numerous well known method of optimization are in disposal.However they were worked out under different assumptions and represent different approches in realization of analysis.Sensitivity of this methods for variation of the given objective functions depend significantly of type of the structure.One can easy notice that optimization problem,especially if the building consist any spatial grid structure,should be formulated as discrete,nonlinear,multicriteria and multilevel problem with possibly deep decomposition.

Application of optimization methods in structural design is not sufficient up to now.This is due to underdevelopment of discrete nonlinear optimization methods.Applications are mostly concerned with structural elements then optimization system.Often it is not necessary to use in optimal analysis contemporary methods perfectly worked out and usually complicated.Fact that is possible oncoming to the optimal solution without its precise evaluation can not be ignored.In many cases it satisfy designer.Nevertheless in such case also correctness of the choice can be confirmed only by employing chosen methods of optimization.

Appendix References

1.Gutkowski W.,Bauer J.and Iwanow Z.,"Support Number and Allocation for Optimum Structure" Proceedings IUTAM Symposium Discrete Structural Optimization,Zakopane,Poland,1993,Springer Verlag(in print)
2.Brandt A.M.(ed),"Criteria and Methods of Structural Optimization", PWN Warszawa,Martins Nijhoff Publishers,The Hague/Boston/Lancaster 1984,
3.Gass S.,"Linear Programming"(in Polish),PWN,Warszawa,1973,
4.Jendo S.,"Multiobjective Optimization,in:Save M.and Prager W.(eds) Structural Optimization, vol. 2: Mathematical Programming, Plenum Press,New York/London,1990,pp 311-342,
5.Karczewski J.A and Paczkowski W.M., "Decompositions principle in discrete multicriterion optimization of the space grid structure", Proceedings of IASS- CSCE International Congress -Innovative Large Span Structures,vol.2,Toronto,July,1992,p.232-243,
6.Karczewski J.A., Niczyj J. and Paczkowski W.M.,"Multicriterion selection of spatial truss crown", Proceedings of International IASS Symposium,Istambul,May,1993,p.461-470,
7.Majid K.I., "Optimum Structural Design" (in Polish), PWN,Warszawa, 1981,
8.Paczkowski W.M.,"Algorythm of Discrete Structural Polyoptimization (in Polish), Proceedings of Mech.Department of WSI Koszalin,13, 1988,
9.Wit R.,"Methods of Nonlinear Programming"(in Polish),WNT,1986.

Field Measurements of Tensile Forces on a
Spherical Dome Sector Caused by Internal Load
of a Granular Product in Storage

Jack L. Brunk, P. E.[1]

[1]Vice President Engineering, Dome Systems, Inc., 171
South Newton Street, P. O. Box 386, Goodland, IN 47948

Introduction

The hemispherical domes, in which the measurements will be made, are constructed using an inflatable fabric form. All the construction work is performed on the inside of the inflated form. (See photo 1). Access is through an airlock. The inside face of the inflated airform is sprayed with foam insulation. Wet mix type shotcrete (4000 psi) is sprayed onto the foam. The reinforcing steel (60 ksi) is tied in place using the vertical dowels which project from the foundation ring, (See photo 2), and special tabs for the horizontal carrier steel (to which the vertical bars are tied). (See photo 3). After the reinforcing steel is in place, additional shotcrete is sprayed on to encapsulate the bar and develop the final shell thickness.

The horizontal load resulting from the stored product can be calculated using any of a number of existing equations developed for application to silos, bins and hoppers. However, there has been very little work done to measure the actual loads in domes. This is due, in large part, to the fact that this type of dome has only recently (past 10-15 years) been applied to industrial and agricultural storage applications.

This description of the field measurements of horizontal product loads must be done in more than one

Anchored and inflated airforms.   (Photo 1)

report. This, the first report, discusses the physical
elements of the dome, the product fill and reclaim
equipment for the first project to be equipped with
measuring devices, the measurement methods, the data to
be recorded and the expected value to be gained from
this effort. It is because of the time necessary to
develop a history that the results will be presented in
segments.

Dome Systems, Inc. has undertaken the task of measuring these loads to further its own knowledge and add
to the information available to others.

Measurements

The method arrived at for measuring the horizontal
loading is very similar to that used by Professor G. E.
Blight[2]. Professor Blight's willingness to provide this
writer with quality responses to our requests has made
this task much easier. We are in his debt.

Tunnels and foundation ring (to accept the airform) under
construction. (Photo 2)

----
[2]Professor G. E. Blight, Civil Engineering Dept.
Witwatersrand University
P. O. Wits, 2050 Johannesburg, South Africa

A pressure sensor is mounted in a housing which is
enveloped in the concrete shell, as shown in figure 1.
In effect, the surface of the pressure sensor represents
the surface of the dome shell. Because tests will be
run on a number of domes, the sensor housing is designed
with a cover plate so that the sensor can be removed and
the cover plate mounted in place. This allows the
housings and cover plates to be installed in each dome
as it is built. The sensors can be installed when needed for testing and removed when convenient.

The first project to be equipped for measurement
consists of four 39.6m (130 ft.) diameter hemispherical
domes, presently under construction in Western Australia.
(See figure 2). Each dome holds 10,000 MT of wheat.
The wheat is discharged into the top dead center of each
dome from belt conveyors. Reclaim is primarily by
gravity, through gates in the floor to a belt conveyor
in a tunnel below. However, should the reclaim system
be down for repair, front endloaders will reclaim the
grain by entering through an access built into the side
of the dome shell. Either method of reclaim will cause
eccentric loading, although the condition is more severe
when using the endloader.

Horizontal steel beginning to be placed on the
vertical bars. (Photo 3)

The location of the sensors for the initial tests in Australia have been restricted to the lower 15% of the elevation of the shell and at 90 degree increments around the perimeter as shown in figure 2.

The intent is to take pressure readings during the fill process, after 6,000 ton of grain has been discharged into storage. This 6,000 ton will fill the floor area preliminary to the build up of grain on the inside face of the dome shell. Readings will be taken at approximately 300 MT increments (this represents about 0.3m build up on the shell height) as grain enters the dome. The last reading will be taken when the dome is completely full. Also, at this time a measurement will be taken from the concrete pad at the top dead center of the dome to the top of the grain pile.

After a minimum time of two weeks, the pressure will be read again and another measurement will be taken from the top of the dome pad to the top of the grain. Any significant differences will represent the effects from grain compaction while held in storage for

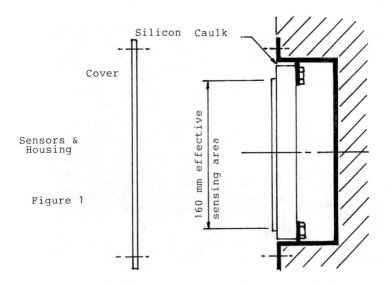

Silicon Caulk

Cover

Sensors &
Housing

160 mm effective
sensing area

Figure 1

the two weeks. Another recording of pressure and change in pile height will be taken after another two weeks. For as long as a significant difference is recorded, the readings will be taken at two week intervals.

When the reclaiming of grain from the dome is begun, readings will be taken at 300 MT increments until grain is no longer in contact with the sensors.

The recorded data will be compared with the design criteria to determine what correlation exists. After a significant number of readings are accumulated, it is hoped that the data either justifies the present assumptions made regarding the horizontal load, or indicates the adjustements to be made so that the design criteria becomes more representative of the real conditions.

The design and analysis was based on the following:

Bulk density for wheat = 800 kg/cu. m

Angle of repose = θ = 24 degrees

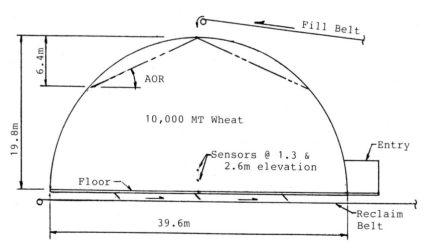

Cross Section of Dome Showing
Uniform Loading of the Filled Storage

Figure 2

Filling the dome completely results in the grain being in contact with the shell to a height of 13.4m. Using an analysis based on the CRSI Handbook chapter on retaining walls with a surcharge (class B material - granular with mixed grain sizes), the pressures on the shell at the elevations of the sensors are 46 kPa at 1.3m and 41.6 kPa at 2.6m.

$$P_h = c \cdot w \cdot h \quad \text{where}$$

c is a constant (.52 in this instance) for a
  sloping backfill condition.

w is the bulk density.

h is the vertical distance from the elevation
  where the grain contacts the shell (13.4m)
  to the elevation at which the horizontal
  load is being measured (in this case 13.4m-
  1.3m or 12.1m and 13.4m-2.6m or 10.8m).

The field test results will be compared to these values.

The next report will contain the field test results from filling and emptying of two or three domes and a discussion of the design value/test value comparison.

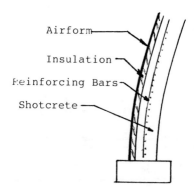

Airform
Insulation
Reinforcing Bars
Shotcrete

Typical Section

Figure 3

References

1) Blight, G. E.: Measurements on Full Size Silos;
    Bulk Solids Handling, Vol. 8, No. 3, 1988,
    pp 343-346.

2) Committee on Design Aids, CRSI Handbook; Chapter 14,
    1984.

Non-Linear Solution Methods for Shell and Spatial Structures

M. Papadrakakis[1]

Abstract

Non-linear solution methods are investigated for the analysis of shell and spatial structures based on limited memory quasi-Newton updates. Under the proposed implementation the preconditioned truncated Lanczos method is used for the solution of the linearized problem in each non-linear iteration. The complete factorization of the stiffness matrix is avoided and large-scale problems can be solved efficiently both in terms of computing time and storage. The non-linear iterative scheme is properly modified to account for loading variation inside the increment in order to trace post-critical equilibrium paths.

Introduction

One of the methods most commonly used for solving large systems of non-linear equations in structural mechanics is Newton's method and its variations. In quasi-Newton (QN) or variable metric methods the costly evaluation of the tangent matrix is replaced by some economically obtained approximation based on rank-one or rank-two correction schemes. The rank-one updates involve the generation of updates using one correction term, while the rank-two updates are generated using two correction terms and provide many potential forms. The most widely used rank-two updates, belong to the Broyden β-family, where the most important members of this family are the DFP (Davidon-Fletcher-Powell) and BFGS(Broyden-Fletcher-Goldfarb-Shanno) updates. Much work has been devoted to the development of this type of methods in the fields of mathematical programming and unconstrained

[1] Institute of Structural Analysis & Aseismic Research, National Technical University of Athens, Athens,Greece 15773

180

optimization. Ref.1 was influential in attracting attention to the application of quasi-Newton methods in non-linear finite element analysis. Some of the first implementations in structural mechanics were reported in Ref.2-6. Extensive studies on the performance of quasi-Newton methods were presented in Ref.7,8. In this work a family of methods is investigated based on the limited memory QN updates. Under the proposed implementation the complete factorization of the stiffness matrix is avoided. It is subsequently replaced by a sparsely populated preconditioning matrix. The solution of the linearized problems is performed by a truncated preconditioned Lanczos iterative procedure. The non-linear iterative scheme is properly modified to account for loading variation inside the increment in order to be able to trace post-critical equilibrium paths.

Constrained Quasi-Newton Methods

Discrete equilibrium equations arising from finite element non-linear formulations may be presented in the general compact form $g(u,p,\theta) = 0$, where u denotes the unknown vector of generalized displacements, p is an array of control parameters, θ is a functional of past history of the generalized deformation gradients, and g is the residual vector of out-of-balance generalized forces. If the system is conservative, g is the gradient of the total potential energy. In many applications the state and control parameters may be segregated and under fixed loading may be expressed as

$$g(u) = F(u) - P = 0 \tag{1}$$

leading to

$$K_i x_i = -g_i \tag{2}$$

where $F(u)$ is the internal force vector, P is the applied external force vector and K_i is the Jacobian of g or the tangent stiffness matrix evaluated at u_i.

Most path-following solution strategies are based on a combination of incremental steps followed by equilibrium iterations in the non-linear space. If we assume that the structure is subjected to a loading λP, where the vector P is a fixed external loading and the scalar λ is a load-level parameter, eq.(3) defines a state of proportional loading in which the loading pattern remains fixed

$$P_{i+1}^m = P_1^m + \delta\lambda_{i+1}P \tag{3}$$

The superscript m denotes the load step number and subscript i the non-linear iteration. The solution procedures described in the previous section may constitute a part of an overall incremental iterative solution strategy in which the fundamental equilibrium equation (2) is given by $\alpha_i K_i^m x_i = P_{i+1}^m - F_i^m$ and the total displacements inside the increment are given by $\Delta u_{i+1} = \Delta u_i + \alpha_i x_i$, in which α_i is the step length parameter.

Non-linear iterations must now be performed in the load-displacement space, where the load factor is considered as a new variable in conjunction with the displacement vector. A constraint of the form $f(\delta\lambda, \Delta u) = 0$ must therefore be added to the equilibrium equations, where f represents the"hypersurface" in the load-displacement space on which the iteration process is constrained. As a result of this additional equation, the load varies inside the increment and iterations are performed under variable loading. The augmented Jacobian is neither symmetric nor banded, and instead of solving the augmented coefficient matrix in one step, a two step procedure is usually adopted (Ref.9,10). Denoting the gradient or the unbalanced force vector $g(u,\delta\lambda)$ as $g_i = F_i^m - P_i^m$ the solution vector is obtained from $\alpha_i x_i = \delta\lambda_{i+1} x_i' + x_i''$ with $x_i' = K_i^{-1}P$ and $x_i'' = -K_i^{-1}g_i$, where the superscript m has been omitted for clarity.

The incremental-iterative procedure of an implicit method under variable loading may be briefly described in the following algorithmic form:

> a. Incremental phase
>   a1. Compute $\delta\lambda_1$, $\Delta u_1 = (K_0^m)^{-1}\delta\lambda_1 P$ (predictor step)
> b. Nonlinear iterative phase (corrector steps), (i=1,2,...)
>   b1. Compute F_i^m, g_i
>   b2. Test for convergence: $\| g_i^m \|/\| P_1^m \| \leq \varepsilon$
>   b3. Define K_i (or K_i^{-1})
>   c. Solution phase   (4)
>   c1. Solve for x_i' and x_i''
>   b4. Compute $\delta\lambda_{i+1}$ from the constraint equation
>   and set $\Delta\lambda_{i+1}=\Delta\lambda_i+\delta\lambda_{i+1}$
>   b5. Compute α_i from a line search routine
>   b6. Update displacements: $\Delta u_{i+1}=\Delta u_i+\alpha_i x_i$
>   and $u_{i+1}^m=u_i^m+\Delta u_{i+1}$
> a2. Update the external load: $P_0^{m+1}=P_0^m+\Delta\lambda P$

One of the most effective quasi-Newton method is considered to be the BFGS update. A sparsity preserving BFGS update can be expressed as

$$K_i^{-1} = \left(I - \varrho_i s_i y_i^T\right) K_{i-1}^{-1}\left(- \varrho_i y_i s_i^T\right) + - \varrho_i s_i s_i^T \quad , \quad \varrho_i = \frac{1}{s_i^T y_i} \qquad (5)$$

or

$$K_i^{-1} = UBFGS(K_{i-1}^{-1},si,yi) = UBFGS(K_0^{-1}, s1,y1, ..., si,yi) \qquad (6)$$

Similar relations can be expressed for other types of QN updates.

A computational efficient variable loading implementation of BFGS is presented thereafter. The two solution vectors using eq.(6) become

$$x_i' = U_{BFGS}\left(K_0^{-1}, s_1, y_1, ..., s_i, y_i\right)P \qquad (7)$$

$$x_i'' = -U_{BFGS}(K_0^{-1}, s_1, y_1, ..., s_i, y_i)g \qquad (8)$$

Further elaboration on the eqs.(7,8) may produce a recursive formula for x_i', which is computational more efficient. Introducing the auxiliary vector w_i and the scalar parameter γ_i, with

$$w_i = K_{i-1}^{-1}(g_i - g_{i-1}') = K_0^{-1}g_i - \sum_{j=1}^{i-1} (\varrho_j s_j w_j^T + \varrho_j w_j s_j^T) g_i + \left[\sum_{j=1}^{i-1} \gamma_j s_j s_j^T\right] g_i + \frac{s_i}{\alpha_{i-1}} \qquad (9)$$

$$\gamma_i = \varrho_i[\varrho_i w_i^T(g_i - g_{i-1}') + 1] \qquad (10)$$

the two solution vectors now become:

$$x_i'' = -w_i + \frac{s_i}{\alpha_{i-1}} + \varrho_i s_i w_i^T g_i + \varrho_i w_i s_i^T g_i - \gamma_i s_i s_i^T g_i \qquad (11)$$

$$x_i' = x_{i-1}' - \varrho_i s_i w_i^T P - \varrho_i w_i s_i^T P + \gamma_i s_i s_i^T P \qquad (12)$$

Both implementations require the storage of a third update vector, namely z_i or w_i, per iteration in addition to s_i, y_i. Despite the efficient storage handling properties of the vectorized BFGS update, the accumulation of two vectors per iteration that have to be stored may substantially increase the storage requirements for the implementation of the method. To further reduce the storage requirements, limited memory quasi-Newton updates have been proposed which occasionally discard all vectors and replace them with new ones. It was found in Ref.11 that the Special Quasi-Newton method (SQN) proposed in Ref.12, in which after M iterations each new update replaces the earliest one, is the most efficient. Iteration control inside an increment is achieved with the constraint equation $f(\delta\lambda, \Delta u)=0$ which permits the determination of the load parameter $\delta\lambda_{i+1}$ (Ref.9,10). The arc-length hypercylindrical constraint $\Delta u_{i+1}^T \Delta u_{i+1} = \Delta l^2$ has been used in this study for testing the performance of the non-linear algorithms, in which Δl is the arc-length radius. This constraint results in a very successful general purpose iterative technique with large domain of attraction at critical points.

The conventional line search approach proceeds first with the bracketing of the solution of the directional derivative $x_i^T g(u_i + \alpha x_i) = 0$ along the search direction x_i. Once the interval is determined an iterative scheme is applied based on curve fitting (quadratic or cubic) or a linear interpolation (or extrapolation) scheme. The regula-falsi approximation is a linear method which has been found in a number of cases to perform efficiently compared with curve fitting techniques. A second approach replaces the one-variable minimization routine by a stability test which insures convergence of the method. If the stability criterion $\left| x_i^T g(u_i + \alpha_i x_i) \right| < \mu \left| x_i^T g_i \right|$ is satisfied, then α_i is accepted as the step-length parameter. Otherwise, depending on the sign of the directional derivative, either α_i is doubled or a regula-falsi step is performed and a stability check is repeated with the new α_i. The value of μ is taken in the range (0-1). In the vicinity of the solution, the step-length should have a value close to unity. When α is too small, the iteration is not effective because the solution vector retracts to the old value, while an excessively large value of α may cause numerical instability. In both cases, the energy function is not well approximated by a quadratic function involving the old stiffness matrix and thus, either a restart procedure is used or a new tangent stiffness matrix is evaluated.

Truncated Newton-Like Methods

The asymptotic convergence rate of the resulting algorithm is controlled by a user-specified parameter which specifies the accuracy of the linear (inner) solution. In this context $x_i^{(j)}$ is acceptable as the solution of the system $K_i x_i = -g_i$, after j iterations of the linear solver, if the truncated termination condition criterion $\|r^{(j)}\| / \|g_i\| < \eta_i$ is satisfied with $r^{(j)} = K_i x_i^{(j)} + g_i$. The parameter η_i is the forcing sequence that controls the required accuracy of non-linear (outer) step i. An expression for η_i is adopted in this study (Ref.13)

$$\eta_i = \min \left\{ \eta_0, \left(\frac{\| g_i \|}{\| g_0 \|} \right)^t \right\} \quad \text{with } 0 < t < 1 \tag{13}$$

which is motivated by similar ones proposed in Ref.14,15 and is scale invariant. The optimal choice of the parameters η_0 and t is problem-dependent and can be adjusted accordingly. This forcing sequence will result in an adaptive algorithm for solving the non-linear problem. When far away from the solution, $\|g_i\|$ is large and hence minimal work is required to obtain a direction which satisfies the truncated termination condition. As the non-linear solution is approached $\|g_i\|$ becomes small and the linear solution is obtained with increased accuracy.

In order that a solution algorithm may allow us to recognize the types of critical points and follow the corresponding paths, it is necessary to investigate the lower eigenvalue spectrum of the tangent stiffness matrix at the predictor step. Whenever a critical point is passed, a change of sign in the determinant or in an eigenvalue of K_0^m is observed. The calculation of $\det K_0^m$ is possible without additional effort if a complete factorization of the matrix is performed. It is well known that $\det K_0^m = d_1 d_2 ... d_N = \prod_i d_i$ where d_i are the factorized diagonal elements of LDL^T factorization. From the Sturm sequence property, the inertia of K_0^m (number of positive, zero, and negative eigenvalues), is reflected in the factorized diagonal elements d_i. In other words, the number of positive, zero and negative d_i corresponds to the same number of positive, zero and negative eigenvalues of K_0^n.

If the Lanczos method is used for the linearized equations, then the required eigenvalue information is automatically extracted. The main feature of Lanczos method is that certain properties of K are gradually transferred to a tridiagonal T_j and an orthonormal basis Q_j as iterations proceed. Since the Lanczos algorithm may also be described in exact arithmetic as the process of factorizing K into the product of an orthonormal matrix multiplied by a tridiagonal multiplied by an orthonormal,

$$K = QNTNQN^T \quad , \quad QNQN^T = I \tag{14}$$

a combination of eqs. 14 gives a similarity transformation which states that K and T_N have the same eigenvalues. Thus, by monitoring the sign of the diagonal factors d_j of T_j during the predictor step, as computed at iteration j of the Lanczos algorithm (Ref.16), we can monitor automatically the change of sign of the detK and we therefore, can trace the appearance of critical points on the load-deflection path.

Numerical Examples

In this section the behaviour and efficiency of different non-linear methods are examined for three benchmark examples: (i) a 168-member articulated dome with N=147 d.o.f.; (ii) a 48-member frame dome with N=222 d.o.f.; and (iii) a clamped quadratic shell with 3x3 elements in each quarter and 129 d.o.f (see Ref. 8,11 for more details). In the first two examples elastic behaviour is assumed, while in the shell example both geometric and material non-linearities are taken into account. Non-linear iterations are terminated when $\|g^m\| / \| P_1^m \| < 10^{-4}$. The required accuracy of the normalized residual vector for the Lanczos algorithm in the solution phase is specified to $\varepsilon = 10^{-4}$. When the truncated version is used then $\eta_0 = 0.1$ and $t=0.2$ of the forcing sequence (13).

Owing to the large number of parameters involved in each method, the following appellation is used to describe the methods:

$$AAA(M)\text{-}BB(\mu)\text{-}TR$$

The symbol $AAA(M)$ corresponds to the type of non-linear iterations used. The symbol $BB(\mu)$ corresponds to the line search used and it takes the form CLS if the conventional line search is used with bracketing followed by regula-falsi iterations or $ST(\mu)$ if the stability criterion is applied with tolerance μ. The symbol TR denotes the application of a truncated version of the method. Figs. 2-4 depict the performance of the truncated versions of the methods in relation to the three examples considered, where MNR and CNR are the modified and conventional Newton-Raphson schemes, respectively.

Concluding Remarks

It is apparent that comparing a group of non-linear algorithms is not an easy task. The variability of the methods, combined with the fact that the performance of the methods is dependent on the type of problem and the non-linearities involved, make it difficult to draw universal conclusions about the behaviour of non-linear methods. However, certain conclusions, or trends, can be drawn on the basis of experience and the numerical results presented in this paper and in Refs.(8,11). Since the solution of linear equations solving represents a major cost of implicit non-linear analysis, the use of iterative solvers, as opposed to direct solvers, for the solution of the linearized equations, allows the efficient solution, in both computing time and storage, of large-scale non-linear problems. Iterative solvers can also be combined with the concept of truncation leading to a further improvement in the efficiency of the methods. The use of truncation may increase the total number of non-linear iterations, but the decrease of effort of the linear solver has an overall beneficial effect on the performance of the non-linear solution scheme. For problems with an expensive gradient evaluation and strong non-linearities, it is advisable to use a more strict forcing sequence parameter in order to reduce the number of additional gradient evaluations.

Secant-type implicit methods with tangential predictors can be combined with inexpensive slack line searches and increase the efficiency of the methods, particularly for highly non-linear problems. The more significant gain from line searches is that they increase the robustness of the iterative procedure. It seems, however, that the choice of a proper iteration matrix is clearly more important than an accurate line search for achieving overall computational efficiency.

It was also observed that some of the methods may run into difficulties in trying to locate an equilibrium configuration, because the iteration path, followed by this particular method, is disturbed by the existence of multiple neighbouring equilibrium states. For these reasons, the computational strategy must be equipped with restart procedures and with an adaptive strategy in the implementation of the non-linear methods. Thus, according to the degree of non-linearities and the peculiarities of the specific problem, the accuracy of the line searches may be adjusted, the number of updates may vary and a recomputation of the tangent stiffness matrix may be necessary inside each increment.

References

1.   H. Matthies and G. Strang, "The solution of nonlinear finite element equations", Int. J. Num. Meth. Eng., 14, 1613-1626 (1979).

2. K.J. Bathe and A.P. Cimento, "Some practical procedures for the solution of nonlinear finite element equations", Comp. Meth. Appl. Mech. Eng., 22, 59-85 (1980).

3. M. Geradin, S. Idelsohn and M. Hogge, "Computational strategies for the solution of large nonlinear problems via quasi-Newton methods", Comp. Struct., 13, 73-81 (1981).

4. A. Pica and E. Hinton, "The quasi-Newton BFGS method in the large deflection analysis of plates", in C. Taylor, E. Hinton and D.J.R. Owen (eds.), Numerical Methods for Non-linear Problems, Pineridge Press, Swansea, U.K., 1980, pp. 355-365.

5. M.A. Crisfield, "Accelerating and damping the modified Newton-Raphson method", Comp. Struct., 18, 395-407 (1984).

6. E. Ramm and A. Matzenmiller, "Large deformation shell analysis based on the degeneration concept", in T.J.R. Hughes and E. Hinton (eds.), Finite Element Methods for Plate and Shell Structures, Pineridge Press, Swansea, U.K., 1986, pp. 365-393.

7. K. Schweizerhof, "Quasi-Newton verfahren und kurvenverfolgungsalgorithmen für die lösung nichtlinearer gleichungssysteme in der strukturmechanik", Report, Institüt für Baustatik, Universität Fridericiana Karlsruhe (TH), Germany, 1989.

8. M. Papadrakakis, "Solving large-scale linear problems in solid and structural mechanics", in M. Papadrakakis (ed.), Solving Large-Scale Problems in Mechanics, John Wiley & Sons, 1993, pp. 183-224.

9. M.A. Crisfield, "A fast incremental/ iterative solution procedure that handles snap-through", Comp. Struct., 13, 55-62 (1981).

10. E. Ramm, "Strategies for tracing nonlinear responses near limit points", in K.J. Bathe, E. Stein and W. Wunderlich (eds.), Nonlinear Finite Element Analysis in Structural Mechanics", Springer-Verlag, 1981, pp. 63-89.

11. M. Papadrakakis and G. Pantazopoulos, "A survey of quasi-Newton methods with reduced storage", Int. J. Num. Meth. Enging., 1573-1596 (1993).

12. J. Nocedal, "Updating quasi-Newton matrices with limited storage", Math. Computation, 35, 773-782 (1980).

13. M. Papadrakakis and V. Balopoulos, "Improved quasi-Newton methods for large nonlinear problems", J. Eng. Mech., ASCE, 117, 1201-1219 (1991).

14. R.S. Dembo and T. Steihaug, "Truncated-Newton algorithms for large scale unconstrained optimization", Math. Program., 26, 190-212 (1983).

15. B. Nour-Omid, B.N. Parlett and R.L. Taylor, "A Newton-Lanczos method for solution of nonlinear finite element equations", Comp. Struct., 16, 241-252 (1983).

16. M. Papadrakakis, "A truncated Newton-Lanczos method for overcoming limit and bifurcation points", Int. J. Numer. Meth. Eng., 29, 1065-1077 (1990).

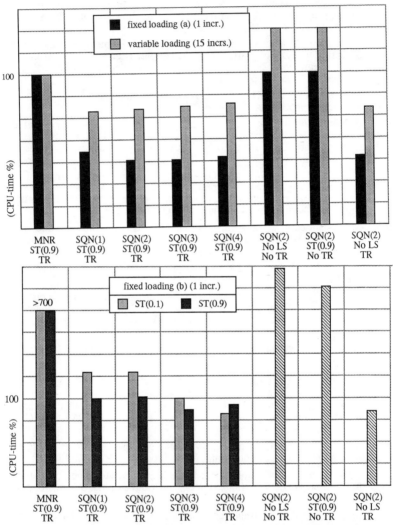

Figure 1. Truss example. Relative performance of SQN(M) and MNR with
and without truncation

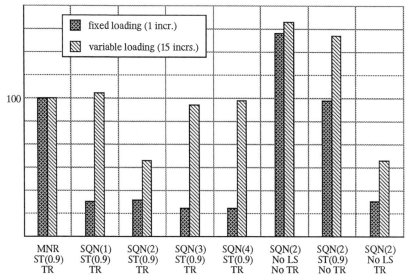

Figure 2. Frame example. Relative performance of SQN(M) and MNR
with and without truncation

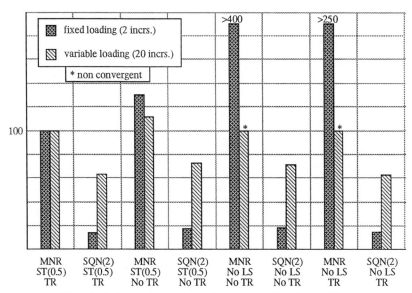

Figure 3. Shell example. Relative performance of SQN(M) and MNR
with and without truncation

AWS versus AIJ Tubular K-Connection Design Rules

Yoshiaki Kurobane, M. ASCE and Kenshi Ochi[1]

Abstract

Among various existing design rules for circular tubular connections, the AWS rules are compared with the AIJ rules because of differences between the two. Subjects of discussion include accuracy in prediction, effects of size and material properties, premature failures due to local buckling of braces and due to fracture at weld toes, and applications to multi-planar connections, focussing them on the ultimate behavior and design of K-connections for the sake of conciseness. Many points required to enhance the reliability of the AWS rules are suggested.

Introduction

The AWS Code provisions for circular tubular K-connections (American Welding Society 1992) are compared with similar provisions proposed in the AIJ Recommendations (Architectural Institute of Japan 1990). Although a great difference in scope exists between the two codes, the both codes recommend ultimate limit state design criteria for the same tubular connections. The two criteria are considerably different from one to the other in spite of the fact that the existing database is common between the two. These differences are discussed hereafter.

Direct comparison of the design rules between the two codes should complicate the points in question because different safety considerations are adopted owing to differences in structures, loads and materials to which the codes are applicable. A comparison is made between the ultimate limit state equations on which the design rules were based. The ultimate resistance equations for AIJ Code are found in an American journal (Kurobane et al. 1984), while those for AWS Code are described with their background in Marshall's book (1992) as well as in the code itself.

The AIJ design equations for tubular connections are essentially identical to those in the IIW design recommendations (1989) and Eurocode 3 (1992) except for a few differences that mainly arose from earthquake resistant design requirements to which Japanese codes always have to pay a special attention.

----
[1]Professor and Associate Professor, Faculty of Engineering, Kumamoto University, Kurokami 2-39-1, Kumamoto 860, Japan

$$P_u = 6\pi\beta\left(\frac{1.7}{\alpha_0} + \frac{0.18}{\beta}\right)Q_\beta^{0.7(\alpha-1)} Q_f \frac{F_y T^2}{\sin\theta} \quad (1)$$

$$\alpha = 1.0 + 0.7 \frac{\displaystyle\sum_{\text{all braces}} P\sin\theta\cos 2\phi\exp\left(-\frac{z}{0.6\gamma}\right)}{[P\sin\theta]_{\text{reference brace}}}$$

or

$$\alpha = 1.0 + 0.7 g/d \quad 1.0 \le \alpha < 1.7$$

$$Q_\beta = 1.0 \text{ for } \beta \le 0.6$$

$$Q_\beta = \frac{0.3}{\beta(1-0.833\beta)} \text{ for } \beta > 0.6$$

$$Q_f = 1.0 - 0.030\,\gamma\,n^2$$

$$z = L/\sqrt{RT}$$

$$P_u = f_0 f_1 f_2 f_3 f_4 f_5 T^2 F_y \quad (2)$$

$$f_0 = 2.11(1 + 5.66\beta)$$

$$f_1 = (2\gamma)^{0.209}$$

$$f_2 = 1 + \frac{0.00904(2\gamma)^{1.24}}{\exp\left(0.508\dfrac{g-3.04}{T}-1.33\right)+1}$$

$$f_3 = \frac{1 - 0.376\cos^2\theta}{\sin\theta}$$

$$f_4 = 1 + 0.305n - 0.285n^2$$

$$f_5 = \left(\frac{F_y}{F_u}\right)^{-0.723}$$

(a) AWS Equations

(b) AIJ Equations

Figure 1 Ultimate Resistance Equations (Note: See Fig.6 for symbols)

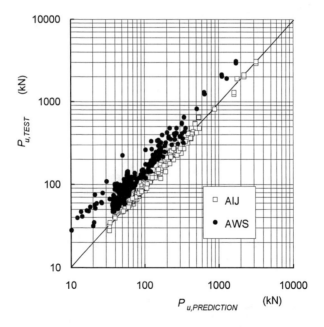

Figure 2 Ultimate Resistances : Test Results Compared with Predictions
for All Non-Overlapping Connections

Accuracy in Prediction

The ultimate resistance prediction equations for K-connections are reproduced in Fig. 1 from both the AWS and AIJ code provisions and are designated as Eqs. 1 and 2. The both equations predict ultimate limit state loads when K-connections fail by localized plastic shell bending deflection of the chord walls, although the AWS equations are not applicable to connections with overlapping braces. In overlapped K-connections shell bending deflection is observed also in the brace walls. The AIJ equations give the resistance of overlapped K-connections as well by a continuous curve applicable also to the range $g<0$. The primary concern should be how accurate these equations are.

Predicted resistances are compared with existing test results in Fig. 2, where the exponential chord ovalizing parameter is used for the AWS equations. The database used here is reliable not only because it omits inappropriate tests but also it includes ALL the reliable test results collected from open literature (Ochi 1984). The mean values and coefficients of variation (COV) of test to prediction ratios are summarized in Table 1. Clearly the AWS predictions scatter much more than the AIJ predictions.

Table 1 Ultimate Resistances : Means and COV's of Test to Prediction Ratios

		Mean	COV
AWS: non-overlapping connections with D>139mm	exponential α	1.74	0.282
	simple α	1.70	0.243
AWS: all non-overlapping connections	exponential α	1.60	0.219
	simple α	1.56	0.206
AIJ: all non-overlapping connections		1.00	0.094
AIJ: all connections		1.00	0.104

Note: COV was computed using degrees of freedom equal to the number of specimens minus 1

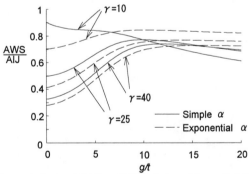

Figure 3 Ratio of AWS to AIJ Predictions (Note : β=0.5 , θ=45°)

The resistance factor for the LRFD is approximately given by

$$\phi = (R_m / R_n)\exp(-0.55\beta V_R) \qquad (3)$$

If the COV, designated by V_R in the above equation, increases from 0.1 to 0.2, the resistance factor decreases by about 15 %, which may not be a serious drawback to overall economy. However, if values of COV become greater than 0.2, it is natural to suspect prediction models to be suffering from a lack-of-fit. To see this, the ratio of AWS to AIJ predictions, tentatively ignoring the effect of material properties (the term f_s in Eq.2), is plotted against g/T in Fig. 3 for typical Warren-type K-connections. This ratio tends to approach to the mean value of $1.8/2.44=0.74$ as g/T gets close to about 12, because the AWS formulae give lower bound predictions of ultimate resistance. However, on the assumption that the AIJ equations are accurate enough to predict the mean resistance of K-connections, the following observations can be made. The errors in AWS predictions possibly contain a systematic component caused by ignoring combined effects of g/T and γ : that is, the connection resistance varies with about 1.8 power of thickness and is more sensitive to variation of g/T as the chord becomes lighter. This was recognized by Marshall as the difficulty that can be traced to the slavish adoption of thickness squared strength formulation in the Yura-based American criteria (Marshall 1992). The systematic errors, unless existing design equations are altered, have to be corrected by applying a constant bias factor, which would make the AWS rules unacceptably uneconomical.

Although accuracy in prediction is one of leading factors affecting the reliability of design, any design equation had better have a simple form and wide range of application so that human errors are automatically avoided. However, an accurate equation is capable of detecting unexpected low-strength failures induced by failure modes different from shell bending failure for which the both equations are provided, as will be discussed later.

Effects of Size and Material Properties

The AWS equations were based on the database in which data for small-size specimens were deleted. Mechanics of continuum shows that no size effect exists if failures are governed by plastic bending deflection of tube walls. If failures are governed by brittle failure under tension, however, there should exist size effects. In the latter case capacity criteria different from those governing the plastic capacity of connections have to be used as will be discussed later. The observed connection resistances governed by these two different failure modes should not be mixed up when formulating resistance prediction equations.

A possible source of size effect on the resistance of tubular connections is the fact that small-size specimens tend to have relatively larger weld size than larger specimens. However, such a size effect can be sorted out in the process of regression analysis. As seen in Eq. 2 in Fig. 1, one of the explaining variables is the nominal gap size minus 3 mm. The value of 3 mm was obtained from the regression analysis. In this way the effect of a larger weld size on the gap size is eliminated. As seen in Fig. 2, ratios of observed to predicted resistances distribute randomly about the diagonal over the whole range varying from 30 kN to 3000 kN when the AIJ equations are used, which proves that errors are independent of the resistance and, therefore, of the size of the connections. On the contrary, the AWS equations underpredict resistances of small-size specimens, showing a significant size effect.

The K-connections included in the database are made of various steels ranging from hot-finished mild steel to high-strength steel with specified minimum tensile strength of 785 MPa. The effect of material properties is explained by a function of the yield stress ratio F_y/F_u of the chord materials in the AIJ equations. In the AWS equations F_y is limited not to exceed $2F_u/3$. The both equations tend to reduce predicted resistances of connections made of materials with high yield stress ratios. However, the AWS 2/3 limitation is an historic provision common to most design codes and does restrict the use of modern high-strength steels. The AIJ equations can be adapted for any steel including high strength steel having a yield stress ratio greater than 2/3.

Local Buckling of Compression Braces

Local buckling of compression braces in areas adjacent to connections is one of typical failure modes observed in both gapped and overlapped K-connections as shown in Fig. 4. When the brace local buckling occurs in overlapped K-connections, the failure mode varies with the diameter to thickness ratio of the brace. Namely, the failure mode varies from dominating local buckles in the compression brace wall (See Fig. 4 (b)) to a mixed failure mode accompanying shell bending deflection in both the chord and brace walls as the brace becomes heavier. Nevertheless, the local buckling and shell bending failure modes should be strictly distinguished between the two because the local buckling strength can be significantly lower than the shell bending resistance as well as the local buckling strength observed in stub-column tests.

(a) Gapped Connection   (b) Overlapped Connection
Figure 4 Local Buckling of Compression Braces

The local buckling strength is given as a function of the nondimensional local buckling parameter $\alpha = (E/F_y)(t/d)$ of the compression brace and the joint resistance to shell bending failure (Kurobane et al. 1986), which can be rewritten in a simplified form as

$$F_{lb} = 0.217 \; \alpha^{1/4} \; (F_y + F_{sb})$$ (4)

in which F_{lb} signifies the local buckling strength and F_{sb} signifies the shell bending strength, both being represented in terms of the average stress on a cross section of compression brace, while F_y used here designates the yield stress of compression brace.

The strength interactions between F_{lb} and F_{sb} are explained as follows. The compression brace sustains a secondary bending moment with greater compression on the toe side than on the heel side owing to uneven support given to the brace by the chord and tension brace. When the load approaches the brace local buckling load, the bending moment at the brace end decreases because the moment is redistributed owing to local buckling or yielding, and then the compression brace carries a further increase of compressive load. When the shell bending failure mode begins to appear, however, the redistribution of bending moment is prevented.

The above equation is the basis for an additional limitation imposed on the design rules for K-connections in the AIJ design rules. The AWS design rules do not contain such a safeguard against the brace local buckling except for ordinary limiting diameter to thickness ratios. As can easily be understood from Eq. 4, this may endanger K-connections with thick-walled chords (F_{lb} does not increases in proportion to F_{sb}) or thin-walled braces designed according to the AWS code.

Fracture at Weld Toes

Cracks are frequently observed along the toes of welds between a tension brace and a chord in the gap region of the chord wall. Although in most cases these cracks extend slowly with no significant effect on load-deformation curves of connections until the chord wall sustains large plastic deflection, there are a few examples in which connections lose stability, even before plastic deflection of the chord walls becomes extensive, as soon as the first cracks are found.

One example of such premature tensile failure is two replicate tests of identical specimens performed at the University of Texas (Yura et al. 1978). Specimen configuration, loading arrangements and a load-deformation curve reproduced from the original report on these tests are shown in Fig. 5. These specimens are subjected to relatively high tensile stress in the chords and have an angle of intersection of 30 degrees between the tension braces and the chords. Both the specimens showed a sudden reduction in stiffness as soon as cracks were found at the weld toes (at the load step 8 in Fig. 5). Although both the specimens withstood a further increase in load after large plastic deflection of the chord walls, the load-deformation curve clearly indicates that the first peak load (the first peak load corresponds to the ultimate resistance according to the definition in both the AWS and AIJ equations) was controlled by a tensile failure.

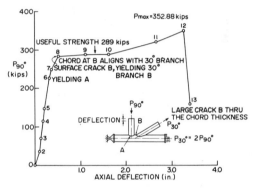

Figure 5 Test Results for K-Connection according to Yura

The AIJ equations suggest that these specimens sustained a premature low-strength failure. The predicted capacity according to the AIJ equations is equal to 364 kips (1620 kN) as compared with 289 kips (1291 kN) observed in the test. The test to prediction ratio is 0.8, which shows a significant reduction in ultimate resistance. On the contrary, the predicted capacity according to the AWS equations is 139 kips (619 kN), giving the test to prediction ratio of 2.09. Thus, the AWS equations cannot capture an occurrence of premature failure, which may result in overlooking this important difference in failure mode.

Another example of tensile failure includes a test of large-sized K-connections performed at Sumitomo Metal Industry (Ohtake et al. 1978). In this test cold-bent high-strength steels with the measured yield stress varying from 739 to 833 MPa were used for the chords.

A very simple analysis was attempted to predict the critical load at which tensile failure of this nature started (Kurobane et al. 1989). Namely, a component of axial load in the tension brace was assumed to be carried by an imaginary tensile strip accommodated in the gap region. A tentative conclusion drawn from this study was to limit the minimum gap size to prevent low-strength tensile failure. A typical value of the limiting gap size for large-sized K-connections like those tested at Sumitomo and at the University of Texas was found to be g/T=5. This result implies that an American practice of using the minimum gap size of 2 inches (API 1991) could incur a danger of low-strength tensile failure in large-sized K-connections.

Many factors are involved in early developments of cracks at the weld toes. Initiating cracks were due to ductile tensile failures in all the existing test results, although these cracks may change to fast unstable cracks if adverse effects are combined (See e.g. Machida et al. 1987). The use of clean, low-carbon and fine-grained steel may be effective to prevent crack extension (Packer 1992). No definite criteria to prevent these cracks are shown in both the AWS and AIJ code provisions except for general cautions on material selection.

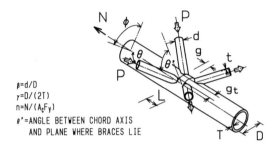

$\beta=d/D$
$\gamma=D/(2T)$
$n=N/(A_cF_y)$
θ'=ANGLE BETWEEN CHORD AXIS
       AND PLANE WHERE BRACES LIE

Figure 6 Typical Warren Type Double-K Connection

Application to Multi-Planar Connections

Most of ultimate resistance or design equations for multi-planar connections, including those in the AIJ code, have a form of uni-planar connection capacity multiplied by correction terms. The AWS equations (Eq. 1) are the only exception that shows general design criteria applicable to any type of non-overlapping multi-planar connections without a need of joint classification. Since this equation was originally developed via a theoretical analysis with no calibration with experimental results (Marshall 1992), it is interesting to see how this proposal compares with the results of tests on double K-connections recently performed at Kumamoto University (Makino et al. 1984, Paul et al. 1993, Makino et al. 1993).

The double K-connection consists of two uni-planar K-connections with a common chord as shown in Fig. 6. Ultimate resistance equations for double K-connections under symmetrical axial loads (Symmetrical loading is shown in Fig. 6) were derived from the three series of tests (Paul 1992). The test results are compared with the predictions in Fig. 7, which shows that the reliability of the proposed equations is excellent at least within variation ranges of experimental variables.

The ratio of double K to uni-planar K-connection resistances, which is called the multi-planar coefficient μ, is calculated by the proposed prediction equations and plotted against the out-of-plane angle ϕ in Fig. 8. Double K-connections selected here are typical Warren type connections, of which dimensions nearly correspond to central values in the variation ranges of geometrical variables for the tests. The values of the multi-planar coefficient calculated from the AWS equation are also shown in the same figure. Further, the AIJ design equations proposed a constant value of 0.9 as the multi-planar coefficient based on the early test by Makino et al.(1984), which is shown by a dashed line in the same figure.

The AWS equations capture well a general trend of multi-planar coefficient: the coefficient first increases with ϕ and then starts to decreases at a certain value of ϕ. However, significant differences are found between the AWS predictions and test results when one looks into details of these curves. The multi-planar coefficient becomes greater than 1.0 in the range where ϕ varies from 45 to 135 degrees according to the AWS equations whereas it usually becomes lower than 1.0 unless both β and g/T are large in tests. This is because a different failure mode starts to appear as ϕ increases: the chord wall sustains radial deflection between the two compression braces eventually creating a fold between them, which is called the failure mode 2 (Fig 9), in contrast with the AWS prediction which speculated that chord wall ovalization was suppressed as ϕ increased showing the maximum resistance at ϕ=90 degrees. The value of ϕ at which the multi-planar coefficient becomes maximum is given as a function of β and g/T according to Paul's formulae. When ϕ decreases from this value, the failure mode 1 occurs as shown in Fig. 9: the two compression braces act as one member and penetrate the chord wall together. When the

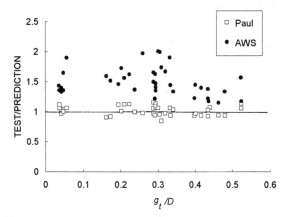

Figure 7 Ultimate Resistances of Double K-Connections: Test to Prediction Ratios

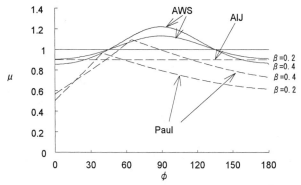

Figure 8 Multi-Planar Coefficient (Note : g/T=10, θ '=45°)

failure mode 1 appears, the multi-planar coefficient given by Paul's formulae decreases with ϕ eventually approaching a value of 0.5 at ϕ=0 degrees, which corresponds to an physical interpretation that each compression brace carries a half of the uni-planar K-connection capacity when the two compression braces overlap completely. Such a trend is not seen in the AWS curves, although this is outside the applicable range for the AWS equations (overlapping braces).

In actual structures the value of ϕ for double K-connections under symmetrical axial loads varies in a range between 45 and 120 degrees. In this range the simple AIJ multi-planar coefficient is even more accurate than the AWS equation.   Errors due to AWS prediction appear on the unsafe side.   Nevertheless, the mean of test to prediction ratios is still greater than 1.0 as shown in Fig. 7 because the AWS equations give lower bound predictions. When the small in-plane gap correction factor proposed by Lalani et al.(1989) is applied to the AWS equation, correlation between test and prediction improves appreciably. But this is because most of specimens have relatively thin-walled chords (γ>18).

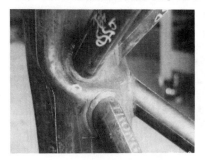

(a) Failure Type 1   (b) Failure Type 2
Figure 9   Failure Modes for Double K-Connections

Applying the same factor to double K-connections with heavy chords leads to an overprediction of connection capacity as predicted by an FE analysis (Wilmshurst et al. 1993).

Study on double K-connections under antisymmetrical loading is now in progress at Kumamoto University. Only one existing data showed a trend opposite to AWS prediction (Makino et al. 1993). Certainly a reexamination of the AWS chord ovalizing factor is unavoidable.

Conclusions

The both AWS and AIJ ultimate resistance equations for K-connections were compared with existing test results. The AWS equations showed errors significantly greater than the AIJ equations suggesting presence of systematic errors in the former equations. These errors bring about relatively unsafe design for connections with heavy chords (especially for double K-connections) and too conservative design for connections with thin-walled chords and small gaps. The AWS design equations, however, have a format so that the design resistance is automatically calculated for any non-overlapping connection that has many members framing into it by using its unique chord ovalizing parameter, whereas, in the AIJ format, engineering judgments have to be exercised when designing complex connections other than those specified in the code. An improvement of the chord ovalizing parameter in accuracy will be effective to enhance the reliability of the AWS design rules.

Low-strength failure events induced by failure modes other than chord wall plastification were found to be hidden behind scatter of data when test results were compared with the

AWS equations. Additional safeguards required of the AWS design rules are those against local buckling of compression braces in areas adjacent to connections and against failure due to tension from cracks at weld toes. Since the latter failure mode is difficult to analyze, no definite criterion has been established in both the AWS and AIJ design rules. This is an issue requiring intensive study in the future.

Space is insufficient to refer to interactions between connection behavior and frame behavior. This subject affords a basis for establishing sound design rules for connections and was dealt with in some detail in a recent study (Kurobane et al. 1993), which demonstrated that the AIJ equations were able to exactly predict the ultimate behavior of connections in trusses both before and after buckling of members. A similar assessment is required to be achieved with regard to the AWS equations as well.

References

American Petroleum Institute (1991)."Recommended practice for Planning, Designing and constructing fixed offshore platforms," API PR2A
American Welding Society (1992)."Structural welding code/steel," ANSI/AWS D1.1
Architectural Institute of Japan (1990)."Recommendations for the design and fabrication of tubular structures in steel," AIJ (in Japanese)
Comité Européan de Normalisation (1992)."Eurocode 3: Design of steel structures," ENV 1993-1-1, British Standard Institution
International Institute of Welding (1989)."Design recommendations for hollow section joints-predominantly statically loaded, 2nd ed.," IIW Doc.XV-701-89
Kurobane,Y., Makino,Y. and Ochi,K. (1984)."Ultimate resistance of unstiffened tubular joints,"J. Struct. Engrg., ASCE, 110(2), 385-400
Kurobane,Y.,Ogwa,K.,Ochi,K. and Makino,Y. (1986)."Local buckling of braces in tubular K-joints," Thin-Walled Structures, 4, 23-40
Kurobane,Y.,Makino,Y. and Ogawa,K. (1989)."Further ultimate limit state criteria for design of tubular K-joints," Tubular Structures (eds. Niemi and Makelainen), Elsevier, 65-72
Kurobane,Y. and Ogawa,K. (1993)."New criteria for ductility design of joints based on complete CHS truss tests," Tubular Structures V (eds. Coutie and Davies), E & FN Spon, 570-581
Lalani,M. and Bolt,H. (1989)."Strength of multiplanar joints on offshore structures," Tubular Structures (eds. Niemi and Makelainen), Elsevier, 90-102
Machida,S.,Hagiwara,Y. and Kajimoto,K. (1987)."Evaluation of brittle fracture strength of tubular joints of offshore structures," Proc. 6th Int. Offshore Mechanics and Arctic Engrg. Symposium, Vol.3, 231-237
Makino,Y.,Kurobane,Y. and Ochi.K. (1984)."Ultimate capacity of tubular double K-joints," Proc. Int. Conf. IIW on Welding of Tubular Structures, Pergamon, 451-458
Makino, Y.,Kurobane,Y. and Paul,J.C. (1993)."Ultimate behavior of diaphragm-stiffened tubular KK-joints," Tubular Structures V (eds. Coutie and Davies), E & FN Spon, 465-472
Marshall,P.W. (1992)."Design of welded tubular connections: basis and use of AWS code provisions," Elsevier
Ochi,K.,Makino,Y. and Kurobane,Y. (1984),"Basis for design of unstiffened tubular joints under axial brace loading," IIW Doc. XV-561-84
Ohtake,F.,Sakamoto,S.,Tanaka,T.,Kai,T.,Nakazato,T. and Takizawa,T. (1978). "Static and fatigue strength of high tensile strength steel tubular joints for offshore structures," Proc. Offshore Technology Conf., 1747-1755
Packer,J.A. (1993)."Overview of current international design guidance on hollow structural section connections," Proc. 3rd Int. Offshore and Polar Engrg., Conf., 1-7
Paul,J.C. (1992)."The ultimate behavior of multiplanar TT and KK-joints made of circular hollow sections," Ph.D Thesis, Kumamoto Univ., Kumamoto
Paul,J.C.,Makino,Y. and Kurobane,Y. (1993)."The ultimate capacity of multiplanar TT and KK-joints: comparison with AWS and API design codes," Proc. 3rd Int. Offshore and Polar Engrg., Conf., 183-191
Wilmshurst,S.R. and Lee,M.M.K. (1993)."Ultimate capacity of multiplanar double K-joints in circular hollow sections," Tubular Structures V (eds. Coutie and Davies), E & FN Spon, 712-719
Yura,J.A. and Frank,K.H. (1978)."Ultimate load tests on tubular connections," CESRE Report, 78-1, Univ. of Texas

The Behaviour and Design of Stressed-Arch (Strarch) Frames

Murray J. Clarke[1]   Gregory J. Hancock[2]

Abstract

The paper presents an overview of the stressed-arch structural system and the associated research program undertaken at the University of Sydney between 1988 and 1992. The objectives of the research were to investigate the strength and behaviour of stressed-arch frames through experimental and theoretical studies, culminating in the development of simple, yet rational and economical, design procedures for the initially curved and yielded top chord.

Introduction

The stressed-arch ("Strarch") system is a unique and innovative structural form comprising prefabricated plane truss frames which are erected by a post-tensioning stressing procedure rather than by the more traditional techniques involving cranes and scaffolding. Cold-formed tubular sections are usually used for the chord and web members of the stressed-arch frames.

The construction procedure for stressed-arch frames initially involves the assembly of the primary load carrying steel trusses of a building at ground level (Fig. 1a). Purlins, roof sheeting and services are also fixed to the trusses while the structure is at ground level so as to form a complete roof system. The truss frame itself consists of two basic components which are termed the *flexible truss* and *rigid haunch* regions (Fig. 1). The top chord in both of these regions consists of a continuous cold-formed square hollow section. Over the flexible truss region of a stressed-arch frame, the bottom chord comprises tubular sections with sliding joints or gaps that lock after a predefined axial shortening. The erection of stressed-arch frames into the final configuration is achieved by tensioning the prestressing cables passing through the bottom chord, thereby causing the bottom chord to shorten at the gap locations in the flexible truss and the complete structure to lift into an arch configuration (Fig. 1b). The erection process induces bending and compression in the top chord, often to the extent that plastic straining occurs. Fixing the sliding columns in position and grouting the bottom

---

[1]Lecturer, School of Civil and Mining Engineering, University of Sydney, N.S.W. 2006, Australia

[2]BHP Steel Professor of Steel Structures, School of Civil and Mining Engineering, University of Sydney, N.S.W. 2006, Australia

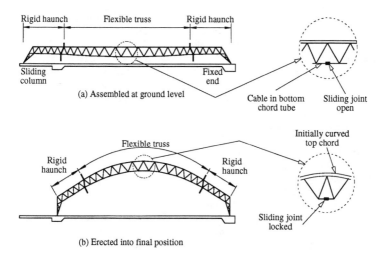

(a) Assembled at ground level

(b) Erected into final position

Figure 1: Stressed-arch (Strarch) frame concept

chord prestressing cables complete the erection procedure. Stressed-arch frames with clear spans of the order of 100 m have been built successfully in Australia, U.S.A., Japan and South-East Asia (ENR 1988, 1993).

During the post-tensioning erection process, the tubular top chord (typically a square section) becomes curved, often causing yielding over a substantial portion of its length. The moment induced in the top chord as a result of the erection procedure alone therefore approaches the section strength. The capacity of the top chord to resist further axial and bending action effects arising from service loads is therefore difficult to quantify in a rational manner using the member design rules available in steel design specifications.

In order to address the significant design issue pertaining to the top chord, and to investigate more generally the strength and behaviour of stressed-arch frames, an extensive research program was conducted at the University of Sydney between 1988 and 1992 (Clarke 1992). This paper summarises the major aspects of this research, which was both theoretical and experimental in nature and resulted in the recommendation of simple yet effective design procedures for the initially curved and yielded top chord.

Statics of Stressed-Arch Frames and Fundamental Top Chord Behaviour

During the post-tensioning erection process, the support at one end of the structure is free to slide and the support reactions are therefore statically determinate. To equilibrate the tension in the prestressing cable, the top chord develops an axial thrust (P) which is statically determinate to within a small error. The moment (M_T) induced in the top chord by the erection procedure is not statically determinate but is a function of the initially applied curvature (κ_0) and the effective flexural rigidity of the top chord. As a consequence of the erection process, the combined tube/cable bottom chord always remains in *net tension* over its full length under the action of dead load alone and will therefore not buckle laterally. The force distribution in the erected stressed-arch frame is therefore different from that which would have resulted had the structure been built in-place to its final geometry. This can be regarded as an advantageous feature of the stressed-arch structural system over conventional systems built in-place where the dead load produces compression forces in the lower chord members near the eaves.

After erection of the structure, the sliding column base is fixed in position so that the frame becomes statically indeterminate under the action of live load. For the indeterminate structure, a uniformly distributed downwards live load over the full span induces superimposed compression in the *bottom chord* which increases towards the eaves. Uniform downwards live load also results in superimposed compression in the *top chord* that is a maximum at the apex of the frame and decreases towards the eaves.

In stressed-arch frames, the lateral restraint provided to the top chord of each frame by the purlins and roof sheeting, in conjunction with the net tension that exists over the full length of the bottom chord from the erection procedure, results in each frame exhibiting predominantly in-plane behaviour under load. In addition, the top chord is usually a square tubular section which does not undergo flexural-torsional buckling between panel points. The nonlinear response of the top chord can therefore be classified realistically as in-plane beam-column behaviour.

Traditional studies of beam-column behaviour usually assume proportional loading. In stressed-arch frames, the top chord is subjected to different loading and restraint conditions in the erection and service loading stages and so the loading is very much non-proportional. The simplified behaviour of a segment of top chord between panel points from the commencement of erection to the resistance of superimposed load is illustrated in Fig. 2. At the conclusion of the erection process, the top chord is acted on by a compression force P and is in approximately uniform bending by the end moments M_T, as shown in Fig. 2a; the resulting curvature (or bending moment) distribution in the top chord is illustrated in Fig. 2b. Under superimposed downwards loading, the top chord segment can be modelled as shown in Fig. 2c, in which the interaction with adjacent segments is now important and is represented by the nonlinear rotational restraints, and P is the axial force (including the compression force from erection) developed in the top chord. The variation in curvature due to the axial load is indicated in Fig. 2d, in which it can be seen that the curvature tends to concentrate near the centre of the segment, but reverses near the restrained ends. The corresponding bending moment variation in the segment is shown in Fig. 2e. Of interest in this figure is the fact that under an increasing axial load P, the bending moment *decreases* both at the centre and ends of the segment. The reduction in moment at the ends is to be expected because of the corresponding reduction in curvature due to the restraint of adjacent segments (Fig. 2d). The reduction in moment at the centre of the segment, however, is a consequence of the fact that, at the conclusion of the erection process, the top chord has a large bending moment which is so close to the section capacity that any *increase* in axial force due to downwards load *must* be accompanied by a simultaneous *decrease* in the bending moment if the cross-sectional strength surface is not to be violated (Clarke & Hancock 1992a, 1992b).

The performance of the top chord according to the model of Fig. 2 is dependent on the end rotational stiffnesses representing the restraint provided by adjacent segments. The only totally rational way to investigate the top chord behaviour theoretically is to

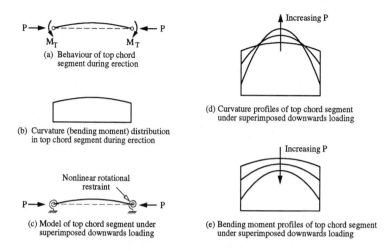

(a) Behaviour of top chord
segment during erection

(b) Curvature (bending moment) distribution
in top chord segment during erection

(c) Model of top chord segment under
superimposed downwards loading

(d) Curvature profiles of top chord segment
under superimposed downwards loading

(e) Bending moment profiles of top chord segment
under superimposed downwards loading

Figure 2: Simplified model of top chord behaviour using isolated segment

perform second-order inelastic analyses of subassemblages or complete stressed-arch frames, where the interaction between adjacent segments is incorporated automatically in the analysis.

Subassemblage Tests

The first two experimental programs performed as part of the research into stressed-arch frames, details of which can be found in Hancock et al. (1988) and Clarke & Hancock (1991), involved the testing of frame subassemblages comprising two top chord segments between pinned ends and a single bottom chord segment (Type A panels, see Fig. 3). The main purpose of the tests was to investigate the nonlinear behaviour of the top chord, and the degradation in top chord strength, as the level of initially applied curvature, and the geometrical slenderness of the top chord, is increased. In the first experimental program (Hancock et al. 1988), tests were performed on panels (denoted SP1 to SP11) of top chord slenderness $\ell_s/r = 33.5$, in which ℓ_s is the length of the top chord segment between adjacent panel points (Fig. 3), and r is the radius of gyration of the cross-section. The mean initial curvature of the top chord from the erection process (κ_0), non-dimensionalised with respect to the curvature to cause first yield of the cross-section (κ_Y), varied between 0.08 and 0.99 (assuming a yield stress $F_Y = 350\,\text{MPa}$). In the second experimental program, four tests (the specimens being denoted panels SP12 to SP15) were performed on subassemblages which were half-scale of the Strarch frames considered for use in a 94 m span aircraft hangar. Two different top chord geometrical slendernesses were considered, corresponding to $\ell_s/r = 46.2$ and 62.9, with the non-dimensional initial curvature κ_0/κ_Y varying between 0.14 and 0.79.

The test rig and Type A panel configuration is shown in Fig. 3. The post-tensioning

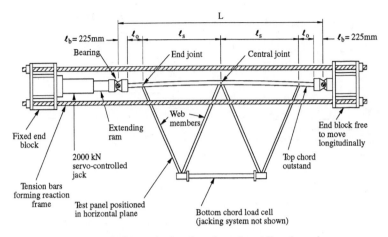

Figure 3: Schematic plan view of test rig and Type A panel

erection process was simulated by curving the top chord a predetermined amount by jacking the bottom chord nodes together. The bottom chord was then locked into position and the top chord loaded in axial compression. This test procedure simulated approximately the behaviour of the top chord in a prototype Strarch frame.

The results of the subassemblage tests are summarised in Fig. 4, in which the results of finite element nonlinear analysis simulations (Clarke & Hancock 1991) are also given for comparison and the degradation in strength of the top chord as both its geometrical slenderness and initial curvature is increased can be seen. In Fig. 4, the axial strengths P_u of the subassemblages have been non-dimensionalised with respect to the experimental stub column strength P_{Su}, and the initial curvature κ_0 at the midpoint of each top chord segment has been non-dimensionalised with respect to the nominal yield curvature κ_Y. The satisfactory agreement achieved between the numerical and experimental results shown in Fig. 4 served to verify the accuracy of the finite element nonlinear analysis to model the behaviour of stressed-arch frames.

Although constituting useful preliminary data for design purposes, the top chord strength estimates obtained from the subassemblage tests were a lower bound to the strength of the top chord of the same slenderness and initial curvature in prototype frames. This is because the pin-ended support conditions and presence of the top chord outstand (ℓ_o) and bearing length (ℓ_b) (see Fig. 3) in the test panels are all weakening effects compared to the continuous nature of the top chord in a complete frame.

Small-Scale Frame Tests

Following the subassemblage tests, a third experimental program involving the testing of small-scale but complete single stressed-arch frames under simulated gravity (vertically downwards) loading was undertaken (Clarke & Hancock 1992a). The aims of this second program were to determine the top chord strength and nonlinear behaviour in a

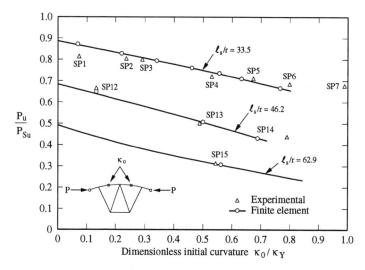

Figure 4: Results of Type A subassemblage tests and theoretical analyses

complete system as opposed to an isolated subassemblage, and also to investigate the behaviour of a complete frame under overload conditions and to ascertain the mode of failure. The test frames, having a span of 15.25 m and an apex height of 4.62 m in the erected configuration (Fig. 5), represented half-scale models of prototype 30 m span structures designed and constructed previously by Strarch International Limited.

Two nominally identical small-scale stressed-arch frames, denoted Strarch frames SF1 and SF2, were tested. The geometrical slenderness ℓ_s/r between panel points of the most critically loaded top chord segment (Fig. 5) was 40.6. The initial curvature κ_0 (the curvature in the top chord after erection but before application of vertical load) of this segment corresponded to 1.70 times the yield curvature κ_Y, based on the measured 0.2 % proof stress of the flats of the square hollow section top chord. A restraint against lateral displacement was provided at every panel point along the top chord to ensure in-plane behaviour. Vertical load was applied to the frames at 8 panel points either side of the apex as shown in Fig. 5.

At the conclusion of the erection process, the experimental curvature profile along the top chord corresponded fairly well to that assumed in the design of the frames. Following erection of the frame, the loading rig was attached (Fig. 5) and the sliding support fixed in position. With the bottom chord remaining ungrouted, vertical load was applied using hydraulic actuators in conjunction with gravity load simulators.

Although the two frames tested did not actually fail in the top chord as was hoped (Clarke & Hancock 1992a), the tests provided substantial information on the *nonlinear behaviour* of the initially curved top chord when subjected to axial compression. The top chord segment for which the largest strains were recorded was the segment second from the apex on the southern side of the frame, between load points L1 and L2 shown

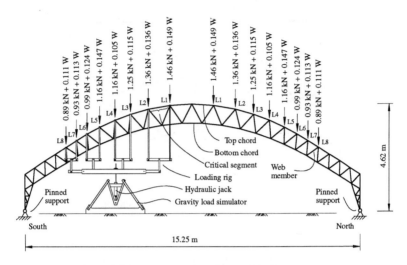

Figure 5: Schematic view of test frame and loading apparatus

in Fig. 5, which contained strain gauge pairs S7, S8 and S9 (see the inset to Fig. 6). The experimental and theoretical load-curvature responses at S7 and S8 are shown in Fig. 6, in which it is evident that the finite element analysis modelled the nonlinear behaviour of the top chord well (the response of S9 is similar to S7 and is not shown on Fig. 6 for clarity). The load-curvature behaviour shown in Fig. 6 is typical of all top chord segments in the flexible truss: the curvature increases at the centre of the segments due to the $P-\delta$ effect, but reverses at the ends due to the restraint afforded by the adjacent segments (see Fig. 2). All segments therefore behaved as effectively built-in at the ends.

Design Recommendations for the Top Chord[3]

The irrationality of applying traditional steel design methods, based on the use of member strength interaction equations, to stressed-arch frames arises because the top chord is subjected to distinctly different loading and restraint conditions in the erection and service loading stages (see Fig. 2). The proposed design procedure for the top chord is not, therefore, based on conventional elastic methods involving the use of interaction formulae for combined compression and bending, but rather utilises the *advanced analysis* provisions of the Australian Standard AS4100–1990 *Steel Structures* (SA 1990). If advanced analysis is performed, AS4100–1990 explicitly waives the requirement to perform independent member capacity checks since the effects of inelasticity and instability have been incorporated *a priori* in the analysis. In this respect, AS4100–1990 is believed to be unique amongst steel design specifications worldwide.

The basis of the proposed design procedure is the quantification of top chord strength

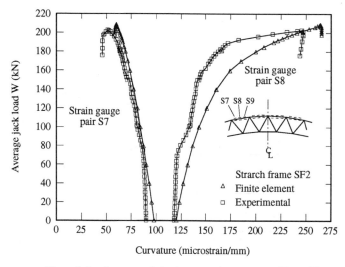

Figure 6: Load-curvature behaviour at strain gauge pairs S7 and S8

as a function of two fundamental parameters: (1) the level of initial curvature, defined non-dimensionally as κ_0/κ_Y, and (2) the geometrical slenderness λ_s of the top chord segment, expressed in normalised form as

$$\lambda_s = \frac{1}{\pi}\frac{\ell_s}{r}\sqrt{\frac{F_Y}{E}}$$

The top chord strength curves are based on finite element nonlinear ("advanced") analyses of stressed-arch frame subassemblages comprising three top chord segments and two bottom chord segments (termed Type C panels, see the inset to Fig. 7). The three-segment Type C panel provides a more realistic model of top chord behaviour in prototype frames than does the two-segment Type A panel, while providing estimates of top chord strength which remain conservative (but not overly so) due to the pinned end conditions. The results of the parametric investigation, summarised in Fig. 7, are most conveniently expressed in a format similar to that of multiple column curves (SSRC 1976), whereby a family of curves, each curve corresponding to a different value of initial curvature κ_0/κ_Y, defines the strength of the top chord as a function of its normalised geometrical slenderness λ_s. In Fig. 7, the top chord strength is expressed non-dimensionally as P_{Tu}/P_Y, in which P_{Tu} is the top chord ultimate strength and P_Y is the squash load of the top chord section.

The philosophy of the design procedure is that the primary distribution of axial forces in the top chord can be obtained with sufficient accuracy by conducting a *first-order elastic analysis* of the erected frame, and then adding these forces to those induced by erection. The strength of individual members can then be assessed by reference to the strength curves given in Fig. 7, which are used in lieu of the conventional member

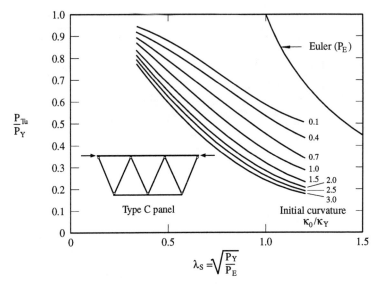

Figure 7: Strength curves for the top chord of stressed-arch frames

design rules for combined compression and bending and which account for the strength degradation caused by the initial curvature from erection, yielding, and the second-order effects and interaction local to a single panel segment and its immediate neighbours.

The rational and conservative nature of the analysis/design procedure has been verified through its application to predict the strength of the small-scale (15.25 m span) test frame SF2 subjected to vertically downwards load (Clarke & Hancock 1992b). It was shown that conservatism exists both with respect to the member strength curves (Fig. 7) and the global analysis. The conservatism of the member strength curves arises because the top chord end-restraints implicit in the use of the Type C panel underestimate the actual restraints in prototype frames. The conservatism of the global analysis arises because the use of first-order elastic analysis overestimates the top chord axial force compared to a rigorous nonlinear (advanced) analysis.

When the proposed analysis/design procedure was applied to Strarch frame SF2, the predicted maximum load was approximately 25 % conservative compared to both the experimental and nonlinear analysis ultimate loads. In view of the complex behaviour of stressed-arch frames, and the inability of conventional methods of steel design to rationally determine the top chord strength, this degree of conservatism is believed to be within acceptable limits.

Although not practicable for routine design application, the conservatism resulting from the use of first-order elastic global analysis can be removed through the use of rigorous nonlinear (advanced) global analysis. When the latter method of analysis was employed in conjunction with the top chord strength charts, the predicted maximum capacity of Strarch frame SF2 was only 4 % below the experimental strength.

Conclusions

The stressed-arch (Strarch) frames described in this paper are a comparatively recent development in steel structures, the unique feature of the system being the post-tensioning erection process resulting in substantial bending and yielding of the cold-formed square hollow section top chord. From a design perspective, a fundamental issue is that the strength in axial compression of the initially curved and plastically deformed top chord cannot be quantified rationally using conventional steel design methodologies.

The research on stressed-arch frames comprised three experimental programs—two initial programs on subassemblages of the structural system and a third program on small-scale but complete single frames—in conjunction with the development of a finite element nonlinear analysis facilitating rigorous theoretical study. The experimental and theoretical studies culminated in the development of a simple, rational and economical design procedure for the top chord, which is currently being used to design stressed-arch frames throughout the world.

Acknowledgements

The authors are grateful to Strarch International Limited for providing the test frames and funding for the research program, and for permission to publish the results. In particular, the contribution of Dr Peter Key, Engineering Manager, Strarch International Limited, to the research program is much appreciated.

References

SA (1990). *AS4100–1990, Steel Structures*. Standards Australia, Sydney, 1990.

Clarke, M.J. (1992). "The behaviour of stressed-arch frames.", *Ph.D. Thesis*, School of Civil and Mining Engineering, University of Sydney.

Clarke, M.J. & Hancock, G.J. (1991). "Finite-element nonlinear analysis of stressed-arch frames." *Journal of Structural Engineering*, ASCE, 117(10), 2819–2837.

Clarke, M.J. & Hancock, G.J. (1992a). "Tests and nonlinear analyses of small-scale stressed-arch frames." *Research Report*, R659, School of Civil and Mining Engineering, University of Sydney.

Clarke, M.J. & Hancock, G.J. (1992b). "On the design of the top chord of stressed-arch frames." *Research Report*, R668, School of Civil and Mining Engineering, University of Sydney.

ENR (1988). "Cabled truss curves as it rises." *Engineering News-Record*, 221(15), 14–15.

ENR (1993). "Down Under roof system goes up fast in Portland." *Engineering News-Record*, 230(10), 22.

Hancock, G.J., Key, P.W. & Olsen, C.J. (1988). "Structural behaviour of a stressed arch structural system." *Recent Research and Developments in Cold-Formed Steel Design and Construction: 9th International Specialty Conference on Cold-Formed Steel Structures*, University of Missouri–Rolla, St. Louis, MO, 273–294.

SSRC (1976). *Guide to stability design criteria for metal structures*. 3rd ed., Johnston, B.G., Ed., Structural Stability Research Council, Wiley.

DESIGN OF HALF-THROUGH OR "PONY" TRUSS BRIDGES
USING SQUARE OR RECTANGULAR HOLLOW STRUCTURAL SECTIONS

STEVEN J. HERTH, P.E.[1]

ABSTRACT

The initial part of this paper will outline some of the research, testing, and investigation which has been done on half-through truss bridges. This research is primarily concerned with two items:

1. Design of the top chord of the truss considering out-of-plane buckling problems.

2. Design strength and stiffness requirements for the "U-frame" formed by the truss web members and the bridge floor beams.

The second part of the paper will outline Continental Bridges' design approach to "pony" trusses. Referencing the above-mentioned research findings, the paper outlines Continental's approach to determining appropriate K-factors for design of the top chord as well as strength and stiffness design of the "U-frame" members.

The final part of the paper is a discussion of some of the connection design ramifications of a half-through truss structure fabricated with square or rectangular hollow structural sections. The connection primarily discussed here will be the one between the truss web members and the floor beams. This connection has design strength requirements for bending moments due to lateral support requirements of the top chord in a "pony" truss bridge.

---

[1] Bridge Engineer, Continental Bridge
     8301 State Hwy 29 N
     Alexandria, MN 56308

RESEARCH AND FINDINGS

The out-of-plane buckling problem of the compression chord of a "pony" truss can be equated to a column supported by elastic restraints at the truss panel points. The lateral support for the top chord is provided by the truss transverse frames or "U-frames" (floor beams and truss verticals). The "U-frames" must be adequately designed for both strength and stiffness to provide the lateral support needed for top chord stability.

The problem of the top chord buckling of a half through truss was brought to light in the late 1800's after a series of "pony" truss failures. The first successful attempt to explain these failures and to provide a method of analysis was done by Engesser (1884, 1893). Since that time a number of people have investigated "pony" truss bridges.

Some of the most extensive research and testing of "pony" truss bridges was done by Edward C. Holt (1951, 1952, 1956, 1957) at the Pennsylvania State College. With sponsorship from the Column Research Council of Engineering Foundation and the Pennsylvania State Highway Department, Holt researched and conducted full scale testing on "pony" truss bridges and wrote a series of four reports on his findings. His final report (1957) gives recommendations for design of bridge chords without lateral bracing.

In 1989, the DeBourgh Manufacturing Company, a manufacturer of pedestrian "pony" truss bridges of primarily tubular construction conducted strain gage tests on a full scale (80' long x 10' wide) "pony" truss bridge made from square and rectangular tubing [Weinstein, James (1989)]. Their findings indicate slightly more stringent requirements for tubular "pony" truss bridges than were dictated by the Holt reports.

The buckling of the top chord of the "pony" truss has two limiting bounds:

1. If the lateral restraint provided by the "U-frames" is very flexible, the chord will tend to buckle in one single half wave over its entire length.
2. If the lateral restraints provided are infinitely stiff, the chord will tend to buckle between the truss panel points.

Neither one of these extremes is seldom if ever reached in practice as either would be uneconomical. The actual buckled shape used in design is somewhere between these two extremes: some number of half-waves less than the total number of bays in the truss.

"U-FRAME" STIFFNESS REQUIREMENTS

Holt's approach will be utilized here for the determination of top chord K-factors used in design. This approach determines the K-factor for out-of-plane buckling of the top chord based on the stiffness provided by the "U-frames". Holt's solution for the allowable buckling load of the top chord of a "pony" truss assumes the following conditions:

1. The transverse frames (U-frames) at all panel points have identical stiffness.

2. The radii-of-gyration of all top chord members and end posts are identical.

3. The top-chord members are all designed for the same allowable unit stress (A's and I's are proportional to the compressive forces).

4. The connections between the top chord and the end posts are assumed pinned.

5. The end posts act as cantilever springs supporting the ends of the top chord.

6. The bridge carries a uniformly distributed load.

The results of Holt's investigation are presented in Table 1, which gives the reciprocal of the effective length factor K as a function of n (the number of panels) and of Cl/Pc where:

C   is the furnished stiffness at the top of the least stiff transverse frame. (See figure 1)

l   is the panel point spacing of the truss

Pc   is the maximum design chord stress multiplied by the desired factor of safety.

Note: Because of uncertainties involved in the analysis of the top chord of a "pony" truss, it is reasonable to require a factor of safety for overall top chord buckling greater than that used when designing typical columns; However, since each member in the continuous top chord of a "pony" truss with parallel chords cannot be simultaneously stressed to its critical buckling load, it is reasonable to use a safety factor of 1.5 for this situation.

Various secondary effects on top chord buckling such as the lateral support given to the chord by the diagonals,

TABLE 1 - 1/K FOR VARIOUS VALUES OF Cl/Pc and n

1/K	n						
	4	6	8	10	12	14	16
1.000	3.686	3.616	3.660	3.714	3.754	3.785	3.809
0.980		3.284	2.944	2.806	2.787	2.771	2.774
0.960		3.000	2.665	2.542	2.456	2.454	2.479
0.950			2.595				
0.940		2.754		2.303	2.252	2.254	2.282
0.920		2.643		2.146	2.094	2.101	2.121
0.900	3.352	2.593	2.263	2.045	1.951	1.968	1.981
0.850		2.460	2.013	1.794	1.709	1.681	1.694
0.800	2.961	2.313	1.889	1.629	1.480	1.456	1.465
0.750		2.147	1.750	1.501	1.344	1.273	1.262
0.700	2.448	1.955	1.595	1.359	1.200	1.111	1.088
0.650		1.739	1.442	1.236	1.087	0.988	0.940
0.600	2.035	1.639	1.338	1.133	0.985	0.878	0.808
0.550		1.517	1.211	1.007	0.860	0.768	0.708
0.500	1.750	1.362	1.047	0.847	0.750	0.668	0.600
0.450		1.158	0.829	0.714	0.624	0.537	0.500
0.400	1.232	0.886	0.627	0.555	0.454	0.428	0.383
0.350		0.530	0.434	0.352	0.323	0.292	0.280
0.300	0.121	0.187	0.249	0.170	0.203	0.183	0.187

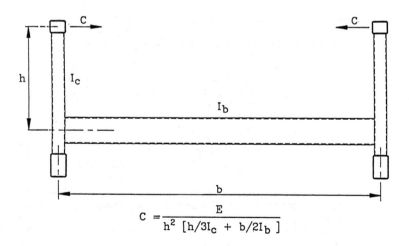

$$C = \frac{E}{h^2 \left[h/3I_c + b/2I_b \right]}$$

FIGURE 1

effects of floor beam deflections due to live loads, etc. have been studied by Holt and others. A full discussion of all aspects influencing the top chord stability of a "pony" truss bridge is prohibited here by the length limit of this paper. I recommend obtaining the "Guide to Stability Design Criteria for Metal Structures" (Galambos, 1989). Much of the information on "pony" truss design presented here is contained in Chapter 15 of that reference. Table 1 and Holt's assumptions are reprinted from that source with the permission of John Wiley and Sons, Inc.

"U-FRAME" STRENGTH REQUIREMENTS

     Strength requirements for the "U-frame" members vary from source to source (research findings, design codes, etc.). Most approaches require an additional moment capacity in the truss verticals, floor beams and their connections. This moment is over and above the moment determined by classical analysis and is calculated assuming the vertical is a cantilever, fixed at its base, which carries a transverse force at its upper end. It is the opinion of this author that the most rational "pony" truss design approach equates the required out-of-plane bending strength of the "U-frame" to the top chord compression and to the K used for top chord design. (If K out-of-plane equals the number of bays, the chord would be designed to buckle in one long half wave. In this case, no out-of-plane bending strength would be required in the "U-frames" for lateral support of the top chord).

The strength requirements suggested by Holt (1957) are:

1.   The end post is a cantilever which carries, in addition to its axial load, a transverse force of 0.3 percent (.003) of its axial load at its upper end and

2.   The moment at the lower end of each vertical may be approximated satisfactorily by applying a transverse force at its upper end equal to 0.2 percent (.002) of the average of the axial loads in the two adjacent top chord members.

While never going less than Holt's suggested requirements, Continental Bridge has adopted the following guide lines based on the more conservative "German Buckling Specifications" (DIM 4114) which are now out of print:

1.   For the interior "U-frames" use 1/100K times the average compressive force in the two adjacent top chord members as the force applied at the top of the truss verticals. NOTE: We have chosen to limit K for uniformly loaded pony truss bridges of tubular construction to a maximum value of 2.5. This gives a

minimum out-of-plane force of 0.004 (1/100K) times the
top chord compressive force. This minimum is in close
agreement with the 1989 strain gage testing of tubular
"pony" truss bridges done by DeBourgh Manufacturing
(Weinstein, 1989) which found for the structure tested
that an average of 0.0027 times the top chord axial
load was transmitted as a lateral load to the center
vertical member.

2.   For end frames, the same applies except K is omitted
      (0.01 agrees with the recommendations of the "Guide to
      Stability Design Criteria for Metal Structures").

DESIGN APPROACH

       The economical design of a "pony" truss bridge using
hollow structural sections is an iterative process. There
exists an almost infinite number of solutions to the design
of the top chord and its lateral bracing system (U-frames).
The best top chord tubular section for a "pony" truss is
rectangular with a wide horizontal face. This section has
a good radius-of-gyration for out-of-plane buckling.
Directly opposed to this in regards to economics will be
the requirements of this face for connection strength
design (simple tubular connections are more economical when
the chord face is narrow and thick, having a low width to
thickness ratio).   While the most economical design for
large heavily loaded structures may be to size the truss
members for strength and stiffness requirements, then
design connections as required, most structures least cost
alternative will be determined by considering steel cost
verses the cost of the tubular connections.

       Following is the design approach adopted by
Continental Bridge for uniformly loaded simple span bridges
utilizing simple welded tubular truss connections (tubular
members are miter cut and welded directly to the face of
the framed to member). These bridges will have their floor
beams welded directly to the truss verticals (See Figure
1).

1.   Determine the truss configuration required based on
      span, deflection limits, aesthetic considerations,
      etc.

2.   Analyze the bridge structure for all applied loads.

3.   Using a K factor of approximately 1.5 for out-of-plane
      buckling (1.3 to 2.0 is typically an economic range
      for tubular structures) and 1.0 for in-plane buckling,
      determine a tube size required for the top chord based
      on the design loads.

4. Design the truss web members and floor beams for their design loads, including the out-of-plane bending moment required for top chord stability. Keep in mind that the vertical's dimension perpendicular to the chord face, must be equal to or less than the width of the chord face.

5. Calculate the spring constant (C) furnished by the "U-frame" having the least transverse stiffness (See figure 1).

6. Calculate the value Cl/Pc.

7. Enter table 1 with n (the number of bays in the truss) and Cl/Pc and find the correct l/K valve for a compression-chord panel, interpolating as necessary.

8. Determine the actual K value and:

    −   If the calculated K is less than the K value initially assumed, check the "U-frame" for the new out-of-plane bending moments based on the lower K value; however, it may be possible to reduce the size or thickness of the top chord based on a lower Kl/r value.

    −   If K calculated is greater than the K initially assumed in sizing the top chord you must either:

        a. Check the top chord for a higher Kl/r value and if necessary, increase its size,

        b. Increase the stiffness of the "U-frame" members to achieve a lower K value, or

        c. Some combination of a and b above.

9. Check tubular connections as outlined in the next portion of this paper.

10. Iterate steps 4 through 9 to final solution.

Bear in mind that while the "pony" truss considerations and the connection design criteria are kept separate here for simplicity, the economic design of a "pony" truss fabricated from tubular members will consider both "U-frame" requirements and tubular connection efficiencies simultaneously.

CONNECTION DESIGN

As stated above, the economical design of tubular structures is highly dependent upon connection design. The most cost effective design is usually some middle ground between the least weight alternative and the least fabrication cost alternative.

If you are doing tubular connection design, I would highly recommend obtaining the "Design Guide for Hollow Structural Section Connections" (Packer, Henderson, 1992) published by the Canadian Institute of Steel Construction. This guide is an excellent source of current design information on hollow structural section connections. Portions of this guide are reprinted here with permission.

The connections of primary importance in a tubular "pony" truss are:

1. The main load carrying (vertical) truss connections at each nodal joint where the truss web members attach to the chord members.

2. The joints between the floor beams and the truss verticals.

The design approach for truss nodal joints is well documented in the above-referenced design guide. In the United States, the same design approach found in this guide has also been adopted by the American Welding Society (See Chapter 10 of the 1992 Structural Welding Code: ANSI/AWS D1.1-92). Either of these sources may be used in checking truss joint capacity.

While a full discussion of tubular joint design is limited here by the length of this paper, I would like to make the following points:

1. The vertical members in a tubular Pratt type "pony" truss, because of economics and "U-frame" considerations, are typically very nearly or are the same width as the chord members.

2. The design capacities which have been developed based on full scale testing of tubular joints have a somewhat limited "range of validity".

Based on these two points, I have found that once "U-frame" requirements and validity limits are met, the actual main truss connection resistance provided is in many instances greater than that required for actual member loads; therefore, during the iterative design process, you typically need only consider connection parameters, staying within the appropriate "range of validity" for the connection you intend to use. You can then make final connection capacity checks after all members have been selected. NOTE: If straying outside the "range of

validity" established for tubular connections, the designer
is on his own. While connections outside the validity
range obviously have some capacity, I do not recommend
their use. If using connections outside the appropriate
"range of validity", the designer needs a very good
understanding of the possible failure modes in a tubular
connection (i.e. punching shear, chord shear in gap joints,
chord face plastification, etc.) and how these factors
influence connection capacity.

The second connection of importance, which is
primarily controlled by "U-frame" considerations, is the
one between the truss verticals and the floor beams. Along
with the end shear reaction of the floor beam, this
connection must be capable of resisting the out-of-plane
bending moment induced in the truss verticals (See previous
discussion on strength requirements of the "U-frame").
NOTE: Secondary stresses due to floor beam deflections are
typically quite small in a uniformly loaded bridge and in
most cases can be neglected.

Simple tubular connections have a certain amount of
flexibility due to deformation of the tube face. In a
"pony" truss, the floor beam to vertical connection is
assumed to be rigid in order to provide lateral support to
the top chord. Because of these facts, β (the width ratio
between the floor beam and vertical) should be
approximately equal to one for this connection.

After sizing the "U-frame" members and determining
design loads, the connection must be checked for its
required capacity. Typical tubular floor beam members are
deep narrow sections (TS 8x3's, TS 10x3's, etc.) with a
relatively high bending strength about their strong axis.
These efficient beam sections are usually outside the
"range of validity" currently established for plain T-type
connections with in-plane bending moments (See "Design
Guide for Hollow Structural Section Connections", Chapter
6, Packer and Henderson, 1992). It is still usually more
cost effective to use these efficient beam sections and
design appropriate connections for their use.

In designing tube-to-tube floor beam connections which
are outside the established "range of validity" for T-type
tubular moment connections, one may conservatively treat
the floor beam as you would a wide flange beam framing into
a tubular column. The vertical faces (webs) of the tube
are assumed to carry the shear load in the floor beam to
the truss vertical through the side welds. The end moment
in the floor beam (out-of-plane bending moment in the truss
verticals), as in the case of a w-shape beam, can be
resolved into two equal and opposite flange forces. These
forces are applied at the top and bottom horizontal tube

faces of the floor beam. The top and bottom tube faces can then be equated to a plate welded transversely to a hollow structural section. The "flange" capacities of the tubular floor beam (or w-shaped floor beam) can then be checked using existing design rules for transverse plates welded to hollow structural sections (See Table 2 copied from Packer, Henderson, 1992).

Weld design for both main truss joints and floor beam connections shall be per the applicable design code. Bear in mind that in tubular connections such as these, transfer of load across the weld is highly non-uniform. Welds must be large enough to enable adequate load redistribution to take place within the joint, preventing a progressive failure of the weld and insuring ductile behavior of the joint.

CONNECTION TYPE	FACTORED CONNECTION RESISTANCE
Transverse Plate	$\beta \approx 1.0$ Basis: CHORD SIDE WALL FAILURE
	$N_1^* = Fy_0\, t_0\,(2t_1 + 10t_0)$
	$0.85 \leq \beta \leq 1 - 1/\gamma$ Basis: PUNCHING SHEAR
	$N_1^* = \dfrac{Fy_0\, t_0}{\sqrt{3}}\,(2t_1 + 2b_{ep})$
	ALL β Basis: EFFECTIVE WIDTH
where $\beta = \dfrac{b_1}{b_0}$	$N_1^* = Fy_1\, t_1\, b_e$

FUNCTIONS

N_1^* CONNECTION RESISTANCE, AS AN AXIAL FORCE
Fy_0 SPECIFIED MINIMUM YIELD STRENGTH OF TUBE
Fy_1 SPECIFIED MINIMUM YIELD STRENGTH OF PLATE
$\gamma = \dfrac{b_0}{2t_0}$

$b_{ep} = \dfrac{10}{b_0/t_0}\, b_1$, but $\leq b_1$ $b_e = \dfrac{10}{b_0/t_0}\,\dfrac{Fy_0\, t_0}{Fy_1\, t_1}\, b_1$, but $\leq b_1$

RANGE OF VALIDITY: $b_0/t_0 \leq 30$

TABLE 2
FACTORED RESISTANCE OF PLATE TO RECTANGULAR HHS CONNECTIONS
(LIMIT STATES OR ULTIMATE LOAD FORMAT)

REFERENCES

American Welding Society (1992) "Structural Welding Code"
    ANSI/AWS D1.1-92, Chapter 10.

Engesser, F. (1885), "Die Sicherung offener Brucken gegen
    Ausknicken," Zentralbl. Bauverwaltung, 1884, p. 415;
    1885, p.93.

Engesser,   F.   (1893),   "Die   Zusatzkrafte   und
    Nebenspannungen eiserner Fachwerkbrucken," Vol. II,
    Berlin.

Galambos, T.V. (1988) "Guide to Stability Design Criteria
    for Metal Structures," 4th ED., PP 515-529. Copyright
    © 1988 by John Wiley and Sons, Inc.

Holt, E.C. (1951), "Buckling of a Continuous Beam-Column
    on Elastic Supports," Stability of Bridge Chords
    without Lateral Bracing, Column Res. Counc. Rep. No.
    1.

Holt, E.C. (1952), "Buckling of a Pony Truss Bridge,"
    Stability of Bridge Chords without Lateral Bracing,
    Column Res. Counc. Rep No. 2.

Holt, E.C. (1956), "The Analysis and Design of Single
    Span Pony Truss Bridges," Stability of Bridge Chords
    without Lateral Bracing, Column Res. Counc. Rep. No.
    3.

Holt, E.C. (1957), "Tests on Pony Truss Models and
    Recommendations for Design," Stability of Bridge
    Chords without Lateral Bracing, Column Res. Counc. Rep
    No. 4.

Packer, J.A. & Henderson, J.E. (1992) "Design Guide for
    Hollow Structural Section Connections," 1st ED.,
    Copyright © 1992 by Canadian Institute of Steel
    Construction.

Weinstein, James (1989) "Strain Gage Test, 10 ft x 80 ft
    Pedestrian Bridge," DeBourgh Manufacturing Company.

Research and Application of Various Joints of Space Grid Structures in China

Liu Xi-Liang[1]

Synopsis

This paper introduces the research work results and applications of various joints of space grid structures in China. It also gives out formulas for determining the loading capacity and constructional sizes of various joints. Some support joints types are suggested and some examples of novel projects are presented.

Introduction

There are mainly three types of space structures joints to be used in China. They are bolted ball joints, welded hollow spherical joints and plate joints.

These three kinds of joints have   with st aod in time. Especially the former two sorts have been used popularly all over the whole country.

Space grid structures, adopted in China has only a history of about 30 years. We have more than 30 space frame factories and companies. More than 4000 projects have been built, 4 millions square metre and 100,000 tones of steel. We have compiled the national code for design and construction of double-layer grid structures. Space Structure Research Committee has been established 10 years ago. In this committee we have organised many activities in the fields of design, analysis and construction of space structures. The research of joint system is also important work in this committee.

----

[1] Liu Xi-Liang, member of IASS and ASCE, Professor, Director of Steel Structural Research Section, Department of Civil Engineering, Tianjin University, Tianjin, P. R. China (300072)

1 Bolted ball joints

The bolted ball joint is composed of five elements: Ball joints, bolts, nose-cone, set-screw and sleeve nut (Fig. 1). Nose-cone (or close plate), sleeve-nut and tubular member are made of low carbon steel (A3) or low alloy steel (16Mn), according to national standard; ball joint is made of medium carbon steel No.45; bolt is made of high strength steel.

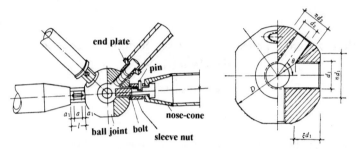

Fig. 1 Bolted ball joint.   **Fig. 2** Ball joint.

The diameter of the ball is determined by the following formula, (Fig. 2):

$$D \geq \sqrt{\left(\frac{d_2}{\sin\theta} + d_1 \mathrm{ctg}\theta + 2\xi d_1\right)^2 + \eta^2 d_1^2} \tag{1}$$

For the purpose of satisfying the contact area of sleeve nut, it also checked by another formula as follows (Fig. 2).

$$D \geq \sqrt{\left(\frac{\eta d_2}{\sin\theta} + \eta d_1 \mathrm{ctg}\theta\right)^2 + \eta^2 d_1^2} \tag{2}$$

where, D is diameter of the ball joint; θ is the minimum angle between two bolts; d_1 and d_2 are the diameters of bolts (mm), $d_1 > d_2$. ξ is proportional volume between the bolt length into ball joint and the diameter of the bolt; η is proportional value of sleeve nut diameter and bolt diameter.

Using the above two formulas to determine the diameter of the ball joint, we take the greater value.

The design value of tensile capacity of the high strength bolts is determined by the following formula:

$$N_t^b = \psi A_{eff} f_t^b \tag{3}$$

where, N_t^b is tensile design value of high strength bolts (N); ψ is a coefficient according to the diameter of bolts; f_t^b is design value of tensile strength through heat treatment (N/mm^2); A_{eff} is effective cross section area of bolts (mm^2).

The length of sleeve nut is determined by the following formula (Fig. 1).

$$l = a + 2a_1 \tag{4}$$

where $a = \xi d_0 - a_2 + d_s + 4$ mm; ξd_0 see the Fig. 2; a_2 is about 4-5 mm; d_s is diameter of set screw.

2 Welded hollow spherical joints (WHSJ) (Fig. 3)

The first project with WHSJ was built in Tianjin Scientific Committee Hall in 1966. Over half of the space grid structures are of the WHSJ grids. Particularly, of all the gymnasium roof structures which were built for the 11th Asian games held in China (1990), more than 90% are the WHSJ grids. This can be seen clearly from Table 1.

Table 1, Gymnasiums for the 11th Asian Games in Beijing (1990)

Steel Space Grid Structure	Welded Hollow Spherical Joints (WHSJ)	Beijing College Student Gymnasium (64m×64m)
		Haidian Gymnasium (48m×52m)
		Yuetan Gymnasium (Irregular Octagon 57m×66m)
		Tennis Gymnasium of BITC (60m×60m)
		Guang Cai Gymnasium (46.8m×67.6m)
		Gymnasium of Fengtai Sports Center (54.6m×76.6m)
	Bolted Joints	Ditan Gymnasiums (Hexagon Side 30m)
Steel Latticed Shells	Welded Hollow Spherical Joints (WHSJ)	Gymnasium of the Olympic Sports Center (70m×83.2m)
		Natatorium of the Olympic Sports Center (70m×100m)
		Shijing Shan Gymnasium (3-Δ99.7m)
		Gymnasium of Beijing Institute of Physical Education (53.2m ×53.2m)
	Bolted Joints	
Cable Structure		Chaoyang Gymnasium (66m×78m)
Plane Truss		Capital Training Gymnasium for Speed Skating (88m×180m)

The WHSJ grids have a series of apparent advantages, first of all is cheap in cost, 25% cheaper than other joint systems. The second is simple in construction. The only disadvantage of WHSJ is that it involves large field welding, although this causes no problem in China because of low labour cost.

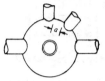

Fig. 3 Welded hollow spherical joint.

Fig.4 Spherical joint with stiffening plate.

The WHSJs are made of two hollow semi-spheres by butt welding (Fig. 4). The semi-spheres can be manufactured by hot pressing. In order to improve the load capacity of WHSJ a stiffening plate can be added into the sphere at the butt welding seam and the three parts are welded into a whole body. The load capacity of the WHSJ can be improved by 15 to 40 % by means of the stiffening plate.

The sphere has two main factors: the outer diameter of the sphere and the thickness of the sphere wall. Usually the diameter is firstly designed according to the structural demands, then thickness can be calculated. The joint diameter is determined by the following formula:

$$D = \frac{d_1 + 2a + d_2}{\theta} \tag{5}$$

where θ is the angle (in rad.) between any two tube lines at the center point of the sphere; a is a constant no less than 10 mm; d_1 and d_2 are the outer diameter of the two tubes respectively.

According to the design requirements, the ratio of the outer diameter to the thickness of the spherical joint can be usually chosen from 25 to 45. The ratio of the sphere wall thickness to the maximum tube wall thickness should be from 1.2 to 2. The thickness of the sphere wall generally should not be less than 4 mm. When the diameter of the sphere is more than 300 mm, stiffening plate can be added into the sphere in order to increase the load capacity of the joint.

On the basis of a vast amount of experiments, some conclusions can be drown[1, 3]. (1) The load capacity in tension is larger than that in compression. (2) The collapse load can be gained according unidirectional load. (3) The stiffening plate will improve the load capacity in large degree. (4) Both thickening the sphere wall and enlarging tube diameter will improve the load capacity of the WHSJ. (5) In case of tension load the joint will usually be destroyed with the tube broken. (6) In case of compression load the joint will usually be destroyed with shell plate buckled along the circumference of the tube. According to the vast experimental results, the distribution diagram of the collapse experimental data of the WHSJ in different cases can be drawn. So the load capacity formula under compression or in tension can be regressed by statistic method.

The formulas adopted by the reference (3):

1) For the compression case

$$N_c = \eta_c \left(400td - 13.3 \frac{t^2 d^2}{D} \right) \qquad (6)$$

where N_c is the design value of the WHSJ under axial compression; D is the outer diameter of the hollow sphere; t is the wall thickness of the hollow sphere; d is the outer side diameter of the tube; η_c is the ratio that improved the load capacity of the hollow sphere when stiffened in compression, $\eta_c = 1.0$ without stiffening plate, $\eta_c = 1.4$ when stiffened.

2) In tension case

$$N_t = \eta_t (0.6td\pi)f \qquad (7)$$

where N_t is the design value of WHSJ in axial tension, f is the design strength of steel in tension, η_t is the ratio that improve the load capacity of the WHSJ when stiffened in tension, $\eta_t = 1.0$ without stiffening plate; $\eta_t = 1.1$ when stiffened. It obviously point out that increase the load capacity only 10 %.

The WHSJ series are listed in Tables 2 and 3.

3. The welded plate joints

The welded plate joints were used very rarely only in several projects in our country. The welded plate joint is composed of cross plate and cover plate (Fig. 5).

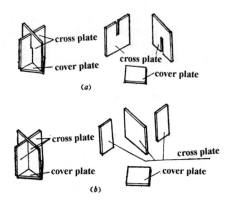

Fig. 5 Welded plate joints.

Table 2, WHSJ Series

No.	Designation	Size and Thickness (mm×mm)	Diameter of Tubes (mm)	Tensile Limit Capacity (kN)		Compressive Limit Capacity(kN)
				16Mn	A3	
1	WS1604	D160 × 4	60	212	145	148
2	WS1606	D160 × 6	60	318	217	216
3	WS1804	D180 × 4	63.5	224	153	156
4	WS1806	D180 × 6	63.5	336	230	228
5	WS2206	D200 × 6	76	403	275	273
6	WS2008	D200 × 8	76	537	366	355
7	WS2206	D220 ×. 6	89	471	322	319
8	WS2208	D220 ×. 8	89	629	429	412
9	WS2406	D240 ×. 6	102	540	369	363
10	WS2408	D240 ×. 8	102	720	492	470
11	WS2410	D240 ×. 10	102	900	615	568
12	WS2606	D260 ×. 6	114	604	412	404
13	WS2608	D260 ×. 8	114	805	550	523
14	WS2610	D260 ×. 10	114	1006	687	632
15	WS2808	D280 ×. 8	133	939	641	603
16	WS2810	D280 ×. 10	133	1174	801	727
17	WS2812	D280 ×. 12	133	1409	962	839
18	WS3010	D300 ×. 10	140	1236	844	767
19	WS3012	D300 ×. 12	140	1483	1012	887

Table 3 WHSJ with stiffening plate series

No.	Designation	Size and Thickness (mm×mm)	Diameter of Tubes (mm)	Tensile Limit Load Capacity (kN) 16Mn	A3	Compressive Limit Load Capacity(kN)
1	WSR3008	D300 × 8	140	1088	742	891
2	WSR3010	D300 × 10	140	1360	928	1074
3	WSR3012	D300 × 12	140	1631	1114	1242
4	WSR3510	D350 × 10	146	1418	968	1142
5	WSR3512	D350 × 12	146	1701	1161	1327
6	WSR3514	D350 × 14	146	1985	1355	1496
7	WSR4012	D400 ×. 12	152	1771	1209	1406
8	WSR4014	D400 ×. 14	152	2067	1410	1591
9	WSR4016	D400 ×. 16	152	2362	1612	1762
10	WSR4018	D400 ×. 18	152	2657	1813	1920
11	WSR4514	D450 ×. 14	159	2162	1475	1689
12	WSR4516	D450 ×. 16	159	2470	1686	1876
13	WSR4518	D450 ×. 18	159	2779	1897	2050
14	WSR4520	D450 ×. 20	159	3088	2108	2210
15	WSR4522	D450 ×. 22	159	3397	2319	2356
16	WSR5016	D500 ×. 16	168	2810	1782	2005
17	WSR5018	D500 ×. 18	168	2937	2004	2194
18	WSR5020	D500 ×. 20	168	3263	2227	2370
19	WSR5022	D500 ×. 22	168	3589	2450	2532
20	WSR5025	D500 ×. 25	168	4079	2784	2749

4. Some types of support joints

Plate type support joints are used in space grid structures with smaller span (Fig. 6).
Single face arc type support joints are used in space structures with medium span (Fig. 7).
Double faces arc type support joints are used in space structures with large span multipoints supports (Fig. 8).
Rubber cushion type support joints are used in space structures with medium and large span (Fig. 9).

Conclusions

In our country, we have also some another new type joints of space structure, but they do not withstand in time. Foe example watermine type bolted ball joints, half bolted ball joints, half and hollow ball joints, improved Triodetic System joints, prefabricated tetrapyramid units and punching plate joint etc..

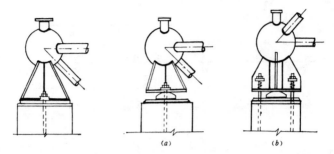

Fig. 6 Plate type support joint.

Fig. 7 Single-face arc type support joints. (a) Two-bolt joint. (b) Four-bolt joint.

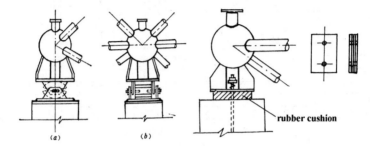

Fig. 8 Double-face arc type support joint. (a) Side view. (b) Elevation.

Fig. 9 Rubber cushion type support joint.

References

1. Liu Xi-Liang, Liu Yi-Xuan "Design, Analysis and Construction of double-layer grids" China Architectural Industry Press, Beijing, China, 1979.

2. Code for the design and construction of double layer grid structures, JGJ 7-91, China, 1991.

3. Commentary on JGJ 7-91, China, 1991.

Spherical Domes Subjected to Horizontal Earthquakes

Haruo KUNIEDA[1], Shigehiro MOROOKA[1] and Koji ONODERA[1]

Abstract

The purpose of this paper is to get a powerful tool for response analysis of a spherical dome subjected to dynamic excitation based on mathematical analytic method, i.e., the Galerkin procedure in modal analysis, with sufficient accuracy and practicality.

At first, this paper provides an approximate solution of eigen modes, which has sufficient accuracy and practicality for response analysis in anti-symmetric($n=1$) and asymmetric($n=2$) states.

In the second stage of this paper, response analysis of a dome subjected to horizontal earthquakes is executed as the application of these approximate modes. Many important response characteristics may manifest themselves through parametric survey of material and geometric properties.

Introduction

When we design large span structures like domes, it is substantial to take into consideration of influence of external forces such as earthquakes and winds. Dynamic excitations such as earthquakes and winds have frequency characteristics, and, therefore, we must take into account of the effect of the natural frequencies of structures in the response analysis.

For a spherical dome, exact eigen modes are already available. In modal analysis, however, the exact modes are of little use, because they are represented in terms of Legendre bi-function with complex or real fractional order. The Galerkin procedure, thus, is impractical, since it involves integrals of the product of deflection modes with respect to spatial variables, which cannot be easily reduced to functions of closed form. From this point of view, one of the authors[Kunieda] has presented approximate eigen modes of a spherical dome in axisymmetric state[3] and showed the response behavior[4] subjected to up-down earthquake. The paper[4] indicates that the response acceleration of a dome subjected to a vertical earthquake (El Centro'40) may reach more than 4000 gals under the condition of 5% damping ratio.

---

1 : DPRI Kyoto University, UJI, KYOTO 611, JAPAN

One of the purpose of this paper is to make the approximate eigen mode in asymmetric state. In this state, there exists the term of inextensional displacement mode with respect to rotation as a whole, which does not exist in axisymmetric state. This term is the function of natural frequency which should be obtained as the result of calculation, and it contains unknown constant. Thence, we use the following procedure to solve this problem. At the beginning of the calculation, we postulate an arbitrary value referring to the exact solution to the initial value of the natural frequency, and then substitute the result to this value, and repeat this procedure. We can see that these values converge to the exact solutions in this procedure. The approximate solution obtained has sufficient accuracy and practicality for use in response analysis in anti-symmetric($n=1$) and asymmetric($n=2$) states. Each deflection mode is set as the sum of Legendre bi-polynomials, satisfying the ordinary boundary conditions and the orthogonality condition. To satisfy the boundary conditions, additional exact eigen modes in circumferential and longitudinal directions are considered. The integrals due to these approximate deflection modes can be reduced to functions of closed form in linear as well as nonlinear state.

In the second stage of this paper, response analysis of a dome subjected to horizontal earthquakes is executed as the application of these approximate modes. Many important response characteristics may manifest themselves through parametric survey of material and geometric properties. We will choose two important examples in the sense that they well illustrate their response characteristics under such excitations.

1. Approximate Eigen Modes of a Spherical Dome in Asymmetric State

1.1. Eigen Modes

We can see in the paper[2] that there is no serious difference of natural frequencies and modes between normal and flexural vibration. On the other hand, the assumption of flexural vibration makes it easy to construct the approximate solution. Thus, we employ the assumption of flexural vibration.

Governing equation and compatibility condition based on the assumption of flexural vibration in the asymmetric state are expressed as

$$H_2 H_2(w) - \frac{a}{D} H_2(\psi) + \frac{ma^4}{gD} \ddot{w} = 0 \tag{1-1}$$

$$H_2 H_1(\psi) - (1-\nu)\frac{D}{a} H_2 H_2(w) + Eha H_2(w) = 0 \tag{1-2}$$

where w is the displacement component in normal direction. Stress function ψ relates to stress resultants and displacements u and v as follows.

$$N_\theta = \frac{1}{a^2}\left[\psi^{oo} + \psi - \frac{D}{a} H_2(w)\right]$$

$$N_\phi = \frac{1}{a^2}\left[\frac{\psi''}{\sin^2\phi} + \psi^o \cot\phi + \psi - \frac{1}{a} D H_2(w)\right]$$

$$u = -\frac{(1+\nu)}{Eha}\psi^o + \frac{1}{\sin\phi}f' \quad, \quad v = \frac{(1+\nu)}{Eha}\frac{\psi'}{\sin\phi} - f^o$$

where f is stress function, which is expressed by Legendre bi-function.

We assume that a dome has one edge and a closed apex and is in an asymmetric deformation state. Each deflection mode is assumed to be set as the sum of Legendre bi-polynomials. Thus, the displacement component w is expressed as

$$w = \sum_{i=1}^{N} B_i \, \tilde{W}_i(\phi) T_i(t) \cos n\theta , \qquad \tilde{W}_i(\phi) = \sum_{j=J_i}^{K_i} F_{ji} P_j^n(\phi) \qquad (1\text{-}3)$$

where N is the number of modes we need, and n is the order of expansion in circumferential direction. P_j^n is Legendre bi-polynomial of order j in the nth rank. $\tilde{W}_i$ is temporary mode. The way of choosing J_i and K_i included in this term has a great influence on the accuracy of the final solution.

Legendre bi-polynomial whose rank $n=1$ and $n=2$ can be expanded in series as

$$P_m^1(\cos\phi) = D_m \sin m\phi + Func.(\sin k\phi ; \quad k=0\sim(m-1))$$
$$P_m^2(\cos\phi) = D_m \cos m\phi + Func.(\cos k\phi ; \quad k=0\sim(m-1))$$

In the anti-symmetric state($n=1$), this implies that totally $m/2$ nodal points exist in the range of $0 < \phi \leq 90°$ in the curve of $P_m^1(\cos\phi)$. We may assume that these nodal points are uniformly distributed between $0°$ and $90°$ in case of sufficiently large m. From the similar consideration of the nodal points in the asymmetric state, we can assume that ith nodal point is founded in $0 < \phi \leq \phi_o$, if we take the i of P_i^n as $(2i \cdot 90° / \phi_o + n + 1)$.

Legendre bi-polynomials of $n=1$, however, are equal to zero when $\phi=90°$, and the nodal points exist at $\phi=0°$ and $180°$ when m has small value. We can see from this point of view that we must modify the preceding term when ϕ_o is close to $90°$. We can recognize from several calculations of approximate eigen modes as

$$K_i = int. \left[\frac{180 \, i}{\phi_o} + n + 1 + 0.3 \right] \qquad (1\text{-}4)$$

To consider all terms of Legendre bi-polynomials of order i under K_N, without lacking any Legendre bi-polynomial in (1-3), the J_i should be taken as

$$J_i = min. \left[K_i - (i+3) , K_{i-1} + 1 \right] \qquad (1\text{-}5)$$

Boundary Condition

Referring to the exact solution[1], the part of the inextensional deflection mode included in u and v are in the following form

$$u = \frac{1+\nu}{Eha} GP_{\mu_4}^n \frac{n}{\sin\phi} , \qquad v = -\frac{1+\nu}{Eha} GP_{\mu_4,\phi}^n$$

where

$$\lambda_4 = -\{2 + 2(1+\nu)\Omega^2\} , \qquad \mu_4 = -\frac{1}{2} + \sqrt{\frac{1}{4} - \lambda_4}$$

Using these equations and eq.(1-3), we can express the boundary conditions. For example, the boundary condition of fixed end $(w = u = v = w^o = 0)$ is represented as

$$
\begin{cases}
\tilde{W}_i = 0 & (w=0) \\[2mm]
-\lambda_i\,\tilde{W}_{i,\phi} + G\dfrac{n}{\sin\phi}P^n_{\mu_4} = 0 & (u=0) \\[2mm]
\lambda_i\dfrac{n}{\sin\phi}\,\tilde{W}_i - GP^n_{\mu_4,\phi} = 0 & (v=0) \\[2mm]
\tilde{W}_{i,\phi} = 0 & (w^o=0)
\end{cases}
\tag{1-6}
$$

where G is the unknown constant, μ_4 of $P^n_{\mu_4}$ is the function of the natural frequency Ω.

Conditions of orthogonality

We can obtain the following $(i\text{-}1)$ conditions of orthogonality from the governing equation(1-1) and compatibility condition(1-2).

$$
\int_0^{\phi_o}[H_2H_2(\tilde{W}_p)-\frac{a}{D}H_2(\Psi_p)]\,\tilde{W}_i\,\sin\phi\,d\phi
\tag{1-7}
$$

$$
= \int_0^{\phi_o}[H_2H_2(\tilde{W}_i)-\frac{a}{D}H_2(\Psi_i)]\,\tilde{W}_p\,\sin\phi\,d\phi \quad (p=1\sim(i-1))
$$

Substituting temporary deflection modes(1-3) into this equation, and executing integrals of the product of deflection modes with respect to spatial variables, we obtain the following equation with closed form functions.

$$
\sum_{q=J_k}^{K_k} F_{qk} \sum_{j=J_p}^{K_p} F_{jp}[\{2-q(q+1)\}^2-\kappa\lambda_q\{2-q(q+1)\}
\tag{1-8}
$$

$$
-\{2-j(j+1)\}^2+\kappa\lambda_j\{2-j(j+1)\}]S^n(q,j)=0
$$

Approximate Eigen Modes are calculated through a two-step process. First, the temporary modes $\tilde{W}_i$ are determined as satisfying boundary conditions and conditions of orthogonality. In the second step, we determine the final approximate solution of eigen modes, using these temporary modes and applying the Galerkin procedure to kinetic equations of free vibration.

The first step (determination of temporary modes $\tilde{W}_i$)

Calculation in the first step falls into two categories. There are two problems in determining $\tilde{W}_i$. The expression of boundary condition contains unknown constant G and $P^n_{\mu_4}$ which is equivalent to inextensional deflection mode. Notice that $P^n_{\mu_4}$ in boundary conditions is the function of Ω which must be obtained as the results.

Thus, we discriminate the first temporary mode from the others as to assume the initial value of Ω. Referring to the exact solution[2], we assume the initial value of natural frequency Ω as $0.7\sim0.9$. We repeat the such routine that we substitute the natural frequencies obtained in the preceding calculation to those in the next calculation. Natural frequencies converge to the exact ones in this method rapidly, i.e., about five times.

Calculation procedures are summerized as

1) In the first mode, assuming $F_{j1}|_{j=J_1}=1$ and satisfying four boundary conditions, we determine $F_{j1}(j\neq J_1)$ and the unknown constant G.

2) In the ith mode, we determine F_{ji} by using constant G determined in the first step, and satisfying four boundary conditions and $(i\text{-}1)$ conditions of orthogonality.

In linear simultaneous equations derived for F_{ji}, the coefficient matrix does not necessarily become square. Usually, the matrix becomes rectangular. Although several solutions for such linear simultaneous equations with rectangular matrices are available, the solution based on the method of minimum square and minimum norm is used in this paper.

The second step (determination of Eigen mode W_i)

Substituting eq.(1-3) into eq.(1-1) and eq.(1-2), we apply the Galerkin procedure to these equations. Using F_{ji} determined in the first step, we can obtain the following equation with respect to B_i with closed form functions.

$$-\Omega^2[m_{ij}]\{B_j\}+[k_{ij}]\{B_j\}=0 \qquad (1\text{-}9)$$

where

$$m_{ij}=m_{ji}=\sum_{r=J_i,s=J_j}^{K_i}\sum^{K_j}F_{ri}F_{sj}S^n(r,s)$$

$$k_{ij}=k_{ji}=\sum_{r=J_i}^{K_i}\sum_{s=J_j}^{K_j}F_{ri}F_{sj}[2-r(r+1)]\left[\frac{2-r(r+1)}{\kappa}-\lambda_r\right]S^n(r,s)$$

where k_{ij} in eq.(1-9) gets to become symmetric, because eq.(1-7) is adopted. Using eq.(1-9), we can easily determine the natural frequency Ω_i and normalized coordinate vector $\{B_j\}_i$ in ordinary method. Substituting $\{B_j\}_i$ into eq.(1-3), the displacement w can be finally represented as

$$w=\sum_{i=1}^{N}A_iW_i(\phi)\cos\theta\ ,\quad W_i(\phi)=\{B_s\}_i^T\{\tilde{W}_s\}=\sum_{j=1}^{K_N}G_{ij}P_j^n(\cos\phi)$$

$$G_{ij}=\sum_{s=1}^{N}\{B_s\}_i^TF_{js} \qquad (1\text{-}10)$$

where $W_i(\phi)$ represents the ith eigen mode, and $A_i\equiv A_i(t)$.

This approximate eigen mode is described in terms of orthogonal Legendre bi-polynomials, and, therefore, W_i is in perfect system. However, there is no guarantee of all eigen values given by eq.(1-9) being correct, since we adopt finite series of Legendre bi-polynomials and adopt the application of Galerkin procedure, et al. There may be some possibility that we get some complex eigen values. Thus, we must check the effectiveness and appropriateness of all eigen values. We can estimate it in the following method.

Calculating $\tilde{\Omega}_i$ from next equation,

$$\tilde{\Omega}_i^2=[\{B\}_i^T[(m_{pq})]\{B\}_i]^{-1}[\{B\}_i^T[(k_{pq})]\{B\}_i]$$

and comparing $\tilde{\Omega}_i$ with Ω_i given by eq.(1-9), we can confirm its usefulness if the difference being small.

1.2. Examples of calculation and conclusion

An example of eigen modes calculated by this method is shown in Fig.1. An example of eigen values is shown in Table 1. Close resemblance of natural frequencies and modes estimated by this approximation procedure to those obtained from the exact solution[2] can be easily recognized.

What should be noticed is that Legendre bi-polynomials in anti-symmetric state $(n=1)$ are exactly equal to zero when the order of the Legendre bi-polynomial is even and $\phi=90°$. When half open angle is as $\phi_o=90°$, such terms are of no effect in the satisfaction of boundary conditions. Thence we use 89.8° in spite of 90° in these examples.

We can see roughness of deflection in Nth mode or the mode in the vicinity of Nth modes as shown in Fig.1. This is because of truncation error due to the finite series of Legendre bi-polynomials up to K_N instead of using the infinite series. When we apply these modes to response analysis, it is necessary to calculate eigen modes by this method more than those we need in response analysis.

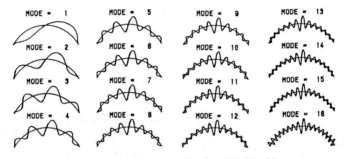

Figure 1. Example of Anti-symmetric Eigen Modes of Dome;
ϕ_o=60, a/h=100, Fixed End

ϕ_o	1st	2nd	3rd	4th	5th	6th	7th	8th	9th	10th	11th	12th
30	1.041	1.208	1.657	2.409								
	1.041	1.212	1.661	2.413								
45	0.990	1.039	1.156	1.392	1.764	2.268		Upper: Exact Solution				
	0.990	1.040	1.158	1.395	1.767	2.270		Lower: Approximate Sol.				
60	0.933	0.955	1.042	1.132	1.287	1.516	1.821	2.199				
	0.933	0.956	1.043	1.134	1.288	1.518	1.823	2.201				
75	0.836	0.974	1.007	1.049	1.120	1.233	1.392	1.600	1.856	2.157	2.250	
	0.836	0.975	1.007	1.050	1.121	1.233	1.393	1.601	1.857	2.160	2.506	
90	0.703	0.955	0.990	1.016	1.060	1.118	1.205	1.323	1.476	1.658	1.884	2.133
	0.704	0.957	0.991	1.017	1.055	1.115	1.204	1.324	1.478	1.666	1.888	2.141
105	0.547	0.931	0.979	1.000	1.021	1.058	1.109	1.180	1.274	1.391	1.544	1.706
	0.548	0.937	0.980	1.000	1.024	1.059	1.110	1.181	1.277	1.397	1.540	1.716
120	0.389	0.903	0.967	0.989	1.007	1.029	1.061	1.105	1.164	1.239	1.354	1.455
	0.392	0.916	0.969	0.990	1.008	1.030	1.062	1.107	1.167	1.244	1.340	1.454

Table 1. Example of Modified Natural frequency Ω of Dome;
a/h=100, Fixed End

2. Application to linear response analysis

2.1. Governing equations on linear response analysis

The spherical dome subjected to horizontal dynamic excitation is now analyzed. The governing equation with the assumption of flexural vibration in anti-symmetric state is given by replacing the right term in eq.(1-1) with the next term of external force.

$$3\frac{ma^4}{gD}\ddot{V}\sin\phi\cos\theta$$

Substituting approximate deflection modes (1-10) into the governing equations, and adopting the Galerkin procedure, we can obtain the following equation which is equivalent to that of one degree of freedom system with respect to time function $T(t)$

$$\ddot{T}_i + {}_i\beta_1 T_i = {}_i\beta_2 \ddot{V} \qquad (i = 1{\sim}N) \qquad (2\text{-}1)$$

where N is the number of modes considered and

$$_i\beta_0 = \sum_{k=1}^{K_N}\sum_{j=1}^{K_N} G_{ji}G_{ki}S^1(j,k)$$

$$_i\beta_1 = \frac{1}{\kappa A\,_i\beta_0}\sum_{k=1}^{K_N}\sum_{j=1}^{K_N} G_{ji}G_{ki}\{2-j(j+1)\}\{2-j(j+1)-\lambda_j\}S^1(j,k)$$

$$_i\beta_2 = \frac{3}{_i\beta_0}\sum_{k=1}^{K_N} G_{ki}S^1(k,1)$$

$$\lambda_j = \frac{(1-\nu)\{2-j(j+1)\}-\kappa}{1-\nu-j(j+1)} \quad, \quad A=\frac{ma^2}{Ehg} \quad, \quad \kappa=\frac{12(1-\nu^2)a^2}{h^2}$$

Response behaviors in the anti-symmetric state are elucidated by solving the eq.(2-1).

2.2. Earthquake Excitation

Two typical important examples are shown, which represent well their response characteristics under excitations. Two earthquake data are used, which are normalized as each maximum acceleration being 100gal. Since these earthquake records are given as digital quantity and interval of time is 0.01s or 0.02s, we cannot determine the functions $T_i(t)$ analytically. We calculate, thence, them by the direct numerical method of Runge-Kutta-Gill method in this paper. The lowest natural frequency of a usual spherical dome, however, is not lower than the predominant frequency of these earthquakes as seen in Table.1., and it is apparent that the use of those data in original form is unsuitable from the point of view of accuracy and convergence in numerical calculation. Accordingly, we divided the original time interval into ten sub-interval and data in sub-interval are linearly interpolated in this calculation.

Effects of radius and radius-thickness ratio are included in material parameter A. Since earthquake data are given in physical quantities, radius is given in physical one in this paper. Taking account of the response spectrum of earthquakes and the usual real domes and materials, we determine physical and material parameter A as shown in the following table.

Earthquakes	El Centro'40-SN, Taft-SN
Poisson ratio (ν)	0.3
Damping ratio	0.05
Half open angle (ϕ_o)	60, 75, 90, 105
Number of Modes (N)	1st ~ 6th
Boundary Condition	Pin support, Fixed end
Radius (a)	3000cm
Radius-Thickness ratio (a/h)	100
Parameter (A)	$0.5 \sim 20.0 \times 10^{-4}$

2.3. Results of Response calculation and conclusion

The first example elucidates the differences of resultant forces between pin support case and fixed end case. Distribution of resultant forces is depicted in Fig.2. In this calculation, $\phi_o=60°$, $A=6.5\times10^{-4}$, and the earthquake El Centro'40 record are used. The maximum displacement appears simultaneously at $t=4.94$ sec., and resultant forces in Fig.2 are those at this time. It can be seen that the horizontal shearing force Q_i in Pin support is larger than that in Fixed end. Taking into consideration the inextensional mode, it is observed that the deflection mode is larger in Pin support.

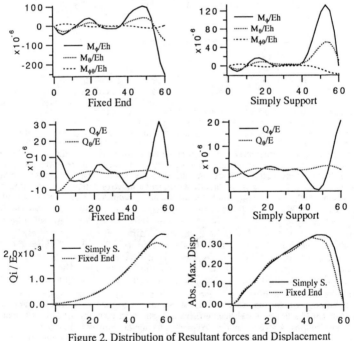

Figure 2. Distribution of Resultant forces and Displacement
($\phi_o=60°$, $A=6.5\times10^{-4}$, El Centro, t = 4.94sec.)

The second example shows the necessity and importance of response analysis of domes. The distribution of the absolutely maximum acceleration is depicted in Fig.3 without regard of the position and the period. This figure shows that the possibility of the appearance of extreme large acceleration, depending on the value of parameter A. An example of distribution of response acceleration is shown in Fig.4. This figure shows that the position where the maximum acceleration appears is inside a little bit from its boundary. It may be possible to specify the position of the maximum acceleration according to the half open angle.

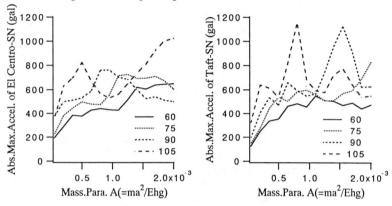

Figure 3. Absolutely Maximum Response Acceleration (gal)

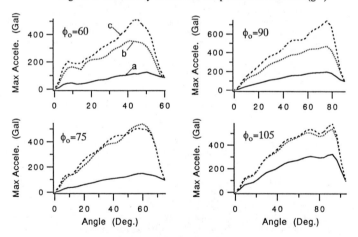

Figure 4. Distribution of Absolutely Maximum Response Acceleration
along the longitudinal direction
(A; a=0.5×10^{-4}, b=5.0×10^{-4}, c=12.5×10^{-4})

Appendix 1. Notation

a = radius of dome
E = Young's modulus
g = acceleration of gravity
h = thickness of dome
m = the mass of the unit area on the neutral surface
θ = spatial variable in circumferential direction
ν = Poisson's ratio
ϕ = spatial variable in longitudinal direction

$$B^o = B_{,\phi} \ , \quad B' = B_{,\theta} \ , \quad \dot{B} = B_{,t}$$
$$H_0(B) = B^{oo} + \cot\phi B^o + \csc^2\phi B''$$
$$H_1(B) = H_0(B) + (1-\nu)B \ , \quad H_2(B) = H_0(B) + 2B$$

Appendix 2. Integrals of Legendre bi-polynomials

$$S^n(r,s) = \int_0^{\phi_0} P_s^n(\cos\phi)P_r^n(\cos\phi)\sin\phi \, d\phi$$

$$= \begin{cases} \dfrac{1}{r(r+1)-s(s+1)}\left[(s-r)\cos\phi P_s^n P_r^n - (s+n)P_{s-1}^n P_r^n + (r+n)P_s^n P_{r-1}^n\right]_{\phi=\phi_0} & (r \neq s) \\ \quad -\left[P_r^n P_r^{n-1}\sin\phi\right]_{\phi=\phi_0} + (r+n)(r-n+1)S^{n-1}(r,r) & (r=s) \end{cases}$$

$$S^0(r,s) = \int_0^{\phi_0} P_s(\cos\phi)P_r(\cos\phi)\sin\phi \, d\phi$$

$$= \begin{cases} \dfrac{1}{r(r+1)-s(s+1)}\left[(s-r)\cos\phi P_s P_r - sP_{s-1}P_r + rP_s P_{r-1}\right]_{\phi=\phi_0} & (r \neq s) \\ \dfrac{1}{2r+1}\left[1 - P_r P_r \cos\phi\right]_{\phi=\phi_0} - 2\sum_{p=1}^{[r/2]}(2r+1-4p)S^0(r,r-2p) & (r=s) \end{cases}$$

Appendix 3. References

[1]   Y. Yokoo, O. Matsuoka and H. Kunieda 1963 *Transactions of the Architectural Institute of Japan No. 83*, 7-14. General solutions of spherical shell in state of free vibration: Part 1. Exact solution (in Japanese)

[2]   H. Kunieda 1983 *Transactions of the Architectural Institute of Japan No. 325*, 57-64. Solutions of free vibrations of spherical shells: Part 3. Natural Frequencies and Modes in Axi- and Anti-symmetric State

[3]   The first in H. Kunieda 1984 *J. Sound and Vibration No. 92*, 1-10. Flexural Axi-symmetric Free Vibration of a Spherical Dome: Exact Results and Approximate Solutions
The revised in H. Kunieda 1992 *Journal of Engineering Mechanics, Vol. 118, No. 8*, August, 1513-1525. Classical Buckling Load of Spherical Domes under Uniform Pre ssure

[4]   H. Kunieda 1986 *Shells, Membranes and Space Frames, IASS Symposium, Osaka, Vol. 1*, 33-40. Responses of Spherical Domes Subjected to Vertical Earthquakes

Dynamic Analysis of Unstable Structural Systems
With Prescribed Velocities

Ken-ichi Miyazaki[1], Yasuhiko Hangai[2] and Ken-ichi Kawaguchi[3]

Abstract

Structural systems that have large inextentional deformation, or in other words, deformation without stain are called 'Unstable Structural Systems.' They include cable structures, membrane structures and movable structures such as foldable space structures and etc. This paper presents a method for the dynamic analysis of unstable structural systems with prescribed velocities. In order to examine the validity of the present method, two illustrative examples are numerically analyzed.

Introduction

Basic equations prescribing the motion of an unstable structural system consist of both kinematic equations of motion (1) of n prescribed particles which represent a configuration of a structure and m geometrically constraint equations (2) describing distance between each particle (see Fig.1).

$$F_i\left(x_1,\cdots\cdots,x_n\right) = f_i \ , \ \left(i =1,\cdots\cdots,n\right) \tag{1}$$

$$C_j\left(\dot{x}_1,\cdots\cdots,\dot{x}_n\right) = b_j \ , \ \left(j =1,\cdots\cdots,m\right) \tag{2}$$

Eq.(2) can be described in the matrix form as follows.

$$C\dot{x} = b \ . \tag{3}$$

This non-homogeneous equation represents the constraint condition for the kinematic equation (1). The $m \times n$ coefficient matrix C is called constraint Jacobian matrix. As some of the variables $x_1,\cdots\cdots,x_n$ in kinematic equations are generally dependent on

[1] Graduate student, [2] Professor and [3] Lecturer,
Institute of Industrial Science, University of Tokyo,
7-22-1, Roppongi, Minato-ku, Tokyo 106, Japan

each other and the constraint equations are singular due to $m < n$, it is hard to get $x_1, \cdots\cdots, x_n$ uniquely. In order to overcome this singularity, many methods have been presented. These methods are slightly different from each other by the formulation of basic equations and the reduction of degrees of freedom(D.O.F.) in kinematic equations.

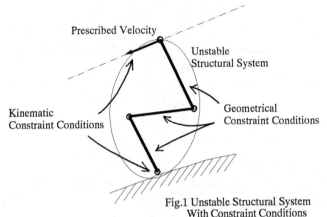

Prescribed Velocity

Unstable Structural System

Kinematic Constraint Conditions

Geometrical Constraint Conditions

Fig.1 Unstable Structural System
With Constraint Conditions

In the following, these methods, in which reduced kinematic equations are derived by the use of the condition that null space of constraint Jacobian matrix signifies the direction of motion of unstable structural system, are introduced.

Method using Kane's equation [1], [2]

Kane's equation is a set of equiliblium equations between externally applied forces and inertia forces in the direction of motion of particles. This method has the advantage that the equilibrium equations have no internal forces explicitly. Kamman et al.[2] derived kinematic equations of motion based on Kane's method. Then they obtained the null-space matrix from eigen vectors corresponding to zero eigen values of constraint Jacobian matrix. By multiplying null space matrix to kinematic equations of motion, they reduced the total number of D.O.F. of kinematic equations to the number of D.O.F. of inextensional motion.

Method using Lagrange multiplier [3]

It is possible to apply the Lagrange multiplier method to kinematic equation of motion with constraint conditions. Kim et al. derived basic equations by using this method. To obtain the null space matrix from constraint Jacobian matrix, they adopted QR decomposition method. It is possible to avoid QR decomposition in every incremental step in the case that the rank of constraint Jacobian matrix is full so that this method is suitable for incremental analysis. Furthermore, they adopted Newton-Raphson method in order to ensure the accuracy of calculation.

Method using the generalized inverse [4]

Kawaguchi et al. directly solved geometrically constraint equation (3) by introducing the generalized inverse. They make use of the condition that the space spanned by homogeneous solution corresponds to the direction of inextensional motion. The null space of constraint Jacobian matrix is automatically obtained from homogeneous solution of eq. (3). By multiplying null space to kinematic equations of motion , they also reduced the number of D.O.F. of kinematic equations.

The method proposed in this paper is based on the generalized inverse theory. However, the reduction by using null space matrix is not used, which is a different point from above papers.

Analytical method

Let us consider a link model which consists of particles with mass connected by massless rigid bars (see Fig.2). Let the coordinate, the mass and the externally applied forces of particle i be represented respectively as

$$\text{particle } i : \left\{x_i, y_i\right\}^{\mathrm{T}}, \ M_i = diag\left[m_i, m_i\right] \text{ and } f_i = \left\{f_{xi}, f_{yi}\right\}^{\mathrm{T}}. \qquad (4)$$

The constraint reaction force applied to particle i which is denoted by g_i takes the form

$$g_i = g\lambda , \qquad (5)$$

where λ is normal direction vector of bar a and g is the bar force.

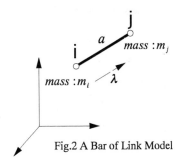

Fig.2 A Bar of Link Model

Then, kinematic equation of motion of particle i is

$$M_i \ddot{x}_i + g_i = f_i. \tag{6}$$

Kinematic equation for a whole link structure can be derived by combining eq.(6) to all particles as

$$M\ddot{x} + G = F . \tag{7}$$

On the other hand, the following geometric constraint equation can be led by the invariability of bar length,

$$\left(x_j - x_i\right)^2 + \left(y_j - y_i\right)^2 = l^2 . \tag{8}$$

The time derivative of eq.(8) leads to the following equation in the matrix form.

$$\begin{bmatrix} -\lambda^{\mathrm{T}} & \lambda^{\mathrm{T}} \end{bmatrix} \begin{Bmatrix} \dot{x}_i \\ \dot{x}_j \end{Bmatrix} = 0. \tag{9}$$

Coefficient matrix on the left hand side of the above equation is constraint Jacobian matrix. From the above equation, geometric constraint equation for a whole unstable structural system is generally described in homogeneous form;

$$H\dot{x} = 0. \tag{10}$$

When the velocities for specified particles are prescribed,then homogeneos geometric constraint equation (10) is changed into the non-homogeneous form;

$$H\dot{x} = b. \tag{11}$$

As the matrix G in eq.(6) can be represented by using the transpose of the constraint Jacobian matrix H as $-H^{\mathrm{T}}n$, the kinematic equation of motion is described as

$$M\ddot{x} - H^{\mathrm{T}}n = F , \tag{12}$$

where M : mass matrix, H : $m \times n$ constraint Jacobian matrix, x : position vector prescribing the configuration, n : constraint reaction force and F : externally applied force.

Therefore basic equations for an unstable structural system consist of kinematic equation of motion (12) with non-homogeneous constraint equation (11).

For the purpose of simplification, change eq.(12) into

$$M\ddot{x} + r = F , \tag{13}$$

where

$$r = -H^{\mathrm{T}}n . \tag{14}$$

In order to change eq.(11) into the homogeneous form by eliminating b on the right hand side, let us introduce following vector in R^n,

$$\dot{u} = \dot{x} - H^+ b \tag{15}$$

where H^+ means Moore-Penrose generalized inverse of H.

Introduction of eq.(15) into eqs.(12) and (15) leads to

$$M\ddot{u} + r = h \tag{16}$$

and

$$H\dot{u} = 0 \tag{17}$$

respectively where

$$h = F - M\frac{d}{dt}(H^+b) . \tag{18}$$

The orthogonal condition of r and $\dot{u}$ can be proved by using eq.(16) as

$$r^T\dot{u} = (-H^Tn)^T\dot{u} = -n^T(H\dot{u}) = 0 . \tag{19}$$

From eq.(19), we have

$$\dot{u} \in L \text{ and } r \in L^\perp \tag{20}$$

where L is a linear subspace in the n dimensional space R^n and $L^\perp$ the orthogonal complement to L. Furthermore, a vector a in R^n, projector matrix P_L to the subspace L and projector matrix $P_{L^\perp}$ to the subspace $L^\perp$ can be defined as

$$u = P_La \text{ and } r = P_{L^\perp}a \qquad (P_LP_{L^\perp} = 0) \tag{21}$$

respectively. Substituting eq.(21) into eq.(16) gives

$$MP_L\ddot{a} + P_{L^\perp}a = h . \tag{22}$$

Multiplying P_L^T to eq.(22) from the left gives

$$P_L^T MP_L\ddot{a} + P_L^T P_{L^\perp}a = P_L^T h . \tag{23}$$

Because P_L is a projector matrix, then

$$P_L^T = P_L . \tag{24}$$

Therefore, eq.(23) becomes

$$M_L\ddot{a} = P_L^T h , \tag{25}$$

where

$$M_L = P_L^T MP_L . \tag{26}$$

When constraint Jacobian matrix is given as

$$B = \begin{bmatrix} I_m & 0_{n-m} \end{bmatrix} , \tag{27}$$

projector matrices are obtained[5] as

$$\overline{P}_L = \begin{bmatrix} 0 & 0 \\ 0 & I_{n-m} \end{bmatrix} \text{ and } \overline{P}_{L^\perp} = \begin{bmatrix} I_m & 0 \\ 0 & 0 \end{bmatrix} . \tag{28}$$

The above condition can also be developed in the case of general constraint Jacobian matrix H. By multiplying appropriate normal matrices from the left and right to constraint Jacobian matrix H, it could be transformed to a diagonal matrix

$$B = UHV,\qquad(29)$$

where $U : m \times m$ normal matrix, $V : n \times n$ normal matrix. By introducing v, q and t in R^n, change u, r and h into

$$u = Vv, \; t = V^T r \text{ and } q = V^T h,\qquad(30)$$

respectively. Substituting eq.(30) into eq.(16) gives

$$\overline{M}\dot{v} + t = q,\qquad(31)$$

where

$$\overline{M} = V^T M V.\qquad(32)$$

Similarly substitute eq.(30) into eq.(17) and multiply U from the left, then

$$B\dot{v} = 0\qquad(33)$$

is obtained. eqs.(31) and (33) are equal to eqs.(16) and (17) respectively because they could be derived from the latter set of equations by elementary transformation for matrices.

In order to reduce the kinematic equation of motion of eq.(31), we can introduce a vector w in R^n that satisfies the following two equations

$$v = \overline{P}_L w \text{ and } t = \overline{P}_{L^\perp} w,\qquad(34)$$

Because B in eq.(33) takes the form of eq.(27), then, eq.(31) can be changed into a 2nd order differential equation with respect to w as follows

$$\overline{M}\overline{P}_L \ddot{w} + \overline{P}_{L^\perp} w = q.\qquad(35)$$

In accordance with eq.(27), eq.(35) can be decomposed into two independent equations as follows.

$$\overline{M}_{12}\ddot{w}_1 + w_1 = q_1\qquad(36)$$

and

$$\overline{M}_{22}\ddot{w}_2 = q_2\qquad(37)$$

where

$$w = \begin{bmatrix} w_1 \\ w_2 \end{bmatrix}, \; q = \begin{bmatrix} q_1 \\ q_2 \end{bmatrix}, \; \overline{M} = \begin{bmatrix} \overline{M}_{11} & \overline{M}_{12} \\ \overline{M}_{21} & \overline{M}_{22} \end{bmatrix}, \; w_1, q_1 \in R^m \text{ and } w_2, q_2 \in R^{n-m}$$

w_2 is a variable which represent the displacement in the direction of inextensional motion and the inextensional displacement is obtained from eq.(30) as follows.

$$u = V \begin{bmatrix} 0 \\ w_2 \end{bmatrix}\qquad(38)$$

w_1 is a variable which represent the displacement in the direction of constraint motion and the constraint reaction force is obtained from eqs.(30) and (14) as follows.

$$n = -(H)^{\mathrm{T}}(V^{-1})^{\mathrm{T}}\begin{bmatrix} w_1 \\ 0 \end{bmatrix} \tag{39}$$

Numerical examples.

Two numerical examples are presented here. The model for one example is a 6bar-link model. The free-fall of this model sustained at both ends is calculated by tracing its time response. Parameters of this model is presented in Fig.3. The model for the other example is a 3bar-link model where one of the ends is fixed and the other is made to move at a prescribed velocity, and two middle particles are free. Parameters of this model is presented in Fig.4. In order to ensure the convergence of geometric constraint equations, Newton-Raphson method is adopted to both examples.

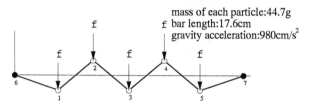

Fig.3 6Bar-Link Model

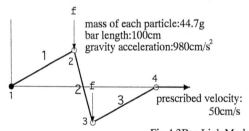

Fig.4 3Bar-Link Model

Fig.5 and Fig.6 show the change of configuration of 6bar-link model. The configuration of at 0.220sec is very close to the configuration at the statical equilibration. Fig.7 shows the kinematic energy and the potential energy for the 6bar-link model. They change periodically and their amplitudes are constant because this model is a conservative system. Though amplitudes of displacement and velocity are not constant, they change periodically in their time responses, as shown in Fig.8 and Fig.9.

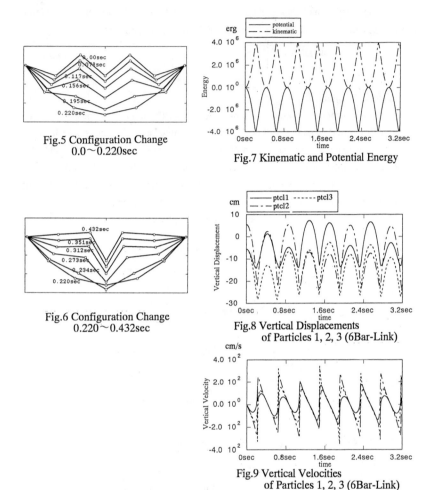

Fig.5 Configuration Change
0.0∼0.220sec

Fig.7 Kinematic and Potential Energy

Fig.6 Configuration Change
0.220∼0.432sec

Fig.8 Vertical Displacements
of Particles 1, 2, 3 (6Bar-Link)

Fig.9 Vertical Velocities
of Particles 1, 2, 3 (6Bar-Link)

Fig.10 shows the change of configuration of 3bar-link model from 0sec to the time when the particle 3 reaches its lowest position at the first time, and Fig.11 shows the change of configuration to the time when the particle 3 reaches its highest position. Vertical velocities of particles is shown in Fig.12. The periods of velocities become shorter in accordance with the displacement of particle 4 at around 2sec. Fig.13 shows the bar forces of bar 1, 2 and the horizontal reaction force of particle 4 with prescribed velocities, which take small

values in almost all time interval as shown in Fig.13, but get very large values at the time
when the configuration of the model become singular. In the case that the prescribed velocities of particle 4 for 3bar-link model is taken as zero, displacements and velocities of 3bar
model are shown in Fig.14 and Fig.15, respectively, which change periodically. This motion is different from 6bar-link model.

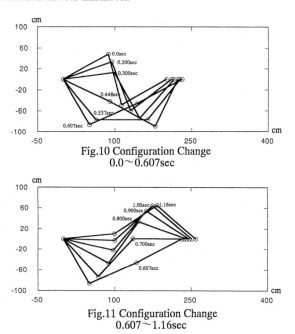

Fig.10 Configuration Change
0.0 ~ 0.607sec

Fig.11 Configuration Change
0.607 ~ 1.16sec

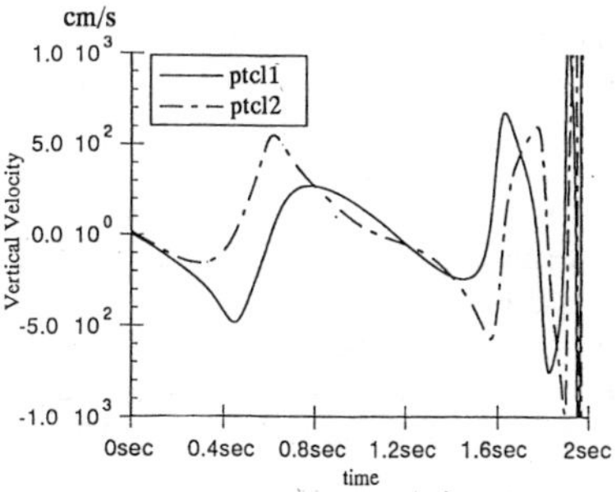

Fig.12 Vertical Velocities
of Particles 1, 2

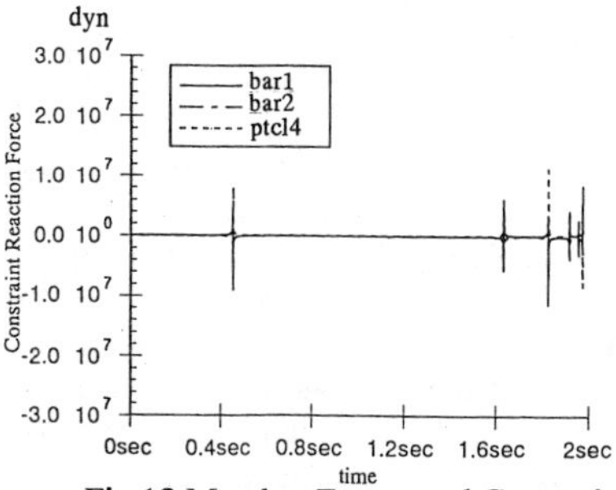

Fig.13 Member Forces and Constraint
Reaction Force of Particles 4

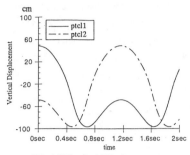

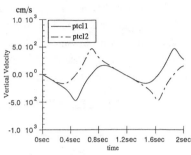

Fig.14 Vertical Displacements
of Particles 1, 2 (3Bar-Link)

Fig.15 Vertical Velocities
of Particles 1, 2 (3Bar-Link)

Conclusion

A method that is available to the dynamic analysis of unstable structural systems is presented in this paper.

The characteristics of the method are

(1) the generalized inverse theory is used,

(2) the reduction based on null space matrix is not used, and

(3) the decomposition method is adopted by orthogonal projection matrices.

Numerical examples show the validity of the present method. This method can also be applied to problems with non-homogeneous constraint conditions.

REFERENCES
1. T.R.Kane, 'Dynamics of Nonholonomic Systems', Journal of Applied Mechanics, Transactions of the ASME, December 1961, pp.574-578.
2. J.W.Kamman and R.L.Huston, 'Dynamics of Constrained Multibody Systems', Journal of Applied Mechanics, December 1984, Vl.51, pp.899-903.
3. S.S.Kim and M.J.Vanderploeg, 'QR Decomposition for State Space Representation of Constrained Mechanical Dynamic Systems', Journal of Mechanisms, Transmissions, and Automation in Design, June 1986, Vol.108, pp.183-188.
4. Ken-ichi Kawaguchi, Yasuhiko Hangai, and Ken-ichi Miyazaki, 'Dynamic Analysis of Kinematically Indeterminate Frameworks', Proceedings of the IASS-MSU International Symposium, May 1993, Istanbul, pp.569-576.
5. Y.Hangai, 'Shape Analysis of Structures, Theoretical and Applied Mechanics', Vol.39, University of Tokyo Press, 1990, pp.11-28

RESPONSE OF TRUSS STRUCTURES SUBJECTED TO DYNAMIC LOADS DURING MEMBER FAILURE

Ramesh B. Malla[1], Member, ASCE and Butchi B. Nalluri[2]

ABSTRACT

The paper presents a method for and results from an investigation of the response of a truss structure during the chain reaction of its member failure while subjected to external static and dynamic loadings. The main emphasis is to include dynamic effects of progressive member failure on the overall behavior of the structure. The member failure is considered to be sudden. The methodology adopted in the study takes into account the member failure effects by applying external forces, that are equivalent to the reduction in the member capacity of the failed members, at the joints where the two ends of the damaged member are connected with rest of the structure. For a compressive member that buckles and snaps through into inelastic postbuckling regime, the material is considered to be elastic-perfectly plastic having reduced yield stress. A three-dimensional double-layer grid truss structure subjected to external static and harmonic dynamic loads is analyzed. Results presented include natural frequencies, displacements, stresses, and/or forces.

INTRODUCTION

As truss-type structures are stronger, lighter, and easier to construct, they are widely used or preferred for large-scale constructions on earth, at sea, and in space. They are normally designed with a large number of redundant members. However, there have been several catastrophes involving large latticed roof collapses (Task Committee 1984).

Reasons encompassing material defects, fabrication errors, construction/connection flaws, impact, accidents, extreme temperature variation, and abnormal loads, can

----

[1]Asst. Prof., Dept. of Civil Engineering, Univ. of Connecticut, Storrs, CT 06269
[2]Grad. Res. Asst., Dept. of Civil Engnrg., Univ. of Connecticut, Storrs, CT 06269

249

cause damage to an individual member or a portion of the structure. Moreover, member behavior under compression such as buckling may also lead to structural failure. Irrespective of the cause, the collapse of large space trusses was observed to be sudden and give little warning of impending events (Schmidt, Morgan, and Clarkson 1976). Buckling of members may cause the structure undergo a "dynamic jump" into the postbuckling regime (Davies and Neal 1959). The member force redistribution during the progressive failure in truss- like structures is thus a dynamic process. Several static analysis studies have been reported on progressive collapse of truss structures (e.g. Murtha-Smith 1988). However, substantial work remains to be done on the dynamic effect of member failure in the progressive failure scenario. Some work on dynamic member failure of truss structures subjected to static (dead) loads have been reported (Malla and Nalluri 1993; Malla and Wang 1993; Malla, Wang, and Nalluri 1993; Morris 1993)

The main objective of the paper is to present a simplified approach for and results from an investigation of the behavior of truss structures subjected to external dynamic loadings during progressive member failure. The study aims to incorporate sudden nature of member failure including the snap-through into inelastic postbuckling after initial buckling. A double layer grid structure is analyzed during consecutive failure of several members.

ANALYSIS METHODOLOGY

Members of a truss system primarily carry axial loads. Failure of members can take place by yielding or buckling. In this study, emphasis is given for member failure under compression. There are two main reasons for this: a tension member normally carry more load after yielding due to strain hardening; where as a buckled compressive member loses strength and shed load on to other members. Furthermore, the critical stress beyond which a compressive member buckles is normally smaller than the yield stress. Throughout the analysis, it is assumed that the member failure is sudden. A damaged or failed member loses its load carrying capability (full or partial), but remains physically attached with the structure (i.e., the mass matrix for the structure remains unchanged, while the stiffness matrix changes). In reality, the structural vibration will be reduced due to the presence of damping in the structural material and/or joints. However, the present study deals with the undamped situation. A small deformation formulation is considered. Material nonlinearity in terms of elastic-perfectly plastic behavior is considered for members that yield in tension or buckle in compression after exceeding the Euler's critical stress and enters into the postbuckling regime. All other members are considered to behave elastically.

Member Failure Representation

The dynamic nature of a member failure is represented as below (Malla and Nalluri 1993; Malla and Wang 1993). While a particular member is intact in the structure, the equation of motion for undamped structural vibration can be written as:

$$[M]\{\ddot{y}\} + [K]\{y\} = \{p_s\} + \{p_d(t)\} \tag{1}$$

where, $\{y\}$ and $\{\ddot{y}\}$ are displacement and acceleration vectors, respectively. The variable t indicates time. The matrices [M] and [K] represent mass and stiffness matrices of the structure before the failure of the member in consideration. $\{p_s\}$ is the static load matrix (normally, dead loads) applied quasi-statically at each joint on the structure. $\{p_d(t)\}$ is the applied dynamic load vector.

If the member is damaged, it is assumed that the reduction in the load carrying capacity of the member can be represented by equivalent external forces $\{f_k\}$ applied at its end joints in the direction of global degree of freedoms. Therefore, for the damaged structure with the additional external joint forces, the equation of motion can be written as:

$$[M]\{\ddot{y}\} + [K']\{y\} = \{p_s\} + \{p_d(t)\} + \{f_k\} \tag{2}$$

Where, [K'] represents stiffness matrix of the structure with the member in consideration being damaged. It should be noted that the underlying principle here to determine $\{f_k\}$ is that Eqn. (1) and Eqn. (2) give the same structural response.

The structural response after the failure of the member is obtained by solving Eqn. (2) with the forcing functions $\{f_k\}$ reduced to zero at a time when the member is determined to fail. An abrupt drop of this force vector to zero represent a sudden failure of the member. The required forcing functions $\{f_k\}$ can be readily obtained by subtracting Eqn. (1) from Eqn. (2):

$$\{f_k\} = ([K'] - [K])\{y\} = [\Delta K]\{y\} \tag{3}$$

where $[\Delta K]$ is the change in stiffness of the structure due to the member damage. Equation (3) gives $\{f_k\}$ given the nodal displacement vector $\{y\}$ of the structure before the member damage. For the structural response from Eqn. (2) to be the same as that from Eqn. (1), the forcing functions, $\{f_k\}$, need to be determined and applied at each time step until a particular member fails.

Member Behavior Under Compressive Load

A compressive structural member buckles after reaching its critical stress and enters into postbuckling. The load carrying capacity of a buckled member is substantially less than that of a prebuckled member. Several analytical and experimental studies have been reported on the behavior of a structural member under the action of cyclic axial load applied quasi-statically (Nonaka 1973; Toma and Chen 1979; Chen and Han 1985; Papdrakakis 1985; Hill, Blandford, and Wang 1989). A typical load versus deflection behavior of a pin-ended member under a cyclic axial load consists of: elastic prebuckling, elastic buckling, inelastic post-buckling, and elastic unloading and tensioning, elastic-plastic tensioning, plastic tensioning, and elastic unloading.

For an abrupt load-shedding of a compressive member after its initial buckling, there occurs an unstable situation; where the member may snap-through characterized by a sudden drop in the load carrying capacity. The structure in this case may experience a dynamic jump. The dynamic jump may take place if the negative stiffness of the member is greater than the stiffness/resistance provided by the rest of the structure (Davies and Neal 1959; Galambos 1968).

In the present study, the member behavior is represented by a simplified model in the postbuckling regime where a member behaves inelastically. Once it is seen that the member enters into the inelastic postbuckling regime, it is assumed to snap-through from its Euler's critical load to a lower capacity. The member then is considered to be completely plastic having a yield stress equal to the predetermined lower capacity. The sudden decrease in the capacity of the member is accounted for by abrupt drop of the equivalent joint forcing functions determined according to the procedure outlined above to zero at the instant when the member is determined to experience the snap-through (Malla and Nalluri 1993; Malla, Wang, and Nalluri 1993).

APPLICATION AND RESULTS

Fig. 1 shows a three-dimensional double-layer grid truss structure made of aluminum circular tubes. The structural geometric parameters and material properties are given in Table 1. All members are made of the same cross-sectional dimensions. The chord members have Euler's critical stress equal to 62.19 MPa and the diagonal members equal to 82.37 MPa. The top layer of the structure has 3x3 bays with total dimension 27.43 m x 27.43 m. The bottom layer has 2x2 bays with total dimension of 18.29 m x 18.29 m. The structure has total of 72 members (36 chordal and 36 diagonals). The structure is pin supported at the four corner points of the top layer (Nodes 1, 4, 13 and 16).

The structure is subjected to uniformly distributed static loads of 3.49 kN/m² applied vertically downward (-Z direction) on the top layer. The static loads are applied quasi-statically on top chord joints (nodes) by a ramp distribution with rise time 1.5 s (approximately 10 times the first natural period of the intact structure). Each of the bottom chord joints (nodes) has additional mass of 8972.4 kg applied. A single harmonic dynamic load is applied at the center of the bottom layer (node 21). The distribution of the dynamic load is given by $p(t) = P_0 \cos\bar{w}t$, where $P_0 = 2.82 \times 10^6$ N and $\bar{w} = 82.738$ rad/s. The dynamic load is considered applied after the static load is fully in place. The load was started from time 1.5 s in +Z direction.

The dynamic analysis was based on step-by-step direct time integration using the Newmark-Beta method. Lumped mass finite element models of the structures are considered for the analysis. The analysis to determine the structural response was performed with the help of the nonlinear module of the COSMOS finite element software package running on a Sun computer system. The response was obtained for a total duration of 3.5 s with 0.002 s time step size for integration.

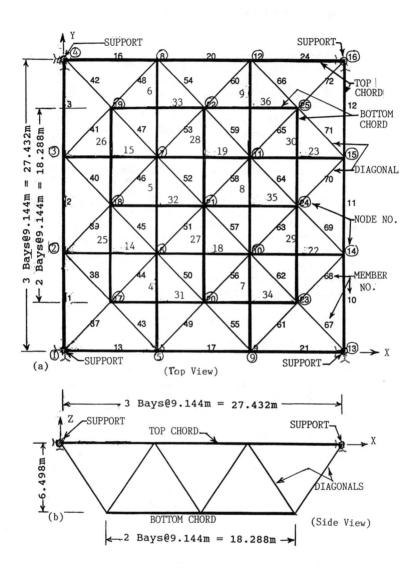

Fig. 1. A Three-Dimensional Double-Layer Grid Structure: (a) Top View; (b) Side View

Table 1. Double Layer-Grid Structure: Material, Geometry and Loadings

(a) Material Properties
Material: Aluminum   Modulus of Elasticity, $E = 72.0$ GPa
Density, $\rho = 2710$ Kg/m³   Yield Stress, $\sigma_Y = 95.0$ MPa

(b) Geometry: All members: circular tubes (Outer diameter = 1.389 x inner diameter)

Member ID	Element No.	Outer Diameter (m)	Length (m)	Area ($\times 10^{-02}$) (m²)	Slenderness Ratio	Euler Crit. Stress (MPa)
Top chord	1 - 24	0.2777	9.144	2.9179	106.89	62.19
Bottom chord	25 - 36	0.2777	9.144	2.9179	106.89	62.19
Diagonals	37 - 72	0.2777	7.945	2.9179	92.87	82.37

(c) Loadings:
Static (dead) load: 3.349 kN/m² applied vertically downward on top chord
Dynamic load: $p(t) = P_0 \cos(\bar{w}t)$; where $P_0 = (2.82 \times 10^6$ N), $\bar{w} = 82.7382$ rad/s
        applied in vertical direction on Node No. 21 (mid bottom chord)
Additional mass on each bottom chord joint = 8972.4 kg

Member Failure Sequence and Structural Response

The axial stress versus axial strain behavior curves (Fig. 2a,b) for pin-ended diagonal and chord members were obtained using the method similar to Nonaka (1973) and Papdrakakis (1985). A parabolic axial load and moment interaction relation was used. Although the member behavior curves show a small elastic postbuckling plateau (AB), it was assumed in this study that as soon as it reaches the Euler's critical stress, a compressive member lost its capacity suddenly to 28.5 MPa and goes into plastic behavior (BC'C). To begin with, all members of the intact structure were designed safe against the critical and yield stresses.

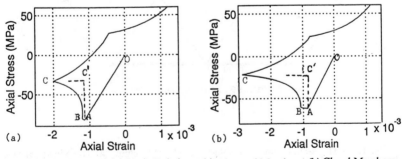

Fig. 2. Axial Stress-Axial Strain Relations: (a) Diagonal Members (b) Chord Members

In order to initiate the member failure process in the structure, the diagonal member 51 was arbitrarily selected to fail first. Although the actual maximum compressive stress of approximately -72.32 MPa at time 1.876 s in member 51 was less than its critical value, the member was assumed to lose all its load carrying capacity abruptly at this time. Failure of the member below its critical stress value may be justified by the fact that in the real-life structures there exist numerous adverse causes such as, material defects and construction and fabrication flaws to cause such damage to the member. To get the dynamic effects of the member failure, the damaged model was subjected to additional joint excitations (forcing functions), which were determined as outlined in the "Analysis Methodology" section; and the forces were dropped to zero at time 1.876 s.

After the failure of member 51, the stress in member 58 exceeded its critical level at time 1.972 s. The member (no. 58) was considered to buckle and snap into the inelastic postbuckling region (and the dynamic jump was said to take place). To facilitate the sudden jump in the load carrying capacity and the perfectly plastic material behavior in the postbuckling regime, the member was modeled to have elasto-plastic material behavior with reduced modulus of elasticity (E) equal to 21.6 GPa and yield stress (σ_Y) of 28.5 MPa. The external nodal forces, which were to be applied on the member 58 damaged structure and removed abruptly at the time the member failed, were determined using the method outlined in "Analysis Methodology" section. Two separate dynamic responses of the structure were required for this purpose, one when the member 58 was intact (original E) and the other when member 58 had elasto-plastic material properties with the reduced E and σ_Y. The sudden reduction on member 58 capacity was obtained by dropping the additional nodal excitation forces to zero at time 1.972 s. In the present analysis, the methodology employed for failure of member 58 (elasto-plastic with the reduced E and σ_Y) was used for all members that reached/exceeded the critical stress.

Failure of member 58 led to tension yielding of members 37 and 72 (both diagonals). The analysis was repeated with these two additional members modeled to have an elastic-perfectly plastic behavior with $\sigma_Y = 95$ MPa and $E = 72$ GPa. It was observed that the failure of member 58 with members 37 and 72 modeled as elasto-plastic material resulted in members 52 and 57 (both diagonals) to approach their critical stresses at time 2.030 s. When the snap-through failure of these two diagonal members was accounted for, members 8 and 19 (top chord) were found to reach their critical stresses at time 2.182 s. Accounting for these member failure, in turn, resulted failure of members 25 and 31 (bottom chord) also by reaching their critical stress at time 2.852 s. After this sequence of member failure, no additional member of the structure was observed to exceed either the critical stress or the yielding stress.

Table 2 shows the range of natural frequencies (lowest and highest) of the structure as the members failed. All sixty-three possible frequencies were obtained for each damaged structure. For the intact structure the frequencies vary from 6.584 Hz to 106.516 Hz. After the above sequence of member failure, the frequencies of the final damaged structure vary from 5.701 Hz to 106.515 Hz. Fig. 3 shows the time

Table 2. Structural Frequencies during Member Failure (Hz)

Members failed	Frequency No.	1	2	3	4	63
Intact structure		6.584	7.642	7.642	9.943	106.516
51^1		6.511	7.499	7.642	8.973	106.516
$51+58^2$		6.459	7.498	7.642	8.821	106.516
$51+58+52,57$		6.417	7.498	7.611	8.821	106.516
$51+58+52,57+8,9$		6.266	7.366	7.582	8.577	106.516
$51+58+52,57+8,9+25,31$		5.701	6.752	7.133	8.493	106.515

[1] $E = 0$ GPa for member 51. For all other failed members, $E = 21.6$ GPa and $\sigma_Y = 21.6$ MPa.

[2] After the damage of member 51, members 37 and 72 have elasto-plastic material properties with $E = 72$ GPa and $\sigma_Y = 95$ MPa .

variations of vertical displacements of nodes 2 and 21 as the member failure progress. The maximum deflection of the mid point of the structure (Node 21) was observed to be 0.117 m in vertical direction. Fig. 4 shows representative stress versus time relationships for four members (members 1, 18, 27 and 47) of the structure. The results show that stresses in members away from the middle region of the structure increase due to the member failure; where as those in members of the middle region decrease.

CONCLUSIONS

The response of a 3-dimensional corner supported double-layer grid, while several members failed sequentially, was investigated. The structure was subjected to a harmonic dynamic loading in addition to the static load. Emphasis has been given to the dynamic (sudden) nature of member failure on overall structural behavior. The study also included the member snap-through to the inelastic postbuckling type failure. A member which reaches its Euler's critical stress during the failure was assumed to snap-through into inelastic postbuckling. The member in the postbuckling regime was considered to behave in an elastic-perfectly plastic fashion. Results delineating the influence of the dynamic member failure included frequencies and time variations of joint displacements and member stresses. For the structural model and loadings considered in the present study, the member failure propagated until certain number of members failed; beyond which the structure appeared to withstand the applied loads. However, considering the failure of a large number of members, it could be inferred that the structure already lost its functional usefulness. It must be pointed out that the study also showed that the member failure path is highly dependent on the frequency and load amplitude of the applied dynamic load.

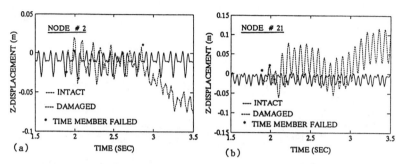

Fig. 3. Time Variations of Joint Displacements in Z - Direction: (a) Node 2 ; (b) Node 21

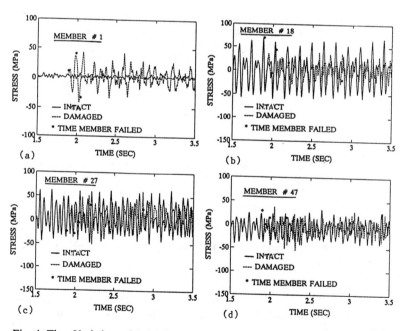

Fig. 4. Time Variations of Axial Stresses: (a) Member 1; (b) Member 18; (c) Member 27: (d) Member 47

ACKNOWLEDGMENT

The authors acknowledge the financial support obtained from the National Science Foundation, Washington, D.C. under Research Initiation Award (Grant No. MSS-9110900).

APPENDIX. REFERENCES

Chen, W., and Han, D. (1985). Tubular members in offshore structures. Pitman Publishing Inc, MA, U.S.A.

Davies, G. and Neal, B. (1959). "The dynamical behavior of a strut in a truss framework," Proc. Royal Soc., A253, pp 542-562.

Galambos, T.V. (1968). Structural Members and Frames, Prentice-Hall, Inc., Chapt.1.

Hill, C., Blandford, G., and Wang, S. (1989). "Postbuckling analysis of steel space trusses," J. of Structural Engnrng., ASCE, Vol. 115, No. 4, April, pp 900-919.

Malla, R. and Nalluri, B. (1993). "Dynamic effects of member failure on the response of truss type space structures," 34th SDM Conference, April 19-21, La Jolla, CA (Paper No. AIAA 93-1434, AIAA, Washington, D.C.)

Malla, R. and Wang, B., "A method to determine dynamic response of truss structures during sudden consecutive member failure," Space Structures-4 (G.A.R.Parke and C.M. Howard, eds.), Vol. 1, Thomas Telford, London, 1993, pp 413-422.

Malla, R., Wang, B., and Nalluri, B. (1993). "Dynamic effects of progressive member failure on the response of truss structures," Dynamic Response and Progressive Failure of Special Structures (R. Malla, Ed.), ASCE, New York, Dec. (in press).

Morris, N., "Effects of member snap on space truss collapse," Jour. of Engineering Mechanics, Vol. 119, No. 4, April 1993, ASCE, New York, pp 870-886

Murtha-Smith, E. (1988). "Alternate path analysis of space trusses for progressive collapse," J. of Struct. Engnrg., Vol. 114, no 9, Sept., pp 1978-1999.

Nonaka, T. (1973). "An elastic-plastic analysis of a bar under repeated axial loading," Int. J. Solids Structures, Vol. 9, 569-580.

Papadrakakis, M. (1985). "Inelastic cyclic analysis of imperfect columns," J. of Struct.Engnrg., ASCE, Vol. 111, No. 6, June, pp 1219-1234.

Salmon, C.G. and Johnson, J.E. (1990). Steel Structures - Design and Behavior, 3rd. Ed., Harper and Row, Publishers, pp 63-64.

Schmidt, L., Morgan, P., and Clarkson, J. (1976). "Space trusses with brittle type strut bucking," J. of the Struct. Div., ASCE, Vol. 102, No. ST7, July, pp 1479-1492.

Task Committee on Latticed Structures under Extreme Loads of the Comm. on Special Structures of the ASCE-Struct. Div., S. Holzer, Chair (1984). "Dynamic consideration in latticed structures," J. of Struct. Engnrg., Vol. 110, No. 10, Oct., pp 2547-2551.

Toma, S. and Chen, W. (1979). "Analysis of fabricated tubular columns," J. of Structural Division, ASCE, Vol. 105, No. ST 11, Nov., pp 2343-2366.

Seismic Behaviour of Tall Guyed Telecommunication Towers

Ghyslaine McClure,[1] Member, ASCE, and Eduardo I. Guevara[2]

Abstract

This paper presents results of a detailed numerical modelling study of two guyed telecommunication towers subjected to seismic excitation: a 107-m (350 ft) tower with six stay levels, and a 342-m (1150 ft) tower with seven stay levels.

Two horizontal accelerograms were used, S00E 1940 El Centro and N65E 1966 Parkfield, with each record being scaled to match the elastic design spectra of the 1990 National Building Code of Canada for the Montréal region. Elements of response generated were: guy cable tensions, horizontal shear forces in the mast, and displacements and rotations at the top of the mast. Effects of surface wave propagation were studied using asynchronous inputs at the ground anchorage points and at the base of the mast. Combined lateral and vertical ground motions were considered for the tallest tower. Results indicate important cable-mast interactions, predominant in the frequency range of the lower axial modes of the mast.

Introduction

Guyed tower design provisions for earthquake effects are still not covered by the Canadian Standard CAN/CSA S-37 *Antennas, Towers, and Antenna-Supporting Structures* (CSA 1986). Such provisions are needed, however, even though seismic risk assessment is highly uncertain throughout Canada. The first motivation is that seismic design is required by the 1990 National Building Code of

---

[1]Asst. Prof., Dept. of Civ. Engrg. and Applied Mechanics, McGill University, Montréal, Québec, Canada, H3A 2K6.
[2]Structural Engineer, P.O. Box 314, San Pedro 2050, San Jose, Costa Rica.

Canada, and seismic design checks are becoming a standard procedure for most constructed facilities. Accordingly, owners of telecommunication towers in areas of high seismic risk (mostly on the West coast) are in a position to ask their designers to evaluate the performance of their towers in the event of a severe seismic excitation. Without proper guidance, there is a danger that tower designers be tempted to directly apply building code procedures to towers, which would be inappropriate. More precise knowledge on the behaviour of guyed towers under seismic excitations is required, however, in order to develop proper design guidelines, and detailed numerical simulations can help get more insight in this problem.

Project Objective

The objective of this project is to investigate the seismic response of tall guyed telecommunication towers using numerical simulations on detailed nonlinear finite element models. The models account for cable geometric nonlinearities and allow for dynamic interactions between the mast and guy cables to take place.

Modelling Considerations

Previous attempts at the numerical modelling of guyed towers were accomplished by simplifications in the model (McCraffrey and Hartman 1972, Augusti et al. 1986, and Ekhande and Madugula, 1988). Often, linear models were used for cables and masts, and cables were even replaced by equivalent springs. More recent studies by Augusti et al. (1990) and Argyris and Mlejnek (1991) have recognized that these simplifications, although not significantly influential when studying the static response of the structure, are no longer appropriate for dynamic analysis. It is the complex behaviour exhibited by the guy cables that has presented difficulties in the modelling of these structures, and with the increased sophistication of the finite element codes available, more realistic models can be studied.

The two guyed towers selected for analysis are real structures used by the industry. They consist of three-legged lattice galvanized steel masts pinned at their foundation and stayed by pretensioned guy wires. Often the guy wires are connected to the mast by outriggers or stabilizers to increase the overall torsional resistance of the structure. The guy cables are attached directly to the mast legs or to stabilizers, and are anchored on the ground on radii oriented at a horizontal angle of 120°.

For the first tower, the mast has a total height of 107 m
and is stayed at six levels; it is tapered at the base and
equipped with outriggers at all cluster levels, with the
exception of the lowest one. Figure 1 illustrates the
geometry of the second tower analysed; it consists of a
non tapered mast of a total height of 342 m stayed by
seven guy wire levels; outriggers are provided only at the
second lowest cluster. Further details can be found in
Guevara (1993).

Mast modelling

    For the shortest tower, the model for the mast is
made of beam-column elements with equivalent properties
(using Timoshenko's beam theory and St. Venant torsion),
whereas the final model for the tallest tower is a
detailed three-dimensional truss. Due to asymmetry in the
pattern of the diagonals for both masts, lateral loading
in a principal direction induces coupled bending and
torsion in the mast. This phenomenon cannot be replicated
in the equivalent 3-D frame models, which do not account
for warping torsion.

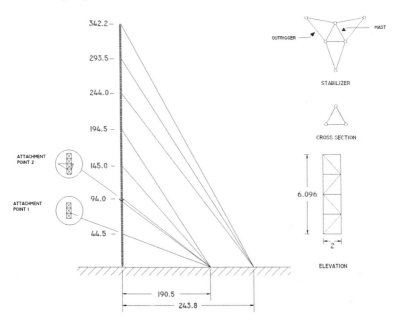

Figure 1. Geometry of 342-m tower (units are meters).

Rotational inertia of individual elements of the mast is neglected in the 107-m tower, considering that translational effects will dominate the response and that the panel width is relatively small (0.6 m). The criterion for mesh selection was to get a good representation of the lowest five flexural modes of the mast, obtained from an eigenvalue analysis in the deformed configuration under self-weight and cable prestressing forces. The natural periods corresponding to these modes are between 0.20 s and 0.38 s.

The equivalent model of the 342-m tower used 60 frame elements. In contrast with the previous model, the panel width is significantly larger (2 m) and contributes to increase the in-plane torsional inertia of the mast, as well as its rotational inertia. This was put into evidence by the frequency analysis, where the corresponding mode of the lowest frequency is a torsional mode in the detailed model, but the first flexural mode in the equivalent one. Furthermore, higher torsional modes of vibration present in the detailed model, are not present in the equivalent frame model. It was therefore decided to use the detailed truss model in this case. The natural periods of the lowest five flexural modes are between 0.92 s and 1.95 s, and the mode shapes are illustrated in Figure 2.

Guy wire modelling

The modelling of the guy wires is of particular importance, especially because of the large geometric nonlinearities they exhibit. These nonlinear effects grow as the cables become slack and the amplitude of motion increases. A sufficiently fine mesh using a large kinematics formulation (but small strains) for the cable stiffness can account for full geometric nonlinearities (ADINA R&D 1987). The selection of the type of truss element (tension-only) and the mesh refinements were based on results from a frequency analysis of the initial configuration of the cable. A parametric study was carried out using three types of isoparametric elements and different meshes. Since transverse vibrations of the guy wires are likely to dominate over the axial high-frequency modes, the criterion was to find a mesh that could give a good representation of the sixth lowest transverse mode of the cable. This was achieved with three-node truss elements (with integration at two Gauss points), and an element size of approximately 1.5 m to 2 m. The lumped mass formulation was retained. The total number of elements required to discretize all the cables was 390 in the 107-m tower, and 1380 in the 342-m tower. All cables are initially prestressed (with tension T_0) to

approximately 10% of their ultimate tensile strength, and
their material properties are assumed linear elastic.

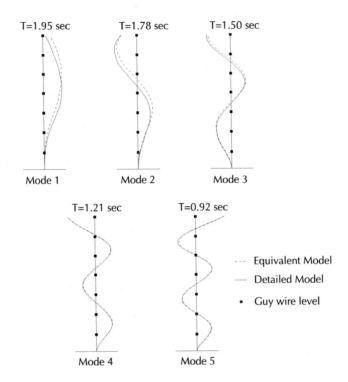

Figure 2.   Five lowest flexural modes of the 342-m tower.

Modelling of damping

Neither cable damping (structural or aerodynamic) nor
structural damping in the mast are modelled. Since the
dynamic analysis is to proceed by direct integration in
the time domain, calibration of a Rayleigh damping model
would have required the calculation of too many mode
shapes in the initial configuration, in order to span
significant frequencies for both the cables and the mast.
Furthermore, due to the nonlinearities in the models,
damping would vary with the frequency and the amplitude of
vibration, and the nature of motion.   In view of the
difficulties associated with realistic modelling of cable

damping for nonlinear analysis, it was decided to rely strictly on the numerical damping of the Newmark-ß integration operator with parameters δ=0.55 and α=0.3, to filter out numerically generated high frequency components of the response.

Response analysis

Shear forces in the mast, axial tensions (T_{dyn}) in the guy cables and displacements parallel to the earthquake direction are taken as the typical elements of response of interest. These response indicators are examined at the guy wire attachment points, while shear forces in the mast are also studied at intermediate points between stay levels. Detailed results and typical time histories are given in Guevara (1993), and only the main observations are summarized in the following paragraphs; a few examples of time histories are also presented.

Response of the 107-m tower

Simulations on this tower were obtained only for the Parkfield ground motion, and time delays (maximum of 0.014 s) were introduced for comparison with the synchronous excitation. Time histories were generated for a total duration of 8 s, considering that the strongest ground shaking takes place primarily in the first 4 s. Results characterizing the peak response of the tower are outlined in Table 1, and examples of time histories are shown in Figure 3.

Response of the 342-m tower

The response of this tower was calculated under the El Centro ground motion for both the synchronous and asynchronous loading conditions (with a maximum delay of 0.065 s). The time histories were generated for a total duration of 8 s, as a compromise between the minimum response time of interest and the computational effort required. In order to test the potential benefits of further investigations of the vertical responses of tall masts, combined horizontal (H) and vertical (V) input ground accelerations were also considered for this tower. Results are summarized in Table 2 and examples of time histories are given in Figure 4.

Summary of observations

The analysis of the 107-m tower indicates that dynamic amplifications in the guy wire tensions (T_{dyn}/T_0) are more important in the bottom and top clusters;

Table 1.   Summary of peak response for the 107-m tower.

	Synchronous	Asynchronous
Bottom Cluster (T_0 = 11.3 kN)		
T_{dyn}/T_0	1.37	1.55
Shear (kN)	1.25	1.25
Top Cluster (T_0 = 41 kN)		
T_{dyn}/T_0	1.67	1.51
Shear (kN)	1.80	2.06

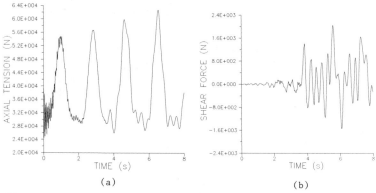

(a)   (b)

Figure 3. Response of the 107-m tower — a) Axial tension
in top guy cluster — synchronous input; b) Shear force in
the mast at top stay level — asynchronous input.

Table 2.   Summary of peak response for the 342-m tower.

	Synchronous		Asynchronous
	H	H+V	H
Bottom Cluster (T_0 = 45.4 kN)			
T_{dyn}/T_0	1.36	1.34	1.36
Shear (kN)	8.0	8.0	10.0
Middle Cluster (T_0 = 56.9 kN)			
T_{dyn}/T_0	2.00	2.14	1.89
Shear (kN)	18.0	18.0	20.0
Top Cluster (T_0 = 76.5 kN)			
T_{dyn}/T_0	1.70	1.85	2.14
Shear (kN)	12.0	12.0	13.5

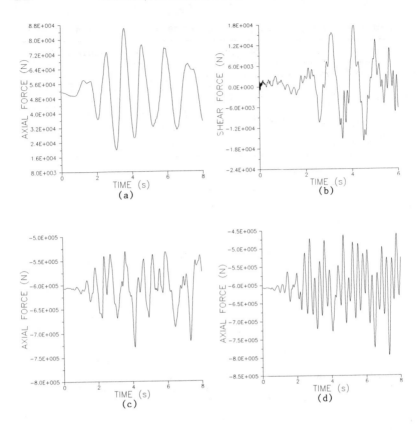

Figure 4. Response of the 342-m tower — a) Axial tension
in middle cluster — synchronous input; b) Shear force at
middle stay level — asynchronous input; c) Axial force in
mast at lowest stay level — synchronous horizontal input;
d) Axial force in mast at lowest stay level — asynchronous
combined horizontal and vertical input accelerations.

however, the shear components in the mast are larger at
intermediate sections. Furthermore, the absolute values
of the dynamic cable tensions and of the shear forces in
the mast are well below strength limits. The tilting
rotations and lateral displacements are also well below
the serviceability limits for this structure.

In general, the analysis presents two interesting

phenomena: the vertical interactions between the mast and
the guy wires, and the tower response under asynchronous
ground motion. The vertical interactions observed suggest
that the inclusion of vertical ground motions in the
analysis could trigger a higher axial participation of the
mast. This was demonstrated for the tallest mast with a
combination of horizontal and vertical excitations. The
vertical effects induced in the mast propagate into the
guy wires, and can generate additional amplifications in
the cable tensions. The participation of the axial modes
is closely linked to the characteristics of the vertical
ground motion. The use of asynchronous motion at the base
of the structure has produced interesting results only in
the extreme clusters for the axial tensions in the guy
wires. However, the difficulties encountered in modelling
the asynchronous motion may not justify such a refinement,
and analysis under synchronous ground motion only should
suffice.

Conclusions

      This study has raised several points that should be
addressed in further investigations. The main ones are:
1) the development and implementation of an accurate
damping model for the cables and the mast, suitable for a
time domain analysis; 2) the investigation of the tower
response to vertical excitations, and the characterization
of the vertical sensibility of the towers to the frequency
content of the input ground motion; 3) the investigation
of the reduction in the bending stiffness of the mast as
a result of higher axial loads induced by vertical
accelerations.

Acknowledgments

We wish to thank Mr. S. Weisman, P.Eng., of Weisman
Consultants Inc., Downsview, Ontario; Prof. M.S. Madugula
of The University of Windsor, Ontario, and members of the
CSA Technical Committee S37 on Antennas, Towers, and
Antenna-Supporting Structures, for providing detailed data
on the towers analysed. The project was supported by
Grant No. OGP0121270 from the Natural Sciences and
Engineering Research Council of Canada.

Appendix. References

ADINA R&D. (1987). "ADINA (Automatic Dynamic Incremental
Nonlinear Analysis) Theory and modelling guide."
Watertown. MA, Report ARD 87-8.

Argyris, J., and Mlejnek, H.P. (1991). "Dynamics of structures." Texts on computational mechanics. Vol. V, North-Holland, New York, 505-510.

Augusti, G., Borri, C., Marradi, L., and Spinelli, P. (1986). "On the time-domain analysis of wind response of structures." J. of Wind Engrg. and Idustrial Aerodynamics, Vol. 23, 449-463.

Augusti, G., Borri, C., and Gusella, V. (1990). "Simulation of wind loading and response of geometrically non-linear structures with particular reference to large antennas." Structural Safety, Vol. 8, 161-179.

CSA (Canadian Standards Association), CAN/CSA-S37-M86, "Antennas, towers and antenna-supporting structures." CSA, Rexdale, Ontario.

Ekhande, S.G., and Madugula, M.K.S. (1988). "Geometric non-linear analysis of three-dimensional guyed towers." Computers & Structures, Vol. 29, 801-806.

Guevara, E.I. (1993). "Nonlinear seismic analysis of antenna-supporting structures." M.Eng. Project Report G93-14, Dept. of Civ. Engrg. and Applied Mechanics, McGill University, Montréal, Québec, Canada.

Guevara, E.I., and McClure, G. (1993). "Nonlinear seismic response of antenna-supporting structures." Computers & Structures, 47(4/5), 711-724.

McCaffrey, R., and Hartman, A.J. (1972). "Dynamics of Guyed Towers." J. of the Struc. Div., ASCE, 98(ST6), 1309-1323.

DYNAMIC CHARACTERISTICS OF TRANSLATIONAL
LATTICE DOMES

Celina U Penalba[1]

ABSTRACT

The feasibility of using lattice domes of various shapes generated by translating a variety of parametric polygons is illustrated and the dynamic characteristics of these lattices determined. The paper discusses the possibility of using energy dissipator devices to control the response of lattice domes generated by translating circular polygons sectors,when subjected to a dynamic environment.

INTRODUCTION

Lattice shell construction has increased considerable during the last years as a response to the needs of spanning large areas aesthetically without obstructions, with engineering efficiency and economy. Their strength inherent from their shapes,relatively high stiffness to mass ratio, easy of construction and prefabrication, makes them attractive to a variety of engineering applications which include, passenger terminals, music and convention centers, theaters, shopping malls, pavilions, sport arenas, hangars, towers, space vehicles and station, and interplanetary reflectors. However dynamic environments, such as wind, earthquakes and equipment vibrations can cause response amplification through interaction with the natural dynamic characteristics of these structural systems and subject these structures to unstable oscillations that could lead to failure.

In this work the feasibility of using Lattice Domes of many shapes which can be generated by translating a variety of polygons is illustrated and the configurations, connections and support conditions required for stability discussed.

The dynamic characteristics of translational lattice domes of different curvatures are presented. To control unstable vibrations the use of energy dissipation devices, located strategically is discussed.

Celina U Penalba
CAL POLY State University
San Luis Obispo, California. U.S.A.

GEOMETRY OF LATTICE SURFACES
The term lattice dome used in this work, referred to a discrete surface generated by two independent sets of space polygons, where the skin connected to the polygonal bars do not participate in the structural action.
A space polygon is defined as a one dimensional sequence of straight elements whose shape is given by the position of its nodes.Its geometrical properties can be conveniently defined as function of a discrete variable a in a similar manner as used in the continuum Figs.1(a,b)

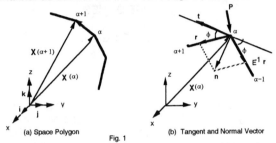

(a) Space Polygon   (b) Tangent and Normal Vector

Fig. 1

A lattice surface is described by appropriate sets of parametric polygons using discrete variables designated by α_1 and α_2 . The position of a typical joint takes the form:

$$\overline{X} (\alpha_1, \alpha_2) = x (\alpha_1, \alpha_2)\, \overline{i} + y (\alpha_1, \alpha_2)\, \overline{j} + z (\alpha_1, \alpha_2)\, \overline{k} \qquad (1)$$

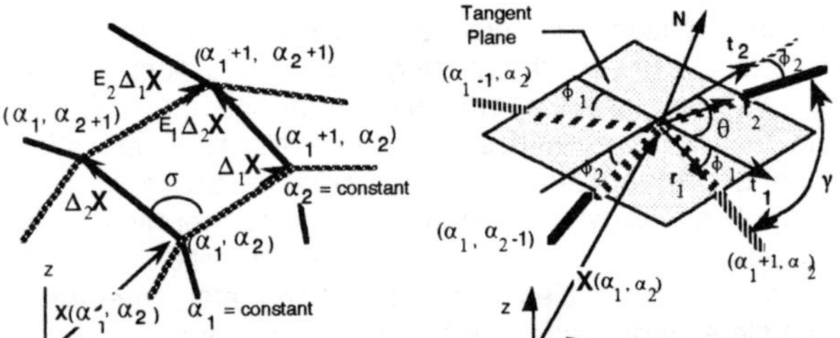

(a) Parametric Polygons   (b) Tangent Plane and Lattice Shell Normal

Fig. 2

To facilitate the establishment of the equilibrium conditions, the force-deformations relations and to gain insight into the behavior of the lattice, a local coordinate system defined by the tangents and normal to the lattice at a joint, is used, Figs.2(a,b).

TRANSLATIONAL LATTICE SHELLS
A translational lattice is generated by effecting upon space polygons with position vectors $X (\alpha_1)$, a translation in which each of its joints describe space polygons defined by the position vectors $X (\alpha_2)$. This definition applies to the converse case, yielding the position of a typical node:

$$\overline{X} (\alpha_1, \alpha_2) = (x_1 + x_2)\overline{i} + (y_1 + y_2)\, \overline{j} + (z_1 + z_2)\, \overline{k} \qquad (2)$$

For translational lattices generated by plane polygons:

$$\overline{X}\,(\alpha_1, \alpha_2) = x_1\,(\alpha_1)\,i + y_2\,(\alpha_2)\,j + \left[z_1\,(\alpha_1) + z_2\,(\alpha_2)\right]\,\overline{k} \qquad (3)$$

Spectacular shapes which may be used to roof large areas, can be generated by the process described above. As an illustration, consider the lattices generated by translating a parabolic polygon over another of the same or opposite curvature, Figs 3 (a,b).

$$\overline{X} = \frac{a}{n_1}\,\alpha_1\,i + \frac{b}{n_2}\,\alpha_2\,j + \left(\frac{c}{n_1^2}\,\alpha_1^2 + \frac{d}{n_2^2}\,\alpha_2^2\right)\,\overline{k} \qquad 4\,(a,b)$$

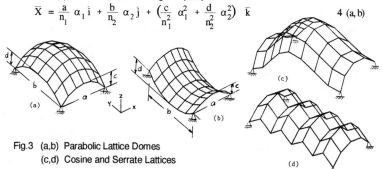

Fig.3 (a,b) Parabolic Lattice Domes
       (c,d) Cosine and Serrate Lattices

Variations of the above lattices are obtained by translating parabolic polygons over circular, elliptical, cosine, catenary and other polygons. When properly supported these lattices provide stable configurations for both, static and dynamic environments.

An interesting lattice results when a cosine polygon is translated over a circular polygon, Fig.3 (c), or a cosine polygon. A non-conventional lattice of potential structural applications, due to its easiness in prefabrication and erection, is generated by translating a serrate polygon over another polygon. When the translation is performed over a circular polygon, the result is the lattice of Fig.3 (d).

The applicability of translational domes for roofing square, rectangular, circular or oval areas, will be illustrated on a lattice generated by two sets of circular polygonal sectors of different curvatures.

The location of the joints as functions of the discrete variables α_1 and α_2 and the parameters indicated in Fig.4, is given by:

$$\overline{X} = a \sin\frac{\beta_1}{n_1}\,\alpha_1\,i + b \sin\frac{\beta_2}{n_2}\,\alpha_2\,j + \left[a\,(1 - \cos\frac{\beta_1}{n_1}\,\alpha_1) + b\,(1 - \cos\frac{\beta_2}{n_2}\,d_2)\right]\,\overline{k} \quad (5)$$

The length of the members of the α_1 and α_2 polygons denoted respectively L_1 and L_2 and the angle γ between these members, of primary importance for the design of connections, are defined by:

$$L_1 = 2a \sin\frac{\beta_1}{2\,n_1} \quad \text{and} \quad L_2 = 2b \sin\frac{\beta_2}{2\,n_2} \qquad 6\,(a,b)$$

$$\cos\gamma = \sin\frac{\beta_1}{n_1}\,(\alpha_1 + \frac{1}{2})\,\sin\frac{\beta_2}{n_2}\,(\alpha_2 + \frac{1}{2}) \qquad (7)$$

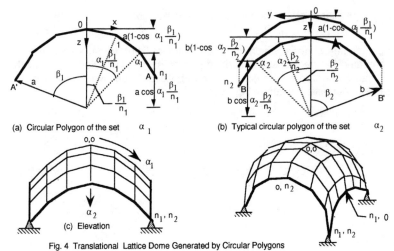

Fig. 4 Translational Lattice Dome Generated by Circular Polygons

Although the above imply non-uniform joints, the double symmetry of the structure and the properties of the planes defining a joint, facilitates the design of the connections.

The model for the analysis depends on the connections and the conditions required for stability. Two sets of parametric polygons will suffice to define a properly supported translational lattice dome with rigid connections, while stable pin jointed lattices will require another set of space polygons and additional members to provide boundary supports.

The design of the lattice of Fig.4, as a pin-jointed structure requires the addition of a set of dependent polygons whose member length L_3 are given:

$$L_3^2 = L_1^2 + L_2^2 - 2L_1 L_2 \sin \sigma_1 (\alpha_1 + \frac{1}{2}) \sin \sigma_2 (\alpha_2 + \frac{1}{2}) \qquad (8)$$

$$\sigma_1 = \frac{\beta_1}{n_1} \quad ; \quad \sigma_2 = \frac{\beta_2}{n_2}$$

Single layer pin jointed lattices have tendency to develop mechanisms due to improper configuration and supports, static and dynamic instability and hazardous dynamic environments. Member and joint buckling, snap through and general buckling will not be discussed here. However, configurations that render stable translational domes will be examined.

A stable configuration for the basic pin-jointed translational lattice ($n_1 = n_2 = 1$), supported to restrain rigid body motion, requires five additional members to those provided by the three sets of polygons, or the replacement of these members by reaction components. Figs.5 (a,b).

A study of the lattice, $n_1 = n_2 = 2$, Fig. 5 (c,d) shows that a stable configuration can be reached when thirteen additional members are provided, two at each polygonal boundary and five at the base, or alternatively providing eight boundary members and replacing the five members at the base by reaction components

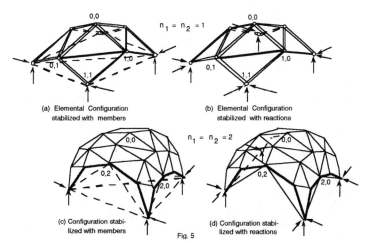

(a) Elemental Configuration
stabilized with members

(b) Elemental Configuration
stabilized with reactions

$n_1 = n_2 = 1$

$n_1 = n_2 = 2$

(c) Configuration stabi-
lized with members

(d) Configuration stabi-
lized with reactions

Fig. 5

The applicability of the concepts of Difference Geometry to establish the governing equation for the static and dynamic analyses of a lattice, will be illustrated on a single layer pin jointed translational dome.The equilibrium of a typical joint, Fig. (6) yields the vector partial difference equation:

$$\nabla_1 (F_1 \bar{r}_1) + \nabla_2 (F_2 \bar{r}_2) + \nabla_3 (F_3 \bar{r}_3) + \bar{P} = 0 \qquad (9)$$

where F_1, F_2, F_3 are scalar forces along the members defined by unit vectors:

$$\bar{r}_1 = \frac{1}{L_1} \Delta_1 \bar{X} \; ; \; \bar{r}_2 = \frac{1}{L_2} \Delta_2 \bar{X} \; \text{and} \; \bar{r}_3 = \frac{1}{L_3} \Delta_3 \bar{X} \qquad 10 \text{ (a, b, c)}$$

where Δ_1, Δ_2, $_1$ and $_2$ are the forward and backward partial difference operators. For the dependent polygons

$$\Delta_3 F_3 \equiv (E_1 E_2 - 1) F_3 \equiv F_3 (\alpha_1 + 1, \alpha_2 + 1) - F_3 (F_3 (\alpha_1, \alpha_2)$$

$$\nabla_3 F_3 \equiv (1 - E_1^{-1} E_2^{-1}) F_3 \equiv F_3 (\alpha_1, \alpha_2) - F_3 (\alpha_1 - 1, \alpha_2 - 1)$$

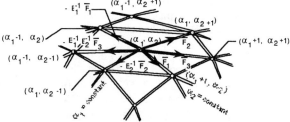

Fig. 6 Forces at a Typical Pin-Jointed Lattice

Statically indeterminate lattices or those whose boundary conditions must be given in terms of joint displacements , require a model for the joint deformations. The joint displacement vector may be expressed in terms of its components along the local coordinate system t_1, t_2, N.

$$\bar{U}(\alpha_1,\alpha_2) = u_1(\alpha_1,\alpha_2)\,\bar{t}_1 + u_2(\alpha_1,\alpha_2)\,\bar{t}_2 + u_N(\alpha_1,\alpha_2)\,\bar{N} \qquad (11)$$

The force-deformation relations are established from the operations:

$$F_i = Q_i\,(\Delta_i\,\bar{U}\cdot\bar{r}_i) \;\; ; \;\; Q_i = \frac{A_i\,L_i}{E_i} \;\; ; \;\; i = 1,2,3 \qquad (12)$$

The substitution into eq. (9), gives a vector difference equation applicable to each node α_1, α_2. Using a matrix notation, the resultant expression becomes:

$$[k]\,[u] = [P] \qquad (13)$$

The elements k_{ij} (i, j = 1,2,3) of the above matrix are composed of partial difference operators related by the geometric properties (coefficients of the First Fundamental Form, tangential and normal curvatures) of the polygons. Eqs.(13) represent a set of three partial difference equations in the unknowns u_1, u_2, u_N which may be solved in close form for specific lattices (conoidal, serrate, parabolic). The general case is solved in open form "walkthrough", numerical techniques, matrix approach. When eqs.(13) are applied at each node, the result is the matrix equation:

$$[K]\,[U] = [P] \qquad (14)$$

where   [K] = stiffness matrix of order N x N, N being the number of free
degrees of freedom.

[U] = Displacement matrix of order N x 1

[P] = Applied load matrix of order N x 1

Dynamic Characteristics
When lattices domes are subjected to a dynamic environment, the model becomes a partial Difference-Differential equation. Using a simplified model obtained by lumping the masses at the lattice joints, the governing equation takes the form:

$$m(\alpha_1,\alpha_2)\,[\ddot{u}] + c(\alpha_1\cdot\alpha_2)\,[\dot{u}] + [k]\,[u] = [P] \qquad (15)$$

The application to each joint gives a standard dynamic matrix formulation.

$$[M]\,[U] + [K]\,[U] = [P] \qquad (16)$$

in which: [M] is the mass matrix of order N x N
[C] is the damping matrix of order N x N
[K] , [U] and [P] as defined before.

Equation (16) is applicable to the case of rigid connections, provided that the effect of rotations and applied joint moments is included.

To study the dynamic characteristics of translational lattices and assess their vulnerability to dynamic vibrations, the natural frequencies, ω_i and modes shapes are determined from the free vibration model:

$$\lceil M\,\omega^2 - K \rceil\,\Phi = 0 \qquad (17)$$

where, ω and ϕ, represent the natural frequencies and mode shapes of vibration. The frequencies ω_i , serve to identify the modes of vibration which are likely to experience resonance during a particular loading condition.

Lattice shells are characterized for their numerous mode shapes of vibration compatible with the number of degrees of freedom exhibited.

These modes are complex but due to the symmetry of these lattices, it is easy to recognize the patterns of "nodal polygons" that designates those polygons that exhibit opposite displacements on opposite sides.

The characteristics of the mode shapes of translational lattice domes, can be inferred by considering the simple and heuristic case of the basic (elemental) translational lattice dome, Fig. 7 (a). By placing the origin at one of the supports, the lattice can be described by the parametric polygons α_1, $\alpha_2 = 1, 2, 3$.

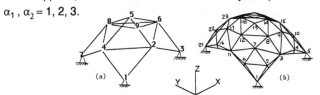

Fig.7 (a) Basic Translational Dome, (b) Circular Translational Dome

The illustration is on a pin-jointed steel pipe lattice with plan dimensions 30' x 24' (9.144 m x 7.315 m) and height of 8' (2.438). A study of the 15 modes of vibration available, the first six shown in Fig. 8, reveals that for the first mode all the polygons develop into nodal polygons, with the central polygon $\alpha_2 = 2$ being determinant in shaping the mode. It is also observed that the 'y' displacements of this polygon are symmetric respect the $\alpha_2 = 2$ polygon and has negligible displacements in the x direction. The second mode exhibits similar characteristics with the nodal polygon $\alpha_2 = 2$, shaping the mode. The "x" displacements of this polygon are symmetric respect to $\alpha_2 = 2$ and take the maximum value, while the "y" displacements are negligible.

The large relative displacements between the joints of the nodal polygons result in sharp modes shapes. The third and fourth modes share the interactive action of the two nodal central polygons. The small relative displacement between the joints of these polygons, provide smoother shape. The characteristics of mode five and six are similar, however the large relative vertical displacements of the joints of edge polygons may lead to mechanism if activated.

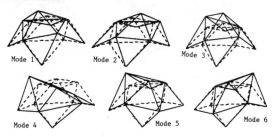

Fig.8 Mode shapes of pin-jointed Lattice Dome

TABLE 1. NATURAL FREQUENCIES ω_i -- Pin Connections

LATTICE	1st	2nd	3rd	4th	5th	6th
Basic 5(a) $n_1, n_2 = 3$	53.81	64.55	66.23	69.64	78.79	93.08
Circ. 5(b) $n_1, n_2 = 5$	28.59	29.21	30.28	30.77	33.70	33.91

As a second illustration, consider the pin-jointed lattice with steel pipe members shown in Fig 7 (b). The lattice has 37 joints, 92 members and the properties: a = 42.25 ft. (12.88 m.), b = 31.25 ft. (9.52 m), β_1 = 45.24°, β_2 = 53.13°, $n_1 = n_2 = 2$, $K_1 = 0.3922$ and $K_2 = 4595$.

A study of the first six modes of the 87 available, Fig. 9 and Table 1 shows the lattice to be more flexible than the basic dome and to have sharper modes making it prone to develop mechanisms.

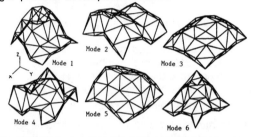

Fig. 9 Mode shapes of circular pin-jointed Lattice Dome

Identifying the nodal polygons facilitate the visualization of the mode shapes and their vulnerability to resonance. The observation of the first mode reveals that the α_1 and α_2 polygons deformed respectively, in one and two waves, and that the polygon $\alpha_2 = 3$ acts as nodal for the vertical displacement and exhibit symmetric 'y' displacement and negligible deformation in the x direction. Similar observation can be made for the other modes.

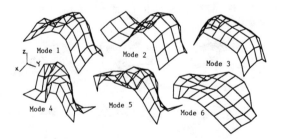

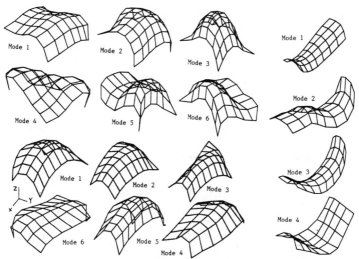

Fig 10 Mode shapes of Translational Lattice Domes rigid connected

TABLE 2. NATURAL FREQUENCIES ω_i -- Rigid Connections

LATTICE	1st	2nd	3rd	4th	5th	6th
Basic $n_1, n_2 = 3$	2.842	3.093	4.828	11.589	12.659	15.701
Circular $n_1, n_2 = 5$	0.876	1.045	1.423	1.486	1.685	1.922
C. Parab. (-) $n_1, n_2 = 7$	0.615	1.154	1.162	1.671	1.842	1.928
C. Parab. (+) $n_1, n_2 = 7$	0.949	1.048	1.317	1.383	2.042	2.065
S.S. Conoidal $n_1, n_2 = 7$	0.855	1.227	1.900	2.030	2.085	2.428
C. Parab. (-) $n_1, n_2 = 9,5$	0.248	0.554	0.615	0.881	1.137	1.179
C. Parab. (+) $n_1, n_2 = 9,5$	0.336	0.441	0.641	0.925	1.026	1.238
S.S. Conoidal $n_1, n_2 = 9,5$	0.725	1.148	1.719	2.046	2.073	2.374
Clamped Conoidal $n_1, n_2 = 9,5$	1.652	2.045	2.689	3.208	3.851	4.178
Cosine C. $n_1, n_2 = 9,5$	1.119	1.501	1.778	1.934	2.021	2.565

Table 2 lists the first six natural frequencies of lattices of different shapes and number of joints. The mode shapes for some of these lattices are shown in Fig 10. Further analysis show that the mode shapes and frequencies are influenced by the number and type of joints, the curvature of the polygons and the dimension and shape of the roofed area. Furthermore, larger translational domes of rectangular plan, tend to be more flexible and develop shaper modes of vibration.

Structural Application of Energy Dissipation Devices

The next step toward the control of dynamic vibration, consists in determining the most vulnerable modes of excitation and the location where energy dissipation devices are more effective. The natural frequencies, ω_i, obtained from the free vibration analysis will be used to identify the modes of vibration that may experience resonance during a particular loading condition. To illustrate the procedure, the Spectral Density Function of the loading is represented by the plot of Fig. 11 (a). From the graph, the intensity of the loading function S_i for each mode of vibration can be determined. The modes which fall in the high intensity region of the loading function, are identified as the more likely to get excited.

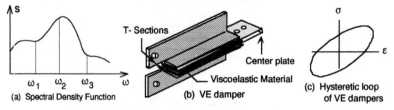

(a) Spectral Density Function   (b) VE damper   (c) Hysteretic loop of VE dampers

Fig. 11 Chracteristics of VE dampers

Many types of dissipation devices have been proposed recently, among them the viscoelastic dampers (VE) seems to be practical and very effective in reducing structural vibrations due to wind, traffic, or seismic excitations.

A typical VE damper, Fig.11 (b), consists in a double layer viscoelastic material placed between two axial members. When a relative motion between the end joints occurs, and forces are transferred from one axial member to another, the viscoelastic material deforms in shear dissipating energy. The behavior of the viscoelastic material is such that the internal stresses and strains are always out-of-phase.The performance of the damping devices is most effective when acting through large relative displacements. Thus, the identification of the most effective location, is based on selecting from the vulnerable modes the joints that experience significant relative motion.

The energy dissipation capacity of a VE damper is a function of G' the shear modulus which determines the stiffness of the damper k' and the shear loss modulus G". The ratio of G' and G" defines the loss factor $\eta = G'/G"$. The stress strain relation of the VE damper can be expressed as:

$$\gamma = \gamma_0 \sin \omega\tau \qquad\qquad \tau = \gamma_0 (G' \sin \omega\tau + G" \cos \omega\tau)$$

where ω is the frequency of the applied harmonic motion.

Recent publications by Chang, Soong, Lai, Nielsen and Oh, show that the Modal Strain Energy Method, can be used to predict with reliability, the equivalent structural damping and the dynamic response of a multi degree of freedom viscoelastically damped structure.

Once the vulnerable modes and the location for the inclusion of the V E dampers have been identified, the adequate loss factor η and the desired level of the viscous damping ratio, ξ_i , for the i mode of vibration, selected, the effective stiffness of the lattice with the supplemental damping, k_{vi}, can be computed from the relation:

$$\xi_i = \frac{\eta}{2} \left(1 - \frac{\Phi_i^T K_s \Phi_i}{\Phi_i^T K_v \Phi_i}\right) = \frac{\eta}{2} \left(1 - \frac{k_{si}}{k_{vi}}\right) \tag{18}$$

where K_v and K_s are the stiffness of the lattice with and without the damper, k_{vi} and k_{si} the effective stiffness in mode i , with and without the damping device. The contribution of the damping element k_{di} can be calculated from the relation : $k_{vi} = k_{si} + k_{di}$ and the k' corresponding to the location of the dampers computed from this value.

The design of the damper follows by selecting the dimensions that satisfy the relation:

$$k' = G' \frac{A}{\eta}$$

CONCLUSION

The versatility of using single layer translational lattice domes for roofing large areas with aesthetic and structural practicality has been illustrated.

A study of the dynamics characteristics of this class of lattice shells, reveals a wide range of dynamic responses determined by the dimensions, geometry, connections, number of joints and material used. For those lattices that exhibit vulnerability to unstable oscillations, the inclusion of viscoelastic dampers is considered.

Although the identification of the modes that are more likely to get excited and the location where the damping devices are most effective, seems to be a monumental task due to the proliferation of the degrees of freedom of these structures; the availability of computers and analytical techniques , should be helpful in reducing their excessive vibrations due to dynamic excitations.

The technique of providing supplemental damping could be particularly useful in the retrofit and rehabilitation of lattice domes subjected to seismic or wind environments.

REFERENCES

[1] Dean, D.L. "Discrete Field Analysis of Structural Systems", Courses and Lectures No.203, International Centre for Mechanical Sciences, Italy ,Springer -Verlag Wien - New York,1976.

[2] IASS Working Group on Spatial Structures, "Analysis, Design and Realization of Space Frames. A State-of- the- Art Report Bulletin of IASS, No.84/85, 1984

[3] K.C.Chang. T.T. Soong, S.T. Oh, M. Lai (1992) "Seismic response f steel- frames structures with added viscoelastic dampers " Vol.19 Proc.10th World Conference on Earthquakes Engineering., Madrid, Spain.

[4] K.C.Chang. T.T. Soong, S.T. Oh, M Lai and E.J, Nielsen. " Development of a Design Procedure for structures with Added Viscoelastic Dampers.

Damage Prediction in
Reinforced Concrete Structures
under Cyclic Loading

B. Garstka[1], W.B. Krätzig[2], F. Stangenberg[3]

Abstract

Assessment of the stiffness, resistance and damage behaviour of reinforced concrete members under high cyclic loading due to earthquake actions
must consider material nonlinear effects very accurately.

In the presented paper hysteretic models after the lamina method
for flexure and axial force and after the truss analogy for shear have been
derived based on uniaxial material laws for concrete and steel. These models
have then been implemented in a physically nonlinear reinforced concrete
beam element numerically described by the Finite–Element–Method. The
theoretical models were calibrated with experimental data obtained by own
large–scale tests and from the literature.

The damage indicator D_Q for reinforced concrete sections has been
evaluated and verified by test results in the course of the "Special Research
Centre of Structural Dynamics" at Bochum (SFB 151). It allows a meaningful
quantitative damage assessment under both monotonic and cyclic loads
between 0% and 100% (virgin state to collapse). This damage indicator is
implemented in the above described nonlinear beam element to predict
damage evolution in cyclically loaded reinforced concrete members.

Introduction

In modern design of structures under extreme loads (e.g. earthquake)
a certain amount of inelastic damage may well be accepted as long as
overall structural integrity is maintained. Seismically excited structures
should withstand excessive local accumulation of inelastic damage or dete-

[1]Dr. of Civil Engineering, Hochtief AG, Ingenieurbau und Konstruktions-
software Essen (IKSE), Rellinghauser Str. 53–57, D-45128 Essen
[2]Prof. Dr.-Ing., Institute for Statics and Dynamics, Ruhr–University Bochum,
D-44780 Bochum
[3]Prof. Dr.-Ing., Institute for Reinforced and Prestressed Concrete, Ruhr–
University Bochum, D-44780 Bochum

rioration of their load–carrying capacity in the sense of low–cycle fatigue without collapse.

This design philosophy requires new physical values to quantify damage in structures and more precise material models for reinforced concrete under cyclic loading. The ductility balance method is an analysis method for assessing a structural system considering its local weak zones and so it can serve as a valuable tool in the design process, if assessment of planned, existing or damaged buildings is desired. The principle is simply a comparison of local nonlinear damage with the available ductility reserves of the system.

Rational prediction of the non–linear cyclic behaviour of reinforced concrete members is complicated by the variety and complexity of the physical phenomena involved. But modern computer technology and the improvement of numerical calculation algorithms offer a good platform to develop better material models for describing the stiffness, resistance and damage evolution of reinforced concrete members.

A special problem is the unfavorable behaviour of short columns. The study of buildings after earthquakes in the last two decades shows that a lot of damage is caused by the failure of short columns. These short columns fail in a brittle manner, because of a combination of axial and shear forces. The shear capacity is less than the flexural capacity. A lot of models for the flexural behaviour of reinforced concrete have been developed, but usually the influence of shear is taken into account approximately. In this presentation, modelling of axial force, flexure and shear is carried out on the same level of precise approximation.

Cracked Reinforced Concrete Beam Element Model

To describe nonlinear behaviour of reinforced concrete up to collapse, many numerical models on different approximation levels have been developed. In Fig. 1 the relationships between the levels of approximation in the course of a finite–element procedure are illustrated. Following this concept a finite element, called BREL2 (beam crack element; version 2 with shear modelling), has been developed for plastic regions in linear reinforced concrete elements under high cyclic loading.

The basis of this element model is the assumption that cross sections with primary flexural cracks are known. As the observation of tests establishes, cracks firstly occur in cross sections which are weakened by stirrups. *In* these known cracked sections the hysteretical behaviour of reinforced concrete can be predicted by cross sectional models. *Between* two cracks, models for tension stiffening can be implemented. As shown by experimental evidence in Bochum rebar strains between two cracks are well correlated with strains in the adjacent crack sections. Mathematical models of this dependance can be applied, but empirical approximations of rebar strain distributions can also furnish the desired information about bond slip, stress reduction and thus stiffness variation. So the tangential stiffness coefficients can be derived in the element.

To introduce this mechanical concept in a numerical model, a physically nonlinear algorithm has been developed on the basis of the finite element displacement method. Central point of this model is the determination of internal forces, which will be an indicator for the validity of the whole model.

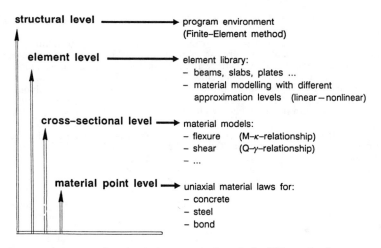

Fig. 1: Approximation levels in structural analysis (FE–method)

Internal Forces

● Flexural and Axial Force Modelling:

Procedures which are based on rational models rather than empirical equations enable the engineer to develop a better understanding of actual structural behaviour. A widespread and very clear assumption is the "plane section theory" for flexure and axial load as the basis for the lamina model (Fig. 2). Since it is assumed that plane sections remain plane, only two variables (e.g. concrete strain in section middle and curvature) are required to define the longitudinal strain distribution over the cross section. Knowing this strain distribution, the cross section may be divided into any number of laminas, in which it is valid to use uniaxial material laws for concrete, steel and for bond. Thus the corresponding stress distribution of the exa-

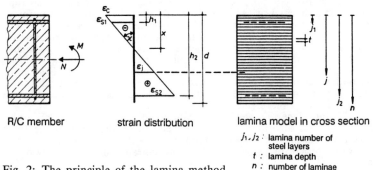

R/C member   strain distribution   lamina model in cross section

j_1, j_2 : lamina number of
          steel layers
t : lamina depth
n : number of laminae

Fig. 2: The principle of the lamina method

mined cross section can be determined and the inner forces — moment and axial force — follow by integration.

- Shear Modelling:
  On the course of looking for an adequate numerical description of linear reinforced concrete elements under shear, the "compression field theory" as the basis for the truss model from Collins and Mitchell (1980) has been transformed for the displacement method and extended for the case of hysteretic loading. With compatibility and equilibrium conditions in the assumed truss the hysteretical behaviour of the reinforced concrete member under shear can be simulated by using the same uniaxial concrete and steel material law for the struts in the truss as in the lamina model (Fig. 3).

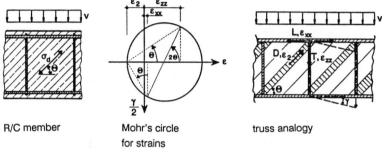

R/C member   Mohr's circle   truss analogy
                          for strains

Fig. 3: The principle of the truss method

Central parameter in this model is the angle of inclination of diagonal compression θ, which is equivalent to the direction of the diagonal strut in the truss according to the assumption of the "compression field theory". With the help of an algorithm of iteration to get this angle the geometry of the assumed truss is determined and shear force and shear rigidity is evaluated. This model may easy be implemented in a numerical procedure.

The solution begins by assuming a value of θ. Knowing θ, the tensile strains in the longitudinal and transverse steel and the diagonal compressive strain in the concrete strut can be determined from the following truss kinematics relationships:

$$\varepsilon_d = \left(\frac{\gamma_{xz}}{2} - \tan \gamma_{xz} \right) \cdot \tan \theta \tag{1}$$

$$\varepsilon_{xx} = \left(\gamma_{xz} - \tan \gamma_{xz} \right) \cdot \tan \theta \tag{2}$$

$$\varepsilon_{zz} = \frac{\dfrac{\gamma_{xz}}{2} - \tan \gamma_{xz} \cdot \sin^2 \theta}{\sin \theta \cdot \cos \theta} \tag{3}$$

with: ε_{xx} : strain in longitudinal reinforcement,
         ε_{zz} : strain in stirrups,
         ε_d $= \varepsilon_2$: strain in main compressive direction
                          respectively in the diagonal strut,

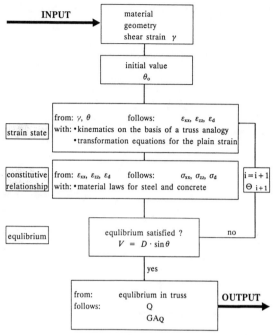

Fig. 4: Algorithm for determinig the Q–γ–relationship by the truss analogy (displacement method)

γ_{xz} : shear strain *(known from FE–computation)*,
θ : angle of inclination of diagonal compression.

Knowing the stress–strain characteristics of the reinforcement and of the diagonal concrete strut, stresses in the truss can be determined. The calculated values of forces then can be used to check the initial assumption of θ by an equilibrium condition evaluated for the assumed truss. If the equilibrium condition is satisfied, then the solution would be correct. If it does not agree, then a new estimate of θ could be made and the procedure repeated.

● Interaction Effects:
The behaviour of linear reinforced concrete elements depends on an interaction between bending moment, axial force and shear force values. It has been demonstrated, that the ordinary uncoupling of flexural / axial force effects and shear force effects holds true for relatively high values of the "shear ratio":

$$\alpha_s = \frac{M}{Q \cdot h} := \frac{a}{h} \qquad (4)$$

with: α_s :   shear span ratio,
          Q :   shear force,
          a :   shear span and
          h :   effective depth.

On the other hand, for low a/h values this uncoupling is no longer valid and ultimate moment M_u values calculated on the basis of the capacity of the ordinary flexural theory overestimate the bearing capacity of the reinforced concrete element. Using the described cross–sectional models for flexure and shear, interactions between the computed inner forces have to be considered especially for lower α_s values (see Garstka 1993).

• Modelling of Uniaxial Material Laws:
          An important aspect of the successful use of the above described models on cross sectional level is the quality of the uniaxial material laws. In the course of the Research Project A4 of the "SFB 151", very detailed uniaxial material laws for steel and concrete under cyclic excitation have been derived (Meyer 1988, Garstka 1993).
          In Fig. 5a the hysteretic behaviour of steel is shown. The primary curve is divided into three branches (elastic – yielding – hardening) and is valid for the tensile and the compressive zone. Unloading is given by a linear relationship. Reloading is given by an estimate from Ramberg/Osgood

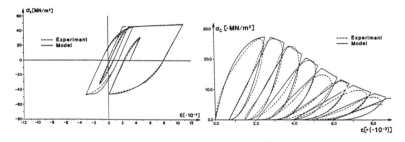

Fig. 5: Uniaxial material law for steel and concrete under cyclic excitation

for Bauschinger–effects and from Ma/Bertero/Popov for cyclic strain–hardening.
          Fig. 5b illustrates the cyclic behaviour of concrete. Three further effects are considered:
– contact effects by closing of cracks during cyclic loading,
– influence of reinforcement ("confinement") and
– characteristics of diagonally cracked concrete in the truss model.

Damage Model D_Q

The damage indicator D_Q is based solely on absorbed energy, with maximum energy E_u under monotonic loading up to failure as a normalizing factor (Fig. 6). The energy absorbed is divided in two parts, E_{si} and E_i. Their physical meaning is described below, employing the concepts of "primary half–cycle" (PHZ) and "follow half–cycle" (FHZ). After Ötes (1985) "primary half–cycle" is the name for any half–cycle with maximum amplitude, followed by a certain number of "follower half–cycles" of smaller amplitude. Whenever a certain deformation maximum v_i, corresponding to the primary half–cycle PHZ_i is exceeded, a new primary half–cycle PHZ_{i+1} is established. Every PHZ corresponds to a certain damage degree.

The first energy part, E_{si}, belongs to the energy absorbed during PHZ_i, the second, E_i, to the energy absorbed during FHZ_i. Mathematically D_Q can be expressed with the help of the intermediate variables D_Q^+ and D_Q^- as it is shown in Fig. 6. Properties of this damage model can be summarized in the following way:

- The energy absorbed during monotonic loading up to failure (corresponding to a simple primary half–cycle) has to be furnished by a certain number of primary half–cycles in the case of cyclic loading.
- The energy absorbed by "following half–cycles" is reflected in a relatively small increase of the damage indicator.

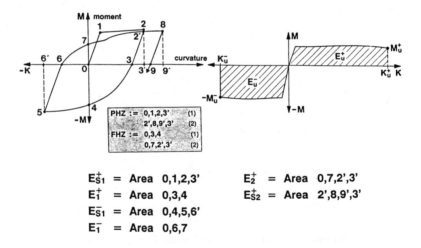

$$E_{s1}^+ = \text{Area } 0,1,2,3'$$
$$E_1^+ = \text{Area } 0,3,4$$
$$E_{s1}^- = \text{Area } 0,4,5,6'$$
$$E_1^- = \text{Area } 0,6,7$$

$$E_2^+ = \text{Area } 0,7,2',3'$$
$$E_{s2}^+ = \text{Area } 2',8,9',3'$$

Fig. 6: The D_Q damage indicator

Because of its energy based equation, D_Q doesn't depend on the level of structure (e.g. lamina–, cross–section– or storey–level). But for considering interaction effects between combined loading by flexure and shear, a failure condition has been derived by evaluating data gained by tests on short columns (Fig. 7). With the help of this interaction relationship failure under combined loading can be predicted (see also Garstka 1993).

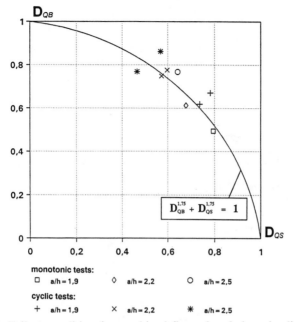

Fig. 7: Failure condition for combined flexural and shear loading on short columns

<u>Large-Scale Tests</u>

During the last 10 years several large–scale tests on reinforced concrete members were carried out by the Research Project A4 of the "SFB 151". An exemplary designed building was divided into simple reinforced concrete members, which were examined carefully. The experimental program included tests of reinforced concrete beams, columns and joints, which were subjected to a simulated earthquake loading history. The results of these tests are the basis of modelling the hysteretic behaviour of reinforced concrete members under high cyclic loading.

In 1992 short columns with different values α_s have been tested monotonically and cyclically. A specimen with $\alpha_s = 1,9$ is shown in Fig. 8. Horizontal load was applied by a 1000 KN–cylinder and was equivalent to a linearly increasing, sinusoidal force excitation.

For the analysis by the above described reinforced concrete beam element under high cyclic loading, this structure has been discretized from stirrup to stirrup. Cross sections have been divided into 9 laminas for the application of the lamina model. The comparison between experiment and model is represented in Fig. 9 and proves the good quality of the described material models.

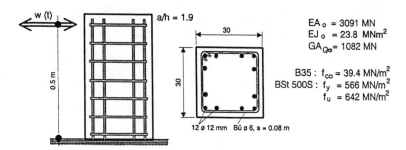

Fig. 8: Material and geometry of short columns [Bochum 1992]

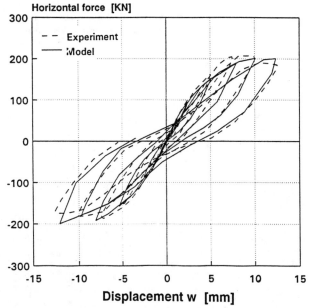

Fig. 9: Horizontal force–displacement relationship of short R/C columns
with a/h = 1,9

Acknowledgement

    This research project has been financed by the DFG (German Research Foundation) in the course of the "Special Research Centre of Structural Dynamics" at Bochum (SFB 151); this support is gratefully acknowledged by the authors.

References

CEB Bulletin d'Information N° 161, 1983. Response of Structural Concrete Critical Regions under Large Amplitude Reversed Actions.

Chung, Y.S., Meyer, C., Shinozuka, M. 1987. Seismic Damage Assessment of Reinforced Concrete Members. National Center for Earthquake Engineering Research, State University of New York at Buffalo, Technical Report NCEER-87-0022

Collins, M.P., Mitchell, D. 1980. Shear and Torsion Design of Prestressed and Non–Prestressed Concrete Beams. Journal of Prestressed Concrete Institute (PCI Journal), Band 25 (5), Chikago, 32–100

Eibl, J., Keintzel, E., Charlier, H. 1988. Dynamische Probleme im Stahlbetonbau. Teil II: Stahlbetonbauteile und –bauwerke unter dynamischer Beanspruchung. Deutscher Ausschuß für Stahlbeton, Heft 392

Garstka, B., Krätzig, W.B., Stangenberg, F. 1993. Damage Assessment in Cyclically Loaded Reinforced Concrete Members. 2nd European Conference on Structural Dynamics (EURODYN '93), Trondheim/Norway, A.A. Balkema, Rotterdam (NL), Vol. I, page 121–128

Garstka, B. 1993. Untersuchungen zum Trag– und Schädigungsverhalten stabförmiger Stahlbetonbauteile mit Berücksichtigung des Schubeinflusses bei zyklischer nichtlinearer Beanspruchung – Investigations on Resistance and Damage Behaviour of Reinforced Concrete Linear Elements Considering Shear Effects under Cyclic Nonlinear Loading. Ruhr–Universität Bochum (Dissertation). Technisch–wissenschaftliche Mitteilungen des Instituts für Konstruktiven Ingenieurbau, Nr. 93–2

Hanskötter, U. 1993. Strategien zur Minimierung des numerischen Aufwands von Schädigungsanalysen seismisch erregter, räumlicher Hochbaukonstruktionen mit gemischten Aussteifungssystemen aus Stahlbeton. (Dissertation) Ruhr–Universität Bochum

Ma, S.M., Bertero, V.V., Popov, E.P. 1976. Experimental and Analytical Studies on Hysteretic Behaviour of Reinforced Concrete Rectangular and T–Beams. Earthquake Engineering Research Center, University of California, Berkeley: Report EERC 76–2, PB–260 843

Meyer, I.F. 1988. Ein werkstoffgerechtes Schädigungsmodell und Stababschnittselement für Stahlbeton unter zyklischer nichtlinearer Beanspruchung. Technisch–wissenschaftliche Mitteilungen des Instituts für Konstruktiven Ingenieurbau, Nr. 88–4, Ruhr–Universität Bochum

Ötes, A. 1985. Zur werkstoffgerechten Berechnung der Erdbebenbeanspruchung in Stahlbetontragwerken. Mitteilungen aus dem Institut für Massivbau der TH Darmstadt, Heft 25, Ernst & Sohn Verlag für Architektur und technische Wissenschaften, Berlin

Park, R., Kent, D.C., Sampson, R.A. 1972. Reinforced Concrete Members with Cyclic Loading. Journal of the Structural Division, ASCE, Vol. 98, No. ST7, p. 1341–1359

Blast demolition of reinforced concrete
industrial chimneys

Friedhelm Stangenberg[1]

Abstract

On some of recently demolished chimneys, transient
measurements were taken, in the vicinity of the tilting
hinge, during the process of falling down. Thus, corresponding dynamic computational models could be verified
and adapted.

Computational result variations due to input variations are studied. The analytical models, on the combined basis consisting of three different branches:
1. full scale in-situ tests, 2. smaller scale laboratory
detail tests, 3. computational models verified on both,
are a well proved tool for solving practical problems
and for explaining the occurring phenomena. The investigations are aiming at an artificial intelligence system,
in future to be used as expert system for preparing
chimney blast demolition based on the available knowledge in this field and using computer simulations.

Introduction

This contribution does not deal with traditional
structural safety problems, rather than with structural
collapse and removal (fig. 1). Not construction rather
than destruction, in a controlled way, is demanded here.

---

[1]Prof. Dr.-Ing., Institute for Reinforced and Prestressed Concrete, Ruhr-University Bochum, D-44780
Bochum

Fig. 1 Blast collapse of a chimney

The demolition of industrial chimneys becomes necessary when the concerning facilities are no more up to
date or no more in acceptable condition. The standards
are set by intensified environmental directions.

Formerly, chimneys had to smoke, visibly proving
wealth, but, nowadays, smoke must be extremely clean,
invisible, unsmelling, colourless and without poisonous
or polluting contents, as far as possible.

Consequently, lots of old bad chimneys have to
disappear and to be replaced by new, modern, efficient,
non-polluting chimneys.

Applying blast technics for demolition is more
economic than cutting piece by piece, as far as the
blast demolition is kept controlled within tolerable limits with acceptable rest risk.

Analytical studies and experimental validations

By blast demolition, the chimney is aimed to fall
into a longitudinal horizontal position (fig. 2). For
reducing the induced soil vibrations, generally an earth
bedding dam is prepared. This is comparable to wood cutting where notches are used.

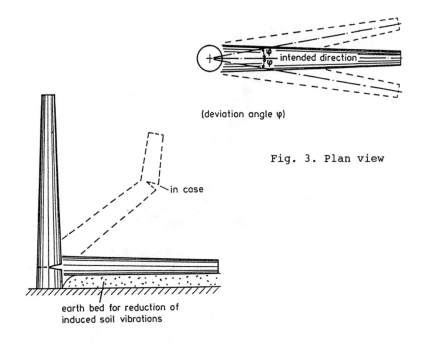

(deviation angle φ)

Fig. 3. Plan view

in case

earth bed for reduction of
induced soil vibrations

Fig. 2. Chimney layed down
by a locally blasted notch

The main blast notch - called blast mouth - is generated on that chimney side where the intended falling
direction is. Casually, a partial incision is placed on
the opposite side (backward incision) e.g. as deep as to
cut the outer vertical steel reinforcement.

While a wood cutter can gradually approach to the
critical state of beginning instability and can escape
in time from the arising dangerous situation, the blast
mouth comes up abruptly and therefore must be in the
right form and sufficiently exact, at the first attempt.
Additional failure hinges can be welcome or even wanted,
as far as pieces do not drift away.

The right form and the right details were realized, when the final results are okay: the intended collapse mechanism must have been realized, and the chimney
wreck must have the intended final horizontal position
without spreading wreck parts beyond the target area.
Such deviations must not occur by drifting apart due to

centrifugal forces in longitudinal direction, as well as
by exceeding the angular tolerances of the aimed position on the ground (fig. 3).

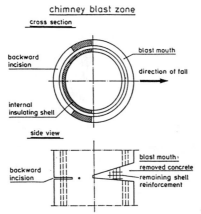

Fig. 4. Details of the results to be obtained by blast

Areas in the vicinity which have to be protected
must not be affected. Special emphasis must be given to
the detail design of the notch to be blasted (form,
height, circumferential section, etc., fig. 4) and to
the backward incision detailing, as well as to the determination of the result sensitivity to be expected;
the blast procedure also has its own tolerances. Before
a decision to use the blast technics, the realizable accuracy must be clarified.

If the blasted mouth turns out to be too small and
the remaining concrete cross section, thus, is too large
(fig. 5), a sudden collapse is risked not to be enforced. Then any imponderables influence when and where the
chimney falls; perhaps a slight gust then is responsible
for an uncontrolled failure. In such an instable situation, subsequent corrections are highly dangerous. The
blast operators avoid these situations (psychologically
understandable!) by dosing the explosives rather too
high than too low.

Thus, the opposite danger can be provoked: a too
large blast mouth is generated, leading to a compressive
failure of the rest cross section, initiating an over-
pronounced vertical motion of the total shaft, before
the tilting motion is started. This also can lead to an

uncontrolled wrecking and uncontrolled tilting direction.

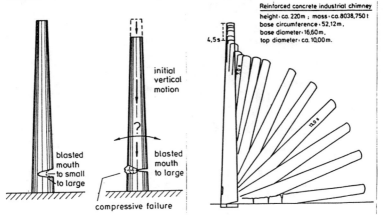

Reinforced concrete industrial chimney
height·ca. 220m ; mass·ca.8038,750 t
base circumference·52,12m.
base diameter·16,60m.
top diameter·ca.10,00m.

initial
vertical
motion

?

blasted
mouth
to small
to large

blasted
mouth
to large

compressive failure

Fig. 5. Extreme cases of   Fig. 6. Collapse phases of
size of blasted mouth   a chimney after blasting

Fig. 6 refers to such an example of a too large
blast mouth where, nevertheless, the failure mechanism
fortunately turned out as planned; just by chance, not
forced. The first 4.5 seconds, here, show the typical
vertical initial motion, due to the too large blast
mouth. But afterwards - in spite of this uncontrolled
initial phase - the falling chimney takes the direction,
almost as planned.

Fig. 6 shows several phases of a falling chimney.

On some of recently demolished chimneys, transient
measurements were taken, in the vicinity of the tilting
hinge, during the process of falling down. Thus, knowledge about the occurring dynamic strains in this region
was obtained and corresponding computational models
could be verified and adapted. Realistic analyses have
to start from the present material and geometric data,
imperfections, tolerances, in case: uncertainties etc. -
all that in such a statistic form that statistically
formulated quantities of deviations can be related to.

The stress resultants M (moment), N (normal
force), H (horizontal force), in the tilting hinge, are
of special interest (fig. 7). H results from the rotational acceleration and the lateral acceleration of the
total mass of the tilting chimney. In the blast hinge, a
reinforced concrete problem of a combined dynamic ben-

ding-compression-shear failure with the consequent problem of decreasing rest resistance exists.

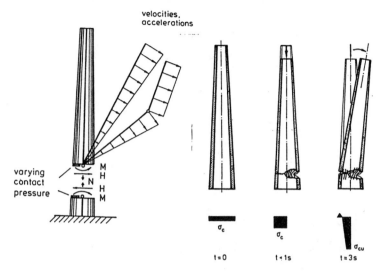

Fig. 7. Dynamic forces in the   Fig. 7a. Collapse phases
blast hinge of the collapsing   of a concrete chimney
chimney   under bore holes
                                 blasting

The nonlinear dynamic analysis using 3-dimensional
finite elements focusses the local stress and strain
concentrations, in the hinge vicinity. Fig. 7a gives the
concrete stress distributions in the hinge section, in
an idealized presentation, in order to illustrate the
typical changes during the first seconds of the collapse
process.

Without going into analytical details, some computational results are discussed. Result variations due to
input variations, within the statistic frame mentioned
above, can be studied. Thus, a solid basis can be generated, from which recommendations, e.g. also for special
measures for influencing the collapse mechanism, can be
derived for improving the probability of success.

Fig. 8 refers to in-situ measurements. Above the
blast mouth, the initial vertical motion can be noticed,
and, on the opposite side, the initial uplifting effect.
2 seconds later, the downward motions begin, according
to the measured results, and in conformity with the ana-

lyses. The inbetween curves refer to intermediate points.

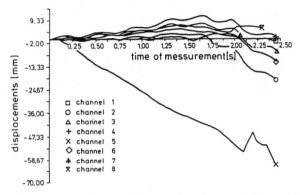

Chimney of HEW power plant Hamburg - Neuhof 20.July 1991, rel.vertical displacements in the tilting hinge

Fig. 8. In-situ measurements during the collapse

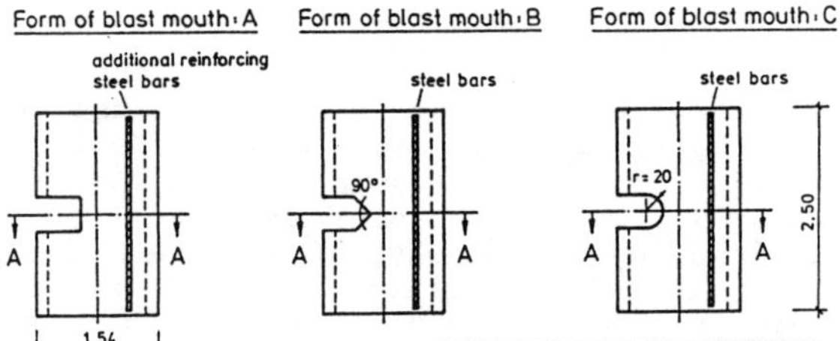

Fig. 9. Laboratory test program

Fig. 9 refers to laboratory tests for investigating the dynamic response details in the blast mouth vicinity, also with the intent to develop and improve computational models. The specimen are tubular concrete sections, 2.5 m high and some 1.5 m in diameter. The reinforcement and the form of the anticipated open

mouths are varied. The testing machines have a limited
velocity due to the high load level (corresponding to
the vertical chimney dead load), and therefore it is
difficult to simulate the highly dynamic change effect
of the structural behavior due to the blast mouth generated abruptly in reality. For some of the tests the
static preload situation was realized by a supporting
element to bridge the gap of the preformed mouth. Thus,
the static equilibrium under vertical dead load was the
initial state. By suddenly removing the supporting element, the dynamic fracture mechanism is started. This
turned out to be an effective test improvement.

It must be remarked that the horizontal (shear)
force due to inertia effects as mentioned above cannot
be included in these laboratory tests (restriction to
bending-compression failure without shear component).
Nevertheless, these experimental laboratory results are
very useful, particularly in combination with additional
parallel computational studies bridging the discrepancy
between smaller scale laboratory tests and the full
scale, shear force including, reality.

Tube 1, right side strains, vertical

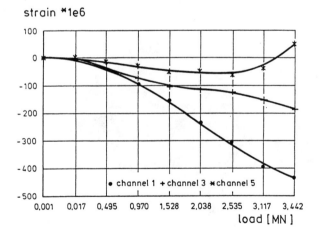

Fig. 10. Load dependent measured strains

The coincidence between laboratory tests and corresponding computational analyses of the same tests is
excellent, fig. 10. Otherwise, there would not be enough

confidence to the analytical models, which, in this verified form, are applied for extrapolation on full scale
reality problems of collapsing chimneys. The combined
basis consisting of three different branches:

1. full scala in-situ tests,
2. smaller scale laboratory detail tests,
3. computational models verified on both,

is a well proved tool for solving practical problems
and for explaining the occurring phenomena.

    Fig. 11 presents further results - here some time
history phases of special computational stress distributions. Because of 2 symmetric planes (both through the
centre of the blast mouth), one quarter of the test specimen is sufficient for calculating results. The blast
mouth vicinity has high strain concentrations.

    Finally, some remarks should be added concerning
the general frame where these studies belong. Just the
reinforced concrete part was dealt with. There exists
another part dealing with the informatic point of view.
Both are aiming at an artificial intelligence system, in
future to be used as expert system for preparing chimney
blast demolition based on the available knowledge on
this field and using computer simulations.

Fig. 11. Computational studies on the laboratory tests

Conclusions

By means of the above described tools, the risks for future chimney blast demolitions can be

- lowered
- estimated more reliably,
- and their cost accounting can be predicted more realistic.

References

Freund, H.U. 1992. Dynamic forces acting on material and structure during concrete blasting, First Inter-national Concrete Blasting Conference, Copenhagen

Lee, E.H. 1952. Symonds, P.S. & Providence R.I Large plastic deformations of beams under transverse impact, Journal of Applied Mechanics, pp 308-314

Melzer, R., Blum, R., Hartmann, D., Stangenberg, F. 1991 Expertensystem für den Sprengabbruch von Stahlbeton-Industrieschornsteinen, 25. DAfStb-Forschungskolloquium, Ruhr-University, Bochum

Stangenberg, F. 1992. Sprengabbruch von Stahlbeton-Industrieschornsteinen, SFB 151, Berichte Nr. 23, Ruhr-University Bochum

Sarma, B.S., Stangenberg, F., Melzer, R. 1993. Analysis of reinforced concrete chimney structures being demolished by local blast, 2nd Eurodyn, Trondheim, Norway

CONTINUOUS PULSE CONTROL OF STRUCTURES WITH MATERIAL NONLINEARITY

CHRIS P. PANTELIDES*
Dept. of Civil Engineering, University of Utah, Salt Lake City, Utah 84112, U.S.A.

AND

PAUL A. NELSON+
Reaveley Engineers & Associates, Inc., Salt Lake City, Utah 84104, U.S.A.

ABSTRACT

A control method is presented for reducing the dynamic response of structures in the inelastic material range using a control force from an active control system. The proposed method of continuous pulse control uses closed-loop feedback control as a combination of two algorithms. The first is the instantaneous optimal algorithm based on linear material behavior, and the second is pulse control which applies a corrective pulse when a prespecified structural displacement threshold is exceeded. Results of the analysis are compared to both a nonoptimal pulse algorithm and an instantaneous nonlinear optimal algorithm. Comparisons between continuous pulse and nonoptimal pulse control for seismic structures in the inelastic range show that the continuous pulse control uses less control energy and reduces the response better than the nonoptimal pulse control. Comparisons between the continuous pulse and instantaneous nonlinear optimal algorithm show that the continuous pulse uses a larger control force but is more effective than the optimal algorithm, in the sense that it can reduce the response of a given structure to any probable earthquake. The optimal algorithm is more effective than the continuous pulse for a single specific earthquake but is not as effective for other earthquakes which may occur in the life of the structure.

INTRODUCTION

Passive, semi-active, and active control devices are placed in civil structures to improve their performance under dynamic loading conditions. Usually, a structural system is designed to deform inelastically during severe loading conditions in order to dissipate the earthquake input energy more efficiently and thus be designed for lower seismic forces. The use of energy absorbing devices, such as active or passive control, can reduce the response of a structural system designed to deform in the inelastic range. In certain cases, use of an active system may reduce the response considerably to a point where

---

*Assistant Professor
+Project Engineer

300

inelastic response may not occur at all. In addition, active control devices can respond to any type of loading at any range of frequencies. In contrast, some passive control devices, such as viscoelastic dampers, are frequency dependent.

Three control algorithms are compared which control a structure in the inelastic material range. The first algorithm is a nonoptimal algorithm that applies a corrective pulse. This algorithm has been studied by Reinhorn et al. [1]. Early efforts on the application of active pulse control to civil structures, including experimental results, were reported by Masri et al. [2], [3]. The second algorithm is the instantaneous closed-loop nonlinear optimal control which has been studied by Yang et al. [4]. The third algorithm is the proposed continuous pulse algorithm which combines an instantaneous closed-loop algorithm based on linear behavior with a pulse control which applies a corrective pulse when a prespecified structural displacement threshold is exceeded. The active bracing system [5] is used in all of the following, although the algorithms can be used for any active control device, such as an active mass damper with some modifications. The active bracing system considered in this paper is shown in Fig. 1. It consists of a sensor, computer, servovalve, actuator, and a conventional steel brace member. Full-scale experimental studies of the active bracing system in Japan have proven its effectiveness during recent moderate earthquakes [6].

EXISTING CONTROL ALGORITHMS
Pulse Control

The pulse control algorithm is a nonoptimal algorithm that applies a corrective pulse when an established threshold of structural response has been surpassed [1]. The pulse control algorithm employs the linear acceleration method to solve the equation of motion. The secant method is used to model the inelastic behavior of the system [7]. The equation of motion for a multiple-degree-of-freedom (MDOF) system can be written as:

$$[M]\{\ddot{y}(t)\} + [C]\{\dot{y}(t)\} + [K]\{y(t)\} = [D]\{u(t)\} + \{F(t)\} \tag{1}$$

where [M], [C], and [K] = mass, damping, and stiffness matrix, respectively; $\{y(t)\}$ = displacement vector; [D] = control force location matrix; $\{u(t)\}$ = control forces vector; and $\{F(t)\}$ = externally applied forces vector due to earthquake acceleration. An overdot denotes time derivative.

In order to evaluate the displacement, velocity and acceleration at each time interval, the linear acceleration method has been used [1], to implement the following algorithm:

$$\{y_{i+1}\} = \{y_i\} + \{\dot{y}_i\}\Delta t + \frac{1}{2}\{\ddot{y}_i\}(\Delta t)^2 + \frac{1}{6}\{\Delta\ddot{y}_i\}(\Delta t)^2 \tag{2a}$$

$$\{\dot{y}_{i+1}\} = \{\dot{y}_i\} + \{\ddot{y}_i\}\Delta t + \frac{1}{2}\{\Delta\ddot{y}_i\}\Delta t \tag{2b}$$

$$\{\ddot{y}_{i+1}\} = \{\ddot{y}_i\} + \frac{6}{(\Delta t)^2}\{\Delta y_i\} - \left(\frac{6}{\Delta t}\right)\{\dot{y}_i\} - 3\{\ddot{y}_i\} \tag{2c}$$

When the structural response exceeds a predetermined threshold, a corrective control force, $\{u(t)\}$, is calculated at each instant of time and applied to the system. The corrective pulse force is proportional to the mass of the structure and the difference between the threshold displacement and response displacement. The corrective pulse applies an initial velocity to the structure. This velocity is added directly to the velocity of the system. The pulse force is applied over a time duration which ranges between 10-20 percent of the time increment Δt. The corrective control force is found in reference [1] as

$$\{u(t)\} = \frac{[M](\{y_{lim}\} - \{y_{i+1}\})}{[Q]\Delta tp}$$

(3a)

$$[Q] = \left(\frac{6[M]}{(\Delta t)^2} + \frac{3[C]}{\Delta t} + [K]\right)^{-1} * \left(\frac{6[M]}{\Delta t} + 2[C]\right)$$

(3b)

where Δtp = pulse duration. An upper bound on the control force magnitude is checked at each pulse so that the control force remains below it. If the required control force calculated from Eq. (3) exceeds the prespecified upper bound, then the value of the upper bound force is applied. This satisfies the physical requirements of a realistic actuator.

Nonlinear Optimal Control

Under severe dynamic conditions, a structural system may experience permanent deformations and the material of the structure enters the inelastic range. This type of behavior allows the structure to dissipate more energy and consequently, the designer can design the structure for lower seismic design forces. The nonlinear optimal control algorithm proposed by Yang et al. uses an instantaneous closed-loop control algorithm which was developed based on inelastic material behavior [4]. The algorithm is optimal for controlling structures which exhibit inelastic behavior. The nonlinear optimal control algorithm uses the Wilson-θ method for time integration and the secant method to model the nonlinear stiffness.

The optimal control force $\{u(t)\}$ is found by minimizing the following time-dependent performance index as shown in reference [8]

$$J(t) = \{z(t)\}^T [Q]\{z(t)\} + \{u(t)\}^T [R]\{u(t)\}$$

(4)

where $J(t)$ = instantaneous performance index, $\{z(t)\}$ = state vector, and $[Q]$, $[R]$ = weighting matrices. The state vector is defined as $\{z(t)\}^T = \{\{y(t)\}^T \mid \{y(t)\}^T\}$. The resulting optimal control force which minimizes the performance index of Eq. (4) subject to the constraint of the equation of motion Eq. (1) is shown in reference [4] to be

$$\{u(t)\} = -\theta^{-2}[R]^{-1}[A_2]^T[Q]\{z(t)\}$$

(5)

where $[A_2]$ is a function of $[M]$, $[C]$, $[K]$, $\theta\Delta t$, and $[D]$, and θ is the parameter used in the Wilson-θ method. Note that the matrices $[Q]$ and $[R]$ must be fixed before the control system is operational. Earthquakes are very random in both frequency and magnitude. Therefore, once the peformance of the control system is fixed by choosing certain fixed values of $[Q]$ and $[R]$ for a specific earthquake, the efficiency of the algorithm may not be as good if a completely different earthquake occurs from the one used to fix the $[Q]$ and $[R]$ matrices.

CONTINUOUS PULSE CONTROL

The continuous pulse control algorithm presented in this paper combines an instantaneous closed-loop control algorithm which was developed based on linear material behavior [8], with the pulse control of Eq. (3) when a prespecified structural displacement threshold is exceeded. If the displacement threshold is not exceeded but the structural response is not zero, the instantaneous closed-loop control algorithm is applied. The algorithm is used to obtain the control force which is applied to the structure in either the linear or inelastic material range. In this sense, the algorithm is considered to be suboptimal for controlling structures which exhibit inelastic behavior. The continuous pulse control uses the linear acceleration method for time integration and the secant method to model the nonlinear stiffness.

The control force $\{u(t)\}$ for the instantaneous closed-loop algorithm is found by minimizing the time-dependent performance index of Eq. (4) subject to the constraint of Eq. (1). The resulting optimal control force from minimizing the performance index for linear structures is found in reference [8] as

$$\{u(t)\} = -\left(\frac{\Delta t}{2}\right)[R]^{-1}[B]^{T}[Q]\{z(t)\}$$

(6a)

$$[B] = \left[\frac{0}{[M]^{-1}[D]}\right]$$

(6b)

Note that [Q] and [R] are fixed matrices and hence they must be set at certain fixed values when an active system is installed, before an earthquake event occurs.

The control force of Eq. (6) in the continuous pulse method is replaced by the pulse control given in Eq. (3) when the structural displacement exceeds a predetermined threshold. An upper bound on the control force magnitude is checked at each pulse so that the control force remains below it. If the control force required is larger than the prespecified upper bound, then the value of the upper bound force is applied. In this respect, the control force is continuous for most of the earthquake and pulsed only near a sudden peak of the response. As will be seen in the numerical examples, this helps to reduce the required number of pulses as well as their duration. In addition, it creates an adaptive feature which the nonlinear optimal control algorithm does not possess.

COMPARISON OF PULSE, OPTIMAL CONTROL, AND CONTINUOUS PULSE

The material model implemented for the pulse, optimal control, and continuous pulse algorithms is either an elastoplastic or bilinear model to represent inelastic behavior. Comparisons are made for earthquake excitations in terms of response reduction, control force, and hysteretic behavior.

Example 1: SDOF Elastoplastic Model Under Earthquake Excitation

A structural model of a SDOF system is considered with the following properties: M = 38.5 tons,   C = 2.42 kN sec/m, K = $1.52*10^3$ kN/m. The structure is assumed to behave in an elastoplastic fashion with a yielding limit of 160 kN. The threshold displacement for activating the control force for the pulse and continuous pulse algorithms is |0.045 m|. The upper bound control force for the pulse and continuous pulse algorithms is limited to 113 kN. The weighting factors are chosen as follows. For the optimal algorithm, the R factor in Eq. (5) is defined as $R_o = 4.64 * 10^{-6}$. For the continuous pulse algorithm, the R factor in Eq. (6) is defined as $R_c = 1.00 * 10^{-5}$. The [Q] matrix is the same for the optimal and continuous pulse algorithms and is given as

$$[Q]_c = [Q]_o = \begin{bmatrix} 320.0 & 0.0 \\ 0.0 & 8.2 \end{bmatrix}$$

It should be noted that the pulse control algorithm does not require any weighting matrices for its operation. The structure is subjected to the N-S component of the 1940 El-Centro earthquake, a portion of which is shown in Fig. 2.

Numerical results are presented in order to compare the three algorithms which include displacement, control force, and hysteresis loops. As can be seen from Figure 3, the continuous pulse algorithm reduces the response better than the pulse control (Fig. 3(c)) while using less control force (Fig. 4(b)). However, for practically the same response reduction (Fig. 3(b)), the continuous pulse algorithm requires a larger control force (Fig. 4(a)) than the optimal algorithm. It is also interesting to note that in all three controlled cases the structure remains in the linear material range, with a maximum displacement of 0.049 m for the continuous pulse and optimal algorithms, whereas for the pulse control the maximum displacement is 0.086 m. The uncontrolled structure remains in the inelastic range with a maximum displacement of 0.172 m.

COMPARISON OF OPTIMAL AND CONTINUOUS PULSE CONTROL ALGORITHMS

Certain control algorithms have been developed specifically for controlling structures which are always in the inelastic material range. Yang et al. [4], have developed the nonlinear optimal control algorithm based on a nonlinear equation of motion which minimizes the performance index of Eq. (4). The algorithm assumes that the structural system will behave inelastically and the control force is therefore optimized in the inelastic range. In the present paper, a different philosophical approach is used. Structures in earthquake-prone areas will experience many more moderate earthquakes than severe ones. It is assumed that in their lifetime, most structures will not be loaded with severe loadings. Therefore, it is possible that with an active control system in place, the structure may not deform inelastically. If in a given earthquake the structure remains in the linear range, then the continuous part of the continuous pulse algorithm is optimal [9].

In the likely event that a different earthquake occurs from the specific earthquake that was used to determine the $[Q]_o$ and $[R]_o$ matrices for the optimal algorithm, the structural response reduction may not be as effective. On the other hand, in the case of the continuous pulse control the different earthquake can be handled by the pulse component of the algorithm given by Eq. (3), which responds to any structural response over and above what the continuous component of Eq. (6) can handle. This pulse force is kept within the upper bound limits of the controller.

Example 2: SDOF Under Earthquake Excitation Using Optimal and Continuous Pulse Algorithms

This example presents a comparison of the continuous control developed herein with the nonlinear optimal control algorithm of reference [4]. The structure is assumed to behave inelastically in a bilinear fashion. The structure is modeled as an SDOF system and has the following properties: M = 345.6 tons, C = 54.29 kN· sec/m, linear stiffness = 85.3×10^3 kN/m and bilinear stiffness = 9.7×10^3 kN/m. The yielding limit is 2,046 kN. The threshold displacement for the continuous pulse control is set at |0.024 m|. The upper bound control force for the continuous pusle algorithm is limited to 1,017 kN. The R weighting factor is defined as follows: $R_c = 5 * 10^{-8}$ for the continuous pulse and $R_o = 2.54 * 10^{-8}$ for the optimal control. The [Q] matrix is the same for both algorithms and is shown below

$$[Q] = \begin{bmatrix} 788.5 & 0.00 \\ 0.00 & 3.20 \end{bmatrix}$$

Figures 5-7 show the behavior of the structure subjected to the N-S component of the 1940 El-Centro earthquake, and to two times the N-S component of the 1940 El-Centro earthquake. In this respect, the latter case represents a different, and much stronger earthquake. Note that the [Q], R_c, and R_0 weighting matrices are fixed and remain the same for both the El-Centro and two times El-Centro case.

Figure 5(a) shows that for the El-Centro earthquake the structural response using either the optimal nonlinear algorithm or the continuous pulse control is practically the same for the chosen weighting matrices. However, for the two times El-Centro earthquake the response using the continuous pulse is clearly better than the optimal algorithm as shown in Fig. 5(b). This is to be expected since the $[Q]_0$ and $[R]_0$ matrices used for the two times El-Centro earthquake are the same ones that were chosen for the El-Centro earthquake and hence can not adapt to the new situation.

From Figs. 6(a) and 6(b) it can be seen that the continuous pulse control applies a larger control force than the optimal algorithm for both earthquakes, but it reduces more effectively the response in the two times El-Centro case as was seen in Fig. 5(b). Comparing Figs. 7(a) and 7(b) one can see that the continuous pulse reduces the inelastic response better than the optimal control for the two times El-Centro excitation. The maximum displacement in the latter case is reduced from 0.058 m for the optimal control to 0.043 m for the continuous pulse control. Thus, the continuous pulse control is more adaptable to different earthquake excitations when the [Q] and [R] weighting factors are kept constant.

CONCLUSIONS

The continuous pulse control algorithm has been presented which illustrates the use of active structural control when it is applied to seismic structures in the inelastic range. The method is compared to a nonoptimal pulse control algorithm and a nonlinear optimal control algorithm. The structures in the numerical examples were analyzed in both the linear and inelastic material range, and in the latter case both elastoplastic and bilinear models were used.

Numerical simulations have shown that the continuous pulse algorithm controlled the displacement of the structure more effectively than the pulse control algorithm. In addition, the continuous pulse algorithm controlled the structural response by using fewer pulses of smaller duration than did the pulse control algorithm. Comparisons between the continuous pulse control and the nonlinear optimal algorithm show that even though the continuous pulse control applies larger control forces, it can reduce the response more effectively than the optimal algorithm when the earthquake excitation is not the same as the one used to determine the weighting matrices. Its adaptability makes it a good candidate for active control applications in civil engineering structures.

ACKNOWLEDGEMENT

The writers acknowledge the financial support provided through a Faculty Research Grant by the University of Utah Research Committee.

REFERENCES

1. Reinhorn, A.M., Manolis, G.D., and Wen, C.Y., "Active Control of Inelastic Structures", J. Engineering Mech., ASCE. 113, No. 3, pp 315-333, 1987.

2. Masri, S.F., Bekey, G.A., and Caughey, T.K., "On-Line Control of Nonlinear Flexible Structures", Journal of Applied Mechanics, ASME, Vol. 49, No. 4, pp 871-884, 1981.

3. Masri, S.F., Bekey, G.A., and Caughey, T.K., "Optimum Pulse Control of Flexible Structures", Journal of Applied Mechanics, ASME, Vol. 48, pp 619-626, 1981.

4. Yang, J.N., Long, F.X., and Wong, D., "Optimal Control of Non-Linear Flexible Structures", Report NCEER-88-0002, Jan. 22, 1988.

5. Soong, T.T., Active Structural Control: Theory and Practice, John Wiley & Sons, Inc., New York, 1990.

6. Reinhorn, A.M., Soong, T.T., Lin, R.C., Riley, M.A., Wang, Y.P., Aizawa, S., and Higashino, M., "Active Bracing System: A Full Scale Implementation of Active Control", Report NCEER-92-0020, Aug. 14, 1992.

7. Paz, M., Structural Dynamics: Theory and Computation, Litton Educational Publishing, Inc., New York, 1980.

8. Yang, J.N., Akbarpour, A., and Ghaemmaghami, P., "New Optimal Control Algorithms for Structural Control", J. Engineering Mech., ASCE 113, pp 1369-1386, 1987.

9. Pantelides, C., and Nelson, P., "Suboptimal Linear Control of Structures with Material Nonlinearity", Proceedings 63rd Shock and Vibration Symposium, Vol. 1, SAVIAC, Las Cruses, NM, pp 182-190, 1992.

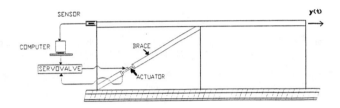

Figure 1. Active Bracing System

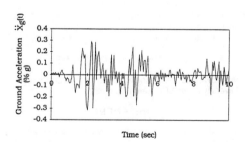

Time (sec)

Figure 2. N-S Component of the 1940 El-Centro Earthquake

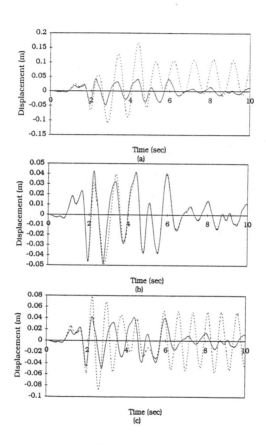

Figure 3. Displacement of Ex. 1 Structure: (a) – – = No Control,
___ = Continuous Pulse, (b) – – = Nonlinear Optimal,
___ = Continuous Pulse, (c) – – = Pulse Control,
___ = Continuous Pulse

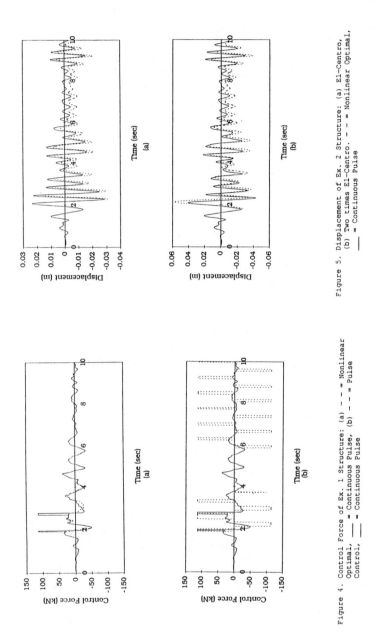

Figure 4. Control Force of Ex. 1 Structure: (a) - - = Nonlinear Optimal, ———— = Continuous Pulse, (b) - - = Pulse Control, ———— = Continuous Pulse

Figure 5. Displacement of Ex. 2 Structure: (a) El-Centro, (b) Two times El-Centro. - - = Nonlinear Optimal, ———— = Continuous Pulse

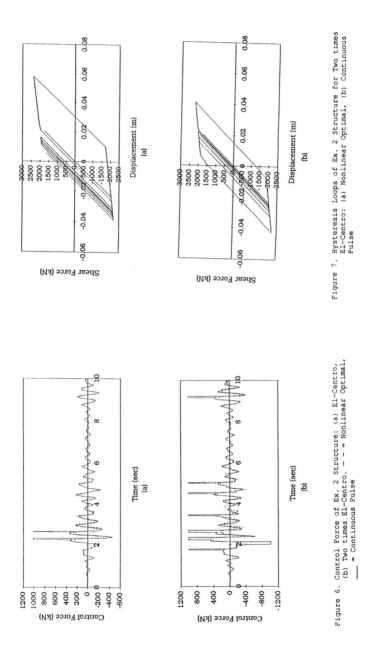

Figure 6. Control Force of Ex. 2 Structure: (a) El-Centro, (b) Two times El-Centro. - - = Nonlinear Optimal, ___ = Continuous Pulse

Figure 7. Hysteresis Loops of Ex. 2 Structure for Two times El-Centro: (a) Nonlinear Optimal, (b) Continuous Pulse

Imperfections and Thin-Walled Bars Behaviour

Professor J.B.Obrębski[1]
Doctor M.E. El-Awadi[2]

Abstract

Experimental and theoretical observations allowed us to present some new conclusions concerning the proper, and safe designing of the structures composed of thin--walled bars (TWB). These structures are very sensitive to small changes of cross-sections (CS) - their dimensions, boundary conditions (BC), material properties etc. . Almost all contemporary structures may be considered (as comprising the elements) of a such type. It makes the effects of this work so important. Below we quoted only some most significant conclusions.

Introduction

The present authors investigated last time an influence of some constructional solutions of thin-walled bars on its behaviour (Obrębski 1986,1988,1990,1991,1993, Obrębski and El-Awadi 1991, Obrębski et al 1992, Obrębski and Urbaniak 1990,1992). It is connected with influences as follows: type of CS, its proportions, bar length - slanderness, boundary conditions, material properties etc. For each of the cases there were observed curves describing of: warping function, moments of inertia, sectorial moment of inertia, critical forces, curves of critical stresses due to axial force, bending moments and

---

[1] Warsaw University of Technology, Institute of Structural Mechanics, Al. Armii Ludowej 16, 00-637 Warsaw, Poland

[2] Zagazig University, Faculty of Engineering, Department of Structural Design, Zagazig, Egypt

bimoments. Among others were also considered characters of proper diagrams, its slope etc. . The authors elaborated the proper lists of CS showing the rank of their strength when applying different criteria of the utility (El-Awadi 1993). Such observations were presented on some international conferences, (Obrębski 1990, Obrębski and El-Awadi 1991, Obrębski at al 1992, Obrębski and Urbaniak 1990, 1992), etc., and were enriched by pointing a method of determination of the CS proportions, BC, bars walls thickness δ, etc., which brings about the negligible sensitivity of the TWBs to the small differences of dimensions (El-Awadi 1993).

Basic Assumptions and Some Relations of Applied Theory

All comparisons were computed applying theory elaborated by Obrębski (1982-1991). It is partially very close to Vlasov's (1940) and Rutecki's (1957) theories. Many examples for particular bar, formulae and describing them proper program for determination of critical forces were performed or derived by El-Awadi (1993). Each time the only one parameter was changed.

□ The considered bars are straight, prismatic, made of elastic material, and may have every type of CS. There were also compared bars which lengths of its CS middle line are identical. In stability problems the length of the bar can vary from zero up to infinity.
□ Comparisons concern the bar with different kinds of CSs: open, closed; and different types e.g. I, T, C, regular polygon, tubular etc., Figure 1a.
□ Two functions describing two deflections in princpal planes and function of rotation of longitudinal bar axis may have independently 3 different BC, Figure 1b-e.

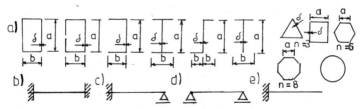

Figure 1. Applied bar CSs and boundary conditions

□ The curve was applied for short bars, for stability problems, for non-linear buckling accordingly to the hypothesis used in Polish Standard 1980. It might be changed by any other.
□ Some reduced geometrical characteristics of CS:
   - moments of inertia $\bar{I}_2 = \oint \eta_3{}^2 \, d\bar{A}$, $\bar{I}_3 = \oint \eta_2{}^2 \, d\bar{A}$

- sectorial moment of inertia $\bar{I}_\omega = \oint \bar{\omega}^2 d\bar{A}$

Here δ is thickness of bar walls. In the manufactured regular space bar structures it may be of importance the designing of typical bars regarding the proportions of the CS, which assure the highest, but identical moments of inertia for both principal axis.

□ Torsional bar rigidity $K_s = \beta\frac{1}{2}\Sigma Gb\delta^3$

□ Torsional equilibrium eqns (here linear) $E\bar{I}_\omega\theta^{IV}-K_s\theta''=m_1$

□ Differential definitions of internal forces:

- longitudinal force $T_1=\bar{E}\bar{A}v_1'$
- shear force $T_2=-\bar{E}\bar{I}_3v_2''', \quad T_3=-\bar{E}\bar{I}_2v_3'''$
- torsional moment $M_1=M_s+\bar{M}_\omega=K_s\theta'-\bar{E}\bar{I}_\omega\theta'''$
- bending moments $M_2=-\bar{E}\bar{I}_2v_3'', \quad M_3=\bar{E}\bar{I}_3v_2''$
- bimoment $B=-\bar{E}\bar{I}_\omega\theta''$
- bending-torsional moment $M_\omega=-\bar{E}\bar{I}_\omega\theta'''$

All above quantities depend on δ, shape of CS and material properties, because $d\bar{A}=E_1\delta ds/\bar{E}$. E_1 & $\bar{E}$ are a modified & general modulus of elasticity (e.g.Obrębski 1991).

□ Stresses:

- normal stress

$$\sigma_1=\frac{E_1}{\bar{E}}\ (m_b^1\frac{T_1}{\bar{A}}-m_b^3\frac{M_3\eta_2}{\bar{I}_3}+m_b^2\frac{M_2\eta_3}{\bar{I}_2}+m_b^4\frac{B\ \omega}{\bar{I}_\omega})\qquad(1)$$

- shear stress

$$\tau=\tau_o+\frac{(T_2+m_3)\ \tilde{S}_3}{\bar{I}_3}+\frac{(T_3-m_2)\ \tilde{S}_2}{\bar{I}_2}-\frac{(M_\omega-b)\ \tilde{S}_\omega}{\bar{I}_\omega}\qquad(2)$$

where m_ω^i are proper buckling coefficients for all internal forces (Obrębski 1990a-e). Two last stresses above depend additionally on the length of the bar (slanderness).

Character of Some Typical Diagrams

- Warping function, Figure 2a.
- Moments of inertia, Figure 2b.
- Sectorial moment of inertia, Figure 2c.
- Critical stresses for longitudinal force, Figure 2d,
  where: BB - bending type buckling, BTB - bending
  torsional buckling, TB - torsional type buckling.
- As above when $\bar{I}_\omega=0$, Figure 2e.
- Critical stresses for bending moments, Figure 2f.
- Critical stresses for bimoments, Figure 2g.
- Critical stresses for different materials, Figure 2h.

Determination of the Bar Sensitivity on Imperfections

When observing the diagrams quoted above, and its slope (derivative), we can conclude, that for some theirs parts small changes of particular parameters may bring about to quite rapid changes of computed quantities e.g. critical forces and then stresses.

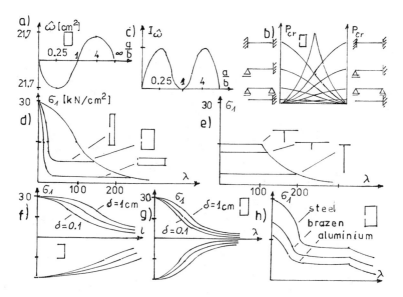

Figure 2. Typical diagrams

We considered two kinds of problems:
□ We assume required tolerance of values counted on ordinate axis and we look which is the influence of the change of parameters given on abscissa axis, e.g. Figure 3a.
□ We assume required tolerance of values counted on abscissa axis, and we look which is admissible change of parameter given on ordinates axis, e.g. Figure 3b.

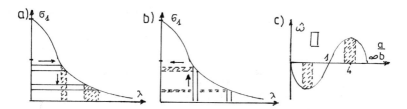

Figure 3. Different cases of CS sensitivity

In both cases above we assume the required accuracy (sensitivity) of the bar performance where e.g. the stresses can vary ±2% from expected value. Than we predict the bar CS proportions for which such sensitivity will exist, Figure 3c, (shadowed area).

Sensitivity of the Bar on CS Dimensions Imperfections

This time the attention is paid especially to the character of curves mentioned above, which allow to foresee, which bar type and its proportions are very sensitive to small imperfections of dimensions occuring at the manufacturing bars. Especially for some types of CSs were investigated:
- proportions which a small alteration gives high or small changes of moments of inertia, and the stresses,
- thickness of bar walls (see e.g. Fig.2f,g),
- the bar length (see e.g. Figure 2f),
- slanderness (see e.g. Figure 2d,e,f,h).

Such sensitivity is very important at automatic dimensioning of large space bar structures (kind of optimisation), e.g. Obrębski 1984,1985,1990a.

Sensitivity on Type of Boundary Conditions

Other problems concern the boundary conditions, which simultaneously can be different for three of the bar axis displacements (two deflections and angle of torsion). The smallest bending-torsional rigidity of the bar (deciding very often about bar capacity, Obrębski and Urbaniak 1988, 1990,1992a,b, or about direction of buckling of investigated bar), strongly depends on the applied type of construc-tional solution. The BC does not change the character of proper critical stresses curve, but lets to use only other its part, e.g. Figure 3d.

Imperfections Obtained by Production

Causes of imperfections, are originated at the sheets production in foundry. In practice the strength and capacity of proper structure elements can significantly be different from the designed in project.There are described below some results, obtained on the basis of the authors' own measurements, done on the series of specimens prepared for the experiments arranged especially by them, Obrębski and Urbaniak 1988-1992. The observations concern:

- Small differences of the dimensions appearing by cold forming of TWBs and differences of material properties, and their influence on bar behaviour. There were coldly formed TWBs with intended dimensions 196x20x10 cm. For these models were drawing up inventory in 5 CSs $\eta=0$, $\eta=1/4$, $1/2$, $31/4$ and 1. In each of them were observed the real length of middle line, which totals L_m were different from intended L_a!, see Table 1.

Table 1

Model No	11	12	13	14
L_m	599.88	618.05	600.04	619.62
L_a	592.0	612.0	592.0	612.0

- Small differences may concern the thickness of metal sheets from which were formed proper bars with determined CS. There were measured the brazen sheets from which were performed 14 examined by TWB models. For 7 sheets declared as $\delta=0.5$mm - real values are shown in the Table 2. Dependently on the particular sheet the deviations ranged from δ_{min} to δ_{max}, Table 2.

Table 2

Model No	1	2	3	4	5	6	8
δ_{min}	0.47	0.47	0.47	0.461	0.50	0.49	0.48
δ_{max}	0.50	0.50	0.50	0.545	0.53	0.54	0.51
δ_{mean}	0.481	0.480	0.477	0.485	0.509	0.505	0.496

More Important Observations

□ All geometrical characteristics of bar CS, in some particular cases, are very sensitive to small imperfection of such parameters as: proportions and dimensions of bar CS, thickness of the walls δ, and material properties - E,G,ν, (Eqns 1-3).

□ For open CS torsional rigidity K_s nonlinearly depend on changes of δ and θ', Obrębski and Urbaniak 1992. Here can be $0.95<\beta<2.5$ for similar bars!

□ Torsional bar rigidity K_s is of strong influence on disposition of torsional forces M_1, M_s, $M_{\hat{\omega}}$, B, along the bar length and at all on its maximum values, e.g. Obrębski 1990,1991. For open CSs when δ increase - quickly are increasing the values of torsional forces. For closed CSs only changes of area of closed circumferences (internal tube area) have impact on K_s and further on torsional forces, Obrębski 1990,1991.

□ The changes of parameters mentioned above and then internal forces, first of all M_s, $M_{\hat{\omega}}$, M_1, B, strongly influence on stresses (1,2).

□ By some unfortunately given BC and bar CS, the warping stresses $\sigma_{\hat{\omega}} = B\hat{\omega}/I_{\hat{\omega}}$ can reach about 50% of normal

stresses from nonsymmetrical bending! (Obrębski 1993). In such case TWB is very sensitive on changes of ω diagram.
□ As a rule the change of δ, CS proportions and material properties, gives new shape of critical curves. They can be significantly different, e.g. Figure 2d.
□ The change of BC has no influence on shape of critical stresses curves, but only on the range of the last and value of critical force. Simultaneously, bar with BC from Figure 1b has the highest critical forces and is the more sensitive on changes of planned BC (e.g. partially flexible). The bar with BC from Figure 1e has the smallest critical forces and the smallest sensitivity.
□ Influence of bar instability should be taken into consideration for all 4 internal forces: T_1, M_2, M_3, and B (bimoment), too! Obrębski 1990a-e,1991.
□ When determining the best CS proportions a/b (the highest values of $\bar{I}_i$ and small sensitivity) is useful diagram of type shown in the Figure 2b. The safe zone is below two curves answering given BCs in two principal axes. The proportions of a/b giving the highest values $\bar{I}_i$, shows e.g. Table 3. The ranges of sensitivity were pointed by Awadi 1993.

Table 3

Internal force	proportions of proper CS a/b						
	rectangular	open rect.	[	I	Z CS	T	poligonal
T_1	1	1.5	1.71	0.55	1.26	0.52	0
M_2 & M_3	1	1	0.73	0.55	1.07	0.71	0

□ Curves for critical stresses for bending moments, with regard to compressing stresses, for nonsymmetrical CS also are nonsymmetrical, Figure 2f.
□ The safe zone always is below of critical curve.
□ Changing of δ especially for higher values, sensitivity of the bar on its bending-torsional capacity strongly increases (Obrębski and Urbaniak 1988-1992), accordingly to 3d power.
□ Stronger material is more useful for the reason of higher strength and critical forces, but also it strongly increases its own weight of structure, Table 4, what does not need further comments. There are shown some comparative multiplyers E/E_{steel} and ρ/ρ_{steel}.
□. Tubular bars with regular poligonal CSs or circular ones, or composed of plane walls connected only in one point, e.g. L, I, [, have $\omega=0$ ($\sigma_\omega=0$ & $\tau_\omega=0$). The unexpected changes of proportion a/b, especially for open CS, may give unexpected warping and then warping stresses, too.

Table 4

Material	E [GPa]	E/E_{steel}	ρ [g/cm^3]	ρ/ρ_{steel}
steel	205	1.0	7.85	1.0
brazen	110	0.536	8.39	1.068
aluminium	70	0.341	2.79	0.355
timber	9	0.044	0.5	0.064

Conclusions

The wide knowledge of TWB behaviour is useful for good choice of constructional solutions, giving high capacity and safety of designed bars. Designing the structures composed of TWBs is - in a fact - a certain kind of the art!

There are recommended following parameters of TWB:
1. Proportions of CS given in Table 3, for large repeatable structure, where bars can be loaded by any type of internal moments and forces. Some of these CS are very sensitive to the changes of CS dimensions.
2. BC fully fixed at both ends, Figure 1b.
3. Closed CS of regular polygonal or the better circular shape.
4. The bars with closed CS have much smaller sectorial stresses than similar one but with open CS.
5. Bars with planary constrained displacements at both ends, even with open CS!
6. The CS with smaller δ are less sensitive to thickness variation.
7. The shorter bars are more sensitive to length differences, but have much higher capacity (strength).

Observations pointed in this paper may be used as basis to optimisation process, for separate TWBs or for space bar structures composed of TWBs, with regard to strength criteria, and parameters such as δ , a/b , bar length etc. . Bar   BC and type of   CS should be rather assumed by designer.

References

El-Awadi   M.E., 1992, Influence of chosen structural parameters on stability and strength of thin-walled bars. Doctor thesis, Warsaw, (supervisor J.B.Obrębski), pp.166.

Obrębski J.B., 1982, Dynamics and stability of perforated thin-walled bars with open cross-section and including in it closed circumferences, (In Polish). 28 Sc.Conf. KILiW PAN i KN PZITB, Krynica, Poland pp.113-120.

Obrębski J.B., 1984, Application of the WDKM Program System to Analysis of Space Structures. 3d Int.Conf. on Space Structures (poster), Guildford, UK.

Obrębski J.B., 1985a, Application of the WDKM Program System to Analysis of Space Structures. Space Structures, v.1. Elsevier Appl. Sc. Puubl. UK,1985, pp.93-98.

Obrębski J.B., 1985b, Numerical algorithm of computat. the structural sets composed of thin-walled bars, (In Polish). 31st Sc.Conf.KILiW PAN & KN PZITB, Krynica, pp.149-154.

Obrębski J.B., 1986a, Influence of buckling on strength of straight thin-walled bars, (In Polish), 2d Sc. Conf. of Facullty of Civil Engin. of Warsaw Univ. of Technology, Warsaw, pp.178-185.

Obrębski J.B.,1986b, Second order and second approximation theory in statics and dynamics of thin-walled straight bars. Proc. 1st Int.Conf. on Lightweight Structures in Architecture, Sydney, pp.81-97.

Obrębski J.B., Urbaniak Z., 1988, Experimental analysis of non-linear effects of bending-torsion loaded thin-walled bars. 18th Yugoslav Cong. of Theoretical and Applied Mechanics. Vrnjacka Banja, pp.141-144.

Obrębski J.B., 1988, Menace of thin-walled structures by defects in the light of theory and experiments, (In Polish). 10th Symp. on Investigations of reasons and protections against damages of civil engineering structures. Szczecin, pp.299-306.

Obrębski J.B., 1989, Second-Order and Second-Approximation Theory in the Statics and Dynamics of Thin-Walled Straight Bars. Thin-Walled Structures, Appl. Sc.Publ., pp.81-97.

Obrębski J.B., 1990a, Numerical dimensioning of large steel space frames.Int.Conf. on Steel Struct. and Space Frames, Singapore, CI-PREMIER.

Obrębski J.B., 1990b, Menace of thin-walled structures by defects. Int. Conf. on Modern Techniques in Construction, Singapore, CI-PREMIER.

Obrębski J.B., 1990c, Selected problems of stress analysis for thin-walled bars. 19th Yugoslav Congress of

Theoretical and Applied Mechanics. Ohrid, Yugoslav.

Obrębski J.B., 1990d, Influence of buckling on strength of straight thin-walled bars. Int. Conf. on Applied Stress Analysis, University of Nottingham, UK, (poster and oral presentation only).

Obrębski J.B. 1990e, Influence of constructional solutions on stress distribution in a bars of thin-walled space structures. Int. IASS Conf., Dresden, Germany, pp.23-32.

Obrębski J.B., El-Awadi M.E., 1991, On the optimal designing of some families of thin-walled bars. Int.IASS Symp., Copenhagen, Denmark, pp.237-244.

Obrębski J.B., 1991, Thin-Walled Elastic Straight Bars, (In Polish). Printed by Publishers of Warsaw Univ. of Techn., (lecture notes), Warsaw ,pp.452.

Obrębski J.B., Urbaniak Z., El-Awadi M.E., 1992, On designing of the thin-walled bars in the light of theory and experiments. Int.IMEKO-GESA Symp. on Risk Minimisation by Experimental Mechanics, Dusseldorf, FRG, pp.109-116.

Obrębski, J.B., and Urbaniak, Z., 1990, Investigations of thin-walled bars behaviour at growing bending-torsional loading. (In Polish). Proceedings, 14th Polish Symposium on Experimental Mechanics of Solid, Jadwisin, pp.201-204.

Obrębski J.B., Urbaniak Z., 1992a, Experimental and numerical investigations of certain class of thin-walled bars. NUMEG'92 (Numerical methods in geomechanics),Prague, CSFR, pp.57-62.

Obrębski J.B., Urbaniak Z., 1992b, Selected results of experimental investigations of certain class of thin--walled bars. IV Int. Konf. on Safety of Bridge Structures, Wroclaw, Poland. pp.395-400.

Obrębski J.B., 1993, Thin-walled failure in the light of experiments. Int. Conf. on Lessons from structural failures. Prague, Czech Republic, pp.135-142.

Rutecki J., 1957, Strength of thin-walled structures, (In Polish). Warsaw, PWN, pp.343.

Vlasov V.Z., 1940, 1959, Thin-walled elastic bars. (In Russian). Moscow, Gosstojizd. 1940 & Gos.Izd. Fiz-Mat. Lit., 1959, pp.368.

Polish standard PN-90/B-03200: Steel structures, design and analysis.

COMPARISON OF TWO DIFFERENT MODELS OF A SHELL ROOF

Oscar A. Andrés [1], Member, IASS, Néstor F. Ortega [2] and
Carlos A. Schiratti [3]

Abstract

The main purpose of this study is to compare a
physical model of a shell roof with a theoretical model
of the same shell roof in order to verify the reliability
of the Homeostatic Model Technique as a tool for the
design of structural shapes. The first model is generated
by the HMT while the second is determined following the
Korda's theory and methodology. The analysis is performed
from the geometrical and mechanical points of view,
comparing shapes and structural behaviours. Diagrams and
a photograph illustrate and synthesize results.

Introduction

This paper is an analysis of the reliability of the
Homeostatic Model Technique (HMT) as a tool for the
generation and determination of structural shapes.

Previous papers of the authors (Andrés 1989, Andrés
and Ortega 1991) have described the HMT method and
explained the advantages of the method from functional,
aesthetic and modelling viewpoints.

1 Prof. Dept. of Engrg., Universidad Nacional del Sur,
Av. Alem 1253, 8000 Bahía Blanca, Argentina.
2 Res. Asst. Dept. of Engrg. Universidad Nacional del
Sur, 8000 Bahía Blanca, Argentina.
3 InPro Engineering Consulting, Belgrano 587, 8000 Bahía
Blanca, Argentina.

In this study attention will be focused on the mechanical behaviour of structures designed using the HMT models. It will be analysed how do they perform and how do they compare with structural shapes designed by a different method.

In order to achieve these purposes, two different models have been generated and analysed:

a) A physical model using HMT
b) A theoretical model according with Korda's theory
   (Korda, 1965)

Both models represent the same structure: a square plan concrete shell roof with side length L = 20 m supported at each corner, having the four edges free and a uniform load applied over the surface.

FEA (Finite Element Analysis) has been used to analyse both models.

Homeostatic model (HMT)

A homeostatic model is a physical model obtained by the simultaneous action of heat and loads over a thermoplastic plate. Details about the method can be found in the above mentioned papers. The characteristics of the HMT analysed model are:

- Material: Acrylic
- Initial side length of the plate: 360.00 mm.
- Thickness: 2.40 mm.
- Load: 0.01 N/cm^2
- Maximum oven temperature: 125°C.

Once the model reached its final equilibrium shape it was taken out from the oven.

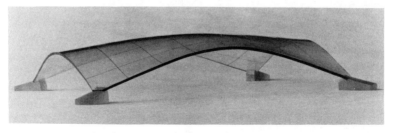

Fig. 1 - Homeostatic model.

The model showed a smooth surface, having no discontinuities (Fig. 1). The shape was measured taking an XYZ coordinate system with a grid of 20x20 points on the XY plane (Fig. 2). Coordinates have been measured with an accuracy of 1/100 mm. Final measured side length was 352.2 mm., thus giving a scale factor of 20000/360 = 56.7666

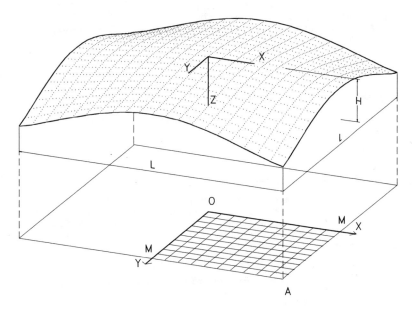

Fig. 2 - Structure, axes and grid.

Coordinates of the real structure were calculated applying this factor to the coordinates previously measured on the model. For example, at Support A:

$$z_{model} = 53.73 \text{ mm, giving a final}$$

$$z_{calc} = 3.05 \text{ m} = H_{str}$$

All coordinates of the mesh were calculated following this procedure.

Fig. 3 and 4 show Z coordinates for vertical sections along planes Y=0, Y=2, Y=5, Y=8, Y=10 and diagonal sections.

Korda's Model

Korda's model is an ideal model; it is theoretically perfect. Following this ideal, the analysed model was calculated for the above mentioned boundary and loading conditions.

Its geometry was obtained using the Undefined Shape Method based on Pücher equation for the equilibrium of membrane structures:

$$\frac{\partial^2 z}{\partial x^2}\frac{\partial^2 F}{\partial y^2} - 2\frac{\partial^2 z}{\partial x \partial y}\frac{\partial^2 F}{\partial x \partial y} + \frac{\partial^2 z}{\partial y^2}\frac{\partial^2 F}{\partial x^2} = V(x,y) \quad\dots\dots\dots (1)$$

Where:

  $Z(x,y)$: Shape function.
  $F(x,y)$: Stress function.
  $V(x,y)$: Load over the surface measured in $Z=0$

Taking into account boundary conditions Korda uses the following stress function $F(x,y)$:

$$F = C[(1-n^2)\ (1-m^{\frac{2}{1-n^2}}) + (1-m^2)\ (1-n^{\frac{2}{1-m^2}})] \quad\dots\dots\dots (2)$$

Where:

  C: Constant to be determined once H is choosen.
  $m = \dfrac{2x}{L}$; $n = \dfrac{2y}{L}$

For a given $V(x,y)$ it is possible to solve (1) and calculate Z coordinates.

In this case the model was calculated at real size (L= 20.00 m) and loaded with 3.4 kN/m^2 over the surface. By choosing H = Z = 3.05 m at point A, the model was made compatible with the HMT model, thus allowing a comparison.

To calculate Z coordinates Korda's tables were used, firstly taking V(x,y) constant (surface projected on XY), but finally adjusting it for its real value. In this way both models were loaded with the same uniform load over the surface, making models also compatible from this point of view. Fig. 3 and 4 show also Z coordinates for this model.

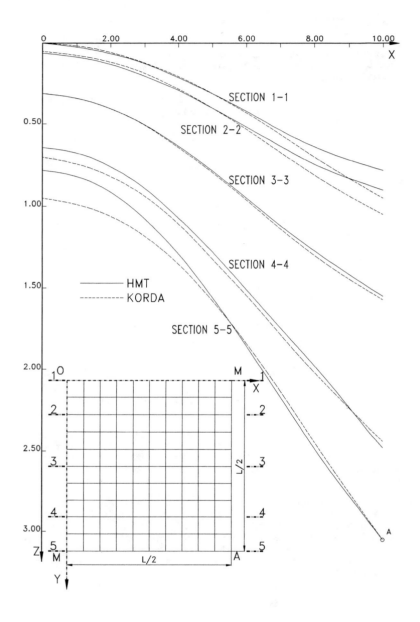

Fig. 3 - Vertical sections

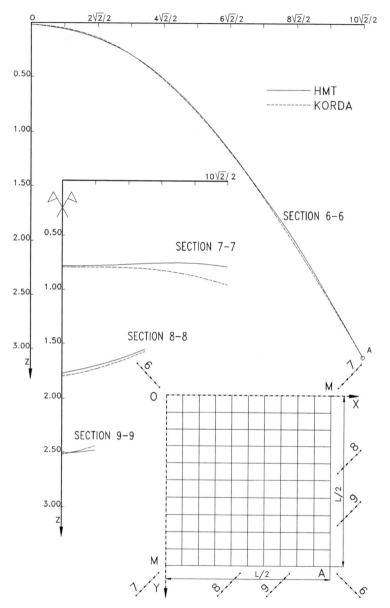

Fig. 4 - Diagonal vertical sections

Geometric analysis

a) In Fig. 3 and 4 it can be seen that coordinates of both structures are almost coincident, even when they were obtained following quite different methods.

b) Differences are generally less than 2% of Hmax

c) The maximum difference is at point M, at the middle of the edge, where it reaches 5.5%.

d) The diagonal section 6-6 is almost identical for both structures, showing perfect coincidence in four points.

e) Even in curvature change zone, a good convergence is shown in Fig. 4.

Stress analysis

Using FEA two real size structures were modelled from HMT and Korda methods. The following general parameters were used:

- Material: Reinforced concrete
- Young module: E=30000 MPa
- Thickness: 0.1 m.
- Load: 3.4 kN/m^2
- Poisson module: 0.2

A square mesh of 120 triangular an quadrilateral elements were used, taking into account symmetry conditions and using a linear static solver. The following parameters were calculated:

- Principal inner and outer surface stresses.
- Principal membrane stresses.
- Principal bending stresses.
- Vertical (Z) displacements.

Fig. 5 and 6 summarize the most representative values. The following can be noted:

a) <u>Principal minimum stresses</u>: Membrane stresses are always compression. Surface stresses (membrane plus bending stresses) are compression over the outer surface and over the 95% of the inner surface.
In any case tractions can be neglected, because they are in the order of 0.1 MPa.

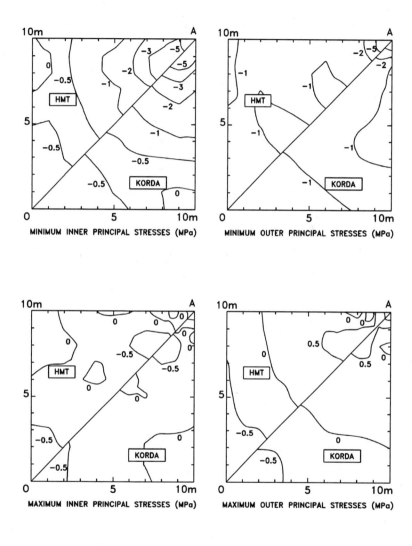

Fig. 5 - Surface principal stresses

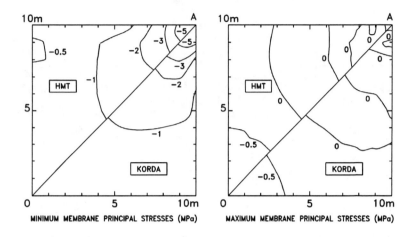

Fig. 6 - Membrane principal stresses

b) <u>Principal maximum stresses</u>: Membrane traction stresses exist in both structures with a magnitude less than 0.7 MPa, distributed over the 37% of the structure for HMT model and over 27% for Korda's model. Considering outer surface stresses those percentages increase to 47% for HMT model and to 46% for Korda's model. They decrease to 20% for both models considering the inner surface. In limited small zones stress magnitude reaches 1.5 MPa.

It is convenient to state that in the above results the zone near the support, where stress gradients are very high, have been excluded. In practice, thickness needs to be increased in this zone.

c) <u>Vertical (Z) displacements</u>: They are almost coincident for both structures, being somewhat smaller for HMT model. For example the maximum displacement (at point O) is 0.59 cm for HMT model and 0.60 cm for Korda's model. In any case they keep small over the whole structure as one can expect for an almost membrane structure.

Conclusions

a) It is evident that coincidence exists between both models from geometric and structural behaviour. The interest and importance of this coincidence must be considered taking into account the different principles from which each model have been generated: the HMT model starting from a biological principle and following a physical way for its implementation, and Korda's model from membrane theory and a mathematical way for its application.

b) Even if HMT structure does not behave as a perfect membrane, which is reasonable since it has been obtained from a physical model, it is shown that a quite good approach exists to the ideal Korda's model.

c) From the point of view of practical applications the HMT obtained shape is highly recommended since it leads to a very efficient dimensioning of the structure, being compression stresses dominant, while bending and traction keep small and limited to small zones.

d) This study confirms the reliability of HMT as a tool to find out and design new shapes. Moreover, being the HMT a general method, without limitations about geometry, loading and boundary conditions, it is very suitable for the design of free form structures.

References

Andrés, O.A. (1989) "Homeostatic Models for shell Roof Design" Proc. of IASS Congress, Madrid, 1989, Vol 1.
Andrés, O.A. and Ortega, N.F. (1991) "Experimental Design of Free Form Shell Roofs" Proc. of IASS Symposium, Copenhagen, 1991, Vol. 2, 69-74.
Korda, J. (1965) "Ribless Membrane Shell with Point Supports at the Corners" Proc. IASS Symposium, Budapest, 1965, 179-190.

Sensitivity and Optimum Structural Design

Witold Gutkowski[1] and Jacek Bauer[2]

Abstract

Modern design requires more advanced considerations leading to light and safe structures. One of the way to reach this aim is through structural optimization. Today's computational possibilities give us very powerful tools allowing to optimize many types of structures subjected to different types of loading and constraints and with different objective functions. However real structures differ from their design. This is caused by technological tolerance in the process of manufacturing and erection. It is then the aim of the paper to consider the very important relation between an optimum structure and sensitivity of active constraints imposed on state variables.

Introduction

There is an increasing number of methods and available computer programs for structural optimization. However the number of application of optimum designed structures is still relatively small. One of the main reason of this limitation is designers concern about the safety of a structure which has to be erected. This is caused by the fact that real, erected structure differs from the optimally designed by some values. These differences result from technological imperfections in the process of manufacturing structural elements and assembling of the whole structure. Other words the real structure differs from the designed one due the fact that prefabrication and erection are carried out with some tolerance. It is then important to know at least with some approximation what will happen to constraints imposed on some state variables (displacements, stresses, eigenvalues) if the variation of design variables (cross section area or moment of inertia of beams, thickness of plate or shell etc.) will vary by small numbers arising from the tolerance of manufacturing them. To make the problem more clear let look at an optimum problem of cantilever beam subjected

[1]Professor, [2]Doctor at the Institute of Fundamental Technological Research
ul. Świętokrzyska 21, 00-049 Warszawa, Poland

to follower force (fig. 1a) [3]. At fig. 1b we see the variation of the radius of the circular beam cross section in the case when the column is divided in to ten finite elements. In this case changing the volume only by one percent of one of the element we are causing the change of critical force by more than 10 percent. The example is showing that in this case even very small variation of design variable may have dramatic influence on the behaviour of the total structure. Other words the particular structure is very sensitive.

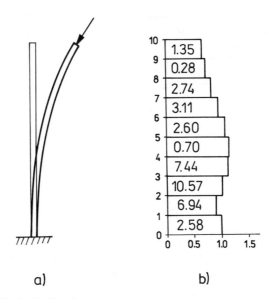

a)   b)

Fig.1 a. Beck column
b. Variation of critical force P in percent
with 1% variation of each cross section separately.

Let remind in few lines the notion of sensitivity and its calculation which will be used in following parts of the paper. The objective of the analysis of design sensitivity consists in finding total dependance of state variable constraints on design variables. If we denote constraint function by Ψ_k and design variable vector by $\mathbf{A} := [A_1, A_2, ..., A_q]$ the sensitivity is then defined in this case as total derivative of Ψ_k with respect to A_j i.e. $d\Psi_l/dA_j$. Assuming that the discussed constraint may depend on A_j and components of displacement vector $\mathbf{u}_i$ we may find the total derivative as:

$$\frac{d\Psi_k}{dA_j} = \frac{\partial\Psi_k}{\partial A_j} + \frac{\partial\Psi_k}{\partial u_1}\frac{du_1}{dA_j} + \frac{\partial\Psi_k}{\partial u_2}\frac{du_2}{dA_j} + ... + \frac{\partial\Psi_k}{\partial u_n}\frac{du_n}{dA_j} \tag{1}$$

or in vector form

$$\frac{d\Psi}{dA} = \frac{\partial\Psi}{\partial A} + \frac{\partial\Psi}{\partial u}\frac{du}{dA} \tag{2}$$

knowing or predicting the tolerance δA_j of j-th structural element, we may find the variation of the Ψ_k constraint

$$\delta\Psi_k = \frac{d\Psi_k}{dA_j}\delta A_j \tag{3}$$

Below it will be shown how optimizing the structure we may at the some time find sensitivity of active constraint. It will be also shown that there is a possibility to include among constraints on state variables also sensitivity of these variables with respect to design variables.

One of the method of finding sensitivity is based on notion of adjoint variable Φ[5] which may be found from relation

$$K(A)\Phi = \frac{\partial\Psi^T}{\partial u} \tag{4}$$

$$\Phi = K^{-1}(A)\frac{\partial\Psi^T}{\partial u} \tag{5}$$

Taking derivative of equilibrium equations $K(A)u = P$ with respect to A we finally obtain

$$\frac{d\Psi}{\partial A} = \frac{\partial\Psi}{\partial A} + \Phi^T[\frac{\partial P(A)}{\partial A} - \frac{\partial}{\partial A}K(A)\cdot u] \tag{6}$$

Sensitivities of active constraints for an optimum structure

Let consider a minimum weight truss subjected to two loading conditions for which constraints are imposed on displacement. We are limiting our attention to this relatively simple problem because it is illustrating the general approach without loosing generality. More detailed information on sensitivity analysis for sensitivities of other state variable may be found in monograph [5]. Additionally more elaborated problems of optimum structures with the approach presented in this paper may be found in earlier Authors paper [4]. It is obvious that more than two loading conditions may be considered in the some way.

The objective function for minimum volume design of a truss may be given as

$$f = l^T A \longrightarrow min \tag{7}$$

where l is vector of structural members length and A vector of their cross section areas.

The problem is subjected to following constraints. First of all let write down equality conditions in terms of equilibrium equations for both loading conditions

$$K(A)u^1 = P^1 \quad K(A)u^2 = P^2 \tag{8}$$

The state variable constraints are imposed in our problem on displacements for both loading conditions

$$\mathbf{u}^o - \mathbf{u}^1 \geq 0 \quad \text{and} \quad \mathbf{u}^o - \mathbf{u}^2 \geq 0 \qquad (9)$$

where $\mathbf{K}$ like usually denotes stiffness matrix $\mathbf{u}^1$, $\mathbf{u}^2$ and $\mathbf{P}^1$ and $\mathbf{P}^2$ displacement vectors and external load vector respectively for both loading conditions. Let now write Lagrangean for our problem

$$L = -\mathbf{l}^T \mathbf{A} + \boldsymbol{\Lambda}^{e1^T}[\mathbf{K}(\mathbf{A})\mathbf{u}^1 - P^1] + \boldsymbol{\Lambda}^{e2^T}[\mathbf{K}(\mathbf{A})\mathbf{u}^2 - P^2] + \boldsymbol{\Lambda}^{d1^T}[\mathbf{u}^o - \mathbf{u}^1] + \boldsymbol{\Lambda}^{d2^T}[\mathbf{u}^o - \mathbf{u}^2] \qquad (10)$$

and Kuhn-Tucker necessary conditions assuring the optimal solution

$$\frac{\partial L}{\partial A_j} = -l_j + \boldsymbol{\Lambda}^{e1^T}[\mathbf{K}(A_j)/A_j \cdot \mathbf{u}^1] + \boldsymbol{\Lambda}^{e2^T}[\mathbf{K}(A_j)/A_j \cdot \mathbf{u}^2] = 0 \qquad (11)$$

where $\mathbf{K}(A_j)$ denotes the stiffness matrix with all terms equal zero except those which are containing A_j. Due to the linear dependence of $\mathbf{K}$ on A_j, differentiation is equivalent with respect to A_j is equivalent to the division by A_j. This may be done because there are not terms in stiffness matrix which are free from the cross section area. Relation (11) may be also presented in a simpler way

$$-l_j + El_j\varepsilon_j^1 e_j^1 + El_j\varepsilon_j^2 e_j^2 = 0 \qquad (12)$$

where $\boldsymbol{\varepsilon} = \mathbf{B}^T \mathbf{u}$; $\mathbf{e} = \mathbf{B}^T \boldsymbol{\Lambda}^e$

$$\frac{\partial L}{\partial \mathbf{u}^1} = \boldsymbol{\lambda}^{e1^T} \cdot \mathbf{K}(\mathbf{A}) - \boldsymbol{\lambda}^{d1^T} = 0; \quad \frac{\partial L}{\partial \mathbf{u}^2} = \boldsymbol{\lambda}^{e2^T} \cdot \mathbf{K}(\mathbf{A}) - \boldsymbol{\lambda}^{d2^T} = 0 \qquad (13)$$

$$\mathbf{K}(\mathbf{A})\mathbf{u}^1 - \mathbf{P}^1 = 0; \quad \mathbf{K}(\mathbf{A})\mathbf{u}^2 - \mathbf{P}^2 = 0 \qquad (14)$$

$$\boldsymbol{\Lambda}^{d1^T}(\mathbf{u}^o - \mathbf{u}^1) = 0; \quad \boldsymbol{\Lambda}^{d2^T}(\mathbf{u}^o - \mathbf{u}^2) = 0 \qquad (15)$$

$$\boldsymbol{\Lambda}^{d1}; \quad \boldsymbol{\Lambda}^{d2} \geq 0 \qquad (16)$$

Multiplying (11) by A_j and adding all equations from $j = 1$ to $j = p_o$ (number of structural elements) we obtain

$$f = \boldsymbol{\Lambda}^{e1^T} \cdot \mathbf{K}(\mathbf{A})\mathbf{u}^1 + \boldsymbol{\Lambda}^{e2^T} \cdot \mathbf{K}(\mathbf{A})\mathbf{u}^2 \qquad (17)$$

Next multiplying (12) by $\mathbf{u}^1$ and $\mathbf{u}^2$ respectively and substituting the results into we obtain

$$f = (\boldsymbol{\Lambda}^{d1^T} + \boldsymbol{\Lambda}^{d2^T})\mathbf{u}^o \qquad (18)$$

Components of $\boldsymbol{\Lambda}^{d1}$ and $\boldsymbol{\Lambda}^{d2}$ differ from zero only for active constraints. In an iterative process of solution the system of equations (11)-(14) we may consider at each iterative step [2] that only one inequality constraint is active which means that a n-th step we have

$$f(n) = \Lambda_i^{dk}(n)u_i^o \qquad (19)$$

From above we find the Lagrange multiplier $\Lambda_i^{dk}(n)$ where k is equal 1 or 2.

If we look at equation (4) defining adjoint function Φ we observe a certain similarity between Φ and Λ^{e1} or Λ^{e2} as Lagrange multipliers associated with equations of equilibrium. Both function are defined then by

$$\mathbf{K}(\mathbf{A})\Phi^1 = \frac{\partial \Psi^T}{\partial \mathbf{u}^1}; \quad \mathbf{K}(\mathbf{A})\Phi^2 = \frac{\partial \Psi^T}{\partial \mathbf{u}^2} \qquad (20)$$

$$\mathbf{K}(\mathbf{A})\Lambda^{e1} = \Lambda^{d1^T}\frac{\partial \Psi}{\partial \mathbf{u}^1}; \quad \mathbf{K}(\mathbf{A})\Lambda^{e2} = \Lambda^{d2^T}\frac{\partial \Psi}{\partial \mathbf{u}^2}$$

The above (19) equations are given in vector form. It has to be pointed out that each element of the matrix $\frac{d\Psi}{d\mathbf{A}}$ has to be calculated from the following relations

$$\frac{d\Psi_i^k}{dA_j} = \frac{\partial \Psi_i^k}{\partial A_j^i} + \Phi_i^k[\frac{\partial \mathbf{P}^k}{\partial A_j} - \frac{\partial}{\partial A_j}(\mathbf{K}(\mathbf{A}) \cdot \mathbf{u}^k)] \quad k = 1, 2 \qquad (21)$$

Assuming that at each iterative step one constraint $\Psi_i = u_i^o - u_i$ [..] may be considered as an active one, we may find simple relation between Φ_i^k and Λ^{e1^T}

$$- \Lambda_i^{dk} \cdot \Phi_i^k = \Lambda_i^{ek} \qquad (22)$$

It means that we are able to define easily the sensitivity of "most active" constraint and then "most dangerous" constraint from the designers point of view. Substituting in our particular case for $\Psi_i = u_i^o - u_i$ and noting that A_j and u_i are in our consideration independent and that both $\mathbf{P}^1$ and $\mathbf{P}^2$ do not depend on $\mathbf{A}$ we find

$$\frac{du_i^k}{dA_j} = \Lambda_i^{ek^T}\frac{\partial}{\partial A_j}[\mathbf{K}(\mathbf{A}) \cdot \mathbf{u}^k]/\Lambda_i^{dk} \qquad (23)$$

in this case as we are considering one active constraint only the relation (21) holds only for k=1 or 2 (not for both at once).

In the case of a truss, as we said before, the derivative of stiffness matrix with respect to j-th design variable allows to write

$$\frac{du_i^k}{dA_j} = \frac{\Lambda_i^{ek^T}\mathbf{K}(A_j) \cdot \mathbf{u}^k}{\Lambda_i^{dk} \cdot A_j} = \frac{E\varepsilon_j e_{ij}}{\Lambda_i^{dk}} = \frac{l_j}{\Lambda_i^{dk}} = \frac{l_j}{f}u_i^o \qquad (24)$$

where $\boldsymbol{\varepsilon}$ and e are defined by known kinematic relations. The relation (24) holds only in the case when there is no active constraint on minimum or maximum value of A_{j_o}.

Number of design variable and sensitivity relation

Let consider again a truss of given lay out, composed of j_o element. The structure is loaded with forces independent of design variables which we assume to be cross section areas of structural members. We will discuss two design of the structure. In the first optimum design we assume that each of j_o structural member may have different cross section area. In the second optimum design we assume that all structural members are divided in to q_o linking groups. This means that members belonging to one linking group have the same cross section area. Both structures are subjected to the same constraint imposed on u_i component of displacement vector. This last assumption is making our considerations more clear without loosing generality.

The main assumption in our consideration is that the structure with larger number of design variables has smaller volume that in the case when members are linked in groups. We denote the volume of the first structure by

$$f_1 = l_1 A_1 + l_2 A_2 + ... + l_{j_o} A_{j_o} \tag{25}$$

and of the second one

$$f_2 = (l_1 + l_2 + ... + l_{k_o})A_1 + ... + (l_{j_o-m} + l_{j_o-m+1} + ... + l_{j_o})A_{q_o} \tag{26}$$

Under assumption that $f_1 \leq f_2$ taking in to account (17) we find that $\Lambda_i^{d1} \leq \Lambda_i^{d2}$. If we want to compare the influence of sensitivities in both cases we have to compare variation of our constraint by changing design variable. Other words we have to compare for instance δu_i^1 with δu_i^2 when changing A_1 to A_{k_o} in the first case, and A_I in second case. This may be formulated as follows

$$\delta u_i^1 = \frac{du_i^1}{dA_1}\delta A_1 + \frac{du_i^1}{dA_2}\delta A_2 + ... + \frac{du_i^1}{dA_{k_o}1}\delta A_{k_o} \tag{27}$$

$$\delta u_i^2 = \frac{du_i^2}{dA_I}\delta A_I$$

Recalling (20) and (22) we can find (23)

$$\delta u_i^1 = [\varepsilon_1^1 e_{i1}^1 l_1 \delta A_1 + \varepsilon_2^1 e_{i2}^1 l_2 \delta A_2 + ... + \varepsilon_{k_o}^1 e_{ik_o}^1 l_{k_o} \delta A_{k_o}]\frac{E}{\Lambda_i^{d1}} \tag{28}$$

$$\delta u_i^2 = [\varepsilon_1^2 e_{i1}^2 l_1 + \varepsilon_2^2 e_{i2}^2 l_2 + ... + \varepsilon_{k_o}^2 e_{ik_o}^2 l_{k_o}]\frac{E}{\Lambda_i^{d2}}\delta A_I \tag{29}$$

and with (12)

$$\delta u_i^1 = (l_1 \delta A_1 + l_2 \delta A_2 + ... + l_{k_o}\delta A_{k_o})\frac{E}{\Lambda_i^{d1}} \tag{30}$$

$$\delta u_i^2 = (l_1 + l_2 + ... + l_{k_o})\delta A_I \frac{E}{\Lambda_i^{d2}} \qquad (31)$$

assuming the same variations of design variables for all structural elements $(\delta A_1 = \delta A_2 = ... = \delta A_I = \delta A)$, and noting that $\Lambda_i^{d1} \leq \Lambda_i^{d2}$ we find that

$$\delta u_i^1 \geq \delta u_i^2 \qquad (32)$$

which means that changing variables by same value δA, possible violations of active constraint is larger for an optimum structure with larger number of design variables.

Optimum structural design with constraints imposed on sensitivities

Let us come back to the problem described in the previous chapter adding to the inequality constraints some limitation on sensitivity. This may be presented as follows

$$\mathbf{s}^o - \frac{d\mathbf{u}^k}{d\mathbf{A}} \geq 0 \quad k = 1, 2 \qquad (33)$$

$$\text{with} \quad \frac{d\mathbf{u}^k}{d\mathbf{A}} = \mathbf{\Phi}^k [\frac{\partial}{\partial \mathbf{A}} \mathbf{K}(\mathbf{A}) \cdot \mathbf{u}^k]$$

where $\mathbf{s}^o$ a matrix of given number of proposed largest sensitivities allowed in our problem. In order to define the sensitivities we need to introduce the adjoint function $\mathbf{\Phi}^k$ by relation

$$\mathbf{K}(\mathbf{A})\mathbf{\Phi}^k = \frac{\partial \mathbf{\Psi}^T}{\partial \mathbf{u}} = -I \qquad (34)$$

remembering that in our case $\mathbf{\Psi}^T = (\mathbf{u}^o - \mathbf{u}^k)^T$. Extending our Lagrangean (10) we find the necessary conditions given by Kuhn-Tucker theorem.

The conditions and solution algorithm are presented in earlier Authors paper [4]. Rao and others [6] have presented recently multicriteria approach to this problem.

Examples

Let consider a simple example of a truss (fig. 2) commonly used as banch mark problem in structural optimization. The structure is subjected to constraints on displacements ($u^o = 2in$) and stresses ($\sigma^o = 2.5 \cdot 10^4 psi$). On fig. 3 a curve representing the relation between sensitivity and number of iteration during the process of optimizing the structure is given. The curve shows the increasing sensitivity as we approach to an optimum structure. Next let find variation of constraint imposed on stress in 5-th number σ_5 for four different number of design variables.

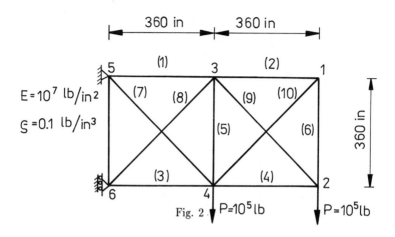

Fig. 2

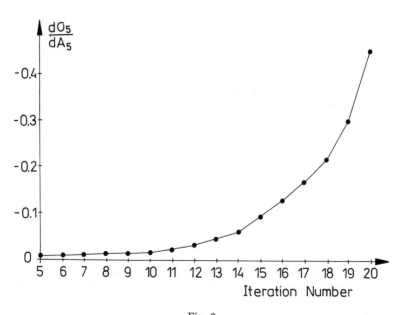

Fig. 3

The considered cases are as follows:

Case A all structural members are
having the same cross section
$\delta\sigma_{5A} = -0.0036\delta A$

Case B $A_1 = A_2 = A_3 = A_4 = A_I,$
$A_5 = A_6 = A_7 = A_8 = A_9 = A_{10} = A_{II}$
$\delta\sigma_{5B} = -0.0063\delta A$

Case C $A_1 = A_2 = A_3 = A_4 = A_I,$
$A_5 = A_6 = A_{II}$
$A_7 = A_8 = A_9 = A_{10} = A_{III}$
$\delta\sigma_{5C} = -0.0157\delta A$

Case D all structural members are
different.
$\delta\sigma_{5D} = -0.6270\delta A$

Conclusions

An important problem of relation between optimum structure and sensitivity of active constraint is solved. Designer can evaluate possible violation of most important constraint knowing manufacture tolerance of design variables. This can be done in the process of optimization without need of additional sensitivity analysis.

The paper discuss a very important relation between the number of design variables and variations of active constraint caused by variations of design variables.

Finaly the paper gives the possibility to include sensitivity of state variables as constraints.

Acknowledgement The paper was sponsored by Polish Committee for Scientific Research grant No 3 0939 91 01.

References

1. J. Bauer, W. Gutkowski, Structural Optimization with Sensitivity Constraints and Multiloading Conditions; Structural Optimization 93, The World Congress on Optimal Design of Structural Systems, Rio de Janeiro, August 2-6, 1993, Ed. J. Herskovits, Vol. I, 65-74.

2. W. Gutkowski, J. Bauer, Z. Iwanow, Explicit Formulation of Kuhn-Tucker Necessary Conditions in Structural Optimization, Comp. Struct. Vol. **37**, No5, p.753-758, 1990.

3. W. Gutkowski, O. Mahrenholtz and M. Pyrz, Minimum - Weight Design of Structures under Non - Conservative Forces, [in] NATO ASI Series,

Optimization of Large Structural Systems, Ed. G.I.N. Rozvany, Kluwer, 1993, Vol. II, 1087-1099.

4. W. Gutkowski, J. Bauer, Structural Optimization with Sensitivity Constraints; Statics, Comp. Struct. (accepted for publication).

5. E. J. Haug, K. K. Choi and V. Khomkov, Design Sensitivity Analysis of Structural Systems, Academic Press, Orlando, (1986).

6. S. S. Rao, K. Sundararaju, C. Balakrishna and G. Prakash, Multiobjective Insensitive Design of Structures, Comp. Struct., **45** p. 349-359 (1992).

Research of strength and deformations with
long-time loading and the use of glassfiber
reinforced concrete in shells

G.K.Khaidukov[2] , M.M.Lachinov[1]

Introduction

Experimental research has been carried out on the
properties of behaviour on compression and tension of
glassfiber reinforced concrete of alkali-resistant glassfiber under long-time load and higher air humidity. Joint
work of glassfiber reinforcement with rod steel in bent
elements is also being considered. On small samples and
thin-walled foulded beams the appearance and development
of cracks and tense deformed state under short-time and
long-time load have been researched and the way of strength
evaluation of glassfiber concrete elements has been suggested. The samples were made of glassfiberconcrete mixture prepared in spiral-vortex or vibroscrew mixer placed
in a mould with vibration and pressure (1)(2). Foulded
plate beams were formed by bending method. The structure
and construction of fiberconcrete shells of 30 mm width
and 12 m length by spraying of glassfiberconcrete on the
pneumatic decking is described. The consumption of material per 1 m² of the building floor area is given.

---

[1] Professor, Managing Director, Scientific and Educational
Company of the Intourreklama Corporation, Russia, Moscow,
103031, Neglinnaya street, 8, fax 923-7057, telex 411211

[2] Head of laborotary, Professor, Doctor of technical sciences, Science Research Institute of Reinforced Concrete,
Russia, Moscow, 109428, 2-ya Institutskaya street, 6,
fax 200-2217

Materials and Methods of Research

    In NIIZHB (Scientific Institute of reinforced concrete
Moscow) systematic research of physical-mechanical properties of glassfiber reinforced concrete on tension and
bending by short-time and long-time (2) load has been
carried out. Samples are in both cases similar by dimensions and the technology of manufacture. For long-time
tests slabs and foulded-plate beams were manufactured of
fine-grained reinforced concrete with the strength on
compression R_b = 22 MPa and on tension R_{bt} = 1,8 MPa. Fiber
was made of alkali-resistant glass filament with the diameter 12-16 mkm and strength for rupture R_f = 1600 MPa.
The length of fiber was 40 mm, the volume contents amounted to 2% and 2,8% as most adequate according to the data
of short-time research (1).

    As rod steel-reinforcement of beam samples smooth
wire with 4 mm diameter was used. On axis tension and
bending glassfiberconcrete plates of 45 mm width and
400mm length and 10-12 mm thickness corresponding the
thickness of the walls of thin-walled beams with "П "
section tested on bending (Figure 1) were researched.

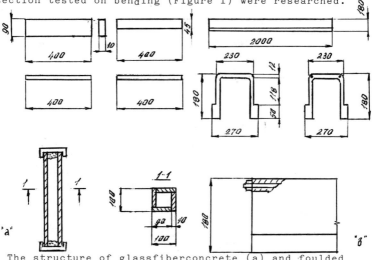

The structure of glassfiberconcrete (a) and foulded
plates samples with "U" section by combined reinforcement (b) (μ_{fv} = 2%, μ_{fv} = 2,8%, μ_s = 0,55%)

Figure 1

The strength of glassfiberconcrete on compression
was researched according to the methodology of Prof.
Khaidukov on hollow prisms made of plates with 90x400 mm
dimensions and 10 mm thickness.

Research Results

Relative humidity of environment where the samples
were put during long-time tests influences substantially
strength and deformations of glassfiber reinforced concrete. Tensile strength of glassfiberconcrete samples put in
the laboratory chamber with the humidity W=96-98% with
t=20-22°C decreases most intensively during the first 18
months of storage and is stabilizing towards the end of
2 years.

Though at the age of 2 months a certain increase of
tensile strength was noticed towards 170 days of humid
storage the strengh of glassfiber concrete decreased to
0,75-0,8 from R_{fbt} (28). The reduction of R_{fbt} can be
explained by mechanical micro damage of the surface of
glassfiber and further (effect) influence of alkali environment. In table 1 recommended values of coefficient
γ_{fb1} decrease in R_{fbt} for glassfiber reinforced concrete
with μ_{fv} from 1,5% to 3% during humid storage are given.

	Storage conditions (relative air humidity)			
	W = 80%	W > 80%	in water	in the open air
γ_{fb1}	0,9-1,0	0,75-0,8	0,7-0,75	0,8

The highest values of γ_{fb1} correspond to μ_{fv} =1,5%
short-time and long-time tests have shown different behaviour of glassfiberconcrete on tension and bending after
the formation of cracks. With short-time load foulded-
plate beams with the contents of glassfiber exceeding the
critical point (with μ_{fb}=1,5%, $R_{fbt} > R_{bt}$) after the
achievement of the load microcracking (P_{fcrc}, relative
deformations of concrete ε_{fb} =10.10^{-5}) continued to carry
the load till the values which were 1,5-2 times more.
With short-time load the relative deformations were from
25 to 30.10^{-5} with long-time bending loads the beam samples after the appearance of tiny cracks (5-10 mkm).

Plastic deformations of tension (up to $40\text{-}45.10^{-5}$mm) were observed and the destruction of samples with M_{fv} = 2,0% on 7-15 days and with M_{fv}=2,8% on 30-60 days took place. In this case the load was approximately equal to the short-time load (Figure 2) in the moment of cracks formation which is shown in Figure 2.

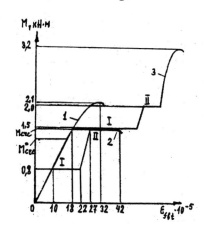

Fig. 2

Moment (M)-strain ($\mathcal{E}$) diagram by bending (M - bending moment in beams sample, $\mathcal{E}$ - strein in border layer at glassfiberconcrete beams speciments.
1. Short-time tests M_{fv} = 2,8%.
2. Long-time tests (M_{fv} = 2,8%) on 1 stage with M = 0,4 M_{lim} and with M = 0,7 M_{lim} by 2 stage.
3. Test "U" samples with combined reinforcement by long-time load on 1 stage with M=M$_{crc}$ and on 2 stage with M = 0,65 M_{lim} and to structure of reinforced rod.

For the samples with combined reinforcement (M_{fv} = 1,5-2,8% and rod steel M_s=0,55%) exposed under load till 270 days in time mergence of microcracks into cracks with the width up to A$_{crc}$=0,2 mm was noticed, glassfiber ruptured or slipped and only rod steel worked while the load corrsponded to the limit for reinforced concrete section. In this case the sagging of glassfiber reinforced concrete beams increased by the moment of destruction 1,2-1,7 times and with combined reinforcement no more than by

20% which is shown in Figure 3.

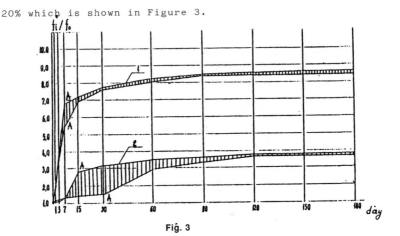

Fiģ. 3

Deflections in the middle part of span of beams speciments with combined reinforcement by long-time load with
$M = 0,95 \ M_{crc}$
1. with $\mu_{fv} = 2\%$,
2. with $\mu_{fv} = 2,8\%$ and rod reinforcement with $\mu_s = 0,55\%$.

    It should be noted that fiber reinforcement restrai-
ned non-elastic deformations both in compressive and tensile zone. For the evaluation of the strength of normal
sections of glassfiberconcrete bent element the limit of
long-time tensile strength can be identified (found) by
the formula:

$$R_{fbt} = R_f \cdot \mu_{fv} \cdot m \cdot \varphi \cdot \eta_o \cdot \eta_1 \quad (1)$$

where

R_f – conventional limit of fiber yield under long-
            time load depending on the strength of fiber R_f

$R_f = 0,7 \ \gamma_{fbl} \cdot R_f$

μ_{fv} – coefficient of volume fiber reinforcement

m – coefficient taking into account the strength of
            matrix and equal 0,4-0,8 (die).

η_o – coefficient taking into account fiber orientati-
            on for used concrete method equal to 0,3-0,375
            (the highest value in the ratio of the wall

thickness to the width equal to 0,5).

η_1 — coefficient taking into account the influence of the fiber length and equal to 0,9.

φ — coefficient taking into account the aggregate state of fiber equal to 0,35-0,42.

The values m, η_o, η_1 are the same as on short-time loading.

The experiments have shown that for the evaluation of the strength of glassfiberconcrete bent elements (bending moment to normal section) one may use the methods of evaluation for ultimate state (3) taking for the compressed zone the triangular strength epure and in tension zone - rectangular one with the coefficient of epure completeness

$$W = 0,5 - 2,5 \; \frac{R_{fbt}}{R_b}$$

where R_b - prism strength of matrix.

For the long-time calculation of glassfiberconcrete bent elements with the ultimate state 2 conditions should be observed:

a) on crack formation $M_{fcrc} \leq 1,1 \, W_{bpl}$ (4)

b) on strength $M_p \leq M_{fb}$

where

W_{pl} — plastic resistance moment of reduced (taking into account the reinforcement) concrete section,

M_p — bending moment of load action,

M_{fb} — moment inner forces taking into account tension resistance of glassfiberconcrete no more as conventional "yield" strain $0,7 \, R_{fb}$.

For elements with combined reinforcement the crack formation moment it can be find on (4) too, taking into account rod reinforcement and correspondent initial modulas of elasticity.

Limit moment in normal section at glassfiberconcrcte with rod reinforcement bent element can be calculated with add tension strain taking into account limit deformation of glassfiberconcrete (Figure 2). Strain in tension that of "I" section must take equally R_{fbt} =0 but for strain epure of wall one must take coefficient W on (3).

The calculation of glassfiberconcrete bent beams with combined reinforcement can account add up, the curvatures $(1/\rho)$ by short-time $(1/\rho_1)$ and long-time loading $(1/\rho_2)$.

$$1/\rho = 1/\rho_1 + 1/\rho_2 ,$$

where

$$1/\rho_1 = \frac{M_1}{B_{fb1}} \quad \text{and} \quad 1/\rho_2 = \frac{M_2}{0,85\ B_{fb1}} ,$$

$$B_{fb1} = (0,85 + \alpha \mu_{fv})\cdot E_b \cdot J_{red},$$

$$\varphi = 0,9(2,5 + \alpha \mu_{fv}); \quad \alpha = E_f/E_b$$

E_b, E_f - modulus of elasticity of concrete and fiber.

J_{red} - reduced inertia moment of element section with taking into account fiber and rod reinforcement on corresponding modulus of elasticity.

This research made one's contribution to greater accuracy description of characteristic and design method of glassfiber reinforced concrete structures.

The appliance of glassfiber concrete in spatial structures

For more than 10 years the world practice has known the appliance of glassfiber reinforced concrete in panels for walls and coatings, floor slabs, spatial houses, shell structures etc.

In Studgart (Germany) in the middle of 70-ies glass-fiberconcrete hyperbolic shell structure of a pavilion coating with 30 m diameter and abput 15 mm thickness was built (4).

The thin-walled structure was assembled of 8 previously made by spraying of glassfiberconcrete on the wood-steel shell centering sectorial hypar shells with further making monolithic joints and placing rod steel in them.

In Russia NIIZHB together with a number of other organizations have elaborated, tested on models and fragments and built in Voronezh in 1979-81 glassfiber-concrete wave-type the Vault shell of 12 m span and

30 mm thickness (Figure 4).

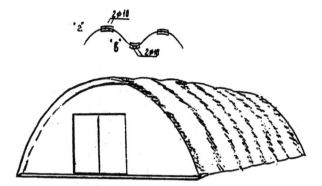

Fig. 4

View of glassfiberconcrete shell 30 mm thickness and 12 m span
a) rod reinforcement above the vaults,
b) rod reinforcement between the vaults.

The surface of the earth-earth-vault waves of 3 m length and circular outline and parabolic outline in the cross section of the building.
In accordance with the Project Report on the crest of a wave and in the joint between waves several rods of 10 mm diameter were placed.

Though several experimental shell structures erected on non-shifted foundations have been existing safely for 10 years without having steel reinforcement. Taking into account temperature-humid effects and local loads combined reinforcement in tensile zone of bearing shell structures should be used.

The manufacture of spatial structure was carried out on pneumashuttering the waveness of which was reached by the tensioning of cables placed every 3 meters along the building and fixed to the foundations. In order to secure the outline of a building with a pointed vault, the cables in the lock of the vault structure were raised by a prop by 40-50 cm.

Concreting was carried out by spraying of glassfiberconcrete mixture on the pneumatic centering in two layers.

Excessive air pressure inside of shuttering must be about 150 mm of water column.

The calculation of the structure vault of an unheated building was carried out on the influence of its own weight, snow, wind and changes in temperature typical of the climate region of Moscow.

The identification of strain-stress state of the wavy spatial structure with rigid diaphragms can be performed by the method of finite elements.

However, if the spatial structures are under the influence of a load uniform along the length of the building then the practical dεsign can be performed as for two hinges of a thin-walled arch of the width which is equal to the wave length and the span of 12 m.

For the disign of glassfiberconcrete shell structure the following initial data were taken: matrix of fine-grained concrete of B35 class with water consumption by weight 8%, alkali-resistant glassfiber of fiber roving RCR type length of roving 40 mm appeared to be right on the nozzle and had tensile strength 1000-1200 MPa, fiber consumption 75 kg/m^3, shrinking of glassfiberconcrete was taken into account by the temperature difference equivalent $\Delta t = -20°C$, possible changes in temperature + 20°C and - 40°C in the ratio to foundation, tensile resistance of glassfiberconcrete $R_{fbt} = 2,92$ MPa and on compression $R_{fb} = R_b = 16,9$ MPa, initial modulus of elasticity of glassfiber $E_x = 70$ MPa and concrete $E = 24.10^3$ MPa.

Consumption of materials per 1 m^2 of the spatial structure floor area of the building of glassfiber reinforced concrete amounts to: fine-grained concrete - 0,045 m^3, alkali-resistant fiber - 1 kg, steel reinforcement - 0,7 kg, Portland cement M400 - 30 kg.

Appendix (references)

1. G.K.Khaidukov, V.V.Volkov, A.X.Karapetyn
   "Strength, deformation glassfiberconcrete units"
   Beton and ferroconcrete 1988 N 2 (in Russian).
2. G.K.Khaidukov, V.V.Volkov, M.M.Lachinov
   "Work of thinwalled glassfiberconcrete elements with long-time loading"
   Beton and ferroconcrete 1990 N 9 (in Russian).

3. SNIP203.03-85 Disign rules. Ferroconcrete structures. (in Russian).
4. J.Schlaich Prof.Dr.eng. "The application of Glassfiber reinforced concrete Shell" Proceeding of
   Symposium (June 1980).

Possibilities and Problems of Latticed Structures

Mamoru Kawaguchi[1]

Abstract

Several fundamental aspects of space frames which are sometimes referred to as latticed structures are discussed. Terminology of the structural system is first considered. Special features of the structure in contrast to the other type of structural system, plane structure, are studied. Advantages and problems of the structural system are discussed. Constructional aspects of space frames are treated as one of the most important features of space frames. A patented structural system named Pantadome System is presented as a method to solve the constructional problems of space frames.

1. Introduction - Terminology

Nearly three decades have passed since a structural system which is sometimes called a "latticed structure" has become familiar in the field of building technology. It seems no longer necessary to ask the question 'What is a latticed structure?' Nevertheless, the meaning of the term is not always clear, and there are many occasions when structural engineers and architects seem to fail to convey with it what they really want to communicate. Thus it may be appropriate to review the terminology of this kind of structure here.

The definition of a latticed structure as understood in U.S.A. may be represented by an ASCE report(ASCE,

----

[1]Professor, Department of Architecture, Hosei University, Koganei, Tokyo 184, Japan

ST11, 1976) which states: ".... a latticed structure is a structural system in the form of a network of elements (as opposed to a continuous surface).... Another characteristic of latticed structural systems is that their load-carrying mechanism is three-dimensional in nature." The structural system which is defined in the same way as the above is often referred to as a "space frame" in some other countries including Japan.

In the present paper the term "space frame" is used to denote that structural system, since the author is more familiar to the term than to "latticed structure".

It may be of interest here to compare the above definition of a space frame to the definitions proposed elsewhere. In a number of cases the definition of a space frame is the same as ours (Gaylord,1968; Mainstone, 1975; Merritt,1975; Salvadori,1963). Some authors define space frames only as double-layered grids(Morgan,1964; Merritt,1972; McGraw-Hill,1977). A single -layered space frame which has the form of a curved surface is termed by them a "braced vault or a "braced dome".

A definition which we find quite impossible to accept is given in the Uniform Building Code used in much of the United States. It states: "Space Frame is a three-dimensional structural system without bearing walls, composed of interconnected members laterally supported so as to function as a complete self-contained unit, with or without the aid of horizontal diaphragms or floor bracing systems". Clearly, this definition refers only to a framing system which can transfer horizontal forces acting in any direction so as to make a multistory building earthquake or wind resistant -- a very narrow definition, indeed, which has almost nothing to do with a great majority of correctly defined space frames.

2. Constituent Elements of A Space Frame

As stated in the previous section, a space frame is an assembly of smaller members, linear or two-dimensional, arranged in three dimensions. To make a structure of this type a reality, it is indispensable that joints be introduced which firmly connect the elements to one another, resulting in a single, integrated structural entity. Thus the members and the joints are the two essential constituents of a space frame.

Linear members are usually straight, but there are cases where curvilinear members have been used. Mild steel is the most common material for the elements, but aluminum space frames are encountered as well. Though not very often, wood is also used for linear members from time to time (Medwadowski,1981). Tubes are the most common section used for linear steel elements, because of their structural efficiency in compression, but open section such as angles, channels, and I or H sections are also utilized when their efficiency or their ability to transfer intermediate loads in flexure is required.

Though the joints are also essential in framing systems other than space frames, their role in the latter type of structure is more important by far because, usually, more members are connected to them. Further, the members are located in a three-dimensional space, and hence the force transfer mechanism is more complex. The role of joints in a space frame is so important that most of the successful commercial space frame systems utilize patented joint systems. Thus the joints in a space frame are usually much more sophisticated than the joints in other structures, such as plane trusses, where simple gusset plates typically suffice.

3. Joint Systems

Materials of which space frame joints are made are steel or aluminum. Among the popular commercial systems one can find joints of cast or forged steel, pressed steel plates (sometimes partially welded), and extruded aluminum. Space frame joints are usually machined to a high degree of precision. Typically, the ends of the elements are also carefully machined, so that their lengths and angles fit the joints to which they are to be attached. On occasion, linear elements are continuous through the nodal point, in which case simpler joints such as straight or U-shaped bolts which clamp the inter-secting members are encountered. In the analysis, the mutual connection of linear elements at the joints is considered pinned, unless the elements are continuous through the nodes, and this continuity is intentionally taken into account in the analysis. However, there are many jointing systems where the ends of the linear ele-ments are clamped by the joints in such a way that they can resist a certain amount of rotation, thus resulting in an extra flexural and buckling capacity.

Studies of joint systems constitute the most impor-tant part of the design of a space frame. And in design of joint systems one of the most important problems is accumulation of errors at the connecting points. In an assembly system where connection of several new linear members to the points which have been set by a preceding assemblage determines the positions of new points. errors in fabrication of the members and in their assemblage are liable to accumulate as the work in the site proceeds. In this kind of system the members of the space frame should be fabricated to extremely accurate dimensions to avoid the accumulation of errors. This is the case in most of the commercial space frame systems.

There is another system of assembly where every point of linear members is located independently in the site, and the linear members are set between the determined connecting points. This system has an advantage that errors do not accumulate as far as the joints have a capacity of absorbing them between the two adjacent connecting points. In this system it is not necessary to impose any strict requirements for accuracy on manufacturers or contractors of the space frame. The system is not always applicable to every type of space frame, because the independent setting of the location of connecting points in the air sometimes may produce difficulties.

The author adopted this system for design of the space frame for the grand roof of Expo'70 to make use of the advantage stated above. Since the roof structure was assembled on the ground level and lifted to the specified height afterward, the location of joints was easily determined from the ground.

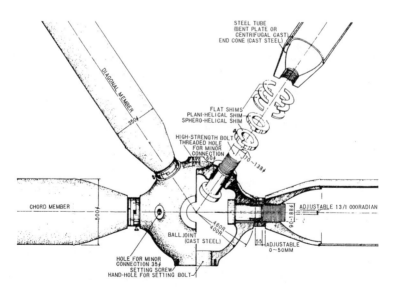

Figure 1. Mechanism of Joint

Possible errors in fabrication of members and in their assemblage in the site were carefully studied, and reasonable tolerances of the members were set on agreement with the fabricators and the contractors. These tolerances were evaluated in the design to determine the required error-absorbing capacity of the joint system. The mechanism of the joint and the evaluated errors in the study to be absorbed by the joint are shown in Figure 1 and Table 1.

Table 1. Factors and Amounts of Errors

Factors	Errors in Angle (1/1 000 rad.)	Errors in Length (mm)	Remarks
Setting Joints			*
Positional*	0.9	20	
Directional**	1.0	0	**
Length of Tubes	0	3	
End of Cones***	1.1	0	***
Ext. Dia. of Joints	0	2×2	
Holes of Joints			
Positional****	0.8	0	****
Directional*****	0.8	0	
Radii of Bolts	0.3	0	*****
Shim Plates	0	2×1	
Deflection of Tubes	1.0	0	
Thermal Deformation			
Assembled Part	1.1	2×6	(±20°C)
Member to be set	1.9	3	(±15°C)
Slope for Camber	3.2	0	
Total	13/1 000 rad.	44 mm (22 mm for One End)	

Today commercial space frame systems are popular all over the world. There are lots of patented systems used often for lighter roofs. Most of the joints in such commercial systems look very much like each other, and the mechanism of their connecting mechanisms is also similar to each other. It may be of interest, however, to review the special features of the various joint systems in market. Typical joint systems currently available in Japan are shown in Figure 2 and their features are listed in Table 2.

4. Special Features of Space Frames

As noted in the preceding section, the principal characteristics of a space frame (as compared to those structures which are based on a plane frame system) are its three-dimensional features, and the manner in which it is assembled. This is illustrated in the following examples.

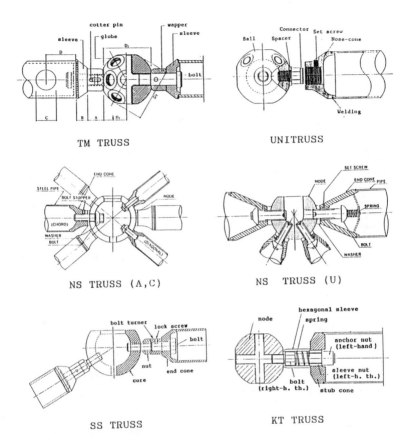

Figure 2. Typical Commercial Joint Systems in Japan

Figures 3(a) and (b) show two different ways of framing a dome. The dome shown in (a) is a complex of elements, or frames, which lie in a plane: arches, primary and secondary beams and purlins. Each of these elements constitutes a system which is, of itself, stable. In contrast, the structure shown in (b) is not an assembly of stable plane subassemblies. Rather, it is a system whose stability is assured only through its integrated action as a whole. This is seen clearly when its central section is compared to the central section of dome (a).

The difference between an assembly of plane frames and a space frame noted above can be understood also from

Table 2. Features of Various Joint Systems

System	T M (M E R O)	TOMOE UNITRUSS	NS TRUSS (A,C TYPES)	NS TRUSS (U TYPES)	SS TRUSS	KT TRUSS
Maker	TAIYO KOGYO	TOMOEGUMI IRON WORKS	NIPPON STEEL	NIPPON STEEL	SUMITOMO METAL IND.	KAWATETSU KENZAI
Bolts Advance from	end cone	ball joint	end cone	end cone	end cone	end cone
Bolts Advance to	ball joint	end cone	ball joint	ball joint	ball joint	ball joint
Driving Work Operated from	out side	out side	inside	inside	out side	out side
Bolts Projected by	─────	turning bolts	(not necessary)	spring	─────	spring
Prestressed Connection	y e s	n o	y e s	y e s	n o	y e s (strong)
Loosening Prevented by	prestress	set screw	ser screw	set screw	n u t	prestress
Other Features	long experience	turnbuckle effect, adjustable length	concentric fit	direct connection	simple application	water tight, high torque

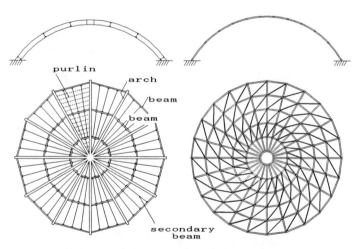

purlin
arch
/beam
beam
secondary beam

(a)Plane-Frame Complex   (b)Space Frame
Figure 3. Plane-Frame Complex and Space Frame

the point of view of the sequence of flow of forces. In system (a), the sequence of transfer of the load applied to the roof is determined beforehand in such a way that the force is transferred successively through the pur- lins, secondary beams and primary beams to the arches,

and through them finally to the ground.   In each case,
members of the framing system transfer loads from the
members of the lighter class to the members of the heavi
er class.   As the sequence progresses, the magnitude of
the load to be transferred increases, as does the span of
the element.

Accordingly distinct ranks are produced among the
elements of different classes, not only from the point of
view of the size of their cross sections, but also from
the point of view of the importance of the tasks assigned
to them.

In contrast, in the structure of system (b) the
sequence of load transfer is not set from the beginning,
and all elements contribute to the task of supporting the
applied load in accordance with the three-dimensional
geometry of the whole structure.   For this reason, the
ranking of the constituent members similar to that of
system (a) is not necessarily observed.   From this it
follows that a space frame can be characterized as a
'spatially framed structure without appreciable rank' (or
hierarchy) among its constituent elements.

This characteristic may not be so rigorous as to be
used as the definition of a space frame, but it is intui-
tively well understood, and it accords with the visual
impression made by a space frame.   Moreover, this characteristic seems to give a direct hint as to the possible
advantages and disadvantages -- or problems -- which
space frames may possess.   This topic will be discussed
in the following sections.

5. Advantages of A Space Frame

One of the most important advantages of a space
frame is its lightness.   It is due mainly to the fact
that material is distributed spatially in such a way that
the load transfer mechanism is primarily axial -- tension
or compression -- so that in any given element all material is utilized equally.   Further, the constituent
elements 'without rank' are arranged more or less uniformly, bracing each other so as to prevent buckling of
the individual elements in compression.   In large span
roofs, where the self weight of the structure constitutes
an important part of the total load, the lightness of the
constituent elements tends to magnify cyclically to a
considerable extent.

Another advantage of space frames may be due to the

production and construction techniques which are industrialized to an extent greater than is the case in conventional systems. The linear elements and joints are
both prefabricated, so that the jointing work at the site
is relatively simple. Thus a space frame can be erected
by semi-skilled workers. The light weight of individual
elements also makes the handling and assembly work easier.

A space frame is usually sufficiently stiff in spite
of its lightness. This is due to its three-dimensional
character and to the full participation of its constituent elements 'without ranks', adapting themselves equally
well to almost all types of loading. This feature of a
space frame is most symbolically shown by a design example where this system is employed in a three-dimensional
structure for a big telescope (Medwadowski,1981) which
demands very high rigidity as well as lightness.

A functional advantage can be observed in that space
frames require only the addition of quite light cladding
elements to enclose space. The same functional need in
conventional structures may result in an elaborate system
of secondary elements in addition to the cladding.

A space frame is often very attractive from the
architectural point of view, as its regular pattern with
taut, incisive lines creates the atmosphere of aesthetic
beauty.

6. Problems in Space Frames

While a space frame offers a number of advantages
noted above, some problems arise as well. These problems
should be given careful consideration at the time of
design. In the following, these problems are discussed
on the basis of a space frame considered in section 1.3.
The discussion is quite general, and it is entirely
possible that some of these problems may have been solved
in a specific established system. Nevertheless, it is
felt that the following review will prove useful.

6.1   Problems of Production and Over-design

As noted before, a special feature of a space frame
in its modern sense is that it is a spatial assembly of
structural elements 'without ranking', with as homogeneous a pattern as possible. Clearly, the concept of mass
production has been more or less a prerequisite to a
successful realization of space frames. The elimination
of ranking of the structural elements implies a minimum

of variation in their cross sections. It further means a
departure from the basic objective of the classical
structural design, namely that each element should have
the minimum section required for the performance of its
task. It is conceivable that, in an extreme case, only
one element of a large space frame has just the adequate
section, and all the remaining elements are over-de-
signed.

In the early stages of development of space frames
it was believed that the waste of materials due to the
over-design of elements would be compensated for by the
expected benefits resulting from mass production. At
this time, it is not clear whether this expectation has
been fulfilled. Buildings are usually custom-produced.
It appears that no effective mass production techniques,
such as those used in car making, have been introduced as
yet into the construction industry.

It appears that it may be difficult to rule out completely the ranking of elements if space frames are to
compete successfully with conventional structures in the
economical arena. It is common now to permit ranking of
the elements to a considerable extent, in terms of the
cross sectional areas, so as to keep the quantities of
materials used as low as possible.

Recently, great progress has been made in the development of automatic tools -- robots -- made possible in
turn by the phenomenal advancements in the field of
electronics and computers. Such robots make possible
efficient production of small quantities of a great
variety of accurately finished parts. It is expected,
and hoped, that this technology will provide in the very
near future the solution to the problem of space frame
design described above.

6.2 Problems of Accuracy

A structure which consists of a number of plane
frames exhibits several levels of ranking among its
constituent elements. It should be noted that, on occasion, this ranking proves convenient in solving various
problems that occur during the process of construction.

As an example, consider that dome (a) shown in
Figure 3. The most important elements of the system are
the arches placed along the diameters of the plan, and it
is obvious that the dimensions of these arches should be
given with the highest accuracy. However, this means
simply that it is the accuracy relative to the other
elements of the structure, since the arches are the

biggest in size among the constituent elements of the
dome. Other elements are constructed with an accuracy
appropriate to their size.

It should be noted that the influence of errors in
the dimensions of the elements is 'directed' quite distinctly from the higher to lower ranks. In the example
of dome (a), for instance, the errors may be dealt with
in turn, beginning with the arches, through primary and
secondary beams, and ending with the purlins. The completion of erection of the arches means that the outline
of the dome has been set, and it is followed by the
erection of the minor elements - the beams - and attaching them to the arches.

Should dimensional difficulties arise in attaching
the beams to the arches, one attempts to deal with them
by introducing modifications to the beams, or to their
connections to the arches. Almost never does one re-
erect the arches, even when the difficulty might have
been caused by some inaccuracy in the dimensions of the
arches. An exception might be made if an extraordinarily
large discrepancy due to some mistake has been found in
the fabrication or erection of the arches. The same
phenomenon can be observed successively with regards to
the connections of primary to secondary beams, and the
secondary beam to a purlin.

Thus in every step of erection of a structure which
is an assembly of plane frames efforts are concentrated
on attaching elements of lower rank to those of higher
rank. When the work of attachment has been completed,
the erection of the structure proceeds to the next phase
of work by attaching elements of still lower rank to
those erected in the previous phase. In this way the
erection of a structure which is an assembly of plane
frames proceeds very steadily, generally without the need
for retracing steps.

This cannot be done in the case of a space frame
which is an assembly of elements without rank. All the
elements have to contribute to the constitution of the
structure with equal accuracy. It is very difficult to
predict in advance how an error in a certain element of a
space frame might influence the subsequent erection work,
and it is almost impossible to judge which errors might
have caused some difficulty in assembling a certain
portion of the total structure. It is not uncommon that
a successful assembly of a space frame becomes altogether
impossible for the reason discussed above.

To prevent this from happening, high standards of
accuracy may have to be imposed on the length and angular

dimensions of elements. However, it should be noted that
the significant efforts expended at the time of fabrication and erection of the elements of a space frame are
never appreciated by those who see it or who use it after
completion, and to whom a dimensional error of a few
centimeters is always tolerable.

The problem of accuracy seems to have been solved in
many space frame systems thanks to the developments in
the field of machine tools. Today, many tools are fully
automated, and joints and linear elements can be machined
with very high precision without a corresponding appreciable increase in costs.

Similar observations can be made concerning the
influence of unintentionally loose connections.   In a
plane frame assembly the extent to which a loose fit in a
connection influences the deformation of the whole structure varies with the rank of the elements which are being
connected. In general, the range of influence of a loose
fit which has occurred in some element is only a fraction
of that of an element one rank higher. Thus the problem
of loose connections can be effectively avoided by concentrating more attention on the connections involving
elements of higher rank.This approach is very practical
as well, since in general the higher the rank of elements
the smaller is the number of their connections.

In a space frame, on the other hand, there exists no
such rule regarding the influence of a loose connection
of an element on the total deformation of the structure.
Consequently, equal attention has to be paid to all
connections in a space frame. The most promising solution of this problem seems to lie, again, in the rapid
development of the computer-driven automatic machine
tools. One such tool is the NC processing machine which
may accomplish highly accurate production of elements
with differing dimensions at a reasonable cost.

6.3 Problems of Falsework

The problems of falsework are important in the erection of every structure, not only because it is sometimes
the largest part of the total cost of construction, but
also because falsework determines the accuracy of the
structure which it supports, and to which it gives form.
In this connection a plane frame assembly has again the
advantage. In such structures the constituent elements
are assembled in the order of ranking, and the attachment
of elements is completed rank by rank. Therefore, at
every stage of erection the elements of higher rank which
have been already installed in their proper position in

the structure play the role of a very accurate falsework
for the assembly of the elements of lower rank.

A space frame does not have such a sequential order
of erection and partial completion. For this reason, a
space frame structure requires special consideration of
its method of erection, or else a complete falsework has
to be provided.

The problems of falsework have been solved in most
cases by adopting the technique of lifting the whole
structure, or its large part, after an assembly on the
ground. Double-layered grid systems seem to have been
most successful in this regard, since the system as a
whole is flat and well suited to assembly on the ground.
Such solutions have been made possible through the developments of the lifting systems. The lifting technique
of erection is now almost routine for double-layer roof
grids. For other types of space frames, domes and vaults
for instance, it does not appear that such routine procedures have been found as yet. The manner in which the
erection of such systems has been tackled by engineers is
discussed in the following.

7. Constructional Aspects of Space Frames

7.1 Problems

A domical space frame, once completed, is one of the
most efficient spatial roof structures capable of covering a very wide area. It is not always efficient, however, from the viewpoint of construction, because it requires big amount of scaffoldings, labor and time and
often encounters difficulties in terms of accuracy,
reliability and safety of work during its erection.
Modern erecting methods such as lifting systems which are
very often adopted in erection of double-layer grids of
plate type can not equally be applied to a domical space
frame.

Buckminster Fuller tried to solve this kind of
problems in a few ways when he encountered them in building some of his geodesic domes. For construction of one
of his domes in Honolulu in 1957 he adopted a system in
which a temporary tower was erected at the center of the
dome from top of which concentrically assembled part of
the dome was hung by means of wire ropes. As assembly of
the dome proceeded the dome was gradually lifted, enabling the assembling work to be done along the periphery
of the dome always on the ground. He also adopted another method when he built a huge dome of 117m in diameter
at Wood River, U.S.A., in 1959, where the assembled part

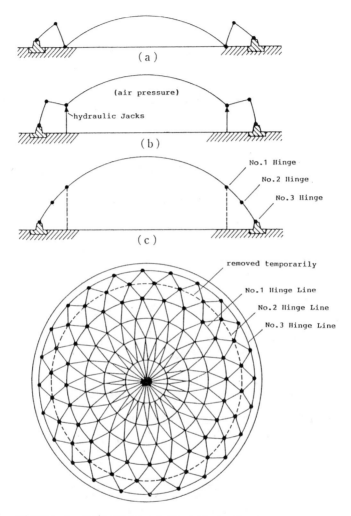

Figure 4. Principle of Pantadome System

of the dome was raised on a balloon-like enclosure. Some
other cases have also been reported where different
lifting methods have been applied to different domes.
However, none of the above methods for lifting domes have
become popular unlike many lifting methods which became
widely used to raise plate-type space frames.

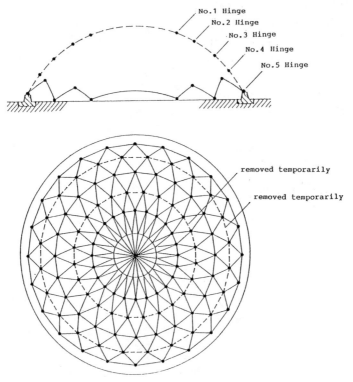

Figure 5. Doubly Folded Pantadome System

7.2 Principle of Pantadome System

A patented structural system called 'Pantadome System' which had been developed by the author for a more rational construction of domical space frames was successfully applied to the structure of a sports hall completed in Kobe 1984. Pantadome System was then applied to the Sant Jordi Sports Palace in Barcelona and the National Indoor Stadium of Singapore and are now being applied to a few other wide-span structures.

The principle of Pantadome System is to make a dome or a domical space frame geometrically unstable for a period in construction so that it is 'foldable' during its erection. This can be done by temporarily taking out the members which lie on a hoop circle. Then the dome is given a 'mechanism', that is, a controlled movement, like a 3-D version of a parallel crank or a 'pantagraph' which is generally applied to a drawing instrument or a power collector of an electric car (hence the name, 'Pantadome') (Figure 4).

Since the movement of a Pantadome during erection is 'controlled one' with only one freedom of movement (vertical), no means of preventing lateral movement of the dome such as staying cables or bracing members are necessary during its erection. The movement and deformation of the whole shape of the Pantadome during erection are three dimensional and may look spectacular and rather complicated, but they are all geometrically determinate and easily controlled. Three kinds of hinges are incorporated in a Pantadome system which rotate during the erection. Their rotations are all uni-axial ones, and of the most simple kind. Therefore, all these hinges are fabricated in the same way as normal hinges in usual steel frames.

In Pantadome system a dome is assembled in a folded shape near the ground level. As the entire height of the dome during assembling work is very low compared with that after completion, the assembly work can be done safely and economically, and the quality of work can be assured more easily than in conventional erection systems. Not only the structural frame but also the exterior and interior finishings, electricity and mechanical facilities are fixed and installed at this stage. The dome is then lifted up. Lifting can be achieved either by blowing air inside the dome to raise the internal air pressure, or by pushing up the periphery of the upper dome by means of hydraulic jacks. When the dome has taken the final shape, the hoop members which have been temporarily taken away during the erection are fixed to their proper positions to complete the dome structure. The lifting means such as air pressure or hydraulic jacks can be then removed, and the dome is completed.

7.3 Applications of Pantadome System

Pantadome system can be successfully applied to domical frames of various configurations. Figure 6 shows a model for a project of a 200m diameter indoor stadium, in which the principle of Pantadome System stated above is directly applied.

World Memorial Hall in Kobe in which the Universiad '85 was celebrated has an oval plan of 70m x 110m. It was designed by the author in cooperation with Architect Mitsumune, and it was constructed by means of Pantadome System as shown in Figure 7. It was the first dome to which Pantadome System was applied.

Figure 6. A Study Model of Pantadome System

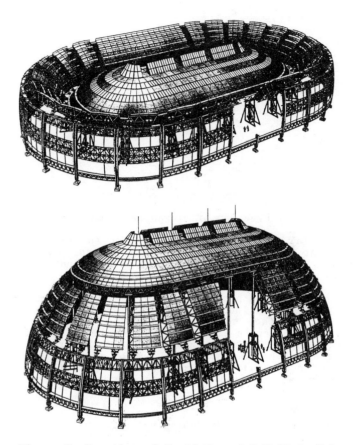

Figure 7. Erection of World Memorial Hall in Kobe

The second example of Pantadome application is the Singapore National Indoor Stadium having a rnombic plan of 200m x 120m in the diagonal directions, the structure of which was designed by the author in cooperation with Architect Kenzo Tange.

National Indoor Stadium has an arena of 3000m^2 and grandstands for 12000 seats. The geometry of the roof is constituted by four cylindrical surfaces, each convex inward, having the axes of the cylinders parallel to the four sides of the rhombic plan. Although the roof surface which is convex inward gives the visual impression

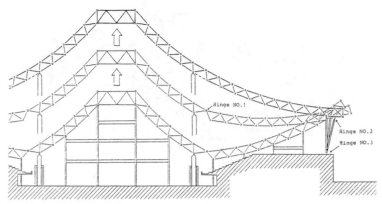

Figure 8. Erection of Singapore Indoor Stadium

of a hanging roof, actually it has a sufficient dome effect. The hinge lines for Pantadome mechanism were set along the straight lines parallel to the boundaries of the roof plan.

The method and processes of erection of the roof of Singapore National Indoor Stadium are shown in Figures 8 and 9. Around the two opposite corners of the rhombic plan the floor levels are elevated, and the height of the roof is lower. So it was considered that the erection of the roof could be easily carried out by means of conventional methods around these areas. This is why the Pantadome System was applied only to the central part of the roof(Figure 9).

7.4 Barcelona Sant Jordi Sports Palace

Sant Jordi Sports Palace in Barcelona was one of the venues for the Olympic Games '92. As a result of an international design competition the author had the opportunity of designing the structure of this important building.

The roof structure which covers an area of 128mx106m for the arena and the grandstands seating 15,000 is constructed by a steel space frame consisting of 9,190 tubes connected by 2,403 joints. The shape of the roof structure is constituted by the central domical part and the four toroidal parts surrounding it(Figure 10). The area of the whole roof surface is 13,460 m^2. The type of the space frame is what is called a double layer grids

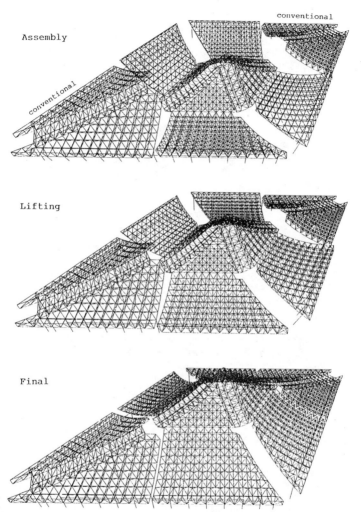

Figure 9. Lift-up Processes of Singapore Indoor Stadium

structure having a depth of 2.5m. The diameter of the
steel tubes of the space frame ranges from 76mm to 267mm
with exceptions of bigger tubes for valley and ridge
members(406mm) and for the peripheral members(508mm).

The roof has a rise of 21m as a dome. The weight of

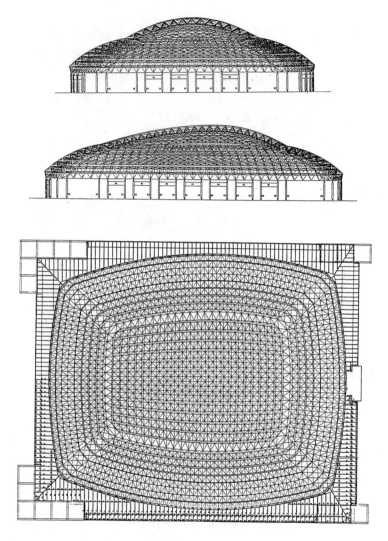

Figure 10. Structural System of Sant Jordi Sports Palace

the whole roof including finishing is about 3,000 tons,
of which 1,000 tons is the weight of the steel space
frame. the roof structure is supported by 60 columns

which stand on the reinforced concrete substructure along the periphery of the grandstand. The height of the roof is 31m above the base of the columns. Since the floor on which the columns stand is 14m high from the arena level, the maximum height of the roof from the arena floor is about 45m.

Each adjacent pair of the 44 columns is rigidly connected to each other at their tops by means of a lattice girder to form a portal frame in the peripheral direction. Since all the columns are kept hinge-connected at their tops and bottoms even after completion so that the roof is free to expand and shrink without producing thermal stresses in the structure when temperature changes, the lateral resistance of the roof structure is exclusively given by those 22 portal frames (14 in the longitudinal and 8 in the transverse directions, respectively) along the periphery of the roof.

Assembly of the roof structure was effected in the site by means of hoisting systems. It consisted of the following two different stages:

a) Assembly of the central dome on the arena, supported by provisional light steel columns.
b) Assembly of the 16 peripheral segments of approximately 400m² weighing up to 40tons.

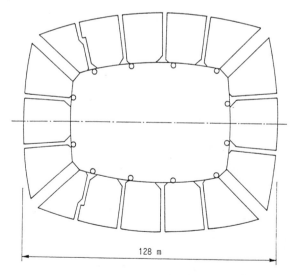

Figure 11. Lifting Segments of the Roof (small circles indicate locations of lifting towers)

The equipment for pushing up the roof was composed of 12 lifting towers(Figure 11). Each tower consisted of(Figure 12):

a) A tetrahedron-shaped spreader truss which transferred the loads from two ball joints of the roof structure evenly and without constrains to the tower.
b) The tower itself (a standard temporary 4-panel tower), which was extendible at its lower end. The height of one section was 3.04m.
c) A lifting frame which was suspended from two lifting units by means of strand cables. Special locking devices provided the connection between the frame and the temporary tower.
d) Two narrower portal frames which supported the lifting units.

When all the specified roof members had been assembled, and necessary lifting equipments had been installed, the lifting operation began(Figure 13). The lifting operation was carried out in stages of 3.04m, which corresponds to the height of one full tower element. The work to accomplish one stage was:

A = Support frame for lifting unit
B = Lifting unit
C = Standard Acrow tower
D = Lifting frame

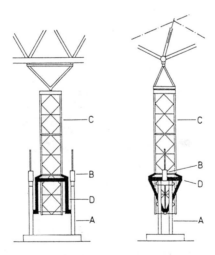

Figure 12. Lifting System

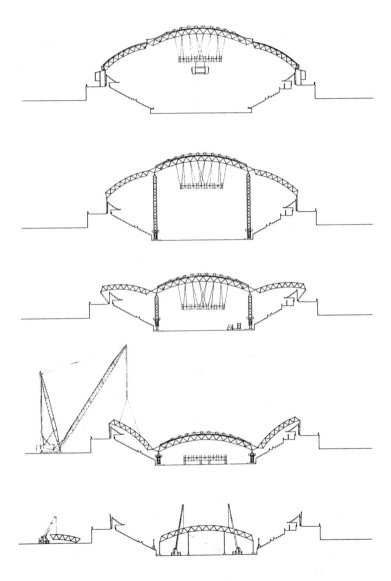

Figure 13. Erection Processes of Sant Jordi Sports Palace

1) Engaging of the mechanical locking devices between lifting frames and towers.
2) Taking over of the load by the hydraulic lifting units, lift off and removal of the four bearing shoes of each tower.
3) Lifting in small steps through a distance of 3.04m with intermediate level readings.
4) After reaching 3.04m, tape reading and complementary measurements were analyzed and adjustments made, where necessary.
5) Extension of the towers by means of 12 new elements and installation of the bearing shoes.
6) Synchronous lowering of all towers, thus transferring the loads from the jacks to the bearing shoes.
7) Opening of the locks between lifting frames and towers and lowering of the frames.

The lifting operation began on November 21st, 1988. The load on the 24 lifting units were initially 760 tons. After 2.74m of lifting, a platform measuring 60x20m and weighing 80tons which carries electrical and ventilation equipment was hung from the central part of the roof by tensioning the suspension cables. In 10 days the roof was pushed up the height of 9 full and one half tower elements, which corresponds to 28.83m. Once the final height had been reached, the towers were set down onto steel plate packers of 16cm height. Thereafter, lifting units, pumps and controls were dismantled and removed.

The members which had been kept away from their positions during the lift were then fitted in and welded to complete the roof structure. At the same time the installation works of insulation, roofing and interior finishing continued. After completion of the structural system the loads on the temporary towers were relieved successively in small steps by jacking up one tower at a time and by removing gradually the packer plates. The vertical displacements of the center of the roof due to removal of the lifting towers was 140mm, showing a close agreement with the result of analysis.

7.5 Other Applications

Three other buildings are now under construction on the Pantadome System. One of them is a multi-purpose building in Fukui to be used for the World Championship Games of Athletics in 1995. The roof covers a circular area of 116m in diameter. Since the site is a heavily snowed area, the weight of the roof structure itself is considerable. The lifting of the roof is scheduled for August 1994. Another application of Pantadome System is being tried for a sports hall to be built in Osaka. It has an oval plan of 110mx125m. The roof is slightly(5 degrees) tilted, and so the lifting direction is not vertical. This roof is to be lifted in November 1994. The last one is a small dome of tensegrity which will be lifted in January 1994.

8. Conclusive Remarks

Several aspects of space frames which are sometimes called latticed structures have been discussed. Special features of space frame in comparison with plane frames have been described. Space frames have lots of advantages on one hand, but they have problems on the other hand. One of the biggest problems of space frames is difficulties in construction. So the construction aspects of space fames have been studied.

As a solution for the construction problems of domical space frames a patented structural system named Pantadome System has been presented. Pantadome System has been applied to large-span domes of a few different geometries. It has so far been successfully applied in a few different countries (Japan, Singapore and Spain) where local building conditions are different from each other. It seems that these examples support rationality of the system in terms of safety, construction speed, quality of built structures and economy of space frames.

References

ASCE-ST11 "Latticed Structures:State-of-the-art",1976

Gaylord,D.H. et al "Structural Engineering Handbook", 1968, McGraw-Hill

Kawaguchi,M. et al "A Domical Space Frame Foldable During Erection" Proc. Third Int. Conf. on Space Structures, 1984

McGraw-Hill "Encyclopedia of Science and Technology",1977

Mainstone,R."Developments in Structural Form", MIT Press, 1975

Medwadowski,S."Modern Spatial Wood Structures in the United States", IASS Bulletin No.76, 1981

Medwadowski,S."Conceptual Design of the Structure of The UC Ten Meter Telescope", Univ. of California, 1981

Merritt,F.S."Building Construction Handbook" McGraw-Hill, 1975

Merritt,F.S."Structural   Steel   Designers'   Handbook",
McGraw-Hill, 1972

Morgan,W."The Elements of Structure", Sir Isaac Pitman &
Sons, 1964

Salvadori,M. et al"Structure in Architecture",Prentice-
Hall, 1963

Project Case Study
Space Frame Roof Structure
Main Stadium
Shah Alam Sport Complex
Selangor, Malaysia

James A. Moses, P.E.

The Shah Alam Sport Complex is a $100-million multi-use sports facility located just outside of Kuala Lumpur, Malaysia, in the city of Shah Alam. The main feature of the complex is an 80,000 seat "Olympic type" Stadium. Seating on each side of the stadium is protected from the weather and the tropical sun by a pair of barrel-vaulted polycarbonate sky roofs. Each roof is supported by an arched Unistrut Space Frame. The sides and back of each space frame are supported by the concrete structure of the stadium. The front arch of each space frame has a free span of 284 m (931 feet). Each space frame cantilevers a maximum of 69 m (226 feet) from back to front, over the seats and raises 61 m (200 feet) above the playing field.

Each of the space frames weighs 700 metric tons, has 10775 tubular members and 2778 nodal joints. The tubes range in size from a diameter of 76.3 mm (3") to 356mm (14"). The surface area of each frame is 17,913 M^2 (185,063 square feet).

The extremely complex and challenging job of designing and constructing these space frame structures has been detailed in other papers. The purpose of this paper is to discuss the management of this international project from its initial developmental stages through completion.

1. Project Development

The major portion of Unistrut's involvement began with the bidding process

Vice President of Engineering and Construction, Unistrut Corporation, 1500 Greenleaf Avenue, Elk Grove Village, IL 60007-4701 in the latter half of 1988 and first half of 1989. The original design of the stadium involved architects from Malaysia, Singapore, and the USA and used a small stadium in Split, Croatia (formally Yugoslavia) as a model. In this case, the term model is appropriate since the Split Stadium is only two-thirds as large as the Shah Alam Stadium. As a coincidence, the general contractor on the project, Mewah Sekitar Sdn Bhd, is a joint venture whose major partner, Energoprojekt, is a Serbian (Yugoslavian) company based in Belgrade.

As is becoming obvious, this project was truly an international affair, in which participants from over a dozen countries eventually became involved. The bidding for the space frame structures alone, attracted companies from Germany, Japan, Malaysia, and the USA, among other places. It is important to note, however, that all foreign entities were required to work with and/or through local Malaysian companies. This requirement not only ensures local economic gain from the project, but also a transfer of technology and experience which will enable domestic companies to maintain the stadium and tackle the additional construction projects the sports complex requires.

2. Project Management

A project of this size and technical complexity creates a unique set of challenges for the subcontractor. Standard procedures for documentation, inventory control, quality assurance, design, manufacturing and construction must all be reviewed, modified and updated. In general, methods and procedures which are usually very effective to provide internal control will not be sufficient to effectively manage a project of this size.

Unistrut reviewed all of its internal processes for project management. In many instances significant changes were required. This paper reviews some of the situations that were encountered and analyzes the changes that were adopted to effectively manage the project.

3. Design Development

A. Project Standards

Malaysia is a former British colony. All of the internal design practices are governed according to British standards. This project was no exception. The project specifications required compliance to British Standards.

For smaller international projects Unistrut would design according to U.S. standards which are generally acceptable in most of the world. Since the primary construction material for space frames is steel - one finds great similarity between the most commonly used specifications (American, British, Japanese, German). Unistrut would normally merely highlight the differences between the two standards and demonstrate that the standard used either "meet or exceeds" the local standard, or that the exception was accounted for.

For Shah Alam, Unistrut acquired a full set of current applicable British Standards. These were used to re-create the basic design standards used to engineer the space frames. All of the submittals fully complied to British Standards.

B. Data Set Development

Each member in a space frame is designed to carry specific loads in both tension and compression. Through the years Unistrut has developed a series of members each of which can meet specific structural requirements. These common members were developed to make the most efficient use of readily available, cost effective materials and manufacturing processes.

If a particular member can accommodate a compression force of 35 kips, the next member in the data set may be designed to a compression strength of 50 kips. The structural analysis of any one space frame may determine that a given member will experience a worst case compressive force of 36 kips. This will cause the selection of the 50 kip member even though its compression value is almost 30 percent greater than what is required. Under normal circumstances it would be more efficient and cost effective to proceed with this member selection.

For a large project, the weight and cost of the structure can be substantially reduced by eliminating the inefficiency described above. This is done by a detailed analysis of the data set and the design of members that meet the most repetitive structural demands on the members. Unistrut estimates that the re-designed data set internally developed by this process reduced the weight of the structures by 8%. This amounted to over 100 metric tons.

C. Special Design Criteria

Conventional published Design Standards are adequate to control the design of the great majority of structures. This project was unique in size, shape and span. The structure had to be studied to develop and/or confirm the design criteria concerning wind and dynamic loading as well as the final deflected shape of the structure.

The impact of wind on the two large crescent shaped arches was analyzed by RWDI,an internationally accepted wind load expert based in Ontario, Canada.

In order to assess the behavior of the structure under various simultaneous load conditions, a dynamic analysis was performed by computer simulation to determine the modal shapes which could occur.

The mid span elevation and deflection criteria for this structure was very precise. In analyzing the impact of temperature expansion, member length change due to buildup of galvanizing or paint or a small variance in the design load versus the actual weight of the cladding, Unistrut determined that the mid-span elevation could be impacted by several inches. To prevent this occurrence, manufacturing tolerances were tightened and design loads very precisely determined.

Each of these situations required unique modification to conventional design practices.

4. Project Control

Once again, strictly because of the size and complexity, conventional methods of project control had to be modified.

Unistrut started at the very top with the creation of a key man Steering Committee chaired by the President of the Corporation. This committee met once a month and conducted a complete review of the overall project status. Task force subcommittees were established to address unique requirements. The individuals in charge of each function reported their activity to the Steering Committee.

The Steering Committee was needed to assure that all activities were coordinated. Without this top management control, two individuals could be working on the same task - or more importantly, a particular task could have been missed under the assumption it was being handled by someone else.

Even the procedures for documentation had to be modified. Existing standard procedures for document control of conventional projects provide central storage cabinet filing of all the documents for one project sequentially by project number in one folder. Chronological document filing in specially designated cabinets was established for Shah Alam. The documents were numbered, titled and entered into a computerized library index to ease future access and retrieval. The amount of documentation was voluminous.

Special Testing and Review

A paradigm could be described as an inability to see clearly because your vision is affected by past experience. Unistrut has been designing, manufacturing and installing space frames for 50 years. Unistrut has never experienced a structural failure. The company takes pride in the thousands of space frames erected under their control.

But when a project comes along that pushes the edge of the design envelope, a company must be careful not to let its past experience create a paradigm ego. For Shah Alam, Unistrut contracted with a well-recognized professional in lattice structure design, Professor Erling Murtha-Smith of the University of Connecticut.

Professor Murtha-Smith reviewed the entire design package and participated in a number of spirited discussions with Unistrut's key technical personnel. Professor Murtha-Smith made several valuable suggestions which Unistrut adopted. In one particular case it was decided to test the structural impact of the hole in the side of the tube used for insertion of assembly hardware. The tests revealed an interesting phenomenon which is discussed in a separate paper.

6. Quality Control

For this international project, Unistrut elected to source the materials from a supplier in Korea. Unistrut had successfully used this company on many prior occasions. Materials when imported into the United States were then inspected to assure correct fabrication and quality.

This project alone represented 1400 metric tons of steel to be produced and shipped over several months. To assure the highest possible levels of quality at the start and through the duration of the process, a completely new quality control manual was developed. Standards for inspection were established. Welds were x-ray inspected by an independent company. All inspection procedures were documented.

Finally, Unistrut hired an independent testing company to inspect all of the materials in each shipping container prior to dispatch. If any discrepancies were encountered, the shipment was delayed until appropriate corrective measures were taken.

7. Inventory Control

Each part in the structure was given a unique part number. Without proper inventory control procedures it would have been impossible to locate the right part at the time it was needed in the construction sequence.

Using computers, Unistrut sorted all the components by size to facilitate efficient manufacturing and handling. The parts were packaged in specific containers designed to carry a specific grouping of parts. The containers were numbered and became the inventory control stock keeping unit. The parts were placed in the containers in the sequence they would be needed.

Each container was double checked to assure it contained all the correct parts. Inventory control and location was so simplified by this process that not one piece of the 27000 nodes and tubes was lost or missing.

8. Communication and Administration

The impact of the fax machine on the global business community was never more clearly demonstrated than in the course of this project.

Kuala Lumpur and Seoul, Korea are 13 hours ahead of Unistrut's headquarters in Michigan. Because of language difficulties and the lack of permanent record telephonic communication was avoided. International courier services could take 7-10 days to deliver. Unistrut relied on the fax machine for daily communication to the factory and the site.

From an administrative and contractual standpoint, Unistrut relied on its experience in the U.S. construction marketplace. The differences in construction practices are dramatic and could be the basis for another paper. U.S. practice is much more "legalized" and specification oriented. Although bureaucratic at times, the benefits of such procedural documentation became very apparent in their absence.

As the globe continues to shrink, as free trade becomes a reality, as technology continues to simplify transportation and communication, many companies should assess their capability to service such projects - so they are prepared to grab the opportunity when it comes.

STRUCTURAL TESTS ON THE "SUSPEN-DOME" SYSTEM

Mamoru Kawaguchi[1]   Masaru Abe[2]
Tatsuo Hatato[3]   Ikuo Tatemichi[3]
Satoshi Fujiwara[3]   Hiroaki Matsufuji[3]
Hiroyuki Yoshida[3]   Yoshimichi Anma[3]

Abstract

The "suspen-dome" system, which the authors have developed, consists of a single-layer truss dome and a cable structure. The present paper deals with the results of structural tests conducted on this suspen-dome system. The test models used were circular, with a diameter of 3.0 m. Brass pipes were employed for the single-layer trusses, and steel rods for the cable systems. The results of the tests, by controlled loading, confirmed that the incremental axial stresses in the dome truss members were smaller in the suspen-dome than in the single-layer truss dome. In the buckling test, the suspen-dome was found to have a buckling load 1.5 to 1.8 times that of the single-layer truss dome.

Introduction

The "suspen-dome" system is a composite structure consisting of a single-layer truss dome, struts and cables.[1] An outline drawing of the structural system is given in Figure 1. The struts are suspended from the nodes in the single-layer truss dome and the lower ends of these struts are bound to the nodes on the outside on the single-layer truss dome by "radial cables." The lower ends of the struts are also bound to each other by ring-shaped "hoop cables."

By applying an appropriate amount of pre-stress to the radial cables, the overall thrust on the dome can be reduced, leading to a reduction in the stresses acting on the outer ring girders and other truss members. At the same time, when extra loads, such as that due to snow, are applied, additional tension is generated in the cables, thereby reducing the increases in the thrust. The cables and struts also provide greater stability against buckling.

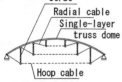

Figure 1 Suspen-Dome
(Structural System)

Single-layer truss domes and cable domes have been the subjects of a large

1. Professor, Hosei University, Dr. Eng., Koganei, Tokyo 184
2. Lecturer, Hosei University, Koganei, Tokyo 184
3. Maeda Corporation, 2-10-26 Fujimi, Chiyoda-ku, Tokyo 102

number of studies (see References 2) and 3)). The suspen–dome is a new dome structure that combines features of these. Model tests were conducted on the suspen–dome to study the effects of the cable tensioning, and the load transmission capacity and structural stability of the suspen–dome.

Outline of Tests

Test Models and their Assembly

The tests were conducted on two test models (A and B). For control, the same tests were also carried out on a single–layer truss dome (one model) without the struts and cables. The test models had a span of 3.0 m and a rise of 0.45 m (rise–span ratio: 0.15). Each suspen–dome model had a mass of approximately 250 kg. The test models are shown in Figure 2 and Photograph 1.

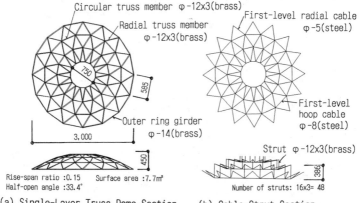

(a) Single-Layer Truss Dome Section   (b) Cable-Strut Section

Figure 2 Test Models

Photograph 1 Test Model

1) Truss and strut members:　Brass rods (ϕ -14) were used for the outer ring girders, while brass pipes (ϕ -12×3) were used for other truss and strut members.

2) Cable members:　Steel rods (ϕ -3 to 8) were used to facilitate assembly and measurement.

3) Bearings:　Flat rollers were used and were made free to move along the radial direction (Figure 3).

4) Connections:　Disk-shaped brass nodes (diameter: 10 cm) were used at the connections for truss members (Figure 4). The truss members were screwed into the nodes via steel bolts (M8). The radial cables were bound with bolts and tensioned by rotation of nuts.

The errors in the lengths of the members after assembly are given in Table 1. The errors in the node elevations were within 1% of the rise in most cases (Figure 5).

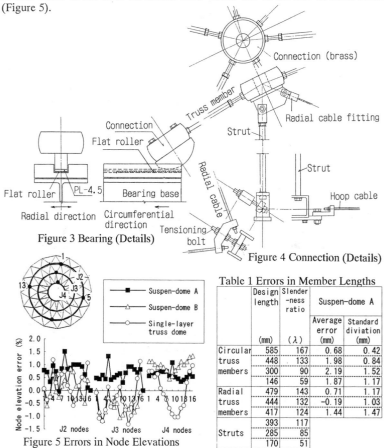

Figure 3 Bearing (Details)

Figure 4 Connection (Details)

Figure 5 Errors in Node Elevations

Table 1 Errors in Member Lengths

	Design length	Slender-ness ratio	Suspen-dome A	
			Average error	Standard diviation
	(mm)	(λ)	(mm)	(mm)
Circular truss members	585	167	0.68	0.42
	448	133	1.98	0.84
	300	90	2.19	1.52
	146	59	1.87	1.17
Radial truss members	479	143	0.71	1.17
	444	132	-0.19	1.03
	417	124	1.44	1.47
Struts	393	117		
	285	85		
	170	51		

Mechanical Properties of Test Model Members and Connections

Tension tests were conducted on the test model members and connections. The connections consist of truss members, connection bolts and a node. The load–elongation curves for the brass members and connections are given in Figure 6. The linearity in the elastic region can be observed, but there were no clearly marked yield points. The connections have a greater elastic rigidity than the truss members, but their yielding strength is smaller. The Young's modulus for the brass members (ϕ -12×3) was 1.00×10^5 N/mm^2, which is approximately a half of the value for the steel members (ϕ -5: 2.11×10^5 N/mm^2).

Bending tests were also conducted on the members and connections. The moment–rotation angle curves obtained are shown in Figures 7 and 8. In the connections, the bending yield of the M8 bolts predominated and non–linearity came into operation once the moment reached around 1.5 kN•cm.

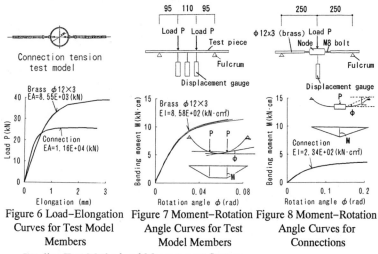

Figure 6 Load–Elongation Curves for Test Model Members

Figure 7 Moment–Rotation Angle Curves for Test Model Members

Figure 8 Moment–Rotation Angle Curves for Connections

Loading Test Method and Measurement System

Controlled loading was implemented using an oil jack. Loads proportional to the dominant area of each node were applied using a "tournament" system of weights as shown in Figure 9. The mass of the tournament system was 716 kg.

The displacement measurement points are shown in Figure 10. The displacement was measured with wires installed between the measurement points and the displacement gauges. The strain was measured using strain gauges. The measurement points for the strain on truss members are shown in Figure 11. The axial strain was measured on all the cable members because of the need for tension control.

Tensioning and Elastic Loading Test

Cable Tensioning

A tension corresponding to around 40% of the design tension (the tension at

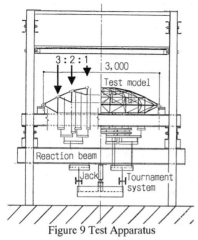

Figure 9 Test Apparatus

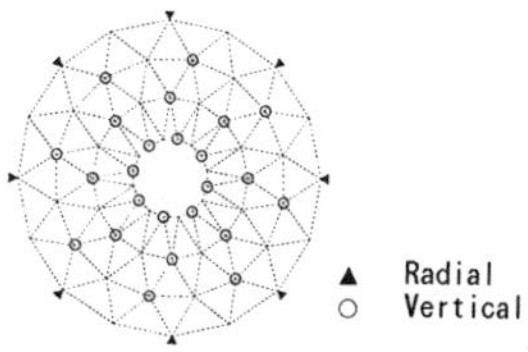

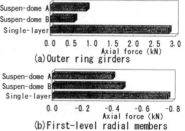

Figure 10 Displacement
Measurement Points

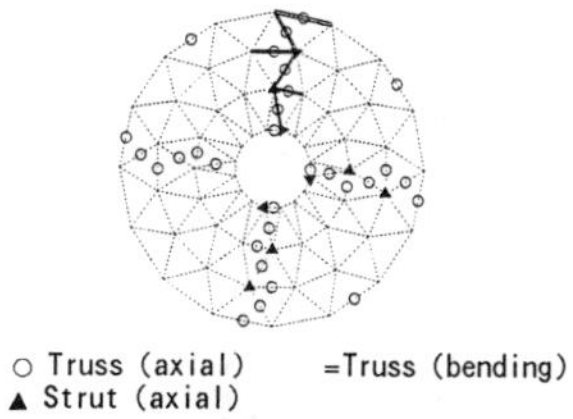

Figure 11 Strain Measurement Points

which the axial stresses in the outer ring girders are reduced, more or less, to zero: 391 N in the first-level radial cables) was applied as the initial tension. Additional tension (30%) was then applied by suspending the tournament system and adjustments were made on the tension by hand. The stresses in the truss members after adjustment are shown in Figure 12. The tensioning had the expected effects, such as the reduction of the stress in the outer ring girders.

(a) Outer ring girders

(b) First-level radial members

Figure 12 Truss Member Stress after Tensioning

Elastic Loading Test

The incremental axial stresses in the truss members are shown in Figure 13. The incremental axial stresses are smaller in the suspen-dome than in the single-layer truss dome. The results of analysis by the finite-element method are shown together in the figure. The equivalent rigidity was applied to the truss members in the analysis. For the cable members, however, the rigidity was made to simulate the test results through a parameter study, as the analysis results varied greatly according to the estimations of the rigidity. The additional tension can be applied with greater effect as the rigidity of the cables increases. In the case of the test models used here, the rigidity as obtained through analysis was significantly smaller than the axial rigidity of the cable members.

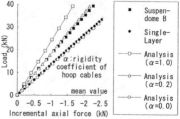

Figure 13 Incremental Axial Stress in First-Level Radial Members

Buckling Test

Rupture of Suspen–Dome

The load–deformation curves are given in Figure 14. The curves begin to diverge in the results for the top section once the load has reached around 50 kN. This reflects the deformation of the top compression ring outside the surface line of the shell. The test models suddenly gave large cracking noises at 93.2 kN (test model A) and 108.9 kN (test model B), and this was followed by a rapid reduction of the load.

Observation of the conditions after rupture revealed large rotation at the connections on the second level from the bearings and deformation of the connecting bolts for the radial members (Photograph 2). The vertical displacement distributions at the maximum load and after rupture are shown in Figure 15. Partial depression can be observed here centring around the nodes on the third level from the bearings. The maximum displacement was 36.7 mm on test model A and 38.0 mm on test model B. On each test model, one tensioning bolt was found to have ruptured at the lower end of a second–level radial cable on one side of the depression zone.

Rupture of Single–Layer Truss Dome

The load–deformation curves are given in Figure 16. The curves begin to diverge in the results for the top section, as in the suspen–dome, once the load has reached around 30 kN. After an appearance of a slight non–linearity in the values for the nodes on the second level from the bearings, a sudden lowering of the load was observed, together with a large noise, at 60.8 kN.

Observation revealed a major depression of the nodes at the second level from the bearings. The first–level radial truss members were found to have undergone major deformation outside the surface line of the shell and the sudden rotation at the bearings had resulted in the dislocation of the flat rollers (Photograph 3). The vertical displacement distribution after the test is shown in Figure 17. The maximum displacement here, at 157.0 mm, was far greater than that observed on the suspen–dome models.

Photograph 2 Deformed Section
(Suspen–Dome B)

Photograph 3 Deformed Section
(Single–Layer Truss Dome)

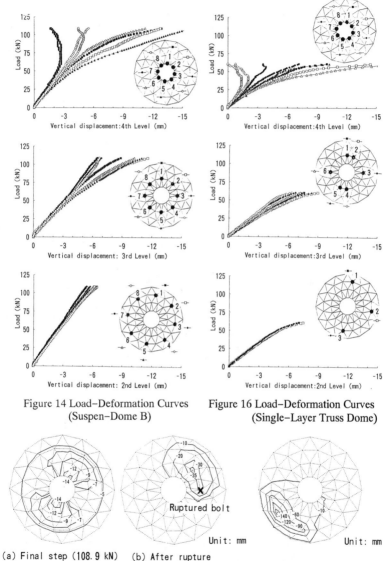

Figure 14 Load–Deformation Curves
(Suspen–Dome B)

Figure 16 Load–Deformation Curves
(Single–Layer Truss Dome)

(a) Final step (108.9 kN)   (b) After rupture
Figure 15 Vertical Displacement Distribution
at Maximum Load and after Rupture
(Suspen–Dome B)

Figure 17 Vertical Displacement
Distribution after Rupture
(Single–Layer Truss Dome)

Investigations on Buckling Test Results

Axial Stress in Radial Truss Members   The load–incremental axial stress curves for the radial truss members are given in Figure 18. These showed linear increases and remained in the elastic region for the materials up to the maximum load. The elastic buckling loads for the truss members with nodes were estimated through analysis and were compared with the test results. As boundary conditions, the members were assumed to have either pin connections at both ends ($\ell_k = \ell$, where ℓ_k: buckling length, ℓ: member length) or fixed connection at both ends ($\ell_k = 0.5\ \ell$).

The analysis results for the truss member buckling loads and the test results for the axial stresses are plotted on the figure showing the tension test results (Figure 19). The maximum axial stresses obtained in the loading tests were slightly greater than the buckling load for a member with pin connections at both ends. As the connections in the test models were not completely rigid, the stress conditions were such as to allow elastic buckling of the truss members.

Bending Stress in Radial Truss Members   The load–bending stress curves for the radial truss members obtained through measurement are shown in Figure 20. It can be seen that the bending stresses show a strong non–linearity under loading, and a maximum bending stress of nearly 2 kN•cm was observed in one member. Since the elastic limit of the connections, according to the bending test results, is approximately 1.5 kN•cm, it is possible that the yielding of the bolts at the connections in certain parts in the tests was caused by the increase in the bending stress. The measurement values obtained in the tests are plotted on a figure showing the results of the bending test for connections (Figure 21).

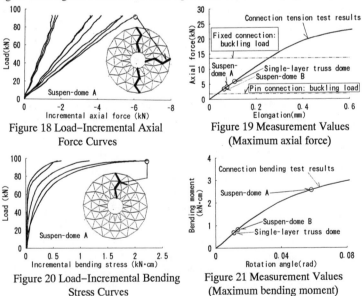

Figure 18 Load–Incremental Axial
Force Curves

Figure 19 Measurement Values
(Maximum axial force)

Figure 20 Load–Incremental Bending
Stress Curves

Figure 21 Measurement Values
(Maximum bending moment)

Suspen-Dome and Single-Layer Truss Dome   The collapse of both the suspen-dome models and the single-layer truss dome model is thought to be due either to the elastic buckling of the truss members or to the member buckling due to yielding of connection bolts. This resulted in depression at the connections immediately above these members and large rotation at the connections immediately below them.

The conceptual drawings for the depression zones are given in Figure 22. On the suspen-domes, the member buckling occurred on members one level further in towards the centres of the domes than on the single-layer truss dome. In other words, the corresponding points have shifted one step towards the centre in the suspen-domes when compared with the single-layer truss dome. This is attributable to the small reinforcement effect of the inner cables, despite the large additional tension in the first-level radial cables.

The bending moments and axial stresses generated in the radial truss members are shown in Figure 23. The values for the suspen-dome are those observed under approximately twice the load for the single-layer truss dome. The stresses are generally kept at a lower level in the suspen-dome. It is to be noted that the nodes subject to the maximum moment are one step further in on the suspendome than on the single-layer truss dome. The buckling, in other words, occurred on the members where both the axial and bending stresses were large.

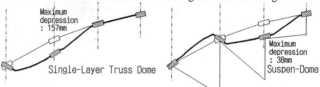

Figure 22 Depression Zones (Conceptual Drawings)

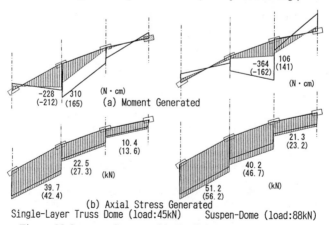

Figure 23 Stresses Generated in Radial Truss Members
☐ and figures without brackets: Measurement values (average)
▥ and figures in brackets: Analysis values (average)

Safety of Suspen–Dome against Buckling   Because of the smaller increases in the member stresses, the suspen–dome is less liable to member buckling and, when buckling takes the form of member buckling, the buckling load is greater than in the single–layer truss dome. Since the member stress increase can be checked by increasing the additional tension in the cables, provision of adequate cable (esp. hoop cable) rigidity will lead to an increase in the buckling strength of the suspen–dome. Furthermore, it is possible to control the buckling load through adjustment of the initial tension.

The axial stress in the truss members at the time of buckling was found to fall between the analysis values for the buckling loads for members with pin connections and those with fixed connections. In the design of a suspen–dome, a conservative buckling load can be obtained by using the buckling load of members with pin connections at both ends ($\ell_k = \ell$). In this case, however, considerations will need to be made on the behavior of the connections, as well as the members themselves in the estimation of the buckling load.

The rupture of the tensioning bolts in the suspen–dome (Photograph 4) is thought to have been caused by the large bending tensile stress generated in the bolts, due to the large difference in the axial stresses in the adjacent radial cables at the edge of the depression zone created by member buckling.

Conclusion

Photograph 4 Rupture of Tensioning Bolt (Suspen–Dome)

The following observations were made in loading tests conducted on structural models of suspen–domes.

1) A significant reduction of the stress in outer ring girders and other members can be achieved through use of the suspen–dome system.

2) The suspen–dome system has a greater buckling strength than the single–layer truss dome.

3) The rupture positions, which are found on the flanks of the domes on single-layer truss domes, are shifted towards the centres of the domes on suspen–domes.

References

1) Kawaguchi, M., Abe, M., Hatato, T., Tatemichi, I., Fujiwara, S., Matsufuji, H.:"On A Structural System 'SUSPEN–DOME'", Proc. of IASS Symposium Istanbul, 1993, pp.523~530
2) Suzuki, T., Ogawa, T.,Kubodera, I., Ikarasi, K.:"Experimental and Theoretical Study of a Single Layer Reticulated Dome", Proc. of IASS Symposium Copenhagen , 1991, Vol III, pp.85~92
3) Yamaguchi, I., Okada, K., Kimura, M., Magara, H., Okamura, K., Ohta, H., Okada, A., Okuno, N., "A Study on the Mechanism and Structural Behaviors of Cable Dome", Proc. of IASS Symposium Peking, 1987

Design of Low Rise Dome for the
University of Hawaii at Manoa

Alfonso Lopez, M. ASCE[1]
Gerald Orrison, M. ASCE[2]

Abstract

The 97.5m span dome covering the new multipurpose sports arena at the University of Hawaii in Manoa is a spherical, triangulated dome structure with a rise of 16.2m. In addition to the loads required by code, the dome was designed for equipment loads totaling 2,000 kN. The center section of the dome is reinforced using a double layer space truss to support the concentrated loads. A description of this new double layer system is presented, as well as a comparison of the behavior of this system with the single layer design.

Introduction

An essential element in the design of a multipurpose arena is the capability of the roof structure to support large amounts of suspended equipment and other concentrated loads. In addition to accommodating major athletic events, large assembly facilities must also be designed for concerts, theatrical productions and other entertainment events that impose extensive rigging and equipment loads on the long-span roof system. Single layer geodesic geometries framed with aluminum structural members have been used successfully for these types of facilities for spans up to 98m.

The design of low rise, single layer geodesic domes subjected to large concentrated loads (and/or severe environmental wind and live loads) is usually controlled by constraints against overall buckling or localized buckling of some portion of the structure. For domes built with aluminum extrusions, buckling is often prevented by increasing the member depth and cross sectional properties. Aluminum structural members up to 460mm in depth have been used in multipurpose sports arenas and large bulk storage dome covers.

[1]Director of Engineering, Temcor, 24724 S. Wilmington Ave., Carson, CA. 90745.
[2]Manager, Technical Services, Temcor, Carson, CA.

This paper presents an alternate solution to the stability problem for the roof systems described above. The concept is a simple one: The geodesic roof structure is reinforced using a double layer space truss system in areas of high load concentration. Stability analyses are used to verify the integrity of the double layer space truss system, and comparisons of the behavior of the single layer and reinforced geodesic geometry are presented.

The design of single or double layer dome structures requires overall stability analysis as well as design of the members and connections for maximum member forces and stresses. In the case of the structure presented here, the structural analysis of the dome considered numerous load combinations of the suspended rigging and live loads, thermal loads, seismic loads, and settlement loads, as well as external live and wind load conditions. The objective of this paper is to demonstrate the effectiveness of the double layer system compared with the original single layer geometry. The results are presented only for the uniform load case (live plus suspended loads). The conclusions arrived at using this load case, however, were found to apply to all load conditions where stability was the controlling mode of failure.

Description Of The Structural System

The 97.5m diameter dome that covers the new University of Hawaii Sports Arena is spherical in shape with a double layer triangulated lattice framework within the center half (55m diameter) and a single layer geometry in the perimeter half (55m to 97.5m diameter). The outer layer of the center half and the single layer triangulated framework are in the same spherical surface and form a continuous modified geodesic geometry (83.2m spherical radius). The reinforcement layer at the center half is in a concentric surface with a spherical radius of 82.3m. The framework, panel covering and battens, as well as the internal suspended rigging grids and catwalks, are all of aluminum construction. Figure No.1 shows the basic dome dimensions and geometric configuration.

The framework of the dome outer surface uses custom aluminum "I" beams connected with gusset plates bolted to the top and bottom flanges. The gusset plates are conically formed to produce the desired spherical shape. A typical joint configuration is shown in Figure No.2. The reinforcement (secondary) layer at the center is framed with aluminum "T" beams which are connected to the main grid with aluminum pipes. A detail of the double layer system is shown in Figure No.2. The dome rests on a support structure comprised of 32 concrete columns and a concrete support/tension ring.

The structure has a total of 1696 joints and 2464 "I" beams on the outer spherical surface, 805 joints and 1540 "T" beams on the inner (second layer) surface, and 1610 pipe connectors between the inner and outer layers.

The dome was designed for a total concentrated load of 2,000kN at the center area of the roof plus the live and wind load conditions and combinations required by local code over the entire roof area.

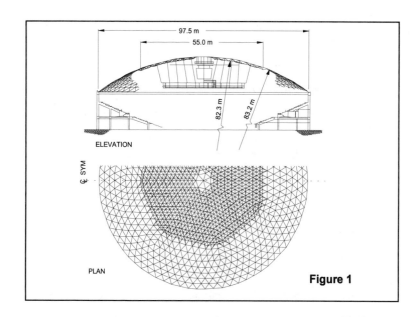

ELEVATION

℄ SYM

PLAN

Figure 1

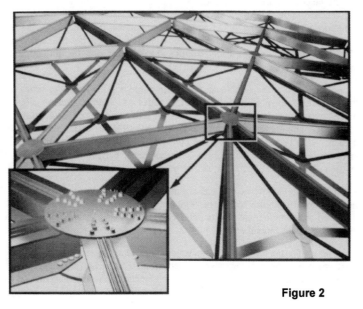

Figure 2

Stability Analysis of the Single Layer Geodesic Dome

In order to determine base line behavior, the single layer dome was first analyzed without reinforcement. The displacement and stress patterns of the single layer geometry were then analyzed to determine the area that should be stiffened using the double layer system.

Several closed form solutions for the analysis of snap-through buckling of triangulated spherical structures are available for structures subjected to uniform load with uniform edge conditions (Wright 1965, Buchert 1970). These formulas can be used to determine approximate collapse loads for lattice structures. Unfortunately, they cannot be used to predict allowable buckling loads for structures subjected to high concentration of loads, nonsymmetrical loads, settlement loads or structures with a non-uniform grid pattern. The critical buckling load for the subject structure, therefore, was obtained using a finite element analysis program capable of modeling nonlinear behavior. For this effort, the nonlinear finite element program NISA was employed (NISA 1991).

Definition of the nonlinear finite element model requires great care and, if possible, preliminary analyses to verify the ability of the model to accurately represent existing physical test data. In order to determine the accuracy of the structural model in predicting the collapse loads for the 97.5m dome, the results from previous load tests performed on a 13.7m dome (Richter 1984) were compared to the numerical results obtained by using several nonlinear structural models.

In the computational model, the beam elements were rigidly connected at the joints. The assumption of rigid joints was based on the following characteristics of the structural system: a) The gusset plates used to connect the beam members at the nodes have a much larger effective cross-sectional area than the beam flanges, and b) The bolt holes on both the extrusion flanges and gussets are only 0.13mm larger than the bolt diameters.

With regard to the actual test, the 13.7m dome was loaded at the nodes using hydraulic jacks. Displacements at all the nodes as well as stresses along the members were monitored for each load increment. Figure No.3 shows the force/displacement curve for the fastest moving point (center node) for the uniform load case. This figure also shows the numerical results obtained from the finite element analysis program using three beam elements per physical member. The beam/truss elements used are 3-D quadratic large deflection elements.

The results for the case of a concentrated load applied near the center of the dome are shown in Figure No.4. This figure also shows the location of the applied load as well as the numerical results obtained when using one, two and three beam elements per physical member.

The results for the 13.7m diameter dome test are presented here to illustrate the performance of the nonlinear model when compared to actual tests and to explain our approach to the selection of the structural elements. The results suggest that three

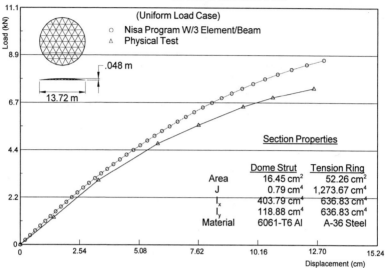

Figure 3: 13.72 m Test Dome Load / Deflection Plot

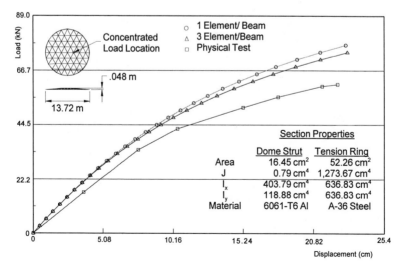

Figure 4: 13.72 m Test Dome Load / Deflection Plot

3-D quadratic beam elements should be used to predict the collapse load of these type of structures when subjected to large concentrated loads. Due to the highly localized deformations of the dome under large concentrated loads, the model with three elements per beam provided answers that are closer to the test data.

For all the load tests performed on the 13.7m diameter dome, the predicted collapse loads (using nonlinear finite element models with three elements per beam) were larger than the actual collapse loads. The difference between the loads predicted by the NISA program and the results of the physical tests can be attributed to relative displacements or yielding at the bolted connections. For the uniform load condition (Figure No.3), the difference between the actual buckling load and the test results is 14%. Similar results were obtained for nonsymmetrical load conditions.

The comparisons using the 13.7m diameter dome, as well as other results using smaller models, provided the authors with confidence in the capability of the selected structural model to predict the nonlinear behavior of the larger structure. Given this experience, however, we were forced to recognize that the complexity of the problem and the computer time required to perform the analysis using three elements per beam would be too great given available resources. As a result, symmetry was used to simplify both the loads and the geometry. In particular, a 1/14 pie sector of the structure was modeled and analyzed for the uniform (controlling) load case.

The load/displacement curve for the fastest moving node of the 97.5m diameter single layer geodesic dome subjected to symmetrical live/suspended loads is presented in Figure No.5. The geometric configuration of the model and the section properties of the extrusions used are also shown in Figure No.5.

Figure 5: Load / Deflection Plot
97.5 m Dome Without Truss

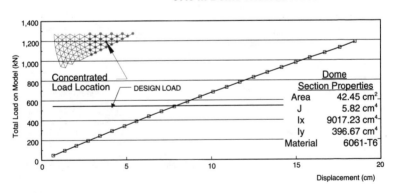

As indicated in Figure No.5, the 355mm deep extrusion used in the analysis of the single layer 97.5m dome provides an adequate solution to the snap-through problem from a structural point of view. This solution, however, is uneconomical because of the high manufacturing cost of the extruded material. The use of a double layer system as an alternate solution was investigated as an option to minimize the main layer extrusion size (and to increase the safety factor against buckling).

Design of the Reinforced Geodesic Geometry

The double layer reinforcement at the center of the dome changes the deflection patterns of the structure and reduces the axial forces and bending moments of the geodesic geometry. The reinforced area, as expected, is much stiffer and capable of supporting larger suspended loads without buckling. As a result of the added stiffness, snap-through buckling is no longer the controlling mode of failure. Instead, bending stresses at the transition area between the single and double layer are the most critical design condition.

The extrusion size of the single layer geodesic geometry was controlled by snap-through buckling. Because this mode of failure is eliminated with the addition of the second layer, the extrusion size of the dome was reduced from a depth of 355mm to 255mm. Figure No.6 shows the deflected shape of the reinforced geometry. The relative displacements and rotations at the center half are much smaller than the displacements of the unreinforced geometry. The rotations shown at the transition area create large bending stresses that control the design of the upper layer extrusion members.

Figure 6: Dome Deflected Shape With Truss (Amplified)

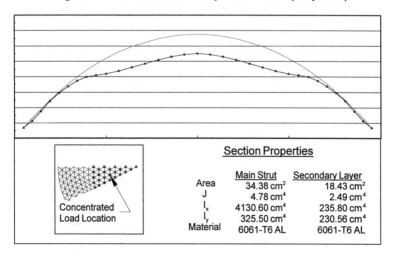

Section Properties

	Main Strut	Secondary Layer
Area	34.38 cm^2	18.43 cm^2
J	4.78 cm^4	2.49 cm^4
I_x	4130.60 cm^4	235.80 cm^4
I_y	325.50 cm^4	230.56 cm^4
Material	6061-T6 AL	6061-T6 AL

Concentrated Load Location

Conclusions

The accurate prediction of the nonlinear behavior of a triangulated dome structure requires an understanding of the behavior of the individual members and connection stiffness. Nonlinear finite element programs can be used to predict buckling modes of these structural systems but must be tested against physical tests of similar structures to verify the accuracy of the modeling assumptions.

The results and tests presented in this paper, therefore, are only applicable to the structural system tested. Different modeling techniques might be required for other structural systems (such as structures with hinged nodes or domes with steel or wood members). The modeling procedure will also depend on other variables such as the type of nonlinear elements used, the size of the model (number of nodes and elements), the framing pattern and the type of load applied.

A double layer space truss system has been designed to reinforce a single layer geodesic geometry without affecting the exterior geometry configuration. This was done at a much lower cost than would have been required for a safe single layer dome. The double layer system eliminated the stability problem of the single layer geometry, reduced the structure weight and increased the overall safety factor of the structure.

References

Buchert, K.P., and Crooker, J. (1972). "Reticulated Space Structures", Journal of the Structural Division, Proceedings of the American Society of Civil Engineers, March, 1970, 7180, pp. 687-700.

NISA II User's Manual; Version 91 (April 1991). Engineering Mechanics Research Corporation, Troy, Michigan.

Richter, D.L. (1984). "Temcor Space Structure Development" Third International Conference on Space Structures, pp. 1019.

Wright D.T. (1965). "Membrane Forces and Buckling in Reticulated Shells", Journal of the Structural Division, Proceedings of the American Society of Civil Engineers, February, 1965, 4227, pp. 173-201.

Analysis and Design of a Double Layer Barrel Vault with a Large Span

Ruo Jun Qian[1]   Zu Yan Shen[2]

Abstract

The roof of coal storage in JiaXin Power Station is a double-layer barrel vault. This roof covers span of 103.5 M and a longitudinal span of 88.0 M. The height of this space truss is 3.5 M and the rise is as high as 37.27 M. The transversal shape of the barrel vault consists of three circles with a radius of 62.957 M and 37.427 M, respectively, This truss has 2553 nodes and 10062 members. The analysis and design of this Barrel Vault are discussed in this paper. The CAD system of Space Structures AADS is used for analysing and designing the structure.

Introduction

The Coal Shed in JiaXin power station has a span of 103.5 M and is 88.0 M long. Fig.1.a. A certain space is demanded for storage of coal and performance of the machine. Fig.1.b. The traditionally adopted structure are plane frames, such as trusses or arc. It manages to cover the span as large as JiaXin power station with those structures, however, it is obviously infeasible in consideration of finance. The advantage of space structure is consequently revealed. The Double-Layer Barrel Vaults is most suitable for Coal Shed. In this paper, having studied the behaviour of the structure, the authors present the suggestions regarding the selection of type of structure and the design of the structure.

Selection of type of Double-Layer Barrel Vaults

Double-Layer Barrel Vaults is finally adopted in JiaXin station Coal Shed, although it seems to be a possibility for selecting other type of space frames.

Structural type

The selection of type of a structure is concerned with the shape, dimension of the structure, supporting condition and loading. The structural behaviour extremely depends on the shape and dimension of the structure. The boundary condition,however, is very important for giving play to spatial behaviour of a structure. It is favourable to the spatial behaviour of JiaXin roof which the plane approximates to a

----

[1]Professor, College of Structural Eng. Tongji University, 1239 SiPin Road,Shanghai, 200092, China
[2]Professor, Vice Chancellor, Tongji University, 1239 SiPin Road, Shanghai, 200092, China

square with a ratio of 1 to 1.176 in length and span respectively. The structures, however, exhibits the behaviour of a 2-D structures due to opening both at two sides.

The pyramid and three way and trusses systems are appropriate for the roof structure. Fig.2. It was noticed that opening is being both at two sides, transverse stiffness of trusses system is relativelly small and an extent accuracy in manufacture and assembly for three way system is required. Having compared to the pyramid system, diagonal system and trusses system, the authors prefer the diagonal system to square on square. The diagonal pyramid system is able to exhibit the spatial behaviour as a stiffened arc are designed at both opening sides.

Geometry configuration

Also, a certain geometry configuration of Barrel Vaults was meticulously designed. The structural behaviour, the stiffness, the bearing capacity of the structure would be unfavourably changed, more materials would be consumed if the geometry configuration of the structure is not appropriate. The determination of a structural geometry usually depends on the space demanded and the topology of the structure and the member type which is required to reduce as less as possible.

In JiaXin Coal Shed. A normal vault and a vault that the section is composed with 3 circles, Fig.1.c, d, are presented. A certain space would be saved, the configuration of the roof approximates the boundary of technological requirements as a 3 circles vault being adopted. It was shown that the height of the building is 37.27 M if 3 circles vault is adopted, while the member in the second layer of a normal vault which is 38.0 M high would be the inside of technological boundary, otherwise, 42. 0 M high is required. The determination of geometry configuration of the Double-Layer Barrel Vault depends on the principle that members respectively located at large and small circle vault must be approximately equal. This principle is presented for convenience in manufacture. An iteration was carried on for adjusting the radius of large circle and grids number from the predetermined reference dimension of a grid, Sy; and determining the coordinates of the centre of small circle. Finally, a transformation of the coordinates of nodes defined on the small circle coordinates system to globe system was performed. Fig.3. The following parameters were derived after calculation:

R_1 = 62.957 M; R_2 = 37.427 M; F = 37.270 M;
S_1 = 76.000 M; F_2 = 24.500 M; F_1/S = 1/2.78;
h = 3.5 M; Sx = 4.0 M; Sy = 4.8 M

The number of grids in longitudinal and transversal direction are 22 and 29, respectively, the length of the arc, Sc = 139.2 M, the ratio of height to span of the structures is 1/29.6 and 1/39.8, respectively. The Barrel Vault has 2553 nodes 10062 bars and 88 supports. The structural projection has a area of 9108.0 M^2.

The shape of generatrix of the vault is more significant for the structural behaviour although the transversal section of the vault had meticulously designed already.

In JiaXin barrel vault, two shapes of generatrix were taken into account. Fig.4. These are no doubt that the barrel vaults with a hyperboloid surface is more favourable. Straight generatrix was finally adopted from the consideration of manufacture and assembly.

Supporting Structure

The barrel vault is directly supported on the foundation. Both horizontal and vertical reaction of the Double-Layer Grids and the action of the piled loading to the ground were taken into account for designing the pile, strip footing foundations. The foundation are built as the pile capping beams so as to be integrated enough. The boundary nodes located on the upper and bottom chord are supported on foundation beams. The boundary nodes are horizontally restricted on the spring supports, therefore, the deformation caused by variation of temperature would be released, then the temperature and other secondary stress are reduced. Besides, the structure behaviour in stability is improved, the inner force is reduced and with the increase of supports.

Type of Loading and Combination of Structural Response

Type of Loading and Standard Loading

The type of loading are as follows :

The self weight of the structure and uniformly distributed load, denoted q_d, q_d = 0.3 KN/M²; uniformly distributed living load or snow load denoted q_s, q_s = 0.3 KN/M²; basic wind load denoted q_w, q_w = 0.6 KN/M²; and the temperature difference denoted Δt, Δt = 30℃.

Boundary Condition

Three boundary conditions are in consideration. The horizontal restraint stiffness of supports are 15 KN/cm and 25 KN/cm along X and Y direction respectively, denoted B.C.1. The forced horizontal displacement in Y direction is ± 1.25 cm, denoted B.C.2. A subsidence of 0.4 cm between two supports is assumed at the corners and the middle of the structure, which is denoted B.C.3 Fig.5.

Wind Load

Wind load is a main load applying to the barrel vault. A wind tunnel test was carried on for measuring the wind pressure on the roof. Fig.6 is a wind diagram tested. These shape coefficients of the structure are provided from table 1. The roughness of the ground is grade A. From the wind diagram and coefficients in table 1, the wind load applying to every nodes is evaluated by AADS.

Table 1. The Wind Pressure Coefficients

Angle	position	a	b	c	d	e
0°	1,2,3	0.7	-0.8	-0.8	-0.5	-0.3
30°	1	1.4	0.8	-0.8 -2.5	-1.4	-1.4
	2	1.0	0.5	-0.8	-1.4	-1.4
	3	0.6	0.4	-0.4	-0.5	-0.2

Combination of Structure Response
The following   Combinations of   structural response   were   performed
from the results of analysis.
(1). Dead load + Living load
(2). Dead load + Wind load (W_1 and W_2)
(3). Dead load + Living load + Temperature difference +
     Subsidence (B.C.2 and B.C.3)

The Analysis and Design
FEM of   pin–joint element system is employed. The fundamental equation
of FEM is
$$KU = P \qquad (1)$$
in which
     K is the stiffness matrix of barrel vault in globe system;
     P is the load vector
In   designing   procedure, it should lay special emphasis   on designing
the stiffness of   a   structure. Because, a structure with   reasonable
stiffness exhibit   the spatial behaviour and reduce the response under the
action   of   different   loading.   Therefore,   it is significant to
redesign the structural stiffness for controlling the force flow so that
the reduction of stress in members will be expected.
Because of multiple loading case and a large number of members and nodes,
a reference   load   vector was predicted for selecting the reference section
area of the   members,   then,the further analysis was carried on for deriving
the displacement of nodes and forces in members,   and reanalysis and
redesign were performed if required.

The Behaviour of the Structures
It is   required to study the behaviour of the structure   for design,
which are the response of the   structure, the force flow   influence   of
boundary conditions to force in bars and displacements of nodes. In this
barrel vault, the dead, living and wind loads are main loads. Fig.7, 8, 9
are the stress diagrams under the dead and living loads, dead   and wind
load with an angle of 90° to the generatrix, and dead load and wind load
with an angle of 0° , respectively.   Fig.10, 11,   12 are the force in
members of the side arc under the loads mentioned above. It may be seen
that the force is more well distributed under dead and living loads, however,
the forces in the members near the side arc become larger under the 30 °
wind loads.   The loads   transfer to the supports and the side arc,
respectively. The longitudinal bars near the boundary and the arc have to
be strengthened. Fig.13 shows the flow.

Design of the Barrel Vault
The section area of the members are selected according to the
combination of the responses under different types of loads, the behaviour
of the structure, and force flows. The shape of the structures , topology,
boundary conditions and the path for transferring forces should be
redesigned as well.   The strength of the members, the stability,   of the
structure, the influence of secondary stresses and imperfection have to be
taken into account for selection the section of members.   It   is   more
significant to determine appropriate theoretical strength of members for

selection of section area. There is a great need for controlling the theoretical strength of the compressed members, which locate at side arc and of the tension members in the boundary region.

A set of reference section area is designed under a reference load vector, the maximum stress in members less than 7.4 KN/cm², the stress of the members in side arc and of longitudinal bars is less than 16 KN/cm².

An adjustment of restricting stiffness of the boundary nodes is performed for reducing bar stress in members produced by the temperature variation instead of increasing the section area. The temperature stress will be released as a certain restricting stiffness of support being reduced, while, rigid body motion of the structure will not occur.

Also, it is not the way to reduce the stress in members, which is produced due to unequal settlement of the foundation. The control of deformation of the arc foundation is more significant, the pre-loading and an appropriate isolation technique are strongly recommended.

The numerical analysis shows that the response of the wind load with an angle of 30° is unfavourable to design, therefore, a suggestion regarding changing the configuration of the roof and opening the windows on the roof is presented.

A nonlinear analysis for tracing the equilibrium path is performed, it was shown that it may not collapse due to instability of the structure. The tension members in boundary region will be yielded after loading.

Conclusion

<1>. The selection of type, shape and boundary condition of the structure is required for designing a Coal Shed. The Double–Layer Barrel Vault with 3–circles is a reasonable structure using in the Coal Shed with a large span structure.

<2>. The structure stiffness and restricting stiffness of supports is more significant in design of Barrel Vault with a large span.

<3>. The control of force flow is another important aspect regarding the design of a large span structure especially in case of multiple loading, therefore, the path of transferring the loads has to be designed.

<4>. The secondary stresses in a traditional structure has to be taken into account in the design of a large span structure.

<5>. The response of wind load is the most significant in designing a Barrel Vault. Some suggestions of changing the configuration of the Barrel Vault and opening some windows on the surface are presented.

<6>. An appropriate isolation technique is suggested for controlling the deformation of the foundation.

<7>. The influence of imperfection and errors produced in installation of the Barrel Vault has to be taken into account.

<8>. It is necessary to trace the equilibrium path of the structure with a nonlinear analysis method. The concept of collapse of the structure would be derived from the numerical result.

Reference :
[1]. The specification of Design and Construction of Double–Layer Grids, JGJ 7–91.
[2]. Z. S. Makowski, Analysis, Design and Construction of Barrel Vaults.

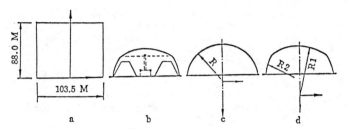

Fig.1 The Shape of Coal Shel in JiaXin Power Station

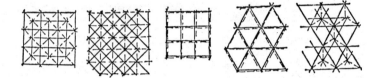

Fig.2 Type of Double-Layer Barrel Vaults

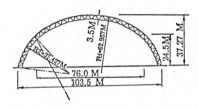

Fig.3 The Co-ordinates of JiaXin Barrel Vault

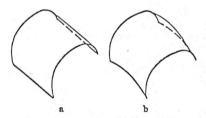

Fig.4 The Shape of JiaXin Barrel Vault

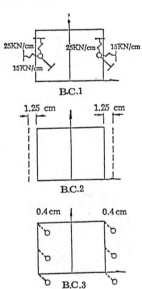

Fig.5 Boundary Conditions

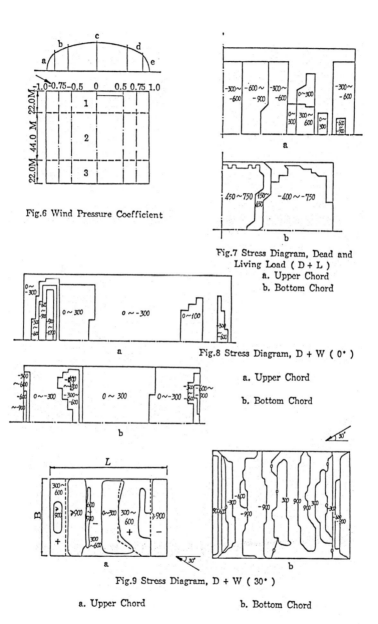

Fig.6 Wind Pressure Coefficient

Fig.7 Stress Diagram, Dead and
Living Load (D + L)
a. Upper Chord
b. Bottom Chord

Fig.8 Stress Diagram, D + W (0°)

a. Upper Chord

b. Bottom Chord

Fig.9 Stress Diagram, D + W (30°)

a. Upper Chord   b. Bottom Chord

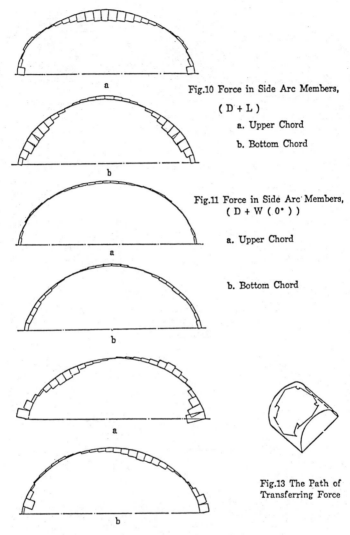

Fig.10 Force in Side Arc Members,

(D + L)

      a. Upper Chord

      b. Bottom Chord

Fig.11 Force in Side Arc Members,
(D + W (0°))

      a. Upper Chord

      b. Bottom Chord

Fig.13 The Path of
Transferring Force

Fig.12 Force in Side Arc Members, (D + W (30°))

    a. Upper Chord   b. Bottom Chord

Structural Integrity of Space Trusses

Erling Murtha-Smith, M. ASCE[1],
Anurag Chaturvedi[2] & Shawn F. Leary[2]

Abstract

Space trusses of the double layer grid configuration can experience rapid propagation of failure in the ultimate load condition and thus are potentially vulnerable to progressive collapse - collapse due to localized damage at loads less than the normal design load values. Structural integrity is a term used to indicate resistance of a structure to progressive collapse following localized damage. Results of a study is reported which used both linear and nonlinear methods of analysis investigating the effect on the structural integrity of the locations of supports and load levels at member loss. The results indicate that space trusses with perimeter supports appear to just have sufficient structural integrity to survive the loss of a critical member whereas space trusses with supports placed at the four corners do not appear able to survive critical member loss at the appropriate load levels for this condition.

Structural Integrity and Progressive Collapse

When a structure is at or below design levels of loading and is then locally damaged due to *abnormal* events, there is a possibility that this localized collapse can propagate and lead to what is known (Breen 1975) as "progressive collapse." Structural integrity can be defined as the satisfactory resistance to progressive collapse. Progressive collapse as a mode of failure received considerable attention following the collapse of the Ronan Point building caused by a "small" gas explosion in one of the apartments in the building (Griffiths et al. 1968). The debris overloaded the floor below which then collapsed and that accumulated debris caused the collapse

---

[1]Professor, Dept. of Civil Engrg, Univ. of Connecticut, Storrs, CT 06269-3037
[2]Research Assistant

to propagate down to the ground. Meanwhile, support was lost to the levels above the explosion and the collapse also propagated upwards to the roof. Changes were immediately made to the UK building regulations (Statutory Instrument 1970). In the years following, other countries amended their building regulations (BOCA 1978, Building 1982, Breen 1975).

Ellingwood and Leyendecker (1978) developed quantitative recommendations and design methods for consideration of progressive collapse. One method, the "alternate path" method considers the ability of a structure to survive the loss of a component. This condition was developed considering the probability of an abnormal event (for example, vehicle impact, explosions, accidental abuse, excessive corrosion, etc), the probability that that event would cause loss of the functioning of a structural component and the probability that this loss would lead to progressive collapse. Thus, when the alternate path method is used, the structure must have sufficient capacity to resist the dead load plus 45% of the live load plus 20% of the wind load when a critical component is not functioning.

The design equation values used by Ellingwood and Leyendecker were developed from loading data for multistory residential and office structures and thus are not strictly applicable to double layer grid roof space trusses subject to light live loading but significant snow loading. Further, roof structures are probably not vulnerable to many of the abnormal events considered by Ellingwood and Leyendecker. On the other hand, since this form of structure is used to cover large public assembly places, the loss of life due to the collapse of such a structure could be very large. Therefore, using a similar rationale to Ellingwood and Leyendecker, a conservative reliability analysis gives that locally damaged double layer grid roof space trusses should be able to survive a load level of the design dead load plus 35% of the (50 year) design snow load plus 20% of the design wind load, the design load values being those for normal design (Building 1982). However, a more detailed and precise reliability analysis could show that these values might be excessively conservative.

Double Layer Grid Space Trusses

A typical double layer grid space truss is shown in Fig. 1. In normal design, the usual loading environment considered is that of gradually increasing applied load until failure occurs. Changes in any one component behavior from linear to nonlinear behavior will lead to nonlinear changes in the forces in other components and the system behavior becomes increasingly complex. Failure can occur for components in tension, by the yield and rupture of members, connectors and joints, and, for components in compression, by yielding, crushing or buckling.

Failure of a component can cause failure to propagate and can lead to geometric instabilities independent of buckling instabilities. For example, the rupture of a tension member can cause major redistributions required of the adjacent members. Similarly, compression members entering their postbuckling regime can require large redistributions. If these redistributions cause adjacent members to rupture or buckle then collapse can easily propagate.

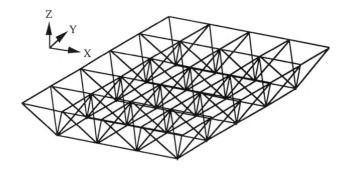

Fig. 1 Double layer grid space truss

Certainly, space trusses must be designed for the monotonically increasing load condition. However, as the Ronan Point collapse demonstrated, structures with inadequate integrity are vulnerable to total collapse due to local damage even at load levels well below the full normal design load. Double layer grid space trusses often experience rapid propagation of failure in the ultimate load condition and thus are potentially vulnerable to progressive collapse following local damage at loads less than the normal design load values.

Previous Studies

The first author (Murtha-Smith 1988) published a study of alternate path analyses of space trusses. The effect was reported of the loss of a single member on the safety of the space truss. Linear analysis was used to determine the resulting factors of safety in the remaining members. In addition, nonlinear analysis was used to determine the system factor of safety of the damaged space truss. Conventionally designed space trusses were analyzed at a load level of the dead load plus 45% of the design live load with a single member removed. Linear analysis of the damaged structures gave the resulting factors of safety, defined as ultimate capacity divided by the force in each member following the loss of the member. The linear analyses showed that when one of the more highly loaded members is removed, a significant number of remaining members become unsafe. One of the worst cases was nonlinearly analyzed and had a very small (6%) margin of safety. It is likely that all the worst cases would collapse under the loading of dead plus 45% of the design live load.

Hanaor and Ong (1988) published a study investigating the structural redundancy of space trusses. They investigated the effect of the loss of one or more members on the load bearing capacity and deflections of double layer grid space trusses. The space trusses had members all of the same length and all of the same cross-section and thus did not represent a practical design condition.

In a recent study (Leary - 1992, Murtha-Smith and Leary - 1993), the following parameters were investigated: support locations - (a) along the entire perimeter, and (b) at the corners, and for both (i) on exterior joints and (ii) on interior joints (one half module in from the exterior joints); span-module size ratios; span-depth ratios; and long span to short span ratios. The space trusses were designed using conventional design methods with members for all the designs investigated selected from a library of tubular members with 44 different members. The dead-live load ratio was 1-1. After design, each truss had a single member removed and the truss was linearly reanalyzed. The resulting member factor of safety, the ultimate member capacity divided by the force in the member following the removal of the single member, was calculated for each member.

Figs 2(a) and 2(c) show typical plots (Leary 1992) of members becoming unsafe (member force following damage exceeds ultimate capacity) following the removal of the member indicated at a load level of the dead load plus 45% of the live load. In those figures, only the safe top (compression) chord members are shown in addition to the removed and the unsafe members. Generally, the unsafe members have factors of safety that increase with distance from the removed member. It can be seen that there is a significant difference in the number of members becoming unsafe between the perimeter and corner supported cases. Plots not shown (Leary 1992) for 12x12 and 16x16 are similar to plots for the 8x8.

a) Linear analysis - unsafe members

b) Nonlinear analysis - nonlinear members

c) Linear analysis - unsafe members

d) Nonlinear analysis - nonlinear members

▨ = removed member

Fig. 2 - Analysis at 100% Dead + 45% Live

For the perimeter supported space trusses (Leary 1992), the maximum number of unsafe members for the worst case is relatively insensitive to the span-module size ratio, span-depth ratio and long span to short span ratio. For the corner supported space trusses, the worst case is when the diagonal member at any one support is removed. This removal causes the support to be ineffective and the structure becomes geometrically unstable because of an inherent internal degree of freedom in the geometry. However, the numbers of unsafe members for the next four worst cases are insensitive to the span-module size ratio and the span-depth ratio. Reviewing the summarized results above indicates that vulnerability to progressive collapse is very sensitive to the location of the supports but is relatively insensitive to the span-module size ratio and the span-depth ratio.

Current Study using Nonlinear Analysis

The current study reported in the following investigated the whether or not the damaged systems would survive using nonlinear analysis methods. The worst cases revealed by the linear study (Leary 1992; and Murtha-Smith and Leary 1993) were reexamined using nonlinear analysis. The nonlinear analysis (Smith 1984) modelled the nonlinear behavior of the members through yield and rupture for tension members and buckling and postbuckling for compression members. Load is monotonically increased as members change state. The member and system force and displacement histories are developed. As a datum, the undamaged structures were nonlinearly analyzed. In addition, these structures were analyzed for the case of losing the largest member at a predesignated load level. For these analyses, a single member is removed by "rupturing" the member. If the space truss can thereafter stabilize then the load is further increased until the structure becomes unstable or develops very large displacements.

The structures were conventionally designed using the AISC specification (American 1986) for load levels of 120% of the design dead load plus 160% of the design live load and capacity reduction factors of 0.9 for the tension members and 0.85 for the compression members. For all the structures, the design dead load was assumed equal to the design live load. For the nonlinear analysis, the capacities of the members was taken as the nominal capacities as given by the AISC specification, that is assuming the capacity reduction factors all to be *unity*.

The undamaged space trusses were nonlinearly analyzed and found to have capacities between 10 and 20% in excess of the ultimate design condition.

Approximately 10% of the excess capacity was due to the design using capacity reduction factors of 0.9 and 0.85 for the tension and compression members respectively whereas the nonlinear analysis assumed the capacity reduction factors were unity. The remaining excess capacity was due to the redundancy in the system.

The structures were also nonlinearly analyzed for the damaged condition. In different nonlinear analyses, the *single* critical member was removed at load levels from 10% of dead, through 100% dead plus 45% live, and up to 100% dead plus 100% live, and the resulting histories compared. Figs 2(b) and (d) show plots for the nonlinear members resulting from being damaged at 100% dead plus 45% live. Comparing these plots with the corresponding results shown in Figs 2(a) and (c) which used linear analysis, it can be seen that there is significant qualitative correlation, particularly for the perimeter supported case. However, greater divergence can be seen for the corner supported case. Further, both the linear and nonlinear analyses show that the extent of the damage is far greater for the corner supported case than for the perimeter supported case.

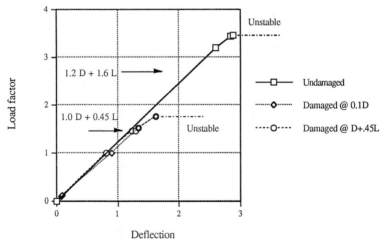

Fig. 3 - Load vs. Deflection - perimeter supported

Fig. 3 plots the load versus central deflection for the perimeter supported truss for the undamaged case in addition to the cases for damage at 10% of dead, and at 100% dead plus 45% live load. First, as explained above, it can be seen that the capacity of the undamaged case exceeds the design capacity. Second, it can be seen that both the 10% dead and the 100% dead and 45% live case survive the loss of the critical member and have reserve capacity up to a load level of 100% dead plus 76% of live load. This survival load is approximately 50% of the undamaged actual capacity. However, not shown are other nonlinear analyses which show that if the member is lost at load levels above the 100% dead plus 76% live, the damaged structure cannot stabilize and survive.

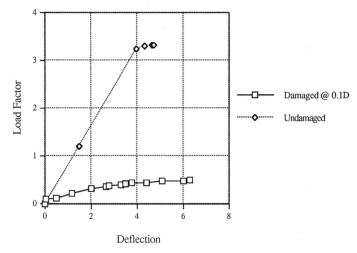

Fig. 4 - Load vs. Deflection -corner supported

Fig. 4 plots load versus deflection for the corner supported case. It can be seen that the case damaged at 10% dead survived but thereafter only achieved 49% dead before the structure became unstable. This damaged capacity is approximately only 15% of the undamaged capacity and approximately only a third of the required capacity of 100% dead plus 45% live. All the cases damaged at loads in excess of 49% dead did not stabilize and survive. Thus it is clear that the corner supported configuration is highly vulnerable to progressive collapse following the loss of a critical member at very low loads and cannot satisfy the safe alternate path requirement.

Summary and Conclusions

The results of both the linear and nonlinear studies indicate that the primary variable of those considered is that of the support location. The perimeter supported trusses do appear to safely survive the loss the largest most critical member.However, when there are only four vertical supports placed at the corner, the structure does not appear to satisfactorily survive the loss of a criticial member at all but very low load levels.

Future Research Needed

The design load levels to be considered for the damaged structure need to be reviewed. For example, the equation of Leyendecker and Ellingwood (1978) was developed considering probablity and structural reliability in the context of reinforced conrete panel type structures. Thus a study is required applying reliability theory to space truss roof structures and the corresponding initiating causes for progressive collapse. It is the opinion of the authors that the Leyendecker and Ellingwood equation is over conservative for space truss roof structures.

In addition, the mobilization of secondary resisting mechanisms should be studied. For example, many space truss systems have continuity at the joints which can slightly decrease the effective length of some of the members and thereby slightly increase the capacity of the members. However, for an efficiently designed space truss this increase would probably be quite modest.

It would be valuable to develop a predictor parameter that could be used in combination with linear analysis to identify critical areas. Such a parameter might compare the sum of the absolute values of the capacities at a joint with the sum of the absolute values of the forces at the joint following loss of the critical member. When this quantity is less than a particular value then special design action is required.

Possible design remedies need to be investigated. For example, factors of safety can generally be increased. Alternatively, the critical member or members can be designed with larger factors of safety. On the other hand, those members failing first as a result of the loss of a critical member could be designed with larger factors of safety to survive the damage.

Acknowledgements

The authors gratefully acknowledge the sponsorship of the National Science Foundation through project MSM - 89-14142. The opinions expressed in this paper are those of the authors and do not necessarily reflect those of the sponsoring organization.

Appendix: References

American Institue of Steel Construction. (1986). "Load and Resistance Factor Design Specification for Structural Steel Buildings" AISC, Chicago, Illinois.

"BOCA: basic building code, 7th edition." (1978). Building Official and Code Aministrators International, Inc. Homewood, Illinois.

"Building code requirements for minimum design loads in buildings and other structures, ANSI A58.1-1982." (1982). American National Standards Institute, New York, NY.

Breen, J E. (1975). "Summary report of research workshop on progressive collapse of building structures". University of Texas at Austin, Austin, Texas.

Ellingwood, B. and Leyendecker, E. V. (1978). "Approaches for design against progressive collapse" Journal of the Structural Division, ASCE, 104 (ST3), March.

Griffiths, H, Pugsley A, and Saunders, O. (1968). "Report of the inquiry into the collapse of flats at Ronan Point, Canning Town" Her Majesty's Stationery Office, London, UK.

Hanaor, A and Ong A-F. (1968). "On Structural Redundancy in Space Trusses" International J. of Space Structures, 3 (4).

Leary, S F. (1992). "A Parametric Study of the Susceptibility of Space Truss Structures to Progressive Collapse" MS thesis, Department of Civil Engineering, University of Connecticut, Storrs, CT.

Murtha-Smith E. (1988). "Alternate Path Analysis of Space Trusses for Progressive Collapse" Journal of Structural Engineering, ASCE, 114 (9), September.

Murtha-Smith, E. and Leary, S. (1993). "Space Structural Integrity" Space Structures 4, Parke, G. A. R. and Howard, C. M. Eds, Telford Press, London UK, Proceedings of the Fourth International Conference on Space Structures, University of Surrey, Guildford, Surrey, UK.

"Statutory Instrument 1970, No. 109, Building and Buildings. The Building Fifth Amendment Regulations" (1970). Her Majesty's Stationery Office, London, UK.

Smith, E. A. (1984). "Space Truss Nonlinear Analysis" Journal of Structural Engineering, ASCE, 110 (4) , April.

OPTIMIZATION OF SPACE LATTICE STRUCTURES WITH MULTIMODAL FREQUENCY CONSTRAINTS

by

J. Czyz and S. Lukasiewicz

Abstract

The paper presents an optimization algorithm for the design of space trusses subjected to constant value constraint. The goal of the optimization is a structure of the highest multiple natural frequency. The optimization algorithm is based on the recently developed optimality conditions which are true for multiple frequency constraints. In all the previous papers in this field, the Kuhn-Tacker conditions were used. However, it was proved that they were not applicable in the case of multiple eigenvalues problems. In the present paper, finite element method was used as a tool to obtain numerical results. The optimization algorithm was implemented in an automated structural optimization computer code.

Introduction

The structural optimization problems with constraints imposed on natural frequencies received great attention in the past several years. The optimization of a structure usually leads in this case to multiple eigenvalues. It is known that Kuhn-Tacker conditions of optimization are valid only for the problems of distinct eigenvalues. In the case of multiple eigenvalues, the eigenvectors are not unique and an infinite number of linear combinations of the eigenvalues satisfies the problem. Therefore the directional derivatives cannot be used in the optimization procedure, Masur (1982). A new optimization algorithm based on the optimality conditions described by Czyz and Lukasiewicz (1993) is presented in this paper. It is proved that the new algorithm is effective and gives a good convergence.

Formulation of the Problem

The present paper is devoted to the optimization of trusses. The method of finite element is applied to solve the problem. The optimization variables are in this case the areas of the cross-section of the members. The optimization is carried out to find the distribution of the parameters h_i for which the fundamental natural frequency reaches the maximum value while the volume remains constant. To find a satisfactory solution from the technical point of view, some additional conditions are imposed. It is assumed that the area of the cross-section of the member is $h_i > h_{min}$. The maximized fundamental frequency can be multiple, i.e. $\omega_1 = \omega_2 = \ldots = \omega_m$, where ω_i is the "ith" natural frequency; m is the number of the multiplicity of the fundamental frequency. The solution of the problem is obtained using the new optimality condition. The modelling of the trusses and the solution of the eigenproblems is performed using ANSYS. The discussed optimization problem can be defined as follows:

The set of optimization variables h_i has to be found for which the lowest natural frequency ω_i reaches the maximum value,

$$\lambda_i(h_i) = \text{max} , \tag{1}$$

and the following constraints are satisfied.

The lowest eigenvalue λ_i is a multiple eigenvalue

$$\lambda_j - \lambda_1 = 0 , \qquad (j = 1, m) , \tag{2}$$

the variable $h_{min} < h < h_{max}$. The volume of the optimized structure does not change during optimization.

$$\sum_{i=1}^{n} V_i - V_o = 0 , \tag{3}$$

V_o -   total volume of elements,
V_i -   volume of "ith" element.

The optimization of the structure requires the solution of the eigenvalue problem.

$$Kx_j = \lambda_j Mx_j , \qquad (j = 1,\ldots,m), \tag{4}$$

where K -   stiffness matrix,
        M -   mass matrix,
        x_j -   eigenvector,
        λ_i - ω^2 is the natural frequency of the structure.

Optimality Conditions

The goal of this paper is to find a point h in the p-dimensional design space which provides the highest value of a multiple fundamental frequency of a structure of given volume V_o. That problem is equivalent to minimization of the weight of the structure with multiple frequency constraints. The detailed derivation of the optimality conditions for multiple frequency constraints were given by Czyz and Lukasiewicz (1993). However, a short summary of these conditions is also presented here. At the point of optimum, besides the condition $V(h) = V_o$, the following condition holds

$$\delta \lambda_1(\delta h) \leq 0 , \tag{5}$$

for any small enough increment δh of the design vector ($\| \delta h \| < \varepsilon$) which does not change the volume of the structure, i.e.

$$\delta V(\delta h) = \sum_{\ell=1}^{n} \frac{\partial V}{\partial h_\ell} \delta h_\ell = 0 . \tag{6}$$

The optimal solution is m - modal if the following additional conditions are satisfied at point h (but not necessarily in its vicinity, even in an infinitesimal region):

$$\lambda_j(h) = \lambda_1(h) , \qquad (j = 2,...m) . \tag{7}$$

If the vector δh is limited to the subspace ΔH in which the modality of the first eigenvalue is maintained i.e.

$$\delta \lambda_j(\delta h) = \delta \lambda_1(\delta h) , \qquad (j = 2,...m), \tag{8}$$

than, as it was proved in the previous papers, $\delta \lambda_j$ is a linear function of δh and the following is true:

$$\delta \lambda_j(-\delta h) = -\delta \lambda_j(\delta h) , \qquad (j = 2,...m). \tag{9}$$

Of course, if the vector δh satisfies condition (6) for keeping the volume constant, the vector $-\delta h$ also satisfies that condition. As there are no additional constraints, admissibility of vector δh also implies admissibility of vector $-\delta h$. Therefore, taking into account Eq. (9), the inequality in Eq. (5) has to be ruled out. Using Eq. (6), the optimality conditions can be presented in the form:

$$\delta \lambda_1(\delta h) = \sum_{\ell=1}^{n} \omega_{ii}^{\ell} \delta h_\ell = 0 , \qquad (i = \ell,...m) , \tag{10a}$$

$$\delta \lambda_{ij} = \sum_{\ell=1}^{n} \omega_{ij}^{\ell} \delta h_\ell = 0 ,$$

with

$$\omega_{ij}^{\ell} = X_i^{\top} \left(\frac{\partial K}{\partial h_\ell} - \lambda \frac{\partial M}{\partial h_\ell} \right) X_j . \tag{10b}$$

for each δh satisfying both constant volume conditions (6) and the constraints (2) which define elements of the subspace ΔH.

$$\sum_{\ell=1}^{n} (\omega_{jj}^{\ell} - \omega_{11}^{\ell}) \delta h_{\ell} = 0 , \qquad (j = 2,...m) , \qquad (11a)$$

$$\sum_{\ell=1}^{n} \omega_{ij}^{\ell} \delta h_{\ell} = 0 , \qquad (i \neq j; i, j = 1,...m) . \qquad (11b)$$

Using Lagrange multipliers k and γ_{ij} ($i, j = 1,...m$, except for $i = j = 1$), the optimality condition (10a), with the constraints (6) and (11) on δh, can be written as the following condition without any constraints on δh:

$$\sum_{\ell=1}^{n} \omega_{11}^{\ell} \delta h_{\ell} + \sum_{i=2}^{m} \sum_{\ell=1}^{n} \gamma_{ii} (\omega_{ii}^{\ell} - \omega_{11}^{\ell}) \delta h_{\ell} +$$

$$\sum_{i,j=1}^{m} \gamma_{ij} \sum_{\ell=1}^{n} \omega_{ij}^{\ell} \delta h_{\ell} + k \sum_{\ell=1}^{n} \frac{\partial V}{\partial h_{\ell}} \delta h_{\ell} = 0 . \qquad (12)$$

Introducing γ_{11} defined as

$$\gamma_{11} = 1 - \sum_{i=2}^{m} \gamma_{ii} , \qquad (13)$$

and taking into account the fact that Eq. (13) must hold for any set of δh_{ℓ} ($\ell = 1,...p$), the following set of p optimality conditions is derived:

$$\sum_{i,j=1}^{m} \gamma_{ij} \omega_{ij}^{\ell} + k \frac{\partial V}{\partial h_{\ell}} = 0 , \qquad (\ell = 1,...n) , \qquad (14)$$

where ω_{ij}^{ℓ} is defined by Eq. (10b).

The above equation represents general optimality conditions which take into account the fact that the multimodal frequencies are not differentiable in the design space. All other papers, except the paper by Masur (1982), assumed existence of those derivatives. The optimality conditions derived using Kuhn-Tacker conditions which do not contain factors ω_{ij}^{ℓ} for $i \neq j$ are generally false.

Optimization Algorithm

The solution of the presented problem can be achieved by an iterative process. One iteration step requires solution of the eigenvalue problem (4) for current set of the design variables h_{ℓ}, and determination of their change δh_{ℓ} which drives them towards the optimum. The resizing of the design variables is based on the optimality criterion approach, in which some special features are implemented to provide convergence in the case of multiple frequencies. The multimodal optimality criterion given in the general form in the previous paragraph can be used to compute the increments of the design variables.

$$\sum_{\substack{i \geq j=1}}^{m} \bar{\gamma}_{ij} \, e_{ij}^{\ell} - 1 = 0 \; , \qquad (\ell = 1,...) \; , \qquad (15)$$

where
$$\bar{\gamma}_{ii} = \frac{\gamma_{ii}}{k} \, , \qquad \bar{\gamma}_{ij} = \frac{\gamma_{ij} + \gamma_{ji}}{k} \; .$$

γ_{ij} are the Lagrange multipliers which have to be found during the optimization process.

$$e_{ij}^{\ell} = \omega_{ij}^{\ell} \, / \, \frac{\partial V_i}{\partial h_{\ell}} \, , \qquad\qquad (16)$$

are the energy densities of the element ℓ.

From the multimodal optimality conditions can be presented in the following way:
$$\delta h_{\ell} = \beta \, \varepsilon_{\ell} h_{\ell} \, , \qquad (\ell = 1,...,n), \qquad (17a)$$

where β- step size, ε_{ℓ} - the residual of the optimality Eq. (5) which is satisfied only at the optimum, defined as

$$\varepsilon_{\ell} = \sum_{\substack{p \geq q=1}}^{m} \bar{\gamma}_{pq} \, e_{pq}^{\ell} = 1 \; , \qquad (\ell=1,...,n). \qquad (17b)$$

The above formula has often been used before but never with true optimality conditions for the case of multiple frequencies. Combining Eqs. (15) and (17b) yields:

$$\delta h_{\ell} = \beta \left(\sum_{\substack{p \geq q=1}}^{m} \bar{\gamma}_{pq} \, e_{pq}^{\ell} - 1 \right) h_{\ell} \, , \qquad (\ell=1,...,n). \qquad (17c)$$

In order to use Eq. (17c) the Lagrange multipliers γ_{ij} have to be computed first. It can be done by imposing some constraints on vector δh, which depends on γ_{ij}. The following three types of the constraints are proposed in order to determine the Lagrange coefficients:

I   The condition of constant volume - Eq. (3).
II   The conditions (10b) ensuring that when the first m eigenvalues are equal, formula (10a) can be used to compute the increments $\delta \lambda_i$ of those eigenvalues.
III   The conditions which facilitate the convergence of the first m eigenvalues to a common, multiple value λ_o of the fundamental frequency:

$$\lambda_i + \delta\lambda_i = \lambda_o , \qquad (i = 1,...m). \qquad (18)$$

The value λ_o has to be determined too. When eigenvalues λ_i $(i = 1,...m)$ are equal to each other, the condition (18) is equivalent to Equation $\delta\lambda_j(\delta h) = \delta(\lambda_1)\delta h$, which says that the increment of the multiple eigenvalue should be the same, i.e. that the multiplicity of the eigenvalue should be maintained. That condition together with the condition II confines δh to the subspace ΔH in which the multiplicity of the eigenvalue λ does not change. Now, let us derive the constraints on the Lagrange multipliers from each of the above three conditions.

Condition I

Using Eq. (17c), the constant volume condition (3) is expressed as:

$$\delta V = \sum_{\ell=1}^{n} \frac{\partial V}{\partial h_\ell} \delta h_\ell = \beta \sum_{\ell=1}^{n} \frac{\partial V}{\partial h_\ell} \left(\sum_{p \geq q=1}^{m} \bar{\gamma}_{pq} e_{pq}^{\ell} - 1 \right) h_\ell = 0 , \qquad (19)$$

or, after division by β

$$\sum_{p \geq q=1}^{m} a_{pq} \bar{\gamma}_{pq} = 1 , \qquad (20)$$

where

$$a_{pq} = \frac{\displaystyle\sum_{\ell=1}^{n} \omega_{pq}^{\ell} h_\ell}{\displaystyle\sum_{\ell=1}^{n} \frac{\partial V}{\partial h_\ell} h_\ell} . \qquad (21)$$

Condition II

Substituting Eq. (17c) into Eq. (10b) the following constraints are obtained (after replacing $\omega_{ij}^{'\ell}$ with ω_{ij}^{ℓ}):

$$\sum_{p \geq q=1}^{m} \left(\sum_{\ell=1}^{n} \omega_{ij}^{\ell} e_{pq}^{\ell} h_\ell \right) \bar{\gamma}_{pq} = \sum_{\ell=1}^{n} \omega_{ij} h_\ell , \qquad (i, j=1,...m; \ i>j) . \qquad (22)$$

Condition III

Because conditions II, based on Eq. (10b), are satisfied, the formula (10a) for the increments $\delta\lambda_i$ of the multiple eigenvalue is valid. Substituting Eqs. (10a) and (10b) to Eq. (18) one obtains:

$$\sum_{p \geq q=1}^{m} \left(\sum_{\ell=1}^{n} \omega_{ii}^{\ell} e_{pq}^{\ell} \right) \bar{\gamma}_{pq} - \frac{1}{\beta} \lambda_{o} = -\frac{\lambda_{i}}{\beta} + \sum_{\ell=1}^{n} \omega_{ii}^{\ell} h_{\ell} \sum_{p \geq q=1}^{m} e_{p\nu}^{\ell} . \quad (23)$$

There are m constraints of type (23), $m(m-1)/2$ constraints of the type (22) and one constraint (19), which all together make $m(m+1)/2+1$ equations. Those equations allow to determine univocally $m(m+1)/2$ Lagrange coefficients and the value of λ_o.

The modality m of the problem may change with subsequent iterations, while the frequencies converge. However, at each iteration step the modality can be determined automatically. Starting from a certain predicted (expected) value m, the Lagrange coefficients can be computed using the above presented algorithm. If all the computed values γ_{ii} are positive, the solution is accepted and resizing is performed. If one of γ_{ii} is negative the number m is decreased by 1 and the computation repeated till all γ_{ii} for $i=1,...m$ are positive. The condition that γ_{ii} should be positive was proved by Masur, (1982). It is also confirmed by numerical examples.

EXAMPLE

The presented algorithm has been implemented into a computer program which optimizes the cross section area of truss elements in order to achieve the maximum value of the multiple fundamental frequency of a truss structure of constant volume. The FEM model as well as the modal analysis of free vibrations is performed using ANSYS. The results from ANSYS are used by the optimization program.

The program has been used for optimization of 36 element truss presented in Fig. 1. The initial cross-section of all the elements was 2cm x 2cm. All the longitudinal elements are 0.5m long, but the elements forming a triangle are of different length: 0.4m, 0.45m and 0.5m. There are three masses, 1 kg each at one end of the structure, and three nodes on the other which are fixed. The Young's module is 2.1E-6, Poison's ratio 0.3 and density 7,000 kg/m3. The lowest frequencies of the initial (uniform) structure were: $\omega_1^0 = 55.45 Hz$, $\omega_2^0 = 60.35 Hz = 1.09*\omega_1^0$ and $\omega_3^0 = 101.9 Hz = 1.84*\omega_1^0$. In the result of the optimization, the two lowest frequencies became equal to each other and they reached the value $\omega_1 = \omega_2 = 113.91 Hz = 2.05*\omega_1^0$, which means that the fundamental frequency became bimodal and was over two times higher than the fundamental frequency of the initial structure. The third frequency became close to the bimodal fundamental frequency and reached the value $\omega_3 = 125.63 Hz = 2.26*\omega_1^0$. Fig. 2 shows the results of the optimization for each iteration step, at which only one modal analysis is performed. Fig. 3 displays the ratio of the first frequencies to the initial fundamental frequency. It also shows the error, which represents the maximum error of optimality conditions for all of the elements. As can be seen, the optimum was practically achieved in 6 iterations.

The optimization was also attempted using the assumption that the Lagrange coefficients γ_{ij} are equal to zero for all $i \neq j$. As was pointed out earlier, these generally incorrect optimality conditions were used in all the algorithms published to date. As seen from the optimization history shown in Fig. 3, this algorithm encounters problems when the natural frequencies are close to each other and the bimodal frequency does not fully converge.

Fig. 1   36 element truss after optimization

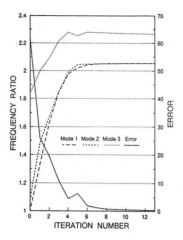

Fig. 2 Optimization history of 36 element truss using correct optimality conditions

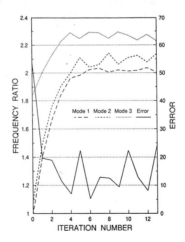

Fig. 3 Optimization history of 36 element truss using incorrect optimality conditions.

REFERENCES

[1]   Berger, M. and Porat, L., *"On Non-Smooth Beam Shapes for Maximal Transverse Natural Frequencies"*, Journal of Sound and Vibration, 132(3), pp. 423-432, 1989.

[2]   Czyz, J. and Lukasiewicz, S.A., *"Optimality Conditions for Multiple Structural Frequency"*, 1993.

[3]   John, K.V. and Ramakrishnan, C.V., *"Discrete Optimal Design of Trusses with Stress and Frequency Constraints"*, Eng. Comput., Vol. 7, March 1990.

[4]   Khan, M.R. and Willmert, K.D., *"An Efficient Optimality Criterion Method for Natural Frequency Constrained Structures, Computers and Structures"*, Vol. 14, No. 5-6, pp. 501-507, 1981.

[5]   Masur, E.F., *"Optimal Structural Design Under Multiple Eigenvalue Constraints"*, Int. Journal, Solids and Structures, Vol. 20, No. 3, pp. 211-231, 1982.

[6]   Sadek, E.A., *"An Optimality Criterion Method for Dynamic Optimization of Structures"*, International Journal for Numerical Methods in Engineering, Vol. 28, pp. 579-592, 1989.

[7]   Szyszkowski, W. and Czyz, J., *"Bimodal Optimization for Maximum Stability and Highest Natural Frequency"*, Seventh International Conference on Mathematical and Computer Modelling, Chicago, August 1989.

[8]   Tze Hsiu Woo, *"Space Frame Optimization Subject to Frequency Constraints"*, AIAA Journal, Vol. 25, No. 10, 1987.

[9]   Vanderplats, G.N. and Salajegheh, E., *"An Efficient Approximation Technique for Frequency Constraints in Frame Optimization"*, International Journal for Numerical Method in Engineering, Vol. 26, 1057-1069, 1988.

Optimal Design Conception of
Grid Domes in Practice

Hu Xue-Ren 1
Luo Young-Feng 2

Abstract

In this paper,the optimal design conception of grid domes in practice is presented. The classical dome shell structures covering some irregular boundaries often seems not to be satisfied by creative architects and engineers. The cable net form finding method controlling lower stress level within the system make it possible to furnish with wide range of optimal solution.
Two examples of such grid dome recently designed and constructed are explained. Model tests as well as the nonliner analysis show that the inversed cable net grid domes are much superior to the mathematical configulated shells.

Introduction

Eduardo Torroja,the prominent pioneer in the field of spatial structures , deeply explained the best structure "which is supported by its shape and not by the hidden strength of its material"(1). Depending on its natural form, shell structure become optimal in comparison with the others , and hence, its variable forms would attract many investigators.

It is believed that the eariest forms of shell are classical domes and barrel vaults which suit to cover some big area with regular contour of circular,

----
1 Professor, College of structural engineering,Tongji University,1239 Siping Rd.Shanghai,200092,China
2 Doctor, College of structural Engineering, Tongji University,1239 Siping Rd.Shanghai,200092,China

polygonal and rectangular plane.

Elaborative modeles made of hinging cloth with water frozen and reversed , pneumatic rubber , soap membrane with rigid edges or metal membrane prestressed are a great help to visualize some special design. Important contribution has been made by many investigators and designers ,especially , a famous Swiss engineer Heinz isler who developed various kinds of reinforced concrete shells which stand unrivalled for its elegance ,efficiency and good behaviour for long term service. E.Ramm (6,1992)(7,1993) analyzed the optmization model, suggested applied procedues and to a great extent will stimulate the discussion on free form shells.

Space frame shell structures, as special types of shell exist also the shape finding and shape optimization problem as the continuous one. Y. Hangai (3,1989) developed the theory for shape finding of unstable link structures in unstable state. The memberane theory as well as the force density method (Scheck,1974) (2,1985) (9,1989) (4,1990) (5,1991) may play an important role in this field.

Fundamental equation

The practical model for running form finding is based on force density equations and is considered as choosing equilibrium position of network joints , thus being much different from ordinary structural analysis calculation.

Figure 1. shows a part of cable network. At each node,single equilibrium equation exists if the following assumption is used:

(1) nodal loads only in Z direction
(2) the nodal joint coordinates projected on XY plane being square grid and in keeping with the original position.

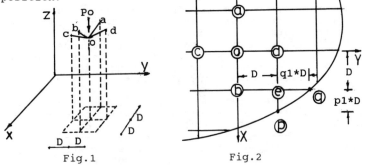

Fig.1   Fig.2

The equilibrium equations in X,Y direction are

$$Toa*Doa/Loa=Tob*Dob/Lob \qquad (1)$$
$$Toc*Doc/Loc=Tod*Dod/Lod \qquad (2)$$

where Toa,Tob,Toc and Tod are axial force in oa,ob,oc,od
Doa,Dob,Doc and Dod are projected grid side(=D)
Loa,Lob,Loc and Lod are the length of oa,ob,oc,od
Defining the force density Q as T/L,and considering the same equations as Eq.(1) and Eq.(2) crossed at node "O"in X,Y direction,we have

$$Qoa=Qob=......=Qx(i) \qquad (3)$$
$$Qoc=Qod=......=Qy(j) \qquad (4)$$

where Qoa,Qob are the force density of oa,ob
Qoc,Qod are the force density of oc,od
Qx(i) are the force density of i row in X direction
Qy(j) are the force density of j column in Y direction
The only equilibrium equation in Z direction at nodal point "O" is:

$$Qx(i)*(Za-2*Zo+Zb)+Qy(j)*(Zc-2*Zo+Zd)=Po \qquad (5)$$

where Zo,Za,Zb,Zc,Zd are coordinate of nodes:o,a,b,c,d.
Po is the nodal load at "O"
It can be found that Eq.(5) is similar to the Pucher's membrane equilibrium equation only neglecting the shear force which does not exist in grid domes.

In the case of irregular boundary,as it is at nodal point "e" (Fig.2), the projected length may be written as
$$Dep=p1*D \qquad Deq=q1*D$$
Where p1 and q1 are length ratio of ep and eq repectively. From the equilibrium condition in X,Y direction at point "e",the force density of ep and eq are

$$Qep=Qx(i+1)/p1 \qquad (6)$$
$$Qeq=Qy(j+1)/q1 \qquad (7)$$

The equilibrium equation at point "e" is

$$Qx(i+1)*[Zd-(1+1/p1)*Ze+Zp/p1]$$
$$+Qy(j+1)*[Zb-(1+1/q1)*Ze+Zq/q1]=Pe \qquad (8)$$

where Zd,Ze,Zp,Zb,Zq are the height coordinates of d,e,p, b and q. If the boundary point p,q are fixed,the known values Zp,Zq must be used for calculation. If p,q are free from support,the unknown values Zp,Zq may be found from solving the simutaneous equation.
It is clear that the unknowns are the numbers of the nodal points and the force density of cables , while the numbers of equation are only equal to the nodal

points.   The unknowns more than equations could be used
to alternate and adjust the equation, and to find the
more forms for choice.   From Eq.4 and Eq.8, it is found
that the ratio of the force density versus the nodal
force can be used as a parameter which control the form
and its optimization.   The higer these values, the bigger
the forces within the elements.   The optimization in
the form with lower forces distribution are predicted
while assuming lower ratios.
        In the case of more flat membrane the equilibrium
equation at nodal point "O" (Fig.1) is

$$Qx(i)=Qy(j)=Q \qquad (9)$$
$$Za+Zb-4*Zo+Zc+Zd=Po/Q \qquad (10)$$

It can be found that Eq.(10) is similar to the Prandtl's
equation which is more easy to control the ratoi of nodal
load versus force density.

Example

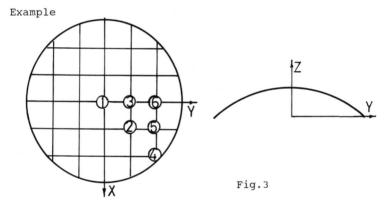

Fig.3

        A 25 nodal points grid dome (Fig.3),span= 6000.0mm,
the grid side length projected in XY plane is 1000.0mm,
the rise at vertax=1000.0mm,compare the possible forms.

        In Tab.1 we use the nodal load Po to conrol the rise
of the structure and the force density Q1,Q2 and Q3 to
adjust forms.   It can be found that the computed height
coordinates is generally near the spherical surface but
differs from it.   It is difficult to predict a spherical
surface by only adjusting the force density except
changing the load Po simutaneously.   We may find that
the grid dome with the second column coordinates is super
to others as the force density is lowest and so as the
forces in cables.   Increasing the force density in the
middle part of the shell may also enlarge the height
coordinates and increase the forces in the members.

Tab.1

No.	Spherical dome	Po=43.90 Q1=1 Q2=1 Q3=1	Po=56.56 Q1=1.5 Q2=1.2 Q3=1.0	Po=59.38 Q1=2.48 Q2=2.27 Q3=-2.21
1	1000.0	1000.0	1000.0	1000.0
2	795.8	781.4	814.7	813.0
3	899.0	890.1	905.8	940.2
4	123.1	128.4	147.2	803.5
5	472.1	453.1	488.1	555.1
6	582.6	558.8	580.1	873.6

Engineering practice

The grid dome may be regarded as one of simplest dome structures as there are only four elements jointed together and the nodes are comparablly smaller. Usually, the rigidly jointed connection at nodes should be secured for preventing from buckling perpendicular to the shell surface. Light weight coloured or penetrative roof covering increases graceful view and the easy connection method seems much favoured.

Multi-Purpose Hall of Qingdo Exhibition Centre (Fig.4)

As it is located in beautiful sea shore of Qingdo scenery area , a sea shell-like dome structure seems much harmonious with surrounding environment, being approved by architects. The plane of the hall is a equilateral triangle with circular curve sides, but the elevation shows one vertex of the triangle is 8m. higher than two others, thus forming an irregular

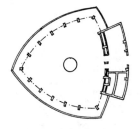

Fig.4

inclined curve beam boundary and requiring a sea shell imitated grid dome on it. The span of the dome is 28.392m. with 1m. square grid and the construction of whole building completed in 1988.

The dome shell surface geometry coordinates were

configurated by using either the Bezier function or
the force density approach under the same boundary
coordinates and the same structural rise at the vertex.
Fig.5 shows the slight difference in XZ and YZ plane of
the cross section.   The solid curve which is found   by
using the force density method seems thinner than   the

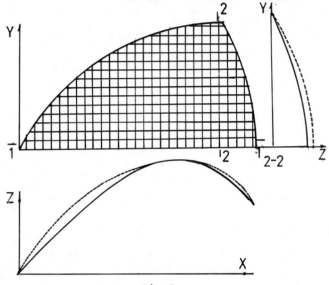

Fig.5

dotted curve of Bezier function made by a mathematician
and both can meet the architectural requirement.
     From the structural design it was found that   the
overall buckling of the shell principally dominate the
element section as the internal forces in elements are
rather small.   The model test comparison of stability
was made between two kinds of different geometrical
surface.   Four elaborative models with 1/30 scale were
manufactoried and tested on a vacuum apparatus.   The
test shows that the buckling mode presents unsymetrical
in all spacimen (Fig.6).   The experimental values
are stable and reliable(Tab.2).

Buckling load of specimen (Mpa)   Tab.2

Bezier Function Model		Force Density Model	
No.1	No.2	No.3	No.4
27.0	26.2	38.1	36.5

The buckling load of real structure was pridicted by an approximate formula (Schmit,1961):

$$q=k(1/RxRy)\sqrt{EtmD}$$

where,q is buckling load of the grid shell
    k is coefficient,relating with forms and grids
    Rx,Ry are the principal curvature of the shell
    tm=A/D1 is the reduced thickness of the shell
    A is the cross section area of the member
    D1 is the grid size
    D=EI/D1 is the reduced cylinderical stiffness
    EI is bending rigidity of the member

Fig.6

The pridict buckling load of real structure is 3.31KN/M2 while using the Bezier function surface and is 4.54KN/M2 using the force density method configurated surface.

The Amphbian Reptile Exhibition Hall (Fig.7)

The Amphbian Reptile Exhibition Hall is a new type of spatial structure group in Shanghai Zoo locating in Hongqiao opening area. The turtle-like entrance hall as a symbol of this exhibition building is a single layer hexagonal grid dome with three elements jointed small nodes. Three cylinderical buildings and corridors are composed of shells and grids.The whole building will be completed in this year.
    Two composed half grid domes with different

diameters supported by a spatial truss arch as a
skylight are finally used in the ecology hall (Fig.8).

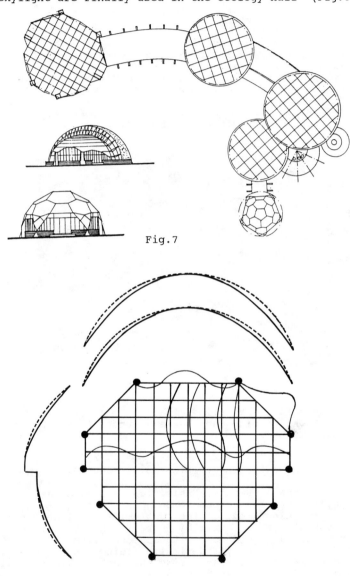

Fig.7

Fig.8

The stress analysis shows that some points as at
"17"and "4" where open the door or window have larger
vertical deflections and the deflections perpendicular
to the door surface plane.

A calculated comparison is made between the
inversed cable net dome and the spherical latticed
shell. The load versus deflection curves at point "17"
from the nonlinear analysis shows that the reversed
smaller deformation and larger ultimate load capacity
(7.7% more).
Four specimen have been tested by using the same
vacuum apparutus and the symmetrical buckling mode is
found in all specimen (Fig.9). The nonlinear analysis
basically proves that the buckling load of specimen
with imperfections approachs to the experimental values
(Tab.3).

Buckling load of specimen (Mpa)   Tab.3

1/25 , D=2.1MM		1/25, D=2.8MM	
No.1	No.2	No.3	No.4
8.10	9.10	18.0	18.0

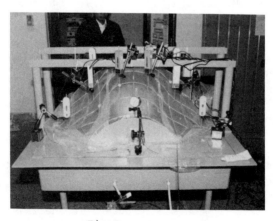

Fig.9

Conclution

The grid dome shell is one of simplest structures
which can be easily jointed and used to cover middle
span area. An optimized form of grid dome may be found
by adjusting the force density.

Acknowlegement

        This research is supported by the natural and
scientific fund of China and the fund of city and
country construction. The author thanks to Mr.Xu P.R.,
Mr.Chen,G.Y.,Mrs.Lin B., Mrs.Wang,Z.S. and Mr. Si R.for
kindly helping in this paper.

Reference

1 Andres,O.A."Homeostatic models for shell roof design"
   10 Years of progress in shell and spartial structures
   30 anniversary of IASS Madrid 1989
2 Grundig,L. " The Force Density Approach and merical
   method for the calculating Networks"   The Third
   International Symposium on Long span Structures.
   Stuttgart.1985
3 Hangai,Y.,Kawaguchi,K.I. "Analysis for Shape-finding
   of Unstable Link Structures in the Unstable State"
   Proceeding of the International Colloquium on Space
   Structures for Sport Building 1987
4 Hu,X.R., Lin,B., et at.   "An Investigation on the
   Optimal Forms of Latticed shell Structures by the
   Force Density Method" Proceedings of Fifth Conference
   on Space Structures(in Chinese) 1990,7.
5 Kaneyoshi,M., Tanaka,H., Furuta,H. "Determination of
   optimun shape for cable net systems based on fuzzy
   set theory"Proceedings of the third East Asia-Pacific
   Conference on Structural Engineering and Construction
   1991
6 Ramm,E. "Shape Finding Methods of shells" Bulletin of
   the International Association for Shell and Spatial
   Structures (IASS) Vol.33 1992,n.2, 89-99.
7 Ramm,E.,Bletzinger,K.-U.,Reitinger,R.   "Shape
   Optimization of Shell Structures"   Bulletin of the
   International Association for Shell and Spatial
   Structures (IASS) Vol.34 1993,n.2
8 Sun,J.H., Hu,X.R. "The Investigation on the stability
   by Geometrical Imperfections in Latticed Shells"
   Proceedings of Sixth Conference on Space Structures
   (in Chinese) 1992,11.
9 Yoshida,A., Majowiecki,M., Tsubota,H.   "Form finding
   analysis for membrane structures using force density
   method"   10 Years of progress in shell and spatial
   structures. 30 anniversary of IASS Madrid 1989

Explorations in Geodesic Forms

Dr. Dimitra Tzourmakliotou[1]

Abstract

Geodesic forms constitute an important family of structural systems. The objective of the paper is to show how the concepts of a mathematical tool called formex algebra and its programming language Formian can be used to represent and generate geodesic configurations with ease and elegance.

The data generation of geodesic forms is solved in two stages. Firstly, a function called the "polyhedron function" is used to generate a configuration modelled on a polyhedron. The resulting configuration is referred to as a "polyhedric configuration". The polyhedron function constitutes the kernel of the problem handling strategy for the data generation of geodesic forms. In the next stage, a transformation referred to as the "tractation retronorm" is employed to obtain the projection of the polyhedric configuration on the required surface(s). The function allows the choice of different types of surfaces such as sphere, ellipsoid and hyperbolic paraboloid. Also, different types of projections may be employed such as central, axial, and parallel.

Introduction

Geodesic forms allow effective use of materials and space and may be employed to create architecturally interesting and economic building structures. They are presently used in a number of specialised areas of construction such as domes for sports arenas, cultural centres, exhibition halls, Olympic facilities and radomes covering radar installations. Their widespread use has been hindered by the difficulty in defining their geometry. This problem has presented a challenge to engineers and architects for decades. To date a variety of piecemeal methods to

----

[1]Tzourmakliotou, Civil Engineer, 16 Larisis Str, 40400 Ambelona-Larisa, GREECE

439

solve this problem have developed, each with its own classification system and suggested tessellation patterns. These methods have tended to use simple spherical or elliptical envelopes, they are difficult to apply and relate only to a small subset of the potential geodesic forms. It is also interesting to note that this problem is not confined to structural engineers. Structural chemists have also faced similar difficulties when investigating the geometry of large carbon molecules such as C60 (which are geodesic forms). However, the approach presented in this paper provides a methodology that allows data generation for geodesic forms of all kind to be handled with ease and elegance. The method is based on the concepts of formex algebra and its programming language Formian. In actually using the method one has to be familiar with the formex concepts. However, the present paper is written in such a way that allows a reader to follow the basic ideas without any knowledge of formex algebra.

Polyhedric and Geodesic Forms

To begin with, consider the configuration shown in Fig. 1. This is obtained by first placing the triangular pattern of Fig. 2 on the top five faces of an icosahedron, as in Fig. 3, and then projecting the resulting configuration on a sphere which is concentric with the icosahedron. The configuration of Fig. 1 is an example of a "geodesic form" and the configuration of Fig. 3 is an example of a "polyhedric form". So, a polyhedric form is a configuration that is obtained by placing one or more patterns on one or more faces of a polyhedron.

The above described procedure for obtaining a geodesic form involves a number of changeable features: Firstly, the pattern that has been mapped onto the faces of the polyhedron may be changed. For instance, the patterns of Figs 4 and 5, used instead of that of Fig. 2, will give rise to the geodesic forms shown in Figs 6 and 7. Secondly, the part of the polyhedron on which the pattern is mapped may be varied. For instance, the pattern of Fig. 8 mapped onto the faces on one side of an icosahedron will give rise to the polyhedric form shown in Fig. 9 and the geodesic form shown in Fig. 10. Thirdly, the type of polyhedron may be changed. For instance, the pattern of Fig. 5 mapped onto the top four faces of an octahedron will give rise to the geodesic form of Fig. 11. Fourthly, the surface on which projection is obtained may be changed. For instance, Fig. 12 shows the result of projection of the polyhedric configuration of Fig. 13 onto a paraboloid.

The method which will be described in the sequel is based on the concepts of formex algebra and its programming language Formian. In a Formian environment, a formex representing a geodesic form may be generated in two stages. In the first stage, a function is used to create a formex representing a polyhedric configuration. This formex is then transformed in the second stage into another formex representing the projection of the polyhedric configuration on a required surface. These two stages are discussed in the sequel.

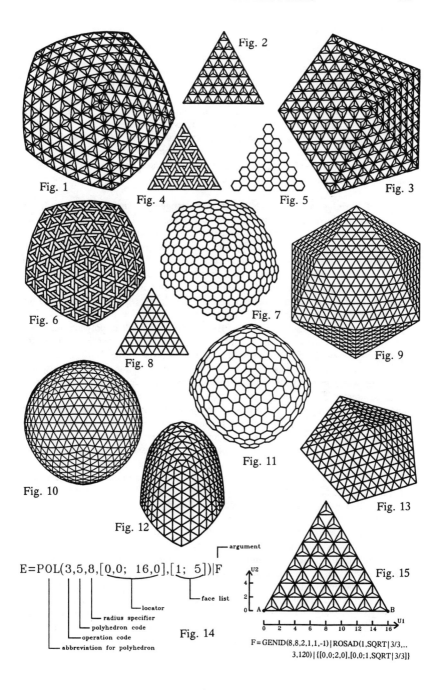

Fig. 2

Fig. 1

Fig. 4

Fig. 5

Fig. 3

Fig. 6

Fig. 7

Fig. 8

Fig. 9

Fig. 10

Fig. 11

Fig. 12

Fig. 13

$E = POL(3,5,8,[0,0;\ 16,0],[1;\ 5])|F$

argument

locator

face list

radius specifier

polyhedron code

operation code

abbreviation for polyhedron

Fig. 14

Fig. 15

$F = GENID(8,8,2,1,1,-1)|ROSAD(1,SQRT|3/3,...$
$3,120)|\{[0,0;2,0],[0,0;1,SQRT|3/3]\}$

Polyhedron Function

A "polyhedron function" may be used to generate a formex representing a polyhedric configuration. For instance, the polyhedric configuration of Fig. 3 can be represented by the "formex variable" E, as shown in Fig. 14.

In actually using Formian for generation of polyhedric or geodesic forms, one must be familiar with "formex algebra". However, reference to formex algebra in this paper is kept to a minimum and a reader who is not familiar with formex algebra can follow the material without any problem. Such a reader may simply assume that a "formex" is a mathematical entity that can represent a configuration and ignore the details of the formulations.

Returning to the discussion of the arrangement shown in Fig. 14, the constituent parts of the formulation are as follows: POL is an abbreviation for "polyhedron" and is followed by a sequence of parameters enclosed in parentheses. The "operation code" specifies the type of action which is required to be performed. The integer 3 given as the operation code in Fig. 14 specifies the operation of mapping a pattern onto the faces of a polyhedron. There are a number of other operations that can be performed through a polyhedron function and these operations may be specified by suitable code numbers. The "polyhedron code" specifies the type of polyhedron which is to be used as the basis for the operation. The polyhedron code for each one of the Platonic and Archimedean polyhedra is given at the top left corner of the squares in Plate 1. The integer 5 given as the polyhedron code in Fig. 14 specifies an icosahedron. The "radius specifier" determines the size of the polyhedron by specifying the radius of the circumsphere of the polyhedron, that is, the sphere that contains all the vertices of the polyhedron. This parameter is given as 8 (units of length) in Fig 14. The "locator" specifies the manner in which a given pattern is to be mapped onto a face of the polyhedron. To elaborate, consider the pattern shown in Fig. 15. This pattern is the same as that of Fig. 2 but is shown here together with a reference coordinate system. A formex variable F representing the pattern relative to the reference system U1-U2 is also given in Fig. 15. Two corners of the pattern are denoted by the letters A and B. The pattern is intended to be placed on a face of the polyhedron in such a way that AB fits an edge of the polyhedron. This intention is conveyed by including the U1-U2 coordinates of A and B in the locator of Fig. 14.

The role of the "face list" in Fig 14 is to specify those faces of the polyhedron onto which the pattern is to be mapped. The specified faces are 1 to 5. The "face numbers" for the Platonic polyhedra are shown in Plate 1. In this Plate, each polyhedron is shown together with the global Cartesian x-y-z coordinate system. The origin of the coordinate system is at the centre of the polyhedron and is indicated with a large dot (The centre of a polyhedron is the same as the centre of its circumsphere). The point where the positive side of the x-axis intersects the polyhedron is indicated by a little circle with an enclosed x. This point is referred to as the "x-point". The positions of the positive directions of y- and z-axes are

indicated by arrows, with z being always vertical. Each face of a polyhedron is
identified with a "face code" which is given at one corner of the face. A face code

Plate 1

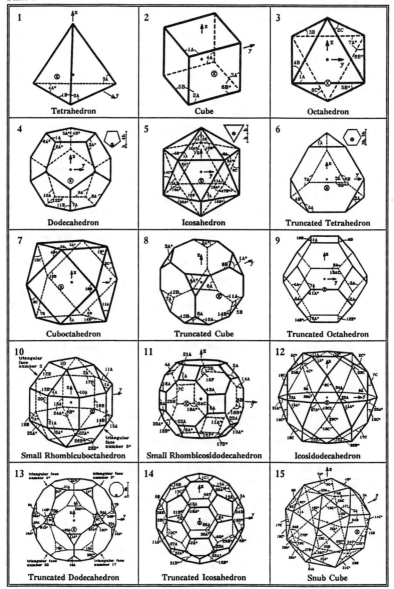

consists of a number followed by a letter and possibly followed by an asterisk. The number in a face code indicates the order in which the face is considered for pattern placement. The letter in a face code determines the points A, B, C,.., etc of the pattern that is to be placed on the indicated corner of the face. These letters for patterns corresponding to different shapes of polyhedral faces are shown in Fig 16. As far as the Platonic polyhedra are concerned, a face is an equilateral triangle, a square or a pentagon. However, Fig 16 also contains information about the corner letters for a hexagon, an octagon and a decagon. This information relates to Archimedean polyhedra, as discussed latter. If a face code has an asterisk, it implies that the pattern placed on the faces is the reflection, with respect to a horizontal plane, of the given pattern. The "argument" in Fig. 14 is the formex variable that represents the pattern of Fig. 15. The symbol preceding the argument in Fig. 14 is used to separate the polyhedron function from its argument.

As the next example, consider the configuration shown in Fig. 17. This is a view of a polyhedric configuration obtained by mapping the square pattern of Fig. 18 onto three faces of a cube. The formulation of the formex variable C1 that represents the polyhedric form of Fig. 17 is given in the same figure where the argument P1 in the formulation represents the square pattern as given in Fig. 18. In creating a polyhedric form, it is possible to use different patterns for different faces of the base polyhedron. For instance, in the case of the polyhedric configuration of Fig. 19, the square pattern of Fig. 18 is placed onto the 2nd and 3rd faces of a cube and the pattern of Fig. 20 is placed onto the 1st face. A pattern that is mapped onto a face of a polyhedron need not necessarily "fill" the face or "match" the boundaries of the face as demonstrated by the polyhedric form shown in Fig. 21. Here, the pattern of Fig. 22 which has a rectangular shape of boundary is mapped onto the 1st face of a cube and extends beyond the edges of the face.

A polyhedric configuration may have more than one layer. For example, consider the configuration shown in Fig. 23. This is a view of a double layer polyhedric configuration which is based on the top five faces of two concentric icosahedra. The double layer pattern used for mapping is shown in Fig. 24. In Figs 23 and 24, the top layer elements are drawn in thick lines and the bottom layer elements as well as the web elements are drawn in thin lines. The new points to be noticed in this example are: (i) In formulating the pattern of Fig. 24, each layer has an identification number. The identification number of the bottom layer is 1 and that of the top layer is 2. These identification numbers appear as "third direction coordinates" in the formex formulation. (ii) The radius specifier in Fig. 23 contains two entries. The first entry, that is, number 10, gives the radius of the circumsphere of the base polyhedron for the first layer (bottom layer) and the second entry, that is number 12, gives the radius of the circumsphere of the base polyhedron for the second layer (top layer). (iii) The locator in Fig. 23 contains two parts. The first part of the locator specifies the coordinates of points A1 and B1 that should fit the end vertices of an edge of the base polyhedron of the first layer (bottom layer). The second part of the locator specifies the coordinates of points A2 and B2 that should fit the end vertices of an edge of the base polyhedron of the

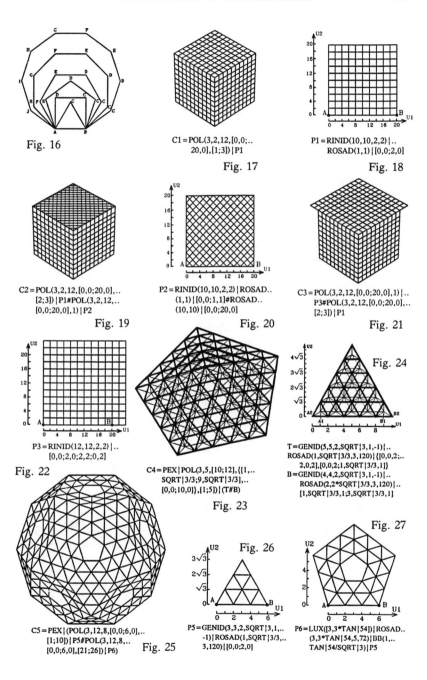

Fig. 16

C1 = POL(3,2,12,[0,0;..
20,0],[1;3]) | P1

Fig. 17

P1 = RINID(10,10,2,2) |..
ROSAD(1,1) | [0,0;2,0]

Fig. 18

C2 = POL(3,2,12,[0,0;20,0],..
[2;3]) | P1#POL(3,2,12,..
[0,0;20,0],1) | P2

Fig. 19

P2 = RINID(10,10,2,2) | ROSAD..
(1,1) | [0,0;1,1]#ROSAD..
(10,10) | [0,0;20,0]

Fig. 20

C3 = POL(3,2,12,[0,0;20,0],1) |..
P3#POL(3,2,12,[0,0;20,0],..
[2;3]) | P1

Fig. 21

P3 = RINID(12,12,2,2) |..
[0,0;2,0;2,2;0,2]

Fig. 22

C4 = PEX | POL(3,5,[10;12],{[1,..
SQRT|3/3;9,SQRT|3/3],..
[0,0;10,0]},[1;5]) | (T#B)

Fig. 23

Fig. 24

T = GENID(5,5,2,SQRT|3,1,-1) |..
ROSAD(1,SQRT|3/3,3,120) | {[0,0,2;..
2,0,2],[0,0,2;1,SQRT|3/3,1]}
B = GENID(4,4,2,SQRT|3,1,-1) |..
ROSAD(2,2*SQRT|3/3,3,120) |..
[1,SQRT|3/3,1;3,SQRT|3/3,1]

Fig. 27

C5 = PEX | (POL(3,12,8,[0,0;6,0],..
[1;10]) | P5#POL(3,12,8,..
[0,0;6,0],[21;26]) | P6)   Fig. 25

Fig. 26

P5 = GENID(3,3,2,SQRT|3,1,..
-1) | ROSAD(1,SQRT|3/3,..
3,120) | [0,0;2,0]

P6 = LUX([3,3*TAN|54]) | ROSAD..
(3,3*TAN|54,5,72) | DB(1,..
TAN|54/SQRT|3) | P5

second layer (top layer). Situations when polyhedric configurations involve more than two layers are dealt with in an analogous manner.

In the case of Archimedean polyhedra the faces are not all of the same shape. In spite of this, one may create a polyhedric configuration by mapping a single pattern on all or some of the faces of an Archimedean polyhedron. However, in creating polyhedric configurations that are based on Archimedean polyhedra, typically, one would use different patterns for different face shapes. For instance, the polyhedric configuration of Fig. 25 is based on an icosidodecahedron and is obtained by mapping the patterns of Figs 26 and 27 onto the triangular and pentagonal faces, respectively. A selection of Archimedean polyhedra together with their face codes are shown in Plate 1.

Tractation Retronorm

A geodesic configuration may be represented by a formex. For example, the geodesic configuration of Fig. 1 may be represented by the formex variable G whose formulation is shown in Fig 28. The constituent parts of this formulation are described as follows: TRAC is an abbreviation for "tractation" and is followed by a sequence of parameters enclosed in parentheses. The construct that consists of TRAC together with the ensuing parameter list is referred to as a "tractation retronorm". The "projection code" specifies the type of projection to be used. The projection code may have the value 1, 2, 3 or 4 indicating central, parallel, axial or radial projection, correspondingly. The projection code in Fig. 28 is given as 4, implying radial projection. The "surface code" specifies the type of surface on which projection is to be made. Table 1 lists a number of surfaces together with their corresponding codes. The surface code in Fig. 28 is given as 1, implying a sphere, and this is followed by three zeros and the number 8 that specifies the radius of the sphere. These zeros are the x-y-z coordinates of the centre of the sphere. The items that should follow the surface code in different cases are listed in Table 1. The "selection code" specifies the course of action to be followed when the projection of a point cannot be determined uniquely. To elaborate, in obtaining the projection Q of a point P on a surface S, the following situations may arise: (i) Q is determinable uniquely, Fig. 29, (ii) Q is nonexistent, Fig. 30, (iii) there is more than one solution for Q, Fig. 31. Various possible courses of action in the case of nonexistent of multiple solutions for Q, together with the corresponding selection codes are listed in Table 2. The "argument" in Fig. 28 is a formex variable representing the polyhedric configuration of Fig. 2. The formulation of this formex variable is shown in Fig. 14.

Application of the tractation retronorm is further exemplified through Figs 32 and 33 that show central projections of the polyhedric form of Fig. 34 on a sphere and an ellipsoid, respectively.

In using the tractation retronorm, it is possible to have more than one surface

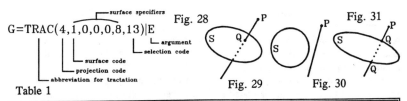

$$G=TRAC(4,\overbrace{1,0,0,0,8,13}^{\text{surface specifiers}})|E$$

- surface specifiers
- argument
- selection code
- surface code
- projection code
- abbreviation for tractation

Fig. 28

Fig. 29

Fig. 30

Fig. 31

Table 1

Surface	Code	Items That Should Follow the Surface Code
Sphere	1	x, y and z coordinates of the centre of the sphere followed by the radius of the sphere
Ellipsoid	2	x, y and z coordinates of the centre of the ellipsoid followed by the three semiaxes of the ellipsoid in the x, y and z directions
Paraboloid	3	x, y and z coordinates of the centre of the paraboloid followed by three semiaxes of the paraboloid in the x, y and z directions

Table 2

Selection Code	Indicated Solution — In the case of multiple solutions for Q choose the one which has the	Indicated Solution — In the case of nonexisting solution	Selection Code	Indicated Solution — In the case of multiple solutions for Q choose the one which has the	Indicated Solution — In the case of nonexisting solution
11	greatest x component		21	greatest x component	
-11	least x component		-21	least x component	
12	greatest y component		22	greatest y component	
-12	least y component		-22	least y component	
13	greatest z component	ignore the cantle (a part of) which represents point P	23	greatest z component	accept point P as the solution
-13	least z component		-23	least z component	
14	greatest distance from the origin		24	greatest distance from the origin	
-14	least distance from the origin		-24	least distance from the origin	
15	greatest distance from P		25	greatest distance from P	
-15	least distance from P		-25	least distance from P	

on which projection is made. For instance, the double layer polyhedric form of Fig. 23 is projected on two concentric spherical surfaces, using radial projection, and the result is shown in Fig. 35. The formulation for this double layer geodesic form is also shown in the figure. Two new points are to be noticed in this formulation. Firstly, the particulars for each surface are given separately, enclosed in square brackets. Secondly, when there are more than one surface for projection, it would be necessary to identify the intended surface for each point to be projected. This identification is achieved through two new items added at the beginning of the set of particulars for each surface. The new items in the particulars of the first surface in Fig. 35 are 4 and 1. These indicate that the first surface should be used for the projection of every point whose fourth coordinate is equal to 1. The new items in the particulars of the second surface in Fig. 35 should be interpreted analogously. Insertion of "layer identification numbers" in fourth coordinate positions is effected through the polyhedric function.

The applications of the tractation retronorm as described so far concern the usage in relation to geodesic forms. However, the tractation retronorm may be employed in any situation when it is required to obtain the projection of a single -or multi- layer configuration on one or more surfaces. An important application of the tractation retronorm is in generation of "grid domes". For example, the grid dome of Fig. 36 is obtained by parallel projection of the pattern of Fig. 37 on a spherical surface. The formulations are included in the figures.

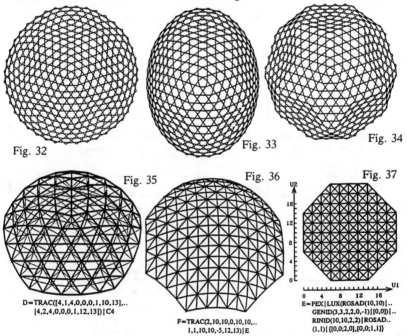

Fig. 32

Fig. 33

Fig. 34

Fig. 35

D=TRAC([4,1,4,0,0,0,1,10,13],..
[4,2,4,0,0,0,1,12,13])|C4

Fig. 36

F=TRAC(2,10,10,0,0,10,10,..
1,1,10,10,-5,12,13)|E

Fig. 37

E=PEX|LUX(ROSAD(10,10)|..
GENID(3,3,2,2,0,-1)|[0,0])|..
RINID(10,10,2,2)|ROSAD..
(1,1)|{[0,0;2,0],[0,0;1,1]}

DISCRETE FIELD STABILITY ANALYSIS OF X-BRACED LATTICES

Raja R. A. Issa[1] and R. Richard Avent[2], Members ASCE

ABSTRACT

The three-dimensional X-braced lattice is used as a compresion member in structures ranging from the long span roof of buildings to radio and television antennas. In such cases the design is often controlled by the stability of the overall system as opposed to local member buckling. The purpose of this paper is to present aplications of discrete field mechanics techniques in the elastic stability analysis of three dimensional X-braced lattices.

The use of discrete field mechanics offers several advantages over alternative approaches for the structural analysis of latticed systems. The foremost of these advantages is that the discrete mathematical model is usually "exact" in the sense of linear elastic, small deflection theory, thus yielding very accurate results. Moreover, the solution to the resulting difference equation model quite often lends itself to a field solution which usually leads to closed form analytical formulas. Finally, the solution forms are independent of the number of joints in the system thus allowing the user to avoid having to solve large systems of equations for eigenvalues as is usually the case with other elastic stability analysis methods. Thus, the critical buckling load for any given X-braced lattice can be determined without a huge computational effort.

---

[1]Associate Professor, M.E. Rinker, Sr. School of Building Construction, Univ. of Florida, Gainesville, FL 32611-5703. e-mail: rrissa@eagle.circa.ufl.edu

[2]Professor and Chair, Department of Civil Engineering, Louisiana State University, Baton Rouge,LA 70803.

449

Literature Review

Dean (1969,1976) has published several papers on the analysis of various types of latticed structures. Dean and Jeter (1972) used the discrete field stability approach to analyze a planar X-braced truss subjected to equal axial load couples causing bending at the ends. Avent et.al. (1991) extended the Dean-Jeter concept to include pure compression on a planar X-braced truss. This study shows the application of the discrete field stability approach to three dimensional X-braced lattices. The resulting formulation is used to study the elastic stability of two typical X-braced lattice structures. The lattice sections are treated as ideally, simply supported axially loaded members with no applied lateral loads and all panels are considered to be "regular", i.e. the panel spacing is constant and the panel properties do not vary with length. In addition the lattice is treated as an ideally, simply supported, axially loaded member with no applied lateral loads.

Development of Mathematical Model

A regular three-dimensional X-braced lattice is shown in Fig. 1. The lattice is assumed to have two equal top chords (lines 2 and 3 in Fig. 1) but the bottom chord (line 1) may be different. All posts are identical and similarly all diagonals are equal. The member properties are constant from panel to panel with A_t, A_b, A_c, and A_d representing the area of the top chord, bottom chord, post and diagonals, respectively; and E = the modulus of elasticity; and ℓ = the panel length. The mathematical

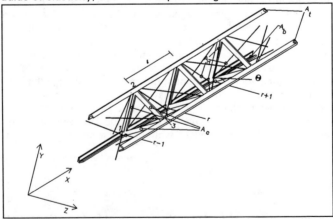

Figure 1. Three-Dimensional X-braced Lattice

model for such a regular three-dimensional X-braced lattice is obtained by performing a routine stiffness analysis (Issa and Avent 1984) for a typical interior and boundary joint lines. A typical joint line will consist of 3 joints, designated 1, 2, and 3, with three degrees of freedom per joint (since all joints are assumed to be hinged). Joint lines in the X direction are designated by the discrete variable r and the lattice has $n+1$ panels. The horizontal rollers at both ends imply that the applied horizontal joint loads, $P_x^1(r)$, $P_x^2(r)$ and $P_x^3(r)$, have to be self-equilibriating for geometric stability. These boundary conditions are required to affect the finite Fourier series solutions to be presented. The bar forces are $F_t(r)$, $F_b(r)$, $F_c(r)$, $S_r(r)$, and $S_l(r)$ representing the top chord, bottom chord, post, and right and left diagonals of the rth joint, respectively.

Nonbuckling Analysis

The mathematical model for regular X-braced planar and space trusses subjected to static loads have been derived by Avent and Issa (1981,1982) and Issa and Avent (1984), respectively. For a three-dimensional X-braced lattice (Fig. 1), the general form of the joint equilibrium equations written in difference operator notation is:

$$C_r X(r) = \frac{-4\ell}{\alpha EA_d} F(r) \tag{1}$$

in which C_r = the structural stiffness matrix operator; $X(r)$ = the static joint displacement matrix; $F(r)$ = the joint loading matrix; and r designates the X or longitudinal direction (see Fig. 1).

Buckling Equations

The modified Dean-Jeter (1972) matrix expression for truss buckling in general is:

$$\sum_m [(\frac{F}{L})_{km}(R_{km} - I_{km}) - K_{km}](U_k - U_m) = -P_k \tag{2}$$

in which the subscripts k and m refer to a member connecting joints k and m, the summation on m refers to all members connected to joint m, and F_{km} the forces due to the applied axial load, P, as found from considering statics only; L_{km} = the length of a member between joints k and m; R_{km} = the rotational matrix that transfers the member from local to global coordinates; K_{km} = the global stiffness matrix of the member, of length L and area A, connecting joints k and m, i.e. $E(\frac{A}{L})_{km}R_{km}$; I_{km} = the identity matrix corresponding to K_{km}; U_k, U_m = the joint displacement vectors at joints k and m due to the applied joint loads; and P_k = the load vector applied at the joint k.

The second term in Eq. 2 represents the initial deformations of the undeformed structure, due to applied static loads, using the undeformed truss geometry, that is,

$$\sum_m K_{km}(U_k - U_m) = P_k \tag{3}$$

in which $K_{km} = \frac{\alpha EA_d}{\ell} C_r$; $U_k - U_m = X(r)$; and $P_k = F(r)$

Thus, Eq. 4 is converted to the mathematical model for the three dimensional X-braced lattice subjected to static loads that has been derived by Issa and Avent (1984). The bar forces obtained from this solution can be used in Eq. 3 as part of the stability analysis.
The first term of Eq. 3 represents the effects of the axial forces on the deformed shape and is represented by:

$$\sum_m S_{km}(U_k - U_m) = 0 \tag{4}$$

where,

$$S_{km} = (\frac{F}{L})_{km}(R_{km} - I_{km}) \tag{5}$$

The general form of the governing elastic buckling equations for the three-dimensional lattice in difference operator node is derived from Eq. 2 as:

$$A_r X(r) = \frac{\ell}{\alpha EA_d} F(r) \tag{6}$$

in which $A_r = \overline{C}_r - C_r$ with $\overline{C}_r = -(\frac{\ell}{\alpha EA_d}) S_{km}$; $X(r)$ = the joint displacement matrix accounting for deformed geometry; $F(r)$ is the joint loading matrix; and the discrete variable r designates joint lines in the X or longitudinal direction (see Fig. 1). For the three-dimensional X-braced lattice with three degrees of freedom per joint, Eq. 6 becomes:

$$
\begin{bmatrix}
A_{11} & 0 & 0 & A_{14} & A_{15} & A_{16} & A_{17} & A_{18} & A_{19} \\
0 & A_{22} & 0 & A_{24} & A_{25} & A_{26} & A_{27} & A_{28} & A_{29} \\
0 & 0 & A_{33} & A_{34} & A_{35} & A_{36} & A_{37} & A_{38} & A_{39} \\
A_{41} & A_{42} & A_{43} & A_{44} & 0 & 0 & A_{47} & 0 & A_{49} \\
A_{51} & A_{52} & A_{53} & 0 & A_{55} & A_{56} & 0 & 0 & 0 \\
A_{61} & A_{62} & A_{63} & 0 & A_{65} & A_{66} & A_{67} & A_{68} & A_{69} \\
A_{71} & A_{72} & A_{73} & A_{74} & 0 & A_{76} & A_{77} & 0 & 0 \\
A_{81} & A_{82} & A_{83} & 0 & 0 & A_{86} & 0 & A_{88} & 0 \\
A_{91} & A_{92} & A_{93} & A_{94} & 0 & A_{96} & 0 & 0 & A_{99}
\end{bmatrix}
\begin{Bmatrix}
U^1(r) \\
V^1(r) \\
W^1(r) \\
U^2(r) \\
V^2(r) \\
W^2(r) \\
U^3(r) \\
V^3(r) \\
W^3(r)
\end{Bmatrix}
= \frac{-4\ell}{\alpha E A_d}
\begin{Bmatrix}
P_x^1(r) \\
P_y^1(r) \\
P_z^1(r) \\
P_x^2(r) \\
P_y^2(r) \\
P_z^2(r) \\
P_x^3(r) \\
P_y^3(r) \\
P_z^3(r)
\end{Bmatrix}
\qquad (7)
$$

in which the stiffness matrix elements, C_r, are defined in the first two columns of Table 1. The boundary conditions, $r = 0$ and n, are shown in columns 3 and 4, respectively. The joint deflections are designated as U^i, V^i and W^i in which the superscript, i, designates either joint 1, 2 or 3 at joint line r. The geometric and material variables used in Table 1 are defined as follows: $\alpha = \cos \Theta$, $\beta = \sin \Theta$; Θ = angle of diagonal to chord members; r = joint-line number; $a_1 = \dfrac{A_b}{\alpha A_d}$; $a_2 = \dfrac{A_c}{\beta A_d}$; $a_3 = \dfrac{A_t}{\alpha A_d}$;

$b_1 = \dfrac{R_b}{\alpha R_d}$; $b_2 = \dfrac{R_c}{\beta R_d}$; $b_3 = \dfrac{R_t}{\alpha R_d}$, where R_b, R_c, R_d, R_t = the ratio of the member forces to the axial load P in the bottom chords, the posts, the diagonal members and the top chords respectively.

These equations include the member forces generated from the non-buckling analysis. Since the individual member forces vary from member forces vary from member to member, a variable coefficient set of difference   equations is produced.   However, an excellent approximation (for $n > 4$) is to take the force in each type of member as a constant equal to the member force at the truss center.

Solution to the Mathematical Model.-- Avent et al. (1991) have shown that for certain classes of boundary conditions, finite Fourier series can be used to solve Eq. 6 for critical buckling loads. For the purpose of this study, only simply supported lattice sections are considered, i.e. horizontal rollers at all but one boundary node, which is hinged. The general form for the finite Fourier series expansion of the joint deflection matrix (Issa and Avent, 1984) is:

$$X(r) = \sum_{k}^{n} [X_k G_k(r)]^T \tag{8}$$

and $X(r)$ and X_k represent the displacements and the corresponding Euler coefficients. The summation index, k, is to be taken over the range $k = 0, (1), n$ for cosine series and $k = 1, (1), n-1$ for sine series. The joint loading matrix, $F(r)$, is expanded as an appropriate series. The general form of the series expansion is:

$$F(r) = \sum_{k}^{m} [F_k G_k(r)]^T \tag{9}$$

Substituting the series solution form of the deflections (Eq. 8) and the loadings (Eq. 9) into the governing equations (Eq. 6) and matching coefficients will yield a set of algebraic simultaneous equations (Eq. 10) relating Euler coefficients of displacement to Euler coefficients of applied loadings. The general form of that equation for the buckling analysis is:

$$A_k X_k = \frac{4\ell}{\alpha EA_d} F_k \tag{10}$$

The elements of A_k are shown in Table 1, column 5. The general criteria for stability can be obtained by setting the determinant of Eq. 10, i.e. A_k, equal to zero and solving for the corresponding eigenvalues.

Numerical Examples

Example 1.-- Consider a guyed tower section of $n = 25$ panels, each of length $\ell = 1.52$ m (5ft) and angle $\Theta = 30°$. Each of the three faces of the equilateral triangle forming the crossection is 0.88 m (2.9 ft) wide. The tower leg area $A_t = A_b = 27.9$ cm^2 (4.33 in^2) and the post and diagonal areas are $A_c = A_d = 16.1$ cm^2 (2.5 in^2), respectively. Using the discrete field techniques presented in this paper, the critical buckling load is determined to be 657 kN (147.7 k).

Example 2.-- Consider a lattice column section of $n = 10$ panels, each of length $\ell = 1.22$ m (4 ft) and angle $\Theta = 45°$. Each of the three faces of the equilateral triangle forming the crossection is 1.22 m (4 ft) wide. The tower leg area $A_t = A_b = 36.5$ cm^2 (5.66 in^2) and the post and diagonal areas are $A_c = A_d = 25.8$ cm^2 (4 in^2), respectively. Using the Discrete Field techniques presented, the critical buckling load is determined to be 25421 kN (5715.2 k).

Table 2. Non-Zero Elements for Operator Matrix $A_r = \overline{C}_r - C_r$.

Matrix Element (1)	A_r (2)	A_0 (3)	A_n (4)	A_k (5)
A_{11}	$8\overline{P}(2\beta^2+b_2)-4(a_1\diamond,-4\alpha^2)$	$8\overline{P}\beta^2+8\alpha^2-4a_{1\Delta r}$	$8\overline{P}(8\beta^2+b_2)+8\alpha^2+4a_{1\nabla r}$	$8\overline{P}(2\beta^2+b_2)+8(a_1\sigma_k+2\alpha^2)$
A_{22}	$2\overline{P}(2b_1\diamond,-6C^2-b_2+4)+12\beta^2+6a_2$	$2\overline{P}(3C^2+b_2-2b_{1\Delta r}-2)+6\beta^2+6a_2$	$2\overline{P}(2b_{1\nabla r}+3C^2+b_2-2)+6\beta^2+6a_2$	$-2\overline{P}(4b_1\sigma_k+6C^2+b_2-4)+12\beta^2+6a_2$
A_{33}	$2\overline{P}(2b_1\diamond,-2C^2-3b_2-4)+4\beta^2+2a_2$	$2\overline{P}(C^2+3b_2-2b_{1\Delta r}+2)+2\beta^2+2a_2$	$2\overline{P}(2b_{1\nabla r}+C^2+3b_2+2)+2\beta^2+2a_2$	$2\overline{P}(4b_1\sigma_k+2C^2+3b_2+4)+4\beta^2+2a_2$
$A_{44}=A_{77}$	$8\overline{P}(2\beta^2+b_2)-4(a_3\diamond,-4\alpha^2)$	$8\overline{P}\beta^2+8\alpha^2-4a_{3\Delta r}$	$8\overline{P}(8\beta^2+b_2)+8\alpha^2+4a_{3\nabla r}$	$8\overline{P}(2\beta^2+b_2)+8(a_3\sigma_k+2\alpha^2)$
$A_{55}=A_{88}$	$-\overline{P}(4b_3\diamond,-6C^2-5b_2-4)+6\beta^2+3a_2$	$-\overline{P}(4b_{3\Delta r}-3C^2-5b_2-2)+3(\beta^2+a_2)$	$\overline{P}(4b_{3\nabla r}+3C^2+5b_2+2)+3(\beta^2+a_2)$	$\overline{P}(8b_3\sigma_k+10C^2+3b_2-4)+6\beta^2+3a_2$
$A_{66}=A_{99}$	$-\overline{P}(4b_3\diamond,-10C^2-3b_2+4)-10\beta^2-5a_2$	$\overline{P}(4b_{3\Delta r}-5C^2-3b_2+2)-5(\beta^2+a_2)$	$-\overline{P}(4b_{3\nabla r}+5C^2+3b_2-2)-5(\beta^2+a_2)$	$-\overline{P}(8b_3\sigma_k+10C^2+3b_2-4)-10\beta^2-5a_2$

Table 2. Non-Zero Elements for Operator Matrix $\boldsymbol{A}_r = \overline{\boldsymbol{C}}_r - \boldsymbol{C}_r$ (Continued).

Matrix Element (1)	A_r (2)	A_0 (3)	A_n (4)	A_k (5)
$A_{25} = -A_{52} =$ $= A_{28} = -A_{82}$	$-(3\beta^2+\overline{P}(3C^2-2))(\square_r+2)$ $-3\overline{P}b_2-3a_2$	$-\overline{P}(3\alpha^2+1)(\triangle_r+1)$ $-3\beta^2(\triangle_r+1)-3a_2-\overline{P}b_2$	$-\overline{P}(3\alpha^2+1)(\triangledown_r-1)$ $-3\beta_2(\triangledown_r-1)-3a_2-\overline{P}b_2$	$-\overline{P}(6\alpha^2+b_2+2C_k)$ $-6\beta^2C_k-3a_2$
$A_{36} = A_{63} =$ $= A_{39} = A_{93}$	$-[\beta^2+\overline{P}(\alpha^2+3)][(\square_r+2)$ $-3\overline{P}b_2-a_2$	$-(\beta^2+\overline{P}(\alpha^2+3))(\triangle_r+1)$ $-3\overline{P}b_2-a_2$	$(\beta^2+\overline{P}(\alpha^2+3))(\triangledown_r-1)$ $-3\overline{P}b_2-a_2$	$-\overline{P}(2(\alpha^2+3)C_k+3b_2)$ $-2\beta^2C_k-3a_2$
$A_{56} = A_{65} =$ $-A_{68} = -A_{86}$	$\sqrt{3}(2(\overline{P}-1)\beta^2+\overline{P}b_2-a_2)$	$\sqrt{3}((\overline{P}-1)\beta^2+\overline{P}b_2-a_2)$	$\sqrt{3}((\overline{P}-1)\beta^2+\overline{P}b_2-a_2)$	$\sqrt{3}(2(\overline{P}-1)\beta^2+\overline{P}b_2-a_2)$
$A_{58} = A_{85}$	$-4\overline{P}(b_2+2+\square_r)$	$-4\overline{P}(b_2+\triangle_r+1)$	$4\overline{P}(\triangledown_r-b_2-1)$	$-4\overline{P}(b_2+2C_k)$
$A_{69} = A_{96}$	$-4(\beta^2+\overline{P}\alpha^2)(\square_r+2)-4a_2$	$-4(\overline{P}\alpha^2+\beta^2)(\triangle_r+1)$ $-4a_2$	$-4(\overline{P}\alpha^2+\beta^2)(\triangledown_r-1)$ $-4a_2$	$-4\overline{P}\alpha^2C_k-8\beta^2C_k$ $-4a_2$
$A_{15} = A_{18} =$ $-A_{42} = -A_{72}$	$4\sqrt{3}\alpha\beta(\overline{P}-1)\square_r$	$2\sqrt{3}\alpha\beta(\triangle_r+1)(\overline{P}-1)$	$2\sqrt{3}\alpha\beta(\triangledown_r-1)(\overline{P}-1)$	$4\sqrt{3}\alpha\beta(\overline{P}-1)B_k$

Table 2. Non-Zero Elements for Operator Matrix $A_r = \overline{C}_r - C_r$ (Continued).

Matrix Element (1)	A_r (2)	A_0 (3)	A_n (4)	A_k (5)
$A_{24}=A_{27}=$ $-A_{61}=A_{81}$	$4\sqrt{3}\alpha\beta(\overline{P}-1)\square_r$	$2\sqrt{3}\alpha\beta(\triangle_r+1)(\overline{P}-1)$	$2\sqrt{3}\alpha\beta(\triangledown_r-1)(\overline{P}-1)$	$-4\sqrt{3}\alpha\beta(\overline{P}-1)B_k$
J1	$-4(\alpha^2+\overline{P}\beta^2)(\square_r+2)-4\overline{P}b_2$	$-4(\alpha^2+\overline{P}\beta^2)(\triangle_r+1)-4\overline{P}b_2$	$-4(\alpha^2+\overline{P}\beta^2)(\triangledown_r-1)-4\overline{P}b_2$	$-4\overline{P}(2\beta^2 C_k+b_2)-8\alpha^2 C_k$
J2	$4\alpha\beta(\overline{P}-1)\square_r$	$2\alpha\beta(\triangle_r+1)(\overline{P}-1)$	$2\alpha\beta(\triangledown_r-1)(\overline{P}-1)$	$4\alpha\beta(\overline{P}-1)B_k$
J3	$4\alpha\beta(\overline{P}-1)\square_r$	$2\alpha\beta(\triangle_r+1)(\overline{P}-1)$	$2\alpha\beta(\triangledown_r-1)(\overline{P}-1)$	$-4\alpha\beta(\overline{P}-1)B_k$
J4	$-\sqrt{3}\beta^2(\overline{P}-1)(\square_r+2)+\sqrt{3}(a_2-\overline{P}b_2)$	$-\sqrt{3}\beta^2(\overline{P}-1)(\triangledown_r-1)-\sqrt{3}(a_2-\overline{P}b_2)$	$-2\sqrt{3}\beta^2 C_k(\overline{P}-1)+\sqrt{3}(a_2-\overline{P}b_2)$	

Note: $A_{14}=A_{41}=A_{17}=A_{71}=A_{47}=A_{74}=$J1; $A_{43}=A_{19}=-A_{16}=-A_{73}=-\frac{1}{2}A_{49}=-\frac{1}{2}A_{76}=$J2; $C^2=1+\alpha^2$;

$A_{61}=A_{37}=A_{34}=A_{91}=-\frac{1}{2}A_{67}=-\frac{1}{2}A_{94}=$J3; $A_{26}=A_{62}=A_{35}=A_{53}=-A_{29}=-A_{92}=-A_{38}=-A_{83}=$J4

$\square_r=X(r+1)-2X(r)+X(r-1)$; $\square_r'=\frac{1}{2}[X(r+1)+X(r-1)]$; $\triangle_r X(t)=X(r+1)-X(t)$; $\triangledown_r X(t)=X(t)-X(r-1)$; $C_k=\cos\frac{k\pi}{n}$; $B_k=\sin\frac{k\pi}{n}$; $\sigma_k=1-C_k$

Summary

The techniques of discrete field analysis were used to investigate the elastic stability of the X-braced, three-dimensional lattice. The solution form for the critical buckling load thus determined, is independent of the number of joints in the structure. Accordingly, the user can avoid having to solve large systems of equations for eigenvalues as is usually the case with other elastic stability analysis methods. Consequently, the critical buckling load for any given X-braced lattice can be determined without a huge computational effort.

References

Avent, R. R., and Issa, Raja R. A. (1981). "X-Braced Truss Stiffnes Matrix Analysis of Roofs," *Proc.*, Symposium on Long Span Roof Trusses, ASCE National Convention, St. Louis, MO, Oct. 1981.

Avent, R. R., and Issa, Raja R. A. (1982). "Beam Element Stiffness Matrix for X-Braced Truss," *J. Struct. Engrg.*, ASCE, 108(10), 2192-2210.

Avent, R. R., Issa, Raja R. A., and Chow, Man L. (1991). "Discrete field stability analysis of planar trusses," *J. Struct. Engrg.*, ASCE, 117(10), 423-439.

Dean, D. L. (1969). Discussion of "Behavior of Howe, Pratt, Warren Trusses," (by John. D. Renton), *J. Struct. Engrg.*, ASCE, 95(9), 1997-2000.

Dean, D. L., and Avent, R. R. (1976). "Design Formulas for Deep Space Trusses," *Proc.*, IASS Congress on Space Enclosures, Concordia University, Montreal, Canada.

Dean, D. L., and Jeter, F. R. (1972). "Analysis for truss buckling", *J. Struct. Engrg.*, ASCE, 98(8), 1893-1897.

Issa, Raja R. A., and Avent, R. R. (1984). "Superelement stiffness matrix for space trusses," *J. Struct. Engrg.*, ASCE, 110(5), 1163-1179.

Simulation Procedure for Buckling Behaviors of a Reticular Dome

Shiro Kato[1]
Hideyuki Takashima[2]
Isao Kubodera[3]
and
Takashi Ueki[4]

Abstract

Complex behaviors of the ball-joint connection, especially used in steel space structures, are expressed in the numerical model to confirm the estimative ability of such a model by comparing the analytical results with experimental ones for realistic scale single layered reticular domes being built on a hexagonal plan.

Suitable material non-linearities are established for the fiber elements constituting the connection section, to express the special effects of the present connection.

Reasons about the differences between numerical analyses and experiments are investigated and some conclusions for them are drawn.

Introduction

When one follows the structural behaviors of a reticular dome analytically, the estimation of the connection stiffness has to be noticed. This paper describes about the numerical simulation of the steel reticular dome considering features of the connection. To observe the actual characteristic behaviors of the reticular dome, the analytical results are compared with the experimental ones. In several studies[1,2], such a system identification has been carried out by replacing the connections to rotational springs. However, there are several restrictions for the spring model to follow

1 Professor, Department of Architecture/Civil Engineering, Toyohashi University of Technology, Tempaku-cho, Toyohashi 441, JAPAN
2 Research Associate, Department of Architecture/Civil Engineering, Toyohashi University of Technology
3 Department of Technical Development, TOMOE CORPORATION, 4-5, Toyosu 3-chome, Koto-ku, Tokyo 135, JAPAN
4 Department of Technical Development, TOMOE CORPORATION

precisely the elasto-plastic behaviors. In Ref.(3) , the authors have tried to establish the analytical model of the ball-joint connection and obtained some good agrees with actual connection tests. The present study is placed on the next phase of the above-mentioned study.

Description of Objective Ball-Joint

Currently, there are some types of ball-joint system. Among them, the present authors select the connection whose properties are described in Fig.1 and as followings:

(i)   the member is connected to the ball as a spherical node, by screw
(ii) there is a pressure, caused by inserting and tightening connector, in contacted surface between the ball and the nose of the member

From owning the second feature, relatively higher bending rigidity is kept for the present connection and the decrement of the dome's buckling strength caused by relaxation at the connector is avoided to some extent.

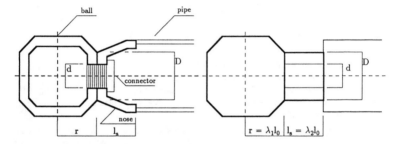

Fig.1 Objective Ball-Joint Connection   **Fig.2** Finite Element Model

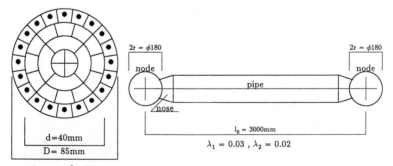

● : nose element
Fig.3 Fiber elements in
connection section

Fig.4 Constitutive Member

Analytical Model for Ball-Joint

The finite element model to follow the actual behaviors of the objective ball-joint was established in the former paper.[3] In which, the load-displacement curves obtained from the bending tests of a simple beam, were followed by the present analytical model. The degree of comparison between analyses and experiments could be judged to be admissible. The explanations for analytical model are illustrated in Figs 2 and 3. The section of the connection is divided into small fiber elements, in which stress-strain relation is assumed as material non-linearity to describe each characteristic behavior. The stress-strain relationship for the connector includes the relaxation at the initial state, and the nose elements behave effectively on compression field only. Both relations are illustrated in Fig.5.

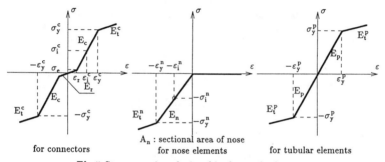

for connectors A_n : sectional area of nose
for nose elements   for tubular elements

Fig.5 Stress-strain relationship for each element

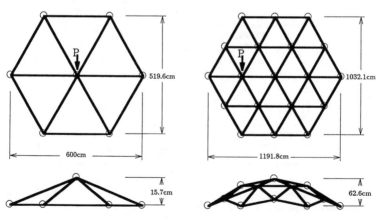

Fig.6 Geometry of the unit dome   **Fig.7** Geometry of the 4-span dome

Fig.5 also includes an explanation of bi-linear relation for tubular elements.

Fig.4 describes the analytical. model for member-unit being composed of the present simulated ball-joint connection and a tubular element.

Brief Explanations for Experiments of Single Layered Domes

The experiments whose results will be compare with results of the present simulation, have been already carried out and the contents have been also presented in Refs.(2) and(4). So, in this section, the details will not be described, but brief explanations are attached to this section.

Geometries of the realistic scale domes for loading tests are shown in Figs. 6 and 7. Fig.6 describes the basic dome being constitutive unit of the large retucular dome and Fig.7 also expresses the reticular dome whose diagonal line includes 4 members and the dome is named by "4-span dome" in this paper. The boundaries of both domes are supported as roller in the radial direction from the apex, but can not move vertically. The specifications of the experiments for them are also summarized in Refs.(2) and(4) , respectively.

Comparison between Analytical and Experimental Results

In this section, the load-displacement curves obtained from previous experiments will be compared with the analytical ones to observe the applicability of the present numerical modeling.

Table 1 Material properties

material properties	pipe $\phi139.8\times4$	connector $\phi40$	nose $\phi85\times22.5$
moment of inertia (cm^4)	393.73	11.46	243.67
sectional area (cm^2)	17.07	12.57	44.18
yield stress (tf/cm^2)	3.66	9.94	3.13

Table 2 Coefficients in stress-strain relationships

E_c, E_p	$2,100(tf/cm^2)$	En	$2,100\times0.25(tf/cm^2)$
E_r	$210(tf/cm^2)$	Etc, Etp, Etn	$21(tf/cm^2)$

Objective domes are illustrated in Figs.6 and 7, as already explaining in previous section and the constitutive member-unit is also shown in Fig.5. Table 1 introduces the properties of the member-unit, reflecting the material parameters described in Fig.4. The calculations are performed based on changing the amount of relaxation at the connector and the initial pressure between the ball and nose section.

Unit dome: First of all, the investigations for the unit dome are discussed. Fig.8 shows the relationship between the external load subjected in vertical direction and the vertical displacement at center nodal point. Solid lines mean analytical results and blank square marks denotes experimental one. Analyses are performed under changing relaxation $d_r = \varepsilon_r \times l_s = $ 0.0, 0.005, 0.010, 0.015, 0.020, 0.0225 and 0.025 cm, and the initial compression N_i varies 4.0tf which is introduced on the connection at actual construction. From 7 analytical curves, the effect of the relaxation on the ultimate strength can be found as that the strength decreases approximate propotionally with respect to amount of relaxation. It is confirmed that the curve in the case of d_r = 0.0225cm is closest to the experimental one. From this fact, the amount of existing relaxation can be predicted as 0.0225cm, but a few remarks will be appeared to discuss about this prediction. These will be described after seeing Fig.9.

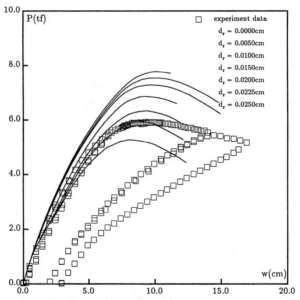

Fig.8 Load-displacement curves for the unit dome
under changing relaxation in the case of N_i = 4tf

Fig.9 shows the load-displacement curves which are calculated under N_i = 0tf being different from above-mentioned case, and changing amount of relaxation d_r. These calculations are carried out to expect the variable range of relaxation with regard to the initial axail force N_i which is considered to be decreased under cyclic loading. From the both results of Figs.8 and 9, it is confirmed that the existing relaxation will vary in the range from $d_r = 0.015$ to 0.0225cm in spite of changing initial axial force N_i during cyclic loading. However, if in the experiment, the initial fluctuation which means an initial vertical displacement or imperfection, was occured at the apex by the cyclic loading until loading over non-linear range not being started, the predicted existing relaxation will have to be refined from further calculations. And from seeing these load-displacement curves, a slight difference between analysis and experiment in the post buckling range (after maximum load) are found. To investigate the reasons, some added analyses are executed, these results will be explained in Chapter "Discussion".

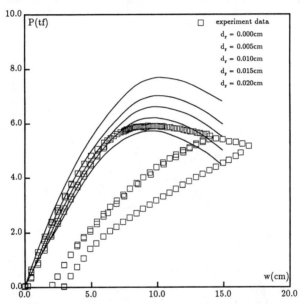

Fig.9 Load-displacement curves for the unit dome
under changing relaxation in the case of $N_i = 0$tf

4-span dome: Parametric analyses are also executed for 4-span dome being illustrated in Fig.7. The material properties of the constitutive members of the dome are simlar to the unit dome's.

Fig.10 shows the load-displacement curves obtained from numerical analyses, and the experimental plots. The load-displacement curves are followed under changing relaxation d_r and fixed value of $N_i = 4$tf. In this case, the closest curve to the experimental result is in the case of $d_r = 0.0125$cm.

From this result, the existing relaxation of the member composed of 4-span dome is expected as d_r = 0.0125cm, but it is different from the case of unit dome. Of cource, one of the considerable reasons for such a difference is due to the individual accuracy of experiments. The considerable reasons from another view will be discussed in next chapter.

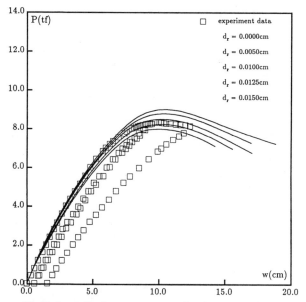

Fig.10 Load-displacement curves for the 4-span dome under changing relaxation in the case of N_i = 4tf

Discussions

In this chapter, the authors will discuss about two problems being appeared from this study. One is that by what the differences are caused between the results of analysis and experiment of the unit dome in the post buckling range, and the other is to investigate the reasons for the inconsistency of the magnitudes of the predicted relaxation between the unit and 4-span domes. These problems will be solved reasonably by one series of analyses for the unit dome, but the authors also feel that other discussions for this solving will be needed in the future study.

Fig.11 shows the results of calculations for the unit dome under initial axail compression N_i = 4tf at the nose of the member and the relaxation at the connector d_r = 0.01cm. In this figure, w_i means initial fluctuation at the apex of the dome and it is introduced as initial imperfection which may be caused by cyclic loading in the range that the linear relation still being kept between the load and the vertical displacement.

By analyzing the curves in Fig.11, the gradient of the curve in the post buckling range is closer to the experimental plots with incresing the fluctuation w_i. This fact suggests that a deviation from the exact position of the apex is existed during cyclic loading. Meanwhile, in the experimental load-displacement plots, such a magnitude of the deviation (about 2cm) is not observed. However, in Ref.(2) , it is confirmed that an initial gap 0.6cm in unloading state was existed, because the initial height of the objective dome was measured as 15.1cm. With consideration of that there was a possibility to increasethe fluctuation after the measurement, a little deviation might be added to the initial gap. Furthermore the calculations for $N_i = 2tf$ which is considered as the case of that the initial tightening is leaved partly, are also executed and from the results, the initial gap can be estimated as $w_i = 1-2cm$. Through the above-mentioned analyses, if such a magnitude of initial gap was existed, the relaxation at the connector for the unit dome would be expected as around $d_r = 0.01cm$. The quantity is relatively closer to the expected relaxation for the 4-span dome.

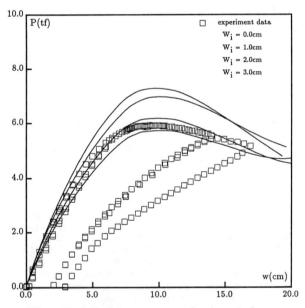

Fig.11 Load-displacement curves for the unit dome
under changing the vertical fluctuation
of the dome height in the case of $N_i = 4tf$

Conclusions

Conclusions obtaind from the prensent study are drawn as follows:

(1) analytical finite element model for the ball-joint connection is introduced and the sections of the elements are divided into small fiber elements which behave under characteristic functions as a material non-linearity, respectively;

(2) relaxation at the connector and initial tightening axail force at the nose of the member are taken into account as typical parameters to describe the behaviors of the reticular dome which is constructed by the present ball-joint system;

(3) by using above-mentioned elements, the load-displacement curves obtained from realistic scale loading tests are followed to confirm the ability of the present model;

(4) from comparisons between the load-displacement curves obtained from the analysis and experiment, the reasons of the differences between analyses and experiments are discussed.

Acknowledgment

The present authors gratefully express their thanks to Mr. Minoru Iida, a graduate student of Toyohashi University of Technology, for his precise calculations.

Notational table Conversion of unit:[tf] into SI

1 [tf]	9.80665×10^3 [N]	1[tf/cm^2]	9.80665×10^7 [Pa]

References

1. Fujimoto,M., Saka,T., Imai,K., and Morita,T., "Experiment and numerical analysis of the buckling of a single-layer latticed dome," *SPACE STRUCTURES 4, Thomas Telford*, vol. 1, pp. 396-405, Sept.,1993.

2. Shibata,R., Kato,S., Yamada,S., and Ueki,T., "Experimental study on the ultimate strength of single-layer reticular domes," *SPACE STRUCTURES 4, Thomas Telford*, vol. 1, pp. 387-395, Sept.,1993.

3. Takashima,H. and Kato,S., "Numerical simulation of elastic-plastic buckling behaviour of a reticular dome," *SPACE STRUCTURES 4, Thomas Telford*, vol. 2, pp. 1314-1322, Sept.,1993.

4. Ueki,T., Mukaiyama,Y., Shomura,M., and Kato,S., "LOADING TEST AND ELASTO-PLASTIC BUCKLING ANALYSIS OF A SINGLE LAYER LATTICED DOME(in Japanese)," *Transactions of AIJ*, no. 421, pp. 117-128, Mar.,1991.

Effect of Joint Rigidity on
Buckling Strength of Single Layer Lattice Domes

Shiro KATO[1], Itaru MUTOH[2] and Masaaki SHOMURA[3]

Abstract

The present paper discusses the effect of the joint rigidity on the buckling strength of single layer lattice domes, with regular three way grids, under vertical loading. The effect of bending rigidity of joints on the overall srength of domes is investigated from three viewpoints; (1)how the joint rigidity contributes to the reduction of buckling loads, (2)how the reduction can be interrelated to compressive strength curves in terms of the generalized slenderness for the member most relevant to the overall buckling of domes, and (3)whether the strength curves will resemble those for beam-columns in tall buildings and for continuum shells, or will not. For present semi-rigid cases, by reflecting the reduction of buckling loads for perfect rigid cases on a specific modification of the generalized slenderness, the member strength curve is available in a similar way for the rigid-jointed lattice domes.

Introduction

The buckling of single layer lattice domes has been analysed and studied from a view point of geometrically and materially nonlinear problems. The results have been often arranged according to a concept of buckling loads per unit area or per nodes taking a shell-like imperfection sensitivity into consideration. On the other hand the results have been reported only in a way of load-displacement relations, buckling modes, buckling loads and so on, for each problem.

In the previous studies the discussion has been made for the rigidly jointed reticular domes with hexagonal planform (Kato et al. 1992; Kato et al. 1993). When we focus axial compressive strength of the member most relevant to the overall buckling then it is recognized that the buckling loads can be interrelated

---

[1] Prof.,Dept.of Regional Plannnig and Civ.Engrg.,Toyohashi Univ.of Tech., 441Toyohasi,Japan

[2] Assoc.Prof.,Dept.of Architecture and Civ.Engrg.,Gifu Nat'al Coll.of Tech., Motosu,501-04Gifu,Japan

[3] Struct. Engr.of Naka-Nihon Consultant,450Nagoya,Japan

468

to the member strength curves in terms of the generalized slenderness for that member. At that time the buckling characteristics of the lattice domes are included into the member strength curves by adopting an imperfection sensitivity from shell-like behaviours. The applicability of the member strength curves in design of lattice domes has been verified under several analytical restrictions: (1)that the domes are rigidly jointed at intersecting nodes, (2)that the joints do not yield and members may yield at both the ends and the center portion, (3)that a boundary condition is pin-supported at dome periphery, (4)that the individual slenderness ratio λ_0, measured using the member length as buckling length, of constituent members are less than about 150, and (5)that the half-subtended angle θ_0 of member at dome apex as a measure of dome shallowness is not greater than 2 degrees (Kato et al. 1993). In the present paper the first restriction, (1), is removed and the joints in lattice domes are presumed as semi-rigids.

In the case that the concept of generalized slenderness is applied to estimate buckling loads of reticular domes under a set of prescribed design loads P_d, the specific member denoted as m in this study is first selected by finding a member with the largest ratio of $N_{cr}^{lin}(m)$ to its yield axial force N_y. The linear buckling axial force $N_{cr}^{lin}(m)$ derived from linear eigenvalue analysis under the same design loads P_d is used to calculate the generalized slenderness Λ,

$$\Lambda = \sqrt{\frac{N_y}{N_{cr}^{lin}(m)}} \tag{1}$$

in a similar way for the procedures in struts in tall steel buildings. The equivalent buckling length ℓ_{eq} can be also obtained for the member with bending rigidity of EI_p as follows. Also using Euler buckling force N_E of struts, ℓ_{eq} relates to a member length ℓ_0.

However a column (member) strength curve different from those for struts in

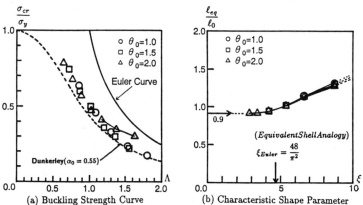

(a) Buckling Strength Curve   (b) Characteristic Shape Parameter

Figure 1. Buckling Strength of Rigidly Jointed Reticular Domes ($n = 12$) and Relationship between Characteristic Shape Parameter and Euivalent Buckling Length Ratio (hexagonal planform)

$$\ell_{eq} = \sqrt{\frac{\pi^2 E I_p}{N_{cr}^{lin}(m)}} \qquad (2)$$

tall buildings is applied. The curve, $\chi(\Lambda)\sigma_y$, for axial strength σ_{cr}, for rigidly jointed domes, for instance, is shown as in Fig.1(a) (Kato et al. 1993) and is given by the dotted curve based on a semi-quadratic Dunkerley formula with a knockdown factor $\alpha_0 = 0.55$ for the initial geometric imperfection assumed at one node as shown also in Fig.2(a) and (b), with a magnitude of

$$w_i = 0.2t_e \qquad (3)$$

Here the configuration of dome is of hexagonal planform subdivided by number of frequency $n = 12$ with length of members $\ell_0 = 4000mm$, and t_e is the equivalent thickness as an isotropic spherical shell for single layer lattice domes.
   The buckling load P_{cr}^{est} may be estimated accordingly as follows.

$$P_{cr}^{est} = P_d \frac{(\sigma_{cr} \cdot A_m)}{N_d} \equiv \frac{P_d}{N_d} \cdot \chi(\Lambda) \cdot N_y \qquad (4)$$

Here N_d means the axial force of the member in dome subjected to P_d, and A_m is the sectional area of that member m.

   The concept hitherto described is extended to the reticular domes with semi-rigid connections at nodes and the validity is discussed in the followings.

Buckling Analysis of Single Layer Lattice Domes

Joint Rigidities and Yield Conditions

   Analytical dome configuration is shown as in Fig.2(a). Bending springs are assumed at both ends between constituent members and joints acting against

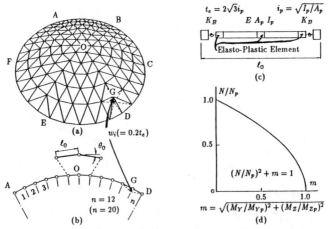

Figure 2. Analytical Model : (a) Dome Configuration ; (b) Half-Subtended Angle θ_0 and Section along A-O-D ; (c) Member Model ; and (d) Sectional Strength Interaction Curve

TABLE 1. Member Properties and Bending Rigidities of Joints

Member Slenderness (1)	I_p ($\times 10^4 mm^4$) (2)	A_p ($\times 10^2 mm^2$) (3)	κ	1.0	2.0	4.0	10.0	100.0
				K_B ($\times 10^6 Nmm/rad.$) (4)				
30	10531.27	59.24		542.0	1084.0	2168.0	5420.0	54200.0
60	1316.41	29.62		67.8	135.0	271.0	678.0	6780.0
90	390.05	19.75		20.0	40.0	80.0	200.0	2000.0
120 ·	164.55	14.81		8.5	17.0	34.0	85.0	850.0
150	84.25	11.85		4.3	16.8	33.6	43.0	430.0

Note. $E = 206 kN/mm^2$. $\ell_0 = 4000mm$.

two rotations as shown in Fig.2(c), since we suppose the domes with screw-type ball joints, for example. Here the torsional rigidity of the springs is assumed negligible. The joint rigidity parameter κ is intoduced in a form of

$$\kappa = K_B \frac{\ell_0}{EI_p} \quad (5)$$

for the bending rigidity K_B (Ueki et al. 1989) of the spring elements. ℓ_0 is the member length, being same for each member on the meridional line A-O-D, and the reticular mesh is given by the subdivision frequecy number n as shown in Fig.2(b) in which the length of the members on a same circumferece are made equal each other.

EI_p and EA_p are the bending and axial rigidities for a tubular member, being assigned commonly to every member in a dome. It is assumed for the constituent members of the domes that the springs do not yield and the members may yield at both the ends and the center portion, and also the yielding condition for members is represented by perfect elasto-plastic hinges with the interaction of strength for axial and bending capacities as shown in Fig. 2(d). In the study the yield stress σ_y and the Young's modulus E are assumed $235kN/mm^2$ and $206kN/mm^2$, respectively.

Geometry of Domes and Loading Condition

As shown in Fig.2(a) and (b), the half-subtended angles θ_0 are 1.0, 1.5 and 2.0 degrees, the individual slenderness ratios λ_0 for the members along the meridian A-O-D are 30, 60, 90, 120 and 150, and the joint rigidity parameters κ are assumed as 1, 2, 4, 10 and 100. The basic data for numerical analyses are shown as in Table 1. Thus the following equation holds.

$$I_p = A_p (\frac{\ell_0}{\lambda_0})^2 \quad (6)$$

The case of $\kappa = 1$ corresponds nearly to pin-connected nodes and the case of $\kappa = 100$ stands for almost rigid-jointed domes. One could point out that for real-sized model tests of screw-type ball joint systems with $\ell_0 = 3000mm$ the degree of semi-rigidity might be the range of κ from 4 to 10, or $K_B \approx 2000 \times 10^6 Nmm/rad.$, for instance (Ueki et al. 1991).

The domes are assumed to be subjected to uniform downward loading at every node, and to be pin-supported at periphery on a circular planform (see, Fig.2(a)). The domes with number of subdivision frequency of two values for

n, 12 and 20, are analysed. For the case of $n = 12$, we discuss the effect of joint rigidity on the buckling strength, mainly.

Analytical Procedure

For linear (eigenvalue analysis) buckling, the axial forces are presumed to be proportional to those obtained by a linear elastic analysis, and the incremental elastic stiffness matrix is derived from the slope-deflection method using stability functions. For elasto-plastic buckling analysis, plasticity is considered at hinges by flow theory using the strength interaction as shown in Fig.2(d). The influence of initial geometrical imperfection is taken into consideration just by assuming 20% of amplitude to t_e, as given by Eq.(3) at the single node G with a half wave-length of $2\ell_0$ as shown in Fig.2(a) and (b). Although this might give a restriction to the results, we assume as a prcatical design limit for designers in order to check the elastic buckling.

Buckling Strength Curves for Rigid-Jointed Lattice Domes

The buckling axial strength for domes with $n = 20$ is shown in Fig.3(a) based on a series of both the linear eigenvalue and the elasto-palstic nonlinear analysis in case of rigidly connected joints. The imperfection, $w_i = 0.2t_e$, also is assumed at the node G shown in Fig.2(a). In every case of analysis, member GD in Fig.2(a) or (b) is the member with the largest ratio of $N_{cr}^{lin}(m)/N_y$. Thus the generalized slenderness Λ is calculated after this member. There seems to be only a little difference between the results for $n = 12$ on hexagonal plan as shown in Fig.1(a) and the results for the present case. The equivalent buckling length ℓ_{eq} for the member GD is also given in terms of ξ (Ueki et al. 1989; Kato et al. 1993), in Fig.3(b). There are no visible differences from those in Fig.1(b) for domes with $n = 12$ on hexagonal plan.

$$\xi(\kappa) = \frac{12\sqrt{2}}{\theta_0\lambda_0(1 + 2/\kappa)} \; ; \; \eta = \frac{P_{cr}}{EA_p\theta_0^3} \qquad (7)$$

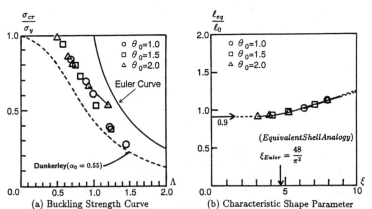

(a) Buckling Strength Curve   (b) Characteristic Shape Parameter

Figure 3. Buckling Strength of Rigidly Jointed Reticular Domes ($n = 20$) and Relationship between Characteristic Shpae Parameter and Equivalent Buckling Length Ratio (circular planform)

where $\kappa = \infty$ means the perfect rigid case. Here the linear buckling load $P_{cr}^{lin}(\kappa)$ is given from $\eta = \xi(\kappa)$ as follows.

$$P_{cr}^{lin}(\kappa) = \frac{12\sqrt{2}}{\theta_0\lambda_0(1+2/\kappa)}(EA_p\theta_0^3) \tag{8}$$

Relation of Elastic Buckling Loads between Rigid and Semi-Rigid Domes

Figs.4(a) to (e) show the ratio of elastic buckling load $P_{cr}^{el}(\kappa)$ to the linear buckling load $P_{cr}^{lin}(\kappa = \infty)$ for domes without imperfections. The reduction due to the effect of semi-rigidity appears in a characteristic manner depending

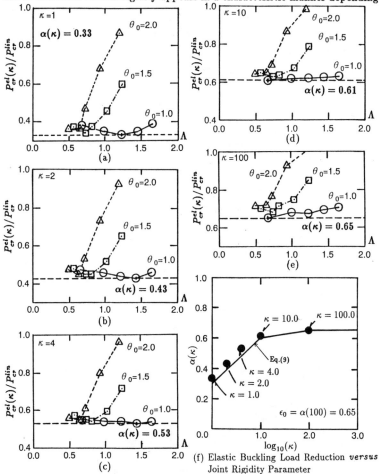

(f) Elastic Buckling Load Reduction *versus* Joint Rigidity Parameter

Figure 4. Effects of Joint Rigidity on Elastic Buckling Load

on Λ and θ_0. Here Λ is the non-dimensional slenderness for domes with perfect rigid connections. The ratio of buckling loads for semi-rigid cases is plotted at the same points on abscissa for rigid domes. The results are also shown in Fig.5(a) in the form of (column) member strength curve in terms of the generalized slenderness. The tendency of reduction is different depending on the magnitude of θ_0, however, the ratio seems to tend towards constant values depending on each value of κ as Λ becomes to zero. The lower bound values of the ratios are denoted by $\alpha(\kappa)$ as shown in Fig.4(f), for example $\alpha(1) = 0.33$ and $\alpha(100) = 0.65$. The trace of the lower bounds is approximately given by Eq.(9).

$$\alpha(\kappa) = \begin{cases} 0.195\log_{10}(\kappa) + 0.195, & 1 \le \kappa \le 10.0 \\ 0.0325\log_{10}(\kappa) + 0.3575, & 10.0 < \kappa < 100.0 \\ 0.65, & \kappa \ge 100.0 \end{cases} \quad (9)$$

Here a new parameter $\beta(\kappa)$ is introduced as,

$$\beta(\kappa) = \frac{\alpha(\kappa)}{\epsilon_0} \text{ where } \epsilon_0 = 0.65 \quad (10)$$

to represent the reduction of elastic buckling loads for domes without imperfections, namely considering the effect of joint rigidity. Thus the elastic buckling load $P_{cr}^{el}(\kappa)$ and accordingly the elastic buckling stress $\sigma_{cr}^{el}(\kappa)$ may be approximately expressed as follows.

$$P_{cr}^{el}(\kappa) = \beta(\kappa) \times P_{cr}^{lin}(\kappa = \infty) \quad (11)$$

$$\frac{\sigma_{cr}^{el}(\kappa)}{\sigma_y} = \frac{\sigma_{cr}^{el}(\kappa)}{\sigma_{cr}^{lin}(m)} \times \frac{N_{cr}^{lin}(m)}{N_y}$$

$$= \frac{\beta(\kappa) \cdot \sigma_{cr}^{el}(\kappa = \infty)}{\sigma_{cr}^{lin}(m)} \times \frac{N_{cr}^{lin}(m)}{N_y}$$

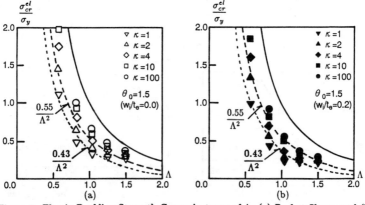

Figure 5. Elastic Buckling Strength Curves in terms of Λ: (a) Perfect Shape; and (b) Imperfect Shape ($w_i = 0.2t_e$)

$$= \frac{\sigma_{cr}^{el}(\kappa = \infty)}{\sigma_{cr}^{lin}(m) \cdot \Lambda_{mod}^2} \tag{12}$$

where $\sigma_{cr}^{lin}(m)$ is the linear buckling stress in perfect rigid case, given by $\sigma_{cr}^{lin}(m) = N_{cr}^{lin}(m)/A_m$, and Λ_{mod} is the modified slenderness in the semi-rigid case for the member m as follows.

$$\Lambda_{mod} = \sqrt{\frac{N_y}{\beta(\kappa) \cdot N_{cr}^{lin}(m)}} \tag{13}$$

The ratio, $\sigma_{cr}^{el}(\kappa = \infty)/\sigma_{cr}^{lin}(m)$, may be approximated by a knock-down factor for elastic buckling, for example by adopting IASS Recommendation (for shells and folded plates) (IASS 1979). One may express a knock-down factor by α_0 as given below.

$$\alpha_0 = \frac{\sigma_{cr}(\kappa = \infty)}{\sigma_{cr}^{lin}(m)} \tag{14}$$

The value α_0 can be assumed as constant, however, with restrictions that the individual slenderness ratio λ_0 is less than about 150, and the half-subtended angle θ_0 is not greater than 2 degrees (Kato et al. 1993). Also we may observe such a constant reduction from Fig.5(a) in case of $\kappa = 100$ for perfect shape as well as from Fig.5(b) for imperfect shape. From Eqs.(12) and (14), the following equation holds even in semi-rigid cases.

$$\frac{\sigma_{cr}^{el}(\kappa)}{\sigma_y} = \frac{\alpha_0}{\Lambda_{mod}^2} \tag{15}$$

where a notice is given that α_0 is the value calculated for rigidly jointed reticular domes depending on the amplitude of assumed imperfections. Finally the elastic buckling load $P_{cr}^{el}(\kappa)$ is approximately expressed as follows.

$$P_{cr}^{el}(\kappa) = \frac{P_d}{N_d}(A_m \cdot \sigma_y)\frac{\alpha_0}{\Lambda_{mod}^2} \tag{16}$$

If Eqs.(10) and (16) are used, elastic buckling loads for reticular domes with semi-rigid connections may be estimated based on the data for rigidly jointed domes once linear eigenvalue analysis is only performed.

Buckling Strength Curve for Semi-Rigid Jointed Domes

As might be expected from modification of the slenderness parameter from Λ to Λ_{mod} as shown in Fig.6, the results shown as in Fig.8 for σ_{cr} in terms of Λ_{mod} are almost on a single curve. Namely, from Fig.6(a), $\bar{\sigma} = \sigma_{cr}/\sigma_y$ for $\kappa = \infty$ with Λ reduced to $\beta_1\bar{\sigma}$ by the influence of joint rigidity. Simultaneously, we can obtain the reduced buckling strength $\beta_1\bar{\sigma}$ through the curve $\chi(\Lambda_{mod})$ against $\beta_2\Lambda \equiv \Lambda_{mod}$. The curve which might be drawn by fitting the results is along with and is a little higher than the curve by Eq.(17) with $\alpha_0 = 0.55$ as well as by Eq.(18) with $k_1 = 1.0$, $k_2 = 3.0$ and $k_3 = 0.5$.

According to IASS Recommendation (IASS 1979), α_0 will be around 0.43 for $w_i = 0.2t_e$, if we adopt the knock-down factor of $\alpha_0 = 0.43$ in Eq.(17) as inserted in Fig.7 and Fig.8, the curve given by Eq.(17) will provide a safe side estimation for σ_{cr}. Here in Fig.7 the results for elasto-plastic buckling loads are plotted for rigidly jointed domes with $n = 12$ on hexagonal plan, and in Fig.8, for semi-rigidly jointed case with $n = 12$ on circular plan, the results obtained from this study. Also some design curves (e.g., Bornsheuer 1985) for shells are inserted in those figures. Insofar as the value of κ ranging 4 to 100 might be restricted, proposed member strength curves derived from Eqs.(17) and (18) are good candidates for evaluation of buckling strength in view of shell-like buckling characteristics (e.g., Mutoh and Kato 1993).

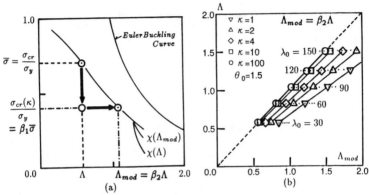

Figure 6. Relationship between Λ and Modified Slenderness Λ_{mod} : (a) Buckling Strength Reduction Scheme ; and (b) Λ_{mod} *versus* Λ

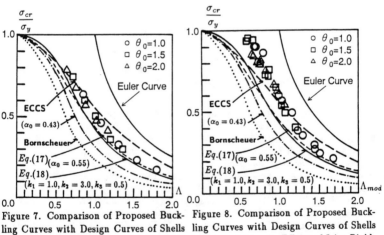

Figure 7. Comparison of Proposed Buckling Curves with Design Curves of Shells in terms of Λ for Rigidly Jointed Reticular Domes

Figure 8. Comparison of Proposed Buckling Curves with Design Curves of Shells in terms of Λ_{mod} by Effects of Joint Rigidity on Elastic Buckling Load

$$\left[\frac{\sigma_{cr}}{\sigma_y \alpha_0 / \Lambda_{mod}^2}\right] + \left[\frac{\sigma_{cr}}{\sigma_y}\right]^2 = 1 \qquad (17)$$

$$\frac{\sigma_{cr}}{\sigma_y} = \frac{k_1}{[k_2 \Lambda_{mod}^4 + k_3 \Lambda_{mod}^2 + 1.0]^{1/2}} \qquad (18)$$

From above discussions, the buckling loads for single layer lattice domes with semi-rigid connections can be approximately estimated as follows.

$$P_{cr}(\kappa) = \frac{P_d}{N_d}(A_m \cdot \sigma_{cr}) = \frac{P_d}{N_d}\{A_m \cdot \sigma_y \cdot \chi(\Lambda_{mod})\} \qquad (19)$$

where σ_{cr} is the (column) member strength and should be calculated by using Eq.(17) or Eq.(18), which considers influencing factors on buckling strength of reticular domes. One possible selection of the knock-down factor is to adopt from IASS Recommendation reflecting the important phenomena of shell buckling.

Conclusion

The effect of semi-rigidity at joints on the buckling strength of single layer lattice domes is studied and the elasto-plastic buckling loads are interrelated with the member strength σ_{cr} of a particular constituent member most relevant to the overall buckling strength of domes. It is confirmed that the member strength curve can be expressed in terms of the modified generalized slenderness Λ_{mod} which reflects the reduction $\beta(\kappa)$ of elastic buckling loads due to the semi-rigidity parameter κ at connections, and that the buckling loads can be effectively estimated by using the proposed strength curve. Also the strength curve is proved to be almost similar to those for spherical shells mainly because of both the nonlinearity before buckling and the imperfection sensitivity.

Appendix. References

Bornscheuer, B.-F. (1985). "Stabilitätsnachwiese für Platten und Schalen in gleicher Darstellung wie für Stäbe." *Der Stahlbau*, 54(12), 364-368.

IASS Recommendations for Reinforced Concrete Shells and Folded Plates. (1979). Int. Assoc. for Shell and Spatial Structures, S. J. Medwadowski, et al. , Eds. , Madrid , 55pp.

Kato, S., Shibata, R., Takashima, H., and Ukeki, T. (1992). "A computational procedure to evaluate buckling strength of reticular domes based on a concept of column buckling." *Proc., IASS-CSCE Symp.(Canada)*, N. K. Srivastava, A. N. Scherbourne and J. Roorda, Eds., The Canadian Society for Civ. Engrg., Montreal, Canada, 552-563.

Kato, S., Shibata, R., Ueki, T., and Matsushita, F. (1993). "Buckling load estimation for single layer domes by the concept of column strength curves" *Proc., IASS-MSU Symp.(Istanbul)*, I. Mungan, Ed., Mimar Sinan University, Istanbul, Turkey.

Mutoh, I., and Kato, S. (1993). "Comparison of buckling loads between single-layer lattice domes and spherical shells" *Space structures 4, Proc., Fourth Int. Conf. Space Structures (Guildford)*, G. A. R. Parke and C. M. Howard, Eds., Thomas Telford Services Ltd., London, UK., 176-185.

Ueki, T., Mukaiyama, Y., Kubodera, I., and Kato, S. (1989). "Buckling behavior of single layered domes composed of members with axial and bending springs at both ends" *Proc., IASS Symp.(Madrid)*, F. del Pozo and A. de las Casas, Eds., Cedex-Laboratorio Central De Estructuras Y Materiales, Madrid, Spain.

Ueki, T., Kato, S., Kubodera, I., and Mukaiyama, Y. (1991). "Study on the elastic and elasto-plastic buckling behaviour of single layered domes composed of members having axial and bending springs at both ends" *Proc., IASS Symp.(Coppenhagen)*, T.Wester, S.J. Medwadowski and I. Mogensen, Eds., Kunstakademiets Forlag Arkitektskolen, Coppenhagen, Denmark, 93-100.

Experimental Study on the Buckling Behavior of
a Triodetic Aluminum Space Frame

- No.2 Ultimate Bearing Strength of a Single Layer
Space Frame -

Kenichi Sugizaki[1] and Shigeru Kohmura[2]

Abstract

The objective of the current study was to see if
the Aluminum Triodetic system was suitable for large
span structures. The authors now report on how they
gained a better understanding of its structural
characteristics through basic testings such as
tensile, compressive and bending tests using
fundamental elements and the loading test on a 10
meter model. In these tests, we can see that the
bearing strength limit was determined by hub's
rotation and was about four times the value of the
design load. The current experimental study enabled
the authors to understand the basic buckling behavior
of the Aluminum Triodetic system.

1. Introduction

As truss system structures using insertion joints
are well suited to secondary processing methods for
aluminum alloy they have been long used in
construction of large-span aluminum alloy space
frames. The triodetic system is a major type of
Aluminum space frame. Previous studies [1] aimed to
clarify experimentally the phenomena governing the
ultimate bearing strength of a single layer Aluminum
space frame with semi-rigid joints and, on the basis
of these phenomena derive formulae for calculating the

----
1. Structural Chief Engineer, Technology Development Div. Shimizu
Corporation, Seavans South, No.2-3, Shibaura 1-chome, Minato-ku,
Tokyo, JAPAN
2. Structural Chief Engineer, Public & Engineering Div. Nippon
Light Metal Co., Ltd., No.13-12, Mita 3-chome, Minato-ku, Tokyo,
JAPAN

ultimate bearing strength. As a result of these studies, it was discovered that the ultimate load is determined by the rotation of the nodes (hubs) from evenly distributed total loading and centrally concentrated loading experiments that were conducted on a shallow small-scale dome model with a diameter of 4.2 meter.

The objective of the current study was to see if the Aluminum triodetic system was suitable for large span structures. The authors now report how they gained a better understanding of its structural characteristics through basic testings such as tensile, compressive and bending tests using fundamental elements and loading tests on a 10 meter model.

2. Features of Aluminum Trusses and the Material Properties

Figure 1 shows the insertion joints and how tubes are joined to hubs. The hubs have slots cut with different internal shapes.

Hubs are formed by making long metal pieces by an extrusion process and then cutting to the prescribed length. The extrusion process makes the slot angles the same in all places. The tube ends are therefore formed in a press for insertion into the different internal configurations of the slots as shown in Fig. 2 in order to make space frames of any shape or mesh pattern. The minimum specifications for the hub material (BS 6082-T6) and tube material (JIS A6063) are shown in Table 1.

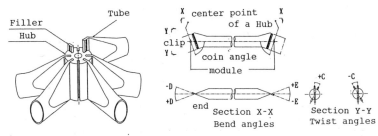

Fig.1 Connection of Hub   Fig.2 Outline of a Tube end
and Tubes

Table 1   Strengths of Hub and Tube Materials

	Tensile Strength (N/mm^2)	o.2% Bearing strength (N/mm^2)	Strain (%)
Hub material	310	270	8
Tube material	206	177	10

3. Fundamental Element Tests (tensile, compressive and bending tests)

The tensile, compressive and bending (out-plane and in-plane) characteristics were confirmed on tube and hub specimens as shown in Figs. 3 through 7. The tubes used in the tests were 66 mm in outer diameter, 3 mm thick and 1.3 m long. The hubs had an outer diameter of 68 mm, a center hole diameter of 21 mm and 6 slots each with a depth of 22.5 mm. These specimens were identical to the members used in the 10 meter model described later.

3.1 Tensile Test

The testing procedure and the load-displacement relationship are shown in Fig.3. The tensile rigidity was approximately 80% of the calculated value 1 (tube overall cross section valid). From this result, we obtained calculated value 2, the tensile rigidity in consideration of the reduction in the insertion points, and applied it to the 10 meter analysis model. The ultimate break point was at the edge of the irregularly processed part of the tube ends.
The average breaking load for three specimens was 117.4 kN. This value is virtually the same as that obtained by multiplying the tube cross sectional area by the tensile strength (206 N/mm^2) so the joint efficiency can be said to be nearly 100%.

3.2 Compressive Test

The test procedure and the load-displacement relationship are shown in Fig.4. The compressive rigidity was approximately 80% of the tensile rigidity and the compressive bearing load limit was determined by the rotation of the end hub. The average ultimate compressive load was 53.9 kN, approximately 80% of the Euler buckling load with a condition which can be assumed to approximate a hinge joint in the in-plane direction.

Figure 5 shows a stress-strain relationship for the irregularly processed part of tube ends. From Fig. 5, it can be predicted that a bending moment will arise along the in-plane axis of the processed part of tube ends causing rotation in the hubs.

3.3 Bending Test (out-plane direction)

The test procedure and load-displacement relationship are shown in Fig.6. The direction of bending is along the out-plane axis for the tube ends and the displacement measured experimentally was close to that calculated for a tube with a circular cross section. The out-plane direction could thus be considered as a rigid joint. When the maximum load had been reached, although the displacement was increased, no tubes broke neither were they pulled out of hubs. The average maximum load for the three specimens was 4.7 kN.

3.4 Bending Test (in-plane direction)

Figure 7 shows the load-displacement relationship and the test procedure. The direction of bending is along the in-plane axis of tube ends. In considering the bending rigidity, it is necessary to think of ends as being members with a section inertia that is little different from the central part of tubes so it can be assumed from the results of both bending tests that the ratio between the out-plane bending rigidity and in-plane ones is approximately 40:1.

3.5 Conclusion from Fundamental Element Tests

(1) The characteristics of the tensile and compressive rigidities were understood for the reduction at the insertion joints.

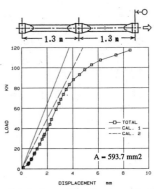

Fig.3 Tensile Test : Test
Procedure and Results

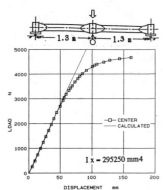

Fig.6 Bending Test (out-plane):
Procedure and Results

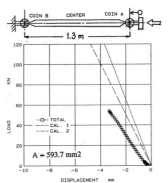

Fig.4 Compressive Test : Test
Procedure and Results

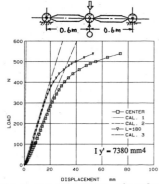

Fig.7 Bending Test (in-plane):
Procedure and Results

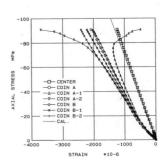

Fig.5 Compressive Test : Test
Results (stress-strain)

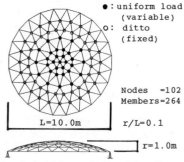

Fig.8 Outline of the 10 meter
model and loading points

(2) From the results of the tensile, compressive and bending tests (out-plane and in-plane directions), the authors confirmed that a semi-rigid joint truss system with out-plane rigid joints and in-plane hinged joints could be configured for a triodetic aluminum space frame.

(3) The ratio between the bending rigidities out-plane and in-plane was seen to be a major factor affecting the rotation of hubs.

4. 10 m Model Experiment

4.1 Details of Experimental Model and Loading Procedure

The shape of the single-layer truss dome model used for this study and the configuration of members and loading procedure are shown in Fig.8 and photos 1, 2. It had a span (L) of 10 meter and the rise to span ratio was 0.1. The load in peripheral areas was uniformly and it was varied in the center with that for the former applied by means of sandbags and for the latter by means of hydraulic jacks via tournament devices. The supporting condition of the model is assumed to be a on-way roller in the meridian direction.

The properties of the members used are listed in Table 2. The Young's modulus of the aluminum alloy was taken as: E = 68.6 GPa (7,000 kgf/mm^2).

Photo 1 Perspective view of the 10 meter model

Photo 2 Outline of loading procedure

Table 2   Properties of Members

Part	Members (material)	Section areas A (mm²)	Section inertia Ix(mm4)	Section inertia Iy(mm4)	Average length of members l(mm)	0.2% bearing strength σy (MPa)
Tube	φ 66.0x3.0 (A6063-T6	593.7	466	295250	630	177
Hub	φ 68.0 (A6082-T6)	--	--	--	--	270

4.2 Experimental Results

After applying an uniform load of 490 N/m2 - assumed to be the finishing load- in the peripheral area, load was added gradually in the center as shown in Fig. 8. Loading could be carried out smoothly up to around 3900 N/m2 but above 4150 N/m2 it became difficult to balance the load and, several hubs were observed to be rotating when the uniform load was around 4400 N/m2 (130 kN), as shown in Fig.9 and photos 3 (A) and (B).

Figure 10 (A) through (C) shows the load-vertical displacement relationship for the main nodes. Figure 11 shows the deformation modes for each step. Figure 12 shows the relationship between load and axial force for the main horizontal members. The stress-strain relationship for the tube ends in Fig.13 (A) and (B) help explain the characteristics of the bending moment due to the hub rotation which arises in the tube ends. Although there is a large bending moment in the tubes around R5 and R6 where rotation is very pronounced, it can be clearly seen that there is not much bending moment at R4 where there is little rotation.

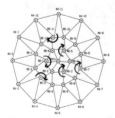  

Fig.9 The state of   (A) R6 (rotated)   (B) R4 (not rotated)
hub's rotation   Photo 3 The state after the occurrence
                                  of hub's rotation

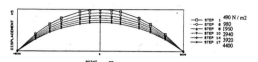

Fig.11 The deformation mode

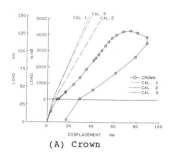

(A) Crown

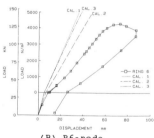

(B) R6-node

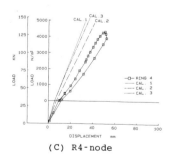

(C) R4-node

Fig.10 The relationship between load and vertical displacement for the main nodes

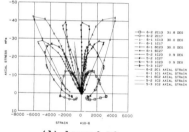

Fig.12 The relationship between load and axial force for main horizontal members

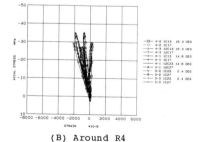

(A) Around R6

(B) Around R4

Fig.13 The relationship between stress and strain for the tube ends

4.3 Discussion of the Results

In the above experiment, it was found that the bearing strength limit of the experimental model, an evenly distributed load of 4400 N/m², was determined by the rotation of the hubs. This is approximately four times the design load. As shown in Figure 13, the bending moment produced by hub's rotation and the axial force in structural members cause a large amount of strain in the tube ends. At the bearing load limit of 4400 N/m², there many parts where the strain exceeded the yield strain for aluminum (3,000 x 10-6) so it can be considered that the bearing strength limit was reached as a result. Further, looking at Fig. 11, there were no instances where local perpendicular deformation was seen to be proceeding at a rapid rate, even when the limit state was reached as a result of hub's rotation.

In Figs. 10 and 12, a comparison is made of values obtained from linear (Cal.1) and non-linear analysis (Cal.3) and those obtained from the analysis in consideration of the reduction in rigidity at the insertion points (Cal.2). The experimental values for stress compare well with those from the analyses but in the case of deformation, the experimental values were much larger than those from the analyses. This is thought to be due to the effect of differences in the mode of deformation due to partial loading and that of local deformation due to rotation and is thus a major area for future study.

5. Conclusion

(1) The bearing strength limit of this model with insertion joints was determined by hub's rotation and was about four times the value of the design load.
(2) The limit state of the truss model occurs because several truss ends become yielding as a result of excessive bending moment produced by the hub's rotation.
(3) In the limit state, this system, which can be thought of as being a truss with overall torsion buckling, is relatively stable despite a sudden increase in the amount of deformation. It can thus be

considered to be better than one in which there is snap through buckling when deciding the design criteria.

(4) In the values obtained from the analysis in consideration of the reduction in rigidity at the insertion joints assumed from the results of the basic tests, there was relatively close conformity between that obtained for the stress and the experimental result. For deformation, however, the value obtained from the analysis was smaller than that achieved experimentally so the authors feel it necessary to study methods for analyzing local deformation due to hub's rotation.

(5) The current experimental study enabled the authors to understand the basic buckling behavior of the triodetic system. In the future, we will do more research on the effect of asymmetric loading, span-rise ratio, truss configuration methods, changes in the mode of buckling due to increased in-plane rigidity and bearing strength limits in order to establish design formulae for determining bearing load limits for the triodetic system.

[Acknowledgment]

The authors wish to express their great thanks to Professor Dr. Yasuhiko Hangai of Tokyo University for his useful advice and helpful suggestions regarding the current studies.

[References]

[1] K.Sugizaki, S.Kohmura, "Experimental Study on Buckling Behavior of A Triodetic Aluminum Space Frame",Proceeding of SEIKEN-IASS 1993,pp.205-212.
[2]Subcommittee on Space Frames,Committee on Shell and Spatial Structures of Architectural Institute of Japan,"Stability Analysis of Single-Layer Lattice Shells, - the State of the Art Report -", 1989.
[3] K.Oda,Y.Hangai,S.Ohya, "Loading Tests of Torsional Buckling of Joint of Six-Member Unit Dome", Proceeding of SEIKEN-IASS 1993, pp.189-196.

A study on Buckling Characteristics
of Single-Layer Latticed Domes
with the Initial Geometrical Imperfection

Hwan-Mok Jung [1]
Young-Hwan Kwon [2]
Moon-Myung Kang [2]

Abstract

The geometeical initial imperfection is generally described that a dome digresses from ideal shape by design and construction error.

The main purpose of this paper is to clarify the effects of initial geometrical imperfection on the buckling characteristics of the single-layer latticed domes with triangular network.

Analysis is undertaken by using two independent methods, namely: one, shell analogy method, and the other, the frame analysis method which is based on the finite element method dealing with geometrically nonlinear problem.

1. Introduction

In recent years, large-scale single-layer latticed domes have attracted many designers and researchers's attention in the world, because single-layer latticed domes as space structure are of great advantage in not only mechanical rationality but also function, fabrication construction and economic[1].

But single-layer latticed domes are apt to become the unstable phenomena that are called "buckling" just as a load comes up to a critical level.

----
[1]Researcher, Department of Architecture Kyung-Pook National University, Taegu, Korea.
[2]Professor, Department of Architecture Kyung-Pook National University, Taegu, Korea.

If these domes have the initial geometrical imperfection, it is expected that buckling strength of overall domes is more decreased.

The object of this study is to investigate the buckling characteristics of single-layer latteced domes subjected to initial imperfection, and to get the data that are to formulate the general equation of buckling strength, which takes the effects of initial imperfection into consideration.

2. Analytical Models

2.1 Geometrical shape model

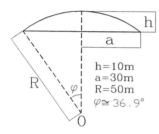

h=10m
a=30m
R=50m
$\varphi \cong 36.9°$

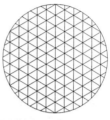

5-division network

Fig.1   Network model

Fig.1 shows the network model in this study. It is composed of the triangular network, and it has circular boundary. In Fig.1, R, a, h, and φ are the radius of curvature, the base radius, the apex height and the half open angle, respectively. In the 5-division network, 5 is the number in division of their meridians from the apex to the boundary.

2.2 Geometrical shape factor S and analytical models

The study of geometrical shape factor for latticed domes is carried out by Forman & Hutchinson[2], Y.Hangai[1], Kollar[3], and M.Yamada[4] etc.

In this study, the geometrical shape factor of latticed dome introduces S which was proposed by Yamada[4] to distinguish overall buckling and member buckling. S is as follow.

$$S = \frac{L}{\sqrt{R}} \left(\frac{K}{D} \right)^{1/4} \qquad (1)$$

where, L, K and D is the member length, the values of in-plane stretching rigidity and out-of-plane bending rigidity respectively. The analytical model is composed on the basis of shape factor S. The section dimension of each models shown table 1. is designed to equal section area of model 5, and the range for values of S is choiced as widely as possible.

Table 1. Geometrical shape factor and analytical models

MODEL	ϕ (mm)	t (mm)	I ($\times 10^4$ mm^4)	i (mm)	S
D1	600	2.45	20527	237	1.9
D2	500	2.95	14024	197	2.1
D3	400	3.69	9030	157	2.3
D4	300	4.96	5006	117	2.7
D5	250	6.00	3425	97	2.9
D6	200	7.61	2131	76	3.3
D7	150	10.49	1125	55	3.9
D8	125	13.08	730	45	4.3
D9	100	17.81	407	33	5.0

In table 1, ϕ, t, I and i are the diameter of the steel pipe, the thickness of steel pipe section, the moment of inertia and the radius of gyration respectively. The Young's modulus, E, and the Poisson's ratio, ν, of members are 2100 t/cm^2, and 0.3, respectively.

3. Analytical Methods

3.1 Shell Analogy Analysis

The typical formulas of buckling strength for shell analogy are proposed by K.Kloppel, K.P. Buchert ,Write.D.T. , del pozo, Y Hangai, M. Yamada etc. [1,5-9]

In this study, the formula applies Yamada's method of shell analogy which is formulated to predict the overall and asymmetrical buckling strength of orthotropic shallow spherical shells. [9] The estimation of equivalent rigidity for present network pattern is applied by Heki's method. [10]

In this paper, the shell analysis plays a fundamntal part in the estimation of buckling strength, i.e., all of the results obtained by the frame analysis are estimated as the ratios to that by the shell analysis; which is based on the fact that the shell analysis is of great advantage to a systematic research and to its application to the practice.

3.2 Frame Analysis

In this study, the frame analysis on geometrical nonlinear deflection problems is formulated based on the combined use of the finite element method and the moving coordinates.

Loading conditions are 2-types, that is, pressure-type-uniform loading and snow-type-uniform loading. Junction's conditions are 3-types, that is, rigid-joint, semi-rigid-joint and pinned-joint. Where semi-rigid-joint means the medium-joint condition of rigid-joint and pinned-joint. In this study, joint flexibility, K_J, is defined by K_θ as follow.

$$K_J = K_\theta \times \frac{6EI}{L} \qquad (2)$$

where, the rigid-joint, the pinned-joint and the semi-rigid-joint condition is defined as $K_\theta = 10^5$, $K_\theta = 10^{-5}$ and $K_\theta = 10^{-1}$ respectively.

4. Mode and Amplitude of Initial Imperfection

The imperfection mode is given by the corresponding buckling mode that will give the largest effects on the buckling strength.

In the cases of rigid-joint and semi-rigid-joint, the nondimentional expression of imperfection amplitude, ξ, is defined by using the equivalent radius of gyration, i_e, as follow.

$$\xi = \frac{\Delta \max}{i_e} \qquad (3)$$

$$i_e = \sqrt{\frac{D}{K}} \qquad (4)$$

In the cases of rigid-joint and semi-rigid-joint, the 4-types ξ, that is, $\xi = 0.2, 0.4, 0.8, 1.2$ are introdeced as analytical examples.

In the case of pinned-joint model, the nondimentional expression of imperfection amplitude, ξ_p, is defined by using the height, h_u, of unit network dome.

$$\xi_p = \frac{\Delta \max}{h_u} \qquad (5)$$

$$h_u = \frac{L^2}{2R} \qquad (6)$$

where, L is the member's length of unit network dome. In the case of pinned-joint, the 4-types ξ_p, that is, $\xi_p = 0.05, 0.1, 0.15, 0.2$ are introdeced as analytical examples.

In the case of the models of the member buckling, it is reasonable that we make an analysis by introducing buckling mode of the models occuring overall buckling which is similar to characteristics of member buckling. But in this study, we introduce the member buckling modes itself as imperfection mode in the the models of member buckling for model's simplification.

5. Results and Discussions

5.1 Buckling strength

Table 1 shows buckling strength for rigid-joint in each model. where LP, LS represents pressure-type-uniform loading and snow-type-uniform loading, respectively. q_{cs}, q_{cf} and q_{imp} is the value of buckling strength for shell analogy method, for frame analysis method and for the model with initial geometrical imperfection by

geometrical imperfection by using frame analysis method. λ_c, α_{is} is the value of q_{imp} divided by q_{cs} and q_{cf}, respectively.   Table 2 shows buckling strength for pinned-joint in each model.

Table 1 Buckling strength for rigid-joint

MODEL (S)		1 (1.9)	2 (2.1)	3 (2.3)	4 (2.7)	5 (2.9)	6 (3.3)	7 (3.9)	8 (4.3)	9 (5.0)
LOAD		LP LS	LP LS	LP LS	LP LS	LP LS	LP LS	LP LS	LP LS	LP LS
q_{cs}		5.16	4.30	3.42	2.55	2.11	1.66	1.21	0.97	0.73
q_{cf}		5.75 5.95	4.55 5.12	3.57 3.86	2.77 2.75	2.19 2.30	1.63 1.65	0.93 0.93	0.61 0.62	0.33 0.33
$\xi=$ 0.2	q_{imp}	5.24 4.73	3.89 3.93	2.82 3.08	2.21 2.20	1.78 1.69	1.31 1.39	0.72 0.80	0.49 0.62	0.31 0.32
	λ_c	0.91 0.80	0.86 0.77	0.79 0.80	0.80 0.80	0.81 0.73	0.80 0.84	0.78 0.86	0.80 1.00	0.95 0.97
	α_{is}	1.02 0.92	0.90 0.91	0.82 0.90	0.87 0.86	0.84 0.80	0.80 0.83	0.60 0.66	0.51 0.64	0.42 0.44
$\xi=$ 0.4	q_{imp}	4.91 4.09	3.49 3.35	2.44 2.59	1.93 1.84	1.57 1.38	1.18 1.23	0.66 0.73	0.45 0.58	0.30 0.30
	λ_c	0.85 0.69	0.77 0.66	0.68 0.67	0.70 0.67	0.72 0.60	0.72 0.74	0.71 0.78	0.74 0.94	0.90 0.90
	α_{is}	0.95 0.79	0.81 0.78	0.71 0.76	0.76 0.72	0.74 0.65	0.71 0.74	0.55 0.60	0.46 0.60	0.41 0.41
$\xi=$ 0.8	q_{imp}	4.41 3.32	2.97 2.67	2.03 2.03	1.58 1.38	1.28 1.00	0.98 1.02	0.57 0.64	0.40 0.55	0.28 0.27
	λ_c	0.76 0.56	0.65 0.52	0.57 0.53	0.57 0.50	0.58 0.43	0.60 0.62	0.61 0.69	0.65 0.89	0.84 0.83
	α_{is}	0.85 0.64	0.69 0.62	0.59 0.59	0.62 0.54	0.61 0.47	0.59 0.61	0.47 0.53	0.41 0.57	0.38 0.37
$\xi=$ 1.2	q_{imp}	4.05 2.88	2.67 2.29	1.90 1.79	1.42 1.27	1.09 0.77	0.83 0.87	0.50 0.57	0.36 0.52	0.26 0.26
	λ_c	0.70 0.48	0.59 0.45	0.53 0.46	0.51 0.46	0.50 0.34	0.51 0.53	0.54 0.61	0.59 0.84	0.80 0.77
	α_{is}	0.78 0.44	0.62 0.53	0.56 0.52	0.56 0.50	0.52 0.36	0.50 0.52	0.41 0.47	0.37 0.54	0.36 0.36

Table 2 Buckling strength for pinned-joint

MODEL (S)		1 (1.9)	2 (2.1)	3 (2.3)	4 (2.7)	5 (2.9)	6 (3.3)	7 (3.9)	8 (4.3)	9 (5.0)
LOAD		LP LS	LP LS	LP LS	LP LS	LP LS	LP LS	LP LS	LP LS	LP LS
q_{cs}		5.16	4.30	3.42	2.55	2.11	1.66	1.21	0.97	0.73
q_{cf}		0.96 1.43	0.96 1.43	0.96 1.44	0.96 1.44	0.96 1.44	0.96 0.96	0.63 0.50	0.43 0.32	0.25 0.18
$\xi_p=$ 0.05	q_{imp}	0.57 0.72	0.55 0.73	0.55 0.73	0.55 0.72	0.55 1.19	0.55 0.92	0.49 0.50	0.33 0.32	0.25 0.18
	λ_c	0.59 0.50	0.57 0.50	0.57 0.50	0.57 0.50	0.57 0.83	0.57 0.96	0.78 1.00	0.77 1.00	1.00 1.00
	α_{is}	0.11 0.14	0.13 0.17	0.16 0.21	0.22 0.28	0.26 0.56	0.33 0.55	0.40 0.41	0.34 0.33	0.34 0.25
$\xi_p=$ 0.1	q_{imp}	0.37 0.47	0.35 0.47	0.35 0.47	0.35 0.47	0.35 0.85	0.35 0.70	0.40 0.44	0.27 0.32	0.24 0.18
	λ_c	0.39 0.33	0.36 0.33	0.36 0.33	0.36 0.33	0.36 0.59	0.36 0.73	0.63 0.88	0.63 1.00	0.96 1.00
	α_{is}	0.07 0.09	0.08 0.11	0.10 0.14	0.14 0.18	0.17 0.40	0.21 0.42	0.33 0.36	0.28 0.33	0.33 0.25
$\xi_p=$ 0.15	q_{imp}	0.25 0.31	0.23 0.31	0.23 0.31	0.23 0.31	0.23 0.58	0.23 0.52	0.33 0.35	0.24 0.28	0.22 0.15
	λ_c	0.26 0.21	0.24 0.22	0.24 0.22	0.24 0.22	0.24 0.40	0.24 0.54	0.52 0.70	0.56 0.88	0.88 0.83
	α_{is}	0.05 0.06	0.05 0.07	0.07 0.09	0.09 0.12	0.11 0.27	0.14 0.31	0.27 0.29	0.25 0.29	0.30 0.25
$\xi_p=$ 0.2	q_{imp}	0.17 0.21	0.16 0.21	0.16 0.21	0.16 0.21	0.16 0.40	0.16 0.36	0.29 0.26	0.21 0.23	0.21 0.11
	λ_c	0.18 0.15	0.17 0.15	0.17 0.15	0.17 0.15	0.17 0.28	0.17 0.38	0.46 0.52	0.49 0.72	0.84 0.61
	α_{is}	0.03 0.04	0.04 0.05	0.05 0.06	0.06 0.08	0.08 0.19	0.10 0.22	0.24 0.21	0.22 0.24	0.29 0.15

Fig. 2 shows the effect of initial imperfection on α–S relation for rigid-joint model under pressure-type-uniform loading.   Dotted lines in Fig.2 indicate the curves of the buckling strength proposed by Yamada[11], namely, α=1.0 indicates overall buckling of rigid-joint framework, $\alpha=\alpha_1$ indicates overall buckling of pinned-joint framework $\alpha=\alpha_m$ indicates member buckling independent of junction's rigidity. Fig. 3 shows λ_c–ξ curves for rigid-joint model under pressure-type-

uniform loading.
    As shown In Fig. 2 and Fig. 3, for $2.3 \leq S \leq 3.9$, the effects of
initial imperfection on the buckling strength are very sensitive,
but for $S \leq 2.1$ or $S \geq 4.3$, the effects of initial imperfection on the
buckling strength are less sensitive than them.

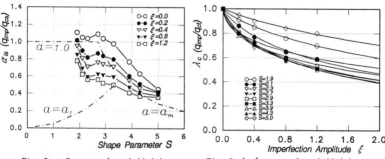

Fig. 2 α-S curves for rigid-joint
under pressure-type-uniform loading

Fig. 3 λ_c-ζ curves for rigid-joint
under pressure-type-uniform loading

    Fig. 4 shows the effect of geometrical imperfection on α-S
relation for rigid-joint model under snow-type-uniform loading.
    Fig. 5 shows λ_c-ζ curves for rigid-joint model under snow-type-
uniform loading.
    As shown In Fig. 4 and Fig. 5, for S=2.9, the effects of initial
imperfection on the buckling strength is the largest, and then the
value of λ_c is 43% in ζ=0.8.   The value of λ_c according to
imperfection amplitude has a tendency to decrease with increasing S,
that is, for the effect of imperfection on buckling strength, it
means that the overall buckling is more sensitive than the member
buckling.

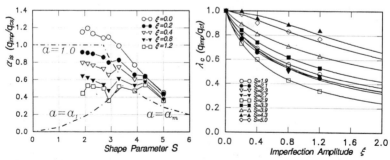

Fig. 4 α-S curves for rigid-joint
under snow-type-uniform loading

Fig. 5 λ_c-ζ curves for rigid-joint
under snow-type-uniform loading

    Fig. 6 shows the effect of geometrical imperfection on α-S
relation for pinned-joint model under pressure-type-uniform loading.

Fig. 7 λ_c-ζ curves for pinned-joint model under pressure-type-uniform loading.

Dotted lines in Fig.7 indicate the curves of the buckling strength on the imperfection amplitude in the case of unit network dome. As shown In Fig. 6 and Fig. 7, for $S \leqq 3.3$, the effects of initial imperfection on the buckling strength is the largest, and then the values of λ_c, independent on the value of S, are constant by about 17% in ζ=0.2. The effect of imperfection on buckling strength is far larger than that of unit network dome. For S=3.9 and S=4.3, the value of λ_c has a similar tendency with it of unit network dome.

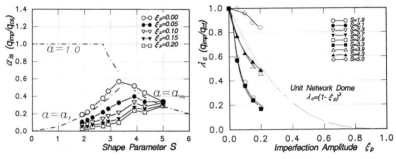

Fig. 6 α-S curves for pinned-joint under pressure-type-uniform loading

Fig. 7 λ_c-ζ curves for pinned-joint under pressure-type-uniform loading

Fig. 8 shows the effect of geometrical imperfection on α-S relation for pinned-joint model under snow-type-uniform loading.

Fig. 9 λ_c-ζ curves for pinned-joint model under snow-type-uniform loading.

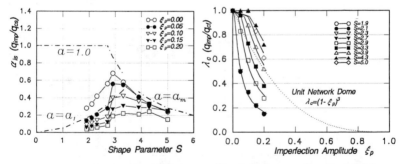

Fig. 8 α-S curves for pinned-joint under snow-type-uniform loading

Fig. 9 λ_c-ζ curves for pinned-joint under snow-type-uniform loading

As shown In Fig. 8 and Fig. 9, for $S \leqq 2.7$, the effects of initial imperfection on the buckling strength is the largest, and then the values of λ_c, independent on the value of S, are constant by about 15% in ζ=0.2. The effect of imperfection on buckling strength is far larger than it of unit network dome. For $S \geqq 2.9$, the value of λ_c has tendency to increase with increasing S except for S=5.0.

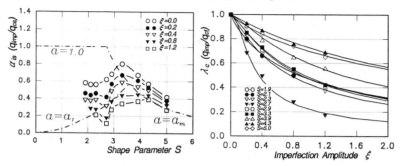

Fig. 10 α-S curves for semi-rigid-joint   Fig. 11 λ_c-ξ curves for semi-rigid-joint
         under pressure-type-uniform loading   under pressure-type-uniform loading

Fig. 10 shows the effect of geometrical imperfection on α-S relation for semi-rigid-joint model ($K_\theta=10^{-1}$) under pressure-type-uniform loading.   Fig. 11 λ_c-ξ curves for semi-rigid-joint model under pressure-type-uniform loading.

As shown in Fig. 10 and Fig. 11, in the case of semi-rigid-joint, the effects of initial imperfection on the buckling strength is more sensitive than them of rigid-joint, and is less sensitive than them of pinned-joint.

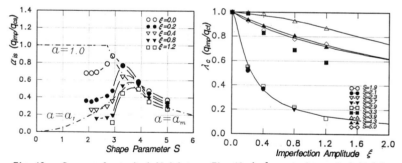

Fig. 12 α-S curves for semi-rigid-joint   Fig. 13 λ_c-ξ curves for semi-rigid-joint
         under snow-type-uniform loading   under snow-type-uniform loading

Fig. 12 shows the effect of geometrical imperfection on α-S relation for semi-rigid-joint model under snow-type-uniform loading. Fig. 13 λ_c-ξ curves for semi-rigid-joint model under snow-type-uniform loading.

As shown In Fig. 12 and Fig. 13, in the case of semi-rigid-joint, the effects of initial imperfection on the buckling strength is more sensitive than them of rigid-joint, and is less sensitive than them of pinned-joint. For $S \leqq 2.7$, the effects of initial imperfection on the buckling strength is more sensitive than them of pressure-type-uniform loading.

5.2 Buckling mode and load-displace relation

Fig. 14 shows the buckling mode for the typical models occuring overall buckling. From Fig. 14, we can find buckling modes of domes with geometrical initial imperfection almost similar to them of perfection domes.

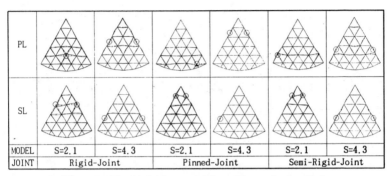

MODEL	S=2.1	S=4.3	S=2.1	S=4.3	S=2.1	S=4.3
JOINT	Rigid-Joint		Pinned-Joint		Semi-Rigid-Joint	

Fig. 14 The buckling mode for the typical models

Fig. 15 shows load-displacement curves in buckling point for the typical models. The nonlinear quality of deflection curves has the tendency to increase with increasing ξ.   These results are well coincide with general deflection characteristics of 3-dimentional structures, i.e., shell etc., with initial imperfection.

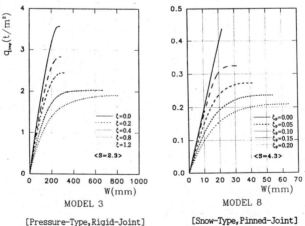

MODEL 3   MODEL 8

[Pressure-Type, Rigid-Joint]   [Snow-Type, Pinned-Joint]

Fig. 15 load-displacement curves for the typical models

6. Conclusion

1. Regardless of the loading conditions and junction's conditions, the effect of initial geometrical imperfection on the buckling strength is much more sensitive in the models of the overall buckling than in the model of the member buckling.
2. From the viewpoint of loading conditions, the effect of initial geometrical imperfection on the buckling strength is generally more sensitive in the models under the vertical-type-uniform loading than in the models under the pressure-type-uniform loading.
3. From the viewpoint of junction's condition, the effect of initial geometrical imperfection on the buckling strength, regardless of loading conditions, is most sensitive in pinned-joint, the second most in semi-rigid-joint and the third most in rigid-joint although they leave a little difference on the model.

Reference

1. Architectural Institute of Japan, STABILITY OF SINGLE-LAYER LATTICED DOMES STATE-OF-THE-ART, edited by Heki, K. August, 1989
2. Forman, S. e. and Hutchinson, J. W., Buckling of Recticulated Shell Structures, Int. j. Solids Structures, Vol.6, 1970, pp.904-932
3. Kollar,L. and Dulacska, E., Buckling of Shells for Engineers, John Wiley & Sons, chichester, 1984
4. Yamada, M., Uchiyama, K., Yamada, S. and Ishikawa, T., Theoretical and Experimental Study on the Buckling of Rigidly Jointed Single Layer Latticed Spherical Shells under External pressure, Proceedings of the IASS Symposium on Membrane Structures and Space Frames, Osaka September, 1986, vol.3, SHELLS, MEMBRAMES & SPACE STRUCTURES, Elsevier, Tokyo, Japan, 1986, pp 113-120
5. Kloppel, K., Beitrag zum Durchschlag problem du nnwandiger wandiger versteifter und unversteifter Kugelschalen fur Vollund halbse itige Belastung, Der Stahlbau, vol.25, no.3, 1956, pp.49-60
6. K.P.Buchert, Buckling of Spherical shells under External Pressure Combination of ECCS Mnaual and SSRC Guide, ASCE, 1977, pp.310-315
7. Write,D.T., Mumbrame Forces and Buckling in Reticulated Shells, Journal of the Structural Division, ASCE, Vol.91, No.ST1, Proc.Paper 4224, Feb, 1965, pp.173-201
8. del Pozo F., and del pozo V., Buckling of Ribbed spherical shells, Preceedings, IASS Congress, Madrid 1979, pp.1.199-1.222
9. Yamada M., An Approximation on the Buckling Analysis of Orthogonally Stiffened and Framed Spherical Shell, Shell and Spatial Structure Enginnering, IASS Symposium, Rio de Janerio, Pentech Press, 1988, pp.177-193
10. Heki, K.: On the Effective Rigidities of Lattice Plates, RECENT RESEARHES OF STRUCTURAL MECHANICS-Contributions in Honour of the 60th Birthday of Prof.Tsubio, Unosheten, Tokyo, 4.1968, pp.31-46
11. Yamada, M., Yamamoto,H., Wang,L.and Jung,H.M., On the effect of joint Flexibility and loading conditions on the buckling characteristics of the single-layer latticed domes,Proceeding of the symposium for 30 Anniversary of Spain, September, 1989, vol.4, pp.503-518

LARGE DEFORMATION ANALYSIS OF LATTICE TRUSS STRUCTURES INCLUDING HYSTERETIC MATERIAL BEHAVIOR

Nayla Khattar[1], George E. Blandford[2], M. ASCE
and Shien T. Wang[2], M. ASCE

ABSTRACT

A methodology is presented to analyze lattice truss structures including both material and geometric nonlinearities. Modeled member material behavior modes include: buckling, yielding, inelastic post-buckling behavior, unloading, and reloading. Geometric non-linearity is based on updating the linearized Lagrangian space truss displacement equations. Interaction of the nonlinear loading behavior results in structure force redistributions which can lead member unloading and reloading response.

INTRODUCTION

The behavior of space lattice systems can only be adequately evaluated with both geometric nonlinearity and member failure considered. To this end, a Space Truss Analysis Program (STAP) has been developed based on the initial development work of Hill, Blandford and Wang (1989). The program traces the inelastic, load-displacement response of a structure in the post-critical range. Constitutive relationships provide the fundamental model of elastic and inelastic member behavior coupled with a geometrically nonlinear finite element model. Geometric nonlinearity is modeled using an iterative updated Lagrangian scheme, i.e., a second-order displacement model which updates the geometry after each iteration. Constitutive details and some analysis details of STAP are presented.

Numerical results are presented for a circular dome truss. These results demonstrate static force redistribution in the members caused by individual member failure modes and the resulting unloading-reloading response of the structure.

---

[1]Graduate Student

[2]Professor, Department of Civil Engineering, University of Kentucky, Lexington, KY 40506-0046

CONSTITUTIVE MODELING OF ELEMENT BEHAVIOR

Stress-strain relationships describing the material behavior are the basis of all structural response. Modeling the individual modes of member behavior with a stress-strain relationship provides a consistent and effective modeling scheme. Behavior modes which are modeled include: elastic, post-yield (tension or compression), elastic and inelastic member post-buckling, stress reversal in buckled or yielded members, and reloading. All of the member behavior models are based on engineering stress-strain relationships.

A bilinear stress-strain relationship is used to model linear elastic, post-yield, and elastic post-buckling behavior of a member:

$$\sigma = E \varepsilon \quad \text{for} \quad \sigma \leq \sigma_k \tag{1a}$$

$$\sigma = E_p \varepsilon \quad \text{for} \quad \sigma > \sigma_k \tag{1b}$$

in which E is the initial elastic modulus; E_p is the "post-elastic" modulus; ε is the engineering axial strain; σ is the engineering stress; and $\sigma_k = \sigma_y$, the yield stress, if yielding is the limiting elastic behavior or $\sigma_k = \sigma_{cr}$, the buckling stress, if buckling is the limiting compression elastic behavior (see Fig. 1).

Equations (1a) and (1b) assume that the transition from the elastic region is abrupt. However, several phenomena give rise to nonlinear chordal stiffness of a truss member. For example, residual stresses in steel members, initial member curvature, load eccentricity, etc. will result in nonlinear member stiffness. A Ramberg-Osgood equation (Richard and Blalock, 1969) is used to provide a smooth stress-strain curve:

$$\sigma = \frac{E \varepsilon}{\left\{ 1 + \left[\dfrac{E \varepsilon}{(1 - E_p/E) \sigma_k + E_p \varepsilon} \right]^n \right\}^{1/n}} \tag{2}$$

in which n is a shape parameter obtained from $n = \ln(2)/\ln(\sigma_k/\sigma_o)$; σ_o is the stress level at the end of the elastic region; and the other symbols are as previously defined.

A stress-strain relationship describing inelastic post-buckling or strain-softening was developed by Hill, Blandford and Wang (1989)

$$\sigma = \sigma_\ell + \left(\sigma_{cr} - \sigma_\ell \right) e^{-\left(X_1 + X_2 \sqrt{\varepsilon'} \right) \varepsilon'} \tag{3}$$

in which σ_ℓ is the asymptotic lower stress limit; ε' is the strain

measured from the beginning of the inelastic post-buckling range, as
shown in Fig. 1; and X_1 and X_2 are parameters calculated using regression analysis techniques to fit the available experimental data.
The initial slope of the inelastic post-buckling curve is a function
of parameter X_1. Parameter X_2 influences the rate of change of the
inelastic post-buckling modulus.

Unlike the compression loading path, which is the same for all
members with the same slenderness ratio, unloading paths are dependent on the state of stress in each member at the instant of unloading (Papadrakakis, 1983). Stress reversal in an inelastically buckled member is represented using a hyperbolic curve

$$\sigma_h = \frac{\varepsilon_h}{a_1 + b_1\,\varepsilon_h} \tag{4}$$

in which σ_h and ε_h are the engineering stress and strain measured
from the intersection of the unloading curve with the initial elastic tension loading curve (i.e., point 1 in Fig. 1); and a_1, b_1 are
interpolation constants obtained by specifying the slope at the
beginning (point 1) and matching the stress-strain values at the
end (point 2) of the unloading curve. A strain value ε_1, point 1 on
the hyperbolic curve, is assigned such that

$$\varepsilon_1 = C_\varepsilon\,\varepsilon_y \tag{5}$$

where $0 \leq C_\varepsilon \leq 1$. Constants a_1 and b_1 are evaluated as

$$a_1 = \frac{1}{C_E\,E} \tag{6}$$

$$b_1 = \frac{1}{\sigma_2 - \sigma_1} - \frac{a_1}{\varepsilon_2 - \varepsilon_1} \tag{7}$$

in which $0 \leq C_E \leq 1$; $\sigma_1 = E\,\varepsilon_1$; and end point (point 2) stress and
strain values σ_2 and ε_2 are known (predicted by STAP).

Experimentally observed responses of steel struts to cyclic
loads involve several complex physical phenomena. One primary observation is that during consecutive inelastic cycles, the maximum
compressive loads tend to decrease. This is in sharp contrast with
the ability of a member to resist tension, which remains essentially
constant regardless of previous cyclic history. Another significant
observation is the progressive lowering of the tangent modulus upon
repeated cycling. This phenomenon, the well-known Bauschinger
effect, holds true regardless of the initial sense of the applied
stress (Black, Wenger and Popov, 1980).

Kahn and Hanson (1976) experimentally investigated a load cycle
involving post-buckling response followed by tension loading. Subsequent loading led them to the observation of two different types

of hysteresis loops. They found that members subjected to cycles of
loading where the maximum tensile force was maintained below a certain value of the yield force, referred to herein as the "elastic
limit", had an "approximately elastic" behavior. Their experimental
results are shown in Fig. 2. One can see that a member experiencing
a stress reversal from inelastic post-buckling unloading follows a
curved path stretching between the point of reload and the beginning
of inelastic post-buckling unloading. Such behavior is modeled by
the hyperbolic stress-strain relationship of (4). The beginning
point (point 1 in Fig. 1) on the hyperbolic curve is redefined as
the point at which stress reversal or reloading occurs. The end
point (point 2) does not change and a new value for b_1 is calculated. This procedure produces stress-strain curves which agree with
the experimentally observed behavior of Fig. 2.

Substantial hysteresis loops were found by Kahn and Hanson
(1976) when the load exceeded the "elastic limit" and they observed
that all specimens exhibited a net elongation, also called cyclic
growth. Some recent experimental results are shown in Fig. 3. To
model such behavior, the Ramberg-Osgood stress-strain relationship
of (2) is used. The stress at the end of the elastic region σ_o is
redefined and it is set equal to the stress value at which inelastic
post-buckling unloading begins (i.e., point 2 in Fig. 1). A new
value of stress at the beginning of the post-buckling region (σ_{cr})
is assigned such that

$$(\sigma_o/\sigma_{cr})_r = S_r \qquad (8)$$

in which $(\sigma_o/\sigma_{cr})_r$ is the ratio used in the Ramberg-Osgood reloading
model to calculate the value of the shape parameter n; $0 \le S_r \le 1$;
and the new value of n for reloading is calculated as n =
$\ln(2)/\ln(\sigma_{cr}/\sigma_o)_r$. The new critical stress is used to define the
initiation of strain-softening (3) for the reloading curve.

If the strain at the new critical stress is less than the value
of strain at which initial inelastic post-buckling occurs (ε_i), an
elastic post-buckling plateau is defined until the strain ε reaches
ε_i. In the event that the strain is greater than ε_i, ε_i is set to
be the value of strain at the redefined critical stress. Cyclic
growth is considered by redefining the beginning point (point 1 in
Fig. 1) of the unloading curve. The strain ε_1 is increased as

$$\varepsilon_1 = E_\varepsilon * \varepsilon_1 \qquad (9)$$

where $E_\varepsilon \ge 0$. Stress σ_1 is unchanged and b_1 is recalculated.

FINITE ELEMENT EQUATIONS

A linearized updated Lagrangian formulation is used to represent the nonlinear, second-order analysis of space trusses. The
starting point involves using the Green-Lagrange strain $\mathcal{E}_x$:

$$\varepsilon_x = \frac{du}{dx} + \frac{1}{2}\left(\frac{du}{dx}\right)^2 + \frac{1}{2}\left(\frac{dv}{dx}\right)^2 + \frac{1}{2}\left(\frac{dw}{dx}\right)^2 = e_x + \eta_x \qquad (7)$$

where e_x is the linear strain; and η_x is the nonlinear strain. Using the principal of virtual work consistent with the linearized updated Lagrangian theory (e.g., Bathe, 1982), the local coordinate (denoted by " ' ") element stiffness equations can be expressed as

$$\{F'\} = ([k'_E] + [k'_G]) \{p'\} \qquad (8)$$

in which $\{F'\}$ is the element force vector; $[k'_E]$ is the element elastic stiffness matrix; $[k'_G]$ is the element geometric stiffness matrix; $[k'_E] = \dfrac{E_T A}{l}\left[\begin{array}{c|c} I_1 & -I_1 \\ \hline -I_1 & I_1 \end{array}\right]$; $[k'_G] = \dfrac{F}{l}\left[\begin{array}{c|c} I & -I \\ \hline -I & I \end{array}\right]$; I_1 is a

3x3 matrix in which all coefficients are zero except the 1,1 coefficient which equals 1; I is the 3x3 identity matrix; $E_T = d\sigma/d\varepsilon$ $(1/L)^3$ is the tangent modulus $(d\sigma/d\varepsilon)$ multiplied by the large strain transformation; L is the undeformed element length; l is the current element length; A is the cross sectional area; F is the current element axial force; $\{p'\}$ is the element displacement vector; and $\{\ \}$, $<\ >$, $[\ ]$ symbolize column, row, and rectangular matrices. Direct stiffness is used to construct the structure stiffness equations.

NONLINEAR SOLUTION ALGORITHM

Evaluating space truss behavior with both geometric and material nonlinearities included requires the use of an incremental/iterative solution scheme. Effective solution of the nonlinear equations, particularly in regions associated with "limit-points" or zero structure stiffness points, requires a proper selection of the incremental step length to reflect the degree of nonlinearity in the equations and the stress history. The incremental/iterative solution scheme consists of stepwise linearization of the nonlinear structural behavior. An updated version of the Hill, Blandford and Wang (1989) computational details associated with modeling second-order geometric nonlinearity and the implementation of the explicit iteration on spheres algorithm (Forde and Steimer, 1987), a modified version of Crisfield's (1983) arc-length algorithm, is given in Blandford and Wang (1993).

NUMERICAL RESULTS

The circular dome truss geometry and loading are shown in Fig. 4 with P = 5 kN; A = 0.10 cm^2; I_y = 4.17x10^{-3} cm^4; E = 2x10^4 kN/cm^2; E_p = 0.001E; σ_y = 40 kN/cm^2; X_1 = 500 and X_2 = 100 (L/r_y ≈ 120); σ_1 = 0.4σ_{cr}; ε_o = ε_{cr} and $(\sigma_o/\sigma_{cr})_r$ = 0.75.

Results presented in Fig. 5 are based on complete inelastic post-buckling (IPB) response using an initial bilinear stress-strain model. The maximum predicted load carrying capacity of the truss is

0.874 P. Figure 5 shows that the IPB results exhibit snap-back response as a member experiences inelastic post-buckling behavior with subsequent reloading.

Figure 6 shows the IPB member force-displacement response for member 7, the first member to buckle. The results show that the member experiences several "approximately elastic" cycles of: (1) inelastic material softening, (2) unloading from inelastic post-buckling behavior, and (3) "elastic" reloading from the unloading curve. The oscillatory nature of the response is due to the force redistribution in the highly indeterminate structure.

Figure 7 shows the symmetric member failure pattern sequence associated with the results of Figs. 5 and 6. The numbers inscribed in the buckling, unloading, and reloading symbols correspond to the load level sequences shown in Figs. 5 and 6, after first buckling. Fig. 7 shows that member 7 buckled first with subsequent buckling of the neighboring members (members 6 and 8). After buckling of these three members, member 7 first unloads and then reloads after several additional analysis steps. Members 5 and 9 buckle next, which results in immediate unloading of members 6 - 8. As shown in Fig. 7, the reloading - member buckling - unloading sequence continued to spread through the circular dome structure until eventual structure failure would occur.

SUMMARY AND CONCLUSIONS

The developed space truss analysis strategy utilizes a constitutive model to define material behavior at the element level. Various stress-strain relationships have been developed. Elastic member behavior is either linear (straight line) or nonlinear and described by a Ramberg-Osgood equation. Elastic post-buckling behavior is also represented by a linear model. An exponential equation models strain-softening, whereas a hyperbolic curve is used for unloading from the inelastic post-buckling range. The hyperbolic and Ramberg-Osgood models are also used to model "approximate elastic" and cyclic hysteretic behavior, respectively.

An accurate finite element analysis procedure to access the behavior of truss systems in their failure modes has been presented. The developed analysis scheme enables a quick and easy analysis for a wide range of truss systems.

Considerably more experimental and theoretical work is needed to define hysteretic behavior. Theoretical methods should be developed to predict the "elastic limit". Reduction in maximum compressive stress with successive cycles needs further analytical development. Research work by Black, Wenger and Popov (1980) discuss the effects of cyclic loading on steel struts and presents an analytical expression to predict cyclic buckling loads. Implementation of such an expression into STAP is necessary. Further experimental work is required for the determination of the "elastic reload limit".

REFERENCES

Bathe, K.J. (1982), *Finite Element Procedures in Engineering Analysis*, Prentice-Hall, Inc., Englewood Cliffs, NJ, Chapter 6.

Black, R.G., Wenger, W.A. and Popov, E.P. (1980), "Inelastic Buckling of Steel Struts Under Cyclic Load Reversals," *Report No. UCB/EERC - 80/40*, Earthquake Engineering Research Center, University of California, Berkeley, CA.

Blandford, G.E. and Wang, S.T. (1993), "Response of Space Trusses During Progressive Failure," *Dynamic Response and Progressive Failure of Special Structures*, ASCE, Ramesh B. Malla (Ed.), 16 pp.

Crisfield, M.A. (1983), "An Arc-Length Method Including Line Searches and Accelerations," *Int. J. Num. Meth. Engrg.*, **19**, 1269-1289.

Forde, B.W.R. and Stiemer, S.F. (1987), "Improved Arc Length Orthogonality Methods for Nonlinear Finite Element Analysis," *Computers and Structures*, **27**, 625-630.

Hill, C.D., Blandford, G.E., and Wang, S.T. (1989), "Post-Buckling Analysis of Steel Space Trusses," *J. Struct. Engrg.*, ASCE, **115**, 900-919.

Kahn, L. and Hanson, R. (1976), "Inelastic Cycles of Axially Loaded Steel Members," *J. Struc. Div.*, ASCE, **102**, 947-959.

Papadrakakis, M. (1983), "Inelastic Post-Buckling Analysis of Trusses," *J. Struct. Div.*, ASCE, **109**, 2129-2147.

Richard, R.M., and Blalock, J.R. (1969), "Finite Element Analysis of Inelastic Structures", *AIAA Journal*, **7**, 432-438.

Sorouchian, P. and Alawa, M.S. (1989), "Hysteretic Modeling of Steel Struts: A Refined Physical Theory Approach," *J. Struc. Engrg.*, ASCE, **116**, 2903-2916.

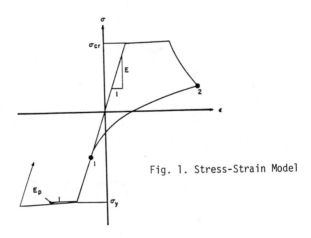

Fig. 1. Stress-Strain Model

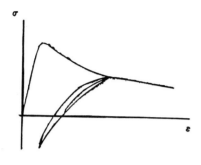

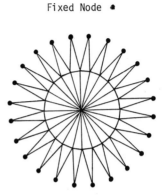

Fixed Node ●

Fig. 2. Experimental Results for
Approximately Elastic
Hystretic Behavior (Kahn
and Hanson 1976)

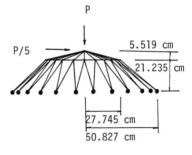

P

P/5

5.519 cm

21.235 cm

27.745 cm

50.827 cm

Fig. 4. Circular Dome Truss
Geometry and Loading

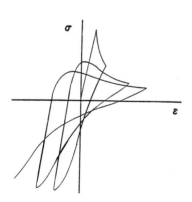

Fig. 3. Experimental Results for
Cyclic Behavior
(Sorouchian and Alawa
1989)

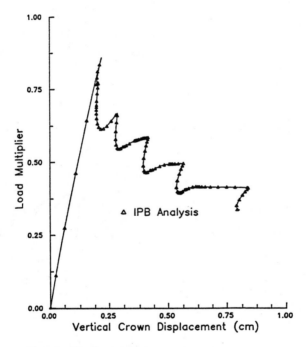

Fig. 5. Vertical Displacement at Crown

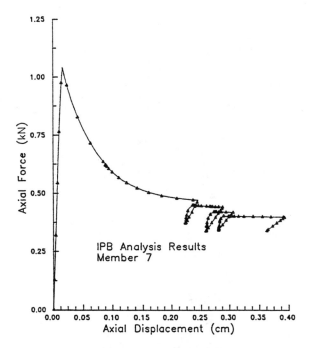

Fig. 6. Axial Force - Displacement Response of
         Member 7

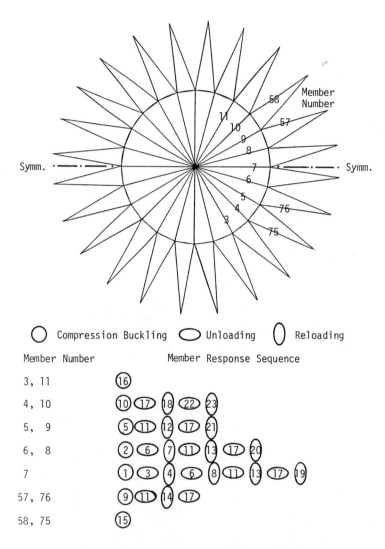

Fig. 7. Member Failure Sequence - Buckling, Unloading and Reloading

Buckling of Single Layer Latticed Dome
under Uniform Gravity Load

Masumi Fujimoto[1], Member, IASS, Katsuhiko Imai[2],
and Toshitsugu Saka[3], Member, IASS

Abstract

This paper reports on experiment and numerical
analysis relating the buckling of a single layer
latticed dome under uniformly distributed nodal gravity
load at the dome central portion. The effect of a half
open angle subtended by a member on buckling load and
buckling behavior of single layer latticed dome is
discussed. The applicability of the discrete treatment
method adopted in this study is also discussed, with
regard to calculating the buckling behavior of a single
layer latticed dome occurring the member buckling and
dimple buckling.

Introduction

With advances in steel structures, many kinds of
joint systems and mesh patterns have been devised for
single layer latticed domes. These domes have been used
as practical space structures. To grasp the buckling
behavior of a single layer latticed dome and obtain
useful information on its structural design, various
experimental studies and theoretical studies (AIJ

[1]Lecturer, Osaka City University, Sugimoto 3-3-138,
Sumiyoshiku, Osaka 558, Japan
[2]Manager, Kawatetsu Steel Products Corporation,
Uozakiminamimachi 3-6-24, Higashinadaku, Kobe 658, Japan
[3]Professor, Osaka City University, Sugimoto 3-3-138,
Sumiyoshiku, Osaka 558, Japan

1989; McConnel et al. 1986; Ueki et al. 1991; Rothert
and Gebbeken 1992) have been conducted. However, few
reports are available on the single layer latticed dome
composed of a standardized truss system.

A previous paper (Fujimoto 1993a) treated single
layer latticed dome composed of a standardized truss
system. The buckling behavior of the single layer
latticed dome under nodal gravity load at the dome crown
was discussed in terms of experiment and numerical
analysis. The validity of the discrete treatment method
was verified with reference to calculating the buckling
behavior of a shallow single layer latticed dome
undergoing snap through buckling.

The present paper also treats the single layer
latticed domes composed of a standardized truss system.
The buckling of such a dome under uniform nodal gravity
loads at the dome central portion is examined by both
experiment and numerical analysis. The effect of half
open angle on the buckling load and buckling behavior of
a single layer latticed dome is discussed in regard to
load displacement relation. The applicability of the
discrete treatment method (Fujimoto 1993b) is also
discussed with comparison of numerical analysis and
experimental results.

Dome Experiment

The parallel lamella is used as the dome mesh type.
The slenderness ratio of a member in the meridian
direction from the dome crown, λ, the half open angle
subtended by that member, ϕ and the dome division number,
n, are used to determine dome dimension. In this study,
two domes are used. One, a Type 3 specimen, has λ =100,
n=3 and ϕ=3°; the other, a Type 5 specimen, has n=3 and
ϕ=5°. The dome spans of both Type 3 and Type 5 are equal
to 6.654 m. The supporting condition of a node on the
peripheral ring is assumed to be a one way roller in the
meridian direction. For Type 5, dome overall
configuration and elevation are shown in Figs. 1 and 2,
respectively.

Circular hollow tubes are used as dome constituent
members. The diameter of member at connection with the
inner nodes is 34 mm; that of member in the peripheral
ring is 60.5 mm. Member sectional properties are shown

in Table 1. As member joints, ball joints composed of
solid spherical node and joint assembly, KT truss system,
are used (Imai et al. 1989). The material of members is
STK400, that of nodes is SCM435. The material properties
of member and of joints obtained by experiment are shown
in Tables 2 and 3, respectively.

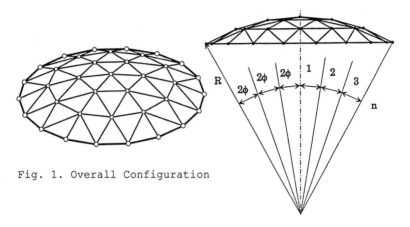

Fig. 1. Overall Configuration

Fig. 2. Elevation

Table 1. Member Sectional Properties

ϕ	A(cm^2)	Zp(cm^3)	I(cm^4)
34	2.07	2.10	2.64
60.5	5.76	10.5	23.7

Table 2. Member Material Properties

Elastic Modulus(kN/cm^2)	2.111x10^4
Yield Stress(kN/cm^2)	38.4
Tensile Strength(kN/cm^2)	47.1

Table 3. Joint Material Properties

Rotational Spring Coef.(kNcm/rad)	9.03x10^3
Yield Bending Moment(kNcm/rad)	64.44

The specimen is loaded using an oil jack. The load
is transmitted uniformly to the dome crown and first
ring nodes by the tournament system. The experimental
setup is shown in Fig. 3. Items and nodes of measurement
are shown in Fig. 4.

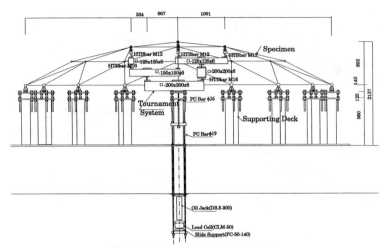

Fig. 3. Experimental Setup

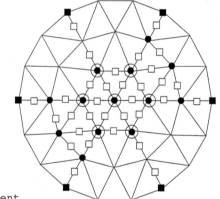

○ Loading Point
□ Axial Strain
● Gravity Displacement
■ Horizontal Displacement
Fig. 4. Items and Nodes of Measurement

Numerical Analysis of Discrete treatment Method

The discrete treatment method considering both geometric and material nonlinearity (Fujimoto 1993b) is used in this study. To consider both joint size and rigidity, ball joints and junctions between tubular members and ball joints are modeled as rigid zone and rotational spring, respectively (Saka and Heki 1984). The member model is shown in Fig. 5. For tubular members, the incremental elasto-plastic stiffness matrix is determined using both the tangent stiffness matrix (Oran 1973) and the generalized hardening hinge method (Inoue and Ogawa 1978). To consider both joint effect and compound nonlinearity simultaneously, a member model introducing the rigid zone and joint rotational spring to the incremental elasto-plastic stiffness matrix (Murakami 1992) is adopted. Moreover, the tubular member expected to be subjected to large compressive axial force is divided into two equal parts, as shown in Fig. 5. As yielding criteria for tubular members, the spherical surface expressions of the interaction equations between biaxial bending and axial force are used. For rotational springs, the relation of moment to rotational angle is assumed to be bilinear and the plastification is determined from the interaction between biaxial bending and axial force. The method tracing the equilibrium path, including the post buckling range, is shown in Fujimoto (1993b).

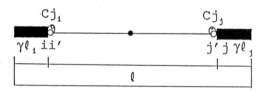

Fig. 5. Member Model

As to the nodal coordinates of the analyzed latticed dome, the design values and measured values are used as the horizontal and vertical components, respectively. For the material properties of tubular members and those of joints, the values obtained in the experiment are adopted. The member strain hardening factor is assumed to be 0.03, and the ratio of rotational spring coefficient of second stiffness to first stiffness to be 0.10. The yield axial force of the rotational spring is assumed to be 60(kN).

Results

The results of experiments are as follows. Buckling load is shown in Table 4; The relation of load to loading point gravity displacement as obtained by experiment is shown in Fig. 6. The experimental and numerical analysis results are compared below. The relation of load to dome crown gravity displacement is shown in Fig. 7. The relations of load to first ring node gravity displacement are shown in Fig. 8. In these figures, the lines with the mark and the lines without denote the results of experiment and analysis, respectively. The thick and thin lines denote results with Type 3 and Type 5, respectively. The final deformation after loading is shown in Photo 1 and buckling mode is shown in Fig. 9.

Table 4. Buckling Load

Specimen	Experiment	Numerical Analysis
Type 3	64.2 (kN)	77.3 (kN)
Type 5	127.1 (kN)	163.1 (kN)

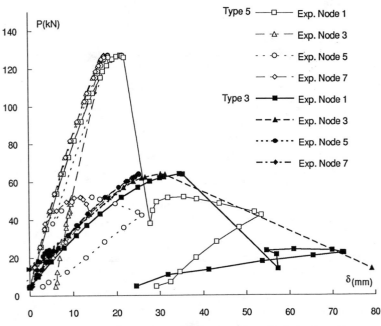

Fig. 6. Relation of Load to Loading Point Gravity Displacement

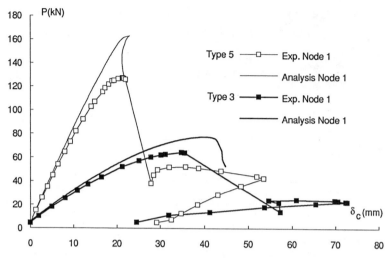

Fig. 7. Relation of Load to Dome Crown Gravity
Displacement

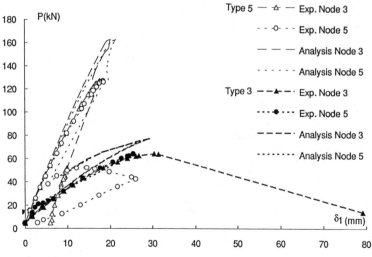

Fig. 8. Relation of Load to First Ring Node Gravity
Displacement

Photo. 1. Final Deformation   Fig. 9. Buckling mode

Consideration

Gravity Displacement of Loading Point   The slope of the
equilibrium path is decreasing due to the influence of
the geometric nonlinearity from initial state to
buckling point. The mode of member buckling appears
slightly in the meridian direction just before the
buckling point. At the buckling point, the member
buckling grows and dimple buckling occurs on both the
dome crown and the one node in the first ring. The
gravity displacement of other nodes in first ring
reverses after the buckling point. The load becomes
about 70 % decrease in buckling load. The process from
buckling point to load decrease is instantaneous. For
the equilibrium path in the range of both initial
loading and post buckling, the slope of Type 5 is larger
than that of Type 3. For gravity displacement at both
buckling point and after dimple buckling, the value of
Type 5 is about 50 % of that of Type 3.

Comparison of equilibrium path of Experiment and that of
Numerical analysis   As to the relation of load to dome
crown gravity displacement, numerical analysis clearly
shows the equilibrium path characteristics obtained in
the experiment. For the relation of load to gravity
displacement of first ring nodes, numerical analysis
shows that gravity displacement reverses after the
buckling point. Thus, the results of numerical analysis
are in good agreement with those of the experiment,
except in regard to the node gravity displacement which
involved dimple buckling. Therefore, the discrete
treatment method adopted in this paper is applicable to
calculation of the equilibrium path of the dome from the
initial state to the range of post buckling.

Buckling Load   The experiment and numerical analysis results show that the buckling load is nearly proportional to the half open angle subtended by member. The numerical analysis result is about 20-30 % larger than the experimental result, because for analysis the member elastic modulus is assumed to be linear, the generalized hardening hinge method is used and the relation of bending moment to the rotational angle of the rotational spring is assumed to be bilinear.

Final deformation of Experiment and Buckling Mode   For dome deformation after the experiment, the mode of member buckling appears at the member in the meridian direction and that of dimple buckling appears at both dome crown and one node in the first ring. For buckling mode obtained by the discrete treatment method, the dome crown gravity displacement appears larger than that of other nodes.

Conclusion

This paper treated a single layer latticed dome composed of a standardized truss system. The buckling of the single layer latticed dome under uniform nodal gravity loads at the dome central portion was examined experimentally and numerically. The effect on buckling behavior of half open angle subtended by a member in the meridian direction was discussed with regard to buckling load and overall load displacement relation. The discrete treatment method, based on geometric and material nonlinearity using a member model, considering the rigid zone and rotational spring of the joint, was used to calculate dome buckling behavior. The applicability of the discrete treatment method presented here was discussed with comparison between the results of numerical analysis and of the experiment. Main conclusions are as follows.
1)   It is confirmed experimentally that the buckling load is nearly proportional to the half open angle subtended by a member.
2)   Comparison of experiment and numerical analysis results shows that the discrete treatment method presented here is applicable both to tracing the fundamental equilibrium path and calculating the buckling load of the single layer latticed dome treated here.

Acknowledgment

The authors wish to thank Mr. Asayama and Mr. Ichikawa for their assistance during the experiment.

Appendix. References

Architectural Institute of Japan (1989). "Stability of single layer latticed dome(in Japanese)", Edited by Heki, K.

Fujimoto, M.,Saka, T., Imai, K.,and Morita, T. (1993a). "Experimental and numerical analysis of buckling of single layer latticed dome", *Proc. 4th Int. Conf. Space Struct.*, Surrey, United Kingdom, 1, 396-405.

Fujimoto, M., Imai, K., and Saka, T. (1993b). "Nonlinear buckling analysis of a single layer latticed dome composed of a standardized truss system under gravity loads at dome central portion", *Proc. SEIKEN-IASS Symp. 1993*, Tokyo, Japan, 175-182.

Imai, K., Wakiyama, K., Tsujioka, S., and Yamada, Y. (1989). "Proposing a new joint system (KT-system) of space frame with threaded spherical nodes and its fatigue characteristics", *Int. Conf. IASS*, Madrid, Spain, 4.

Inoue, K., and Ogawa, K. (1978). "A study on the plastic design of braced multi-story steel frames part 2 on the overall static and dynamic behaviors of plastically designed multi-story braced frames(in Japanese)", *Trans. AIJ*, 268, 87-98.

McConnel, R. E., Fathelbab, F. A., and Hatzis, D. (1989). "The buckling behavior of some single layer, shallow lattice domes", Shells, Membranes and Space Frames, *Pro. IASS Symp.*, Osaka, 3, 97-104

Murakami, M. (1992). "Numerical analysis of elastic buckling of single-layer latticed domes under gravity load", *Proc. IASS-CSCE Int. Cong. 1992*, 2, 576-586.

Oran, C. (1973). "Tangent stiffness in space frame", *J. Struct. Div.*, ASCE, 99(6), 987-1001.

Rothert, H., and Gebbeken, N., (1992). "On numerical results of reticulated shell buckling", *Int. J. Space Structures*, 7(4), 299-319.

Saka, T., and Heki, K. (1984). "The effect of joints on the strength of space trusses", *Proc. 3rd Int. Conf. Space Struct.*, Surrey, United Kingdom, 417-422.

Ueki, T., Mukaiyama, Y., Shomura, M. and Kato, S. (1991). "Loading test and elasto-plastic buckling analysis of a single layer latticed dome(in Japanese)", *J. Struct. Constr. Engrg. AIJ*, 421, 117-128.

Nonlinear Buckling Response of Single Layer Latticed Barrel Vaults

Seishi Yamada [1]   and   Takashi Taguchi [2]

Abstract

Elastic overall buckling of single layer latticed barrel vault roof structures under static vertical loading is investigated using a fully nonlinear finite element solution procedure and a simple reduced stiffness bifurcation analysis. The present rigidly-jointed latticed vault consists of 212 straight members. Two types of boundary support system, a roller support system and a pinned support system, have been adopted. The distributions of displacement and internal axial force are determined by boundary conditions. It has been observed that the nonlinear numerical experiments are very sensitive to geometric imperfections. A correlation between the scattered nonlinear buckling response and the reduced stiffness lower bound load spectrum, has been discussed.

Introduction

The elastic nonlinear buckling of single layer latticed barrel vault roof structures has the complex behavior for overall (general) buckling form due to imperfection sensitivity in contrast with member or local buckling form; resistance to the member buckling is simply estimated by the bending stiffness of the individual members just similar to for column buckling, and resistance to the local (nodal) buckling can approximately be obtained through the analysis of the individual frame units.

For the design of these shell-like structures, it would be very important to ensure adequate resistance to the overall buckling under self-weight, imposed snow loading, or possibly an effectively amplified gravity force arising from seismically induced response. A number of recent studies (Goncalvas and Croll 1992; Yamada and Croll 1989/1993; Yamada et al. 1993) have shown that for initial imperfection sensitive shell structures, a reduced stiffness analysis gives simple, theoretical lower bound estimates of experimental elastic buckling loads. The first author (Yamada 1991) has shown the good agreement between a nonlinear Ritz analysis for the equivalent stiffness continuum analogy and a fully nonlinear finite element

1 Associate Professor, Dept of Architecture and Civil Engrg, Toyohashi University of Technology, Toyohashi 441, Japan
2 Graduate Student , Dept of Architecture and Civil Engrg, Toyohashi University of Technology, Toyohashi 441, Japan

analysis. Also given in the literature are the loss behaviors of incremental membrane strain energies, associated with the loss of stiffness, along its nonlinear equilibrium path .

The present paper seeks to make clear the effects of boundary constraints, initial geometric imperfections, and a geometry (central angle ϕ) on the nonlinear buckling responses. The relationships between the nonlinear detailed analysis and the linear bifurcation analyses for the buckling of latticed vaults are discussed in order to bring the potential advantages of these approaches together for understanding such difficult problems.

The Latticed Vault Model

The present vaults, of central angle ϕ, are adopted as rigidly jointed single layer latticed ones having an isosceles triangle configuration. Each constitutive member is a straight steel tube with Young's modulus E=206GPa, Poisson's ratio 0.3, a 165.2mm outside diameter and a 7.0mm thickness. The nominal length of the members (for initially perfect models) l is fixed as 3.5m; consequently, the member slenderness ratio λ is 62.5. The network pattern has six spans in ridge direction and ten spans in circumferential direction as shown in Fig.1. The joint nodes are on the cylindrical surface defined by the radius of curvature $R= \sqrt{3}l/[4\sin(\phi/20)]$. In the present vaults it can be seen that the length in ridge direction L=6l is fixed as 21m and that in span direction L_Y=2$R\sin(\phi/2)$. Also superimposed upon Fig.1 are uniform vertical loading system and boundary support systems. The external loads P are assumed to be applied at 49 mid-shell joint nodes only. The roller support points are restricted to move in only vertical (Z-) direction but can move horizontally (in both X- and Y-directions); the pinned ones are restricted to move in any direction.

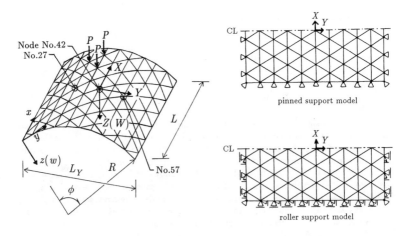

pinned support model

roller support model

Fig.1 A latticed vault model

Nonlinear Numerical Solution Procedure

The geometrically nonlinear numerical analysis is formulated based on the combined use of the finite element method and the method of moving coordinates, and is performed applying the modified incremental method with respect to the load P or to one of the unknown vertical nodal-displacements W_i (Yamada et al. 1986). In the present study the increments ΔP and ΔW_i have been adopted as round 0.25kN and 0.02cm, respectively. A finite element modeling within which line elements are interconnected at nodes is used. The linear and geometric stiffness matrices of the element are of type (12×12) by application of the Bernoulli-Euler hypothesis and the Saint-Venant theory considering torsional rigidity (Sumec 1990). One element in each member is used in the present study in order to save computational time. It has preliminarily obtained that it has only a few percent difference in buckling load in comparison with two elements per member discretization in the case of the present vault model.

Effects of Boundary Conditions on Linear Bending Solutions

Figure 2 shows the linear vertical deflection profiles and the distributions of the associated internal axial force of members in the case of $\phi=\pi/6$. For the pinned support model (Fig.2a) it can be seen that the almost diagonal members at mid-shell are in the uniform compressive state, which would be estimated from a statically determinate truss unit model (Yamada 1991) as

$$N_L{}^d = -\frac{2PR}{3l} \tag{1}$$

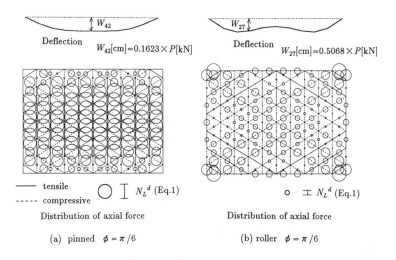

Fig.2 Linear bending analytical results for perfect geometry

In Fig.2b, for the roller support model, diagonal members having $N_L{}^d$ are limited to be in a smaller center region; the compressive force of corner members near the four fixed corner nodes is round three times as large as $N_L{}^d$. Consequently the maximum deflection point for the roller support model moves into near the boundary apart from the center of vault. These results suggest that the 'loading imperfection' (Yamada and Croll 1989) of the roller support model is much larger than that of the pinned support model.

Assumption of Geometric Imperfections

In the design of this kind of shell-like structure, it is very important to estimate the effects of imperfections on the elastic buckling behavior. Many measurements for shell domes by Yamada et al. (1983) have suggested that a longer wavelength mode component is dominant in the initial imperfection and contributes to an extra reduction of buckling load.

In the present imperfect model, the node coordinates are removed from those of the perfect model (X,Y,Z), for the global Cartesian coordinate system as shown in Fig.1, to new positions $(X,Y,Z+W^0)$. Let us define the imperfection profiles as superimposed upon Table 1, referring to a linear bifurcation analysis for the idealized orthotropic continuum analogy which has been previously proposed by Yamada (1991) and also will be reviewed later. The imperfection

$$W^0 = \left[-W_s{}^0 \cos\left(\frac{n_s\pi Y}{L_Y}\right) + W_a{}^0 \sin\left(\frac{n_a\pi Y}{L_Y}\right) \right] \cos\left(\frac{\pi X}{L}\right) \tag{2}$$

Table 1 Adopted imperfections for nonlinear numerical experiments

No.	α_s	α_a
F	0	0
$S_8{}^-$	−8	1
$S_4{}^-$	−4	1
$S_1{}^-$	−1	1
S_0	0	1
$S_1{}^+$	1	1
$S_4{}^+$	4	1
$S_8{}^+$	8	1
A_4	1	4
A_8	1	8

consists of two types: a symmetric mode for circumferentially odd half-wave number n_s and an anti-symmetric mode for circumferentially even half-wave number n_a. Listed in Table 1 are the coefficients, α_s and α_a, which represent the associated component amplitudes with the rate per thousand to the representative length of structure L; that is

$$W_s^0 = \alpha_s L/1000 = 2.1\alpha_s(\text{cm}) \; ; \quad W_a^0 = \alpha_a L/1000 = 2.1\alpha_a(\text{cm}) \qquad (3)$$

Nonlinear Finite Element Analytical Results for Imperfect Geometries

The present results of nonlinear numerical experiments shown in Figs.3−6 are for the following cases:

support	ϕ	R	$L/(R\phi)$	L_Y	rise
(a) pinned	$\pi/3$	28.96m	0.693	28.96m	3.880m
(b) pinned	$\pi/6$	57.90m	0.693	29.97m	1.973m
(c) roller	$\pi/6$	57.90m	0.693	29.97m	1.973m

Figure 3 represents the load versus the vertical deflection at sampling points No. 27, 42 and 57 superimposed upon Fig.1.   The dots indicate the first critical buckling points. The deflection response can be observed to have a dramatically different form depending on the level of the imperfection W^0. The buckling loads are also affected both by the geometric parameter, central angle ϕ (see Figs.3a and 3b) and by the variation of boundary support system (see Figs.3b and 3c).

The dramatic changes of deflection mode are shown in Figs.4 and 5: Fig.4 indicates the distribution of the total vertical displacement W along the Y-direction of $X=0$ at the buckling point: Fig.5 shows the associated incremental displacement ΔW at the same equilibrium state as that of Fig.4. It has experimentally been clarified that the development of buckling lobes for imperfect shells relates to a classical critical mode (Yamada et al. 1983). Also Yamada and Croll (1989) has suggested that it is necessary to consider the development of buckling lobes that allow an essentially non-integer form of buckling mode. The closed dots in Fig.6 for the present imperfect geometries show the relationships between the nonlinear buckling loads and the observed dominant non-integer circumferential half-wave numbers n_N. It is also apparent that the observed modes n_N for the roller models are smaller than those for the pinned models and the lower bounds to the scatted buckling loads are seemed to be related to n. In other words, the increases of buckling load by changing the roller support system into the pinned one are due to the occurrence of a circumferentially higher buckling mode.

Linear Bifurcation Buckling and Reduced Stiffness Analyses

If an analyst were to limit the investigation of the buckling of this class of shell-like structure, it is likely that a designer would be left in a state of some confusion. It would for example, be difficult to understand why the deflection mode undergoes such large changes, and why at a certain level of imperfection the system undergoes a seemingly discontinuous change in qualitative behavior. Even if a lot of non-linear analyses were carried out varying the imperfection modes and levels, the results would be of little direct benefit if another geometry or member arrangement was to be considered. An alternative is to examine the relationship between the linear bifurcation and nonlinear analyses.

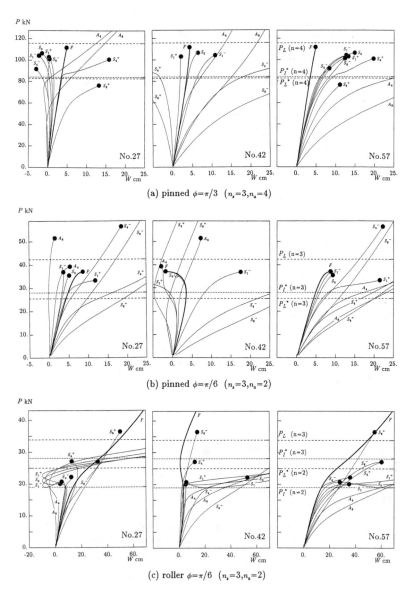

(a) pinned $\phi=\pi/3$ $(n_s=3, n_a=4)$

(b) pinned $\phi=\pi/6$ $(n_s=3, n_a=2)$

(c) roller $\phi=\pi/6$ $(n_s=3, n_a=2)$

Fig.3 Load versus vertical deflection curves

(a) pinned $\phi=\pi/3$

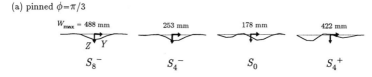

S_8^- S_4^- S_0 S_4^+

(b) pinned $\phi=\pi/6$

A_8 S_1^- S_0 S_1^+

(c) roller $\phi=\pi/6$

S_8^- S_4^- S_0 S_4^+

Fig.4 Total vertical deflection distribution at the buckling point

(a) pinned $\phi=\pi/3$

S_8^- S_4^- S_0 S_4^+

(b) pinned $\phi=\pi/6$

A_8 S_1^- S_0 S_1^+

(c) roller $\phi=\pi/6$

S_8^- S_4^- S_0 S_4^+

Fig.5 Incremental vertical deflection mode at the buckling point

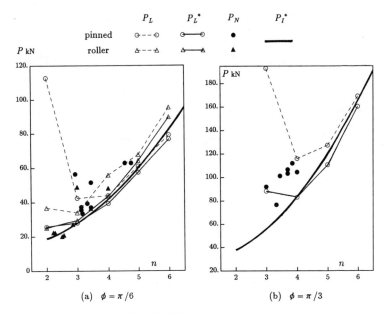

$$(a) \quad \phi = \pi/6 \qquad\qquad (b) \quad \phi = \pi/3$$

Fig.6 Buckling load spectra

The geometric, nonlinear, stiffness matrix $\mathbf{K_G}$ associated with internal axial force vector $\mathbf{N_L}$ has been preparedly computed from the linear bending analysis. Through a stationarity of the total potential energy to represent incremental displacements about this linear bending path, it is possible to derive the linear bifurcation buckling loads P_L as eigenvalues and the associated buckling modes $\Delta\mathbf{u_L}$ as eigenmodes. Consequently, the quadratic form of the change of total potential energy for each buckling mode is obtained as

$$\Pi_2 = \frac{1}{2}\sum(\Delta\mathbf{u_L})^T\mathbf{K_L}\Delta\mathbf{u_L} + P_L\frac{1}{2}\sum(\Delta\mathbf{u_L})^T\mathbf{K_G}\Delta\mathbf{u_L} = 0 \qquad (4)$$

where the superscript T is a transposition operator, and $\mathbf{K_L}$ represents the linear stiffness matrix with respect to the global Cartesian coordinate system (X, Y, Z).

A simple means of predicting reduced stiffness critical loads is to assume that the incremental out-of-plane deformation is largely unchanged as compared with the linear bifurcation buckling mode. Because the incremental bending energy remains unchanged, a reduced stiffness eigenvalue analysis would therefore predict a reduced stiffness critical load as

$$P_L^{*} = P_L \frac{\sum(\Delta\mathbf{u_L})^T\mathbf{K_L^b}\Delta\mathbf{u_L}}{\sum(\Delta\mathbf{u_L})^T\mathbf{K_L}\Delta\mathbf{u_L}} \qquad (5)$$

where $\mathbf{K}_L^b$ expresses the bending, so-called quasi-inextensional, stiffness matrix in which the linear membrane stiffness in each member is eliminated.

The analytical results of the linear bifurcation buckling loads P_L and the reduced stiffness critical loads P_L^* are depicted in Figs.3 and 6. The linear bifurcation buckling analysis provides an unreliable upper bound to the scatter of the buckling loads obtained by the nonlinear numerical experiments P_N. It is also shown in Fig.6 that P_L^* associated with the smaller n, for the roller support model, gives a non-conservative estimation of the lower bound: to examine its improvement consider the following idealized shell model.

<u>Lower Bound Analysis for Idealized Orthotropic Continuum Shells</u>

In the pressure buckling problem of a partial cylindrical shell (Yamada and Croll 1989/1993) or a shallow spherical shell (Yamada et al. 1983; Goncalvas and Croll 1992), which is much similar to the present overall buckling problem, an intrinsic loading imperfection affects the nonlinear buckling behavior. Consequently the buckling mode has at times different lobe rather than that of the linear buckling estimation. It has been suggested in Yamada (1991) that the application of a prebuckling uniform idealized shell model is very convenient to break such a difficult problem. For non-zero buckling mode

$$w = w_{mn} \sin(m\pi x/L) \sin[n\pi y/(R\phi)] \tag{6}$$

in the orthogonal curvilinear coordinate system (x,y,z) as shown in Fig.1, the classical eigenvalue equations of a Donnell-type give the following reduced stiffness critical load P_I^* (Yamada 1991)

$$P_I^* = \left[\frac{\sqrt{3}l^2}{2}\right]\left[\frac{\pi^2 D_y}{RL^2}\right]\frac{B^4 + 2(D_2 + 2D_3)B^2 m^2 + D_1 m^4}{B^2} \tag{7}$$

For the present isosceles triangle latticed vault, the coefficients in Eq.7 are

$$D_y = 49\sqrt{3}EI/(52l) \qquad D_1 = 1 \qquad D_2 = 0.061 \qquad D_3 = 0.469$$

$$B = nL/(R\phi) = 0.693n \qquad\qquad m = 1$$

where $I \equiv$ centroidal moment of inertia of the member cross section. By substituting these into Eq.7 and by using the following hinged-member buckling load predicted from a simple Euler column analysis based upon the uniform axial force N_L^d (Eq.1)

$$P_E = \frac{3l}{2R}(-N_L^d)_{Euler} = \frac{3\pi^2 EI}{2Rl} \tag{8}$$

we obtain

$$P_I^* = P_E \frac{49}{52}\left[\frac{l}{L}\right]^2 \frac{0.480n^4 + 2n^2 + 2.08}{n^2} \tag{9}$$

Superimposed upon Figs.3 and 6 are the present idealized reduced stiffness predictions P_I^* which are governed by the geometric parameter, l/L, and by the load-carrying capacity of diagonal member, P_E. As illustrated in Fig.6 the idealized reduced stiffness model can be seen to yield a reliable lower bound to the scatter of P_N.

Conclusions

Through a selected parametric study the complex nature of the imperfection sensitive buckling of single layer latticed barrel vault roof structures has been demonstrated. It has been shown that the elastic buckling load-carrying capacity varies strongly with geometries: the boundary condition affects the dominant circumferential buckling mode. Also it has been suggested that the present idealized reduced stiffness approach would provide a safe, simple realistic basis to the elastic overall buckling design of latticed vaults containing moderate and practically relevant levels of geometric imperfection.

Acknowledgements

The authors wish to thank Miss Noriko Naruse, undergraduate student of Toyohashi University of Technology, for her kind assistance.

Appendix. References

1. Goncalvas, P.B., and Croll, J.G.A. (1992). "Axisymmetric buckling of pressure-loaded spherical caps." *Journal of Engineering Mechanics*, A.S.C.E., Vol.118, 970–985.

2. Sumec, J. (1990). *Regular lattice plates and shells*, Elsevier, 298–302.

3. Yamada, M., Uchiyama, K., Yamada, S., and Ishikawa, T. (1986). "Theoretical and experimental study on the buckling of rigidly jointed spherical shells under external pressure." *Proc. IASS Symp.*, Osaka, Vol.3, 113–120.

4. Yamada, S., Uchiyama, K., and Yamada, M. (1983). "Experimental investigation of the buckling of shallow spherical shells." *Int. Journal of Non–Linear Mechanics*, Vol.18, 37–54.

5. Yamada, S., and Croll, J.G.A. (1989). "Buckling behavior of pressure loaded cylindrical panels." *Journal of Engineering Mechanics*, A.S.C.E., Vol.115, 327–344.

6. Yamada, S. (1991). "Relationship between non-linear numerical experiments and a linear lower bound analysis using finite element method on the overall buckling of reticular partial cylindrical space frames." *Computer Applications in Civil and Building Engineering, Proc. 4th I.C.C.C.C.B.E.*, Kozo System Inc., Tokyo, 259–266.

7. Yamada, S. and Croll, J.G.A. (1993). "Buckling and postbuckling characteristics of pressure loaded cylinders." *Journal of Applied Mechanics*, A.S.M.E., Vol.60, 290–299.

8. Yamada, S., Uchiyama, K., and Croll, J.G.A. (1993). "Theoretical and experimental correlations of the buckling of partial cylindrical shells." *Proc. SEIKEN–IASS Symp.*, Tokyo, 151–158.

Collapse Behavior of Double-Layer Space Truss Grids with Member Buckling

Toshitsugu Saka[1] and Yoshiya Taniguchi[2]

Abstract

The results of collapse tests of over twenty space truss grids with elastic buckling of members are reviewed. The space truss grids include three types, that is, square-on-square, square-on-diagonal and diagonal-on-square double-layer grids which have a 1320 x 1320 mm square plan. Brass tubular members, ball joints and a bolted jointing system are used. The individual collapse behavior of each type is described, so that designers grasp the differences and the structural characteristics of these grids.

Introduction

Double-layer space truss grids often collapse by buckling of members, and the collapse behavior of grids may differs from continuum plates such as reinforce concrete slabs, in viewing them as the plates macroscopically. In designing a grid, it is important for designers to grasp collapse behavior concerned with collapse modes and load-deflection relationships, etc.

In this paper, individual collapse behaviors of space truss grids of the three typical types, as shown in Fig. 1, are described by comparing the each result of model tests and the theoretical results which were carried out by the present authors. The initial elastic buckling loads, post-buckling behavior and collapse modes are particularly focused on. Moreover, the analytical methods to predict a load carrying capacity of each type are introduced, and the comparisons between the analytical results and the test results are performed.

[1] Professor, Faculty of Engineering, Osaka City University, Sugimoto 3-3-138, Sumiyoshi-ku, Osaka 558, JAPAN
[2] Research Associate, Faculty of Engineering, Osaka City University, Sugimoto 3-3-138, Sumiyoshi-ku, Osaka 558, JAPAN

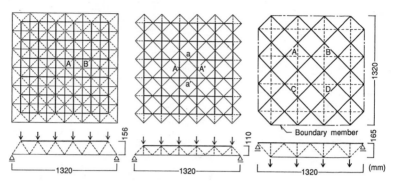

Square-on-Square Square-on-Diagonal Diagonal-on-Square

Fig. 1 Space Truss Grids

Truss Grid and Testing Method

The details of truss grids tested are shown in Table 1. The side length of grids with a square plan is 1320 mm. There are two types which consist of 4 × 4 and 6 × 6 structural units, respectively. The length of upper chord members is 220 mm or 330 mm. That of diagonal-on-square grids is 233 mm. The bolted jointing system shown in Fig.2 is used. The mechanical properties of the jointing system are given in Table 2.

The truss grids are supported at the all perimeter nodes by roller bearings and subjected to a uniform vertical load at the all nodes of the upper layer by a tournament system, as shown in Fig. 3. Applied loads were measured with a load cell, and deflections were done with displacement meters of a strain gage type transducer.

Table 1 Details of Truss Grids

Truss type	Number of truss grid	Lattice pattern	Number of structural unit	Depth of truss grid (mm)	Boundary member	Slenderness ratio of upper member
SS4	3	Square-on-square	4 × 4	233	–	205
SS6	3		6 × 6	156	–	137
SD4	3		4 × 4	165	No	205
SD4EB	3	Square-on-diagonal	4 × 4	165	Yes (Brass tubular)	205
SD6	3		6 × 6	110	No	137
DS4	2		4 × 4	165	No	145
DS4EB	2	Diagonal-on-square	4 × 4	165	Yes (Brass tubular)	145
DS4ES	2		4 × 4	165	Yes (Steel solid rod)	145

Length of span = 1320 mm; Diameter of steel solid rod = 10 mm

Table 2 Mechanical Property of the Bolted Jointing System

Brass tubular member		Joint
Extensional rigidity EA	1.11×10^6 N	Length of rigid end part λ_j 11.9 mm
Flexural rigidity EI	2.87×10^6 Nmm^2	Rotational connecting rigidity C_j 9.22×10^4 Nmm
Yield bending moment M_p	6.37×10^3 Nmm	Yield bending moment of the connection M_{jp}
Yield tensile strength N_p	5.1×10^3 N	1.77×10^3 Nmm

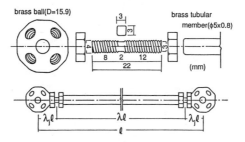

Fig. 2 Bolted Jointing System

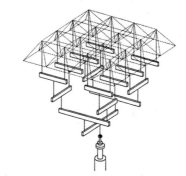

Fig. 3 Tournament Loading System

Analytical Method

The constituent member constructed by the bolted jointing system is idealized to the member model shown in Fig. 4. The rotational spring constant represents the connecting rigidity between a member and a ball joint, and the rigid end part corresponds to a ball joint.

For an elastic buckling analysis, the slope-deflection equation for members under axial forces is used (Saka and Heki 1984).

For post-buckling behavior, a simplified nonlinear analysis which can take into account the effect of sizes and connecting rigidities of joints and initial lateral deflections of compressive members are used (Saka 1989; Saka and Taniguchi 1991). The nonlinear characteristics of buckling members are theoretically derived from the effective strength of grids and a plastic hinge model. The nonlinear analysis is based on a limit state analysis, which is useful in the case of space trusses with a great number of members (Supple and Collins 1981).

In this paper, the analytical methods to predict a load carrying capacity are presented for each grid.

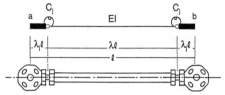

Fig. 4 Member Model

Square-on-Square

The load-deflection curves of SS4 and SS6 grids, at the central node, are shown in Figs. 5 & 6, respectively (Saka and Heki 1984). In the figures, the broken lines are the stiffness of an elastic analysis. The elastic buckling loads by the analysis and load carrying capacities of model tests are given in Table 3.

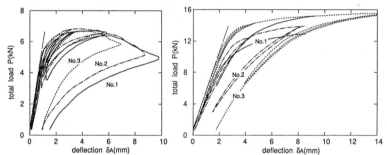

Fig. 5 Experimental Results   Fig. 6 Experimental Results
         for SS4 Grid   for SS6 Grid

The results of the simplified nonlinear analysis, presented by the author, are shown in Fig. 7 (Saka 1989). This method can take into account the stiffness, forces and configuration of the adjoining members, that is, the stress-redistribution after member buckling is considered. This is different from a common simplified nonlinear analysis presented by some authors, for example, Schmidt (1976), Smith (1984) or Hanaor (1985).

In Fig. 7, Model(I) & (II) represent the two kinds of

piece-wise linearization in the unstable range of member characteristics. The load carrying capacities by the present nonlinear analysis are given in Table 3.

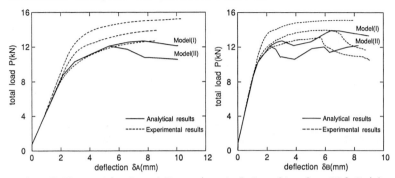

Fig. 7 Theoretical and Experimental Results for SS6 Grid

Table 3 Elastic Buckling Load and Load Carrying Capacity

Truss type	SS4			SS6		
	No.1	No.2	No.3	No.1	No.2	No.3
Load carrying capacity (kN)	6.68	6.83	6.72	13.16	13.95	15.13
	6.74(1±0.01)			14.08(1±0.06)		
Theoretical elastic buckling load (kN)	5.16 (77%)			10.23 (73%)		
Theoretical load carrying capacity (kN)	−			13.90* (99%)		

* Not considering initial lateral deflections of upper compressive members

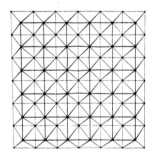

Fig. 8 Residual Deformation for SS6 Grid No.3

In the load-deflection curves of both the theoretical and experimental results, the load bearing capacity increased from the initial buckling load to the maximum load is large, and the plateau is seen. The post-buckling behavior is, therefore, very ductile. The collapse mode was a cross line pattern, as shown in Fig. 8. As for square-on-square grids, a collapse pattern is usually a cross line type in macroscopic point of view, even if there are initial imperfections in members.

The collapse modes and load-deflection curves obtained by the nonlinear analysis are in good agreement with the experimental results.

Square-on-Diagonal

The load-deflection curves at the central nodes A and A' are shown in Fig. 9 for SD4 grids and in Fig. 10 for SD6 grids, respectively (Saka and Heki 1986). The results of the simplified nonlinear analysis, which considered the effect of the initial lateral deflections of compressive members, for SD6 grid are shown in Fig. 11 (Saka and Taniguchi 1991). The elastic buckling loads and load carrying capacities are given in Table 4.

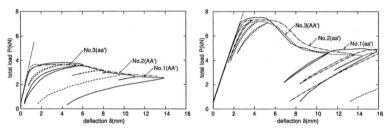

Fig. 9 Experimental Results   Fig. 10 Experimental Results
      for SD4 Grid   for SD6 Grid

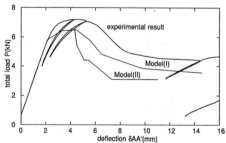

Fig. 11 Theoretical and Experimental Results
for SD6 Grid No.3

In the load-deflection curves, the load bearing capacity does not remarkably increase after the initial buckling, limited plateau is seen. Therefore, square-on-diagonal grids have limited ductility, and the elastic buckling load is considered to be a load carrying capacity.

Table 4 Elastic Buckling Load and Load Carrying Capacity

Truss type	SD4			SD4EB			SD6		
Load carrying capacity (kN)	No.1	No.2	No.3	No.1	No.2	No.3	No.1	No.2	No.3
	3.68	3.66	3.66	3.74	3.70	3.69	7.52	7.31	7.21
	3.67(1±0.003)			3.71(1±0.007)			7.35(1±0.022)		
Theoretical elastic buckling load (kN)	3.43 (94%)			3.60 (97%)			7.05 (96%)		
Theoretical load carrying capacity (kN)	–			–			6.61 (90%)	6.64 (91%)	6.60 (92%)

The collapse modes are a straight line pattern as shown in Fig. 12, which is different from square-on-square grids. However, if square-on-diagonal grids have no imperfection, the collapse mode become a cross line pattern, that is found by the nonlinear analysis.

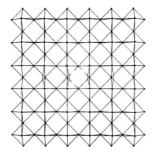

Fig. 12 Residual Deformation for SD6 Grid No.3

The theoretical maximum load is about 90% of the experimental results, however the nonlinear characteristics of post-buckling behavior are in good agreement with the experimental results. The present nonlinear analysis is enough for designers to predict the practical post-buckling behavior of the grid. The load carrying capacity by the analysis would be close to the experimental results if the number of structural units would be larger.

Diagonal-on-Square

The load-deflection curves of DS4, DS4ES grids are shown in Figs. 13 & 14, respectively (Taniguchi and Saka 1991b). The elastic buckling loads by the analysis and load carrying capacities of model tests are given in Table 5.
Since the diagonal-on-square grid is unstable in the pin-jointed case, the simplified nonlinear analysis can not be used, and a method considering materially and geometrically nonlinearity for the member model should be utilized (Taniguchi et al. 1993). However, from the

experimental results, a load carrying capacity can be
estimated as an elastic buckling load.

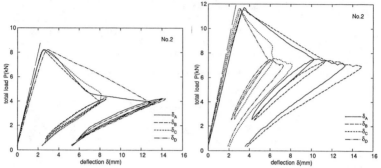

Fig. 13 Experimental Results   Fig. 14 Experimental Results
        for DS4 Grid   for DS4ES Grid

Table 5 Elastic buckling load and load carrying capacity

Truss type	DS4		DS4EB		DS4ES	
Load carrying capacity (kN)	No.1	No.2	No.1	No.2	No.1	No.2
	7.18	8.19	9.46	10.08	10.77	11.55
	7.69(1±0.066)		9.77(1±0.032)		11.16(1±0.035)	
Theoretical elastic buckling load (kN)	7.24 (94%)		10.52 (108%)		13.60 (122%)	

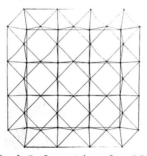

Fig. 15 Residual Deformation for DS4ES Grid No.2

        In the load-deflection curves by tests, the remarkable
decrease of load bearing capacity is seen after initial
buckling, and any plateau can not be found. The initial
buckling load is, therefore, considered to be a load-
carrying capacity, and the deformation capacity in the
ultimate load state is not expected. For diagonal-on-square
grids, the initial stiffness by tests is remarkably smaller
than the one by an elastic analysis of perfect grids. This
is probably because the grids possess the unstable mode due

to the relative rotation of pyramidal units.
The residual deformation is shown in Fig. 15. The two
members which were adjacent to the center ones showed large
buckling deformations, and the another two members near by
the edge also did. A clear line of collapse pattern was not
seen.

Summary and Conclusions

21 space truss grids has been tested, varying the
lattice pattern and the number of structural units. The
individual collapse behavior of each grid was clearly shown
experimentally and theoretically. As the theoretical
estimation, the elastic buckling loads were calculated by a
conventional matrix method using the member model proposed
by the first author. Both the load carrying capacities and
post-buckling behavior were predicted by the simplified
nonlinear analysis, for square-on-square and square-on-
diagonal grids. The present method can take into account the
effect of sizes and connecting rigidities of joints and
initial lateral deflections of compressive members. Within
the limits of the present tests and theoretical results, the
conclusions are as follows:
For square-on-square double-layer grids, the load
carrying capacity is higher than the initial buckling load
and the defamation capacity in the ultimate state is large,
and the collapse mode is a cross line pattern, even if there
are initial imperfections in members. The simplified
nonlinear analysis was found to predict the load carrying
capacity and the post-buckling behavior of the grid.
For square-on-diagonal double-layer grids, the load
carrying capacity is slightly higher than the initial
buckling load. The elastic buckling load is considered to be
a load carrying capacity of the grid. There is limited
ductility in the load deflection relationship. In the case
that the constituent members possess initial imperfections,
the collapse pattern is a straight line type, which is
different from square-on-square grids. The collapse mode and
load carrying capacity can be predicted by the simplified
nonlinear analysis.
For diagonal-on-square double-layer grids, the load
bearing capacity remarkably decreased after initial
buckling. Therefore, the elastic buckling load is considered
to be a load carrying capacity of the grid. The boundary
members restrain the unstable mode due to the relative
rotation of pyramidal units and increase the buckling load.
In comparing the experimental results of SS6, SD6 and
DS4 grids, which have the slenderness ratio of the similar
value for upper compressive members, the post-buckling
behavior and the collapse mode of each type are very different from each other. The collapse behavior is found to be
dependent upon the lattice pattern of each grid.

Acknowledgements

    The authors wish to thank Prof. K. Heki for his valu-
able guidance and advice. Thanks are also necessary for
Mr. K. Kanata, Mr. H. Tsuji and Mr. Y. Sakamoto for their
valuable assistance during the experimental work.

References

Hanaor,A. (1985). "Analysis of double layer grids with
    material nonlinearities - a practical approach." Space
    Structures 1, Vol.1, Elsevier Applied Science Publishers
    LTD, England, 33-40.
Saka,T., and Heki,K. (1984). "The effect of joints on the
    strength of space trusses." Third International Conference
    on Space Structures, H.Nooshin, ed., Elsevier Applied
    Science Publishers LTD, London, 417-422.
Saka,T., and Heki,K. (1986). "The load carrying capacity of
    inclined square mesh grids constructed by a bolted
    jointing system." Proceeding of the IASS Symposium, Vol.3,
    K. Heki, ed., Elsevier Applied Science Publishers LTD,
    Tokyo, Japan, 89-96.
Saka,T. (1989). "Approximate analysis method for post-
    buckling behavior of double-layer space grids constructed
    by a bolted jointing system." 10 Years of Progress in Shell
    and Spatial Structures, Vol.4, F.del Pozo, and
    A.de las Gasas, eds., Madrid.
Saka,T., and Taniguchi,Y. (1991). "Post-buckling behavior of
    square-and-diagonal double-layer grid." Proceeding of the
    IASS Symposium, Vol.3, Kunstakademiets Forlag
    Arkitektskolen, Denmark, 199-206.
Saka,T., and Taniguchi,Y. (1991b). "On the buckling strength
    of diagonal-on-square double-layer grid constructed by a
    bolted jointing system." Summaries of Technical Papers of
    Annual Meeting of Architectural Institute of Japan, 1257-
    1258 (in Japanese).
Schmidt,L.C., Morgan,P.R., and Clarkson,J.A. (1976). "Space
    trusses with brittle-type strut buckling." Journal of the
    Structural Division, ASCE, Vol.102, 1479-1492.
Smith,E.A. (1984). "Space truss nonlinear analysis." Journal
    of the Structural Engineering, ASCE, Vol.110, 688-705.
Supple,W.J., and Collins,I. (1981). "Limit state analysis of
    double-layer grids." Analysis, Design and Construction of
    Double-layer Grids, Z.S.Makowski, ed., Applied Science
    Publishers LTD, London, 93-117.
Taniguchi,Y., Saka,T., and Shuku,Y. (1993). "Buckling
    behaviour of space trusses constructed by a bolted
    jointing system." The Fourth International Conference on
    Space Structures, Space Structures 4, Vol.1, Thomas
    Telford, London, 89-98.

Ultimate Loading Capacity of Braced Domes

X. Chen,[1] N. Wang[2] and S.Z. Shen[3]

Abstract

The stability behaviors of a practical latticed dome with 40 metres span are sutdied in this paper. The geometric and material nonlinear behaviors, as well as the effects of the initial geometric imperfections, are cosidered in the analysis. Some valuable results for the strucutral design are obtained.

Introduction

A braced dome strucutre with 40 metres span was designed for the roof of a car test room. The authors were responsible for the nonlinear stability analysis of this strucuture. Two domes with different net systems, the lamella dome and the Kiewitt dome (parallel lamella dome), were considered for comparision. It was said that from the consideration of stiffness and strength, Kiewitt dome is better than lamella dome (Makowski 1984); but from the point of view of aesthetic feelings, the designers prefer the later one. In this paper, a finite element computer program compiled by the authers is used to analyze the nonlinear behaviors of these two structures. The complete load – deflection responses with

[1]Assoc. Prof. of Civ. Engrg., Harbin Archit. & Civ. Engrg. Inst. , Harbin. China.
[2]Lecturer of Civ. Engrg., Harbin Inst, of Tech., Harbin, China.
[3]Prof. of Civ. Engrg., Harbin Archit. & Civ. Engrg. Inst., Harbin, China.

consideration of initial imperfections and material nonlinearity are studied in detail. Some suggestions for the structural design are made.

Elastic analysis

The geometry of the two structures analyzed which have a span of 40 metres and a height of 12.192 metres are shown in Figs 1 and 2 , respectively . The members are from steel tubes, rigidly

Fig. 1. Lamella Dome.

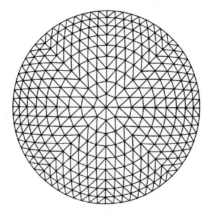

Fig. 2. Kiewitt Dome.

connected to each other by hollow welding ball joints, and the boundary nodes are rigidly surported on a circular reinforced concrete beem. The vertical dead load is exerted on the nodes. Three Kinds of tubes are selected, for the lamella dome, accouding to the force distribution in members: $d = 133mm$, $t = 4mm$ for the two outside radial elements; $d = 76mm$, $t = 3mm$ for the four outside ring elements; and $d = 102mm$, $t = 3.5mm$ for the rest. In this case, the total weight of the steel tubes is $16.5kg/m^2$. For fair comparision, the steel weight of the Kiewitt dome must be as the same as the lamella one. For this purpose, all elements of the Kiewitt dome are selected as $d = 102mm$ and $t = 3.5mm$, providing a total steel weight of the dome of $16.7kg/m^2$.

In stability analysis of latticed dome with initial geometric imperfections, two methods were suggested by the authors (Chen and Shen 1994). The first method is called "random imperfection mode method". In this method, the install errors of each node is considered to comform with the normal distribution over the interval of permitted errors, therefore, the install error of the whole structure is a muti−dimensional random variable, and each sample piece of the sample space corresponds to a kind of imperfection mode of the structure. A sample with a volume of N is taken for analysis, i.e., N imperfection modes are taken at random to trace the load−deflection path for each of them, from which some numerical charactoristics of the sample can be obtained to evaluate the stability behavior of the imperfect structure. This method requires N times of load−deflection analysis and, hence, needs a big consumption of computer time. Then the second method as called "conformable imperfection mode method" was suggested. It is imaginable that the buckling load would depend not only on the size of imperfections but also, in more degree, on the patern of imperfection distribution. We know that the buckling mode of a perfect strucutre represents the deflection tendency, i.e. a potential deflection mode of the structure at buckling point. The real (imperfect) structures will most likely deflect along the lowest buckling mode from initial loading stage; therefore, it would be most disadvantageous if the imperfection distribution of the strucuture is in conformity with the buckling mode. According to this principle, we take the lowest buckling mode as the initial structural imperfection, and the possible smallest buckling load could be obtained by one calculation.

In this paper, a finite element computer program compiled by the authors based on precise theory of element stiffness (Chen

1990) and arc – length method (Crisfield 1983) for nonlinear analysis is used to analyze the complete load – deflection response and conformable imperfection mode method mentioned above is adopted for imperfection analysis. The load – deflection curves and the buckling mode obtained for the two domes are shown in Figs 3 and 4, respectively, in which, ±R denotes the permitted install error of each node. The buckling mode corresponding to the first

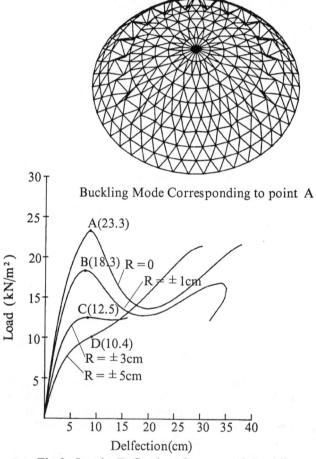

Fig.3. Load – Deflection Curves and Buckling Mode of the Lamella Dome.

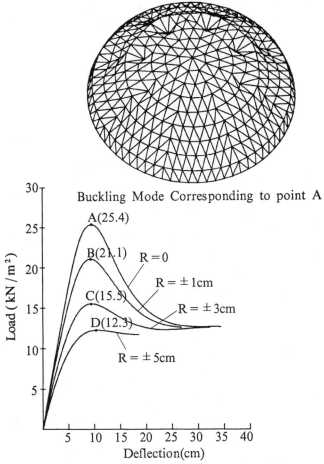

Fig.4. Load – Deflection Curves and Buckling
Mode of the Kiewitt Dome.

critical point (point A) of the perfect structures is used to model
the structural initial imperfections. The maximum value of this
mode is ±R.

It can be seen that the buckling modes of two strctures are all
symmetric. The critical points in perfect and imperfect conditions for
lamella dome A, B, C and D, correspond to critical loads

23.3kN/m², 18.3kN/m², 12.5kN/m² and 10.4kN/m², respectively. It's pointed that, when R = ±5cm, the first limit point becomes a inflection point (point D), which can be regarded as a critical point. The comparable critical loads of the Kiewitt dome are 25.4kN/m², 21.1kN/m², 15.5kN/m² and 12.3kN/m², respectively. It can be seen that the critical loads of the Kiewitt dome are about 2 ~ 3kN/m² higher than that of the lamella one. However, for this design, the stability capacity of the lamella dome was considered to be good enough, and from a aesthetic consideration, the lamella dome was finally selected.

In the above analysis, only geometric nonlinear behaviors of the structure are considered. For the deep dome as studied, part of the sections of some members have been found to have some plasticity developing before reaching critical point. Therefore, it is thought necessary to make further analysis, taking material nonlinearity into account. Elastic — plastic nonlinear behaviors of the selected lamella dome will be discussed in the next part.

Elastic — Plastic Analysis

Finite segment finite element method (Meek and Loganathan 1989; Wang et al. 1993) is used for the elastic — plastic analysis of the lamella dome. The yield stress of the steel tube elements is $f_y = 235N/mm^2$. The initial imperfection mode is the same as that of elastic analysis shown in Fig.3. The complete load — defletction curves obtained for perfect and imperfect structure are shown in Fig.5, in which, point A to point D are the critical points of the structure with different imperfection value, the corresponding critical load are 12.2kN/m², 11.9kN/m², 10.0kN/m² and 7.7kN/m² respectively. All the curves in Fig.5 are corresponding to the node with largest deflection.

It can be seen that the smallest critical load of the dome structure analyzed, 7.7kN/m², is obtained when the install error equals ±5cm and material nonlinearity is taken into account. On the other hand, the design dead load of the structure is 1.9kN/m². Compare it with the critical load obtained above, we can get the conclusion that when install error of each node of the structure is not more than ±5cm, the stability safety factor is at least 4.

Conclusions

1. It is quite necessary to make a complete load — deflection

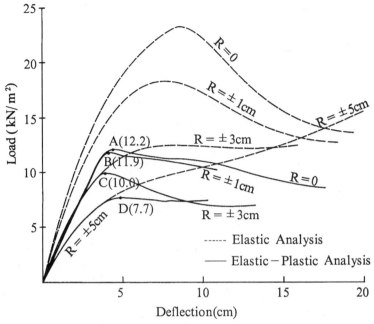

Fig.5. Load – Deflection Curves of Lamella Dome

response analysis if we want to get a overall understanding of stability behaviors of the structure. The classical method of predicting stability load – capacity by dividing the "buckling load" obtained from linear analysis by an empirical safety factor seems to be not accurate, and would sometimes lead to a totally wrong conclusion due to the different geometrical characters of different structures.

2. Plasticity may develop in part sections of some elements of the structure before the first critical point is reached. As the results, the critical load from elastic – plastic analysis is lower than that of elastic analysis. Therefore, the material nonlinearity must be taken into account in the stability analysis of latticed dome.

3. Initial geometric imperfection is an important factor in the stability analysis of latticed dome. Install errors can greatly reduce the stability capacity of the structure. Effective measures must be taken to ensure the accuracy of the installation.

4. The theoretical stability safety factor of the lamella dome analyzed in this paper is at least 4 if install error of each node is

Chen et al.

not more than ±5cm.

5. Stability capacity of Kiewitt dome is higher than that of lamella dome. Therefore, Kiewitt dome will get more economic advantage for larger span structures.

Appendix. References

Chen, X.(1990). "The complete load–deflection analysis of latticed space structures and the study on nonlinear behaviors of latticed shells of negative Gaussian curvature." Ph. D. thesis, Harbin Archit. & Civ. Engrg. Institute.

Chen, X. and Shen, S. Z. (1994). "Complete load–deflection response and initial imperfection analysis of single–layer latticed dome." Int. J. Space Structures.

Crisfield, M. A. (1983). "An arc–length method including line searches and accelerations." Int. J. Num. Meth. Engrg., 19, 1269–1289

Makowski, Z. S. (1984). "Analysis, design and construction of braced domes." First published in Great Britain.

Meek, J. L. and Loganathan, S. (1989). "Theoretical and experimental investigation of a shallow geodesic dome." Int. J. Space Structures. 4(2), 89–105

Wang, N., Chen, X and Shen, S.Z.(1993). "Geometric and material nonlinear analysis of latticed shell." China Civ. Engrg. J., 26(2).

Membrane Structures : Technological Innovation and
Architectural Opportunity

Professor Edmund Happold[1]

Perhaps the most powerful description of civil
engineering was by Thomas Tredgold in 1828 when
supporting the foundation of the Institution of Civil
Engineers he wrote "..... the profession of civil
engineering being the art of directing the great sources
of power in nature for the use and convenience of man".
Civil engineering constructs a new nature, a supernature,
between mankind and original nature.

And to do that it learns from nature, the great
optimiser. Failures become redundant.

When looking one day at why the human cell divides I was
shown the hogweed[1], a plant that visibly grows and

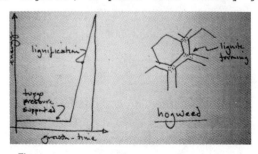

Figure 1.

achieves a great height in one year. For most of that
height it is supported by a turgo system, liquid pressure

[1]Professor Edmund Happold, Senior Partner, Buro Happold,
Camden Mill, Lower Bristol Road, Bath BA2 3DQ

keeping up a soft cellular form and using little, but continuous and quantifiable, energy input. Then the wind suddenly gives it a shake and this time for a quite large quantifiable energy input it produces stiff fibres, lignification, at the junctions of the cells thus converting the soft turgo system requiring low continuous energy to a stiff system requiring no further energy. A trigger in its DNA spiral. Much the same seems to have developed in the tension structures field – and no review should ignore the analogy.

The weight of structural material used in an air house is but a small proportion of that used in a tent and, as the span and volume of the building increases, that difference increases exponentially. And though their first design was by Lanchester, the pioneer of flying in Britain in 1918 and there were unsung but very imaginative developments in tents supported by pneumatic tubes by Kaneshiga Nohmura in Japan in 1929, it was really Walter Bird of Birdair who developed the modern airhouse in the 1950s in the USA, first with radomes, then with a whole range of small commercial buildings and later, when ptfe coated fibre glass was invented, over a series of very big sports stadia. It is important to mention materials fairly early in this paper because recent developments have not been very significant. I show a list from an article on tensile structural materials drawn up by Tony Read and Turlogh O'Brien 13 years ago because it basically shows what we still have today[2].

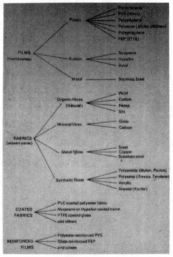

Figure 2.

Getting a body of knowledge organised in the airhouse field was slow to happen. Academics, who develop most theory, had trouble in moving from solutions to rigid shell theory carried out in the early 70s to solutions where deformation had to be resisted by internal pressures. I think there is no doubt that the start of the solution of the problem, the concentration on developing methods, came when money was found to send a young engineer round to enquire and see how existing airhouses were behaving and what were the causes of sudden or long term failure[3]. That defined clearly that catastrophic failure usually came from cuts in the

Figure 3.

fabric, either from being struck by an object or from striking an obstruction during deflection in a gale, and long term usually either from excessive bending of the fabric leading to cracking in the coating and biological attack on the base fabric or a failure of the stitching or welding between the fabric panels. A very broad range of work was carried out into tear propagation, the development of form and patterning in relation to deflection, design loads and load analysis, detailing, the environment inside and so on by a broad group of

researchers, drawing on work by others elsewhere in the
world but the whole regularly being subjected to review
and criticism from Wally Bird and being also presented
regularly to anyone in the industry who wanted to
attend.[4]

The lessons from this, and subsequent study, are clear.
The form of an air house structure influences the wind
pressures on the surface and the distribution of snow on
the surface which determines the peak fabric stresses.
Airhouses on the ground will experience positive wind
pressures, constructed on a wall or beam the frequency
and magnitude of positive pressure can be minimised and
the internal pressure required for stability can be
reduced.   The form of an airhouse is determined by the
patterning of the cloths so the method of defining the
patterning is rather important.

.   At one level, patterning is empirical, the cloths
      are cut to geometrical rules developed by trial and
      error or by cutting paper over stiff models by the
      fabricator.

.   a computer programme fits panels to a geometrically
      defined surface made up of sections of spheres and
      cones.

.   or using programmes which carry out both load
      analysis and equilibrium patterning.

The process is necessarily non-linear; large deflections
must be taken into account. The material properties used
should reflect the stiffness of the fabric in the warp
and fill directions with low shear stiffness.   The
analysis is usually run with quasistatic wind loads taken
from Code-type data. Because of the low fabric shear
stiffness, the warp or fill stresses vary only a little
along a yarn line. This means that a gust averaging time
appropriate to the width of the fabric field can be used.
The results of the load analysis give fabric stresses and
anchorage loads. Normally, the analysis is run with
characteristic loads, and an essentially permissible
stress approach is adopted. This interacting with the
development of the fan of the airhouse gives the best
control of fabric stresses. The cloth seam lines are
included as part of the process. The programme in our
practice is of this type[5].

Figure 4.   Stress

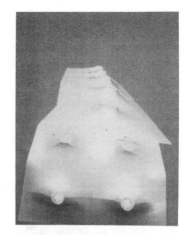

Figure 5.   Form

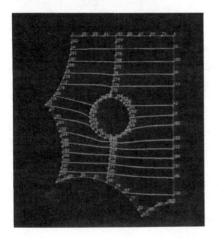

Figure 6.   Patterning

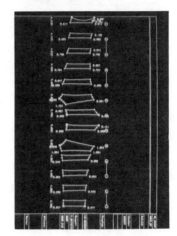

Figure 7.   Cutting Patterns

The fabric used has a length of cut it will allow to propagate and a factor of safety has to be provided to this. The basic strength of fabric is defined by a 50mm strip tensile strength, a 40mm cut will propagate at about 25% of the strip tensile strength so it is necessary to have a factory of safety of 6 to have a reliable product. A conscious choice is useful to the designer.

Detailing of boundaries, connections and foundations is important and has to be well understood, as stress concentrations or gaps can become tears, and must be reduced or reinforced with multilayer patches. When the main fabric meets door frames or cable ends smoothly connecting flexible balloon pieces are required.

There are plenty of safe and reliable airhouses. Our practice has not had any problems. But there have been, in many countries, an unacceptably large number of failures and such failures tend to be highly dramatic as one day there is a huge building and the next, because there is so little material, there is a very noticeable gap. It is important to examine why.

The major problem lies with the cheaper and more common end of the industry: tennis halls, swimming pool covers and the like. In the USA there are an amazing number of failures. A recent meeting was told that of over 200 structures some 70% were compromised as regards performance. The number of failures in Europe due to the gales of recent years have caused catastrophic failure to over 300 in Germany, who have founded a review board and a comparable number in Britain, where the Lawn Tennis Association has also mounted a thorough review. Basically the reasons are very simple.

.   Airhouses are sold by a wide range of firms and buildings of very low cost. Any fabricator who tries to set reasonable standards is immediately undercut.

.   The air houses are sold as engineered when they often have no engineering input. A tarpaulin cover manufacturer can get cutting patterns free from Hoechst with the fabric, buy in doors and ready packaged inflation and emergency fan units - and instal and start running.

.   There are problems of fabrication, installation, operation and maintenance even when the structure has been engineered.

It is the bottom end of the market, they are cheapo short life buildings - of enormous visual impact - and maybe failure does not matter. It is a socio-political problem.

The problems of the larger airhouses however are extremely interesting to an engineer. Mainly they centre on the northern USA wet snow belt region and were to structures now of some age. The Minnesota Metrodome and the Pontiac Stadium developed excessive snow drifting on the roof which led to ponding and to deflation. At Pontiac a deflation was followed by strong winds for three days and the fabric flogged up and down uncontrollably until the panels of the roof were destroyed. Strong winds can also lead to instabilities of the whole roof. During a thunderstorm at the Metrodome the pressure surge from a wind gust caused the internal pressure to rise and the effect was magnified by the shape of the roof and the position of the sensors. The rise in pressure shut off the fans and opened the relief vents and the roof fell allowing very large oscillations. This type of instability has occurred on other stadia roofs at Indianapolis and Dakota.[6]

Some of these problems have already been addressed. Whiplash of panels as a mode of failure was missed in "the examination as to causes of death". A member of the original research group has subsequently addressed it.[7] The snow melt problem was examined very thoroughly with regard to the Minnesota Metrodome and can generally be dealt with.[6]

It is the design of the operating of the airhouses which was the problem area and one which the original research group did not really address. The control systems for the fans operated on pressure gauges; providing extra controls reacting to roof deflections and oscillations seems sensible. Like the modern jumbo aircraft, several parallel and different control systems are required, in effect considerable redundancy. It is in the engineering of resource management that the problems have arisen.

On the whole the actual performance of engineering airhouses has been extremely good. To get ptfe coated glass fibre accepted Birdair had to offer 20 year guarantees on it and time is running out and their problems are non existent. Chemfab still kept the Radome section when they sold Birdair to Taiyo Kogyo. They set the radomes up in extremely remote and exposed positions - on rocks in the north Atlantic, glaciers in the Antarctic - they must be 100% reliable - the worlds aircraft traffic depends on them for their location. It is a tribute to their control engineering. And the example of the most recent stadium roof, the Tokyo dome which has cables at closer centres can be operated at higher pressures and has a much more sophisticated control system than the earlier structures. The airhouse is a machine and has to be designed as such.

While the number of airhouse projects has declined, cable trussed roofs and tent roofs have prolificated. Architects are now interested in them and there are a large number. At this conference many of them will be described. They really divide into two separate types of structure with an area where both systems are used. Perhaps the division really comes from the type of cladding used. Chur Transport Station roof in Switzerland, the National Aquatic Centre, Sydney, Australia are examples of compression shell glazed roofs dependent on tension members. There are many other examples of both horizontal and vertical glazed "skins" with either cable stayed or cable trussed supporting systems. Tensile supports for buildings or bridge structures are a current engineering aesthetic now easy to achieve because of advances in structural analysis.

Figure 8.   Sydney Athletics Stadium
Architect: Philip Cox
Engineers: Connell Wagner

Figure 9.   Imagination Headquarters
Architect: Ron Herron
Engineers: Buro Happold

Cable (and mast) support systems for fabrics are an advance because they provide a softer method of concentrating force from a relatively low level in a roof to a high level in a mast or a foundation. Our roof at the Imagination Building in London is an example. The big roof at Atlanta for the Olympics, the airport terminal at Denver are the current big examples which could push the development of this type in a new direction including covering of intermediate space between a closed, traditional inside and a fairly hostile – hot or cold – external climate.

But as well as the successes there are some major problems in this tenting field. The first is with our understanding of available materials. Apparently an ever increasing choice since offering a new material is how fabricators attempt to get market (and financial) advantage.

Because we need some shear freedom in order to be able to produce the structurally necessary double curvature we are largely dependent on woven fabrics and we control its anisotropy, and protect the base cloth, by coating it. Now an improvement in one property means a deterioration in another, there is no such thing as a perfect material, the trick is to optimise. But even this can be difficult because, in most cases, the properties are not definable by a single valued function – joinability just depends on the right give, weldability requires an autogenous joint and that involves a whole army of materials properties. Add to these fire resistance, insulation values, acoustic performance, optical properties, manufacture, prime cost, maintenance cost etc and we are into an optimisation jungle. And despite the previously mentioned apparently wide range of possible materials, we are limited to three or four which are readily commercially available:-

. PVC coated polyester (PVC/PES)
. PTFE coated glass (PTFE/GS)
. Silicone coated glass (VESTAR)

And it is quite difficult to make comparisons, often for reasons of commercial confidentiality, and in spite of PVC's fiftieth birthday it is especially true of it!

There is not space to discuss all the properties engineers are interested in but I will touch on some mechanical ones. Firstly if you look at the literature all the stress-strain curves appear to be linear – which is a nonsense. Secondly it leads to the supposition that an unloading curve is the reverse of a loading curve. Which is not true. Where does that lost energy go? And thirdly, we lack specific properties – unit weight or

properties per unit of energy consumed in production -
and so on.

Bill Biggs has been talking to me about fracture
toughness[8]. In an isotropic elastic solid it is easy
enough to show that the fracture stress all depends upon
the balance between the strain energy applied and the
energy required to create two new surfaces - the surface
energy S. This leads to the familiar 'Griffith' equation.

$$\sigma_f \sim (ES/\pi a)^{1/2}$$

Where a is the length of the initial crack. Plotting the
two components as a function of crack length shows that
there is an initial crack length a_o, and, once exceeded,
fast fracture will follow under a diminishing or even a
negative stress (windscreens or Comet aircraft). The
equation needs considerable modification for
elastic/plastic materials (since plastic work must be
added to the surface energy term and it is more usually
written in the form

$$\sigma(\pi a)^{1/2} = (E\sigma_c)^{1/2}$$

Where G_c is now the total energy absorbed in making unit
area of crack - a high value as in copper (G_c $10^6 J/m^2$)
means that it is difficult to make a crack propagate $-G_c$
for glass is only - 10 J/m^2. The left hand term crops up
so often that it is generally abbreviated to a single
symbol K called (rather unclearly) the fracture
toughness.

Fabrics invariably tear along the warp or weft directions
- the diagonal directions lack any shear connection and
because of their extensibility it is difficult to release
enough strain energy to propagate the tear. Animal
membranes such as skin, cuticle or artery adopt a similar
strategy and their stress-strain curves are J shaped. In
the early part of the curve strain is almost independent
of stress due to the lack of shear connection in the
material and makes tearing difficult (though of course,
they can be cut or pierced because the deformation is
highly localised). It is when attempts are made to feed

energy into the crack site from remotely loaded regions that difficulties occur.

There we have yet another trade-off problem.   If for respectable engineering reasons one would like a high modulus and some semblance of isotropy we open the door to fast fracture.   A J-shaped curve would give resistance to fracture but may well be unacceptable for other reasons.   On the whole we have probably got it about right but it is interesting that in some ways we are now having more problems with materials than much earlier on.

Which raises the question of what next?   Would it make any difference if the modulus of steel was 231 instead of 210?   An increase of 10% in any property is rarely worth having, a decrease of 10% in cost is usually really worthwhile.   Polymers are energetically expensive materials.   What is the energy cost over the life of the building?   Joint behaviour - methods of jointing are unsophisticated and quality assurance expensive.   We try to apply plane strain fracture mechanics to an isotropic, very thin, material.   Mould growth, wicking, abrasion, cleanability are all problems of the coating and are all limiting life cycle factors.   These are further performance properties that are nearly all non quantifiable in a single term so we like to pretend they do not exist because we cannot put them into our sums - yet!

But like the life history of concrete and steel these are not questions we can ignore.   We are already very skilled though and, at least in the tenting field, architects have now taken them up and some are designing imaginatively with them.   Unfortunately, at least in my country, there is not a body of precedents which have been widely subjected to discussion and criticism so that paradigms are understood and taught to architectural students.

It is Japan who is leading in so many aspects in these fields.   Firstly the Japanese government has embraced such structures into their building codes, designating three categories.

Class I permanent and semi-permanent structures made with PTFE coated fibreglass.

Class II structures manufactured for PVC coated fibreglass

Class III small span, essentially non-permanent construction fabricated with PVC coated polyester.

Secondly the manufacturers joined together to form the Tent Structures Research Group at much the same time as we started here and now the most rigorous codes and much of the best research work comes from them. More than that they have supported teaching membrane structures technology in many Japanese architectural schools and competitions are held to encourage students to experiment.

There are other moves in tensile structures. My own interest at present is in ETFE foils because of light transmission/cost qualities and other environmental possibilities. Moveable roofs are another. Perhaps we will talk about all of them at the conference.

But I return to my analogy at the beginning. Always nature is reacting to need. And engineers just limiting themselves to determining the strength of structures, like the instrument makers of the seventeenth century, will become redundant. We have to see the whole picture. Be concerned not just with the form, but with the function. And that means returning to a broad view of engineering. It is interesting that of the eight young engineers who founded the Institution of Civil Engineers in 1818 for "facilitating the acquirement of knowledge required in their profession and for promoting mechanical philosophy" only one member would be considered a civil engineer today, the remainder being the equivalent of mechanical engineers. The air house, for sure, has a great future - but it is as much mechanical engineering as civil - should we call it building engineeri

As so much more is known we have to broaden our perspective because "....... if the truth is to be told, it is technology that is creative, because it gives us these new opportunities. Historic ideas of art and culture can entrap. It is technology that frees the scene ......"

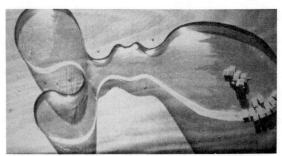

Figure 10.

APPENDIX

References:

1.   by Dr Julian Vincent of the University of Reading
      and author of Structural Biomaterials.   Macmillan
      Press 1982.

2.   "Coated fabrics for lightweight structures" by Tony
      Read and Turlogh O'Brien.   The Arup Journal October
      1980.

3.   The University of Bath and Buro Happold funded the
      first study.   This was followed by an extremely
      generous grant from the Wolfson Foundation and a
      subsequent one from the Science and Engineering
      Research Council.

4.   .   Air supported structures: the state of the art.
           Proceedings of a symposium held in London, June
           1980.   Published by the Institution of
           Structural Engineers, London.

      .   The design of air supported structures.
           Proceedings of a symposium held at Bristol,
           July 1984.   Published by the Institution of
           Structural Engineers.

      .   The Air hall handbook: IL15 .   Published by
           the Institution for Lightweight Structures,
           University of Stuttgart 1983.

5.   My partner, Ian Liddell, has written about this in a
      paper "Air supported structures" The Structural
      Engineer, 20 July 1993. Vol 71. Number 14. I have
      used many of his words.

6.   Surface stressed building structures.   Henderson
      Colloquium.   Pembroke College, Cambridge, July 1992.
      Published by the British Group of the International
      Association for Bridge and Structural Engineering,
      London.

7.   Travelling waves and shading waves on fabric
      structures - C J K Williams. The Structural Engineer
      1990 Vol. 68 No. 21.

8.   This part of the paper owes a lot to the knowledge,
      intelligence and words of Professor Bill Biggs, late
      of Cambridge University and Reading and currently
      consultant to Buro Happold.

The Georgia Dome and Beyond

Achieving Lightweight - Longspan Structures

by Matthys P. Levy

Extended Abstract

The development of longspan roof structures has undergone a quantum leap in the last two decades. Since the introduction of steel structures in the late nineteenth century, exemplified by the great European train stations, roof spans have been continuously stretched, reaching a culmination with the 204m (680 ft) span Louisiana Superdome. That structure, completed in 1975 was, and still is, the largest steel dome ever built. It also achieved a modicum of economy using about 122 kg/m^2 (25 psf) of steel. A parallel accomplishment in Paris was the construction of the world's largest concrete dome for the CNIT, a 216m (720 ft) span exhibit hall, completed in the nineteen sixties. Although it used a unique double-layer shell structure linked by concrete webs, it was many times heavier than the Superdome, since it was built of concrete.

The introduction of fabric structures in the early nineteen seventies completely changed the weight equation. The Pontiac Silverdome, a 235m (770 ft) span air-supported membrane was achieved with less than 15 Kg/m^2 (3 psf) of material. Of course, the roof was supported by air, which some critics might say is not a real structural element.

Nevertheless, this was a significant achievement by the engineer, David Geiger. Air supported membranes can span unlimited distance with virtually no change in unit weight of the material used, a result that is almost as miraculous as perpetual motion.

Not wanting to rely on an air blower, which is a mechanical device, for structural stability, engineers next rediscovered Bucky Fuller's tensegrity structure (he called it an Aspension Dome), in which "islands of compression reside in a sea of tension". Putting aside Fuller's mystical verbiage, this is in reality a prestressed space truss. Geiger developed a radial cable dome version that was used for the 210m (690 ft) span Suncoast Dome and we developed the triangulated Tenstar Dome that covers the 241m (790 ft) span Georgia Dome. These rigid tensegrity structures, although heavier than air-supported membranes, still only weigh an incredibly low $30kg/m^2$ (5.9 psf).

Tenstar domes are characterized by a triangulated cable network that is adaptable to numerous plan configurations: the circle, the oval, the ovalized triangle, the ovalized square. The triangularization is organized in such a way as to define a number of concentric rings, more for larger domes, fewer for smaller domes. This arrangment of elements leads to another possible plan configuration, namely a canopy roof with an open center. In cross section, Tenstar domes are deformed by displacing alternate nodes to create hyperbolic-paraboloid panels with the curvature needed to use fabrics as roof covering.

Perhaps the most exciting possibility presented by Tenstar domes is their use for retractable roofs. Since the basic cable network, even including the rigid posts, is virtually transparent, sections of roof covering can be eliminated to create a window, open to the sky. This window can be covered with a retractable flexible roof section that, although consisting of rigid truss elements, is designed to conform to the deformable structure of the Tenstar dome itself. The result is an economical retractable roof that is unlimited in the spans it can cover. It is truly a new window to the world.

GEORGIA DOME CABLE ROOF

CONSTRUCTION TECHNIQUES

Wesley R. Terry, M.ASCE

SUMMARY

Birdair, Inc. was the roof contractor for the construction of the
Hypar-Tensegrity Cable Roof for the Georgia Dome.   Birdair has
been involved in many of the stadium roofs constructed in the
United States and Worldwide. Some recent examples include: The
Georgia Dome in Atlanta, Georgia, The Rome Olympic Stadium in
Rome, Italy, The Suncoast Dome in St. Petersburg, Florida and The
Riyadh International Stadium in Riyadh, Saudi Arabia.

This paper will provide a case study of the construction of the
Hypar-Tensegrity Cable Roof for the Georgia Dome. The paper will
emphasize innovative computer analysis of the   construction
process and a detailed description of the construction of the
cable roof.

---

1   Principal Engineer, Birdair, Inc., 65 Lawrence Bell Drive,
     Amherst, NY   14221, United States of America

INTRODUCTION

The Georgia Dome is the World's largest cable supported roof with
a clear span measuring 240 m by 193 m, and also being the World's
first large-scale oval domed roof.

Designed by Engineer, Matthys Levy of Weidlinger Associates in
New York City, Mr. Levy calls his design a "Hypar Tensegrity
Dome". The "Hypar Tensegrity Dome" combines Buckminster Fuller's
"Tensegrity" concept with the hyperbolic parabola, a basic
building shape of tensioned membrane structures.

The Georgia Dome, in plan, consists of two circular segment ends
connected by a 56 m long plane tension truss. The cablenet is
anchored to a concrete ring beam supported by 52 columns and is
comprised of a series of three concentric tension hoops, each
stepping inward and upward toward the roof crown. The
triangulated cablenet results in diamond shaped, hyperbolic
parabola fabric membrane panels, which are clad in teflon-coated
fiberglass (Fig. 1).

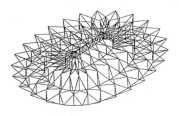 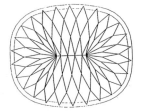

Fig. 1

PRELIMINARY CONSTRUCTION PROCEDURE

The preliminary construction procedure for the cable supported
roof involves not only the roof construction, but the thought
process for the construction of the entire stadium. If the
construction of the roof interferes with or causes delay in the
completion of the seating bowl, then the overall lowest cost for
the stadium construction may not be achieved.

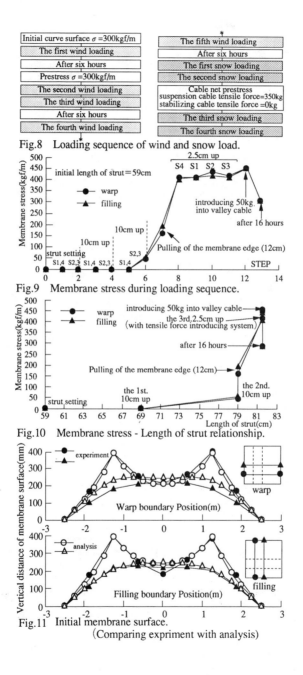

Fig.8   Loading sequence of wind and snow load.

Fig.9   Membrane stress during loading sequence.

Fig.10   Membrane stress - Length of strut relationship.

Fig.11   Initial membrane surface.
(Comparing expriment with analysis)

2.2.   Description of the load tests

Three different load tests were performed to simulate the loading conditions during construction and operation of the membrane structure. To simulate the actual construction situation a test was performed of prestressing the membrane surface. Two tests were performed to simulate wind and snow load conditions in the operational situation. The prestressing method of membrane surface used in the test is based on three operations:

(1)   increasing the length of the struts, thereby inducing compression forces in the struts and tension forces in the pretension cable and the membrane surface,
(2)   pulling the membrane surface edge,
(3)   tensioning the valley cable.

The method of increasing the length of the struts is the most simple system for prestressing the membrane surface in an actual construction situation. The method is illustrated in Figure 6 and shows a strut, consisted of four plates, and the simple equipment needed for elongating the struts. Multiple set of three holes in plates A and C enable the adjustment of the bolted connection of plate B and plates A as well as the connection of plate B and plates C, enabling to change the length of the strut from 59.0 cm to 99.0 cm. This is equivalent to a stroke of 40 cm. The equipment used for prestressing works on the mechanism of a pantograph and supports the membrane surface while changing of the length of the struts (Photo 2).

The test procedure is shown in the flow diagram of Figure 7. First the suspension cables were hanged from a rigid supporting frame. Secondly, the stabilizing cables were put in place, and last the membrane was put on the cable net formed by suspension and stabilizing cables. The tension forces in the stabilizing cables were set to 300 kg. The tension forces in the cables were based on the consideration of decreasing tension loads in the cables under the additional external load, the cable strength, and avoiding initial displacement of the suspension cable. The prestressing stress of the membrane was set to 300 kgf/m, after relaxation. The initial prestressing stress just after completing the final curved membrane surface was 450 kgf/m while the relaxation level was based on results of past experiments.

To simulate the case of full up–lift by wind forces equipment was used to apply an internal pressure of 170 kg/m^2. In order to simulate snow load small sand bags were places on the membrane structure to give a distributed load. In case the total structure was loaded the load was 50 kg/m^2, in case the structure was partly loaded the load was 100 kg/m^2. The loading sequence is shown in the flow diagram of Figure 8.

2.3.   Results of the prestressing test

In the prestressing of the membrane the three components described in part 2.2. contributed to the total prestressing in the following distribution: (1) elongating the struts contributed 60%, (2) pulling the membrane edge contributed 30%, and (3) tensioning the valley cable contributed 10%. The membrane stress history is shown in Figure 9 as function of the loading step. In each of the contributions the warp and the filling direction are equivalent to the membrane tension. The elongation of the struts in the way as described earlier is a very simple erection procedure and has hereby shown to be very effective in prestressing the membrane, confirming a useful practical application of this method.

The relation between the membrane stress and the length of the strut show a nonlinear behavior as shown in Figure 10. In order to give the membrane an effective prestress of 300 kgf/m after relaxation the strut had to be elongated to 81.5 cm, resulting in an axial force of 328 kg.

In Figure 11 a comparison is made between the initial shape of the membrane surface in the test and the initial shape obtained from analysis. The comparison was made for the stage with an effective prestress of 300 kgf/m. The displacements at the top of the struts obtained by the test and analysis compare very well for both the warping and filling directions. However, the test gives a smaller displacement of the point in the middle between the struts. This tendency is especially seen in the warp direction. Looking at these results we believe that it is important to make a careful estimation of the initial membrane stiffness used in the analysis in order to obtain the correct initial shape of the membrane.

2.4.   Results of the wind and snow load tests

In Figure 12 the additional forces in the suspension and stabilizing cables are given as a function of the internal pressure. The influence of the presence of the valley cable is also investigated by considering two support conditions, the presence and absence of the valley cable. For both the support conditions, the force in the stabilizing cable increases with the internal pressure to a maximum after which it decreases. The value of the maximum is found for different values of the internal pressure and depends on the presence of the valley cable. The values of the internal pressure for which the maximum occurs are 60 respectively 80 kg/m^2 in cases with and without the valley cable. For the additional forces in the suspension cable the influence of the presence of the valley cable is much more significant. The presence of the value cable reduces the additional forces in the suspension cables to about a half. The presence of the valley cable influences the change in the shape of the membrane surface thereby influencing the force distribution in the total structure. The behavior of the struts is similar to the behavior of the stabilizing cables; the forces in the struts increase with the internal pressure to a certain maximum after which they decrease as seen in Figure 13. For both the support conditions a maximum is seen for an internal pressure of 80 kg/m^2.

The relation between the force in the strut and its vertical displacement is seen in Figure 14. In the case the valley cable is present the displacement increases almost linearly with the force in the strut. Under increasing displacement the force reaches a maximum and decrease.

Figure 15 shows the stress in the membrane as function of the internal pressure for both the warp and filling directions. Both supporting conditions – with and without the valley cable – are considered. The influence of the presence of the valley cable on the membrane stresses in the valley cable direction are evident and a reduction of about one fourth is observed. In case of the presence of a valley cable the increase rate of the membrane stress decreases after the internal pressure reaches the value of 80 kg/m^2. This is the point were the forces in the stabilizing cable and the struts find their maximum and after which the rate of the displacements with the internal pressure

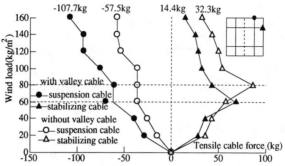

Fig.12   Load - Tensile cable force relationship.

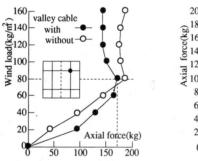

Fig.13   Wind load - Axial force relationship.

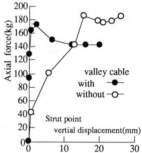

Fig.14   Axial force - Vertical displacement relationship.

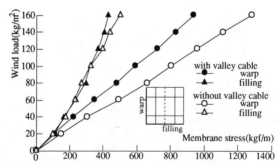

Fig.15   Wind load - Membrane stress relationship.

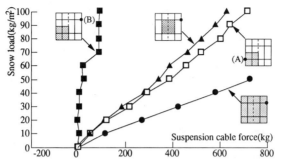

Fig.16   Snow load - Suspension cable tensile force relationship.

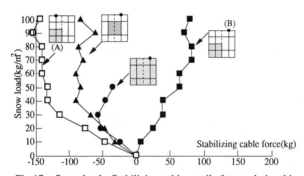

Fig.17   Snow load - Stabilizing cable tensile force relationship.

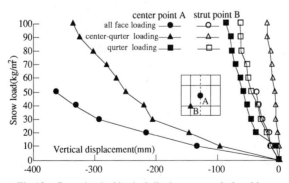

Fig.18   Snow load - Vertical displacement relationship.

increases.

Figure 16 shows the load in the suspension cable with the increase of the snow load. Three loading cases are considered:

(1)   Distributed load over the whole membrane,
(2)   distributed load over the central quarter of the membrane,
(3)   distributed load over the outer corner quarter of the membrane.

In all three cases the loads in the suspension cables increase with the increase of snow load, however the increase is small for the suspension cable under an unloaded part (cable B).

Figure 17 shows the load in the stabilizing cable with increase of the snow load, for the same loading cases as mentioned above. The cable force decreases with the increase of the snow load for both the symmetrical load cases 1 and 2. For the asymmetrical load case 3, the increase or decrease of the cable force depends on the location of the cable relative to the load. If the membrane above the cable is loaded the load in the cable will decrease with the increase of the snow load (cable A), while the cable load will increase when the membrane above the cable is not loaded (cable B).

In Figure 18 the displacements of points A and B is shown as a function of the snow load, for the same load cases. It is seen that the maximum displacement of point A in the center occurs for load for the symmetrical load case 1, while the maximum displacements of the struts occur for the asymmetrical loads case 3.

3.   Conclusions and discussion

The following conclusions result from this investigation:

1.   As only small deformations of the strut are necessary to obtain the desired prestress level, the method of elongating the struts is practical and effective application to induce prestress in the membrane surface.

2.   For design and construction, it is necessary to estimate the interaction between the strut length and the membrane stress.

3.   In the case of wind load the influence of presence of the valley cable on the load interaction of the components is substantial. The presence of the cable reduces the stresses in the membrane, the struts and the stabilizing cable, reduces the displacements while it increases the stresses in the suspension cable.

Although only two different support conditions by the valley cable are investigated (with and without) we believe that the initial shape and stiffness of the valley cable are of prime influence on the structural behavior of this type of membrane structure. However, this subject needs further study.

References

1) Okada, A., Saitoh, M., and Endoh, S. (1992). "Structural Design and Construction of the Tension Lattice Dome". IASS-CSCE International Congress 1992 on Innovative Large Span Structures, Toronto, Canada.

2) Saitoh, M., Kuroki, F., Shimizu, K., Kudoh, K., and Matsuyama, K. (1993). "Study on Structural Characteristics of Suspended Membrane Structures with Cable Net", IASS International Congress 1993, Istanbul, Turkey.

Fabric Structures for Public Buildings

Horst Berger, F.ASCE, M.IASS [1]

The design of the enclosure structure of the new Denver Airport derived from twenty years of experience with the design and engineering of fabric structures for permanent buildings. This paper describes four recent examples illustrating primary forms of fabric tensile architecture, designed by Horst Berger Partners. They include arch supported structures for the University of Wisconsin and for Wimbledon, and tent shaped structures for the Mitchell Performing Arts Pavilion near Houston, Tex., and for the San Diego Convention Center..

Introduction

A number of fabric tensile structures completed over the last few years signal the acceptance of this new technology and architectural art form for use in permanent buildings. Since the structure and the enclosure are provided by the same membrane, and since its form is a critical aspect of its structural design, a totally integrated approach to design is necessary. Architecture and engineering cannot be separated. Structure, enclosure, interior space and exterior sculpture, light, sound, thermal comfort and efficiency, all derive from the same elements. Form and function are one as they were in earlier stages of architectural history. But the technology is new, applying advanced materials and utilizing the power of the computer for developing geodesic shapes and analysing the non-linear behavior.

The fabric roof of the new Denver airport, described in a separate paper, is the latest and most prominent result of a twenty year long series of public building applications of this new tensile architecture. The principal features of four others, completed in the last few years, are described in this paper. Two of them, the McClain Practice Facility of the University of Wisconsin and the Indoor Tennis Facility at Wimbledon, are arch supported structures. The roof over the outdoor exhibition space of the San Diego Convention Center and the roof cover of the Cynthia Woods Mitchell Performing Arts Pavilion each consist of a series of tent-shaped modules with special suspension systems. The main fabric membrane in all cases is an integral part of the structural system and its most visible architectural feature. The Wimbledon project uses a PVC-coated polyester fabric with a Tetlar finish, the other three are made of Teflon-coated fiberglass.

---

[1] Professor, School of Architecture, CCNY; Principal, Horst Berger Consultants; 525 Broadway, No. White Plains, N.Y. 10603

McClain Practice Facility, University of Wisconsin in Madison

This structure covers a football field and serves the football team of the university for indoor practice. The plan dimensions of the roof are 92 m by 67 m. The support system consists of steel arches spanning the 67 m span spaced 18.30m apart. The arches are triangular in cross section, consisting of three continuously curved steel pipes connected by trusslike tubular web members.

This form of steel arch was based on experience with a number of previous applications, primarily using cross-arched configurations as shown in Fig.1. For the larger McClain structure parallel arches were more economical. Their lateral stability was achieved by a set of light steel bracing members which are visible in Fig. 2.

Fig.1: Crossed-Arch Structure

Fig. 2: McClain Practice Facility: Interior View

The fabric membrane was limited to the center half of the roof, with the outer quarters roofed with standing seam metal roofing supported by metal deck framing. This arrangement proved to be the most economical.

It produced the best lighting in the daytime with the light source located centrally and high above the playing field thus avoiding low light angles which are not desirable in a sports facility. The insulation in the opaque portion of the roof helped reduce heating cost. And, finally, the metal roofing at the edge of the building was best suited to handle sliding ice and snow in the heavy winter climate of Wisconsin.

The light steel arches, fabricated in the shop and delivered to the site in 12 m sections, were bolted together on the ground and erected in two half sections each, requiring no scaffolding. Each arch took a day to install. A temporary wood walkway in the 1 m high arches provided access for the installation of the lateral steel and the membrane roof. The fabric was simply draped over the arches, bolted to the steel at the junctures with the rigid roof and stressed with a valley cable which would carry the uplift loads from the wind.

The Architects were Bowen Williamson Zimmerman of Madison, Wisc. Roof Design Consultants and Structural Engineers were Horst Berger Partners of New York. The roof contractor was Birdair Inc. The facility was designed and built within one year and opened in 1987. Its' users have found it to be one of the most pleasant indoor spaces in which to play football.

Wimbledon Indoor Tennis Facility

The functional requirements set by the All England Lawn Tennis and Croquet Club for this building were extremely high, in keeping with the prestigious name of Wimbledon and the objective of having such international events such as Davis Cup competition take place in it. This included high demands of the visual environment. Arches, especially cross arches, were considered visual obstacles on the inside. Therefore the decision was, made early in the design process to work with external arches.

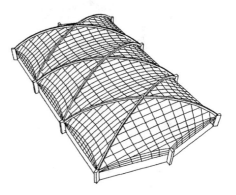

To cover this three-court space precast concrete arches proved economical and had the added advantage of low maintenance when exposed to the elements. A cross-arch arrangement resulted in a system which would be stable and permit a simple construction system using one temporary erection tower at each intersection point.

Fig. 3: Computer Perspective of Wimbledon Practice Facility

Fig. 4: Wimbledon Practice Facility, Exterior View

To speed the erection a steel assembly was designed into the center end of each arch segment which was later imbedded in the cast-in-place keystone section. The latter formed a circular opening for exhaust fans. Pipes cast into the precast arch section provided the location for cable hangers from which the membrane was suspended.

The membrane consists of a pattern of trapezoidal and triangular fabric sections with ridge cables located under the arches. The fabric membrane is one continuous piece with one splice along a ridge cable to make two shippable sections. The peripheral edges are formed by edge catenaries on the three exterior sides of the building and a clamped edge where the tennis hall meets the clubhouse. The space between the edge catenaries and the rigid outside walls is enclosed by fabric infill panels.

The fabric membrane was assembled on the ground and raised to a position several feet below the arches. It was then clamped onto the curb on the side of the clubhouse. Edge catenaries were attached along the other three sides. Stressing of the fabric was accomplished by pulling the suspender cables up through the arches with come-alongs until the proper geometry and stress level had been achieved.

The result is an interior in which the ceiling is a sky which does not visually interfere with playing tennis. In the daytime the courts have excellent natural illumination. In the night floodlights reflecting off the white membrane provide a similarly satisfactory result. On the outside the gentle waves of the membrane suspended from the slender concrete arches form an appealing shape in the valley which is surrounded on three sides by suburban residential buildings and on the fourth is facing the grounds of the famous Wimbledon tennis compound.

The architect was Ian King of London. Roof Design Consultants and Structural Engineers were Horst Berger Partners. The general contractor was Taylor Woodrow of London. And fabric roof contractor was Koit on Germany. Completion 1988.

San Diego Convention Center

The roof over the outdoor exhibition area of the San Diego Convention Center has become a landmark for the city which sees the facility as its "Sidney Opera House". The roof spans 91.5 m between the triangular concrete fins which, like flying buttresses of a medieval cathedral, run the two long sides of the building. The other two sides of the 91.5 m long roof are completely open over their entire width, giving the roof a sense of graceful weightlessness.

Fig. 6: San Diego Convention Center Roof

The roof membrane itself consists of a series of point-supported, tent-like units with two peaks in each of the five 18.3 m wide bays. Main support cables attached to the top of the triangular fins and supporting two flying masts each, near the center, form the main support of the roof system.

Since the flying masts and the main cables are placed on a line located in the midway between two lines of fins, the main cables are split near the supports in a Y- format so that they can be connected to the top of the fins.The membrane structure itself is folded over a series of ridge and valley cables and restrained along all four edges by edge catenaries. Each ridge cable is located vertically above the main support cable and is draped over the two flying masts. V - shaped edge support cables connect each end of each ridge cable to the top of the two adjacent fins. This set of cables and struts carries the down load on the roof, holding up the membrane.

Valley cables located on line with the fins are connecting to them 5 m above the floor bow upwards in the form of an arch. They resist the upwards loads (from wind and prestress) and hold down the roof membrane.

The edge catenaries on the support sides form small drapes, 30 ft. in plan view, between the ends of the ridge and the valley cables. The edge catenaries on the two open sides mirror the shape of the ridge cable in plan projection by swinging from one edge across two projecting support points on line with the flying masts and back to the other support point.

What makes the open ends possible is a 96 m long horizontal floating strut with Y - shaped ends which form the support points for the interior edge catenaries just described. A grid of light cables hanging from the top of the flying masts and restrained against the bottom end of these same masts keeps this truss-shaped strut in position. Its additional function is to support loud speakers, lights and sprinklers.

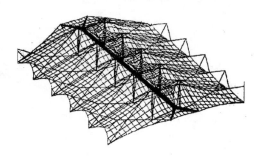

Fig. 7: Computer Graphic showing support system (membrane shown as transparent net)

The main membrane has large openings in the middle of each bay which have the purpose of letting warm air escape at the top and thereby facilitate air flow under the roof. In order to keep rain from coming through these openings a secondary membrane cover is provided above the main roof. This "rain fly cover" is suspended from the tops of the flying masts and held downward by anchorage to the center of the valley cables. This functional feature adds to the visual charm of this membrane roof structure and gives it its special character.

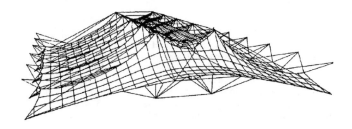

Fig. 8: San Diego Roof: Computer Perspective

The cable system was erected first, independent of the membrane. Fabric panels spanning from ridge cable to ridge cable and from the edge to the center, were then erected and connected to each pair of ridge cables. The valley cables were placed on a wear strip in the middle of each segment. The stressing of the system was achieved by jacks on top of the flying masts and by use of adjustable threaded rods at the upper rings.

The elegance and appeal of this membrane roof structure derives directly from the transparency of the structural order which can be understood and appreciated even by the layman by simply following the lines and shapes of which the structure is composed.

Architects were the Convention Center Architects with Arthur Ericson's office in charge of design. Roof Design Consultant and Structural Engineer for the roof structure was Horst Berger Partners. The roof contractor was Birdair Inc. The Building was completed in January of 1990.

The Cynthia Woods Mitchell Pavilion of the Performing Arts

This facility is a major outdoor performance center located in the new town of Woodlands, 45 miles from the center of Houston. It has a 2,500 sq.m roof, covering 3000 fixed seats, a berm accommodating an additional 7000 people, a stage house capable of providing the backup for classical and rock concerts, theater, ballet, musicals, and opera. It is the summer home of the Houston Symphony.

As principal designer of this facility it was this writer's responsibility to develop the architectural design for the entire facility. It provided the opportunity to give shape not only to the roof but also to the stage house, the fixed and lawn seating and other main features, such as the lobby circles, and to arrange them as an integrated composition of large sculptural volumes.

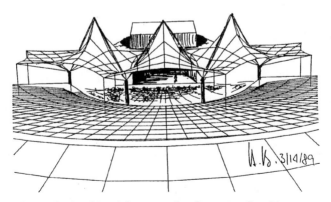

Fig. 9 : Design Sketch Incorporating Computer Graphics

In this composition the fabric roof had the purpose of creating a sense of being inside a protected, friendly space for those under the roof, and forming an inviting proscenium arch for those looking in from the berm. In addition to these primary functions the roof had to produce ample daylight during day performances, act as reflector for artificial light in the night, reduce and scatter reflected sound under the roof, support natural and electronically reinforced sound for the berm.

To achieve these various objectives the fabric roof form best suited and most economical to build was a system consisting primarily of three tentlike membrane units in a fan-shaped arrangement, suspended from three external A-frames. Together with the two wing-shaped end units the roof plan forms approximately a third of a circle.

Fig. 10: The Mitchell Performing Arts Pavilion

The roof structure is supported by four trussed columns located at the spring points of the A-frames and by the stage house. In addition, the four peripheral corner points located radially outside the trussed columns are tied down with thin steel pipe sections.

The horizontal forces of the membrane and cable structure caused by the surface stresses of the fabric are balanced within the structure by trussed struts below the roof surface which, running approximately horizontally on radial lines, connect the peripheral anchor points to the tops of the columns and those in turn to the edge of the stage house. Thus only vertical forces have to be transmitted to the foundation except that overall lateral loads from wind are resisted by the trussing of the stage house walls.

Ridge cables running between rings near the apex of the A-frames, the stage house on the one side and the center of the edge catenaries on the other, hold up the membrane. Valley cables located between the column support points and the peripheral corners on one side and the stage house on the other hold it down. The connection point at the top of the columns is a low point collecting the rain water from most of the roof surface and draining it into a leader inside the column. A gap between the fabric ring and the collection pan below it provides a safety overrun in case a drain or drain pipe is clogged.

Because of the steep curvatures the panels were fabricated in sections bordered by the ridge and valley cables, the edge catenaries, and the stage house edge. They were lifted and clamped onto the cables with the top rings under the A-frame peaks in a lowered position. Stressing was achieved with the help of jacks on top of the A-frames which raised the top rings upward.

Though covering 3000 seats on one level, the roof succeeds in creating a space that feels intimate and festive. On the other hand, the evening photograph shows the spectators on the berm outside are drawn in by the dazzling lights and glowing forms of the soft structure which opens towards them.

The owner of the facility is the Woodlands Corporation. The design was produced by Horst Berger Partners in association with Sustaita Associates, Architects of Houston. The roof contractor was Birdair Inc. The facility opened in April of 1993.

Summary and Conclusion

The four structures presented in this paper illustrate a variety of fabric tensile solutions used in permanent building applications. Together with structures like Canada Place in Vancouver, the Riyadh Stadium, and the Jeddah Airport they were direct forerunners of the tensile structures of the new Denver Airport which are covered in a separate paper.

Fig.11: Construction Photo of the Landside Terminal at the new
Denver Airport

Helium-Supported Hangar for a Small Solar-Powered Airship

Rosemarie Wagner[1], Bernd Kröplin[2]

Abstract

For the IGA Expo '93 the Institute for Statics and Dynamics of Aerospace Structures, University of Stuttgart, together with Stuttgart Solar e.v. developed a solar-powered airship.

For operating the airship during the IGA, a temporary hangar is required to provide safe accommodation during the days not suitable for flighing and during the night. The hangar-structure is closely connected to the ideas and the concepts of the airship. Particular technical challenges are the development of an extreme light-weight and efficient/durable construction of load bearing elements stabilized by internal pressure, and the use of newly developed materials, provided by manufacturers, with a high gas density and tension strength.

The novelty of this construction is the use of pneumatically supported elements, filled with helium to be self-supporting. The loadbearing elements are pretensioned membrane constructions stabilized by internal pressure with the least possible dead load.

The following describes the construction and the loadbearing behaviour of this helium-supported hall.

The Preliminaries

In 1990 an 8m long airship with bio-gas propulsion was built at the Institute for Statics and Dynamics of Aerospace Structures at the University of Stuttgart, Germany.This was followed by an airship, 16 m long and 4 m in diameter, powered by a solar generator followed. It was in operation during the IGA Expo 93 in Stuttgart. The payload of this airship consisted of the remote-control-equipment and a video-camera.

While in operation during the IGA, a temporary shelter was necessary, one which would exceed the possibilities of a hangar. In addition to housing the airship during the IGA Expo 93, several events were to be held and an exhibition showing new technologies in the field of airship-construction, the use of solar energy and new materials was organized.

----
[1] Institute for Structural Design, University of Stuttgart, Germany
[2] Institute for Statics and Dynamics of Aerospace Structures, University of Stuttgart, Germany

The Conception

Despite the manifold requirements this hall had to meet, the idea for the construction was developed from the concepts of the airship. If the airship is able to change its position when parked according to the respective wind direction, optimum starting and landing conditions and the safest possible entry and exit in and from the hall are guaranteed.

The loadbearing structure of the airship consists of elements stabilized by internal pressure. These structural elements, able to withstand stress as well as bending stress by relieving prestress in the membrane, are very efficient and the helium-filling makes them weightless.

This principle was also applied to the design of the hall. The loadbearing elements are prestressed membrane constructions, with the least possible dead load, filled with helium, stabilized by internal pressure. The membrane material has to meet the following requirement: helium density, also at the seams when processed, with the strengths necessary for the load transfer.

The hall is circular in plan with a roof, opening at the top. The main loadbearing element consists of a torus. The torus, stabilized by buoyancy, had to be safe-guarded against lifting off by placing diagonal stays crossing each other at the largest and smallest diameter of the torus and anchoring them in the ground. The roof consists of segments which are also pneumatically supported cushions, filled with helium. Using individual segments allows for a simple opening of the roof, since the buoyancy force places the individual elements in a vertical position (Fig. 1).

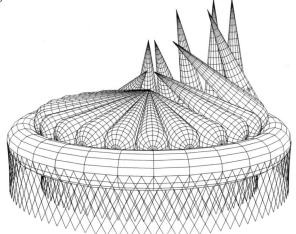

Fig. 1

The rooms resulting from this construction are for one the inner circular are housing the airship with its technical equipment and for the other the annular ambulatory beneath the torus. Aside from housing the airship the inside of the hall provides enough space for holding events. The ambulatory is suitable for exhibitions giving the general public access to the ideas and concepts connected with the construction of the airship.

The Principle

The idea of constructing a hall which has no deadload due to the buoyancy force and therefore does not require any support or stabil sub-structure is fascinating and the realisation is an exceptional challenge. Many aspects of this unusual task require the development of original solutions, since experience is only rarely a factor.

A simple examination of the equlibrium shows the effect of the buoyancy force for the loadbearing ring under horizontal windload. The ring, assumed to be a stiff element and the stays being weightless and stiff, deformes until the resulting load from the horizontal windload and the buoyancy force acts in the direction of the diagonal stays (Fig. 2). This results in a horizontal and in a vertical displacement, since the stays follow the circular movement as stiff pendulums.

An increase in buoyancy leads to smaller displacements, i.e. an increase in volume increases the stability of the ring. Another favourable factor is the linear decrease of the relation between surface, respectively deadload of the foil, to the volume according to the decreasing ring diameter, leading to an increasing buoyancy force. On the other hand, the membrance stresses increase linearly under internal pressure with an increasing ring-diameter.

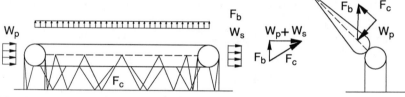

Fig. 2

When placing the roof-segments in a vertical position this rule applies: The buoyancy moment multiplied by the distance between the center of gravity of the volume and the pivot must always be greater than the deadload-moment multiplied by the distance between the center of gravity of the plane and the pivot. There will also be equilibrium between the windload normally affecting the medium-plane and the buoyancy when the roof-segments are opened, thus leading to the resulting load affecting the longitudinal axis of the segments.

The Structure

Due to the size of the airship and the torus-diameter set by the location at the IGA, a 3m-ring-diameter was chosen. The total height of 6m leads to an internal height of 3m in the ambulatory beneath the ring. The stays are most effective when inclined less than 45°, their number - 96 each on the inside and on the outside of the ring - is directly connected to the number of foil-sheets used to form the ring. The three entrances were simply integrated into the anchorings, therefore special elements could be omitted (Fig. 3).

The stay cables are not forming a net, but individually run from the foundation to the ring and back. A translucent foil was placed at the outer anchoring, acting as weather proctection and vertical boundary of the ambulatory. Inside the ambulatory only the stay cables form a separation to the inner room.

The wall-foil is placed between the two inclined cable-shares and attached to the loadbearing ring. Absorbing the windloads affecting the wall-foil leads to great deformations of the cables, spanned straight between the ring and the foundtion. A stiff attachment of the wall-foil to the cables would result in vertical tensile forces, due to deviation, aimed downwards, which would be greater than the total buoyancy force and the wind suction. If the foil is attached only at the upper rim, it is flexible under wind and since the foil is placed between the two cable-sheafs there is no excessive flapping.

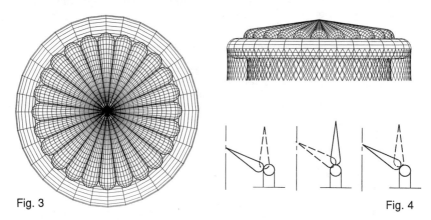

Fig. 3   Fig. 4

Since the main objective of this roof is weather protection, the segments are connected water-tight when closed. Cones stay true to shape even under internal pressure and spherical caps or again cones may be chosen to form the lower boundary. A spherical cap with a tangential connection to the upper cone has the least geometrical discontinuity.

Increasing the number of roof-segments leads to smaller cone-diameters and increases the height of the available inner space, at the same time the relation between the cone-surface, respectively the deadload, to the volume becomes more unfavourable, i.e. less segments result in an increasing buoyancy force. 24 segments were chosen, each 11.5m long, with a maximum diameter of 2.80m.

The attachment at the ring in the area of the inner anchoring would be advantageous, since the stay cables could directly absorb all loads from the roof-segments. This also keeps the total height of the hall small, limits the effect of the horizontal wind-attack, the ring-shape stays dominant and is aerodynamically favourable. The shape of the roof-segments, even when formed out of two connected cones, leads to limited opening-angles. If the ring and the segment are under equal internal pressure, a maximum inclination up to touching the ring is possible. There is another disadvantage: Since rainwater collects in the seam between segment and cone, a drainage system is required (Fig. 4).

If the segments were to be fixed on the ring, all of the rainwater could drain across the ring to the outside and the inner space would increase. The segments could be placed vertically and, when closed, rest on the ring. The individual load induction into the ring are disadvantageous, as well as the larger area subject to wind-attack and the unfavourable aerodynamic shape.

In the realisation, the roof-segments are attached to the ring at a 50° angle from the horizontal direction. The cone-shape in the lower area guarantees that the segments can be placed vertically when opening the roof. The required water-tightness between the segments, the opening and exchanging of segments can be easily achieved using zippers.

Due to its buoyancy, the loadbearing ring is always floating horizontally and the lengths of the stay cables can be adjusted to uneven ground conditions. The area, first intended to be a park, has a roof-profile with a height-difference of about 0.4m. The anchoring of all vertical and horizontal loads from the stay cables is conducted by weight. A 3.5m wide and 0.10m thick concrete slab is placed in the area of the ambulatory, following uneven ground conditions. The stay cables are fixed to concreted twisted-steel mats with adjustable knots.

Aside from the wind-effects the temperature is an essential factor in this type of construction, since a change in temperature leads to a volume increase or decrease, thus affecting the internal pressure and the stiffness of the supporting elements. Based on the absolute pressure relations between surrounding area and internal pressure, a temperature change of 50 K would increase the internal pressure 20 times, thus pressure balance equipment would be required. Three ballonets were placed along the inner anchorings inside the hall for the helium supply and the connections to the moveable roof-segments. Across each Bal-lonett one third of the loadbearing ring, separated by a build-in bulk-head, and 8 roof-segments were controlled. Dividing the loadbearing ring also provides stabi-lity for the entire hall in case of pressure loss in one segment (Fig. 5).

Fig. 5

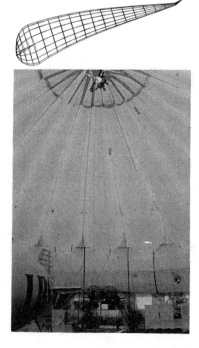

Fig. 8

The Materials

In addition to the wind, rain and exposure to the sun are essential factor affecting the foils during the intended period of operation from April to October 1993. The material used for the loadbearing ring and the roof-segments has to meet further requirements aside from helium density, sufficient tension- and tear-resistance, such as the lowest possible weight, low moisture, water- and UV-resistance. A 3-layer-laminat was used, consisting of 45μm polyester-foil, 50μm polyester-fleece and 45μm polyester-foil (total thickness: d = 0.144mm), with a high helium density, strengths of 115 N/mm^2 (= 16.6 kN/m) across and 90 N/mm^2 (= 13 kN/m) along the fleece, with a deadload of 180 g/m^2 and an available sheet-width of 1.2m.

Sewing the cut segments together could not be considered, due to the comparatively low tear-resistance and the desired helium density. Welding the foils along the curved edges lead to great stresses at the connection of welded and unwelded material and to tears at these points. The individual pieces were glued together to form the ring and the segments with a 2-component-glue used in baloon-construction and, due to the tight schedule, a double-sided adhesive tape without foil was applied. Tests showed that the 2-component-glue possesses the same transversial tension-strength as the foil, and the double-sided .adhesive tape the longitudinal strength of the foil.

During assembly the high tension-resistance of the foil proved to be disadvantageous along the curved edges. The intended helium density could not be achieved, small wrinkles or inclusions lead to an increasing helium loss.

The side-wall had to meet the following requirements: Being non-flammable, crease-resistant, transparent, low weight, sufficient strength and a sheet-width of about 4.5m to avoid additional seams. A transparent 100μm thin hostaphan-foil with a deadweigth of 135 g/m^2 was selected.

The thin stay cables with a diameter of 2.5mm consisted of a polyester coating with a DYNEEMA-fibre-filling and had a ultimate strength of 3 kN.

The Safety Concept

This construction could only be realised with the support of the authorities for supervision of building. Since there are essential differences between the construction of an airship-hall and regular structures, a safety concept was developed for the period of operation. These differences exist in the idea that buoyancy and internal pressure are play an essential part in the loadbearing capability of the helium-filled elements. Therefore the hall needs constant supervision. This includes controlling and adjusting the internal pressure to sustain a constant internal pressure independent from the surrounding conditions (air pressure and temperature). Due to the innovative loadbearing principle and a lack in knowledge about the durability of the foil-material, the seams and the connection-details, no definite statements can be made regarding the durability in general. For this reason, safety precautions were made independent from the strengths of the foil, the seams and the cables, to avoid injury.

Since the foils and the stay cables - both made out of polyester - were flammable, the number of visitors had to be limited to 99 persons. In Germany nonflammability is required only in the case of events with an attendance exceeding 100 persons.

The Loadbearing Behaviour

The windloads, used for dimensioning the foils, the cables and the connections were divided into operation of the airship while the roof is open, use with a closed roof and storm conditions. The airship is operationable up to a wind speed of 10 m/s (36 km/h). Therefore, with an open roof a windload of 0.07 kN/m^2 is assumed. During use with a closed roof, a wind speed of up to 20m/s and a windload of 0.15 kN/m^2 was calculated. A wind speed exceeding 20m/s is considered to be a storm and additional safety precautions take effect. The maximum windload is 0.3 kN/m^2.

In the calculations the buoyancy force amounts to 0.01 kN/m^2 and it is considered to be a load normally affecting the plane and subject to linear changes along the height due to a decrease in pressure. The dead weight of the foil including seams, connecting loops and cables is calculated with 2.3 N/m^2.

A roof-segment weighs 120 N. A volume of 22.5 m^3 and a buoyancy of 10 N/m^3 per segment results in a buoyancy force of 105 N per segment and 2.520 N for the entire roof. The loadbearing ring together with the stay cables and the side-wall foil weighs 2.250 N; the volume is 615m^3, resulting in a buoyancy force of 3.900 N. The total buoyancy force amounts to 6.420 N, corresponding to 82 N/m when applied to the average ring diameter of 25m, the total load is 5.130 N (= 0.83 kg/m^2).

To determine the membrane stresses and cable forces, only the loadbearing ring under construction without the roof was examined. In the final stage, the loads from the closed roof were considered as linear loads affecting the ring.

In the case of cylindrical elements under internal pressure, the ring stresses are twice as big as the tangential stresses. Under wind normally affecting the middle of the segment and affecting the loadbearing ring horizontally, the structural elements act similar to prestressed loadbearing systems under bending stress by relieving the prestress in the stressed compression areas. This means for the loadbearing ring and the roof-segments that without the effect of the windload, the tangential stresses are one half of the ring stresses under internal pressure, with the maximum and minimal stresses occurring under wind in the tangential direction.

In the case of a deformed geomtry due to the windload and a constant surface the volume, the internal pressure and the stresses change. The deformations of the ring under horizontal windload decrease with the increasing internal pressure. The maximum tangential stresses remain constant despite the increased internal pressure. In the case of a lower internal pressure, the larger deformations result in a greater pressure change in the deformed state leading to a smaller increase in stresses with higher internal pressure and the sum of prestress and windload remains constant in this case (Fig. 6).

During use the internal pressure of the loadbearing ring and the roof-segments is estimated to be 0.5 kN/m^2 and may be checked using a rising pipe pressure gauge. During use with the roof there are maximum tangential stresses of 5.1 kN/m and ring stresses of 2.3 kN/m. The internal pressure rises from 0.5 kN/m^2 to 1.6 kN/m^2. During construction without safety precautions the maximum tangential stresses increase in the load case of a storm with 0.3 kN/m^2 horizontal load to 6.3 kN/m and the ring stresses to 3.5 kN/m. The internal pressure rises from 1.5 kN/m^2 to 2.3 kN/m^2. A comparison of the calculated horizontal

deformations of 2.5m of the ring and the horizontal deformations of 1.6m including the roof in the load case of a storm proves the stabilizing effect of the tension stresses acting in a vertical direction. Here only the wind suction affecting the
roof has to be taken in consideration, due to its inclination of about 17 °.

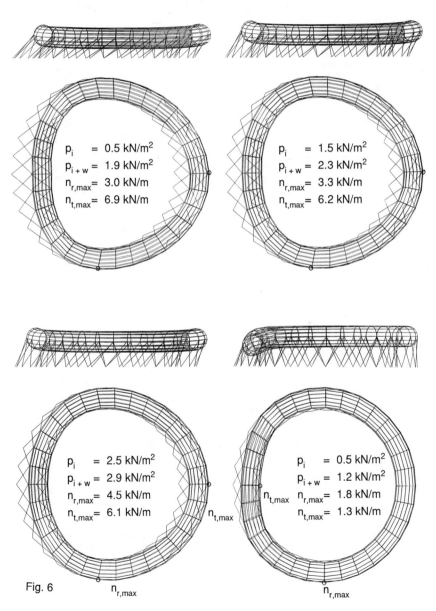

p_i = 0.5 kN/m^2
p_{i+w} = 1.9 kN/m^2
$n_{r,max}$ = 3.0 kN/m
$n_{t,max}$ = 6.9 kN/m

p_i = 1.5 kN/m^2
p_{i+w} = 2.3 kN/m^2
$n_{r,max}$ = 3.3 kN/m
$n_{t,max}$ = 6.2 kN/m

p_i = 2.5 kN/m^2
p_{i+w} = 2.9 kN/m^2
$n_{r,max}$ = 4.5 kN/m
$n_{t,max}$ = 6.1 kN/m

$n_{t,max}$

p_i = 0.5 kN/m^2
p_{i+w} = 1.2 kN/m^2
$n_{t,max}$ $n_{r,max}$ = 1.8 kN/m
$n_{t,max}$ = 1.3 kN/m

Fig. 6 $n_{r,max}$ $n_{r,max}$

Since the area is surrounded by tall trees and buildings, and therefore wind gusts as well as vertical windpressure may occur, the load case of a wind gust with 0.2 kN/m affecting one fifth of the ring in the total diameter and the upper outer quarter were examined. The calculated deformations of 1.2m vertically and 1.0m horizontally with an internal pressure of 0.5 kN/m^2 could be observed at the finished ring without the roof (Fig. 6).

To secure the emergency exits steel-pipe supports were placed over each of the three entries. These supports limit the maximum deformations of the entire hall in the case of wind-speeds exceeding 72 km/h (20 m/s), by placing air-cushions between the supports and the loadbearing ring and prestressing the ring against the supports. This way the diagonal cables also carry the horizontal wind loads and a torsion deformation of the entire ring can be avoided. These three point supports, stationary to the ring horizontally as well as vertically, lead to an increase in tangential stresses to 10 kN/m right next to the anchoring point. Using the supports reduced the maximum deformations by 60 %.

Using three stiff supports offers the possibility of avoiding buoyancy in the ring and fill it with air. The vertical deformations occurring under the total deadweight of the ring of 2.250 N show the influence of the internal pressure on the deformation behaviour. In the case of the estimated operating pressure of 0.5 kN/m^2 the vertical deformations amount to 2cm, if the pressure decreases to 0.3 kN/m^2, 60cm bends occur (Fig. 7).

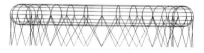

Fig. 7

There is no longitudinal stress in the upper cone-points of the roof-segments under internal pressure. In addition there is hardly any volume, therefore no buoyancy, but a great dead weight. The calculated buckling of the tips also occurred in the constructed segments. There were ruptures and small tears in the foil due to the constant movement of the tips caused by the wind, favouring the helium loss, since helium, being lighter than air, always gathers in the tips (Fig. 8).

In this load case there are only minimum stresses, due to the small windload when the roof is open and the flexibility of the segments. During a storm each segment is additionally supported in the center of gravity of the plane, transferring the loads while acting as a bending element. If the internal pressure is 1.5 kN/m^2, the foil has hardly any slack. Due to the internal pressure maximum ring stresses of about 2.9 kN/m occur.

The closed state, forming a cone-shell is the essential factor in dimensioning the connecting loops at the ring and the connections between the segments. The ring stresses amount to 1.9 kN/m with a windload of 0.10 kN/m^2 and correspond to the tangential support loads. The zippers and the loops between the roof-segments and the connections to the loadbearing ring have to be fashioned according to these stresses.

The Realisation

The cutting, gluing and erecting of the hall were carried out by members and students of the collaborating institute within 8 weeks (Fig. 9). Due to the course of the construction, the manufacture of the concrete foundation, the assembly of the ring, filling it with helium, anchoring it in the in target position, putting up the side-wall foil and finally the fastening of the roof-segments all connecting details had be placed at the ring from the beginning. Even in the case of the roof-segments, consisting of 8 parts glued together, the loops and the zippers for the cables for safety in the open state and during a storm were glued on afterwards.

Unfortunately, experience with the openable roof (Fig. 10) could only be gather during a short period of time. The premature aging of the foil material, the stiff supports unfavourable for the load transfer of the ring, the ring being filled with air and a roof opening of 20 % lead to a failure of the ring during a summer thunderstorm, resulting in the hall being dismantled after only 2 months of operation.

Further cooperation and assistance in the construction was given by the following:
Members of the Institute for Statics and Dynamics of Aerospace Structures, Institute for Structural Design, Institute of Building Materials, Institute of Applications of Geodesy to Civil Engineering.

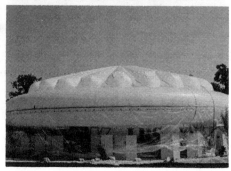

Fig. 10

Fig. 9

Wall Relief and Three Dimensional Suspension Sculpture
Utilizing Tension and Membrane Structures

Warren Seelig   1

Abstract

I am an artist, interested in sharing with engineers, architects and others of the scientific community, my aesthetic concerns and other observations in regard to a series of abstract sculptural forms which involve tension and membrane structures. During the past ten years, I have been constructing forms made of mesh or fabric membranes, which are carried through space on sequential pairs of spokes radiating from axles.   The suspension works began with simple wall reliefs, then evolved into three-dimensional suspensions of radiating lines and arched planes, slicing through space.   The sculptures are dependent upon gravity and a delicate counterbalancing of forces, as the weight of each individual pair of spokes contributes to the subtle but dynamic tension of the whole.   Although not inspired by any particular forms in nature, these constructions have an organic unity and inner life which make them as convincing as certain phenomena found in living things.   The following discussion includes   earlier textile forms, which   strongly   influenced   the   more   recent suspensions sculptures,   and   photographic documentation of each.

- - - - - - - - - - - - - - - - - -

1   Studio Artist and Professor of Art/Textiles, Philadelphia College of Art & Design, University of the Arts, Broad & Pine Streets, Philadelphia, Penna. 19102

Introduction

For about twenty years, I have been constructing flat, relief, and three-dimensional nonobjective forms.   They derive from an intuitive   vision   through   my   observation   of   certain   physical effects caused by tension, compression, gravity and light.   My work has evolved, without any particular knowledge of structural theory or inspiration, from direct observation of nature or the landscape.   I am deeply moved by the mysteries inherent in ordinary realities like the rising of the moon every night, but I have no interest in making such events part of my work.   I am not interested in propaganda and make no attempt to have my work communicate   anything   in   particular;   there   are   no   conscious messages.   As a child, I copied a picture of a seven-masted sailing schooner from the encyclopedia.   I remember becoming so totally absorbed in drawing the rigging, with its complexity of lines, that I completely lost sight of the subject.   I also remember, as a child, looking up to the ceiling for the answer.

Influence of Woven Textiles

Before the radiated steel spoke and skin suspensions illustrated in this paper,   I produced a wide range of woven textile forms. Weaving my first length of cloth on a hand loom in 1971, at   the Philadelphia College of Textiles & Science, proved to be a mesmerizing experience.   There was alchemy in the way that a plane of cloth emerged-------- a gossamer screen built up out of hundreds and thousands of particles of threads, interlacing both vertically (warp) and horizontally (weft). I was mystified by a process and by materials that allowed me to construct a surface with both the hand and the appearance of something organically whole.   This experience has never left me; and in some ways it has informed most of my work, from the early woven constructions to the present.

The experience of weaving led me to a considerable range of textile constructions over the span of about twelve years.   The fan weaving Double Accordian Gesture (1975) and the totem like

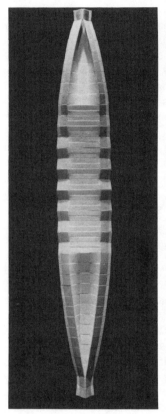 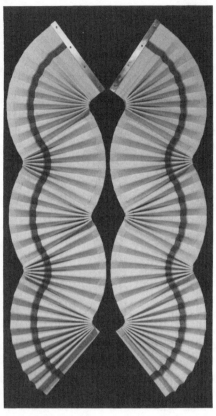

Figure 1.

Vertical Shield #1
14" x 81" x 6" (1977)
Cotton Double Weave and
Rigid Mylar Endoskeleton

Figure 2.

Double Accordian Gesture
40" x 65" x 2" (1975)
Cotton Double Weave and
Rigid Mylar Endoskeleton

Vertical Shield #2 (1977) are typical examples of my woven textile constructions that came out of an invention, utilizing the technique of double-cloth weaving.   The process of interchanged double-plain weave involves weaving two layers of cloth simultaneously on the loom, but in this case the two planes of cloth intersect in patterned squares and rectangles across the width and down the length.   Flexible Mylar segments which were inserted in each "pocket" created by the interchanged layers of cloth segments, became a permanently encased skeleton within the cloth, with built-in natural hinges of warp and weft at each axis of interchange.   This invention created a unique skeleton-skin relationship that allowed the Mylar segments to fold or swing 180 degrees in either direction.   Folding along natural seams to create radiating fans or subtle flutes, as well as long swelling curves, was thereby possible.

Wall Relief and Three Dimensional Suspended Sculptures

Wall suspensions utilizing a ripstop nylon skin, supported by a spoke on axle skeleton construction, eventually replaced my woven textile forms.   Wall-suspended works, such as Gelato (1987) and subsequent fully three-dimensional suspensions: Slice (1991), Double Ended (1992), Oval (1993), were strongly informed by the woven textiles.   The potential for the discovery of new forms was my primary motivation for this change in technique and material,   but my need to make work involving the unified organic construction continued.   As was true of the woven constructions, the technology in fabricating the suspended works is relatively unsophisticated.   Steel spokes are spread at their ends, drilled and threaded, with one spoke of a pair on each side of a horizontal axel. A sleeve is created in the fabric skin or mesh, and a cross rod is inserted.   Pairs of spokes project and hold long slices of various fabric membranes and meshes in suspension.   Spokes energetically radiate from strategically placed axles, projecting the plane of mesh or skin in space, above or below the level of the axle.   The mesh jumps from one pair of spokes to the next, with each set of spokes giving a different load, depending not only on its   relative   length but also on its position, as a spokes radiate from the vertical to the horizontal. The load created by spokes pulling on the fabric skin is practically zero when the spoke is sitting on the axle at 90 degrees and the load is full when spokes are projecting horizontal or at 180 degrees. Spokes collect on either side of an axle, with pairs of outside spokes suspending the mesh at the widest point. As the spokes collect from the outside to the inside on an axle, the distance between pairs of spokes narrows. Consequently the width of the suspended membrane slowly tapers to a point.

In Slice, Double Ended and Oval, the center triangle framework is the stable element that allows the upper and lower wing (Slice, Double Ended) or left and right half (Oval) to pull away in either direction from the top cross member of the triangle. In Slice, the upper half of the serrated wing is held in space by the collection of spokes that draws the Tyvek membrane forward and away from the top of the triangular framework. A second collection of radiating spokes draws the lower wing down with gravity and away from the same cross member as sell, at the top of the triangular framework. The upper and lower collection of pairs of spokes not only holds the white membrane in supremely delicate tension but acts as counterweight to hold the sculpture in balance as it is suspended from two points in space.

Describing the way these works function is like giving insight into the physiology of their form. These are not rigid, welded frameworks; rather, they are collapsible and somewhat yielding structures that behave like organisms in the particular way the parts relate to the whole. During the evolution of this work, when the suspensions moved from wall relief (see Gelato) to three-dimensional space, the perception of the works changed dramatically. The delicate balancing of forces that place tension on the Tyvek skin and the soft radiating "spray" of spokes exude a quiet energy and, ultimately, create a hovering sensation as Slice, Double Ended, and Oval hang in space. These works are not static, but appear to be buoyant or in some kind of suspended animation or stop action. The softly detectable dynamic internal forces, along with the illusion of energy caused by radiating spokes, help to create this sensation. Although the works are

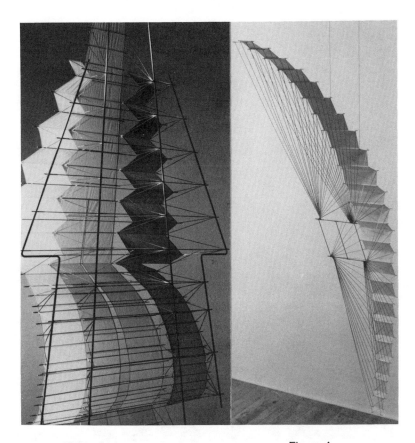

Figure 3.   Figure 4.

Gelato   Slice
Back Detail of Model (1987)   78"x8"x12" (1991)
Ripstop Nylon Membrane   Tyvek (spunbonded olefin
Stainless Steel Spokes   2   fiber skin) Stainless Steel
                                   Spokes and Framework

- -

2 Collaboration with Sheryl Gibson for IBM Corp.
   San Jose, California

Figure 5.

Double Ended
60" x 46" x 12" (1992)
Tyvek (spunbonded olefin fiber
skin), Stainless Steel Spokes

Figure 6.

Oval
63"x 48"x12" (1993)
Tyvek (spunbonded olefin fiber
Skin), Stainless Steel Spokes

stable, they are still closer to the brink of structural stability. Each spoke shares in the load, which, although extremely light, is allowed maximum deflection, thus reinforcing what we can see as dynamic tension.

These forms are not meant to resemble anything, yet as they evolve, some of the pieces take on certain human, animal, or even insectlike qualities.   The center triangle framework, for example, becomes the stable "thorax," while the wings or appendages extend naturally from that region.   The articulation of the spokes and the jointed construction reinforces this feeling of skeletal framework.   The relationship of the hard stainless spokes and the frame to the softer mesh or skin that it supports relates these works as well to the figurative or animalistic.   Most are bilaterally symmetrical, which even furthers the association. The repetitious segmentation or platelet construction of the membrane/mesh over the framework which is basic to all of the suspension works as well as the weavings, evokes images of certain shell formations or translucent carapaces in animals and insects. The three-dimensional works fold up into flat wings, and the upper and lower wing section of Slice collapses into a batlike configuration.   I enjoy the paradox of using industrial materials like stainless steel and spunbonded olefin fibers (Tyvek) or vinyl coated meshes------ to make forms that, strangely, may be as "convincing" as phenomena observed in nature.

Scale is a critical factor when constructing these works.   I usually maintain a size relative to the human figure----- that is, to my own relative dimensions.   Making work that lies within human scale seems to establish a dialogue in which we either consciously or unconsciously compare our own size.   For me, human-scaled work is more likelike and projects viscerally when compared with miniature, handheld work, or work that is larger than life.   As my work grows to human scale, it has a greater propensity to engage the viewer, for its presence becomes a physical challenge, and the dynamic physical forces that are inherent in the sculpture are more potent and readable when there is more to see.   Small-scale or miniature versions (models) of the real   thing   may render   the   form as an   ornamental   or precious object.   As the works grow larger than life or monumental in scale, they begin to reference themselves, to familiar, related forms (Ferris wheel, carnival ride) or appear to become an extension of the architecture.

The way that both wall-suspended works and, especially, three-dimensional suspensions occupy space and define volume may be their most individual quality.   In Oval, for instance, the spokes are actively radiating lines in space that define side walls and, along with the tapering Tyvek membrane overhead, identify the body or shell.   The radiating metallic spokes seem to dematerialize and transform to become lines of energy, while the central triangle framework slices through Oval to outline and define another illusive space.   This is sculpture in which concrete, tangible materials, such as stainless steel and olefin membrane, capture and define a rather intangible space.   I am very interested in making sculpture that creates large volumes of space, yet is ethereal and extremely light in weight.   I enjoy this relationship, in contrast to the more familiar, heavyweight sculpture of often imposing mass.

Gothic architects were strongly motivated to find ways of bringing light into their buildings in an attempt to transcend the mass and material, and ultimately, to dematerialize the building in order to achieve Heaven on earth.   From the beginning, I have been interested in the way that light can transform surface and form.   I have never been interested in re-creating its effects, but, rather, have been drawn to it as a kind of magical material to wash across the fluted planes of the early weavings and, now, to caress the sweeping arcs of recent work like Slice and Oval. Slice is literally a light catcher in that light intermittently brushes the soft, zigzag planes on both the top and underside of the wing.   The form is ever changing, depending on the location of the light source, and a symmetrical form can be put off balance by the weight of shadow concentrating on one side or another. The underside of the upper wing on Slice turns almost black with shadow and then incrementally lightens up to white on the tail.

Various skin or membrane materials, surface treatments, and color have been used in the evolution of constructions suspended from the wall to the freehanging pieces. A durable but translucent woven, ripstop nylon skin was first applied to the wall-suspended works, such as Gelato, so that light coming in from the sides could bounce against the white wall to illuminate the work effectively from behind.

Black and white were used in Double Ended, where flat-black oil stick on Tyvek effectively removed the possibility of shadow and somewhat flattened and weighted out the lower wing. The upper wing is opposite; it catches the light and darkens under shadow as the form seems to balance out on opposites. In a way, the voided black wing tends to reinforce and amplify the illuminated and floating upper wing simply by the dramatic contrast. In Oval, a blue black void is created in the way the interior of the form cuts a black slice out of space, while bright light gently washes the highlighted outer face, then dissapates to dark below.

I am determined to build sculpture that is given life by the effect of both natural forces and the presence of light. I have a strong urge to construct by hand, to invent and to abstract; for me, there is an authentic spiritual element tied to the transformation that occurs when ordinary materials assume an extraordinary form. In his book Abstraction and Empathy, Wilhelm Worringer wrote in 1908, "The urge to abstraction is the outcome of a great inner unrest inspired in man by the phenomena of the outside world." We respond in a manner not unlike the way "primitive" peoples shielded and protected their psyches from what worringer referred to as "the prevailing caprice of the organic." At the end of the twentieth century, our culture is reeling against the overwhelming forces of a world in which technology is encouraging us to experience everyday life as passive participants. For me, working abstractly is an antidote to the rule of blandness, driving me to the sensual, to materiality, to the physical world as a way of coping. The impulse toward abstraction is a way to discover and reveal what is within: the fantastic, the eccentric, the sublime.

Current State of Development and Future Trends in Employment of Air-Supported Roofs in Long-Span Applications

Kris P. Hamilton, M. ASCE[1]; David M. Campbell, M.IASS[1];

and Paul A Gossen, M.ASCE[1]

Abstract

Beginning in 1974 with the Pontiac Silverdome, and progressing through the Korakuen Stadium in Tokyo in 1986, the primary use of long-span low-profile air-supported roofs has been to cover sports stadia. There has been speculation that due to operational expense, problems of crowd control through pressurized exits, and especially problems of some facilities with developing effective snow removal techniques, the day of the air-supported long-span roof is past.

The authors believe that many of the problems that have been encountered in the past have been surmounted, and that air-supported roofs are viable as a structural system. Further improvements that may lower operating costs are also possible, although at higher first cost.

Other conditions indicating the feasibility of an air-supported roof include:

- Relatively low requirements for support of speakers, show rigging, and the like

- Knowledgeable facility management, committed to a program of preventive maintenance

- Budgets placing an emphasis on low initial costs.

In this paper, we will be considering only the low-profile, long span air-supported roof, as opposed to the high-profile" bubble" variety, which has seen a substantially different development and expectation.

Introduction

The initial rush of enthusiasm for the long-span air-supported fabric roof, beginning with the U.S. Pavilion in Osaka in 1970 and continuing through a dozen stadiums and arenas, has worn off in recent years. The advantages of initial economy, low lighting costs, and "open-sky" architectural concepts

---

[1] Principals, Geiger Engineers, 1215 Cornwall Avenue, Bellingham, WA 98225; and, Two Executive Boulevard, Suite 410, Suffern, New York 10901

have seemingly been overwhelmed by problems with ponding, high snow-melting costs, and a much-publicized "unreliability."

The stadia and arenas which have been the major market for this structural type are placing an increasing emphasis on issues such as light and sound control and sub-division into smaller spaces, needs that are not met by the most recent air-roof designs. Other proposed project types using this technology, such as enclosed "moderated weather" campuses (office or educational), and mega-structures for far-north mining camps, have fallen prey to volatile mineral and energy prices, as well as to concerns about reliability engendered by the problems in the smaller facilities.

Historical Development

The first major long-span air-supported roof was the U.S.Pavilion for the 1970 Exposition in Osaka (*Geiger, 1974*). This temporary structure, with major/ minor axis clear spans of 138 m x 78.5 m and a rise of only 6.49 m, had many design features that were not carried through into later structures, because of the increasing number of design constraints encountered in the design of public facilities.

Following the development of a "permanent" fabric for structural use (i.e. Teflon™-coated glass fiber cloth), new long-span air-supported roofs were constructed at a rate of almost one per year from 1974 to 1983. During the years of very high construction cost inflation, the dominant position of the air-supported roof was boosted by the very close attention being paid to first cost for all public facilities.

Table 1 shows the relationship of roof costs between several different stadium structures. The first-cost trend is very clear between the air-roofs and the "conventional" roof structures. These cost savings came from the lower construction costs of the roof system, reduced supporting structure costs, and an overall design attention to economy. There are several areas where the degree of integration of the air-supported roof with lower structure and stadium functions can affect costs:

Roof plan shape: The pavilion at Osaka (Figure 1) was a "super-ellipse" in plan, i.e. having the form $(x/a)^n + (y/b)^n = 1$, where n is a real number greater than two. For the "skew symmetry" used in this structure, such a geometry made the ring funicular for the uniform load case (Geiger 1974). Later roofs approximated this ideal shape, so as to simplify the precast seating structure. The ring beam at Pontiac Silverdome was a doubly symmetric eight-sided polygon. The stadium in Indianapolis (Figure 2) used a dual radius; the curvature of the roof was adjusted to provide horizontal cable reactions giving the lowest bending moments in the ring.

Roof Rise: The clear height of the roof is governed by the functional requirements of the stadium. The spring height of the roof then must be sufficient that the roof in the inverted (construction or deflated) position will have adequate clearance from the seats and

TABLE 1
Costs for North American Covered Stadia

Facility and Location	Roof Type	Roof Cost ($/Ft.2)*
Astrodome, Houston	Steel skew-trussed dome	$74
Superdome, New Orleans	Steel trussed lamella dome	81
Kingdome, Seattle	Concrete radial rib dome	73
Silverdome, Pontiac, Mich	Air-supported fabric	39
Metrodome, Minneapolis	Air-supported fabric	40
B.C. Place, Vancouver	Air-supported fabric	36
Hoosierdome, Indianapolis	Air-supported fabric	39
Thunderdome, St. Petersburg	Cabledome, fabric	49
Skydome, Toronto	Steel Lattice Arch, Movable	262
Georgiadome, Atlanta	Tensegrity dome, Fabric	63

* Cost is per square foot of clear span area, in 1993 dollars U.S.

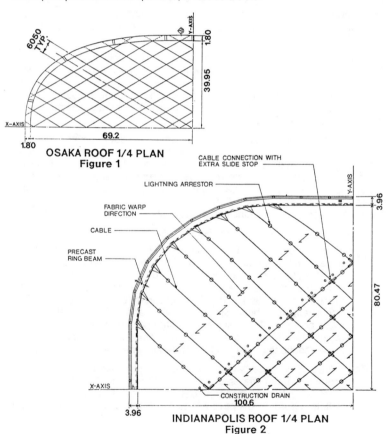

OSAKA ROOF 1/4 PLAN
Figure 1

INDIANAPOLIS ROOF 1/4 PLAN
Figure 2

the events floor. Considerations of drainage, cable length and strength will also influence rise.

Cable spacing: To minimize ring-beam costs (20 - 25% of total roof cost), it has been desirable to avoid having the beam carry large vertical loads. This has in the past dictated that roof cable anchorages should coincide with the supporting columns at the ring beam. The structural column spacing is generally governed by seating width between aisles, which in turn is governed by fire-safety codes. Typical roof cable spacings have thus evolved to about 12-13 meters, compared to 6.05 m in the Osaka structure.

Fabric rise: The rise of the fabric between cables is balanced between various factors. Higher rise permits use of lower strength fabric under uniform loads, and/or wider cable spacing. Lower rise tends to reduce snow-drifting, simplify patterning, and reduce fabric surface area and heat loss. Also, the resulting increase in fabric stress under inflation pressure provides greater resistance to pond initiation.

Fabric clamping: Fire code requirements for stadium roofs have led to the usage of glass-based fabrics, which require clamps at connections, rather than less costly attachment methods. Clamp design is driven primarily by issues of flexibility, installation sequence and serviceability; loading is only a small factor in the design. Clamping costs have served to encourage using maximum feasible cable spacings.

Additional ring-beam functions: Most of the stadia with air-supported roofs have used the ring-beam as a plenum to provide ventilation to the upper seating areas. Some roofs in snow areas have also used this plenum to distribute high-temperature air to the roof surface. As a result, non-structural requirements such as plenum cross-section, guttering for rain and snow run-off, and transportation routes for roof maintenance influence ring-beam dimensions along with structural needs.

Problems to Date

Many different functional problems have been identified with past applications of air-supported roofs for public structures. This paper deals only with the structural problems; architectural problems such as entrance and exiting, acoustic performance, and lighting are equally important in establishing appropriate future usage of this structural type, but are beyond the scope of this paper.

Deflation: This group of problems has received the most publicity, and has certainly been responsible for the public perception of unreliability. Table 2 provides a detailed breakdown of deflation incidents.

TABLE 2
Deflation History of Long-Span Air-Supported Roofs

Facility	Year Inflated	Deflation History
Korakuen, Tokyo	1986	None reported
Hoosierdome, Indianapolis	1983	C, 1988
B.C.Place, Vancouver, B.C.	1982	None reported
Metrodome, Minneapolis	1981	S, 1981 (Twice) (Note 1), 1982
Carrierdome, Syracuse, NY	1980	S, 1982, 1992 (Note 2)
O'Connell Ctr, Gainesville, FL	1979	None reported
Sundome, Tampa, FL	1979	PO, 1983; R,1987,1988 (Note 3)
Dakota Dome, Vermillion,SD	1978	C, 1978 (Note 4); S, 1982
Leavey Ctr, Santa Clara, CA	1976	None reported
UNIDome, Cedar Falls, IA	1975	PO,1975 (Note 4); R, 1976
Silverdome, Pontiac, MI	1975	R, 1975 (Note 5); S,1985 (Note 6)

Key: S = Snow   R = Rain or Lightning   PO = Power Failure
       P = Partial Deflation   C = Control System Failure

Note 1:   Occurred during construction; snow-melting system not functional.

Note 2:   Both were deliberate measures by stadium operators as a means of combatting severe snow accumulations, and did not result in any damage to the roof; re-inflation followed quickly.

Note 3:   Attributed to operator error, in failure to observe decreasing pressures and take appropriate action. System has been replaced by computer management, and problems have not recurred.

Note 4:   Occurred during construction.

Note 5:   Attributed to failure of metal cladding due to wind suction.

Note 6:   About one-half of snow-melt system not functional at time of storm.

Deflation is not in itself a failure; the air roof is designed to go up and down. The use of deflation as a snow control device at the Carrierdome is an example of beneficial controlled deflation. Deflation is only a problem when there is damage to the roof, and time out of service.

In the ten largest collegiate and professional sports stadia with air-roofs, from 1974 to 1993, there have been a total of fifteen full deflation incidents, averaging to approximately one deflation per ten "operating years." However, in the last one hundred operating years, there have been only four deflations, compared to eleven in the preceding fifty-eight operating years. It is also worth noting that of the five accidental snow deflations, the last one occurred at the Pontiac Silverdome in 1985, and none have occurred in the three latest structures. This is a direct result of two improvements which have been made: the introduction of computer patterning and similar design refinements, and greater knowledge and planning on the part of the operators.

Four of the snow-caused roof deflations have shared as a common immediate cause the sudden motion of a pond or drift located at the top of the large rectangular or triangular panels which form the perimeter of the roof. These bodies exist in a state of unstable equilibrium, which can change suddenly as a result of geometry changes in the roof, caused by increasing interior pressure, reducing weight on the roof while maintaining a constant pressure, or similar operational technique. The mass of snow/ice/water then moves rapidly down the roof, generally causing the fabric to tear near the ring beam.

Excessive Operational Costs: Energy costs related to roof inflation, and especially to snow melting, have risen since the first such roofs were constructed. Together with additional operating and maintenance personnel, life cycle costs for these roofs have generally been higher than originally anticipated.

Directions for the Future

Nearly two-thirds of the total operational experience to date with long-span low- profile air-supported roofs has occurred since the construction of the most recent North American facility; there has thus been little opportunity to put into practice the design ideas generated by this experience. Nonetheless, several strategies have been developed by the authors and others to deal with the real problems and the perceived deficiencies.

Of the various problems encountered to date, it is clear that the issues most difficult to deal with from an operator's perspective are those of the structure's response to snow drifting and ponding. (We are not aware of any problems due to wind suction, although down-bursts during thunderstorms have contributed to at least one ponding failure leading to deflation.)

To overcome public and owner resistance, it will be necessary to develop the air-supported roof to a clearly new generation of design. At the same time, it is imperative that any improvements serve to reduce requirements for energy and for operations and maintenance personnel.

Any strategy to increase structural reliability of these roofs must:

A. Reduce the likelihood of ponding or drift formation, and increase ability of the roof to carry ponds

B. Facilitate removal of drifts or ponds if they do occur, to avoid damage

C. If damage does occur, limit so that deflation does not occur, and facilitate repair.

The first of these strategies is the most passive, and is thus by assumption deserving of the greatest attention.

A. Reduce pond formation and size

Both experience and wind tunnel tests have indicated that the worst drifting conditions occur in mild-to-moderate wind conditions, in temperatures of about -4° to 0° C. At a drift, snow accumulates faster than the average ground accumulation rate, and may not be fully melted off. As a drift builds along a cable line, it impedes drainage, and contributes to development of a pond. Ponds due to snow typically occur at the top of a rectangular or triangular panel, initiating at a corner, and moving towards the the center of the panel as they grow. (Ponds caused by rainfall have occurred close to the center of a roof.)

We have identified two aspects of the air-supported roof system that may be manipulated to reduce or eliminate this problem:

1) Modify the detail roof shape so that ponding is prevented or is reduced to such a level that the internal pressure can resist local inversion of the membrane.

   a) Reduce Cable Spacing

      By reducing cable spacing, the rise of the fabric panel between cables will be reduced, thus reducing both the tendency to drift and the size and weight of drifts or ponds that may accumulate.

      Alternately, if the two-way cable grid is maintained over the entire roof surface, as in Figure 1, rather than just over the central area, as in Figure 2, the above benefits will also accrue. The resulting roof shape will have a more uniform curvature, and the resulting fabric panels will be substantially stiffer than the large rectangular and triangular panels of the latest designs.

      As is pointed out above, the current cable spacing standard was developed as a result of cost considerations. Using a layout such as shown in Figure 1 will result in about twice the amount of cable and cable clamping if cable spacings are kept the same; halving the cable spacing will double this again, resulting in a system cost increase of about $12 per square foot for clamping materials, increased fabric complexity and installation. Ring beam costs would also increase somewhat due to the more irregular cable terminations.

      Two important areas for further design development are therefore:

      - in fabric clamp connections, to facilitate installation, reduce material costs, and maintain watertightness; and

      - optimizing cable spacing with respect to ponding potential.

   b) Increase Fabric Operating Stresses

      Increasing the fabric stress under normal operating pressures will result in a higher span to rise ratio, reducing the profile and the tendency to drifting in snow conditions. The membrane will be effectively stiffer, which will also reduce the formation of ponds. Sufficient fabric curvature must remain to permit normal relative displacements of cables during the inflation process.

      Computerized patterning techniques used in the stadia in Tokyo, Vancouver, and Indianapolis have already demonstrated their value over earlier hand methods. By providing a more uniform bi-axial stress field, especially in the panel corners, the tendency to pond is substantially reduced. Any facilities replacing their roof membrane should be re-patterned with these techniques.

2) Improve snow-melting and pressurization systems and controls so that snow drifting and accumulation are prevented by timely delivery of heated air and/or increased internal pressure to drift-prone areas.

   a) Implement Additional Instrumentation into Mechanical Control System

   The early control systems for air-supported roofs used primarily visual observation of the roof to determine its position and monitor the development of any problems or drifts. These methods are dependant on the skill and attentiveness of the operations staff, and do not give sufficient detail to indicate whether the roof is in any distress from a given environmental load.

   Cable force indication (strain gages or load cells) and roof height monitoring together can be used to establish at any time the state of the roof. The readings would be complex, and computerized control systems would be required to correctly and quickly assess the meaning of the readings.

   b) Improving Heat Distribution to Roof

   Providing heated air to the roof to melt snow has been the typical method of dealing with snow on these roofs. However, experience has shown that the current designs do not have the ability to accurately deliver sufficient heat to deal with drifting associated with major storms, unless the entire structure is pre-heated. System capacity and lag time is such that starting to melt snow only after a problem has developed may not be enough to keep the problem from growing. Rising energy costs and practical limitations to heating system size have made it important to be able to deliver only the required amount of heat, and only where needed.

   Many methods have been suggested for such point-delivery of heat, including infrared lamps, vast quantities of heat tape, and vents directly from the roof to the exterior. More practical, we feel, would be to create fabric channels directly at the cable lines that would take hot air directly to where it is needed. Along with this, an additional expenditure on controls and dampers would be required to enable distribution to be limited to affected areas.

   c) Improving Automatic Operation of Mechanical Snow Melt Systems

   The development of fully computerized building management systems has made it much easier to gather information from multiple locations, display it in an understandable manner, and issue commands that can control a large number of operational devices such as fans, dampers, and the like. The Korakuen Stadium in Tokyo has carried this to the furthest extent of any facility; computerized controls permit the display of weather conditions, snow accumulation at multiple points on the roof, temperatures throughout the facility, and roof cable stresses, as well as enable the operators to select and initiate a variety of automated mechanical system sequences in response to developing storm conditions.

The next step in development of such control systems would be the development of algorithms which would accept the various inputs from item 2a) above, and display the shape and stress state of the roof, as a means of establishing which areas need attention in the form of additional heat or snow-removal. These systems could also be used to monitor and control the roof shape when snow is being removed, to reduce the potential of dislodging an unstable mass.

B. Facilitate removal of ponds or drifts

Measures such as outlined above could be reasonably expected to prevent problems with snow and rain accumulation such as have occurred in the past. However, prudence dictates that means should be provided for removing excess loads when they do occur. Stadium Operators and Facility Managers have been actively involved in the development of means to remove excess snow and water, which is the second strategy towards a reliable active structure.

Early methods involved using snow shovels, buckets, pumps and large crews to remove ponds and drifts. Around 1983, operators experimented with using high pressure water to move snow and to serve as a "venturi pump" to empty ponds. This method has grown in acceptance, and is now used throughout the snow belt for these roofs. More permanent installations have been made, and some facilities have used waste heat to warm the water prior to application. These systems have considerable advantages for the roofs, in that they require smaller crews on the roof, and do far less damage to the fabric than earlier methods. Future designs in snow areas should integrate a water distribution system into the roof.

C. Limit damage and facilitate repair

The third part of a strategy for improving reliability is to plan for a failure, and design the system so that at least one component can fail without causing a deflation of the entire roof. The weak member in the system has been the fabric; therefore it is desirable to modify the roof materials and details so as to reduce the likelihood of system failure occurring as a result of failure of any single panel.

Typical experience with previous deflations has been that the initial damage was small enough that the roof support mechanical system could provide the necessary make-up air. However, subsequent tearing of the fabric under snow and/or wind loads progressed more or less quickly until this was no longer possible, and the roof deflated. Ideally, the largest possible opening as a result of a single panel failure should be such that the roof inflation system would be able to maintain a system pressure sufficient to resist all dead loads and a selected amount of snow load.

By using smaller panels, the amount of roof that could be damaged by the failure of any single panel would be reduced. Methods for reducing fabric panel size include reducing cable spacing, as discussed above, and subdividing panels between main cables with an intermediate cable system. Inclusion of rip-stop elements in fabric seams would also accomplish this goal.

Due to their stiffness, currently available glass-based fabrics have very low tear strengths, compared to their tensile strengths. Development of fabrics with a substantial increase in tear strength would be another approach to limiting damage.

Conclusion

The first application of the long-span low-profile air-supported roof in Osaka in 1970 was also the "purest" in following the logic and mathematics of David Geiger's original patents. The roofs that followed responded in different ways to the needs of the stadium building type; i.e. the realities of fitting a roof system to a stadium which has in turn been shaped around a sports field. Since that time, new fabric roof systems have been developed which approach the air roof in life cycle costs.

As currently implemented in the Vancouver, Tokyo, and Indianapolis stadia, the air-supported roof would be a feasible structural solution in areas of light or no snow-fall. With modifications as identified above, some of which return more closely to the 1970 design, it would be a usable system in areas of moderate snowfall. However, the modifications would reduce the first-cost advantages of the air-roof, and careful life-cycle cost study would be required to justify the use of this active structural system over a passive one.

The air-supported roof has limitations which must be recognized when studying potential applications. However, wherever first-cost is an issue, we feel that the air-supported roof should be given serious consideration among the possible options.

REFERENCES

Geiger, D.; U.S. Patents 3,772,836; 3,841,038; 3,835,599 "Roof Construction"

Hamilton, K.; Campbell, D.; and Geiger, D.; *Comparison of Air-Supported Roofs on the Vancouver, B.C. and Indianapolis Stadia*, Space Structures for Sports Buildings, Proceedings of the International Colloquium on Space Structures for Sports Buildings, Science Press, Beijing, 1987

The Use of ETFE Foils in Lightweight Roof Constructions

Craig Schwitter[1]

i. abstract

ETFE foils are polymer film sheets which are enclosed in support frames and inflated to provide stable, insulated and transparent roof cushions for lightweight enclosures. This paper describes foil cushion background with specific attention drawn to material properties of ETFE foil and their implications in structural design of cushions and support structure. Several projects are presented to communicate to the reader the possibilities of this material in construction.

ii. background

Inflatable roof cushions of varying materials have been investigated and used in the construction of lightweight structures since the 1950's. In the 1970's advances were made with polymer films, (most notably polyester films) for inflatable cushion construction. These cushions were for the first time transparent and a suitable alternative to other glazing systems; however, the materials were highly susceptible to UV light, and degraded quickly under sunlight exposure. The development of flouropolymer ETFE foil cushions in the early 1980's resulted in a highly durable material with a wide range of transparencies, from opaque to translucent, and high resistance to UV degradation.

ETFE foils have been commercially developed for construction purposes over the past 15 years, primarily in Germany and the UK by a German Manufacturing firm, Vector Foil Gmbh. Buro Happold have been involved in developing lightweight foil enclosure designs since the early 1980's when cushion construction was first considered for several large enclosure projects including the design of a covering for a township in Northern Canada. Currently, foil cushion roof are being used in applications where controlled environments, but high sunlight transmission is desirable, such as sporting halls, swimming pools, or greenhouses. However, advances in manufacturing processes, detailing and general acceptability have opened up wider possibilities in the use of foil roofs. Due to their high insulation and high light transmission properties, ETFE foil cushion constructions have been

---

[1] Project Engineer, Buro Happold Consulting Engineers, Camden Mill, Bath, U.K.
FTL/Happold, 157 Chambers Street, NY, NY, U.S.A.

increasingly introduced as economical alternatives to glass panel systems and planar glazing for roofing enclosures such as atriums, sports halls, and retail areas.

iii. foil roof systems - rigid and flexible

The ETFE foil cushion assemblies typically consist of two or three layers of ETFE foil extruded into thin sheets. The layers are welded at the perimeter to form air-tight cushions. The cushions are restrained through aluminium edge extrusions or clamping which is then attached to a primary structural system (Figure 1). The cushions are given a constant inflation pressure to stabilize the surface under wind and to offset downward snow loading. Panels widths are restricted by manufacturing and structural concerns to 3.6 m but panel lengths have been designed in excess of 15 m.

Cushion construction over the past 10 years has been primarily concentrated in "rigid" type panels. These are panels which are the direct replacement of a glass or other type rigid panel with shapes of the panels being quite regular, i.e. rectangular or triangular, A considerable advantage of these systems is that the lightweight inflated cushions form the secondary structural roofing system which is normally a purlin, beam or truss system in planar glazing design. However, this elimination means that the cushions edges must take up the horizontal inflation forces. Edge members in horizontal bending can often be sizable. The first two sample projects shown in a following section illustrate "rigid" type foil systems.

Flexible edges on cushions have been considered in the past, and recently progress has been made in more flexible surface forms using ETFE foil cushions. Essentially this is the process of eliminating the rigid edges to be replaced with flexible cable edges and adapting extrusions to clamp the cables and foil panels. Design for movements in this type of system is critical to its success. Nonetheless, it is felt that a flexible system of support and connection will provide for a more uniform stressed surface, particularly regarding edge conditions. The third sample project is a flexible foil enclosure system currently being designed.

iv. relevant design criteria

material properties

ETFE foil consists of a copolymer of linked monomer units of Ethylene and Tetra-Flouroethylene which is processed and extruded into a film of varying thickness and translucency. Material thicknesses are typically in the range of 30 to 200 microns, but can be varied to suit any design. Translucency can be altered by the addition of pigmentation to form a white, milky film of variable light transmission. Alternatively, the light transmission qualities of the film can be altered by printing a matrix of silver dots on the foil sheets, again providing for variable amounts of light transmission.

Testing has been carried out by the manufacturer as well as through a test programme sponsored by Buro Happold (1)(2). Uniaxial tension tests show the basic material characteristics of ETFE foils (Figure 2). ETFE foils display a marked yield point at approx. 3 % strain. After yield, the material shows large elongation to ultimate failure (approximately 200% strain at ultimate). The large amount of energy required from yielding to ultimate failure has several interesting results, most notably that the tear propagation load for the material is extremely high. Thus, small slits in the foil cushion will not tend to widen and lead to possible structural failure as might happen in a stressed skin structure of polyester or glass fibre.

Being a polymer, ETFE foil is susceptible to strength loss at elevated temperatures. At 60° C the strength of the foil is considerably reduced. Thus its use in extremely hot climates or applications may be questionable. However, creep due to extension of foil tends to be elastic and providing that the inflation system is controlled accordingly, panels subject to excess heat will tend to expand and contract when the heat is removed. This action also ensures that air expansion in cushions due to thermal changes can be accommodated.

Weathering tests performed on ETFE Foil cushions have shown that the material has a strong resistance to UV degradation. Less than a 10% decrease in material strengths has been observed after 10,000 hours of artificial weathering. Natural exposure samples (Located in Arizona and Bombay) have performed well and material is expected to have a life-span of 25 to 50 years. Testing has shown that ETFE foil has very low gas and water permeability. Tests on foil samples in heated water showed no signs of strength loss due to water permeability. Most importantly, the low impermeability ensures that dirt and grease particles can not easily be absorbed into the surface, making the cushions highly dirt resistant. Their smooth surface also helps in terms of reduction of wind suction pressures due to surface wind disturbances creating vortices.

Light transmission and insulation properties of foil cushions are two of their strongest attributes. As stated previously options exist for clear, translucent (white pigmented) and shaded panels. The amount of allowable solar gain is typically a design consideration, but transmission values can be anywhere from 20-95%. Insulation is provided through the barrier of air in the cushion and is considerably increased when three layers of foil are introduced.

Fire tests have been undertaken and the material has been approved to both UK (Class O) and German fire codes. ETFE Foil does not support flame spread with an oxygen index of >25% and is thus inherently flame resistant. However, the material has a melting point of approx. 220° C, but does not form into droplets upon melting, preventing dripping during a fire. The melting point does conveniently allow for thermal impulse welding at approx. 300° C. Joints consist of lapped layers of foil, with layers typically limited to 200 microns to prevent thinning of the foils near the welds.

structural design

Cushion panels are formed by placing flat, unpatterned sheets of foil into frames and inflating the panels to the desired inflation pressure. This inflation pressure must be chosen to be suitable for the cushion assembly to perform under all reasonable loading conditions. Loading for the cushions can be considered to be either an initial prestress which is followed by differing combinations of upward and downward loading. For a two way spanning cushion, the internal pressure must be in equilibrium with the principal surface tensions and their respective radii of curvature. However, since the foil elongates to its final position, the two-way relationship must be modeled. Finite element modeling provides accurate results which have shown good comparison to in-service cushion assemblies (3). For a quicker, more approximate result or for a one-way spanning cushion the strain relationship of a curved fabric strip can be utilised where:

S = final length   So = initial length
w = appl. loading/unit width   EA = Stiffness
T = final tension   To = initial tension

Tension is equal to applied loading times radius of curvature (1) and also equal to the stiffness times strain in the material (2):

$$T = w \cdot R \quad (1) \qquad \text{AND} \qquad T - T_o = EA \cdot \left(\frac{S - S_o}{S_o} \right) \quad (2)$$

Final length S is related to the radius of curvature as follows:

$$S = 2R\theta = 2R\sin^{-1}\left(\frac{S_o}{2 \cdot R} \right) \quad (3)$$

Substituting (3) and (1) into equation (2) yields an expression which can be solved interatively for T knowing initial span, prestress, stiffness, and applied loading:

$$T - T_o = EA \cdot \left[\frac{2 \cdot T}{w \cdot S_o} \sin^{-1}\left(\frac{w \cdot S_o}{2 \cdot T} \right) - 1 \right]$$

These equations can be quickly used to find surface tensions and reactions for supporting structure design (Figure 3). Relating the behavior back to the stress-strain relationship of the material one notices that prestress pressures in the cushion push the skin close to the yield point of the material. In analysing load cases wind suction on outer film, internal wind suction on inner layers should be taken in combination.

Snow loading must also be considered on cushion panels. Uniform snow loading will be resisted by the inflation pressure of the cushion. Snow overloads in areas of extreme snow fall can be handled by means of thin stainless steel wires located under the inner film layer. The cushion assemblies invert onto the wires and reinflate after the snow is removed or melted.

Supporting framework for the panels should be designed for the maximum combination of prestress and applied loads. It is important for the designer to realize this particular loading condition for foil cushion assemblies since it implies that stiffness and strength are required in the horizontal plane of the structural system, whereas normally it would be more rational to visualize it in the vertical.

v. a few examples

The following three projects represent a range of projects designed by Buro Happold from 1988-1993. All projects were designed in cooperation with Vector Foil Gmbh and are located in the UK.

westminster chelsea hospital

ETFE foils were used to enclose an enormous atrium space for a new hospital situated in downtown London and completed in 1992. The hospital is organized around a central nave with four crossing transepts (Fig. 4). Considerable cost savings were realized since the cushions could be directly incorporated and supported off the primary structural system and eliminating the need for a secondary system. The overall plan size of the atrium is 116m by 85m. The foil cushion assemblies are fitted to aluminium arched frames with the largest of the foil panels being approx. 4m x 3m. The cushions are inflated to operating pressures of 400PA with a maximum depth of approx. 600mm. A detail of the base structure shows a cross section of the typical foil sections and the incorporation of the air supply valves hidden in the aluminium arch extrusions (Fig. 5).

The atrium is formed from two differing panel constructions. The outer perimeter panels are white translucent film providing shading and controlling direct solar gains while preventing glare in rooms off the atrium. The inner panels are clear transparent allowing for direct sunlight to penetrate the atrium.

schlumberger research atrium

Completed in 1992 as an addition to the existing Schlumberger Research Centre in Cambridge, the new structure had a central atrium with 5 rectangular panels 10m x 3m (Fig 6). The panels were rigidly attached to trusses spanning the atrium with a ventilation area left open at the top. The foil panels were triple layer constructions with a silver dot matrix printed on the inside of the outer layer designed to reduce glare and solar heat gains. It is interesting to note the swaged edge profile developed by the manufacturer which provides a thermal break for the cushion construction designed to reduce the effects of condensation (Fig. 7).

eastleigh tennis halls

Currently being designed, this project covers 10 indoor tennis courts with a plan area of approximately 6000m2. The cushions are supported from a system of highly prestressed parallel cables supported off a ridge cable which is in turn supported by opposed external masts (Fig. 8). Each of the two ridged structures is approx. 78m x 36m and are separated by a steel portal spine. The entire structure is stabilised via a system of ground anchors.

The foil panels are approximately 3m x 18m and have been designed to resist full wind and snow loading. The foil panels are connected to the cables via an aluminium clamping and extrusion system which allows for full movement of the cushions under wind oscillations and thermal variations (Fig. 9). The foil edges at top and bottom are terminated on aluminium edge channels which housing air supply hoses. Differing from the rigid connections of the hospital and atrium roofs, the details for the foil to cable connections have been designed to take into account the movements of the cable net. Rotations in the end connections to perimeter steelwork necessitate the use of sliding panels at the edges to accommodate for in-service movements and easy installation.

vi figures

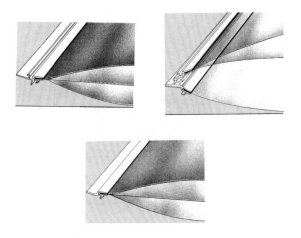

figure 1. cushion and edge extrusion examples

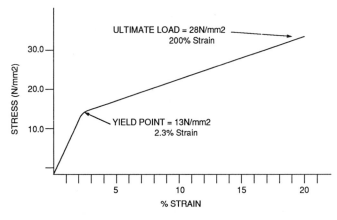

INDICATIVE STRESS/STRAIN DIAGRAM FOR ET FOIL

figure 2. stress/strain diagram

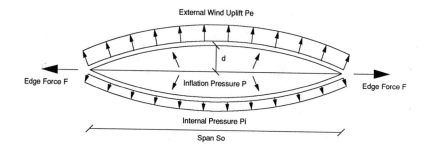

External Wind Uplift Pe

d

Edge Force F

Inflation Pressure P

Edge Force F

Internal Pressure Pi

Span So

figure 3. approximate foil cushion analysis

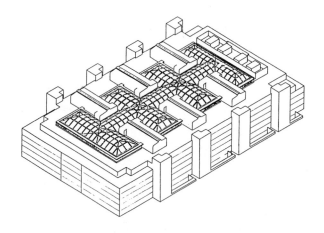

figure 4. westminster chelsea hospital isometric

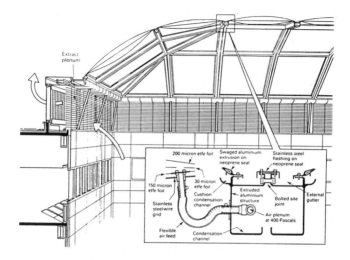

figure 5. hospital roof construction

figure 6. schlumberger atrium interior

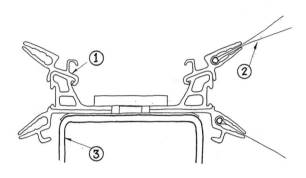

① ALUMINIUM EDGE EXTRUSION
② ETFE FOIL CUSHION
③ PRIMARY STRUCTURAL SECTION

figure 7. schlumberger extrusion detail sketch

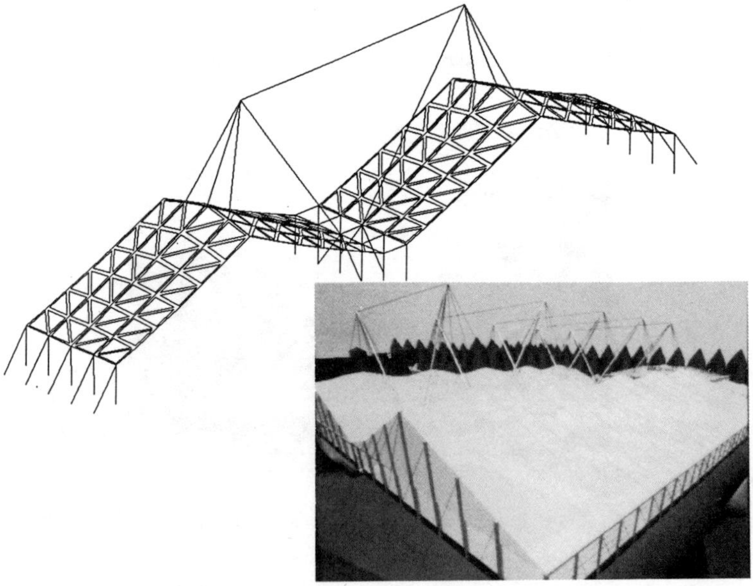

figure 8. computer modeling of tennis halls

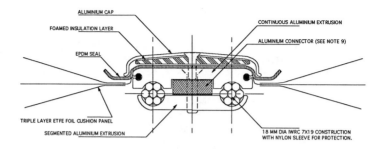

figure 9. flexible extrusion connection detail

vi. references

(1)   Hoescht Plastics. Sales Information VM417 Hostaflon ET 6235, January
         1984.
(2)   Ansell, M.P. *Report of a Test Programme Carried Out for Buro Happold
         Consulting Engineers on the Acceptability of ET Hostaflon Film for Use in
         Buildings.*   University of Bath School of Materials Science, 1985.
(3)   Liddell, W.I. *Westminster and Chelsea Hospital: Atrium Roof Structural
         Design Report*, Buro Happold Consulting Engineers, February 1989.

STUDY ON MECHANICAL CHARACTERISTICS OF A LIGHT-WEIGHT COMPLEX STRUCTURE COMPOSED OF A MEMBRANE AND A BEAM STRING STRUCTURE

Prof. Masao Saitoh, Dr. Eng.[1]
Akira Okada[1]
Katsuo Maejima[2]
Tetsuo Gohda[3]

Abstract

This paper aims at studying a new type of Beam String Structure, in which a membrane is used as finishing material. This paper reports on the mechanical characteristics of this structure, an analytical method, and the execution control method employed during construction.

Introduction

A Beam String Structure, that is BSS, is composed of a structural system which pursues a goal of rationality in transportation, execution and mechanical rationality. In Japan the BSS has been widely applied to the construction of flat frames with a span of 150 m and for buildings located in areas that are subject to heavy snowfall.

The structure reported in this paper constitutes the following features in comparison with the conventional BSS.
(1) A membrane used as finishing material resulted in the realization of a light-weight frame.
(2) The BSS is not connected with a membrane through the use of metallic fittings.
(3) Neither a space for the jack, used for the introduction of prestress, nor such a system as that used as a turn buckle for the regulation of the string, is installed.

This paper will first report on the results of the experiments which were carried out in order to obtain the structural characteristics. Then, through the comparison between the analytical results and the experimental results, the validity of the analytical method will be verified.

---

[1] Department of Architecture, College of Science & Technology, Nihon University, 1-8, Kanda Surugadai Chiyoda-ku, Tokyo, 101, JAPAN

[2] Penta-Ocean Construction CO., LTD., 11-25, Higasi-ooi, 1-Chome, Sinagawa-ku, Tokyo, 140, JAPAN

[3] Ogawa Tent Co., LTD., 10-13, Fuyuki, Koto-ku, Tokyo, 135, JAPAN

Photo.1 Exterior - View and Interior - View of Actual Building

Finally, the construction method and the management method for the actual building will be introduced. At the same time, the actual behavior which was measured during the construction, will be compared with the design value.

Outline of the Actual Building (See Photo.1)

The roof structure of the building is composed of the three pieces of the BSS which are placed with simple support, parallel to each other on a rectangular plan with a span of 22.4 m and a length of 35 m. The BSS is made of a flat arch possessing a rise-span ratio of 0.13, a string with a sag-span ratio of 0.06 and three struts. The BSS members are connected to each other by cross-type stiffening members made of tension materials. The BSS members are covered with a single-sheet membrane. Furthermore, valley cables are installed parallel to the BSS on the membrane. The membrane is not joined with the BSS.

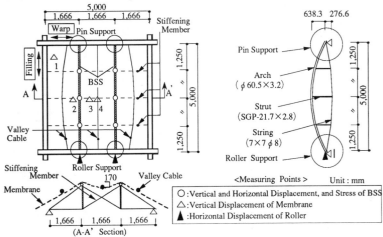

Fig.1 Outline of Scale Model for Loading Test

Table 1 Material Properties of Model for Loading Test

	Arch	Strut	String	Stiffening Member	Membrane	Valley Cable
	$\phi\,60.5\times3.2$	$\phi\,21.7\times2.8$	7×7 $(\phi\,8)$	RB - $\phi\,6$	FGT- 250	7×7 $(\phi\,8)$
A	5.76	1.66	0.31	0.2826	Ext=1.07×10^{2}	0.31
Ix	23.7	0.7586		0.0063	Eyt=1.56×10^{2}	
Iy	23.7	0.7586		0.0063	ν x=0.93	
Z	7.84	0.699		0.0211	ν y=0.64	
E	2.1×10^{6}	2.1×10^{6}	1.6×10^{6}	2.1×10^{6}		1.6×10^{6}

A:Section Area(cm^2)   Ix,Iy:Moment of Inertia(cm^4)   Z:Section Modulus(cm^3)
E:Elastic Modulus(kg/cm^2)   Et:Eloagation Rigidity of Membrane(kg/cm) ν :Poisson Ratio

Outline of the Experiments

Fig.1 shows the outline of the experimental model, which is scaled down to approximately 1/5 from the actual building. The number of the BSS is reduced to two from three for the actual building. Members, as well as the membrane, were selected so that they could satisfy the static scale law as much as possible (Table 1). Two different models were employed: Model A equipped with the BSS only, and Model B, which is covered with a membrane placed over an Model A. The initial tensile force (300 kg/m, 400 kg) imposed upon the membrane and valley cable was set up based on the result of the shape analysis which was carried out beforehand.

Table 2 Case of Loading Test

	Model A	Model B		
	BSS	BSS + Membrane		
	Snow Load	Wind Load (Internal Pressure)		
			With Valley Cable	Without Valley Cable
Uniformly Distributed Load	Maximum Point Load 33.6 kg/m²	Maximum Snowfall Load 100 kg/m²	Initial Membrane Strees : 300 kg/m Prestress of Valley Cable : 400kg	Initial Membrane Strees : 300 kg/m
Partially Distributed Load	Maximal Point Load 24 kg/m²	Maximal Snowfall Load 100 kg/m²	Maximum Internal Pressure 200kg/m²	Maximum Internal Pressure 200kg/m²

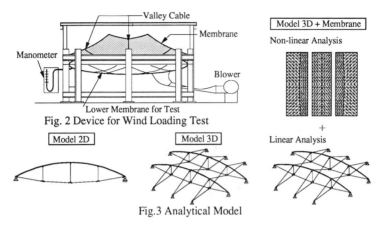

Fig. 2 Device for Wind Loading Test

Fig.3 Analytical Model

Two types of experiments shown in Table 2 were carried out. In the wind loading test, a method in which air pressure was applied to the inside of the model was employed on the assumption that wind force was pushing upwards (Fig.2).

Outline of the Numerical Analyses

In the analyses made in this paper, the three types of models shown in Fig.3 were used. For both the plane model (2D) and the space model (3D), linear analyses were employed. In regard to the space model equipped with the membrane (3D+M), both the geometrical non-linear analysis for the membrane and the linear analysis for the BSS were carried out. The membrane is modeled by separating it into two parts, a central part and an ending part, according to the location of the BSS.

Snow Loading Test

(1) Purpose. Since the shape of the membrane changes due to the load, the stress vector transferred from the membrane to the BSS changes with an increase in the load. This experiment was carried out with the main aim of understanding the interaction between the membrane and the BSS.

(2) Result and study: a) Uniformly Distributed Load. Fig.4 illustrates the relationship between the load and the tensile force for the string. The incremental amount of the tensile force for both models are almost the same. It can be seen that the value obtained from both the space and the plane analyses corresponds quite well to the experimental value.

b) Partially Distributed Load.   In the relationship
between the load and the tensile force for the string, linear characteristics can be found (Fig.5). However, an
increase of the tensile force for string B on the non-
loading side of both model is recognized. It can be interpreted that this is caused by the spatial effect of the
stiffening member.

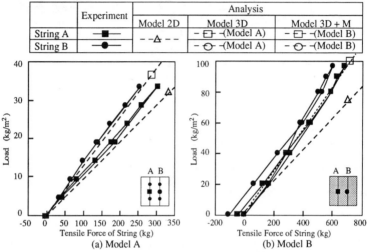

Fig.4 Relationship between Load and Tensile Force of Strings
Obtained from Uniformly Distributed Snow Loading Test

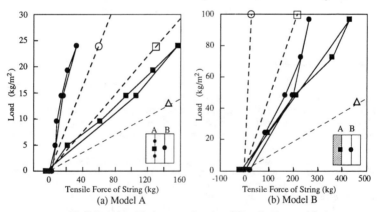

Fig.5 Relationship between Load and Tensile Force of Strings
Obtained from Partially Distributed Snow Loading Test

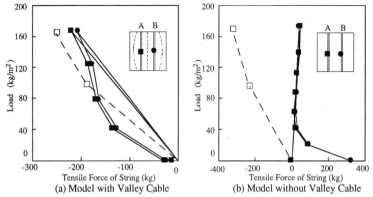

Fig.6 Relationship between Load and Tensile Force of Strings
Obtained from Wind Loading Test

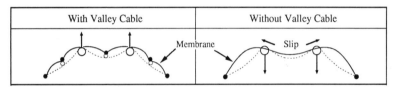

Fig .7 Displacment mode of Membrane Under Wind Load

The ratio of the incremental amount of the tensile force of the string on the non-loading side to that on the loading side is about 20% for Model A and 60% for Model B. This is thought to be caused by the effect of the stress transferred through the membrane. The fact that the analytical value dose not adequately simulate the behavior of Model B shows the necessity of taking into account the effect of the non-connection of the BSS and the membrane.

Wind Loading Test

(1) Purpose. When the structure is subject to the upward wind load, the component of the force in a vertical direction, which is transferred from the membrane to the BSS is determined mainly by the following three elements: The membrane's stiffness against elongation, the elongation stiffness for the valley cable, and the sag-span ratio of the membrane. This experiment was carried out with the aim of understanding the loading transfer mechanism.

(2) Result and Study. Fig.6 shows the relationship between the load and the tensile force of the string obtained from the experiment. When the BSS is equipped with a valley cable, the tensile force of the string tends to decrease. From this phenomenon, it can be explained that the component of the combined stress transferred from the membrane to the BSS shows an upward tendency. On the contrary, when there is no valley cable, the tensile force for the string shows a tendency to increase. From this situation, it is considered that the deformation mode changes according to the existence of the valley cable as shown in Fig.7.

In regard to the case in which the valley cable is installed, the analytical value corresponds quite well to the experimental value. However, when there is no valley cable, the values show no similarity to each other. Consequently, it is thought that the slipping state of the membrane over the BSS must be taken into consideration.

Behavior of the Actual Building During Construction

(1) Construction Procedures. The construction of the roof for this building was conducted according to the following procedures (Fig.8).
1) Erection of the BSS on the ground and the introduction of the prestress to the string.
2) Lift-up and installation for the BSS.
3) Arrangement of the stiffening member and the introduction of prestress to it.

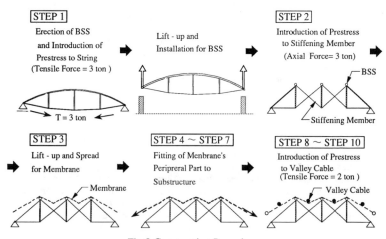

Fig.8 Construction Procedure

4) Lift-up and spread for the membrane and the fitting of the membrane's peripheral part to the substructure.
5) Installation of the valley cable and introduction of prestress to it.

(2) Erection Method of the BSS   The procedure for the erection of the BSS and a control method during erection were determined under consideration of the fact that the shape of the arch is characteristic of the change that takes place when prestress is introduced to the string.

The most distinctive point of the erection method of the BSS in this building, exists in the method of eliminating manufacturing errors for the arch and the string. This method was decided through the investigation carried out with the following three points given as pre-conditions.

1) Both the target shape for the BSS and the set value (3 ton) of the tensile force for the string should be satisfied, while eliminating the errors.
2) Since there is no space for the installation of a jack at the ends of the string, the control value should be based on the shape of either the string or the arch.
3) It is necessary that the length of the cable be estimated under the state in which the tensile force has been introduced to some degree.

(3) Errors Elimination Method and Control Method.   An outline of the method for the elimination of the manufacturing errors and the control method will be described as follows (Fig.9).

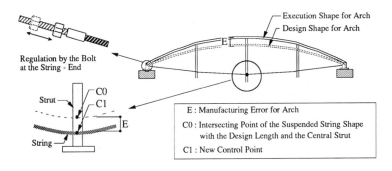

Fig.9 Method for the Elimination of Manufacturing Error and Control Method

1) Through the measurement of a rise of the arch when installed over the place with a prescribed span, the difference in value of E from the rise for the design shape of the arch is measured.

2) The distance between the position of the C0 and the bottom of the central strut is measured and a mark is put on C0. The position of the C0 is an intersecting point of the suspended string shape with the design length and the central strut.

3) A new control value was established as C1. C1, utilized for example when the execution rise for the arch is higher than the design dimension, is set at the point which is below C0 by E.

4) The intersecting point of the actual string shape is regulated by the bolt at the string-end so that point can reach the control value of C1. In this case, the tensile force is introduced to the string to some degree due to the self-weight.

5) The string is pulled down and fixed at the bottom end of the struts. At this time, prestress is introduced to it.

Result of the Measurement During Construction and Study

Fig.10 shows the result obtained from the measurement, which was conducted at each execution step. The value for the abscissa shown in the figure corresponds to the construction step illustrated in Fig.8.

(1) Tensile Force for the String. The tensile force of the string for the three BSS members, when the erection of the BSS has been finished (STEP 1), agrees quite well to the design target value (3 ton). Consequently, the validity of the errors elimination method was confirmed. The tensile force sharply increases up to STEP 6 and a conspicuous change can not be observed afterwards. The analytical value corresponds well to the experimental value.

(2) Axial Force of the Arch. The axial force increases by about 5 ton from STEP 3 up to STEP 5, which is larger than the incremental amount (3 ton) of the tensile force of the string. This is thought to be caused by the effect of friction at the roller support. On the other hand, it is considered that the difference between the decrease in the axial force from STEP 5 to STEP 7 and the incremental amount of the tensile force for the string is caused by resolution of the friction. The analytical value corresponds well to the experimental value.

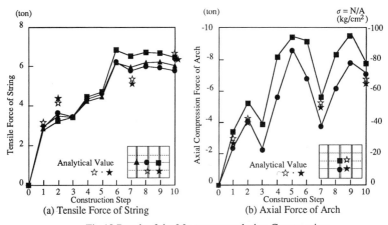

(a) Tensile Force of String   (b) Axial Force of Arch

Fig.10 Result of the Measurement during Construction

Conclusion

From the result of both the experiment carried out using scale models and the measurement during construction of the actual building, the following information was obtained.

(1) The behavior of the BSS during the formation of the shape and during the uniformly distributed loading can be completely understood through the carrying out of the plane and spatial analyses. Furthermore, the upward wind force induced behavior with valley cable can also be simulated by the spatial analysis.

(2) During the partially distributed loading or upward wind loading without valley cable, the slip between the membrane and the BSS must be evaluated.

(3) The validity of the control method, which was employed during the construction of the actual building in order to eliminate the manufacturing errors was confirmed.

References

[1]M. SAITOH, "Recent developments of Hybrid Tension Structures", Proc. of IASS, Copenhagen, Denmark, (1991.9)
[2]M. SAITOH & A. OKADA, "Conceptual Design of Hybrid Tension Structures", Proc. of SEIKEN-IASS, Tokyo, Japan, (1993.10)

Combinatary   Net-Shell   Roofs   of   Elliptic   Paraboloid
of   Zhaoqing   Gymnasium

By   Fa-Kun   Yao[1]

Abstract

Zhaoqing   Gymnasium   is a multi-functional   sports   building,
whose   structural modelling takes Zhaoqing City Flower(Lotus)   as
the main theme; adopts a cambinatory ellipic paraboloid net-shell
roofs.   So   the   entire building structure looks exactly like   a
lotus flower covering on the ground.   This article will summarize
the   building structural modelling and structural design   method;
compare   the   results   of model test   with   that   of   theoretical
calculation;   and finally, will demostrate the performance of the
combinatory   elliptic   paraboloid   net-shell   roofs   and   its
application in sports building construction.

Structural Modelling

Zhaoqing   Sports   Center   is   located   on the beautiful Star-Lake
Scenic Spot, where, according to the city construction program, a
sports park will be built. Xijiang Road, which is the main street
in   this   city,   passes   on its west towards the   North   Door   of
Zhaoqing   (Zhaoqing   Railway   Station).   The   Gymnasium.being   the
principal part of the Sports Center,   should be built as a symbol
of   the city.   This is the peoples great expectation.   So through
careful   and   repeated   demonstration,   the   final   plan   of   the
structural   mould is settled;   the design plan should   imitate   a
lotus   flower(Zhaoqing City Flower),   lotus is the main theme   of
the building.   No doubt, this design is a perfect   combination of
architectural beauty and natural beauty.

----

1.   Fa-Kun   Yao;Director of South-China Space structure   Research
     Center,   Senior Engineer.
     Add;Fenggan 29#,Wufeng,   Guangzhou,China P.C.   510260

The constructional plane of Gym is that of the square of a trancation angle length of 57m, total area of side 8,000 square meter. Four main entrances are set on the four trancation angle sides inside the building, there are 3,000 fixed seats and 400 mobile seats. The building is a multi-functional building, it can be used as a place of holding either sports activities, or singing and dancing parties, or exhibitions, or meetings etc. As the ground water level is comparatively high, the ground elevation is fixed 0.95m higher than that outside. The whole building present a rising structure.

Fig.1 Gymnasium Perspetive   Fig.2 Net-shell Roof
Drawing

Because of this, the main point of structural shape is the roof. Through careful comparison of design plans, we select the plan of composing several ellipse paraboliod net-shell together with the edge of each shell extending to the ground and linking with each other there. This design makes the builtding look like a lotus flower covering on the ground(Fig.1). The design is specific, and is regarded as an exquisite works of architectural art.

The individual net-shell is composed of two orthogonal double-layer parabola arch trusses, with the rises of 10.97m and 5.2m seperately, and the network projection is 2687*2687mm(Fig.3), the thickness is 2m according to architect requirement. Two orthogonal parabola space arch trusses, which are set along the link-line linking mid-points of the opposite side of the building square, are used as the central bearing and also are used to combine the four net-shell of the same shape together(Fig.2). The net-shell's periphery is supported by the rigidity edqe beams of the gymnasium outer-ring, and then continues cantilevering outside of the cantilever part is the shell surface's natural extension, while the lower string is formed by half of the parabola arch. This reduces the thickness from the net-shell to the cantilever angle points to 600mm(Fig.4)

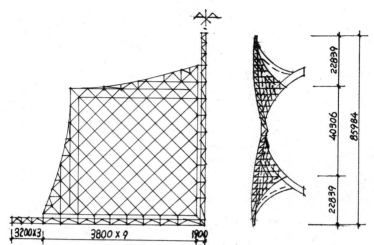

Fig.3 Net-shell Surface   Fig.4 Net-shell Section

The pole used for net-shell structure are no.3 seamless steal tubes. Those poles are connected by bolt ball nodes. The designed steel consumption of the project is 46kg/square meter, as steel replacement will occur during construction, the actual steel consumption will reach 52kg/square meter. The cost of manufacturing and mounting net-shells is 360RMB/square meter.

Structure Design

The discretization pole structure model is a real model of net-shell structure. The usual practice is to adopt the finite element method to process mechanical analysis. The shell surface equation of the elliptic parabotiod net-shell is set up as follows according to a self-decided coordinate(Fig.5).

$$Z= \frac{4(x-ax)f_a}{a} + \frac{(4y-b)f_b}{b} \qquad (1)$$

In the formula;
    a.b--the side lengths of thhe net-shell projection
    f_a --a rise corresponding to side a, f_a=10.97m
    f_b --a rise corresponding to side b, f_b=5.2m

The analysis of the structure model linking to the poles is processed according to hinge. Performed on computers using the finite element program of the space trusses. The structure calculation conducts to the net-shell, with symmetry principle by taking one-fourth or one-eighth seperately from the net-shell.

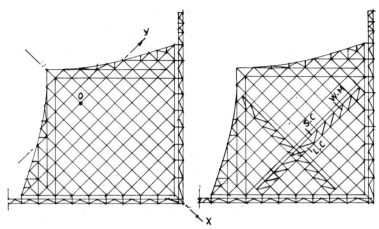

Fig.5 Shell Surface   Fig.6 Pole Layout
     Coordinate System

To calculate load value-obtained, the dead load is q=2,500N/square meter, wind load value is obtained according to the basic wind pressure , W =450N/square meter, vertical earthquake liak is 10% of the dead load, and the thermal stress is calculated based on ±15°C of temperature difference. The calculation results show;

a) The stability of the double-layer net-shell is not the main problem. The structure design doesn't need to be made according to stability control.

b) The calculating internal force of the members is comparatively small, it shows the structure height of the net-shell can be lower than 2m.

C) The cantilever part of the net-shell takes good effect on balancing the internal force around boundary poles and on reducing the horizontal thrust of peripheral bearers.

d) Adopting periphery hinge plan, the thermal stress reduces by 30% comparing with rigidity bearers.

Based on the results calculated above, the net-shell structure is designed completely with the strength calculation. The members layout(Fig.6) is designed along the receiving-force direction of the parabola arch as a double-layer arch truss with the web members. The space area of central bearing can not only suspend the dearing structure of net-shell but also suspend one part of the net-shell, thus assure the deformation harmonizing. The rigidity of the space arch truss is very large, sso it can reduce the internal force of shell's members.

Because the internal force of shell's members is comparatively small, the member section is designed mainly according to the construction, this design makes the members size cut down a lot. Main members use $\phi 60*3.5$ and $\phi 76*3.6$. As the internal force of central bearing arch trusses id very large, the biggest member needs a $\phi 219*12$.

Net-shell joints use bolt balls, but attention should be paid to the layout of slanting web members. The diameters of main joints are $\phi 130$ and $\phi 150$, the diameter of thhe biggest joint are central arch truss is $\phi 230$. In order to reduce the influence of thermal stress, the suport joints are placed at intervals by using rubber suports and flat suports. However, the suport of central bearing atch truss must be rigidity, so as to assure the bulk stability.

The side components of net-shell are curved-beams of multi-column bearing. Structure design should simultaneously consider vertical and horizontal concentrated loads which passed on by net-shell suport jointd, and should also consider the "axial force" parallel to axial line of the beams. Those loads can cause bending moment or eccentric force moment to central-axial line of the beams, therefore the resistance to pressure, the resistance to bending and the torsional rigidity of the outer-ring components, are of the same importance. The keinforced concrete edge beams of box-form section can not only raise the rigidity of the outer-ring components but also decrease their self-weights.

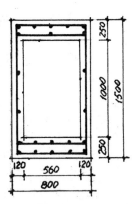

Fig. 7 Edge Member
Components Section

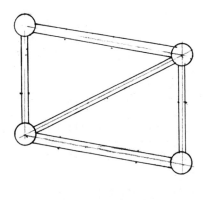

Fig. 8 Shell Surface
Member Nodes

Structure Chart

As the internal force of the members is relatively small, the node diameter may be also short. This certainly will cause difficulties to member-linking. So, reasonably laying out the breast member and possibly increasing the included angles between poles become very important(Fig. 8).

Fig. 9 shows the rubber member. The member itself has enough intensity to bear vertical loads, also can adapt a large shearing deformation so as to satisfy the horizontal displacement caused by various reasons of the upper structure, at the same time, it has a elasticity and can suit the structure revolving on the suport.

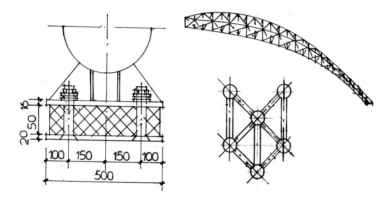

Fig. 9 Rubber Suport Nodes   Fig. 10 Central Bearing Arch Truss Section

As the central bearing of the net-shell, the parabola arch truss is a triangle space arch truss of the net-shell(Fig. 10). The reinforced concrete base which supports the space arch truss is placed into the outside ground to bear the vertical load passed in by the arch truss. The corresponding arch foot bases are connected by drag-poles so as to balance the horizontal thrust of the arch foot.

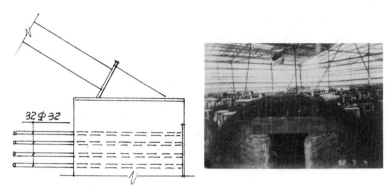

Fig.11 Arch Foot   Fig.12 Model Testing
     Structure   Site

Model Test

        In order to test the structure calculation theory and the
calculation results, a 1:10 net-shell has been model according to
the similarity theory. The model span is 5.7m, and the thickness
is 0.2m, the network projection size is 268.7*268.7mm. The model
members are roundsted of geometric similarity, and the joints are
40 solid balls.   Members and joints aree welded together. And
the suport restraint form mainly corresponds to the prototype
net-shell (Fig. 12).
        Based on the net-shell symmetry, all the measuring points are
distributed within the one-eight of the net-shell.   Among them
are 35 deflection measuring points tested by using percentage
test(Fig.13) and 172 strain measuring points which are used to
measure the strain of 86 member by using 2X3mm resistance   cards
(Fig.14).
        Testing load is appoied to the entire net-shell.   The
procdeure of loading is;
            + 0.2 designed load   stop four hour
    0 ——————————————————► designed load ——————————————►desinged
          5th grade

            + 0.2 designed load
    load ———————————————————► twice of desinged load
          5th grade

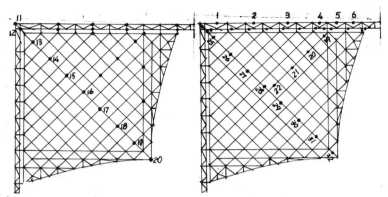

Fig.13 Layout of Deflection
Measuring Points

Fig.14-a Layout of Upper Chord
Strain Measuring Points

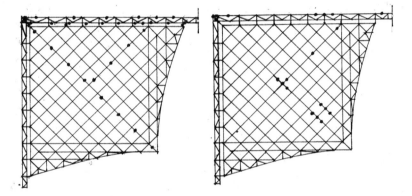

Fig.14-b Layout of Lower Chord
Strain Measuring Points

Fig.14-c Layout of Web Member
Strain Measuring Points

The testing results show:
a) Under twice of the designed load, the biggest stress of net-shell stress is 135.4N/mm², which still belongs to the elastic working period.
b) The ratio of the largest deflection to the span is 1/1662, which is far less than the deflection allowable value.
c) No unstability or other destructions appear.
d) Through the comparison between testing data and calculation data, we know that the structure deflection mainly tallies with the distribution rules of the poles internal force. The fact that the testing data are smaller than calculation data indicates that node rigidity has an internal potential.

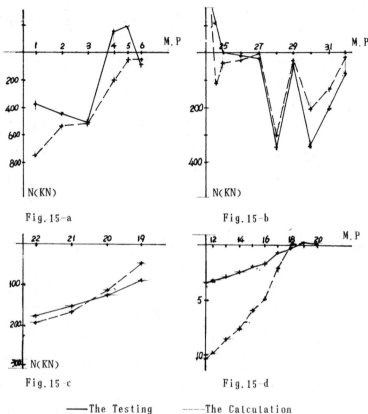

Fig. 15-a   Fig. 15-b

Fig. 15-c   Fig. 15-d

——The Testing   ----The Calculation
Fig. 15 Comparison of Testing Data and Calculation Data
a, b, c--internal Force   d--deflection

Conclusion

Based on the results of theoretical analysis, model test and engineering practice, for the built-up type elliptic paraboliod net-shell roofing, the writer regards as follows:

1. The combinatory elliptic paraboliod double-layer net shell has a good rigidity with the receiving-force well and reasonably distributed. So it is a good structural pattern for a wide span roofs. And the structural style is casual and can meet the needs of sports buliding style varieties.

2. From the model test, we know the limit load of the net-

shell is twice of the designed load. That indicates the current calculating theory is safe and reliable. The structural height is much less than the building heitht. If adopting sigle-layer net-shell or adopting the combiantion of partial single-layer and partial double-layer, material functions may be fully brought into play, and a better economic efficiency may be got, but the stableility of the net-shell will be a real problem.

3. The thermal stress influence to the net-shell can not be ignored. This project has adopted all-steel structure for central arch trusses to harmonize the deformation; also adopted partial rubber suports to loose boundary restraint. Both of them have the effect of reducing thermal stress influence.

4. The net-shell cantilever can partly balance the horizontal thrust of these suports, so it can reduce the loads on beams and pillars and also makes the main structure unusually light and ingenious.

Effects of Spatial Triangulation on the Behavior of "Tensegrity" Domes

David M. Campbell[1] ,MIASS, David Chen[1], MIASS,
Paul A. Gossen[1], MASCE, & Kris P. Hamilton[1] , MASCE

Abstract
A number of long span "tensegrity" dome type structures have recently
been realized pursuant to the inventions of R. Buckminster Fuller and
David H. Geiger. These structures have demonstrated structural efficiency
in a variety of long-span roof applications. The primary attribute that
these structures share which classifies them as tensegrity systems is their
utilization of discontinuous compression elements connected within a
network of continuous tension members. Beyond this common trait the
current built types of tensegrity domes differ somewhat in the geometry of
their networks. These variations in network geometry result in
differences in behavior and structural efficiency.

A significant variable in tensegrity dome networks is the degree of spatial
triangulation, which as a consequence of the geometric non-linear
behavior of these structures is not required for structural stability. The
extent to which spatial triangulation may be advantageous or detrimental
is investigated by comparing the behavior of examples of two tensegrity
dome structural systems. The non-linear behavior of these structures
subjected to various load conditions is compared analytically using
computer modeling techniques.

Introduction
Only two variations of tensegrity domes have been realized in the seven
roof structures built to date, the "Cabledome" (Geiger) and the spatially
triangulated dome first illustrated in 1983 (Fuller). There are currently six
Cabledomes in the world including: the Thunder Dome (formerly the
Florida Suncoast Dome) roof, St. Petersburg Florida and the Taoyuan

---

[1]Principals, Geiger Gossen Hamilton Liao Engineers P.C., 2 Executive
Blvd, Ste. 410, Suffern, NY 10901 & 1215 Cornwall, Bellingham, WA 98225

County Arena roof, Taoyuan, Taiwan ROC. There has been one spatially triangulated tensegrity dome built to date, the Georgia Dome roof, Atlanta, Georgia (Levy 1991).

Many qualitative claims have been made about the relative pros and cons of triangulation of tensegrity dome networks (Geiger 1986), (Rastorfer), (Levy 1991), (Levy 1992). However, the only analytical comparison of the behavior of spatially triangulated and non-triangulated tensegrity dome networks was performed by Yamaguchi et. al. (Yamaguchi), which compared the behavior of the Cabledome with a spatially triangulated tensegrity dome.

All the tensegrity domes built to date are clad with a stressed fabric membrane. There are a number of factors which have lead to the exclusive employment of stressed membranes for these structures, primarily the desire for light transmission through the roof and the economy of low mass membrane structures for long spans. Consequently the two tensegrity structures studied in this paper were configured as required for prestressed roof membranes.

The triangulated dome system has been configured such that the roof surface is primarily comprised of number of quadrilateral anticlastic membrane panels joined at the "ridgenet" cables (Levy 1991). See Figure 1. The roof panels of the Cabledome are "wedge-shaped" in plan. They are joined to adjacent panels at the ridge cables. The membrane is tensioned by valley cables which are arranged radially in each panel between ridge cables. See Figure 2.

Methodology of Structural Model Development
The two roof structures modeled are shown in Figures 1 and 2. The networks were intentionally developed to be as similar as possible while being representative of practical realizable structures. Both structures have a circular plan configuration with a span of 120 m (394 ft.). Both have two tension "hoops" with the same nominal elevation and diameter. The tension hoops are planar. The outer hoop in both structures has a diameter of 90 m (295.22 ft.) at an elevation of -4.45 m (14.61 ft.) below the support. The inner hoop has a diameter of 46 m (150.89 ft.) at an elevation of 0.65 m (2.14 ft.) above the support. Each structure utilizes 16 struts at each hoop. Each structure employs an upper and lower center tension ring with a 5 m (16.41 ft.) diameter.

Each model was developed integrally with its membrane surface. In order to achieve the anticlastic curvature in the membrane panels of the triangulated dome its ridgenet must have a reversal in curvature in the radial direction. As a consequence the minimum overall rise of the roof structure is dictated by the necessity of maintaining positive drainage of the surface to the perimeter. A rise of 15.24 m (50 ft) was judged to be appropriate. This same rise was employed for the Cabledome structure for the sake of comparison. The span to rise ratio for both structures is 7.9, under dead load and prestress. This is considerably higher than is optimal for a Cabledome. Most Cabledomes have been built with span to rise ratios greater than 12.

The structures were analyzed both with and without roof membranes. Although the membrane surface was not modelled in sufficient detail to allow a detailed stress investigation or membrane design, interesting membrane results were obtained and will be presented in a future paper.

The dead load of the two structures was normalized to 0.316 kN/m^2 (6.6 psf). This is significantly heavier than the Cabledomes constructed using cast steel connections and multiple prestressing strand (Tuchman), (Geiger 1986), (Geiger 1988), but is representative of the self weight of a triangulated dome system of this span employing plate weldment connections and socketed cables as employed at the Georgia Dome (Levy 1991).

Load Conditions:
Structural analysis was conducted employing Geiger Engineers' proprietary large deflection nonlinear analysis software. Each structure was analyzed for eight load conditions as follows:

1. **Dead Load and Prestress with Fabric Membrane:** 0.316 kN/m^2 (6.6 psf) This case was used as the input condition for Load Conditions 2 through 6.
2. **Live Load:** Uniform superimposed live load of 0.574 kN/m^2 (12.0 psf) applied to the membrane surface.
3. **Uplift:** Uniform suction of 1.197 kN/m^2 (25.0 psf) applied normal to the membrane surface.
4. **Unbalanced Live Load:** Uniform live load superimposed over 1/2 the roof plan of 0.574 kN/m^2 (12.0 psf) applied to the membrane surface, no live load on other 1/2 of structure.
5. **Unbalanced Uplift:** Uniform suction of 1.20 kN/m^2 (25.0 psf) applied to 1/2 the roof and uniform suction of 0.60 kN/m^2 (12.5 psf) applied to the other 1/2. All pressures are normal to the surface.
6. **Network Dead Load and Prestress without Fabric Membrane:** 0.263 kN/m^2 (5.5 psf). This condition was created by stripping the membrane from the structures as modeled in Load Condition 1. (Campbell 1991) The input condition for Load Conditions 7 and 8.
7. **Network with Concentrated Load:** Single point load of 445 kN (100

kips) applied in the gravity direction (-z direction) at a single outer hoop to strut connection node.

8. **Network with Element Length Error:** Simulation of the result of a - 50 mm (-2.0 in.) fabrication error in the installed length of a single outer diagonal cable, Cable 06.

Results

The circumferential prestress level in Load Condition 1 in the Cabledome is somewhat higher than the triangulated structure. The increased circumferential prestress of the Cabledome in Load Condition 1 is primarily a result of the prestress in the valley cables, Cable 07. See Load Condition 6 member forces.

TABLE 1: TRIANGULATED TENSEGRITY DOME, **MEMBER FORCES**

Members	Load Condition	1	2	3	4	5	6	7	8
CTR ELEVATION (m)		15.24	13.75	15.97			15.39		
CTR TOP (kN)	Max.	342	245	712	302	912	405	414	418
	Min.				240	907		383	405
CTR BOT (kN)	Max.	1,383	1,174	2,709	1,263	2,313	1,437	1,432	1,459
	Min.				1,245	2,264		1,401	1,446
CABLE 01 (kN)	Max.	129	85	342	156	400	165	173	173
	Min.				27	342		133	156
CABLE 02 (kN)	Max.	360	329	729	369	738	391	400	405
	Min.				280	520		356	387
CABLE 03 (kN)	Max.	845	1,005	1,027	987	1,050	850	894	890
	Min.				729	814		654	832
CABLE 04 (kN)	Max.	543	463	1,059	520	970	560	569	578
	Min.				467	814		529	560
CABLE 05 (kN)	Max.	249	391	89	458	160	231	342	258
	Min.				173			120	209
CABLE 06 (kN)	Max.	1,873	2,580	1,254	2,629	1,708	1,766	2,059	2,780
	Min.				1,793	947		1,686	921
CABLE 08 (kN)	Max.	1,139	1,766	360	1,619	498	1,059	1,112	1,081
	Min.				1,259	240		1,005	1,054
CABLE 09 (kN)	Max.	7,126	9,817	4,750	9,408	6,000	6,725	7,339	7,544
	Min.				7,375	4,114		6,628	6,352
STRUT 01 (kN)	Max.	-49	-53	-93	-58	-93	-49	-49	-49
	Min.				-36	-76		-49	-49
STRUT 02 (kN)	Max.	-116	-214	-13	-200	-31	-102	-102	-102
	Min.				-116			-98	-102
STRUT 03 (kN)	Max.	-810	-1,201	-480	-1,148	-654	-747	-787	-787
	Min.				-836	-396		-525	-725

For Load Condition 3, Uniform Uplift, the total loss of circumferential tension in the triangulated dome structure is 38% versus 3% for the Cabledome. The Cabledome is much stiffer in this load condition.

TABLE 2: CABLE DOME, **MEMBER FORCES**

Members Condition	Load	1	2	3	4	5	6	7	8
Ctr Elevation (m)		15.24	14.29	14.90		15.85			
CTR TOP (kN)	Max.	832	405	3,278	596	2,767	805	810	814
	Min.				547	2,593		783	814
Ctr Bottom (kN)	Max.	583	489	1,214	525	1,103	498	503	503
	Min.				498	1,090		485	503
CABLE 01 (kN)	Max.	133	142	9	187	53	316	329	320
	Min.				71			280	320
CABLE 02 (kN)	Max.	387	374	476	414	498	525	547	529
	Min.				342	414		440	534
CABLE 03 (kN)	Max.	1,334	1,548	1,477	1,499	1,535	1,321	1,343	1,339
	Min.				1,370	1,392		1,165	1,330
CABLE 04 (kN)	Max.	236	196	494	245	476	200	209	205
	Min.				178	431		151	200
CABLE 05 (kN)	Max.	823	979	907	894	' 912	698	712	707
	Min.				876	907		636	703
CABLE 06 (kN)	Max.	3,238	4,163	3,025	3,687	3,127	2,811	2,945	2,873
	Min.				3,634	3,122		2,860	2,816
CABLE 07 (kN)	Max.	200	27	1,281	120	1,165			0
	Min.				36	801			0
CABLE 08 (kN)	Max.	2,015	2,384	2,224	2,184	2,228	1,717	1,712	1,730
	Min.				2,148	2,220		1,699	1,730
CABLE 09 (kN)	Max.	7,949	10,222	7,433	9,061	7,691	6,912	7,081	6,970
	Min.				8,923	7,668		7,077	6,970
STRUT 01 (kN)	Max.	-53	-44	-129	-49	-120	-40	-40	-40
	Min.				-40	-107		-40	-40
STRUT 02 (kN)	Max.	-214	-280	-240	-276	-271	-165	-169	-16
	Min.				-200	-214		-142	-169
STRUT 03 (kN)	Max.	-836	-1,148	-765	-1,072	-858	-698	-703	-707
	Min.				-881	-738		-583	-703

The Cabledome structure is considerably stiffer in Load Condition 2, Uniform Live Load. The maximum deflection of the Cabledome system is -951 mm (-3.12 ft) or roughly 1/126th of the span. The maximum deflection of the triangulated system is -1481 mm (-4.86 ft) or roughly 1/80th of the span.

The vertical displacements from unbalanced loads for the two systems are of the same magnitude. Maximum displacement occurs at the top of the inner hoop struts in both structures for Load Conditions 4 & 5. Cable 09 of the triangulated dome in Load Condition 5 has a variation in tension of 1886 kN (424 kips) or 31% of the maximum hoop tension, compared with a variation of 22 kN (5 kips), 0.3%, for the Cabledome.

TABLE 3:   **TRIANGULATED TENSEGRITY DOME**
Ridgenet Node Deflections, Z Displacements (mm)
Top Center Tension Ring

Load Condition	Node: 1	68	176	284	392	500	608	716
2	-1,481	-1,480	-1,479	-1,479	-1,478			
3	1,039	1,039	1,039	1,039	1,039			
4	-512	-541	-611	-681	-711			
5	712	734	788	842	864			
6	395	394	394	394	394	394	394	394
7	-18	-17	-15	-12	-10	-10	-13	-16
8	11	12	12	11	9	8	8	9

Inner Hoop Strut Top

Load Condition	Node: 4	71	179	287	395	503	611	719
2	-795	-795	-795	-795	-795			
3	996	996	996	996	996			
4	98	-31	-365	-692	-769			
5	294	390	732	1,122	1,225			
6	280	280	280	280	280	280	280	280
7	-41	-45	-28	5	23	17	-11	-42
8	22	28	30	18	0	-12	-9	7

Outer Hoop Strut Top

Load Condition	Node: 27	138	246	354	462	570	678	786
2	-212	-212	-212	-212				
3	242	242	242	242				
4	11	-4	-135	-210				
5	54	87	218	273				
6	76	76	76	76	76	76	76	76
7	-168	-5	14	11	10	11	14	-5
8	34	43	13	-1	-5	-7	-11	-15

Cable 06 members of the triangulated dome in Load Condition 7 have a variation in tension of 374 kN (84 kips) or 18% of the maximum tension, compared with a 85 kN (19 kips), 2.9%, for the Cabledome. In both structures the maximum vertical displacement occurs at the point of load application. The maximum Cabledome displacement is -439 mm (-1.44 ft) at node 7 versus -168 mm (-0.55 ft) for the triangulated dome at node 27.

The structures exhibit quite different sensitivities to fabrication error. For the triangulated dome Load Condition 8, resulted in a change in force in Cable 06 of +1014 kN, - 845 kN (+228 kips -190 kips) or +57%,-48% from the initial condition. Similarly the variation in Cable 09 tension is +818 kN, -374 kN (+184 kips -84 kips) or +12.2%,-5.6% from the initial condition. For the Cabledome structure this condition resulted in a change in force in Cable 06 of +62 kN (+14 kips) or +2.2% and a variation in Cable 09 tension of +58 kN (13 kips) or 0.8% from the initial conditions.

Figure 1:   Triangulated Tensegrity Dome Structure

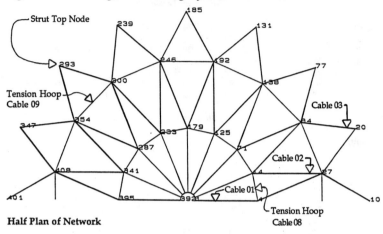

Half Plan of Network

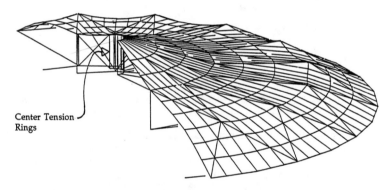

Perspective of Half Model w/Membrane

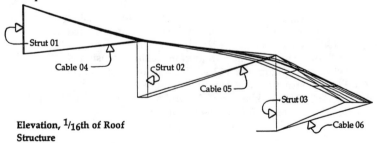

Elevation, $1/16$th of Roof Structure

Figure 2: Cable Dome Structure

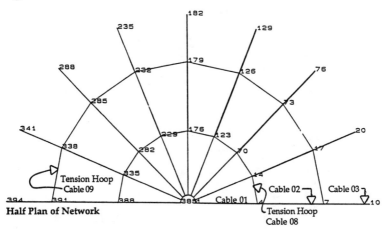

Half Plan of Network

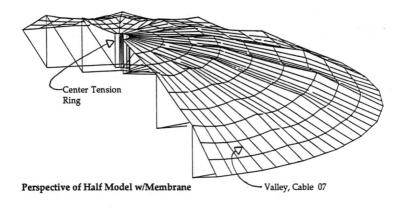

Perspective of Half Model w/Membrane

Valley, Cable 07

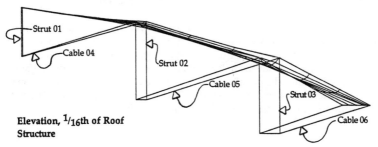

Elevation, $1/16$th of Roof Structure

TABLE 4:   **CABLEDOME**
Ridge Node Deflections, Z Displacements (mm)

Top Center Tension Ring

Load Condition	Node: 1	67	173	279	385	491	597	703
2	-951	-951	-950	-948	-948			
3	-325	-325	-324	-324	-324			
4	-171	-214	-315	-417	-460			
5	-412	-386	-323	-260	-233			
6	847	847	847	846	846	846	847	847
7	-23	-20	-13	-7	-4	-7	-14	-20
8	11	12	12	12	12	12	12	11

Inner Hoop Strut Top

Load Condition	Node: 4	70	176	282	388	494	600	706
2	-695	-695	-695	-695	-694			
3	-8	-8	-8	-8	-8			
4	471	287	-245	-822	-1,011			
5	-545	-401	-38	319	455			
6	532	532	531	531	531	531	531	532
7	-87	-71	-19	46	75	46	-20	-71
8	9	9	9	9	9	9	9	9

Outer Hoop Strut Top

Load Condition	Node: 7	73	179	285	391	497	603	709
2	-194	-194	-194	-194	-194			
3	45	45	45	45	45			
4	219	164	-70	-335	-400			
5	-217	-156	21	211	267			
6	138	137	137	137	137	137	137	137
7	-439	-54	47	87	100	87	47	-54
8	13	6	6	7	7	7	7	6

Discussion

Neither of the tensegrity dome networks studied behave at all like structural domes, rather they are dominated by spatial truss-like behavior. Both structures exhibit non-linear behavior, albeit to different extents. The triangulated system is not a stiff as the Cabledome for uniform loads, but is somewhat stiffer in response to concentrated loads.

The two structures behave differently in response to non-uniform loading, especially with respect to individual member forces. The Cabledome's behavior is unique, member forces simply do not change much under the non-uniform load conditions evaluated. This behavior cannot be explained in classical linear terms as it exploits "inextensional" modes of deformation (Pellegrino). The nonlinear geometric stiffness contribution to the system's overall stiffness is quite large.

The triangulated system also has the potential for this type of response; however, the elastic component of the over-all system's stiffness attracts much of the work done on the structure so that the geometric component of the overall stiffness is not engaged.   Consequently, the triangulated structure is stiffer with respect to concentrated loads, at the expense of relatively large variation in element forces.

Interestingly when both structures are subjected to uniform loads which are resisted primarily by "extensional" (Pellegrino) behavior, ie. by change in the strain energy of the structure's elements, the Cabledome is significantly stiffer than the triangulated dome structure.

The behavior of the triangulated system under uplift is attributable to the reversal in curvature in the ridgenet of the triangulated system.   The result is that the loss of tension in some cable elements is quite large.   As a consequence of this behavior, it may prove to be prudent in designing similar triangulated tensegrity structures to analyze the ultimate condition to insure an appropriate factor of safety against failure (Hamilton).   Under net uplift the Cabledome displays typical geometric nonlinear behavior where, due to the change in the curvatures of both the valley and the ridge cables under load, the center deflection is actually downward.   The Cabledome's membrane valley cables contribute to the structure's stiffness in resisting uplift.

The triangulated tensegrity dome is relatively sensitive to fabrication errors, either in element length or prestress, depending upon the construction technique employed.   The Cabledome is quite insensitive to the same fabrication errors as well as to support movement (Campbell 1992).

Conclusions
In summary, the effects of spatial triangulation of tensegrity dome networks is quite significant, at least as illustrated by the behavior of the two structures evaluated in this study.

The development of suitable free draining membrane form for the triangulated dome system necessitates a significantly greater rise for the roof surface than is required for the Cabledome.   (In reviewing the membrane surface geometry of the triangulated dome in various load conditions, it was clear that the roof rise should be further increased to assure drainage.) The anticlastic curvature of the triangulated dome's ridgenet has a significant influence on its behavior.

The Cabledome is a simpler structure than the triangulated tensegrity dome. Clearly, spatial triangulation of the tensegrity network can only be achieved at increased expense and erection complexity when compared with the simpler Cabledome network. Generally, this added complexity does not seem to yield any direct benefits other than a somewhat increased stiffness in response to load concentrations. Moreover, the Cabledome demonstrates some significant advantages. The Cabledome has demonstrated resistance against progressive collapse (Geiger 1986) (Taniguchi). The Cabledome generally exhibits greater stiffness, much reduced to non-uniform and concentrated loads, an insensitivity to fabrication errors, as well as greater design flexibility of roof form than the triangulated dome system.

References

Campbell, D., "The Unique Role of Computing in the Design and Construction of Tensile Membrane Structures", Proceedings, American Society of Civil Engineers Second Civil Engineering Automation Conference, New York, NY, 1991.

Campbell, D., Chen, D., and Lynch, K., "Effects of Support Settlement on a Long-Span Cabledome Roof Structure ", Innovative Large Span Structures, IASS-CSCE International Congress 1992, published by the CSCE, Montréal, 1992.

Fuller, R., Inventions , The Patented Works of R. Buckminster Fuller, St. Martins Press, New York, NY, 1983.

Geiger, D., "Roof Structure", U.S. Patent No. 4,736,553.

Geiger, D., Stefaniuk, A., and Chen, D., "The Design and Contruction of Two Cable Domes for the Korean Olympics", in Shells, Membranes and Space Frames (Vol 2), Proceedings IASS Symposium, Elsevier Appl. Science, 1986, 265-272.

Geiger. D, Campbell, D., Chen, D., Gossen, P., Hamilton, K., and Houg, G., "Design Details of an Elliptical Cable Dome and a Large-span Cable Dome Under Construction in the United States ", Proceedings, 1st Oleg Kerensky Memorial Conference, London England, 1988.

Hamilton, K., Campbell, D., and Davidson, C., "Comments on Limit States Behavior and Design of Large Deflection Tension Structures ", Proceedings, ASCE Conference, Orlando, August 1987

Levy, M., "Floating Fabric Over Georgia Dome", Civil Engineering, November 1991, Vol. 61, Number 11.

Levy, M., "Triangulated Roof Structure", International Patent Application, Publication Number: WO 982/08015, Application Number: PCT/US91/08096, World Intellectual Property Organization, 1992.

Pellegrino, S., "A Class of "Tensegrity" Domes", International Journal of Space Structures, 1990

Rastofer, D., "Structural Gymnastics for the Olympics", Architectural Record, September 1988, 128-135.

Taniguchi, T., Ishii, K., Toda, I., Komatsu, K., Etoh, T., "Report on Experiments Concerning Tension Dome", Proceedings, IASS Symposium on Space Structures for Sports Buildings, Beijing, Elsevier Appl. Science, 1987.

Tuchman, J., Shin Ho-Chul, "Olympic Domes First of Their Kind", Engineering News Record, March 6, 1986, 24-27.

Yamaguchi, I., Okada, K., Kimura, M., Magara, H., Okamura, K., Ohta, H., Okada, A., Okuno, N., "A Study on the Mechanism and Structural Behaviors of Cable Dome", Proceedings, IASS Symposium on Space Structures for Sports Buildings, Beijing, Elsevier Appl. Science, 1987.

Active Control of Axial Forces in Beam String Space Frames

Shiro Kato [1], Shoji Nakazawa [2],
Yasuo Matsue [3] and Tatsuo Yamashita [3]

Abstract

The present work discusses a formulation for active control of axial forces of strings in beam string structures to dynamically and/or statically reduce the responses, and proves both the effectiveness of the control and the applicability of the formulation by analyzing two types of structures; one is an ordinary, one way beam string structure stiffened by stressing in tendons and the other is a three dimensional beam string structure stiffened by cables underneath. The target function to be structurally optimized is expressed by combination of the followings; strain energy stored in a structure to be stiffened or softened by active control, and quadratic functions in terms of displacements, velocities and accelerations. In static case, a strategy is discussed to reduce the strain energy stored in the stiffened portion by prestressing and in dynamic cases a strategy of instantaneous optimization for kinematic energy is investigated.

Formulation

Beam string structures have been efficiently applied to long span structures and a strategy is adopted to suppress the stresses and displacements under their dead loads and additional loads as snow and winds. Dynamic prestressing in steel space structures will be developed in these ten years since similar systems for active controlling have been installed for high rise buildings.

---

1 Professor, Toyohashi University of Technology, Tenpaku, Toyohashi, Japan 441
2 Graduate Student, Toyohashi University of Technology
3 Chief Engineer, Department of Design, TOMOE CORPORATION, Tokyo, Japan

In the formulation for beam string structures, the structures are decomposed into two portions. One is the part stiffened by prestressing and the other is the part to stiffen the former. The former and the latter might be respectively called a stiffened part and a stiffening part. Fig. 1 shows an illustration of those parts. In the stiffening part, a device to introduce prestressing in strings is assumed to be installed. The behaviors of strings are assumed to be controlled by initial stresses, λ_i, and the equations for axial forces , n_i , are expressed by their initial stresses and the elongations of strings, δ_i , as follows.

$$n_i = k_i \cdot \delta_i + \lambda_i \tag{1}$$

Here λ_i are determined based on undeformed configuration, and k_i is the stiffness of string.

The dynamic equation for motions is expressed in a form of matrix.

$$[M] \{\ddot{D}\} + [C_d] \{\dot{D}\} + [K]\{D\} + \lfloor F \rfloor\{\Lambda\} = - [M]\{\xi\}\ddot{u}_g + \{P_{et}\} + \{P_{eo}\}$$

$$[K] = [K_o] + [K_s] \tag{2}$$

M, C_d, K_o, are the mass, damping, and stiffness of the stiffened part, and K_s and F are the stiffness and force coefficient matrices of the stiffening part. Λ, P_{eo}, P_{et}, and $\ddot{u}_g$ are the initial stress vector composed of λ_i, static load vector, time-dependent load vector, and earthquake accelerations and ξ is the effective matrix for earthquake motions.

The equation (2) holds at the time t_{j+1} .

$$[M] \{\ddot{D}_{j+1}\} + [C_d] \{\dot{D}_{j+1}\} + [K]\{D_{j+1}\} + [F]\{\Lambda_j\}$$
$$= - [M]\{\xi\}\ddot{u}_{gj} + \{P_{etj}\} + \{P_{eo}\} \tag{3}$$

The above equation can be approximately solved by a numerical integration scheme such as Newmark-β method, and the state vector X_{j+1} composed of acceleration, velocity and displacement at time t_{j+1} can be obtained. Here the matrices Λ_j, P_{eo}, and P_{etj} at time t_j and the ground motion $\ddot{u}_{gj}$ are assumed unchanged between t_j and t_{j+1} .

$$\{X_{j+1}\} = [A] \{X_j\} + \{R_L\} + [R_o] \cdot \{\Lambda_j\} \tag{4}$$

where

$$\{X_{j+1}\} = \begin{Bmatrix} \ddot{D}_{j+1} \\ \dot{D}_{j+1} \\ D_{j+1} \end{Bmatrix} , \quad \{R_L\} = \begin{Bmatrix} P \\ P \dfrac{\Delta t}{2} \\ P\beta\Delta t^2 \end{Bmatrix} , \quad [R_o] = \begin{bmatrix} G_o \\ G_o \dfrac{\Delta t}{2} \\ G_o\beta\Delta t^2 \end{bmatrix} \tag{5}$$

$$\{P\} = \{G\} \cdot \ddot{u}_{gj} + \{G_{eo}\} + \{G_{ej}\} , \quad [G_o] = - [\tilde{M}]^{-1} [F]$$

$$\{G\} = -[\tilde{M}]^{-1} [M] \{\xi\} , \quad \{G_{eo}\} = [\tilde{M}]^{-1} \{P_{eo}\} , \quad \{G_{ej}\} = [\tilde{M}]^{-1} \{P_{etj}\}$$

$$[\tilde{M}] = [M] + \frac{\Delta t}{2} [C_d] + \beta \Delta t^2 [K] , \quad \{\tilde{X}_j\} = [A] \{X_j\}$$

$$[A_{11}] = - [\tilde{M}]^{-1} (\frac{\Delta t}{2} [C_d] + (\frac{1}{2} - \beta)\Delta t^2 [K])$$

$$[\mathbf{A_{12}}] = - [\tilde{\mathbf{M}}]^{-1} ([\mathbf{C_d}] + \Delta t [\mathbf{K}]) , \qquad [\mathbf{A_{13}}] = - [\tilde{\mathbf{M}}]^{-1} [\mathbf{K}]$$

$$[\mathbf{A_{21}}] = \frac{\Delta t}{2} [\mathbf{A_{11}}] + \frac{\Delta t}{2} [\,I\,] , \quad [\mathbf{A_{22}}] = \frac{\Delta t}{2} [\mathbf{A_{12}}] + [\,I\,]$$

$$[\mathbf{A_{23}}] = \frac{\Delta t}{2} [\mathbf{A_{13}}] , \quad [\mathbf{A_{31}}] = \beta \Delta t^2 [\mathbf{A_{11}}] + (\frac{1}{2} - \beta) \Delta t^2 [\,I\,]$$

$$[\mathbf{A_{32}}] = \beta \Delta t^2 [\mathbf{A_{12}}] + \Delta t [\,I\,] , \quad [\mathbf{A_{33}}] = \beta \Delta t^2 [\mathbf{A_{13}}] + [\,I\,]$$

The function to be minimized instantaneously at time t_{j+1} is chosen in the present investigation as follows (Yang et al 1987).

$$J_d = \frac{1}{2} \{\mathbf{D}_{j+1}\}^T [\mathbf{K_o}] \{\mathbf{D}_{j+1}\} + \frac{1}{2} \{\mathbf{D}_{j+1}\}^T [\mathbf{Q_D}] \{\mathbf{D}_{j+1}\}$$

$$+ \frac{1}{2} \{\dot{\mathbf{D}}_{j+1}\}^T [\mathbf{Q_V}] \{\dot{\mathbf{D}}_{j+1}\} + \frac{1}{2} \{\ddot{\mathbf{D}}_{j+1}\}^T [\mathbf{Q_A}] \{\ddot{\mathbf{D}}_{j+1}\} + \frac{1}{2} \{\Lambda_j\}^T [\mathbf{C}] \{\Lambda_j\}$$

$$[\,\mathbf{C}\,] = \mu \cdot [\,\mathbf{I}\,] \tag{6}$$

where $\mathbf{Q_D}$, $\mathbf{Q_V}$ and $\mathbf{Q_A}$ are the positive-definite weighting matrices for the responses and $\mathbf{C}$ is also a positive definite matrix for control, defined by a parameter μ . They are to be determined based on a design concept. In this optimization, the first term is the strain energy stored in the stiffened structure. As the function J, other type could be adopted and a set of displacements composed of several nodes and/or the stresses of several members might be set to satisfy a set of given values. The ratio of effectiveness by active tendons depends on both the number and the places of the installation.

Minimization of Eq.(6) with respect to Λ_j after replacing $\mathbf{D}_{j+1}$, $\dot{\mathbf{D}}_{j+1}$ and $\ddot{\mathbf{D}}_{j+1}$ using Eq.(5) leads to determination of Λ_j ;

$$\{\Lambda_j\} = [\,\mathbf{L}\,]^{-1} \cdot \{\mathbf{Z}_j\} \tag{7-1}$$

$$[\,\mathbf{L}\,] = \beta^2 \Delta t^4 [\mathbf{G_o}]^T [\,\mathbf{K_o} + \mathbf{Q_D}\,] [\mathbf{G_o}] + \frac{\Delta t^2}{4} [\mathbf{G_o}]^T [\mathbf{Q_V}] [\mathbf{G_o}]$$

$$+ [\mathbf{G_o}]^T [\mathbf{Q_A}] [\mathbf{G_o}] + [\,\mathbf{C}\,] \tag{7-2}$$

$$\{\mathbf{Z}_j\} = - [\mathbf{G_o}]^T [\,[\,\mathbf{K_o} + \mathbf{Q_D}\,] (\beta \Delta t^2 \{\tilde{\mathbf{D}}\} + \beta^2 \Delta t^4 \{\mathbf{P}\})$$

$$+ [\mathbf{Q_V}] (\frac{\Delta t}{2} \{\dot{\tilde{\mathbf{D}}}\} + \frac{\Delta t^2}{4} \{\mathbf{P}\}) + [\mathbf{Q_A}] (\{\ddot{\tilde{\mathbf{D}}}\} + \{\mathbf{P}\}) \,] \tag{7-3}$$

The procedures in leading Eqs.(7-1,2,3) are ordinary ones and abbreviated in spite that several different schemes lead a same answer. In discussing the procedure, an observer might be formulated in equations from Eq.(3) to Eq.(7-1,2,3), however, the present paper discards this since the present work aims at illustrating the effectiveness of active axial control for beam string structures.

The responses at time t_{j+1} can be calculated by Eq.(4), once the axial forces can be defined by Eq.(7). When an assumption is made as that the controlled responses are stochastically independent of the fluctuating load terms $\mathbf{P_{et}}$ and $\ddot{u}_g$, the term $\mathbf{P}$ in Eq.(5) can be replaced by Eq.(8) .

$$\{\mathbf{P}\} = \{\mathbf{G_{eo}}\} \tag{8}$$

The assumption might hold since the controlled responses are rather static with low varying frequencies for behaviors (Ikeda et al 1992).

Static Control by Prestressing Strings

Static control of a beam string structure of one way is illustrated based on the models shown in Fig.1. The model A is a symmetric one without control for asymmetric loading, and the other model B is a model capable of controlling the asymmetric loading. The dimensions for the stiffened structure is common to the models except the stiffening part. The parameters for both the stiffening and the stiffened are given in Tables 1, 2.

The loads considered in the present example are their gravity loads and additional loads on the half side of the structures, as shown in Fig. 2.

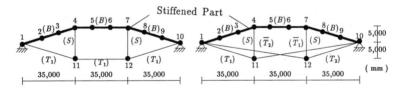

a) Model A   b) Model B

Fig.1 One-way Beam string Structures in Static Control;
the numbers explain the members and nodes.

Table 1 Sections for the Stiffened Part

	A (cm^2)	I (cm^4)
B	127.06	1270600

Table 2 Sections for the Stiffening Part

	A (cm^2)	I (cm^4)
S	30.0	1000
T	30.0	-

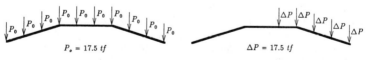

a) gravity load   b) additional load

$P_o = 17.5 \ tf$ $\Delta P = 17.5 \ tf$

Fig.2 Loading Conditions for the Structures in Fig.1

The deformed configurations under their gravity loads are shown in Fig. 3. The static behaviors of both structures are almost same in case of symmetric loadings and the bending moments for beams and the nodal displacements are much reduced than those of uncontrolled structures, depending on the value of the parameter μ in Eq.(6). In case of $\mu = 0$, the structure is completely controlled and on the other hand in case of $\mu = \infty$ the structure is under no control.

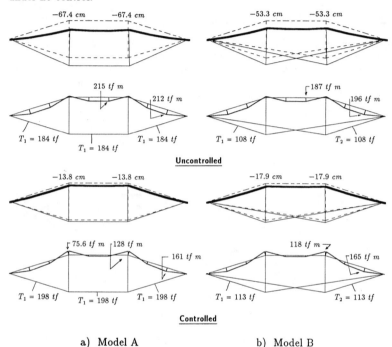

a) Model A   b) Model B

Fig.3 Bending Moments and Displacements under Gravity Loads;
T shows the magnitude of required axial, active force.

Assuming a case that a function of active string control is installed in structures in any time for sudden additional loading, an analysis is conducted for showing the effectiveness still under asymmetric loading. The results are given in Fig.4. In case of the structure without bracing (Model A), it is incapable of controlling the asymmetric additional loading, and on the other hand the structure with bracing (Model B) reveals its ability and the behaviors are controlled symmetric, and, moreover, both the bending moments and displacements are much reduced.

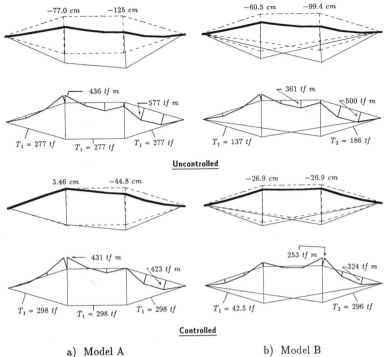

a) Model A   b) Model B
Fig. 4 Responses under Additional Loads on Half Side ;
T shows the magnitude of required axial, active force.

The study is extended to a three dimensional space frame, shown in Fig. 5. The structure stands on a hexagonal plan subjected to both their gravity loads and additional loads on the half of their roof area. The loadings are given in Fig. 6. P_o is the gravity load and ΔP is the additional load.

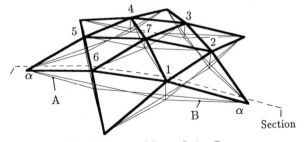

Fig. 5 Hexagonal Beam String Dome

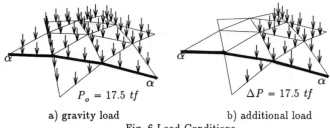

a) gravity load   b) additional load

Fig. 6 Load Conditions

The dimensions for the members and the strings are same as those in the previous examples. The responses are shown in Fig. 7, under the gravity loads. The displacements and stresses are well controlled.

Controlled

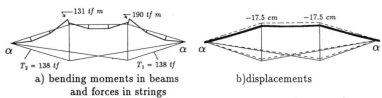

a) bending moments in beams   b)displacements
and forces in strings

Fig. 7 Responses of Stresses and Displacements under Gravity Loads

In case of non-uniform loading due to additional loads, the results are given in Fig. 8. The bending moments in the beams even subjected to the additional loads are much reduced because the symmmetricity of deformation is preserved by control when the control parameter μ is assumed as zero, and the effectiveness can be seen in case of this bracing type beam string structure. The results would be different in case of non-bracing type as seen from the previous example.

Controlled

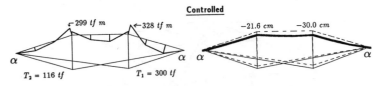

a) bending moments in beams   b)displacements
and forces in strings

Fig. 8 Responses of Stresses and Displacements
under Gravity and Additional Loads

Dynamic Control by Prestressing Strings

Dynamic control of strings of the previous three dimensional space structure is illustrated. The gravity loads first apply to the structure statically and the sudden additional loads follow, as shown in Fig. 9b for time domain and in Figs. 9a in space.

In Eq.(6), the third and fifth terms are utilized for this example in optimization of the target function J. In general it is also possible to include the first term for controlling. The third one corresponds to the kinematic energy of the stiffened part. Thus the masses for the nodes with the number 1 to 7 are considered as the third term and each diagonal term for the matrix Q_V is set equal to the masses for those nodes.

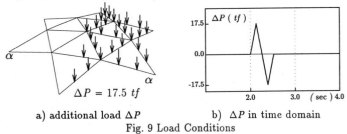

a) additional load ΔP b) ΔP in time domain

Fig. 9 Load Conditions

The effectiveness of the control is illustrated with respect to the parameter μ and the response of the vertical displacement at the node with number 1 is given in Fig. 10. As can be seen, the dynamic response can be controlled completely in case of $\mu = 0$, and the response in case of $\mu = \infty$ is under no control. In this dynamic analysis, the damping matrix C_d is assumed as Rayleigh damping with 2% damping factors for the first and second natural modes.

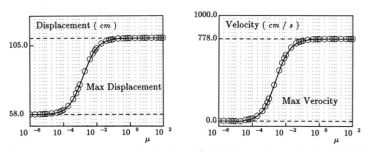

a) vertical displacement at Node 1   b) relative, vertical velocity at Node 1

Fig. 10 Effectiveness of Control Depending on the Parameter μ

Depending on the value μ the effectiveness changes and the value μ might be determined in consideration of design economics for a structure in issue. The time histories for the string tensions are given in Figs. 11 and 12, and for the displacement at Node 1 in Fig.13.

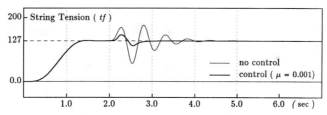

Fig. 11 Time Histories for the A String.

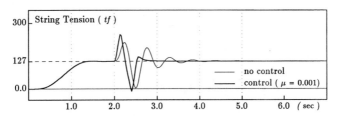

Fig. 12 Time Histories for the B String.

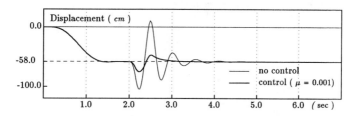

Fig. 13 Time Histories of the Vertical Displacements at Node 1.

Designing with more effects needs greater controlling forces in strings, and needs energy with adaptable high frequencies. This kind of direct control requiring large stresses would be possible, however, an adjustment could be introduced by adopting several passive measures, which might be called as a multi-step control. If a passive system that a series of strings is adjustable just at completion and thereafter unadjustable to additional loads is combined with a series of active strings, the required active forces would be much smaller and accordingly the energy for controlling would be reduced. The structural system could be schematically drawn as in Fig. 14. The formulation for the multi-step control is almost same as the present one except a necessity of two controlling force matrices F and two controlling matrices C .

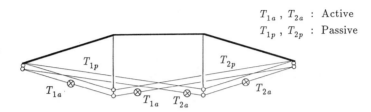

T_{1a} , T_{2a} : Active
T_{1p} , T_{2p} : Passive

Fig. 14 Combined System of Passive and Active Strings

Conclusions

The effectiveness of active control of strings in space structures is illustrated by adopting several, simple beam string structures. The analysis is performed in both static and dynamic cases to investigate their dynamic behaviors. From the results the system has proved to be much effective to reduce the stresses and deformations of those structures, although the ratio of effectiveness depend on the cost invested in those structures. And a possibility of a mixed system is discussed by using a rather simple model to reduce the energy required for these systems.

Acknowledgement

The present work is supported and approved as one of the projects by Toyohashi University of Technology. The first author would like to express his gratitude for the financial support duing the work.

References

Ikeda, Y., Kobori, T. (1992). "Active Variable Stiffness system Based on Instantaneous Optimization for Single Degree of Freedom Structure.", *Journal of Struct. Constr. Engin.*, AIJ, 435(5), 51-59.

Yang, J.N., Akbarpour, A. and Ghaemmaghami, P. (1987). "New Optimal Control Algorithms for Structural Control.", *Journal of Engineering Mechanics,* ASEC, 113(9), 1369-1386.

Fabric Roof Design Responds to New Technologies

Craig G. Huntington , Member ASCE

Abstract

A combination of newer synthetic materials and advanced structural design technologies have given contemporary tensioned fabric roofs vastly improved energy performance and far greater durability and reliability than those built only 20 years ago. Advanced analytical techniques provide reliable form-finding and prediction of fabric stresses, while allowing designers to select unique shapes with assurance of reliable behavior. Structural details have evolved for both improved performance and visually expressive character that fits with their function in the structure.

New Structural Methodologies Improve Performance

Contemporary tensioned fabric roof structures are successors to a tradition of tent building dating back to prehistoric times, and including more recent historical applications for circuses and the military. Low cost and portability gave traditional tents their enormous appeal, while short lived materials, fire hazards, and wrinkled, fluttering shapes gave them their notoriety.

Emerging technologies, however, make these contemporary structures fundamentally different from those created as recently as 20 years ago. Newer materials assure fabric lifespans in excess of 20 years, even in harsh environments, while meeting strict test standards for combustibility, flame spread, and smoke generation. High light reflectivity and natural

---

1 President, Huntington Design Associates, Inc., 4112 Park Boulevard, Oakland, CA 94602

daylighting provide energy efficiency in warm and bright climates, while evolutions in translucent and flexible insulation have improved energy performance in cold climates dramatically, as well. Recent designs have provided R values of 12 or greater while maintaining sufficient translucency to transmit 4% of daylight through the fabric. These developments have given tensioned fabric roofs application in permanent, high occupancy, enclosed structures, and in buildings where energy efficient, conditioned spaces are required.

While materials advances have improved many measures of fabric roof performance, however, it is newer structural design methodologies that have been instrumental in the emergence of tensioned fabric roof structures that are also high grade works of architecture. Evolutions in analytical technique over the past 30 years have been most critical to this change. Up to the 1960's, tent shapes developed through the gradual evolution of previously realized forms, with model studies and simple hand calculations used to evaluate shapes and structural adequacy.

The German designer Frei Otto advanced the use of modeling in membrane roof design, sometimes using scaled assemblies of wire and springs to predict the behavior of full size structures. His best known models, however, were those made using soap bubbles, which have the property of shaping themselves into membranes with equal tension at all points on the soap film and about any axis. By stretching thin soap films between any desired edge condition and graphically recording the resulting shapes, Otto found that he could accurately replicate the shape that a fabric membrane or cable network would take in order to be stable and uniformly prestressed over any proposed system of mast of arch supports.

In the 1960's, the engineers working with Otto began to develop the computer programs that would be used to shape and analyze the Montreal Pavilion for Expo '67 and the Munich Olympic Stadium, structures that were revolutionary in their time for both monumentality and sophistication. While these structures had either polyester fabric or plexiglass roof surfaces supported on steel cable networks, the technology was adopted and further developed by engineers in the Geiger Berger office in New York City for use in designing fabric membranes without cable nets. By modeling a strip of fabric as if it were a cable having the same axial stiffness, they learned, a good approximation of the proper shaping and stress under load of the fabric membrane could be realized.

By the   mid 1970's, these computer programs were as amazing for their relative accuracy and analytical power as they were stupefying for their cumbersome data input and interpretation. The programs were used in the late 1970's to design the Jeddah Hajj Terminal, soon to become the largest fabric   roof   in the world, and analytical results about the forces in various members in the structure were validated within a few percent by test results on a prototype constructed later. As part of the team   of engineers preparing that   analysis, however, the author   spent   many   sleepless   nights preparing input data for computer runs that would cost $5,000 to $10,000 each on the first Cray supercomputers. The programs included no means for automated generation of geometric data; each of several thousand "nodes" defining the shape of the fabric membrane had to be defined and input by hand.   No graphic interface was available to plot this input geometry and assist the engineers in verifying its numerical accuracy.

The refinement of computing technology increased thereafter, until, by the mid 1980's, programmers had automated the input of most data, and could instantly generate plots that showed the structure's geometry and deformation under load.   Engineers at Birdair, America's largest fabric structure contractor, could sit at the computer with the owner or architect of a proposed building and "walk him around" the structure - observing an isometric plot of the shape from any location, even moving "inside" the building in order to get a sense of its interior volume.

A fabric roof is subject to wind flutter unless prestressed over some supporting framework.   With the often complex curvature of these roofs, patterning the fabric into a shape with fairly uniform prestressing over the entire surface area of the roof becomes a daunting task. Reliable structural behavior, though, in which some fibers do not become overly tensioned while others flutter excessively due to a lack of tension, is dependant on providing a regulated level of prestress throughout the surface of the fabric.

The rapid development in computing technology that has taken place since the 1950's has dramatically improved the accuracy and speed with which the shape of the finished structure and the stresses in fabric, cabling, and supporting elements can be predicted under load. Regulated levels of prestress can now be reliably provided, even on structures of unusual shape, and the resulting improvements in structural behavior have been enormous. On old circus tents or other archaic fabric

pretensioning. Moderate winds along the length of the structure created waves in the canopy that rolled repeatedly from one end of the structure to the other until the fabric tattered. Computer analytical techniques have made such disasters completely avoidable, and have created not only the ability to evaluate the stability of a structure throughout the different stages of erection and tensioning of the fabric, but to accurately analyze the secondary stresses that may be introduced by the adjustable cabling or telescoping masts often used to tension contemporary tensioned fabric roofs.

Enriching the Vocabulary of Forms

In addition to improving the accuracy of shaping and stress analysis, new computing technology has created the ability to confirm the viability of unusual shapes and to reliably predict their behavior under load - an enormous boon to designers seeking to make a bold statement or respond to unusual site parameters. Birdair was able to achieve dependable performance with the spherical fabric Radome structures built in the 1950's and 1960's partly because these high profile air-supported bubbles were generally similar in shape and scale. Repeated construction of similar structures allowed gradual refinement in an iterative approach to a nearly perfect design, a process not unlike that of the Gothic cathedrals, or even that occurring in the Darwinian evolution of the natural world.

Contemporary engineers not only have the ability to consider variations in mast location, arch curvature, scale, and other refinements to a design; they also have the capacity for developing reliable, large scale roofs using structural systems with little or no precedent. The recently completed Georgia Dome in Atlanta offers an example. An oval shaped roof with a maximum span of 235 meters, it draws inspiration from both the tensegrity dome concept of Buckminster Fuller and the cable domes created by David Geiger. None of Fuller's domes were built, however, and the radial cable system created by Geiger was substantially different in both shape and structural behavior from the triangulated system created by engineer Weidlinger Associates for Georgia. Yet the analytical tools available to Weidlinger and contractor Birdair allowed them to design and erect one of the world's longest span roofs without the opportunity to build confidence working on smaller roofs of similar design.

The rapidly expanding ability of the structural engineers is, ironically, beginning to give architects a greater role in a medium dominated, until now, by engineers. As long as the number of structural engineers with experience in fabric was very small, and as long as the analytical tools used by them were limited in capability and not publicly available, architects were bound by the skills, intuitions, and preferences of the tension structure specialist with whom they chose to work. As long as the analytical tools available to these specialists were limited, design options were in turn limited, and fabric roof shapes tended to be restricted to those that were familiar and those having simple and readily predictable structural behavior. With contemporary tools, however, architects can ask their structural collaborators to evaluate unusual shapes or those having highly complex behavior, and can receive prompt and frequently favorable evaluations of roof forms that a pragmatically thinking engineer would never seek to use.

As advanced analytical tools and direct project experience have become more widespread, therefore, tensioned fabric roof form has left the world of structural engineering magic to join the mainstream of a construction industry in which architects act as form givers and structural engineers the professionals charged with addressing the practical considerations of bringing those forms to reality.

The maxims followed by structural engineers in tension structure design typically display a desire for simplicity of design, economy of material use, and redundancy that reflect the engineer's training and mind set. As architects have played a more critical role in the design of certain monumental tensioned fabric structures, however, their divergent motives have been expressed. In Vancouver's Canada Place, for example, the peaks of the masts have been lifted to an elevation far greater than that required to attain adequate fabric curvature, thereby increasing the yardage of fabric required and the magnitude of lateral wind loads significantly. The forms give a distinctly sail-like image that architect Eberhard H. Zeidler found appropriate to the waterfront structure. The enormous new airport structure in Denver, Colorado, capped by a repeated module of double masted tents, provides another relevant example. It was structural engineer Horst Berger's natural inclination to simplify design and maximize construction economy by making all 34 masts the same height. However, architect C.W. Fentriss/J.H. Bradburn, chose to provide masts of three different heights, in an effort both to emphasize the junctions

between the different   terminals   and   to   mirror   the
varying heights of   the   snow-capped   Rocky   mountains
visible in the distance.

Evolutions in Detailing

The evolution   in detailing of structural   elements
and their connections   has seen a growth that, while not
as impactful as   the   rapid   expansion   of   analytical
capability, has nonetheless had a significant   effect on
fabric roof design.   A   resurgence   in   the use of cast
steel in place   of   weldments   for   major   structural
connections, for example,   not   only   reduces   stress
concentrations in the   steel,   but   also   gives   these
connections elegant, curving   profiles   that   mirror the
quality of the   fabric   membrane   itself.   Extruded
aluminum fabric edge   clamps,   similarly,   have   both
reduced fabrication costs for   the   clamps   and provided
the possibility for a smooth termination of   the   fabric
at a rounded edge member with no visible bolting.

While steel   casting   and   aluminum   extrusions, as
described above, have   softened   certain   details   and
allowed a purer expression of the natural   beauty of the
curving roof form,   other   details   give a more machine-
like expression of the means   by   which the structure is
erected or tensioned.   Mast jacking mechanisms are often
the most visible of   these,   with   adjusting   bolts   or
turnbuckles typically left exposed to provide a muscular
reminder of the   large   forces exerted to raise the mast
and pretension the fabric.   In   the author's structural
design for a   shade canopy   at a school in   Santa Clara,
California, for example,   the   mast   sits   inside   a
cylindrical pot which   is   raised   into   position   with
portable hydraulic jacks, then secured with   four   large
threaded rods.

In the   tensioned   fabric   roof,   form and structure
are analogous to a degree unprecedented   in the world of
architecture.   In like   manner, structural   details   are
shielded by neither   fascia   or   ceiling,   so that their
elegance and expressiveness   is inextricably linked with
the success of   the   architecture.   It is   no   surprise,
then, that evolutions   in   the technology of shaping and
analyzing these roofs and evolutions in their structural
detailing are having   a   profound   impact   on   the
architecture of the tensioned fabric roof.

structure by contrast, wrinkles (representing a lack of prestress at some location on the structure) and patches (often representing a location where the fabric tore in order to relieve local overstress) were ubiquitous.

Accurately regulated prestress has brought enormous improvement in the visual sophistication of tensioned fabric roofs; so too has the virtual elimination of the visual clutter of turnbuckles and other devices that were used to adjust membrane shape and prestress in the field. With contemporary analytical techniques offering a high likelihood of accurate shaping and prestress, designers often prefer to omit devices that allow field adjustment of cable length or other geometries of the structure, having observed various incidences in which well intentioned field personnel created unintended overstresses or tears in the course of their efforts to perfect the tensioning in a structure.

Tensioned fabric roofs rely upon anticlastic surface shapes, those in which fabric curvatures oppose each other about two axes at any point on the surface, in order to provide stability under load. Saddles, hyperbolic paraboloids, and the characteristic cone shape of traditional tents all have this property. With their opposing curvatures, loading inward or outward against the fabric surface causes an increase in fabric stress about one axis and a decrease about the other. If sufficient prestress and curvature is provided in a properly designed roof, large areas of the fabric are never allowed to go slack.

During erection of a fabric roof structure, however, the membrane may lay draped over the supporting structure without prestress, or a portion of the roof may be tensioned while another is not. Stable shapes with appropriate anticlastic surfaces therefore may not occur throughout erection, and large secondary stresses in supporting members may be introduced by the imbalance of tension throughout the membrane. This may be of little import in roofs of moderate size and simple configuration, where a single crane can lift the entire fabric membrane into place over a supporting framework and prestress it within a matter of hours. In larger contemporary structures like the cable domes, however, it may take several days or even weeks before the fabric is stressed into its final, stable position, during which time the loose fabric is subject to destruction from wind flutter. Several years ago, the author participated in the investigation of the failure of a fabric canopy more than 150 meters in length that was draped loosely over a Hollywood stage set without

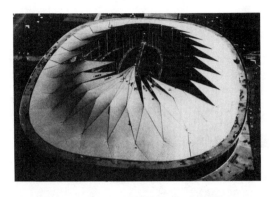

Photo Courtesy
of Birdair

Figure 1.   Georgia Dome; Atlanta, Georgia
Architect: Heery/RFI/TVS Atlanta
Engineer: Weidlinger Associates, New York
Roof Contractor: Birdair, Inc.; Amherst, New York
This stadium is the first to use a triangulated cable
dome. Analysis accounted for wind and live load on both
the completed structure and during cable tensioning and
fabric prestress.

Photo Courtsey
of Birdair

Figure 2.   Canada Place; Vancouver, British Columbia
Architect: Zeidler Roberts Partnership; Toronto
Engineer: Geiger Engineers; Suffern, New York
Roof Contractor: Birdair, Inc.; Amherst, New York
Canada Place, like the Georgia Dome, was dependent on
contemporary computer technology   to achieve reliable
behavior on an innovative   form.   Georgia   Dome   is
emblematic of an engineer's proclivity to enclose space
with maximum efficiency and minimum cost. Exaggerated
mast heights and sail-like fabric shapes at Canada Place
represents the more abstract and symbolic aims of many
architects.

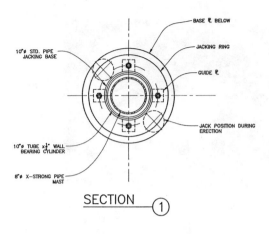

SECTION ①

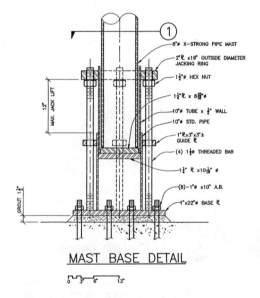

MAST BASE DETAIL

Figure 3.   Buchser School Mast Base
Architect: The Steinberg Group
Engineer: Huntington Design Associates, Inc.
The jacking mast assembly consists of a standard pipe
base welded to the base plate, a cylindrical "pot"
jacked upward inside the base, and a pipe mast that
rides inside the pot.   The mast is raised by hydraulic
jacks and held in position by securing the nuts on the
threaded bars against the underside of the doughnut
plate.

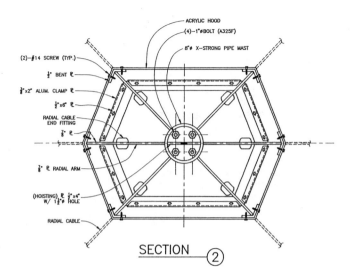

ACRYLIC HOOD
(4)—1"∅BOLT (A325F)
8"∅ X—STRONG PIPE MAST

(2)—#14 SCREW (TYP.)
½" BENT ℓ
⅜"x2" ALUM. CLAMP ℓ
½"x6" ℓ
RADIAL CABLE END FITTING
⅜" ℓ
⅞" ℓ RADIAL ARM
(HOISTING) ℓ ½"x4" W/ 1½"∅ HOLE
RADIAL CABLE

SECTION ②

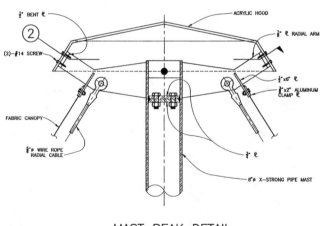

½" BENT ℓ
②
(2)—#14 SCREW
FABRIC CANOPY
⅝"∅ WIRE ROPE RADIAL CABLE

ACRYLIC HOOD
⅞" ℓ RADIAL ARM
½"x6" ℓ
⅜"x2" ALUMINUM CLAMP ℓ
¾" ℓ
8"∅ X—STRONG PIPE MAST

MAST PEAK DETAIL

0 3" 6"   12"

Figure 4. Buchser School Mast Peak
At the start of erection, the mast peak weldment is
connected on the ground to fabric and radial cables,
then hoisted by crane hook over the mast. Besides
serving as a visual focal point and providing the
connections of mast, fabric, and cables, the mast peak
assembly provides venting of heated air daylighting
through the hood.

A PRACTICAL DESIGN AND CONSTRUCTION
OF
TENSION ROD SUPPORTED GLAZING

Toru TAKEUCHI[1], Haruyuki KITAMURA[2],
Kimiaki HARADA[2], Kohei HIGUCHI[1],
Akiho HARADA[1], and Mamoru IWATA[1]

ABSTRACT

In designing a glass skin-covered space, for the atrium, entrance hall or other core part of a building complex, the high openness and quality requirements of the space often lead to the high lightness and refined detail requirements of the glazing support structure. A tension structure composed of rods or cables is a glass skin support system of excellent performance. Following

Fig. 1 Long credit bank HQ./North atrium

- -
[1]Nippon Steel Corp., Otemachi 2-4-6, Chiyoda-ku, Tokyo
[2]Nikken Sekkei Ltd., Koraku 1-4-27, Bunkyo-ku, Tokyo

the results of evaluation by full-size experimentation
and analyses in the previous report,[3] application examples of tension rod structure to support a large glazed
wall area are reported in terms of design, detail, and
construction.

INTRODUCTION

A tensile glazing support structure often becomes a
part of the glass skin design itself as shown in Fig. 1,
because of the glass transparency. In refining and
lightening the design characteristics of the tensile
glazing support system, the following methods are generally effective.

(a) Elimination of some structural elements: The tension
structure is an effective means for alleviating the heav-
iness of the glazing support system while maintaining the
stability and safety. These benefits vary with the arrangement and pretension plan of tensile members. In
some case, a glazing structure of higher transparency can
be obtained by utilizing the large deformation stability
and eliminating some of the structural members necessary
under linear theory.

(b) Utilization of non-structural members: Glass or sash,
which have been traditionally treated as non-structural
members, may be evaluated as structural or reinforcing
members to eliminate redundant structural elements used
there. Since this boundary element have the built-in
function of adjusting assembly accuracy, however, the
redundant may eliminate at the sacrifice of ease of
construction.

(c) Detail refinement of structural parts: Conventional
steel structures are mainly fabricated and elected by
such operations as gas cutting, welding, and friction-
bolt jointing, and their details are often not good
enough for use in the as-finished condition. More refined details can be obtained by combining the forging,
casting and machining of steel for applications other
than building construction and by utilizing mechanical
jointing methods.

A tension structure to support about 3,600 m^2 of
glass skin in an actual project was designed and constructed according to the above methods. The details of
this structure are described below.

GENERAL PROPERTY OF THE PROJECT

The project is outlined in Fig. 2. It comprises a
T-shaped office building about 130 m high and with an

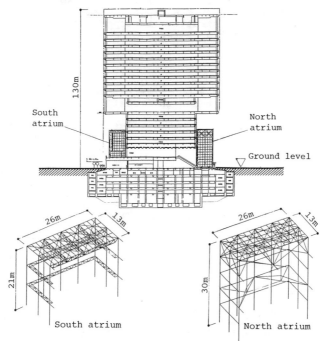

South atrium

North atrium

Ground level

South atrium

North atrium

Fig. 2   Outline of project

Fig. 3   Long credit bank HQ./south atrium

approximately 20 m overhang in front and rear at the 50 m height. Glass boxes, each measuring about 30 m high, 26 m wide and 13 m deep, are built under each overhang at the ground level. They function as an entrance hall and a main sales office, and are hereinafter called the north atrium and south atrium, respectively. Each atrium is covered with a sashless glass panels with DPG (Dot Point Glazing) support. The north atrium is shown in Fig. 1, and the south atrium is shown in Fig. 3. Main structural members are seamless steel pipes measuring 318 mm in diameter and 20 mm in thickness, and are assembled in a grid pattern measuring 6 to 7 m to constitute a box frame supporting the glass skin.

GLAZING SYSTEMS

The glazing support systems of the north atrium and south atrium are schematically illustrated in Figs. 4(a) and (b), respectively.

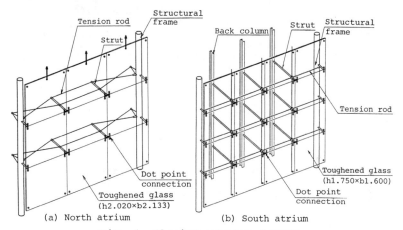

(a) North atrium   (b) South atrium

Fig. 4 Glazing support system

The tensile glazing support system of the north atrium (a) is basically similar with that P. Rice adopted in La Villette in Paris[1] and has a set of three glass panes carried by a frame in curtain form. The wind pressure on the glazing is countered by the tension structure on the inside of the glazing. Strength, stiffness, and economy favored high-strength steel rods of 13-mm diameter over cables as tensile members. Each tension rod is pretensioned to about 30 kN. To allow the tension rods to withstand the pretension and protect the ends of the tension rods from bending stress, special joint blocks with the functions of following rotation and

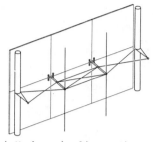 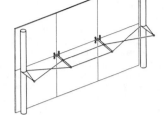

(a) Members by linear theory   (b) Members by nonlinear theory

Fig. 5 Elimination of some structural members by
using large-deformation stability

adjusting length were installed at the points where the
tension rods were fixed.   The structural safety of the
entire glazing support system is verified by the previously reported full-size experiment and analyses.[3]   As
described in the previous report, the stability of the
tension structure under vertical load and asymmetrical
horizontal pressure are guaranteed by large deformation
stiffness with   tension force.   Conventional linear
theory calls for the use of the members shown in Fig.
5(a).   Design, based on geometrical nonlinear characteristics can reduce the use of these members as shown in
Fig. 5(b). Since the boundary condition to guarantee the
vertical stability of the tensile frame is given by the
glazing, and the weight of the lower glazing is carried
by the upper glazing, the glass functions as structure in
this system.

The   south   atrium of Fig. 4(b) is constructed by a
method in rather conventional way.   The glazing is supported by DPG connections at the strut ends.   The strut
ends stabilized in the vertical and in-plane horizontal
directions by tensile members installed in a grid pattern
between the main frames.   The external forces in the out-
of-plane direction are transferred through the struts to
the back columns.   The south atrium is a double-skinned
box having clear sashless glazing on the outside and a
conventional white glass on the inside. The back columns
double as inside glazing sash retainers.   The panes of
plate glass are separate from each other and carry their
own weight alone.   Even when the panes are removed, the
tension structure does not become unstable.   This means
that glass is not considered as structural element.   The
tension frame, including the back columns, is also stable
under linear theory and need not be pretensioned.   Since
the installation accuracy of each pane of glass is determined at the strut ends, however, accuracy must be adjusted after assembly.   For these reasons, the south

atrium used the same joints as the north atrium with length adjusting function.

MATERIALS AND DETAILS

As described at the beginning, the selection of materials and composition of details have a large bearing on the quality of completed products. The materials and details used in the project are reviewed by element below.

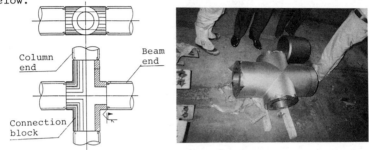

Column end
Beam end
Connection block

Fig. 6   Connection block

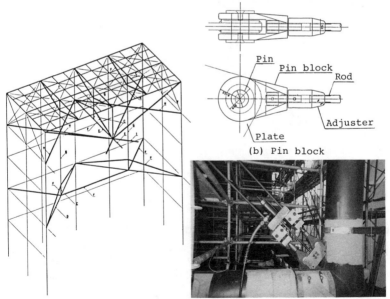

Pin
Pin block
Rod
Adjuster
Plate

(b) Pin block

(a) Tensile brace arrangement (c) Pretensioning of tensile brace

Fig. 7   Tensile braces for main frames

Main frame seamless CHS made of fire-resistant steel
developed by Nippon Steel are used in the column and
roof, in order to omit fire-proof covering.   Since some
part of the roof was found to exceed 600°C according to
the results of fire temperature analysis, a fire-resistant paint that produces an expanded char when exposed to
fire was used in the roof area concerned.   Connection
blocks shown in Fig. 6 were made to ensure rigid column-
beam joints and obtain a continuous intersecting pipe
structure. Since fire-resistant steel for casting was not
developed at the time of their fabrication, connection
block were made from fire-resistant forging steel already
performance proven in the field.

The north and south atriums have a total of 141 tensile ·braces arranged against external forces developed
during strong winds, earthquakes, and snowfalls.   The
arrangement of tensile braces in the north atrium is
shown in Fig. 7(a).   The tensile braces may be divided
into three groups; The first stiffens the horizontal
plane, the second stiffens the vertical external walls,
and the third adds to the vertical load carrying capacity
of the roof.   All types have the function of complement-

(a) Screw joint block

(b) Detail of north atrium   (c) Detail of South atrium
Fig. 8 Glazing support tensile frame joints

ing the buckling capacity of members with a high-slender-
ness ratio. The tensile braces are also made from high-
strength steel rods, measuring 17 to 32 mm in diameter,
and are connected to main frames by forged pin blocks
shown in Fig. 7(b). Some of the tensile braces are
pretensioned to about 50 kN by drawing rod ends into the
pin blocks with oil jacks, as shown in Fig. 7(c).

The tension structures to support the glazing were
also pretensioned, and screw joint blocks as shown in
Fig. 8(a) were fablicated to follow large deformation.
The screw joint block is an improvement on the one used
in the full-size experiment described in the previous report. It is more compact, but retains the length adjusting and rotation following functions. The details of
north and south atrium glazing supports, each constructed
by using this joint, are shown in Fig. 8(b) and (c),
respectively. All screw joint blocks were separately
proof tested to confirm their structural safety.

The materials of steel members used in the main
frames and tension structures of the north and south
atriums are listed in Table 1.

Table 1   Materials of steel members

AREA	PART	STANDARD		MATERIALS
MAIN FRAME	BEAM/COLUMN	JIS	⎧ G3101 ⎨ ⎩ G3106	SS400,SM490,SM490-NFT
	TENSILE BRACE	JIS	G3109	SBPR785/930
	PIN END	JIS	G3106	SM490 (forged)
GLAZING TENSILE STRUCTURE	TENSION ROD	JIS	G3109	SBPR930/1080
	STRUT/JOINT	JIS	G4051	S45C (machined)
	DOT POINT CONNECTION	JIS	G4303	SUS329J1 (casted)
	GLASS HANGING BLOCK	JIS	⎧ G3101 ⎨ ⎩ G4051	SS400,S45C (forged, machined)

ESTIMATION OF PRETENSIONING FORCE

In the north atrium, eight tension structures in the
same level were simultaneously pretensioned with oil
jacks, as shown Fig. 9. The axial pretensioning force
was all controlled by hydraulic pressure. For checking
the residual force of tensile members after completion, a
weight was suspended at the rear of each tension rod by
utilizing the vertical stiffness caused by geometrically
nonlinear deformation. The residual axial force was
checked by measuring the displacement caused by the
weight. The values given for each level are the average
displacement of the rear of the tension structure from
which a 30-kg weight was suspended. The distribution of
the measured values is summarized in Fig. 10. Although
one point registrated an extremely low axial force due to

hydraulic valve clogging, the measured values approximately fell within the control range of 0 to -35%.

When the eight tension structures in the same level are pretensioned, pipe columns at the ends and the corners were expected to move about 2 mm in the analysis, and same order movement were measured in the actual frame. The movement of the tensile member connections in the pretensioning process were expected to half of these boundary movement, which indicates the control of axial force by measuring the displacement of connections are difficult to accomplish in reality. In contrast, the axial force control by hydraulic pressure was confirmed to be practically effective. These variations will be reduced by applying the axial forces while measuring them.

Fig. 9   Pretensioning of glass support tensile members

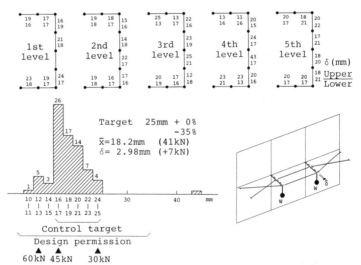

Fig. 10   Measured displacement and axial force

CONCLUSIONS

Following the full-size experiment and analysis described in the previous report, two different types of tensile glazing support structures were applied to a total area of about 3,600 m² in an actual project. Their details were established, and their fabrication, construction and design properties were studied. The following findings were confirmed:

(1) The tension rod structures are effective as both main frames and glass supports. Joint blocks with the function of following rotation and adjusting length are satisfactory as to both fabrication and installation.

(2) The pretensioning method used in the full-size experiment and the method of measuring the axial force by using the stiffness caused by geometrical nonlinear deformation have proved effective in the actual project. The measured axial force are practically within the control target range.

(3) The composition of glass support frame with tensile rods and mechanical joints has provided neat structures benefitting sashless glass skins.

References:

1) P. Rice and H. Dutton, 'Le verre structurel', Architecture Thematique 1991.
2) S. Houghton, 'Marine technologies in architecture', The Journal of SCI, Vol. 6 No. 6, Nov. 1992.
3) Toru Takeuchi et al., 'A basic study of tensile rod supported glazing', IASS-SEIKEN Symposium, 1993.

Structural Shape Analysis under the Prescribed Displacement Mode

Yasuhiko HANGAI [1] , Kazuaki HARADA [2]

ABSTRUCT

An analytical method of structural shape satisfying the prescribed displacement mode which belongs to the inverse problem in the field of structural analysis is presented. In this paper, the following three kinds of analytical methods are presented, and their validity is examined by numerical analyses.
(A) Method by the generalized inverse matrix.
(B) Method by the Bott·Duffin inverse matrix, and
(C) Method by the minimization of the error of the constraint condition.

INTRODUCTION

Many analytical and numerical methods have been developed in order to apply mathematical programming techniques to minimize the weight or volume of structures as well as to maximize the stiffness of structures considering design limit on stress or deflections. In these methods, sizing design variable such as cross sectional area, plate thickness, moments of inertia, etc. have been usually adopted as design variable [1].

Another class of optimization problems is shape optimization in which shape of structures is to be determined[1-3]. The shape optimization is stated mathematically as

$$\text{Find} \quad \min F (x) \quad \text{subjected to } g_i(x) \leq 0, \ i=1\sim N \qquad (1)$$

where F is the objective function, g_i is the constraint function, and x is a generalized coordinate defining the shape of structures and design variable to be determined.

There exists an important class of structural design problems in which the shape of structures is analyzed without the objective function

---

[1] Professor, Institute of Industrial Science, University of Tokyo
7-22-1, Roppongi, Minato-ku, Tokyo 106, Japan.
[2] Research Engineer, Engineering Research Center, Tokyo Electric Power
Company, 2-4-1, Nishi-Tsutsujigaoka, Chofu-shi, Tokyo 182, Japan.

being optimized.   This problem is stated as

Find   x   subjected to $d_i(x) = \bar{d}_i$, i=1~M   (2)

where **x** is a vector defining the shape of the structures, d_i the
displacement components and $\bar{d}_i$ the prescribed displacement mode for
d_i .

In this paper, the shape analysis of structures subjected to
constraints on displacement modes which belongs to the problem given by
Eq.(2), is presented.   The three kinds of analytical methods(ANALYTICAL
METHOD-1, 2 and 3) are presented,   and their validity is examined by   the
numerical examples.   Shape analyses of structure belong to the inverse
problem   in   the structural analysis,   and   show   highly nonlinear
characteristics.   Then,   (a)   the   existence   condition   of   solution,
(b)   the number of solutions   and   (c)   numerical technique of nonlinear
equations   become important research themes.

RESEARCH OBJECT

Let us consider the load-displacement relation

$f = K(x) \, d$ (3)

where f is   n × 1   load vector,   **K**   n × n   stiffness matrix which
is   a   function   of   a   generalized coordinate vector   **x**   representing
structural shape,   d   n × 1   displacement vector and   n   the number
of degrees of freedom.   The generalized coordinate vector   **x** in Eq.(3)
can be divided into three components : x_h the vector   subjected   to
the prescribed displacement mode, x_f the unknown   vector   to be
analyzed and x_c the fixed vector (Fig.1).   The number of degrees
of freedom are   h for x_h and f   for x_f respectively,   and   then
n = h + f .   As a   constraint   condition,   let   us   consider   the
homologous deformation [4]   which   is   given   by   the statement : x_h
keeps the prescribed shape before,   during   and   after the   deformation
under the prescribed load f .   Let   us   consider   an example   of   the
homologous deformation by using a plane truss structure given   in Fig.2.
The homologous deformation   is   that the upper nodes lie on a horizontal
straight line   after   the deformation (DY(1):DY(2):DY(3) = 1. 0:1. 0:1. 0).
We get   the displacement mode shown   in   Fig.3.   This displacement mode
does not satisfy the homologous deformation   stated above,   so that we
change the vertical position Y(4) of node 4.   Fig.4 shows the relation
between Δ (Δ = DY(1)-DY(2)) and Y(4).   The value of Δ becomes zero at
two points,   A and B.   For these two shapes as shown in Fig.5,   we get
tha same vertical displacements for the upper nodes.

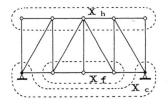

x_h : Prescribed shape
x_f : Unknown or objective shape
x_c : Boundary shape

Fig.1.   Classification of coordinate parameters

ANALYTICAL METHOD-1 [5]

The displacement vector d can be divided into two parts: d_h denoting the displacement vector of prescribed shape, d_f denoting other displacement components in d .   By prescribing the displacement mode d_o , d_h can be expressed by

$$d_h = \alpha \, d_o \,.$$ (4)

where α is an unknown parameter and d_o the prescribed displacement mode.   Based on d_h and d_f , the stiffness matrix $K(x_f)$ and the load vector f can be devided and then Eq. (3) can be expressed in the form

$$\begin{bmatrix} f_h \\ \hline f_f \end{bmatrix} = \begin{bmatrix} K_{hh}(x_f) & K_{hf}(x_f) \\ \hline K_{fh}(x_f) & K_{ff}(x_f) \end{bmatrix} \begin{bmatrix} d_h \\ \hline d_f \end{bmatrix}$$ (5)

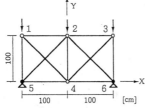

Fig. 2.   Initial shape

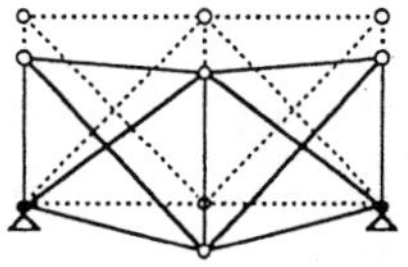

Fig. 3.   Displacement mode

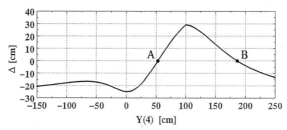

Fig. 4. $Y(4) - \Delta$ relation

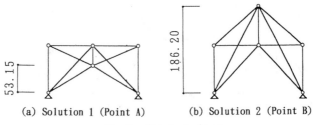

(a) Solution 1 (Point A)   (b) Solution 2 (Point B)

Fig. 5.   Obtained shape

The introduction of Eq. (4) into Eq. (5) leads to

$$\begin{bmatrix} f_h \\ \hline f_f \end{bmatrix} = \begin{bmatrix} k_h & K_{hf} \\ \hline k_f & K_{ff} \end{bmatrix} \begin{bmatrix} \alpha \\ \hline d_f \end{bmatrix} \tag{6}$$

where x_f is omitted and

$$k_h = K_{hh}d_0, \quad k_f = K_{fh}d_0 \tag{7}$$

If the total degree of freedom and the dimension of d_h are denoted by n and m respectively, the following coefficient matrix becomes the $n \times m$ rectangular matrix:

$$L(x_f) = \begin{bmatrix} k_h & K_{hf} \\ \hline k_f & K_{ff} \end{bmatrix} \tag{8}$$

Our problem is stated by "Find x_f, α, d_f subjected to Eq. (6)."

In order to eliminate the unknown of α and d_f, consider the existence condition of solution of Eq. (6). The necessary and sufficient condition for Eq. (6) to have the solution is

$$[L(x_f)L^+(x_f) - I_n] f = 0 \tag{9}$$

where I_n is $n \times n$ identify matrix and $L^+(x_f)$ is the Moore-Penrose generalized inverse matrix of $L(x_f)$. The left-hand side of Eq. (9) denotes the nonlinear function of x_f, which can be expressed in vector form by

$$g(x_f) = [L(x_f)L^+(x_f) - I_n] f \tag{10}$$

If $\overline{x}_f$ is a solution of Eq. (10), $\overline{x}_f$ satisfies the following equation

$$g(\overline{x}_f) = 0 \tag{11}$$

which can be used for the numerical analysis of x_f.

Eq. (11) is a nonlinear equation for x_f and the Newton-Raphson method is adopted for the numerical analysis in the paper.

ANALYTICAL METHOD-2

Let us denote the prescribed displacement mode by

$$A d = 0 \tag{12}$$

where A is $m \times n$ matrix which is called compatibility matrix that denotes the subsidiary condition for the prescribed displacement mode, and m the number of subsidiary conditions. The research object in the section can be stated as "Find x which satisfies Eq. (3) subjected to the subsidiary condition Eq. (12)."

Eq. (3) can be derived by minimizing the total potential energy function

$$\Pi = \frac{1}{2} d^T K d - d^T f \tag{13}$$

where T denotes the suffix for the transposition. In order to analyze Eq. (3) under the subsidiary condition Eq. (12), let us use the Lagrange multiplier method by introducing the potential energy function

$$\Pi_k = \frac{1}{2} d^T K d - d^T f + \lambda^T A d \qquad (14)$$

where λ is a Lagrange multiplier in the vector form. The stationary condition of Eq. (14) are

$$K d - f + A^T \lambda = 0 \qquad (15)$$

$$A d = 0 \qquad (16)$$

If we introduce the notation

$$r = A^T \lambda \qquad (17)$$

then Eq. (15) takes the form

$$K d + r = f \qquad (18)$$

The orthogonality condition of d and r can be proved by using Eq. (16) as

$$d^T r = d^T A^T \lambda = (A d)^T \lambda = 0 \qquad (19)$$

In the result, the minimization problem with the subsidiary condition Eq. (12) given by Eq. (14) results in Eq. (18) with the orthogonality condition of Eq. (17).

From Eq. (19), we have

$$d \in L \qquad \text{and} \qquad r \in L^\perp \qquad (20)$$

where L is a linear subspace in the n dimensional space R^n and $L^\perp$ the orthogonal complement to L. If P_L and $P_{L^\perp}$ are orthogonal projections of a vector a on L and $L^\perp$ respectively, d and r can be obtained by

$$d = P_L a \qquad \text{and} \qquad r = P_{L^\perp} a \qquad (21)$$

In order to determine a, we introduce Eq. (21) into Eq. (18). Then we have

$$[K P_L + P_{L^\perp}] a = f \qquad (22)$$

which gives us

$$a = [K P_L + P_{L^\perp}]^{-1} f \qquad (23)$$

If we introduce the above equation into Eq. (21), we have the solution of Eq. (18) which takes the form

$$d = P_L [K P_L + P_{L^\perp}]^{-1} f \qquad (24)$$

$$r = P_{L^\perp} [K P_L + P_{L^\perp}]^{-1} f \qquad (25)$$

The coefficient matrix of f on the right-hand side of Eq. (24) is called " Bott·Duffin inverse of K " [6], whose automated numerical analysis is presented in [7].

In Eq. (25), the right-hand side is a function of x_f because $K = K(x_f)$. Then r is also a function of x_f as

$$r = r(x_f) \qquad (26)$$

If we can find $\bar{x}_f$ which satisfies $r(\bar{x}_f) = 0$, then $\bar{x}_f$ gives us an shape to be determined because Eq. (18) coincides with Eq. (3) satisfying the subsidiary condition of Eq. (4).

In the result, our problem can be stated as " Find x_f which satisfies the following equation. "

$$r \ (x_f) \ = P_L{}^\perp \ [K \ (x_f) \ P_L + P_L{}^\perp] \ ^{-1} \ f = 0 \qquad (27)$$

Eq. (27) is a nonlinear equation for x_f and Newton-Raphson method is adopted for the numerical analysis in the similar way as ANALYTICAL METHOD-1.

ANALYTICAL METHOD-3

If we consider the structural shape which satisfies $K(x_f) \, d = f$, the shape has the error of the subsidiary condition for the prescribed displacement mode. From Eq. (3), we have

$$d = K^{-1} \ (x_f) \ f \qquad (28)$$

Let us denote the error by

$$e \ (x_f) \ = A \, d = A \, K^{-1} \ (x_f) \ f \qquad (29)$$

The research object in the section can be stated as " Find x_f which satisfies the error of the subsidiary condition given by Eq. (29) is equal to zero." Newton-Raphson method is also adopted for the numerical analysis of the nonlinear Eq. (29).

NUMERICAL EXAMPLES

Consider the truss structure in Fig. 2. For this truss structure, Figs. 6, 7 and 8 show relations of $|g| - Y(4)$, $|r| - Y(4)$, and $|e| - Y(4)$, respectively, and the convergent behavior starting from two points C and D. The values of $|g|$, $|r|$ and $|e|$ become zero at two points A and B which correspond with the shapes given in Fig. 5. Consider the prescribed displacement mode that $DY(1):DY(2):DY(3)$ is equal to $1.0 : 1.2 : 1.1$. In this example, x_f has components of $X(4)$ and $Y(4)$. Figs. 9 and 10 show the contour line of $|r|$ and the surface of $|r|$ by Eq. (26). The value of $|r|$ becomes zero at two points E and F. Figs. 11(a) and 11(b) show shapes at these two points. Figs. 12 ~ 15 show the longitudinal sections at the points E and F in Fig. 9. Consider three cases with the prescribed displacement modes that $DY(1):DY(2):DY(3)$ are equal to $1.0 : 1.0 : 1.0$, $1.0 : 1.1 : 1.0$, $1.0 :1.2 : 1.0$, and $1.0 : 1.3 : 1.0$. Figs. 16 ~ 19 show the relations between $Y(4)$ and $|r|$ under each prescribed displacement mode. Figs. 20, 21 and 22 show the obtained shapes which satisfy the prescribed displacement modes. There exists no solution in the case that $DY(1):DY(2):DY(3)$ is equal to $1.0 : 1.3 : 1.0$.

CONCLUSION

In this paper, three kinds of analytical method of structural shape with the constraint condition of prescribed displacement mode are presented. Numerical examples show the validity and characteristics of the present shape analysis. Shape analysis of the structure belongs to the inverse problem and show highly nonlinear characteristics so that there are important problems to bo solved about the existence condition of solutions, the numbers of solutions and numerical technique of nonlinear equations. Illustrative examples of these three items are shown numerically.

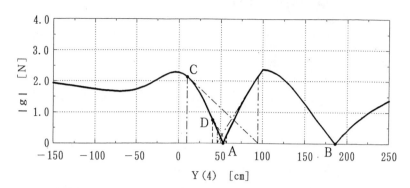

Fig. 6. |**g**| − Y(4) relation and convergent behavior

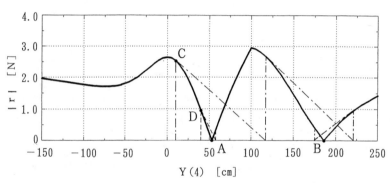

Fig. 7. |**r**| − Y(4) relation and convergent behavior

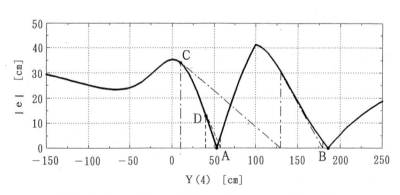

Fig. 8. |**e**| − Y(4) relation and convergent behavior

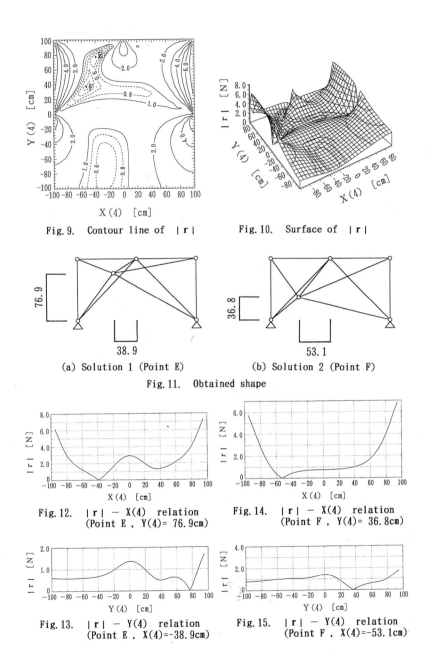

Fig. 9.   Contour line of |r|

Fig. 10.   Surface of |r|

(a) Solution 1 (Point E)

(b) Solution 2 (Point F)

Fig. 11.   Obtained shape

Fig. 12. |r| − X(4) relation
(Point E , Y(4)= 76.9cm)

Fig. 14. |r| − X(4) relation
(Point F , Y(4)= 36.8cm)

Fig. 13. |r| − Y(4) relation
(Point E , X(4)=-38.9cm)

Fig. 15. |r| − Y(4) relation
(Point F , X(4)=-53.1cm)

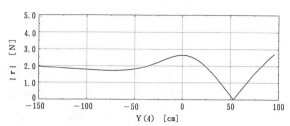

Fig. 16. | r | — Y(4) relation
(ΔY(1):ΔY(2):ΔY(3) = 1. 0 : 1. 0 : 1. 0)

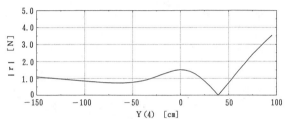

Fig. 17. | r | — Y(4) relation
(ΔY(1):ΔY(2):ΔY(3) = 1. 0 : 1. 1 : 1. 0)

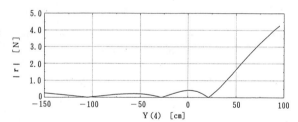

Fig. 18. | r | — Y(4) relation
(ΔY(1):ΔY(2):ΔY(3) = 1. 0 : 1. 2 : 1. 0)

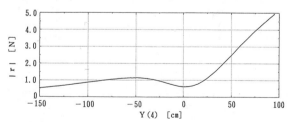

Fig. 19. | r | — Y(4) relation
(ΔY(1):ΔY(2):ΔY(3) = 1. 0 : 1. 3 : 1. 0)

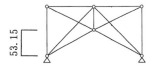

Fig. 20. Obtained shape
($\Delta Y(1):\Delta Y(2):\Delta Y(3)=1.0:1.0:1.0$)

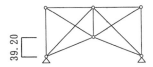

Fig. 21. Obtained shape
($\Delta Y(1):\Delta Y(2):\Delta Y(3)=1.0:1.1:1.0$)

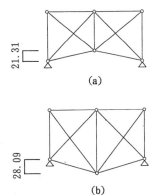

(a)

(b)

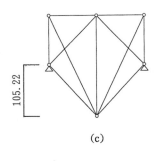

(c)

Fig. 22. Obtained shape
($\Delta Y(1):\Delta Y(2):\Delta Y(3)=1.0:1.2:1.0$)

REFERENCES

[1] Ding, Y. : Shape optimization of structures : A literature survey, Computers and Structures, Vol. 24, No. 6, 1986, pp. 985-1004.

[2] Topping, B. H. V., : Shape optimization of skeletal structures : A review, Journal of Structural Engineering, Vol. 109, No. 8, 1983, pp. 1933-1951.

[3] Hang, E. J., Choi, K. K., Hou, J. W. and Yoo, Y. M. : A variational method for shape optimal design of elastic structures, New Directions in Optimum Structural Design, John Wiley and Sons, 1984, pp. 105-137.

[4] Hoerner, S. : Homologous deformation of tiltable telescopes, Journal of the Structural Division, ASCE, Vol. 93, No. ST5, 1967, PP. 461-485.

[5] Hangai, Y. and Guan, F. L. : Structural shape analysis with the constraint conditions of homologous deformation, Journal of Structural and Construction Engineering, Transaction of AIJ, No. 405, 1989, pp. 97-102.

[6] Domaszewski, M. and Borkowski, A. : Generalized inverse in elastic-plastic analysis of structures, Journal of Structural Mechanics, Vol. 12, No. 2, 1984, pp. 219-244.

[7] Hangai, Y. and Kawaguchi, K. : Generalized inverse and its application to shape analysis. Baifukan, 1991.

[8] Hangai, Y. and Harada, K. : Analytical method of structural shape under the prescribed displacement mode, Journal of Structural and Construction Engineering, Transaction of AIJ, No. 453, 1993, pp. 93-98.

Form Finding Numerical Methods
for
Tensegrity Systems

René Motro Professor[1],
Sihem Belkacem Docteur Ingénieur[2],
Nicolas Vassart, Chercheur[3].

ABSTRACT
The rigidification of Tensegrity Systems is related to
selfstress states which can be achieved only when
geometrical and mechanical requirements are simultaneously
satisfied. This paper deals with several form finding
processes and underlines the interest of the Force Density
method which allows a wider range of forms than other
methods which have been used until now.

1.INTRODUCTION
Tensegrity Systems are reticulated spatial structures
composed of straight members, struts and cables which
define a stable volume in space by the effect of
equilibrium between compression and tension.According to
the definitions given by Fuller and Emmerich , they follow
a principle of compression islands in a sea of tension.

2.FORM FINDING AND SELFSTRESS

2.1 Classification of tension structures
According to the classification which we have established
for "tension structures" these systems belong to the class
of selfstressed systems.Equilibrium equations for
reticulated space structures (with "n" nodes and "m"
members) can be established :

- **in terms of internal stresses**
$[A].\{t\} = \{W\}$ (1)
with [A], "equilibrium" matrix (dimensions 3n x b), {t}

[1] Head of "Structural Design" team.
[2] [3] Members of "Structural Design" team.
Laboratoire de Mécanique et Génie Civil, Université de Montpellier II, 34095,
Montpellier Cedex 5, France.

internal forces vector (dimension b) and {W}, external actions vector (dimension 3n).
To be more precise on the form finding problem it's also useful to split up the equilibrium matrix [A], which can be set in a product of two matrices :

$$[A] = [C*]^t.[F] \tag{2}$$

which leads to :

$$[C*]^t.[F].\{t\} = \{W\} \tag{3}$$

Matrix [C*], whose transpose appears in equation (2), is characteristic of the relational structure of the studied system. Matrix [F] is a square diagonal matrix of dimension "m": its terms are the direction cosines of each member, it depends upon the "3n" nodes coordinates. This matrix might be called "form matrix"

- or in terms of displacements

$$[K_e].\{d\} = \{W\} \tag{4}$$

$$\text{with } [K_e] = [A].[H].[A]^t \tag{5}$$

([K_e] elastic rigidity matrix and [H] "elastic" matrix related to material and geometrical characteristics.)
According to these equilibrium equations we can distinguish:

Funicular systems which satisfy :

$$[C*]^t.[F]_i.\{t\}_i = \{W\}_i \tag{6}$$

each form [F]$_i$ is directly related to each external actions vector {W}$_i$.

Prestressed systems which satisfy :

$$[K_e]_i.\{d\}_i = \{0\} \tag{7}$$

Vector {d}$_i$ includes boundary conditions in terms of some imposed displacements which imply support actions and a state of prestress in the system. [K_e]$_i$ is the sum of the elastic rigidity matrix and of the so-called geometrical matrix taking into account the existing internal stresses. Cable-nets, membranes belong to this class.

Selfstressed systems which satisfy :

$$[C*]^t.[F]_i.\{t\}_i = \{0\} \tag{8}$$

Vector $\{t\}_i$ defines a state of selfstress of the system
and the form characterised by $[F]_i$ for a given connexion
pattern $[C*]$ of the members is the solution of the
formfinding problem. Tensegrity systems belong to this
class.

2.2 Formfinding processes

According to equation (8), formfinding processes can be
achieved in three steps :
    1 Definition of the relational structure given by
matrix $[C*]$.
    2 Calculation of the "3n" coordinates of nodes,
matrix $[F]_i$
    3 Study of the compatibility of the selfstress,
vector $\{t\}_i$ (all cables need to be in tension and all
mechanims have to be eliminated by the selfstress)

This paper deals mainly with step 2.Two types of forms may
be distinguished : monoparameter and mutiparameter forms.
In the first class all struts are of same length "s" and
all cables of same length "c". The resulting form is
defined by the ratio s/c. The second class is
characterised by two vectors $\{s\}$(dimension m_s) and
$\{c\}$(dimension m_c), with m, number of elements equal to m_s+
m_c.

3.FORM FINDING BY GEOMETRICAL AND ANALYTICAL METHODS

3.1 Geometrical methods

A great amount of research has been done on geometrical
basis, mainly by Fuller, Emmerich and Pugh . An intensive
use of regular and semiregular polyhedra led to regular
forms, generally monoparametered, sometimes with several
values of s/c. Assembly of elementary modulus allowed the
design of masts or other structures like torus. Geometry
was regular in general meaning of this word.Formfinding
with geometrical methods doesn't garanty mechanical
equilibrium and solutions are to be checked.

3.2 Analytical methods

When solutions are simple, it's possible to use analytical
methods, by statical considerations or/end by calculation
of a maximum of the ratio s/c.

For prismatic tensegrity systems the ratio s/c can be put
in the form :

$$\frac{s}{c} = \sqrt{\frac{1}{\sin\dfrac{\pi}{p}} \cdot \left[\sin\left(\theta + \frac{\pi}{p}\right) + \sin\frac{\pi}{p}\right]} \qquad (9)$$

p is the number of edges of the top and bottom polygons, and θ the relative rotation of these polygons.The corresponding curve shows (case p = 4) the maximum value which is related to a geometry of selfstress (θ = $\pi/4$).(Fig. 1)

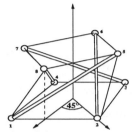
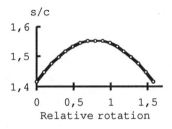

Fig. 1 Selfstress geometry for 4-prismatic tensegrity system

Statical considerations can be achieved on a tensegrity system with 6 struts and 24 cables. We called it "expanded octahedron". The resulting form is reached for a distance between struts equal to their half length. Apices are not those of an icosahedron as it could be anticipated without statical form finding research.

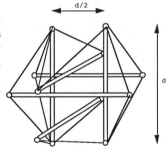

Fig. 2 Expanded octahedron

4.FORM FINDING NUMERICAL METHODS

4.1 Dynamic relaxation with kinetic damping
This method is based on the study of the damped state of a dynamic behavior governed by :

$$[M].\{x''(t)\} + [D].\{x'(t)\} + [K(t)].\{x(t)\} = \{W(t)\} \qquad (10)$$

with [M], mass matrix, [D], damping matrix, [K(t)], rigidity matrix and $\{x''(t)\}, \{x'(t)\}, \{x(t)\}$, respectively

acceleration, velocity and displacement vectors. When
dealing with formfinding external actions vector {W(t)} is
equal to {0}. The resulting form is not dependent from [M]
and it is convenient to choose a diagonal matrix. Damping
is obtained by maximization of kinetic energy, so matrix
[D] disappears. The system to be solved is of the form :

$$[M].\{x"(t)\}+[K(t)].\{x(t)\}=\{0\} \tag{11}$$

According to the diagonality of mass matrix, it can be
solved in incremental way by decoupling the calculation of
nodes position during a time increment Δt from the
evaluation of residual forces corresponding to this
position :

$$\{r(t)\} =- [K(t)].\{x(t)\} \tag{12}$$

A centered finite difference scheme is used to solve eq.
(11). Kinetic energy is computed at each end of time
increment : when a maximum is detected, corresponding
coordinates are chosen as a new basis for calculation, and
velocity vectors are put to 0.
The linearity of the solution scheme is ensured by the
choice of diagonal mass matrix. The movment is started by
a given incremental initial elongation of struts. Cables
have a constant length. Strut's length is increased until
a selfstress state is reached and checked by calculation
of the determinant of rigidity matrix.
This method has been applied for prismatic tensegrity
systems. Numerical precision was checked by comparison
with analytical solution and was better than what can be
reached by achievment by avalaible technology. Convergence
is good for systems with few nodes but more problematic
when the number of nodes increases.

We could verify that for a
truncated tetrahedron nodes of
hexagonal faces were not
coplanar.But the difference with
plan is very slight.

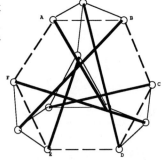

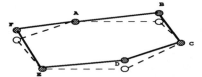

Fig. 3 Truncated tetrahedron

If this method is convenient for only one or two values of
ratio s/c, the monitoring for free forms is not easy since
changes of cable lengths have to be made during the
procedure.

4.2 Force density method

Force density method has been largely used for formfinding of membranes. We applyed it to tensegrity systems. Main features of the method are given in the following sections We consider a tensegrity system with m members and n nodes, coordinates of which are included in three vectors $\{x\}, \{y\}, \{z\}$.
Let $\{t\}$ be the internal stress vector and $\{l\}$ the length vector.The action of a member b (b=1 to m) on a node i (i=1 to n) is given by $s_b = -t_b$ according to classic sign convention with tension being negative value and compression positive value.

* Force density q_b
Force density for a member b is defined by :

$$s_b = -t_b \cdot l_b / l_b = q_b \cdot l_b \qquad (13)$$

and the force density matrix [Q]is a diagonal matrix m,m built with the force density values.

* Connectivity matrix [C].[C] is a diagonal third of [C*]
This m,n matrix is regular and is constructed line by line for each member b whose initial extremity are k and g (with k<g). All terms c_{bi} are equal to 0,except :

$$c_{bk} = -1 \text{ and } c_{bg} = 1 \qquad (14)$$

Components of member lengths along x are given by :

$$\{lx\} = [C] \cdot \{x\} \qquad (15)$$

(Equations along y and z are of the same form and are not listed in the following sections).
According to equation (13), actions of members $\{s\}$ on nodes can be written :

$$\{sx\} = [Q] \cdot \{lx\} \qquad (16)$$

or

$$\{sx\} = [Q] \cdot [C] \cdot \{x\}$$

* Equilibrium of nodes

Each node "i" is in equilibrium under the actions of members which are connected to "i". In case of selfstress no external actions are considered.
Equilibrium equations are :

$$[C]^t \cdot [Q] \cdot [C] \cdot \{x\} = 0 \qquad (17)$$

with $[C]^t$, transpose of [C].

* *Fixed and free nodes*
Nodes n are splited up in two classes, namely n_1 free
nodes and n_f fixed nodes. This leads to a partition of node
coordinates vector and connectivity matrix in the form :

$$[C] = [C_1 : C_f] \text{ and } <x> = <x1:xf> \tag{18}$$
($<x>$ transpose of $\{x\}$)

Equation (17) becomes :
$$[C_1]^t [Q] [C_1] \{x_1\} + [C_1]^t [Q] [C_f] \{x_f\} = 0 \tag{19}$$
Only equations for free nodes are taken into account.

* *Final equilibrium equations:*

With $[D_1] = [C_1]^t [Q] [C_1]$ $\hspace{3cm}$ (20)

and $[D_f] = [C_1]^t [Q] [C_f]$ $\hspace{3cm}$ (21)

equations of equilibrium are
$$[D_1].\{x_1\} = -[D_f].\{x_f\} \tag{22}$$
$$[D_1].\{y_1\} = -[D_f].\{y_f\} \tag{23}$$
$$[D_1].\{z_1\} = -[D_f].\{z_f\} \tag{24}$$

Computation of the three vectors $\{x_1\}, \{y_1\}, \{z_1\}$, for a
force density matrix [Q] and for fixed nodes defined by
$\{x_f\}, \{y_f\}, \{z_f\}$ gives access to the form of the
corresponding system.

4.3 Application to Tensegrity Systems

4.3.1 Force density method and Tensegrity Systems
Force density method allows a multiparametered form
finding process . For a given connectivity matrix [C], the
designer may choose :
    - values of force densities, and the matrix [Q] is
completely defined. In this case force density for a member
b is positive for a cable and negative for a strut.
    - fixed nodes which correspond to geometrical
requirements such as nodes to be in two paralell planes.
The solution is given by :

$$\{x_1\} = -[D]^{-1} [D_f].\{x_f\} \tag{25}$$
When using this method for membranes or cable-nets matrix
[D] is not singular because all the terms of [Q] are of
the same sign. For tensegrity systems two situations may
occur.

a/ Determinant of [D]≠0
The rank of [D] is equal to "1" and the system (25) is a classical Cramer system. The form is governed by the values chosen for [Q] and the monitoring is largely dependent from the experience of the designer.
Attention must be paid to the the specific case which occurs when all fixed nodes $\{x_f\}$ are coplanar: the solution will also be planar and it will be without interest.

b/ Determinant of [D]=0
Let r_d be the rank of [D].System (25) can be solved only when vectors $[D_f].\{x_f\}$ are related to independent vectors of matrix [D].It's necessary to compute all the $1-r_d$ characteristic determinants of [D] for x,y and z directions.
If one of them is equal to zero, there is no solution.Otherwhise there is an infinity of solutions parametered along each axis by the $1-r_d$ components of linked vectors of [D].
Among this infinity,forms may be reached by fixing supplementary nodes. In this case attention must be paid on the fact that force densities have to be chosen such as det[D] and all characteristic determinants must be equal to zero. Nevertheless a wide range of forms is avalaible with this method.

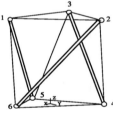

Fig.4 Regular 3-prismatic system
Fig.5 Squew 3-prismatic system

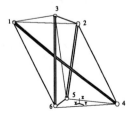

4.3.2 Examples of form finding.
This numerical method was first applied to regular 4-prismatic (Fig.1) and 3-prismatic systems (Fig. 4) to check the validity of the method.

A squew 3-prismatic system was computed (Fig.5).Triangles are in two parallel planes and can move on them. Values of force densities, resulting lentghs and stresses are given in the table.

Member	Force density	Length	Stress
1(c)	1	1	t
2(c)	1	1	t
3(c)	1	1	t
4(c)	1,732	1,302	2,255 t
5(c)	1,732	1,641	2,843 t
6(c)	1,732	1,360	2,356 t
7(s)	-1,732	2,024	-3,505 t
8(s)	-1,732	2,161	-3,743 t
9(s)	-1,732	0,969	-1,679 t

Formfinding was also developed for squew 4-prismatic
systems. These modulus can be use to design single
curvature doube layer tensegrity systems. In this specific
case there are two different strut lengths as shown in the
table. These results have to be compared with those given
in a previous study, where strut's lengths were identical.

Fig.6 Squew 4-prismatic system

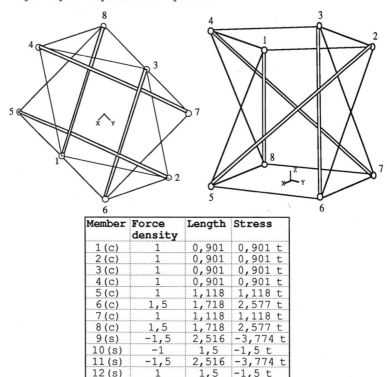

Member	Force density	Length	Stress
1 (c)	1	0,901	0,901 t
2 (c)	1	0,901	0,901 t
3 (c)	1	0,901	0,901 t
4 (c)	1	0,901	0,901 t
5 (c)	1	1,118	1,118 t
6 (c)	1,5	1,718	2,577 t
7 (c)	1	1,118	1,118 t
8 (c)	1,5	1,718	2,577 t
9 (s)	-1,5	2,516	-3,774 t
10 (s)	-1	1,5	-1,5 t
11 (s)	-1,5	2,516	-3,774 t
12 (s)	1	1,5	-1,5 t

5.CONCLUSION
Formfinding is a necessary step in the study of tensegrity
systems   in   order   to   satisfy   mechanical   self
equilibrium.Free forms can only be determined by
multiparametered numerical methods. Force density has been
first applied to prismatic modulus and a wide range of
"irregular" form resulted from this first approach. It can
be anticipated that this numerical method is convenient
for formfinding of more complicated tensegrity systems.

REFERENCES

Belkacem S. " Recherche de Forme par Relaxation Dynamique de Systèmes Réticulés Spatiaux Autocontraints" Thèse DI, UPS Toulouse, juin 1987.

Emmerich D.G. "Structures tendues et autotendantes". Monographie de Géométrie constructive. Editions de l'Ecole d'Architecture de Paris La Villette.1988

Fuller R.B.,Marks R. "The Dymaxion World of Buckminster Fuller". Anchor Books. Ed.1973.

Motro R. "Tensegrity Systems and Geodesic Domes",Special Issue of the International Journal of Space Structures ,"Geodesic Space Structures", Vol.5,n°3,1990,pp.343-354.

Motro R. "Tensegrity Systems : State of Art",Special Issue of the International Journal of Space Structures,"Tensegrity Systems", Vol.7,n°2,1992,pp.75-83

Motro R. "Morphological Aspects of Space Structures requiring Formfinding Processes" Proceedings of 4 International Conference on Space Structures, Parke & Howard ed.,1993, Vol 2, pp 1148-1159;

Pugh A. "An introduction to tensegrity",University of California Press, Berkeley,1976.

Vassart N. "Recherche de forme d'autocontrainte multiparamétrée de Systèmes Réticulés Spatiaux Autocontraints". DEA. Université Montpellier II. Juillet 1993.

Direct and Modal Analyses
for Shape-Finding of Unstable Structures

Tetsuyuki Tanami[1]
Yasuhiko Hangai[2]

Abstract

Shape-finding analysis of unstable structures has been presented by using generalized inverse (Tanaka and Hangai,1986). The "Modal Analysis" using rigid body modes is elegant, but the computer programming for the numerical calculation of the modes is not simple. Therefore we propose another method named "Direct Analysis" without using rigid body modes in the paper.

Introduction

In case of unstable structures composed of cable and membrane materials which can change the shape without strain, shape-finding of the structure under given load is important at design stage. And control of unstable behavior of structure is widely interesting theme, especially, in the field of tension structural engineering.

Shape-finding analysis of unstable structure based on usage of Moore-Penrose generalized inverse is presented. The analysis means to find out transition and final equilibriumed shape to external load acting on the structure. After extracting the rigid body mode components of displacement from general solution of compatibility equation by using generalized inverse, either the work of loading on static problem or the inertia force on dynamic problem has been used in order to decide each unknown coefficient of the normalized modes in the treatment of

[1]Tanami, Center for Space Structural Research, Taiyo Kogyo Corporation, 3-22-1 Higashiyama, Meguro-ku Tokyo153, Japan
[2]Hangai, Institute of Industrial Science, University of Tokyo, 7-22-1 Roppongi, Minato-ku, Tokyo 106, Japan

"Modal Analysis." But getting the rigid body modes in programming would not be easy though the theory of extraction by using generalized inverse is elegant. Moreover there is no advantage in reducing the number of the modes since all of the modes have to be used in the modal analysis. Therefore "Direct Analysis" being in no need of extraction of rigid body modes is presented by focusing on the following concerns in the paper. (1)Rethinking of transition condition having been used in the current modal analysis on static problem. (2)Proposing the direct analysis on static problem. (3)Showing existence of the direct analysis on dynamic problem by solving unstable truss with one rigid mode in comparison with the result of the modal analysis.

Relation between rigid modes and load

Displacement, u, strain, ε, stress, σ, and load, f, have the following relations each other in linear elastic problem.

Compatibility equation	$\varepsilon = A\ u$		(1)
Constitutive equation	$\sigma = E\ \varepsilon$		(2)
Equilibrium equation	$f = B\ \sigma$		(3)

where coefficient matrixes, A and B, have relation of $B=A^t$, in which $(\ )^t$ means transposition. In case of unstable structure with $\varepsilon =0$, the explicit relation connecting the load of Eq.(3) to the displacement of Eq.(1) is impossible. Therefore both the work of loading and the inertia force play important role to connect them on static and dynamic problems, respectively.

Matrices A, E and B are assumed to be full ranks with the sizes of $(m,n),(m,m)$ and (n,m), respectively, with the following relations

rank A = rank B = m   in n>=m,
rank A = rank B = n   in n< m,
rank E = m   in always,

where n=m means statically determinate, and the structure of n<m is statically indeterminate with the order of (m-n). In case of unstable structure with the relation of n>m, the number of rigid modes is (n-m).

Static problem

Transition condition meaning how to get the transition of geometrical behavior to the final stable shape of unstable structure is rethinked in modal analysis. Moreover, after proposing the direct analysis, we can show that both the problems become the same optimal analysis to decide a solution with the indeterminate equation of transition condition by using generalized inverse.

Transition condition

Let us denote incremental strain, $\Delta\varepsilon$, and incremental displacement, Δu , would be related in the form

$$\Delta\varepsilon = A\Delta u \tag{1}$$

In case of unstable structure, the condition of $\Delta\varepsilon = 0$ can be considered into Eq.(1). Then

$$A\Delta u = 0 \tag{2}$$

The general solution has been derived from using generalized inverse in the form

$$\Delta u = \Delta\alpha_1 h_1 + \Delta\alpha_2 h_2 + \cdots + \Delta\alpha_p h_p = H\Delta\alpha \tag{3}$$

in which

$$H = [h_1, h_2, -, h_p], \quad \Delta\alpha = \begin{Bmatrix} \Delta\alpha_1 \\ \Delta\alpha_2 \\ \vdots \\ \Delta\alpha_p \end{Bmatrix} \tag{4}$$

where h_i and $\Delta\alpha_i$, $i = 1 \sim p$ denote normalized vectors of rigid body modes and the corresponding unknown coefficients with the number of rigid body modes, $p(=n-m)$.

Let us now consider the work of loading, $\Delta\Pi$, in the form

$$\Delta\Pi = (\Delta u^t f = \Delta\alpha^t H^t f = \Delta\alpha^t p =) p^t \Delta\alpha \tag{5}$$

in which

$$p = H^t f = \begin{Bmatrix} h_1{}^t f \\ h_2{}^t f \\ \vdots \\ h_p{}^t f \end{Bmatrix} \tag{6}$$

Then Eq.(7) means transition condition where the final stable shape satisfying equilibrium condition is obtained by the expression of $\Delta\Pi = 0$ and the process meaning geometrical orbit from the initial shape to the final shape of unstable structures is given by $\Delta\Pi > 0$.

$$\Delta\Pi \geqq 0 \tag{7}$$

We note that there are infinite sets of solution, $\Delta\alpha$, satisfying indeterminate equation of Eq.(5). Therefore particular solution by generalized inverse is useful to optimally specify the solution in the form.

$$\Delta\alpha = \Delta\alpha\, p \tag{8}$$

in which

$$\Delta\alpha = \frac{\Delta\Pi}{(h_1{}^t f)^2 + (h_2{}^t f)^2 + \cdots + (h_p{}^t f)^2} \tag{9}$$

Eq.(7) is always satisfied under the condition of $\Delta\alpha > 0$.

Substituting Eq.(8) into Eq.(3), rigid body displacement at an incremental step is derived in the form

$$\Delta u = \Delta \alpha H p = \Delta \alpha \{(h_1 f^t h_1) + (h_2 f^t h_2) + \cdots + (h_p f^t h_p)\} \qquad (10)$$

we note again that the solution decided by generalized inverse in Eq.(8) is the same optimal solution corresponding to the method (Hangai and Kawaguchi, 1987) using energy surface of the work of loading.

Direct analysis(Tanami and Hangai, 1993)

Rewrite the compatibility relation with the condition of $\Delta \varepsilon = 0$

$$A \Delta u = 0 \qquad (1)$$

and the work of loading caused by given load, f, is already shown in the form

$$f^t \Delta u = \Delta \Pi \qquad (2)$$

Now, we can make a simultaneous equation composed of both equations of Eq.(1) and Eq.(2) in the following form.

$$G \Delta u = b \qquad (3)$$

in which

$$G = \begin{bmatrix} A \\ f^t \end{bmatrix}, \quad b = \left\{ \begin{matrix} 0 \\ \Delta \Pi \end{matrix} \right\} \qquad (4)$$

Matrix of G in Eq.(3) is always a full rank of the size, $G=(m+1,n)$, with the relations of $m+1 <= n$ and $rankG=m+1$ because in case of unstable truss structure with the number of members, m, and the degrees of freedom, n, for instance, size of the matrix of A is $A=(m,n)$ with the conditions of $m < n$ and $rankA=m$. We note that existence condition of solution on Eq.(3) is always satisfied, in other words, uniqueness of the solution is not always guarantee. Therefore generalized inverse is useful to get an optimal solution in the paper. Then rigid body displacement, Δu, is

$$\Delta u = G^- b \qquad (5)$$

The work of loading, $\Delta \Pi$, can be given at each incremental step in the following two methods. One is simply

$$\Delta \Pi = C o \qquad (6)$$

Another one is given by analogy with the following expression shown in modal analysis.

$$\Delta \Pi \propto (\overline{h}^t f)^2 \qquad (7)$$

Where $\overline{h}^t$ denotes a general term of the normalized rigid

body modes in modal analysis. Therefore in case of direct
analysis, normalized vector of incremental displacement,
$\Delta \overline{u}$, can be used instead of $\overline{h}^t$ in the form.

$$\Delta \Pi = C_o \, (\Delta \overline{u}^t f)^2 \qquad (8)$$

Solving Eq.(3) under an increment given by Eqs.(6) or (8)
is direct analysis for shape-finding of unstable structure in static problem. Both Eqs.(6) and (8) are also
useful in the modal analysis.

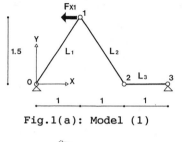

Fig.1(a): Model (1)

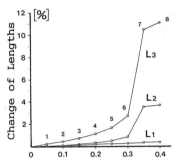

Fig.2: Change of Member's
Lengths in Model(1)

Fig.1(b):Control by Eq.(6)

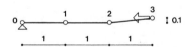

Fig.3(a): Model (2)

Fig.1(c):Control by Eq.(8)

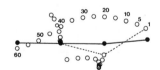

Fig.3(b):Control by Eq.(6)

Fig.1: Example 1   Fig.3: Example 2

Two examples are illustratively demonstrated. In
case of the first example of structure shown in Fig.1(a),
two kinds of control proposed in Eqs.(6) and (8) are
numerically checked as shown in Figs.1(b) and (c) in
which control means how to give the increment at each
step. Control in Eq.(6) has an advantage in getting
transition and final shape roughly by a small iteration.

On the other hand, we can get exact final shape in case
of Eq.(8) based on the consideration of orthogonality of
vectors between load and displacement at the final geometry. Change of the member lengths caused by the linear
analysis is shown in Fig.2 with Co=0.05 in Eq.(6). Another example is applied to the model strongly governed to
the rigid body behavior as shown in Fig.3 in order to
examine the rough behavior of the transition up to the
step number of 60 with each step of Co=0.05 in Eq.(6).

Dynamic problem

Let us show existence of direct analysis by demon-
strating an analysis on unstable truss structure with one
component of rigid body mode and compare the result with
the corresponding modal analysis.

Direct analysis

Neglecting damping term, the equation of motion of
the unstable structure with two degrees of freedom is

$$\begin{bmatrix} m_o & 0 \\ 0 & m_o \end{bmatrix} \begin{Bmatrix} \ddot{u} \\ \ddot{v} \end{Bmatrix} = \begin{Bmatrix} f_x{}^o \\ f_y{}^o \end{Bmatrix} \tag{1}$$

Velocity and displacement components to the x-coordinate
direction are derived from integration of Eq.(1) under
assumption of unit mass matrix in the form, respectively.

$$\dot{u} = f_x{}^o t + C_1 , \quad u = \frac{1}{2} f_x{}^o t^2 + C_1 t + C_2 \tag{2}$$

Integration constants of C_1 and C_2 are decided by the
following initial conditions

$$\dot{u} \mid_{t=0} = \dot{u}_o , \quad u \mid_{t=0} = 0 \tag{3}$$

Note that initial displacement is zero in case of the
motion of rigid body. Then the velocity and the displacement are

$$\dot{u} = f_x{}^o t + \dot{u}_o , \quad u = \frac{1}{2} f_x{}^o t^2 + \dot{u}_o t \tag{4}$$

Similarly deriving the responses to the y-direction, the
following matrix expressions of velocity and displacement
are given with time interval of Δt .

$$\begin{bmatrix} 1 & 0 \\ 0 & 1 \end{bmatrix} \begin{Bmatrix} \Delta \dot{u} \\ \Delta \dot{v} \end{Bmatrix} = \begin{Bmatrix} \dot{u}_o \\ \dot{v}_o \end{Bmatrix} + \Delta t \begin{Bmatrix} f_x{}^o \\ f_y{}^o \end{Bmatrix} \tag{5}$$

$$\begin{bmatrix} 1 & 0 \\ 0 & 1 \end{bmatrix} \begin{Bmatrix} \Delta u \\ \Delta v \end{Bmatrix} = \Delta t \begin{Bmatrix} \dot{u}_o \\ \dot{v}_o \end{Bmatrix} + \frac{1}{2} \Delta t^2 \begin{Bmatrix} f_x{}^o \\ f_y{}^o \end{Bmatrix} \tag{6}$$

Let us now consider compatibility condition. When
the coefficient matrix, **A**, is constant during the time

interval, Δt, stress-strain relations at the time steps of t=n and t=n+1 are respectively

$$A u_n = \varepsilon_n \; , \quad A u_{n+1} = \varepsilon_{n+1} \tag{7}$$

Then the following incremental expression is obtained

$$A\Delta u = \Delta \varepsilon \tag{8}$$

Derivatives of the first and the second orders of Eq.(8) by time are

$$A\Delta \dot{u} = \Delta \dot{\varepsilon} \; , \quad A\Delta \ddot{u} = \Delta \ddot{\varepsilon} \tag{9}$$

In case of behavior of unstable structure, the following constraint conditions have to be considered.

$$\Delta \varepsilon = \Delta \dot{\varepsilon} = \Delta \ddot{\varepsilon} = 0 \tag{10}$$

Direct analysis in dynamic problem resolves itself into the problem which solves three sets of simultaneous equation composed of both the equations of motion and the compatibility with respect to the components of acceleration, velocity and displacement under the corresponding constraint conditions shown in Eq.(10).

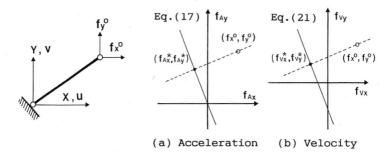

(a) Acceleration   (b) Velocity

Fig.4: Unstable Truss   Fig.5: Acceleration and Velocity
                        Components of Given Load

When direct cosines of member on both axes in Fig.4 are denoted by the notations, l and m, the matrix of **A** is expressed in the form

$$A = [\, l, m \,] \tag{11}$$

In case of this example, the behavior is governed by the following three sets of equation

$$\begin{bmatrix} 1 & 0 \\ 0 & 1 \\ 1 & m \end{bmatrix} \begin{Bmatrix} \Delta \ddot{u} \\ \Delta \ddot{v} \end{Bmatrix} = \begin{Bmatrix} f_{Ax}{}^* \\ f_{Ay}{}^* \\ 0 \end{Bmatrix} \; , \quad \begin{bmatrix} 1 & 0 \\ 0 & 1 \\ 1 & m \end{bmatrix} \begin{Bmatrix} \Delta \dot{u} \\ \Delta \dot{v} \end{Bmatrix} = \begin{Bmatrix} \dot{u}_o \\ \dot{v}_o \\ 0 \end{Bmatrix} + \Delta t \begin{Bmatrix} f_{vx}{}^* \\ f_{vy}{}^* \\ 0 \end{Bmatrix} \tag{12},(13)$$

$$\begin{bmatrix} 1 & 0 \\ 0 & 1 \\ 1 & m \end{bmatrix}\begin{Bmatrix} \Delta u \\ \Delta v \end{Bmatrix} = \Delta t \begin{Bmatrix} \dot u_o \\ \dot v_o \\ 0 \end{Bmatrix} + \frac{1}{2}\Delta t^2 \begin{Bmatrix} f_{Dx}{}^{\scriptscriptstyle I} \\ f_{Dy}{}^{\scriptscriptstyle I} \\ 0 \end{Bmatrix} \tag{14}$$

where we note that unknown load of $(f_x{}^*, f_y{}^*)$ having satisfied existence condition of solution has to be used as the loads in the corresponding acceleration, velocity and displacement equations with the subscripts of A,V and D, respectively, instead of given load of $(f_x{}^o, f_y{}^o)$.

Let us now show the relation between the unknown load and the given load. In case of Eq.(12) relating to acceleration, existence condition of solution is defined by using generalized inverse in the form

$$\left([I_3] - \begin{bmatrix} 1 & 0 \\ 0 & 1 \\ 1 & m \end{bmatrix}\begin{bmatrix} 1 & 0 \\ 0 & 1 \\ 1 & m \end{bmatrix}^{-}\right)\begin{Bmatrix} f_{Ax} \\ f_{Ay} \\ 0 \end{Bmatrix} = \begin{Bmatrix} 0 \\ 0 \\ 0 \end{Bmatrix} \tag{15}$$

Calculating the parentheses, we get

$$\begin{bmatrix} 1^2 & 1\,m & -1 \\ 1\,m & m^2 & -m \\ -1 & -m & 1 \end{bmatrix}\begin{Bmatrix} f_{Ax} \\ f_{Ay} \\ 0 \end{Bmatrix} = \begin{Bmatrix} 0 \\ 0 \\ 0 \end{Bmatrix} \tag{16}$$

Explicit relation of the load components is derived from Eq.(16) in the form

$$f_{Ay} = -\frac{1}{m}f_{Ax} \tag{17}$$

Then the load components, $(f_{Ax}{}^*, f_{Ay}{}^*)$, contributing to the equation of acceleration in rigid body motion is obtained at the point on the line given in Eq.(17) orthogonally intersecting to the given load, $(f_x{}^o, f_y{}^o)$, as shown in Fig.5(a), in the following relations.

$$f_{Ax}{}^{\scriptscriptstyle I} = m^2 f_x{}^o - 1\,m\,f_y{}^o\,, \qquad f_{Ay}{}^{\scriptscriptstyle I} = -1\,m\,f_x{}^o + 1^2 f_y{}^o \tag{18}$$

Substituting Eq.(18) into Eq.(12), the components of acceleration are

$$\begin{Bmatrix} \Delta \ddot u \\ \Delta \ddot v \end{Bmatrix} = \begin{bmatrix} m^2 & -1\,m \\ -1\,m & 1^2 \end{bmatrix}\begin{Bmatrix} f_x{}^o \\ f_y{}^o \end{Bmatrix} \tag{19}$$

In case of the equation of velocity of Eq.(13), the existence condition of solution is

$$\left([I_3] - \begin{bmatrix} 1 & 0 \\ 0 & 1 \\ 1 & m \end{bmatrix}\begin{bmatrix} 1 & 0 \\ 0 & 1 \\ 1 & m \end{bmatrix}^{-}\right)\begin{Bmatrix} \dot u_o + \Delta t\,f_{Vx} \\ \dot v_o + \Delta t\,f_{Vy} \\ 0 \end{Bmatrix} = \begin{Bmatrix} 0 \\ 0 \\ 0 \end{Bmatrix} \tag{20}$$

By carrying out the similar derivation shown in Eqs.(15) to (17), the load relation is

$$f_{Vy} = -\frac{1}{m}\left(f_{Vx} + \frac{\dot u_o}{\Delta t}\right) - \frac{\dot v_o}{\Delta t} \tag{21}$$

Then the load components, $(f_{Vx}{}^*, f_{Vy}{}^*)$, contributing to the equation of velocity are obtained by the relation shown in Fig.5(b).

$$\left\{{f_{vx}}^{\sharp}\atop{f_{vy}}^{\sharp}\right\}=\frac{1}{\Delta t}\begin{bmatrix}-1^2&-1\,m\\-1\,m&-m^2\end{bmatrix}\left\{\dot{u}_o\atop\dot{v}_o\right\}+\begin{bmatrix}m^2&-1\,m\\-1\,m&1^2\end{bmatrix}\left\{{f_x}^o\atop{f_y}^o\right\} \qquad (22)$$

Similarly, the load components, $({f_{Dx}}^*, {f_{Dy}}^*)$, relating to the equation of displacement are

$$\left\{{f_{Dx}}^{\sharp}\atop{f_{Dy}}^{\sharp}\right\}=\frac{2}{\Delta t}\begin{bmatrix}-1^2&-1\,m\\-1\,m&-m^2\end{bmatrix}\left\{\dot{u}_o\atop\dot{v}_o\right\}+\begin{bmatrix}m^2&-1\,m\\-1\,m&1^2\end{bmatrix}\left\{{f_x}^o\atop{f_y}^o\right\} \qquad (23)$$

By substituting Eqs.(22) and (23) into Eqs.(13) and (14), both responses of velocity and displacement of the un-stable truss with two degrees of freedom at a time step are obtained, respectively, in the form

$$\left\{\Delta\dot{u}\atop\Delta\dot{v}\right\}=\begin{bmatrix}m^2&-1\,m\\-1\,m&1^2\end{bmatrix}\left(\left\{\dot{u}_o\atop\dot{v}_o\right\}+\Delta t\left\{{f_x}^o\atop{f_y}^o\right\}\right) \qquad (24)$$

$$\left\{\Delta u\atop\Delta v\right\}=\begin{bmatrix}m^2&-1\,m\\-1\,m&1^2\end{bmatrix}\left(\Delta t\left\{\dot{u}_o\atop\dot{v}_o\right\}+\frac{1}{2}\Delta t^2\left\{{f_x}^o\atop{f_y}^o\right\}\right) \qquad (25)$$

Comparison with modal analysis(Tanami and Hangai, 1991)

Let us rewrite the equations of motion and compa-tibility of the unstable truss

$$\begin{bmatrix}1&0\\0&1\end{bmatrix}\left\{\Delta\ddot{u}\atop\Delta\ddot{v}\right\}=\left\{{f_x}^o\atop{f_y}^o\right\}、\qquad [1,m]\left\{\Delta u\atop\Delta v\right\}=0 \qquad (1),(2)$$

Rigid body mode derived from Eq.(2) is

$$\left\{\Delta u\atop\Delta v\right\}=\alpha\left\{m\atop-1\right\} \qquad (3)$$

After substituting Eq.(3) into Eq.(1), by multiplying that equation by the generalized inverse of the rigid body mode, we can get the following equation with unknown coefficient of α .

$$\ddot{\alpha}=m\,{f_x}^o-1\,{f_y}^o \qquad (4)$$

Integrating Eq.(4) by time

$$\dot{\alpha}=(m\,{f_x}^o-1\,{f_y}^o)\,t+C_3、\qquad \alpha=\frac{1}{2}(m\,{f_x}^o-1\,{f_y}^o)\,t^2+C_3t+C_4 \qquad (5)$$

Integration constants, C_3 and C_4, are obtained under consideration of the following initial condition

$$\dot{\alpha}\,|_{t=0}=\dot{\alpha}_o、\qquad \alpha\,|_{t=0}=0 \qquad (6)$$

where $\dot{\alpha}_o$ is related to the initial velocity in the form

$$\dot{\alpha}_o=m\,\dot{u}_o-1\,\dot{v}_o \qquad (7)$$

Then

$$\dot{\alpha} = (m f_x{}^0 - 1 f_y{}^0) t + (m \dot{u}_0 - 1 \dot{v}_0)\, ,$$

$$\alpha = \frac{1}{2}(m f_x{}^0 - 1 f_y{}^0) t^2 + (m \dot{u}_0 - 1 \dot{v}_0) t \qquad (8)$$

Consequently, the results of acceleration, velocity and displacement at a time step derived by modal analysis are exactly corresponding with Eqs.(19),(24) and (25), respectively, given by the direct analysis as shown in the following expressions.

$$\begin{Bmatrix} \Delta \ddot{u} \\ \Delta \ddot{v} \end{Bmatrix} = \ddot{\alpha} \begin{Bmatrix} m \\ -1 \end{Bmatrix} = (m f_x{}^0 - 1 f_y{}^0) \begin{Bmatrix} m \\ -1 \end{Bmatrix} \Rightarrow \text{ Eq.(19)}$$

$$\begin{Bmatrix} \Delta \dot{u} \\ \Delta \dot{v} \end{Bmatrix} = \dot{\alpha} \begin{Bmatrix} m \\ -1 \end{Bmatrix} = ((m f_x{}^0 - 1 f_y{}^0) \Delta t + (m \dot{u}_0 - 1 \dot{v}_0)) \begin{Bmatrix} m \\ -1 \end{Bmatrix} \Rightarrow \text{Eq.(24)}$$

$$\begin{Bmatrix} \Delta u \\ \Delta v \end{Bmatrix} = \alpha \begin{Bmatrix} m \\ -1 \end{Bmatrix} = \left(\frac{1}{2}(m f_x{}^0 - 1 f_y{}^0) \Delta t^2 + (m \dot{u}_0 - 1 \dot{v}_0) \Delta t \right) \begin{Bmatrix} m \\ -1 \end{Bmatrix}$$

$$\Rightarrow \text{Eq.(25)}$$

Conclusion

By solving illustratively an simple unstable truss with one rigid body mode, existence of direct analysis on dynamic problem is shown with comparison of modal analysis. In contrast with dynamic problem with unique solution, Moore-Penrose generalized inverse is also useful to get the optimal solution on static problem with infinite solutions.

References

Tanaka, H and Hangai, Y., September 1986, "Rigid Body Displacement and Stabilization Condition of Unstable Truss Structures", Proceedings of IASS Symposium, Osaka, Vol.2, pp.55-62

Hangai, H and Kawaguchi, K., November 1987, "Shape-Finding Analysis of Unstable Link Structures (in Japanese)," Transactions of AIJ, No.381, pp.56-60

Tanami, T. and Hangai, Y., July 1991, "Dynamic Analysis of Unstable Structures (in Japanese)," Seisan-Kenkyu, Vol.43, No.7, pp.5-7

Tanami, T. and Hangai, Y., July 1993, "Direct analysis for Shape-Finding of Unstable Structures (in Japanese)," Proceedings of Symposium on Computational Methods in Structural Engineering and Related Fields, Vol.17, pp.151-156

A Shape-Finding Finite Element Analysis for an Equally-Stressed Surface in Membrane Structures

T.J. Kwun [1] , S.Y. Sur [2] and H.W. Cho [3]

Abstract

A finite element method is proposed to obtain an equally-stressed surface in a membrane structure, where the displacements and rotations of a membrane are in general finite and nonlinear. Assuming that the forces of a triangular membrane element can be expressed in terms of generalized plane stresses, nonlinear effect of deformations is considered in the formulation of geometrical stiffness matrix within the framework of displacement-controlled finite element method. In the process of applying incremental steps, unbalanced forces caused by piecewise linearization of displacements are added to the forces of the next step to correct an error. The algorithm is applied to the case of catenoid membrane, for which the results show a close agreement with the exact solution.

Introduction

A membrane, which can only sustain tensile stresses, should be in general prestretched and its materials are easily deformable. If initial element stresses are differently given one another, the shape of the structure will change as the structure gradually evolves to a state of equal stresses. Therefore, in the membrane analysis, an equally-stressed surface, which has the most stable characteristic of a membrane structure, is often used. The surface can be numerically obtained by means of finite difference method[1] or finite element method[2-4], however, these procedures may have some drawbacks, such as unsolvable problems or bad convergences in their iterative processes. As a result, joint deformations may not be compatible with stresses, or in a cutting-pattern analysis unreasonable element meshes can be produced .

Under the assumption that the relationship between stress and strain is linear,

1. Professor, Department of Architectural Engineering, Sung Kyun Kwan University, 300 Chun Chun Dong, Suwon, 440-746, Korea.
2. Full-time Lecturer, Department of Architectural Equipment, Daelim Junior College, 526-7, Bisan Dong, Anyang, Korea.
3. Research Fellow, Institute of Technology, Samsung Construction Co. Ltd., C.P.O. Box 1430, Seoul, Korea.

governing equations at a deformed configuration are formulated and analyzed incrementally within the framework of displacement control method. The gradual drift-off from true solution, which occurs due to the piecewise linearization of displacements, is corrected by adding out-of-balance forces of previous step to the ones of present load level. In the final step, unbalanced forces are reduced by direct iteration method.

Basic Equations

In a local coordinate system, as shown in Fig. 1, it is assumed that a membrane is in a state of generalized plane stresses and the strain-displacement relationship is an ordinary linear one. When in local coordinates, the continuous displacement components of a finite element are approximated by first-order polynomial functions, the strain-displacement relationship may be written as:

$$\{ \varepsilon \} = [A] \{ u \} \tag{1}$$

where $\{ u \}$ is nodal displacement vector in local coordinates, $\{ \varepsilon \}$ is strain vector and $[A]$ is a matrix relating strains and displacements. The geometrical stiffness matrix in a local coordinate system can be expressed in the form of :

$$[k_G] = [\{ k_{G1} \}, \{ k_{G2} \}, \ldots, \{ k_{Gn} \}] \tag{2}$$

$$\{ k_{Gi} \} = [\partial [T]^T / \partial U_j] \{ f \} \tag{3}$$

where $[T]$ is the displacement transformation matrix, U_j's are the displacements in global coordinate system and $\{ f \}$ represents element nodal force vector of the form

$$\{ f \}^T = \{ f_{ix} \ f_{iy} \ f_{jx} \ f_{jy} \ f_{kx} \ f_{ky} \} \tag{4}$$

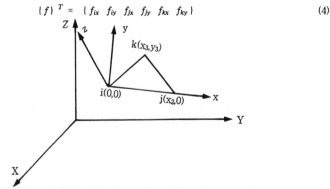

Fig. 1   Global(X, Y, Z) and local(x, y, z) coordinate systems
of a membrane element

In order to obtain the geometrical stiffness matrix in global coordinates, the expression for $[T]$ in Eq.(3) will be described at the current configuration, subsequently.   The matrix $[T]$ relates $\{du\}$ and $\{dU\}$, the small increments of nodal displacements in local and global coordinate systems, respectively, as

$$\{du\} = [T]\{dU\} \tag{5}$$

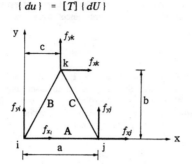

Fig. 2   Fundamental vectors($\mathbf{A}$, $\mathbf{B}$, $\mathbf{C}$) and lengths(a, b, c)
of a triangular element

Introducing fundamental vectors, $\mathbf{A}$, $\mathbf{B}$ and $\mathbf{C}$, as shown in Fig. 2, the fundamental lengths, a, b and c, in Fig. 2 can be described as

$$a = P$$
$$b = (R+S+T)/P$$
$$c = \sqrt{\{(PQ)^2 - (P+S+T)^2\}/P^2} \tag{6}$$

where

$$P = \sqrt{(X_2-X_1)^2+(Y_2-Y_1)^2+(Z_2-Z_1)^2}$$

$$Q = \sqrt{(X_3-X_1)^2+(Y_3-Y_1)^2+(Z_3-Z_1)^2}$$

$$R = (X_3-X_1)(X_2-X_1)$$
$$S = (Y_3-Y_1)(Y_2-Y_1) \tag{7}$$
$$T = (Z_3-Z_1)(Z_2-Z_1)$$

Then, $[T]$ can be expressed in terms of the fundamental lengths of Eq. (6) as follows:

$$[T] = \begin{bmatrix} \lambda & 0 & 0 \\ 0 & \lambda & 0 \\ 0 & 0 & \lambda \end{bmatrix} \tag{8}$$

where the submatrix $[\lambda]$ is

$$[\lambda] = \begin{bmatrix} \lambda_{11} & \lambda_{12} & \lambda_{13} \\ \lambda_{21} & \lambda_{22} & \lambda_{23} \end{bmatrix} \tag{9}$$

Each components in Eq.(9) defined as follows

$$
\begin{aligned}
\lambda_{11} &= 1+(U_2-U_1) & , \quad \lambda_{21} &= (U_3-U_1)/c-b(U_2-U_1)/ac \\
\lambda_{12} &= (V_2-V_1) & , \quad \lambda_{22} &= 1+(V_3-V_1)/c-b(V_2-V_1)/ac \\
\lambda_{13} &= (W_2-W_1) & , \quad \lambda_{23} &= (W_3-W_1)/c-b(W_2-W_1)/ac
\end{aligned}
\tag{10}
$$

with nodal displacements, $U_1, U_2, V_1, V_2, W_1, W_2$ in a global coordinates. Substituting Eq. (8) into Eq. (3) and differentiating each term with respect to U_j yield

$$
[k_G] =
\begin{bmatrix}
k_{11} & 0 & 0 & k_{14} & 0 & 0 & k_{17} & 0 & 0 \\
& k_{22} & 0 & 0 & k_{25} & 0 & 0 & k_{28} & 0 \\
& & k_{33} & 0 & 0 & k_{36} & 0 & 0 & k_{39} \\
& & & k_{44} & 0 & 0 & k_{47} & 0 & 0 \\
& & & & k_{55} & 0 & 0 & k_{58} & 0 \\
& & & & & k_{66} & 0 & 0 & k_{69} \\
& & & & & & k_{77} & 0 & 0 \\
& \text{Symm.} & & & & & & k_{88} & 0 \\
& & & & & & & & k_{99}
\end{bmatrix}
\tag{11}
$$

where

$$
\begin{aligned}
k_{11} &= k_{22} = k_{33} = -f_{ix}/a + (b/ac - 1/c) f_{iy} \\
k_{14} &= k_{25} = k_{36} = f_{ix}/a - (b/ac) f_{iy} \\
k_{17} &= k_{28} = k_{39} = f_{iy}/c \\
k_{44} &= k_{55} = k_{66} = f_{ix}/a - (b/ac) f_{iy} \\
k_{47} &= k_{58} = k_{69} = f_{jy}/c \\
k_{77} &= k_{88} = k_{99} = f_{ky}/c
\end{aligned}
\tag{12}
$$

The geometrical stiffness matrix $[K_G]$ in global coordinates can be obtained as

$$
[K_G] = [T]^T [k_G][T]
\tag{13}
$$

where it is noted that, unlike Eq. (8), the transformation matrix $[T]$ in Eq. (13) should be augmented by the following components of

$$
\begin{aligned}
\lambda_{31} &= \lambda_{12}\lambda_{23} - \lambda_{22}\lambda_{13} \\
\lambda_{32} &= \lambda_{21}\lambda_{13} - \lambda_{11}\lambda_{23} \\
\lambda_{33} &= \lambda_{11}\lambda_{22} - \lambda_{21}\lambda_{12}
\end{aligned}
\tag{14}
$$

The above components represent the transformation equations with respect to normal direction of an element. Therefore the basic equation for a membrane structure is finally expressed as

$$
\{ F \} = [K_G] \{ U \}
\tag{15}
$$

where $\{ F \}$ is a force vector in global coordinates.

Incremental Equations

In order to solve Eq. (15), a displacement control method is used with the introduction of an arbitrary parameter t, and U and F are assumed to be the functions of t, i.e. :

$$U \equiv U(t) \quad , \quad F \equiv F(t) \tag{16}$$

where $t = 0$ represents current configuration, thus, $U(0)$ and $F(0)$ are null vectors. Using a Taylor series expansion for $U(t+dt)$ and $F(t+dt)$, it follows that

$$U(t+\Delta t) \equiv U(t) + \dot{U}(t)\Delta t + \frac{1}{2}\ddot{U}(t)(\Delta t)^2 + \dots$$
$$F(t+\Delta t) \equiv F(t) + \dot{F}(t)\Delta t + \frac{1}{2}\ddot{F}(t)(\Delta t)^2 + \dots \tag{17}$$

Neglecting the higher order terms and substituting $t = 0$ yield

$$U(\Delta t) = \dot{U}(0)\Delta t$$
$$F(\Delta t) = \dot{F}(0)\Delta t \tag{18}$$

In case of a constant load mode, load vector can be expressed in terms of load increment parameter β as

$$F(t) = \overline{F}\,\beta(t) \tag{19}$$

where $\overline{F}$ represents a load mode. Differentitiating both sides of Eq. (19) with respect to t and substituting them into Eq. (18) at $t = 0$, one obtains

$$U(\Delta t) = \dot{U}(0)\Delta t = \dot{U}\Delta t \quad (\dot{U} = dU(0)/dt)$$
$$F(\Delta t) = \overline{F}\dot{\beta}(0)\Delta t = \overline{F}\dot{\beta}\Delta \quad (\dot{\beta} = d\beta(0)/dt) \tag{20}$$

Based on Eqs. (20) with Eq. (15), the basic equations can be written as :

$$[\,\{K_{G1}\}\,,\{K_{G2}\}\,,...,\{K_{Gn}\}\,] \begin{vmatrix} \dot{U}_1 \\ \dot{U}_2 \\ . \\ . \\ \dot{U}_n \end{vmatrix} = \{\overline{F}\}\,\dot{\beta}, \quad or \quad [K_G]\,\{\dot{U}\} = \{\overline{F}\}\,\dot{\beta} \tag{21}$$

If the incremental parameter t is U_1, an arbitrary component of $\{U\}$, i.e. :

$$t = U_1 \quad and \quad \dot{U}_1 = \frac{dU_1}{dt} = 1 \tag{22}$$

then Eq. (21) will become

$$[\ \{ -\overline{F} \}, \{ K_{G2} \}, \ ... \ , \{ K_{Gn} \} \ ] \begin{Bmatrix} \dot{\beta} \\ \dot{U}_2 \\ . \\ . \\ \dot{U}_n \end{Bmatrix} = - \{ K_{G1} \} \qquad (23)$$

Mclaurin series expansion for $U_i(t)$ and $\beta(t)$ produces, neglecting its higher order terms, the followings :

$$U_i(U_1) = \dot{U}_i U_1 \qquad (i = 2,3,...,n) \\ \beta(U_1) = \dot{\beta} U_1 \qquad (24)$$

The substitution of Eq. (24) into Eq. (23) yields

$$[\ \{ -\overline{F} \}, \{ K_{G2} \}, \ ... \ , \{ K_{Gn} \} \ ] \begin{Bmatrix} \dot{\beta} \\ U_2 \\ . \\ . \\ U_n \end{Bmatrix} = - \{ K_{G1} \} U_1 \qquad (25)$$

However, the solution for Eq. (25) may have an error in the process of piecewise linearization at each step. In order to correct the error, a modified incremental method is performed, as shown in Fig. 3, by adding unbalanced forces at current step to the forces of the next step. Then, Eq. (25) is modified as

$$[\ \{ -\overline{F} \}, \{ K_{G2} \}, \ ... \ , \{ K_{Gn} \} \ ] \begin{Bmatrix} \dot{\beta} \\ U_2 \\ . \\ . \\ \dot{U}_n \end{Bmatrix} = \{ R \} - \{ K_{G1} \} U_1 \qquad (26)$$

where $\{ R \}$ is the unbalanced force vector of the previous step. In the final step the out-of-balance forces are evaluated and reduced by direct iteration method.

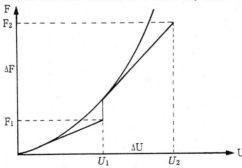

Fig. 3 Modified incremental method

Numerical Example and Discussion

The proposed algorithm is applied to the case of catenoid membrane, as shown in Fig. 4, to find an equally-stressed surface, which can be expressed as

$$Z = a \{ \ln(\sqrt{X^2 + Y^2} + \sqrt{X^2 + Y^2 - a^2}) - \ln(a) \} \qquad (27)$$

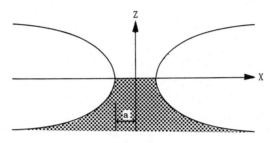

Fig. 4 Catenoid surface

Fig. 5 shows finite element meshes for an initially-undeformed flat membrane, continuously supported along its outer periphery. In that case desired profile may be obtained by forced displacement of inner periphery joints.   The comparison of the numerical results with the exact solution is made in Fig. 6, which shows a close agreement between them. Fig. 7 gives a perspective shape of the surface obtained.

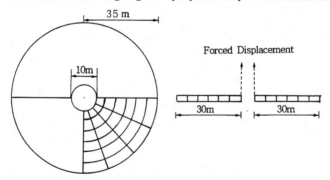

Fig. 5 Finite element meshes for an initially undeformed flat membrane

Conclusion

In the present paper a finite element method was presented to obtain an equally-stressed surface for membrane structures.   Within the simple framework of modified displacement control method using triangular plane stress elements, the effect of large displacement and rotation is taken into consideration in the derivation of displacement transformation matrix.   An error produced at each incremental step

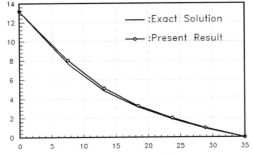

Fig. 6 Comparison of shapes

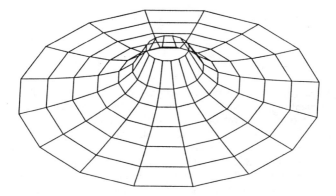

Fig. 7 Bird-eye view of result

can be corrected by taking into account the unbalanced forces in the formulation of force vector at the next step. A good agreement of numerical results with the exact solution for a catenoid membrane demonstrates the validity of the proposed algorithm.

References

(1) Ishii, K. and Suzuki, T., "Shape of Membrane Structures," Proc. of   IASS Pacific Symposium on Tension Structures and Space Frames,   Tokyo, Japan, Oct., 1971.
(2) Haug, E. and Powell, G.H., "Finite Element Analysis of Nonlinear   Membrane Structures," Proc. of IASS Pacific Symposium   on Tension Structures and Space Frames, Tokyo, Japan, Oct., 1971.
(3) Barnes, M.R., " Form-Finding of Minimum Surface Membranes," Proc. of IASS World Congress on Space Enclosures, Montreal, Canada, July, 1976.
(4) Ishii, K., "Analytical Shape Determination for Membrane Structures," Proc. of IASS Pacific Symposium on Tension Structures and Space Frames, Tokyo, Japan, Oct., 1971.

Vibration Characteristics of Suspension Membrane Structures

Norio Masaoka[1], Kazuo Ishii[2] and Ryo Muranaka[3]

Abstract

The present paper specifies fundamental vibration characteristics obtained from experiment on the model of the suspension or the framed membrane structure. The first paragraph describes outline of the free vibration tests on plane quadrilateral and approximate hyperbolic parabolic shaped membrane models. The second paragraph shows influence of vibration amplitude to damping characteristics. Specially, it is clarified that damping force becomes larger according to larger vibration amplitude, and that damping force of the approximate hyperbolic parabolic membrane model is larger than that of the plane quadrilateral membrane model because of different geometrical configuration. In the last paragraph, results of free vibration analysis on the framed membrane structure which was already completed as a roof of the gymnasium is compared with the experimental results of approximate hyperbolic parabolic shaped membrane models. Consequently, fundamental vibration characteristics of the real framed membrane structure is considered to be predicted by assuming stiffness proportional damping or Rayleigh damping.

Introduction

In general, suspension or framed membrane structures are extensively deformed in the out-plane direction because of light weight and applied tensile force,

1 Group Leader, Construction Engineering Sect., Research & Development Dept., TOMOE CORPORATION, 3-4-5, Toyosu, Koto-ku, Tokyo 135, Japan
2 Professor, Department of Architecture,Yokohama National University,156 Tokiwadai, Hodogaya-ku, Yokohama 240, Japan
3 Engineer, Construction Engineering Sect., Research & Development Dept., TOMOE CORPORATION, 3-4-5, Toyosu, Koto-ku, Tokyo 135, Japan

which is necessary to be constructed, introduced from cables or edges of the frame. In the practical design of such the membrane structure, lately it becomes important to figure out its dynamic nonlinear behavior under fluctuating wind load. However, it is thought to be quite difficult because there are many uncertain factors regarding wind load or the material characteristics. To clarify its dynamic behavior, it should be significant to know vibration characteristics of the suspension membrane structures, however, a few of such the investigation has been reported (Benzley et al.1976, Uematsu et al.1986).

In the present investigation, vibration experiment on the membrane model is employed to study its fundamental vibration characteristics. Then, free vibration analysis on the practical example is accomplished and its result is discussed with the experimental result.

Outline of Free Vibration Test with the Membrane Model

Schematic of the present vibration test is shown in Fig. 1. At the experiment, two types of specimens are tested, which are a plane quadrilateral and an approximate hyperbolic parabolic shaped membranes noted here as approx. HP. For the plane quadrilateral type specimen, two kinds of membrane materials are applied to be tested. Each membrane having no welded portion is fixed on the rigid boundary of the steel frame by stretching periphery of the membrane.

On the other hand, for approx. HP type specimen, only single membrane material is tested. Because width of the regular membrane is restricted, this specimen is made of welded three parts of which welded line is determined

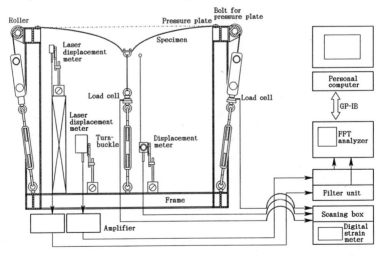

Figure 1. Schematic of the present free vibration test

not to be affect the vibration dominant portion.   It is fixed on the hyperbolic parabolic shaped rigid boundary of the steel frame by forced deforming under stretching at periphery of the membrane which is initially flat plane shaped.

Initial elongation, which causes reduction of the material homogeneity, of each type of the specimen is removed by applying 3% of breaking tensile stress then holding the deformed shape for 40 hours.   Then, it is fixed on the boundary with pressure plates and bolts after applying the regulated tensile force.   Photo 1 shows each specimen after completely fixed.   Regulated tensile forces are applied from the boundary by turnbuckles, which are measured and controlled by load cells.   Fig. 2 and Table 1 show configurations of the specimens and the testing conditions respectively.

Before the free vibration tests, natural frequency and damping factor under

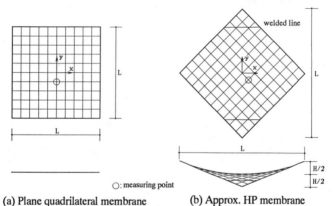

(a) Plane quadrilateral membrane   (b) Approx. HP membrane

Figure 2. Configuration of the specimens

Table 1. Testing condition for the specimen

Model		Plane Quadrilateral		Approx.HP
		PL-A	PL-C	HP-C
Shape of model		L=1.00 m		L=1.41 m H/2=0.14 m
Material type		PTFE-coated glass fiber fabric	PVC-coated polyestetr fiber fabric	
Material properties	Tensile stiffness	Ext = 8112.83 N/cm Eyt = 6052.87 N/cm	Ext = 1600.05 N/cm Eyt = 981.37 N/cm	
	Poisson's ratio	vx = 0.570 vy = 0.425	vx = 0.796 vy = 0.488	
	Shear stiffness	Gt = 501.76 N/cm	Gt = 98.00 N/cm	
	Breaking tensile stress	Tcx = 1568.00 N/cm Tcy = 1306.67 N/cm	Tcx = 463.87 N/cm Tcy = 460.60 N/cm	
	weight	w = 11.76 N/m²	w = 6.47 N/m²	
Regulated tensile force		To = 1960.00 N/m	To = 980.00 N/m	

small vibration are measured by impulse excitation tests using impact hammer. Then, each specimen is free-vibrated by initially applying the several regulated amplitudes at center of the specimen.   Regulated amplitude is given by the following method, shown in Fig. 3: Firstly, center of the specimen is pull down by Kevler strings until the regulated amplitude.   Then, free vibration is initiated by cutting off the strings.

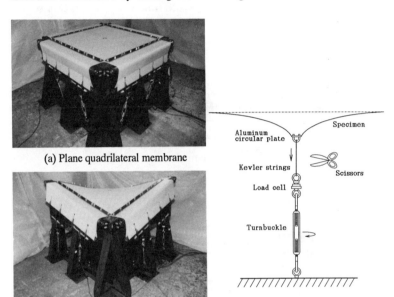

(a) Plane quadrilateral membrane

(b) Approx. HP membrane
Photo 1. Perspective of the specimens   Figure 3. Detail of the loading method

It is noted that pulling point of the specimen is reinforced by aluminum circular plate, ø50x3mm, to reduce stress concentration.   Influence of this additional plate to the vibration is determined by the plane quadrilateral PL-A specimen.   The primary natural frequency of the specimen with the additional plate differs from that without the plate in about 2.5% diversity.

Displacement Response of the membrane is measured by non-contact type laser displacement meter.   And the measured data is analyzed by 4ch FFT analyzer.

Results and Discussion of Free Vibration Test

Vibration Test of Plane Quadrilateral Membrane - Comparing displacement response wave form of specimen PL-A applied δ=50mm amplitude with that applied δ=10mm, shown in Fig. 4(a) and (b),the specimen applied δ=50mm is damped in high gradient from start of the vibration then is transformed to stationary response; therefore, it is considered as stiffness proportional or Rayleigh damping, which is less affected by high order vibration.

Dominant frequency in power spectrum, shown in Fig. 5, obtained from displacement response wave form of the specimen applied the lower amplitude, δ=10mm, is 18.4Hz which is almost equal to the primary natural frequency of 19.2Hz measured by the impulse excitation test.   From this result, although excitation is caused by forced displacement at center of the membrane, first-order vibration is considered as the dominant mode.   The similar

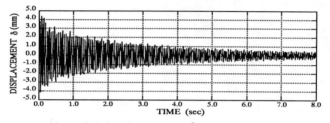

(a) Applied vibration amplitude δ=10mm at the center node

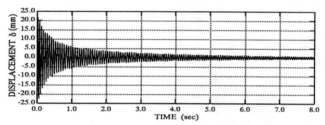

(b) Applied vibration amplitude δ=50mm at the center node

Figure 4. Displacement response wave form of the plane quadrilateral membrane

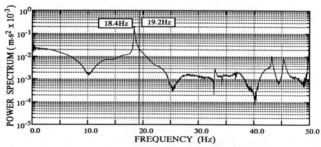

Figure 5. Power spectrum of displacement response wave form at applied vibration amplitude δ=10mm

result is obtained in the different material property type, specimen PL-C. Fig. 6 shows relation between the dimensionless value of vibration amplitude δ divided by the span length L and the damping factor h calculated from the logarithmic decrement of the each enveloped line which is obtained from the displacement response wave form at the each regulated amplitude.   In this figure, the damping factors obtained from the impulse excitation tests are plotted at zero amplitude.   Consequently, from this figure the damping factor h becomes larger according to the larger amplitude.   In region of the smaller amplitude, the damping factor h becomes nearly equal to that obtained from impulse excitation tests.

Free Vibration Test of Approx. HP Membrane - Fig. 7 shows obtained displacement response wave forms of specimen HP-C in case of being applied

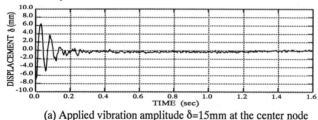

Figure 6. Relationship between damping factor and dimensionless amplitude for the quadrilateral membrane

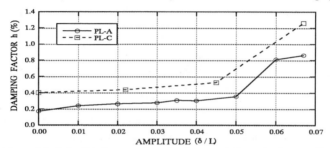

(a) Applied vibration amplitude δ=15mm at the center node

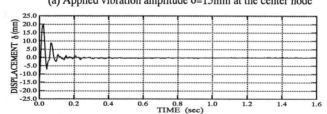

(b) Applied vibration amplitude δ=35mm at the center node

Figure 7. Displacement response wave form of the approx. HP membrane

δ=35mm and δ=15mm amplitude.   Differed from that of the plane quadrilateral
membrane, these wave forms are damped quickly within the initial several
vibration cycles.   Fig. 8 shows relation between the damping factor h and
dimensionless vibration amplitude δ/L which are calculated by the same method
used in case of the plane quadrilateral membrane.   From this figure
the damping factor h becomes larger according to the larger amplitude, however
value of the damping factor is quite larger than that of the plane quadrilateral

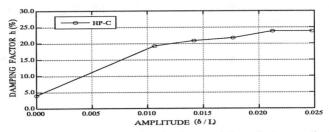

Figure 8. Relationship between damping factor and dimensionless amplitude for
the approx. HP membrane

membrane made of the same material.
     As this reason, geometrical configuration of the approx. HP membrane
is determined as the negative Guassian curved surface, therefore, its vibration
damping mechanism is considered to be different.   In general, restoring
force in out-plane direction of the HP shaped surface is thought to be larger
than that of the plane quadrilateral surface; therefore, if damping force is
assumed to be proportional to the stiffness specified by the membrane structure,
it can be explained that the damping factor of the approx. HP membrane
becomes larger than that of the plane quadrilateral membrane.

Numerical Example

     As the example of numerical analysis, here shows results of the response
analysis on the framed membrane structure which was already completed as
a roof of the gymnasium.   And also vibration characteristics of this
example is discussed concerning with the damping characteristics obtained from
the model vibration experiment. Fig. 9 shows configuration of the structure
and objective of the analysis. Perspective of the structure is shown in
photo 2.   Table 2 shows the analytical condition.   Free vibration
is initiated at the maximum deformed shape which is obtained by applying
uniform suction of 980N/m² as the wind load on the center arched shape
membrane panel.   In this case the maximum displacement of center of
the membrane is about 1/30 of the shorter span length L=3910mm.
     The present analytical method is based on Newmark's ß(=1/6) scheme

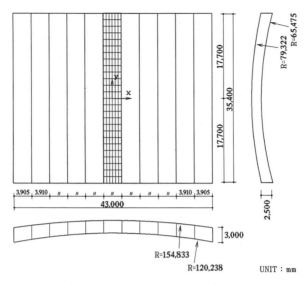

Figure 9. Configuration of the membrane roof of the gymnasium

Table 2. Analytical data of the membrane roof

Material type		PTFE-coated glass fiber fabric
Material properties	Tensile stiffness	Ext = 11597.32 N/cm Eyt = 10074.40 N/cm
	Poisson's ratio	$v_x = 0.769$ $v_y = 0.668$
	Shear stiffness	Gt = 323.40 N/cm
	Weight	$w = 11.76$ N/m^2
	Breaking tensile stress	Tcx = 1568.00 N/cm Tcy = 1306.67 N/cm
Regulated tensile force		To = 1960.00 N/m

Photo 2. Perspective of the membrane roof of the gymnasium

as a numerical time integration, considering geometrical nonlinearity (Masaoka et al.1993).   And also effects of wrinkling is examined, which is considered to be specific phenomena in the membrane structure.   Assuming Rayleigh damping, considering viscous damping and stiffness damping, 3% of the damping factor h1 or h2 is used for the primary and the secondary modes.   Bilinear quadrilateral isoparametric finite elements are used for one-quarter model taking symmetry into account for the purpose of reduction of computation period. And time interval Δt is 0.00075sec.

To assess the fundamental vibration characteristics, eigenvalue analysis is accomplished before the free vibration analysis.   From Fig. 10 showing results of the eigenvalue analysis, natural frequency are considered to be close through the first to fifth mode order.

Fig. 11 shows displacement response wave form at the center obtained from results of the free vibration analysis.   This response wave form

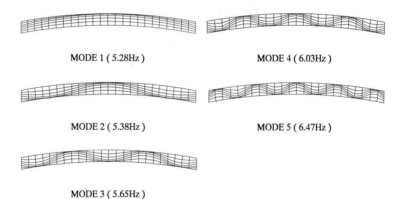

MODE 1 (5.28Hz)   MODE 4 (6.03Hz)

MODE 2 (5.38Hz)   MODE 5 (6.47Hz)

MODE 3 (5.65Hz)

Figure 10. Natural frequency and natural modes of the analytical model

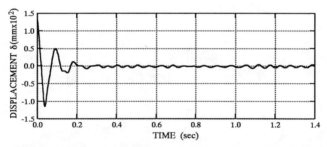

Figure 11. Displacement response wave form of the center node for
the analytical model

is relatively similar to that from the experimental result of approx. HP membrane model, which is damped quickly within the initial several vibration cycles. Calculated damping factor from the same method used in the former paragraph is 19.4%. As this reason, although shape of the present example made of shallow curved surface, it is still negative Gaussian surface, which is the same reason described in the former paragraph.

Conclusions

1. From the free vibration experiment, damping characteristics of the suspension membrane structure is depending on vibration amplitude. Specially, damping force becomes larger according to larger vibration amplitude.
2. On vibration characteristics of approx. HP membrane, consisting of negative Guassian curved surface, strongly damped wave form is observed, which is quite different from that of the plane quadrilateral membrane structure. And this damping characteristics is considered to be predicted by assuming stiffness proportional damping or Rayleigh damping.
3. Response analysis on the practical example of the framed membrane structure shows the similar results obtained from the model vibration experiment of approx. HP membrane.
4. Although vibration characteristics of the real membrane structure has not been measured, vibration characteristics of such the structure will be evaluated by the nonlinear response analysis using with accurately calculated damping factor.

Acknowledgement

The present work is supported and approved as the cooperative project by OGAWA TENT CO., LTD. The authors reserve special thanks for their cooperation.

References

1. Benzley, S. E. and Key, S. W. "Dynamic Response of Membranes with Finite Elements", *Journal of the Engineering Mechanics Div.*, ASCM, EM3, 1976, pp. 447-460.
2. Masaoka, N. and Ishii, K. "Vibration characteristics of suspension membrane structures analyzed with lower-order quadrilateral membrane element ", *Proceedings of the SEIKEN-IASS Symposium on Nonlinear Analysis and Design For Shell and Spatial Structures*, Tokyo, 1993, pp. 393-400.
3. Uematsu, Y. and Uchiyama, K. " Aeroelastic Behavior of an H.P.-shaped Suspended Roof", *Proceedings of the IASS Symposium on Membrane Structures and Space Frames*, Osaka, Vol. 2, 1986, pp. 241-248.

Study on Structural Characteristics of
Air-Inflated Beam Structures

Masaya Kawabata[1] and Kazuo Ishii[2]

Abstract

The purpose of this paper is to clarify the structural characteristics of nonlinear load-carrying
behaviors of air-inflated beam structures in their finite
deflection range and after wrinkles are produced by means
of numerical analysis and experiment. This paper
discusses about the major factors which determine
wrinkling load, collapse load and deflection of those
structures. Also, one of the practical examples is
illustrated.

Introduction

Membrane structures can be divided by their shape
and system into two categories: suspension membrane
structures and pneumatic structures, and the pneumatic
structures can be subdivided into single wall air-
supported structures and dual wall air-inflated
structures. In air-inflated structures and arch, membrane
materials are fabricated into the shape of air-tight bag.
The internal pressure is increased by air supplied to the
bag, and pretension is given to the membrane surface. In
this manner, rigidity is obtained; thus the air-inflated

[1]Assistant, Dept. of Architecture, Yokohama National University., 156 Tokiwadai, Hodogaya-ku, Yokohama 240, Japan
[2]Professor, Dept. of Architecture, Yokohama National University., 156 Tokiwadai, Hodogaya-ku, Yokohama 240, Japan

beam structures can be classified as one of the pneumatic structures. Inflatable structure have a variety of applications where it is desirable to have a small package for transporting, simple erection capability at the destination and self-sufficiency. But deformations of those structures under loads are comparatively large, so nonlinear approach by numerical analysis and consideration of wrinkling are required to grasp the structural characteristics. In this paper, the authors will discuss the fundamental structural characteristics of air-inflated beam structures and introduce a practical example of their applications.

Analysis Method

The solution procedure is based upon the finite element method (FEM) and the followings are particularly taken into account.
1. Geometrical nonlinearlity
2. Orthotropic elastic behaviors of materials
3. Wrinkling of fabrics
The basic theory used as follow: (Ref.1)
The equilibrium conditions can be written as

$$\{\psi(\{\delta\})\} = \int_v [\overline{B}]^T \{\sigma\} dV - \{\overline{R}\} = 0 \tag{1}$$

where $\{\psi\}$: the vector of the sum of internal and external forces
$\{\sigma\}$: the vector of stresses
$\{\overline{R}\}$: the vector of external forces
$[\overline{B}]$: the matrix defined from the strain definition as

$$d\{\varepsilon\} = [\overline{B}]d\{\delta\} \tag{2}$$
$$[\overline{B}] = [B_0] + [B_L(\{\delta\})] \tag{3}$$

$[B_0]$ is the linear part of $[\overline{B}]$ and $[B_L]$ is the non-linear part of $[\overline{B}]$ depending on the displacements. As Newton-Raphson process is to be adopted, the relation between $d\{\delta\}$ and $d\{\psi\}$

$$d\{\psi\} = \int_v d[\overline{B}]^T \{\sigma\} dV + \int_v [\overline{B}]^T d\{\sigma\} dV \tag{4}$$

Using Eq. (2)

$$d\{\sigma\} = [D]d\{\varepsilon\} = [D][\overline{B}]d\{\delta\} \tag{5}$$

therefore

$$d\{\psi\} = \int_v d[B_L]^T\{\sigma\}dV + [\overline{K}]d\{\delta\} \qquad (6)$$

where

$$[\overline{K}] = \int_v [\overline{B}]^T[D][\overline{B}]dV \qquad (7)$$

The first term of Eq. (4) can be written as

$$\int_v d[B_L]^T\{\sigma\}dV = [K_\sigma]d\{\delta\} \qquad (8)$$

where $[K_\sigma]$ is geometric matrix dependent on the stress level. The vectors of internal and external forces, which include inflate pressure were always calculated according to the deformed shape. However the value of inflate pressure was assumed to be kept constant, independent with the variation of volume. Two types of elements, triangular and cable elements, were used for the division of material into elements. (Fig.2) In the analysis using triangular elements, wrinkling of the membrane was accommodated by a combination of the modified stress transfer method (Ref.2) and variable stiffness method. In the case of cable elements the membrane material was assumed as elastic cable members with the direction of warp, weft, and diagonal, and wrinkling was accommodated by making each member effective under the condition of tension only.

Structural behavior

        In order to understand the fundamental structural characteristics of air-inflated beams, we performed loading experiments applying center concentrated load (Fig.1), using simply supported beams as models, and the results were analyzed by means of FEM.(Fig.4 and 5) Also, in this section the features of wrinkling, collapse and deformation are shown about air-inflated beam and arch. Elastic constants for numerical analysis were determined so as to satisfy "the Reciprocal-theory" with considering the results of uniaxial and biaxial loading test of the coated fabric. (Table.1)
        Fig.3 shows the comparison of load-displacement relationships of simply supported beam between the results of experiment and those of FEM.   Strictly speaking, there is a slight difference between the results obtained by FEM and those by experiments. This indicates that in the FEM, in which material linearity is postulated, the relationship between load and displacement is linear in the range of small displacement. However, in experimental results, the properties of

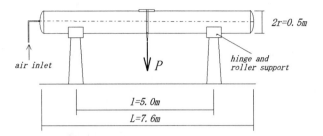

Fig.1. Outline View of Loading Experiment

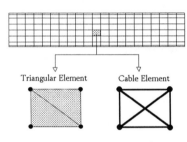

Fig.2. Element Division

Table 1. Elastic
                    Constants

Et_x = 639.45 N/m
Et_y = 426.30 N/m
ν_x =   0.60
ν_y =   0.40
Gt_{xy}= 29.40 N/m

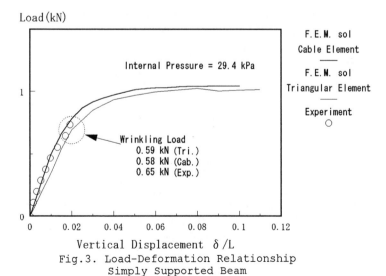

Fig.3. Load-Deformation Relationship
       Simply Supported Beam

material nonlinearity developed already in the small
displacement range. In the case of semicircular arch
nonlinear behaviors due to the geometrical nonlinearity
are shown even in the range of small displacement.(Fig.7)
Development of wrinkling in fabrics and collapse of the
structure is observed clearly in the results of FEM. The
effectiveness of the analysis methods for air-inflated
beam structures are shown by these results.

Internal Pressure = 49.0 kPa

Span = 5.0 m

Radius = 0.25 m

.... Initial Shape

Fig.4. Deformed Shape of Simply Supported Beam

Load(kN)

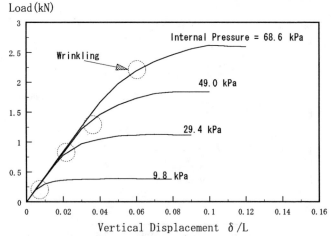

Vertical Displacement δ/L

Fig.5. Load-Deformation Relationship
Simply Supported Beam
(F.E.M Sol., Triangular Element)

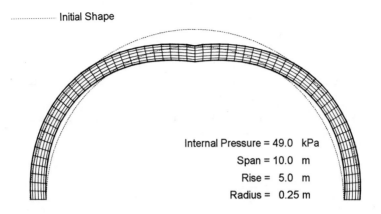

Fig.6. Deformed shape of Semicircular Arch

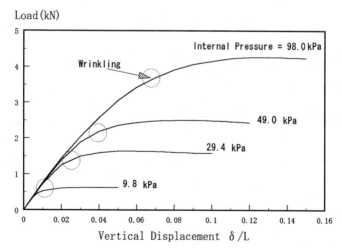

Fig.7. Load-Deformation Relationship
Semicircular Arch
(F.E.M Sol., Triangular Element)

Wrinkling and collapse:
In the range of linear theory tension in the axial direction at compressive side of air-inflated beam with a
constant cross section can be written as

$$N_x = \frac{1}{2}pr - \frac{Mt}{Z}, Z = \pi r^2 t \qquad (9)$$

where Nx, M, t, p, r, and Z represent tension in the axial
direction, bending moment, thickness of membrane, internal pressure, radius of section, and section modules,
respectively. Wrinkling load of simply supported air-
inflated beam which receive center concentrated load can
be obtained by solving Eq.9 for $Nx=0$, $M=Pl/4$ (where l
represents span length) with respect to P.

$$Pcr = \frac{2\pi pr^3}{l} \qquad (10)$$

Wrinkling load of this linear solution corresponds to
that of F.E.M. solution using cable element. In the case
of triangular element the values obtained for wrinkling
loading are greater than those obtained in Eq.10. This
result is based on the assumption that stress is constant
and uniform within elements. Fig.5 and 7 shows that, with
air-inflated beam structures, bending stiffness decreases
due to the development of wrinkling (local buckling), and
the structure does not collapse immediately after the
development of wrinkling. In other words, the structure
has load-carrying capacity arising from redistribution of
stress in the membrane. The similar tendency was reported
for the pure-bending of air-inflated beam. (Ref.3) The
wrinkling and buckling load is usually taken as the proof
load in most practical designs. However this does not
seem appropriate in the light of the fact that those
structures retain load-carrying capability even after
wrinkling or buckling which is produced within a small
area.(Ref.5)
     Deflection:
Deformations of air-inflated structures under loads are
comparatively large so it is important to grasp the
deflection features. The major factors which determine
deflection are the followings.
     1. internal pressure
     2. property of fabric (tensile stiffness and shear
        stiffness)
     3. radius of cross section
The internal pressure mostly exerts its effect in terms
of apparent shear stiffness and change in radius (change
in the moment inertia). Apparent shear stiffness is the
additional stiffness generated by the effect of initial

tension and can be shown by $Gt_a=pr/2$ in the case of inflated cylinder with both ends; for instance $Gt_a=1.225$kN/m when $p=9.8$kN/m^2 and $r=0.25$m.(Ref.4) It differs from material in-plane shear stiffness (represented by Gt_m).

Initial deflection:
In this study, we define the deflection generated before the development of wrinkling as the initial deflection. Fig.8 shows the difference of the relationship between load and deflection for two models, Model- I ($Gt_m=29.4$kN/m) and Model- II ($Gt_m=14.7$kN/m), in order to study the effect of apparent shear stiffness (inflate pressure) and materiel in-plane shear stiffness on initial deflection. In the case of Model- II, the difference in initial deflection due to inflate pressure is greater than that observed in Model- I. Namely, the sum of the material in-plane shear stiffness and apparent shear stiffness is considered to work as the shear stiffness for entire structure. The effect of difference in internal pressure is clearly reflected in the magnitude of initial deflection when the material in-plane shear stiffness is comparatively small.

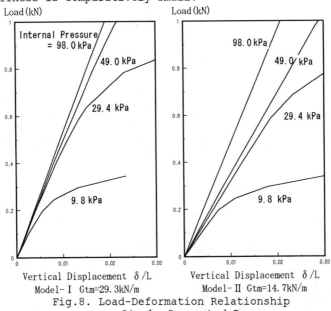

Fig.8. Load-Deformation Relationship
Simply Supported Beam
(F.E.M Sol., Cable Element)

Deflection after wrinkling:
The magnitude of loading at which wrinkling develops is roughly proportional to internal pressure and the magnitude of deflection gradually increases after the development of wrinkling. (Fig.5 and 7) Therefore, after the development of wrinkling, the change in the development of deflection due to the application of different levels of internal pressure becomes great.

Practical Example

In this section as a practical example of air-inflated structures, cross-beam-tents which consists of the combination of air-inflated arches, is introduced. (Photo.1) It offers advantages for some portable or reusable applications such as a pavilion or temporary structure at the site of construction. It takes only about 15 minutes for erection and does not require rigid frame or foundation to fix the fabric and introduce tension. By using the present solution, deformation under wind load (fig.9) and wind velocity in which wrinkles are produced (Table.2), were obtained and the required inflate pressure was determined.

(a)

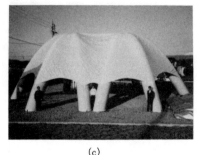

(c)

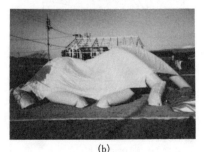

(b)

Photo.1(a)-(c)
Erection of
Cross-Beam-Tents

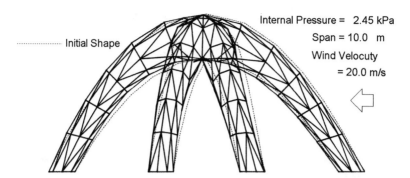

Internal Pressure = 2.45 kPa
Span = 10.0 m
Wind Velocuty
= 20.0 m/s

Fig.9. Deformed Shape of Cross-Beam-Tent

Table.2. Wrinkling Developing Wind Velocity

internal pressure	wind velocity
4.90kN/m	15m/s
2.45kN/m	10m/s
0.98kN/m	5m/s

Conclusion

- The nonlinear load-carrying behaviors of air-inflated beam and arch, which include wrinkling and collapse are clarified through the present method.
- Those structures retain load-carrying capability due to redistribution of tension in the membrane even after the development of wrinkling.
- The effect of inflate pressure on the magnitude of deflection is dependent on the material in-plane shear stiffness.

Reference

1. Zienkiewicz, O.C., "The Finite Element Method in Engineering Science," McGRAW-HILL London, 1971.
2. Zienkiewicz, O.C., Valliappan, S., and King, I.P., "Stress Analysis of Rock as a No Tension Material," Geotecnique, 18:56-66.
3. Stein, M., and Hedgepeth, J.M., "Analysis of Partly Wrinkled Membranes," NASA TN D-813, 1961.
4. Leonard, R.W., "On the Shear Stiffness of Fabrics," J. of the Aerospace Science, Mar., 1962, pp.349-350.
5. Ishii, K., and Aihara, T., "Structural Design and Analysis of Air-Inflated Dual Membrane Structure," Proc. of IASS, Application of Shells in Engineering Structures, Moscow USSR, Vol.4, 25-39 1985.

EXPANDABLE STRUCTURES WITH INCORPORATED ROOFING ELEMENTS.

P. Valcárcel, J.(*); Escrig, F.(**); Martín, E.(*)

Abstract.

In this paper design aspects of expandable structures with incorporated cover plates are studied. These roofing elements can be made by single plates or by self-folding plates. In both cases it is possible for these elements to add to the final resistance of the dome. Also it was studied the problems related to their unfolding, and the effect of these roofing elements on their resistance. At the same time, specific examples of the possibilities of this system will be proposed.

Introduction.

Curre ntly, a revitalization is taking place of transformable structures which can cover an area when necessary, or be removed, thus leaving this enclosure in the open air. This type of roofing is specially useful for sports arenas or auditoriums. In stadiums, football fields or bull fighting arenas in Spain, spectator generally prefer to be out in the open, weather permitting. However, in adverse climatological conditions, it becomes necessary to cover the area, or at least when the audience is seated. Any solution that offers these two options will represent a good solution to the problem.

Recently, significant achievements have dealt with this problem by using large elements witch can be displaced. Important examples, such as the Skydome in Toronto or the Ariake Colosseum in Tokyo come to mind. They are, undoubtedly, exceptional projects which have won great acclaim but, at the same time, there are certain drawbacks to be considered. Since the panels are of great size, the enclosure can only be partially exposed. In addition, the considerable weight of the moving parts makes necessary the use of very complex mechanisms. A curious and not very well known example of this type of systems is the transformable dome, designed by the spanish architect P. Piñero, which never got to the building stage (fig 1).

(*) Departament of Building Technology. E.T.S. Architecture. La Coruña (Spain)
(**) Departamento of Structures. E.T.S. Architecture. Sevilla. Spain.

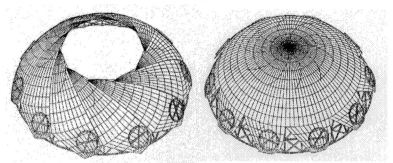

Figure 1.- Retractable dome from Pérez
Piñero.

Another possible solution to this problem lies in the use of expandable bar structures. These consist of modular bar units, made up of mobile bundles or scissors. They can be folded up to form a compact package or unfolded to cover a large enclosure. They are quite simple structures, the main difficulty of which lies in finding an adequate system for covering them.

In 1961, Pérez Piñero proposed the first really effective example of a structure of this kind. The idea was for a dismountable theatre, consisting of a set of three bar bundle modulus. The proposal was straightforward and truly ingenious, to such an extend that Fuller, a member of the jury which awarded Pérez piñero first prize in the U.I.A. contest, was full of admiration. Unfortunately, the structure presented one serious problem, the covering was a textile sheet which folded up, together with the unit of bars. These textile coverings are certainly effective, and have been thoroughly studied by our team, but we have observed that they are only useful for reduced spans. The reference [5] provides a complete study of several of our proposal, in which self folding textile covers are employed.

A truly effective cover must be made with rigid elements, usually plates or sandwich panels. These provide adequate protection against the rain and, at the same time, form heat and acoustic insulation. this aspect is worth noting, since covers made up of stiff textile membranes produce a loud noise under heavy rain.

For this reason our team has carried out a great deal of research into roofing systems consisting of expandable structures with rigid panels which can either fold and unfold together with the structure or in any case, be placed over it easily.

The simplest solution for covering these structures is to unfold the package of bars and, once this is set up and fixed in its final position, place some rigid panels screwed onto the top knots of the truss. This is a straightforward, efficient solution which present only one difficulty: It is necessary to complete the mounting operation at a great height and this could be dangerous for those workmen erecting the structure. Nevertheless, in some interesting types of structures, such as domes, a more efficient strategy is possible and with less risk involved. Expandable domes, based on geometrical partition of the sphere in six sectors (Ref [6]) are characterized by a peculiar expanding process. The package of bars remains in a nearly flat position during the deployment until it is about to cover the enclosure. At this time the roofing plates can be placed in such a way that they overlap one another, just as fish scales do. When the unfolding process continues, the plates slip over one another until the entire structure curves, unfolding completely. At this moment, the plates can be fixed in their final position. this procedure is exceedingly simple and time-saving and can be carried out from a crane, this minimizing the risk for workers.

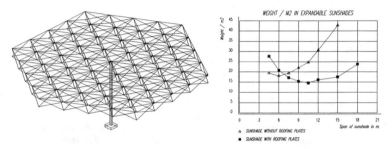

Figure 2.- Comparative weights of a sunshade with and without roofing rigid plates.

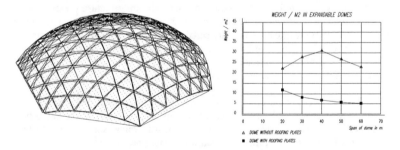

Figure 3.- Comparative weights of a dome with and without roofing rigid plates.

Another important advantage to this system is that when the plates are fixed in place these can add improving the strength of the structures. The procedure is relatively simple and consists of formulating the stiffness matrix by F.E.M. and condense the degrees of freedom which correspond to the vertex of the triangle. This condensed matrix can be added at the stiffness matrix of the bars (References [3], [6]).

In order to test the validity of the system, we have carried out the corresponding calculation for the following structures with the given results (Fig. 2, 3).

This system is straightforward and efficient but one disadvantage lies in that it can only be applied to very specific structures, such as the two shown here, or triangular cylindrical vaults.

Another possibility is to cover the structure with plates which fold up by means of an auxiliary crosspiece structure. This is placed over the holding up structure and merely serves as a guide for the unfolding of the panels. Therefore it can be of reduced cross section.

In figure 4 one can observe the unfolding of a typical module of this system. If the panels are placed so as to fold towards the inside of the truss, and if the entire crosspiece is employed, it is possible to solve the problems of the point of union in order to prevent water from entering. The solution is similar to that used in windows pivoting on a horizontal axis and similar features can be employed.

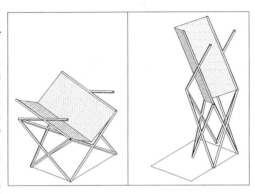

Figure 4

It is worth remarking that the auxiliary structure only needs to support the weight of each of the panels and this, only in the unfolding process, which means that a reduced cross section is required.

The system is directly applicable to flat structures, such as that shown in figure 5 or cylindrical domes, like the one in figure 6. In this case, it is clear that the panels in the upper joint cannot overlap. It is necessary to design a special elastic junction, using a joint widely employed in windows (Fig. 7) or a similar one.

When the structure is made up of triangular modulus, it is possible to use the same system. However, it is not possible to close the structure totally, as is demonstrated in ref. [7], since the panels end up running into one another (see remark zone in fig. 8). Even so, it is possible to carry out a initial reduced unfolding. Then, the plates can be placed over the auxiliary structure and the unfolding can continue without any problems. In fig. 8 the procedure for a sunshade designed by this system is shown.

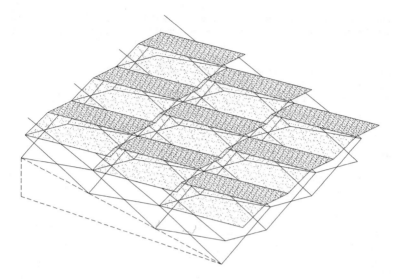

Figure 5.- Flat truss with rigid panels.

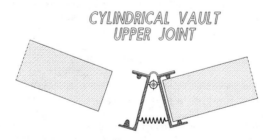

Figure 7.- Detail of upper joint of a cylindrical vault.

If we consider as a unit the set formed by two adjacent modulus, it is possible to design a system which allows for a nearly complete folding. Of course, there is a limit to the degree of folding for real structures which is defined by the thickness of the bars and the roofing plates, In this case the sequence of unfolding would be (fig. 9).

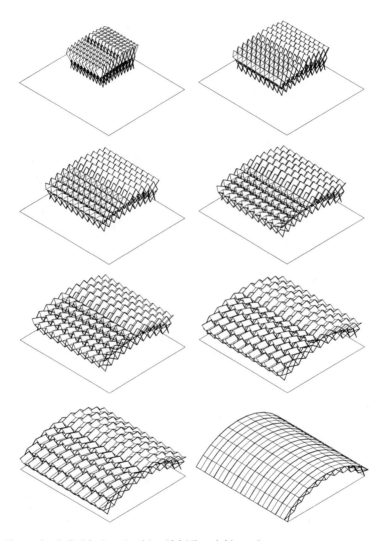

Figure 6.- Cylindrical vault with self-folding rigid panels.

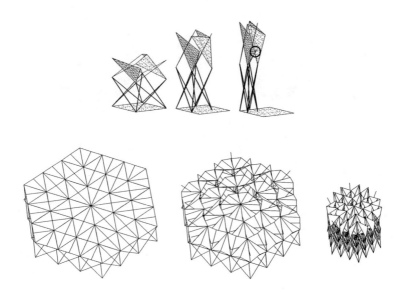

Figure 8.- Unfolding of a sunshade with plates of one modulus.

Figure 9.- Unfolding of a sunshade with plates of two modulus.

If the unfolding structure is not flat, the two adjacent triangles cannot remain covered by a single flat panel. In this case, the plates cannot be unfolded completely. On the contrary, they must remain a bit open in such a way that the panels a-b-c and a-c-d are flat, as is shown in fig. 10.

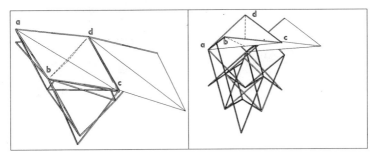

Figure 10.- Detail of the roofing plates for a dome.

This precess can be generalized, even for structures of great complexity, as is the case of an unfolding dome. in this situation, the conditions for geometrical compatibility require all modules of each sector to be different, as shown in ref. [4]. Besides this, it is necessary to take into account that when de truss closes up, one must allow the crosspieces of the auxiliary structure to displace laterally to a slight degree, in order to avoid the lower panels running into the upper ones. In practise, this poses no difficulty, since the auxiliary scissors lack lateral stiffness. It would be unnecessary because its only purpose is to guide the unfolding of the roofing plates. The result of this process is shown inf fig. 11. This has been formulated through a calculus program, the results of which have been translated into an AutoCAD environment, using .SCR files.

CONCLUSIONS.

It is possible to observe that a considerable number of solutions have been proposed for roofing enclosures, using expandable structures with rigid panels. These structures resolve the main problem presented up to now by this type of roofing and one may hope that in the future such structures will be employed as a perfectly viable solution in those situations in which the possibility of covering or leaving exposed an enclosure may provide advantages for its use.

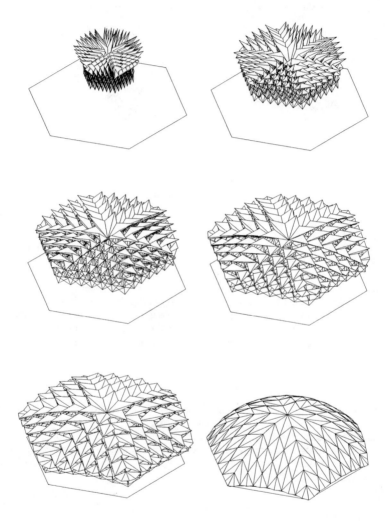

Figure 11.- Unfolding of a dome with roofing panels.

REFERENCES.

1.- Pérez Piñero, E.; Candela, F.; Dalí, S.: "La obra de Pérez Piñero". Arquitectura. Madrid n° 163-164. 1972. pp 1-28.

2.- Escrig, F.; P.Valcárcel, J.B. "Curved Expandable Space grids". Non-conventional Structures '87. London 1987.

3.- P.Valcárcel, J.B.; Escrig, F. "Analysis of curved expandable space bar structures" Int. Symposium on 10 Years of Progress in Shell and Spatial Structures. IASS. Madrid 1989.

4.-Escrig, F.; P.Valcárcel, J.B.; Gil Delgado, O. "Design of expandable spherical grids" Int. Symposium on 10 Years of Progress in Shell and Spatial Structures. IASS. Madrid 1989.

5.- P.Valcárcel, J.B.; Escrig, F. "Expandable Structures with Self-folding Textile Cover". International Conference on Mobile and Rapidly Assembled Structures. MARAS'91. Southampton 1991.

6.- P.Valcárcel, J.B.; Escrig, F.; Estévez, J.; Martin, E. "Large Span Expandable Domes" Int. Symposium on Large Span Structures. IASS. Toronto 1992.

7.-P. Valcárcel, J.B.; Escrig, F.; Martín, E. "Expandable Domes with Incorporated Roofing Elements". Four International Conference on Space Structures. Surrey 1993.

Seismic Performance and Retrofit
of the Golden Gate Bridge

Charles Seim, P.E., F. ASCE[1], Tim Ingham, Ph.D., S.E.[2]
and Santiago Rodriguez, P.E.[3]

Abstract

The Golden Gate Bridge has endured since its opening in 1937 as a symbol of
the City of San Francisco, and is one of the most famous bridges in the world. The
bridge functions as an important transportation link for Bay Area commerce and
commuters. It is owned, operated and maintained by the Golden Gate Bridge,
Highway and Transportation District.

Immediately after the Loma Prieta earthquake in October of 1989, the District
engaged T.Y. Lin International (TYLI) to perform seismic evaluation and retrofit
studies of the bridge. The First Phase evaluation studies concluded that a major
earthquake occurring on the San Andreas Fault and centered near the bridge may
cause severe damage to the bridge and a prolonged interruption of service to the
communities it serves.

The Second Phase retrofit studies included development of measures and cost
estimates to retrofit the bridge against such an occurrence. These studies of the
seismic retrofit of the bridge were divided into five stages: 1) evaluating the seismic
risk and developing site specific response spectra and ground motions, 2) identifying
the seismic deficiencies of the crossing, 3) defining structural performance levels

[1]Senior Principal, T.Y. Lin International, 825 Battery Street, San Francisco, CA 94111

[2]Sr. Proj. Engineer, T.Y. Lin International, 825 Battery Street, San Francisco, CA 94111

[3]Project Engineer, T.Y. Lin International, 825 Battery Street, San Francisco, CA 94111

and developing seismic retrofit design criteria, 4) performing linear and nonlinear and multiple support excitation analysis, and 5) developing conceptual seismic retrofit designs. The work in Stages 1 to 5 was completed in 1992.

Several consultants, in addition to T.Y. Lin International, are now working on the Third Phase which consists of final retrofit designs and preparation of construction plans, specifications and cost estimates. Construction is expected to start immediately thereafter and be completed by the end of 1998.

Introduction

One of the most renowned bridges in the world, the Golden Gate Bridge opened to traffic on May 28, 1937. The design and construction of the Bridge is well documented in the Chief Engineer's final report [3]. The Bridge is operated by the Golden Gate Bridge, Highway and Transportation District (District).

The 2790m overall length of the Golden Gate Bridge consists of a number of different structure types. The bridge's major components are the steel truss approach viaducts, the Fort Point steel arch, the steel cable's concrete anchorages and concrete anchorage housings, the steel suspension bridge, and the art deco concrete pylons which are purely architectural motifs. All of the foundations for these structures except the northern viaduct are supported directly on rock. The northern viaduct piers are founded on spread footing bearing on competent soil.

A lateral bracing system was added to the stiffening truss as wind retrofit in 1954. The vertical suspenders were replaced in the mid 1970s. The original concrete deck was replaced in 1985 with a lightweight orthotropic steel deck with a net reduction in weight of about 11,000 tons.

The Loma Prieta earthquake on October 17, 1989 was rated at a magnitude of 7.1. The Golden Gate Bridge was not instrumented, but appeared to be in an area where the bed rock acceleration was recorded at about 0.08 of gravity. Immediately following the earthquake, the District engaged T.Y. Lin International (TYLI) to perform a seismic evaluation, and about a year later, concept retrofit studies and approximate construction costs. The results of the seismic evaluation were presented in *The Golden Gate Bridge Seismic Evaluation* [1] by T.Y. Lin International in November, 1990. The evaluation revealed that a major earthquake on a nearby segment of the San Andreas or Hayward Faults would likely cause severe damage to the bridge and could cause interruption of traffic and require significant repairs. The *Golden Gate Bridge Seismic Retrofit Studies* [2] in July 1991 included development and evaluation of concept retrofit measures as well as estimation of their construction costs and schedules.

Site Specific Seismic Risk, Response Spectrum and Ground Motions

The determination of site specific seismic risk, response spectrum and ground

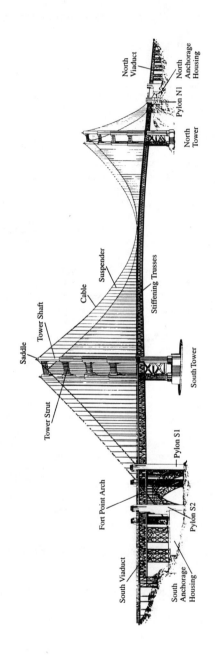

motions was based on evaluations of the geology, the relative location of nearby faults and the reoccurrence interval of the seismic event, seismic risk analyses, and computer simulations of ruptures on nearby faults. The results of these studies are presented in Geological, Geotechnical and Ground motion studies for seismic retrofit of the Golden Gate Bridge in July, 1992 [5] by Geospectra, Inc., Richmond, CA.

Seismic risk was evaluated by applying a probabilistic assessment of ground motion using the maximum expected design earthquake values of about 1,000 years return period for the two main nearby faults, the San Andreas and the Hayward, with a magnitude of 8.25 and 7.3 respectively. The peak ground acceleration estimated for an event comparable to the 1906 San Francisco earthquake on the San Andreas is 0.65g. Site-specific longitudinal, transverse and vertical response spectra were developed on the basis of these seismic risk analyses.

Three different seismic events (San Andreas fault rupture scenario) were developed by Geospectra Inc. to define three independent ground motions. One set of ground motions was developed for the Hayward fault, but the evaluation studies showed that it did not control.

Multiple Support Ground Motion Synthesis

For the suspension spans, six multiple ground motions containing proper estimates of the phase delay and incoherency were used as inputs to the two main towers, the two flanking concrete pylons, and the two cable anchorages for all three sets of ground motions.

Seismic Performance Criteria

The seismic performance criteria for a retrofitted suspension bridge presents a compromise between available retrofit measures, constructibility constraints, user and owner's expectations, and cost. The retrofit measures for the Golden Gate Bridge must preserve life and allow the bridge to be used immediately after the largest expected event, first for emergency vehicles and then for the toll paying general public. To guide the designers, design criteria for the retrofit project was developed by T.Y. Lin International based on the Design Criteria from the new AASHTO Load and Resistance Factor Design.

The Evaluation Studies and the Seismic Retrofit Studies used only dead load and seismic loads were considered. For the final design phase, the Design Criteria requires also using 50% of live load in addition to dead and seismic loads.

Suspension Bridge

The characteristics of the suspension bridge which most significantly influence its seismic response and the analysis requirements related to those characteristics are summarized below.

Large Displacement Effects and Multiple Support Excitation

The stiffness of the structure, which is provided mainly by the shape of the cable adjusted to the applied loads, is very sensitive to its geometry. The distortion of its geometry under applied loads causes its stiffness to change as the loads change. A geometrically nonlinear analysis which considers large displacement effects is required to capture the full behavior under dynamic excitation.

The multiple support excitation imposes relative displacements between the towers and the anchorage blocks that induce both static and dynamic stresses and displacements. The out-of-phase motions were found to give only a minor increase, or in many cases a decrease, in the stresses in the structure, as might be expected considering the flexibility of the bridge. The demands on the expansion joints, however, are generally larger with the multiple support excitation.

Dynamic Characteristics

Due to their dimensions and flexibility, suspension bridges have a long fundamental period of vibration. The seismic response for long period structures is characterized by large displacements and lower seismic forces than for shorter period structures. Although often ignored, higher vibration modes with a shorter period make an important contribution to the seismic forces in a suspension bridge. In a response spectrum analysis based on a mode-superposition procedure, it is necessary to consider a very high number of vibrational modes in order to include the modes with a natural period in the range of the maximum spectral acceleration. The higher modes have a small mass participation when compared with the fundamental modes but may actually have a very large participation in the response of local elements, such as the towers.

For example, in the seismic analysis of the Golden Gate Bridge, it was found that the first longitudinal mode for the towers, with a period of 1.4 seconds, contributed greatly to the towers' seismic forces. However, this mode is the 67th mode for the bridge as a whole and it would be overlooked in an analysis considering less than 67 mode shapes.

Although more difficult to perform, a time history analysis performed by direct integration of the coupled equations of motion avoids the problems associated with the mode-superposition method.

Material Nonlinearity

The main cables and suspenders of the Golden Gate Bridge respond to the maximum credible earthquake in the elastic range. Yielding of structural elements is confined to local areas such as parts of the towers, pylons and some elements in the stiffening truss. Yielding of the stiffening truss and towers may be confined to local areas only, since the strength of the members is not related to the seismic demands. Although only limited areas of the bridge may then need retrofit, the requirement to limit the ductility demand on those local areas is difficult to meet.

Displacement Compatibility

Due to the long natural period suspension bridges have, the seismic displacements are large, on the order of several feet across the expansion joints. In addition to this, different parts of the structure are subjected to different ground motions thus creating large relative displacements at the expansion joints. When the displacement across an expansion joint reaches the maximum capacity, the expansion joint closes and transmits impact forces. A joint closure changes the dynamic properties and the response of the bridge. These effects are considered in the analyses by means of nonlinear gap elements.

Support Conditions

It is also important to consider changes in the support conditions due to seismic forces. In the case of the Golden Gate Bridge, the towers are weakly anchored to the underlying reinforced concrete piers. For service loads, the towers can be considered to be fixed to the piers because the dead load produces compressive stresses which are not exceeded by the bending stresses due to wind loads. However, uplift of the towers can be expected in strong earthquakes. This uplift can significantly change the towers' response characteristics. The rocking motion of the towers actually reduces the seismic stresses at the towers but increases the deflections. The nonlinear rocking behavior is considered in the seismic analysis of the suspension bridge.

Analysis Technology

The analysis of the seismic response of the Golden Gate Bridge requires the following three steps of increasing complexity: definition of the dead load state, modal analysis and response spectrum analysis assuming linear behavior, and nonlinear time history analysis, considering multiple-support ground motion.

Definition of the Dead Load State

The dead load state was defined starting with the geometry and dead load corresponding to the time the construction was completed. The dead load and the construction procedure determined the initial stresses. A static nonlinear analysis considering large displacement theory was performed to check equilibrium and to fine-tune the original geometry. The result of this analysis yielded the dead load state at the completion of construction.

The Newton-Raphson method, based on tangent stiffness iteration, was used for this static load analysis to account for the large displacement effects. For the Design Phase, 50% of liveload is added to the dead stress for combining with the dynamic analysis.

Modal and Response Spectrum Analysis

Once the dead load state was determined, a modal analysis was performed to compute the vibrational properties. The modal analysis assumes linear behavior for small deformations from the dead load state. The natural frequencies and mode shapes computed from the analysis were compared with the ones obtained from ambient vibration tests. This comparison provided a way to verify the computer models.

Following the modal analysis, a response spectrum analysis was performed for the maximum credible earthquake. The response spectrum analysis did not include the effects of the nonlinear behavior or the multiple support excitation. However, it gave an initial insight into the magnitude of the seismic forces, thus redirecting the modeling and analysis efforts towards the problem areas.

Nonlinear Time History Analysis

The analyses of the global response of the bridge to multiple-support ground motion excitations are performed using a dynamic nonlinear finite element computer program, in which large displacement effects are considered by establishing the static or dynamic equilibrium of the structure in its deformed configuration. The effect of a limited displacement capacity at the expansion joints and the nonlinear uplifting behavior of the tower's bases is examined by using "gap" elements.

The nonlinear dynamic analyses are performed by integrating the coupled equations of motion in the time domain by using the acceleration implementation of the average acceleration method. This method is unconditionally stable and has second order accuracy. The ground motion excitation was applied as a time-varying displacement boundary condition at each of the supports. In each time step, the nonlinear system of equations relating the effective dynamic loads and the nodal accelerations are solved by using the Newton-Raphson method, based on tangent stiffness iteration. Rayleigh damping is assumed, in which the damping matrix is proportional to a linear combination of the mass and stiffness matrices. The proportionality factors were computed to yield a damping ratio of 5% at periods of 0.5 seconds and 20 seconds. The damping ratios for periods between those two values are smaller, with a minimum of 1.5% at a period of 3 seconds. These values are consistent with ambient vibration measurements.

Anchorage Housings

The North and South Anchorage Housings are building-type reinforced concrete structures which consist of exterior reinforced concrete walls braced with pilasters, internal reinforced concrete moment frames supporting the roof, and the reinforced concrete roof slab forming the roadway. Linear elastic analysis were performed to assess the ductility demand of the as-built and retrofitted structures.

Vulnerabilities and Retrofit Proposals

The proposed retrofit measures include both tuning the structures, i.e., adjusting the local and overall response of the existing structures so they can better respond to strong ground motions without damage by allowing uplift or addition of damper devices, and strengthening the structures to minimize the damage caused by the structures' responses to strong ground motion.

The vulnerable areas for the suspension bridge and the proposed retrofit are:

1. Connection between Stiffening Trusses and Towers: Impact between the main span stiffening trusses and the towers is predicted since seismic displacements exceed the existing displacement capacity. The proposed retrofit includes the installation of dampers at this location.
2. Reinforced Concrete Pylons: New ductile concrete walls inside the pylons are proposed to increase strength and ductility.
3. Main Towers: Uplift due to rocking motion of the tower would cause high contact stresses at the toe of the uplifting base, requiring strengthening of the steel plates. Because uplifting of the tower bases improves the global seismic performance, it will not be prevented with the retrofit.
4. Side Span Stiffening Trusses: There are high tension and compression stresses in the chords of the side span stiffening trusses. A local nonlinear model of a chord member shows that the required ductility of two is attainable.
5. Piers: The piers supporting the main towers are subjected to high bearing stresses under uplift conditions. The installation of post-tensioning tendons at the top of the piers to provide confinement is proposed.
6. Cable Saddles: The connections of the cable saddles to the tower tops need to be strengthened for shear stresses due to differential cable tension.
7. Deck Panels: The connections of the orthotropic steel plate deck panels to the floor beams need improvement due to high horizontal shear forces.

The anchorage housings have high ductility demands, but they lack the necessary confinement reinforcement to provide ductility. The proposed retrofit measures include installation of new internal framing and shear walls, strengthening the connections between the existing walls and the floor and roof diaphragms, and reinforcing the foundations.

Seismic Deficiencies to Avoid in a New Design

Advances in seismology have also given the engineering community a better understanding of seismic risk. There is more information available regarding ground motions during strong earthquakes and it is thus possible to generate site-specific synthetic ground motions according to local geological conditions. The long term structural behavior may also be affected by unforseen service conditions or by the degradation of structural materials.

In order to increase the overall seismic performance under unforseen circumstances, suspension bridges should be designed taking into consideration their seismic behavior beginning with the conceptual stages. In addition to providing adequate strength for the predicted seismic forces, the design should be based on the following principles:

1. Ductility: The structural elements, connections and components should be detailed to achieve ductility and avoid brittle failure. The connections should exceed the capacity of the members and yielding should occur before local bucking. The hysterectic behavior of a ductile structure provides an energy dissipation mechanism which minimizes the probability of member failure or collapse.

2. Redundancy: Alternative load paths should be provided so that the structure can adjust to loads larger than expected by redistributing the load.

3. Displacement Compatibility: A suspension bridge consists of elements with different vibrational properties such as cables, stiffening girder, towers, and anchor blocks. The *interface* between different elements may have large relative displacements. The design must address this by either leaving enough free space or by keeping the connections between different elements to a minimum. Energy absorption devices such as dampers provide the option of controlling differential displacements but must always include a fail-safe mechanism.

Conclusion

The seismic retrofit studies for the Golden Gate Bridge have shown that it can be adequately analyzed; retrofit measures can be developed, designed and constructed to provide adequate capacity to resist the demands from a major earthquake occurring on the San Andreas fault.

Contract plans and retrofit designs are in preparation and are expected to be completed in late 1994. The three year construction period is scheduled to start in early 1995.

The seismic retrofit studies for the Golden Gate Bridge have also provided a methodology which can be applied to the seismic design of other major crossings. These methodologies consisted of an evaluation of the seismic risk, a generation of site specific ground motions, a definition of performance levels and design criteria, a seismic analysis, both linear and nonlinear where appropriate, and finally a seismic retrofit design.

Important factors that are consider in the analysis of suspension bridges are: large displacement effects, multiple support excitation, dynamic characteristics, material nonlinearity, the displacement compatibility and support conditions.

The retrofit design will provide for adequate ductility, redundancy and will at the same time insure displacement compatibility between different elements.

Acknowledgements

All of the Golden Gate Bridge studies summarized in this paper were made under the direction of the Golden Gate Bridge Highway and Transportation District. The authors wish to thank the General Manager Carney Campion, District Engineer Daniel Mohn, and Deputy District Engineer Mervin Giacomini for their support.

References

1. Golden Gate Bridge Seismic Evaluation, T.Y. Lin International, Golden Gate Bridge, Highway and Transportation District, San Francisco, CA, November 1990.

2. Golden Gate Bridge Seismic Retrofit Studies, T.Y. Lin International, Golden Gate Bridge, Highway and Transportation District, San Francisco, CA, July 1991.

3. Geological, Geotechnical and Ground Motion Studies for Seismic Retrofit of the Golden Gate Bridge, Geospectra, Inc., Golden Gate Bridge, Highway and Transportation District, San Francisco, CA, July 1992.

4. Housner, G.W., et.al., Competing Against Time, Report to Governor George Deukmejian from the Governor's Board of Inquiry on the 1989 Loma Prieta Earthquake, State of California, Office of Planning and Research, May 31, 1990.

5. Strauss, J., The Golden Gate Bridge: Report of the Chief Engineer to the Board of Directors of the Golden Gate Bridge and Highway District, California, 1937, San Francisco, California, Golden Gate Bridge and Highway District, 1938.

Innovations in Major Suspension Bridge Design

Klaus H. Ostenfeld[1]

Introduction

The 1930s was a decade with great development in the area of cable supported bridges, especially in North America. In 1931, the first bridge to overcome one kilometre free span between supports was built: The George Washington Bridge with a main span of 1066 m across the Hudson River. This bridge was quickly followed by the double-suspension bridge, San Francisco-Oakland Bay Bridge, and the most famous Golden Gate Bridge with a main span of 1280 m, fig. 1.

Fig. 1. Golden Gate Bridge.

A few years later, the elegant relatively slender superstructure of the Golden Gate was surpassed in slenderness by the next major suspension bridge, the Tacoma Narrows, which catastrophic destiny has written history in the field of aerodynamic stability for bridge superstructures.

---

[1]Exececutive Director, Cowiconsult Consulting Enginners and Planners, 15, Parallelvej, DK-2800 Lyngby, Denmark.

The trend towards increasing slenderness and lack of understanding for aeroynamic behaviour had gone too far. The Tacoma Bridge failed shortly after the opening due to wind induced oscillations, and the Golden Gate also showed a tendency to wind excited oscillations, but of a more benign nature.

Truss Girders

Common for these historical bridges is the truss girder design which usually can accept quite high wind speeds. As a consequence of the Tacoma disaster, a high degree of aerodynamic stability was built into the Mackinac Bridge, fig. 2, where the stiffening truss is 11.60 m deep. Hardly an optimal solution concerning weight, construction costs and maintenance by today's standards.

The Verrazano Narrows in New York, opened in 1964, followed the then traditional American design. With a main span of

Fig. 2. Mackinac Bridge.

772

1298 m, the bridge was slightly longer than the Golden Gate, and held the world record span for almost 20 years. Aerodynamic stability was ensured by horizontal wind bracing of the bottom chord thus forming a latticed torsion box, and by providing air gaps between the carriageways and footways.

In 1966, the river Tagus Bridge in Lisbon was completed. The design of this bridge followed the tradition with a stiffening truss and a steel grid deck. The grid prevents aerodynamic lift forces altogether.

Trusses can be elegantly designed to provide sufficient torsional stiffness to prevent flutter instability by introducing horizontal top and bottom windbracing. The flutter resistance can be further enhanced by longitudinal open slots in the road deck, a well known feature from post World War II suspension bridges in North America and Japan.

Besides the adoption of the truss girder, other typical features of the 1930-60 suspension bridges are the use of steel towers, large expansion joints at the tower positions, main cables installed by the air spinning method, and the anchor blocks built as dominating monoliths generally filling the first opening of the adjacent approach spans. These characteristics were seriously looked upon in the following two decades.

Box Girders
In 1966 the Severn Bridge, fig. 3, was inaugurated and shortly thereafter a new Little Belt Bridge was built with several distinctive cost saving and advanced departures from tradition. For the main cables, the spun wires were abandoned in favour of prefabricated partial cables made in their full length of

Fig. 4. Little Belt Bridge anchorage.

Fig. 5. Little Belt Bridge box girder.

1.500 m delivered complete with anchorages.

The above-ground anchor blocks were replaced by underground, earth-loaded friction structures with the load carrying cables anchored at ground level, fig. 4, and the main towers constructed in concrete.

Another newcomer was the closed, fully welded box girder of full width, aerodynamically shaped and equipped with wind deflector/guide plates, fig. 5. Dehumidified air was introduced as protection of the interior resulting in initial cost savings and low-cost maintenance.

This new trend was followed in the 70'es by the Bosporus Bridge No 1 and the Humber Bridge also using the steel box girder. The Humber Bridge design also used concrete towers.

The Little Belt Bridge box girder was fabricated in 12 m length sections in a shipyard. From a distance of more than 100 nautical miles, the boxes were transported to the bridge site on barges which were specially built for the purpose. Normally two boxes were transported at a time. The weight

Fig. 3. Severn Bridge.

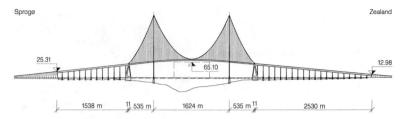

Fig. 6. Great Belt East Bridge elevation.

of a 12 m box was 140 tons. As for the Severn Bridge, direct flotation of the boxes would have been possible but was abandoned because of undesirable salt water contamination of the steel surfaces to be painted.

Continuous Girder Concept

As part of the Great Belt fixed link, the second long span suspension bridge is now under construction in Denmark, fig. 6. This bridge is 6.8 km long and includes a suspension span of 1624 m - the longest slender box girder ever built - with a height of 65 m above sea level, and side spans of 535 m each. The continuous approach bridges with spans of 193 m total approximately 2.5 km and 1.5 km, respectively, on the west and east side.

The suspension span girder is continuous over the full cable supported length of 2.7 km between the two anchor blocks. The traditional expansion joints at the pylons are omitted as are the underlying traditional cross beams on the towers. The cable is fixed to the girder by a central node at midspan, and hydraulic buffers at the anchor blocks control the longitudinal movement. Compared to a system with joints at the pylons, analyses have indicated an approximately 25% reduction in the vertical deflection of the girder from traffic load. This increased rigidity also contributes to a better aerodynamic stability.

If unrestricted movements were allowed, the extreme horizontal movement at the expansion joint from the characteristic traffic load would be up to 1.8 m.

Between the anchor blocks and the girder, hydraulic buffers are arranged which allow for slow horizontal movements up to ±1.0 m and a free angular rotation of the girder, fig. 7. Live load induced rapid movements are restrained by proper tuning of the dampers to increase the fatigue life and reduce the wear on the joints. The movements which are thus

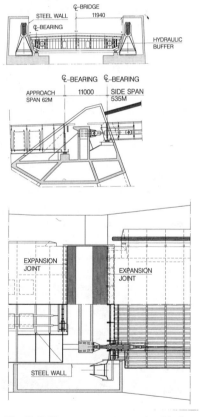

Fig. 7. Buffer arrangement at anchor block.

prevented will of course result in restraining forces, which are easily accepted by the girder and cables via the central nodes.

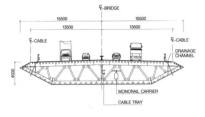

Fig. 8. Great Belt suspension bridge box girder.

The continuous girder concept leads to a very simple arrangement at the pylons, reduced installations and maintenance costs for the joints, improved stiffness of the overall suspension bridge system, and improved aerodynamic stability.

Large capacity expansion joints are arranged at four locations only for the entire 6.8 km long bridge, i.e. at the anchorages and at the abutments.

Joints and Bearings
The choice of bearing types is of great importance for reliability, robustness and durability. Also friction must be small throughout the lifetime of the bearings.

To support the relatively flexible steel superstructure on the Great Belt Bridge piers, spherical bearings with adequate rotational capacity have been selected. The horizontal movements are accommodated by sliding on a stainless steel base plate, and the angular rotation by a spherical lens shaped segment. The sliding medium is PTFE (Teflon) against stainless steel. The advantage of these bearings is the relatively low restraining moments originating from rotation friction in spherical parts.

For the Great Belt suspension bridge, watertight modular expansion joints of the so-called Swivel Joint type have been selected as a result of intensive tests to achieve the required 100 year fatigue life.

Tuned Mass Dampers
The two Great Belt approach bridges are 1,538 m and 2,538 m long, respectively, with spans of 193 m. The superstructures are

closed steel box girders, 6.7 m deep, with a central longitudinal bulkhead. This design was favoured after tendering also for an alternative shorter span concrete box girder solution.

Wind tunnel testing of a full aeroelastic bridge model revealed that the approach bridges might be subject to wind induced oscillations in wind speeds above 20 m/s if the bridges later were equipped with wind screens or congested by heavy queues of traffic. Aerodynamic data derived from wind tunnel testing and structural damping obtained from field tests on an existing bridge were employed in the design of a Tuned Mass Damper system for protection against detrimental wind induced oscillations.

A system which will meet performance requirements for a 5 percent detuning and a 20 percent possibility for variation in the damping of the TMD masses has been selected, fig. 9. The TMD masses which correspond to approximately 0.5% of the modal mass of the bridge structures will be installed in the individual spans which will have the largest motions for a given mode of vibration. The approach spans will be equipped with TMD's to ensure damping of vibrations at wind speeds below 25 m/s.

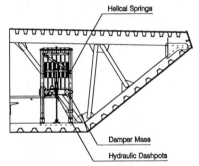

Fig. 9. Tuned Mass Damper for the Great Belt approach bridges.

Environmental Conditions
In Sweden, the Höga Kusten Bridge is now under construction. With a main span of 1210 m and side spans of about 300 m each, the bridge will be one of the largest suspension bridges in Europe. The bridge has a minimum clearance of 40 m under the main span. With an effective width of 17.8 m, the bridge is prepared for a future four-lane

motorway. However, initially the layout will be corresponding to two lanes with emergency lanes. The bridge will be founded on rock.

Two environmental factors have played essential roles in the design: The beautiful landscape surrounding the bridge, and the severe winter climate in the region. The suspension span will be the longest ever built at such a latitude.

Consequently, a conventional three span suspension bridge from coast to coast is chosen, with special architectural treatment of the shapes of the pylons and the cable anchorages. Furthermore, potential snow accumulation has required special attention for aerodynamic stability.

The climate at the site include risk of icing and heavy snowfall. Accumulated snow and ice will - as for aircraft - alter the cross sectional shape of the girder which in turn will alter the bridge's aerodynamic behaviour.

Consequently, the girder cross section is shaped to minimize areas prone to snow accumulation particularly close to the edges, fig. 10. Three different edge configurations have been tested in wind tunnel to select the best shape in terms of aerodynamic stability. There are no horizontal surfaces which are prone to snow accumulation outside the roadway area where removal would be difficult. The sloping girder edges may further be protected against snow accumulation through electrical heating cables installed behind the steel surface inside the box. Further, painting systems are tested full scale in the northern part of Sweden to secure a minimal snow adherence on all inclined surfaces.

The design criteria for the bridge specify that aerodynamic stability must be ensured even under conditions where the openings in the crash barriers are partly snow blocked. Hence the railings are designed with a smooth shape and as open as possible to wind and snow flow. Also monitoring of snowfall on the bridge and subsequent application of special equipment for removing of snow underneath the railings is considered.

Ice may change the shapes of the main cables and hangers and cause wind induced oscillations of the hangers. There may also be a risk to the users due to falling ice. However, expert opinions and icing observations carried out at a measuring station at a nearby mountain have indicated that icing should

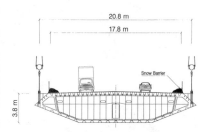

Fig. 10. Höga Kusten Bridge, stiffening girder with potential snow accumulation.

only be a significant problem about 150 m above water level, and hence close to the top of the pylons. Due to a low traffic intensity it is possible to close the outmost roadway areas in cases where falling ice may occur and remove ice prior to reopening.

In the design process many different structural suspension bridge systems have been analyzed with respect to the static and dynamic behaviour of the bridge under these adverse conditions.

The structural configuration has furthermore been optimized to obtain the best behaviour with regard to aerodynamic stability. A girder, 4 m high and 22 m wide has been selected with a distance of 20.8 m between cable planes. The span to sag ratio is 9.5, and the main cable is fixed to the girder by a central node at midspan.

Computer Models
A 3-dimensional computer model facilitates the bridge design, and analysis of many static and dynamic load conditions.

A model of the entire Höga Kusten suspension bridge was analyzed in COWI-consult's in-house developed IBDAS programme (Integrated Bridge Design and Analysis System).

The system is based on two major concepts:
- Integrated solutions of all discipline tasks of a modern bridge design.
- A parametric 3D representation of the bridge.

Parametrical modelling of bridge structures ensures fast operations and easy model changes as e.g. in the conceptual phase of the Höga Kusten Bridge project.

During the detailed design the benefits gained from the IBDAS system mainly relate to the integrated results produced with the computer model. For instance, a full compliance with the computer FEM model and the geometry is ensured, and based on the same simple database, e.g. drawings are automatically produced, static and dynamic calculations are carried out, all relevant geometrical information is determined, and quantities are generated.

The result is a design which minimizes the risk of inconsistency and thereby increases the total quality of the design.

Alternative Development of Main Cables
In Norway, the Askøy Bridge was inaugurated in 1993. The 850 m long main suspension span box girder was installed in 36 m long sections in the record time of a fortnight using a floating crane.

For the cables, an interesting low cost solution is used, fig. 11. Each of the two main cables consists of a bundle of 21 individual cables with a diameter of 99 mm arranged in a rectangular 3x7 cable pattern. They are prefabricated and fully galvanized. Each cable weighs about 71 tonnes and has a guaranteed breaking strength of 9060 kN.

The main cables are anchored in rock. The individual cables in a bundle are spread out using a splaysadle, and anchored in a reinforced concrete block. From this point

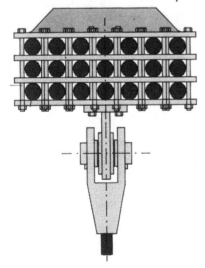

Fig. 11. Cable arrangement, Askøy Bridge.

on, the cable forces are transmitted to prestressing steel tendons passing through predrilled holes in the rock, and anchored in an anchorage chamber, 35 m below the terrain.

The advantage of the system is fast erection of the entire prefabricated cables and no requirement for compacting and wire wrapping. Also cable bands for hangers are greatly simplified. However, the concept is not suitable for very large bridges.

Future Suspension Bridges

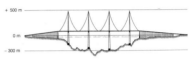

Fig. 12. 3 x 3500 m sspans for a Gibraltar crossing.

Long, continuous superstructures will be generally preferred in cable-supported bridges, because they offer the benefits of better driving comfort and reduced maintenance when the number of expansion joints is limited.

Many alternative superstructure configurations have been proposed in order to improve aeroelastic stability using multiple interconnected box girders, and even laterally connected independent suspension bridges side by side. Such systems will most likely be further developed in the future.

Expansion joint structures and bridge bearings will be developed to permit improved movement capacity for long continuous superstructures.

The still-longer free spans will also require attention to the statical main system to achieve sufficient stability and limit girder deformations to acceptable levels.

Actively Controlled Systems
Simple closed box girder types probably would still be favoured for long spans because of possibility for industrial low-cost fabrication and assembly, and low maintenance costs. However, light structures for very long spans in excess of 2.000 m may become critical to wind induced oscillations.

Traditional methods for assurance of the aeroelastic stability of the basic girder concept are based on passive methods, mostly rigidity of the structural system and proper aerodynamic shaping. Tuned Mass Dampers known from buildings can improve damping and make longer spans possible, as for the Great Belt approaches and the Normandie Bridge.

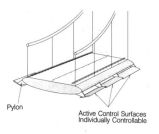

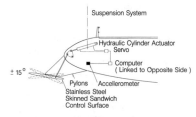

Fig. 13. Active Control System.

Active control systems similar to autopilots in aircraft and more sophisticated systems for control of aerodynamic instability in advanced military aircraft seem promising for the future. Such systems involve active aerodynamic control surfaces acting with opposite phase amplitude relative to the main body movements. They could be operated via hydraulics governed by computerized feedback loop which responds to sensors attached to the main surfaces.

A study conducted by COWIconsult has explored the potential of actively controlled surfaces to moderate flutter instability of long span cable supported bridges, fig. 13. The system, which is patent pending, is based on constantly monitoring movements of the deck and control surface movements to generate aerodynamic stabilizing forces which outbalance destabilizing forces.

In the future, actively controlled systems should be common elements in wind-sensitive bridges and other structures to improve the comfort of users and to reduce fatigue

damage. Similar systems may limit or prevent large deformation of flexible structures subjected to semi static loads.

Development in Cable Technology

Another major development is foreseen for ultra-long suspension spans. The Messina Strait, Gibraltar Strait, Bering Strait, and other potential bridge crossings in Japan may require spans exceeding 3000 m. Since the Brooklyn Bridge was built more than 100 years ago, the dominating material for main cables have been high-tensile steel wires. In the 20th century only modest increases in the tensile strength have been achieved.

In the aerospace industry materials such as carbon fibre reinforced plastics (CFRP) and aramid with improved strength/weight ratios are increasingly used for structural components. Suspension bridge cables and cable stay cables are under rapid development, and price is expected to fall significantly as application and volume increase. However, the lighter carbon cables will require substantially more passive or active damping to eliminate i.a. the risk of wind induced oscillations.

The use of new materials and active systems not presently employed regularly in the industry will lead to considerable reductions in the use of natural resources and new potential for structural development. This will generate a need for highly skilled engineering and labour which should already be accounted for in the future education and training programmes.

Akashi-Kaikyo Bridge Underconstruction

Fumio NISHINO[1], Takeo ENDO[2], Shin KITAGAWA[3] --- Author

Summary of the paper to be presented

This paper presents 1) brief history of the survey work, 2)
brief description of the physical dimensions and structural
system, 3) major studies for the design, 4) development of
higher strength wires, 5) major studies for the construction of the foundation and the erection of the superstructure, and 6) program for the construction work. Emphasis
will be placed more on the items 3) to 5). Some of them
are wind tunnel tests with the 1/100 scale model, development of 180 kg/mm^2 wire, which together with smaller safety
factor adopted for this particular bridge made wires of
single band possible contributing simple design and saving
of cost, construction of 60 meter deep foundation under the
current velocity of 4 m/second, and vibration control of
the main towers during the construction by tuned mass
dumpers.

1) Professor, Department of Civil Engineering, University
of Tokyo
2) Member of Board, Honshu-Shikoku Bridge Authority
3) Director, First Design Section, Honshu-Shikoku Bridge
Authority

Liteature

There are a number of papers already published in English.
They are listed here for those who have interest on Akashi-
Kaikyo bridge. They are:

1) M.Ohashi, T.Miyata, I.Okauchi, N.Shiraishi and
N.Narita, Considerations for wind effects on a 1,990m-
main span suspension bridge, Prerept. 13th cong.IABSE
(Helsinki), 1988.

2) T.Miyata, H.Yamada and H.Akiyama, Codification of wind
effects for a long-span suspension bridge, Proc. Str.
Cong '89, ASCE (San Francisco), 1989.

3) T.Miyata and H.Yamada, Coupled flutter estimate of a
suspension bridge, J.Wind Eng. & Ind. Aero.,33, 1990.

4) T.Miyata, K.Yokoyama, M.Yasuda and Y.Hikami, Akashi
Kaikyo Bridge:wind effects and full model wind tunnel
tests, Aerodynamics of Large Bridges, A.A.BALKEMA,
1992.

5) T.Miyata and K.Yamaguchi, Aerodynamics of wind effects
on the Akashi Kaikyo Bridge, Proc. 2nd Italian Nat.
Conf. Wind Eng. (Capri), 1992, to appeare in J.Wind. &
Ind. Aero.

6) T.Miyata, K.Tada and H.Katsuchi, Wind resistant design
considerations for the Akashi Kaikyo bridge, Proc, Int,
Seminar on Utilization of Large Boundary Layer Wind
Tunnel, 1993. 12 (Tsukuba, Japan).

7) T.Miyata, H.Sato and M.Kitagawa, Design considerations
for superstructures of the Akashi Kaikyo bridge, Proc,
Int, Seminar on Utilization of Large Boundary Lager
Wind Tunnel, 1993. 12 (Tsukuba, Japan).

8) Y. Sumiyoshi, T. Endo, T. Miyata, H. Sato and M.
Kitagawa, Experiments for the Akashi Kaikyo bridge in
a large boundary layer wind tunnel, Proc, Int, Seminar
on Utilization of Large Boundary Lager Wind Tunnel,
1993. 12 (Tsukuba, Japan).

Elasto – Plastic Analysis of Cable Net Structures

J. L. Meek[1]

Abstract

This paper presents a Newton – Raphson interation technique which takes account of plastic cable strains while iterating on the nodal equilibrium equations of cable net structures. The stability of the iteration proceedure is controlled by using a modulus which lies between the cable tangent and secant moduli. In this way the plastic strains in members whose forces have exceeded the elastic limit are built up at a faster rate than simple elastic strain increments. An examle is given which shows how the method controls the number of iterations.

Introduction

High strength cables have been used in recent years in the construction of many large span roof structures (e.g. the shell roofs of the Munich Olympic Stadiums), in which the modern computerized shape finding technique was pioneered by Professor J. H. Argyris (ref(1)). The thrust of the analytical development has been directed toward the determination of the cable structure shapes while the overload characteristics have received less attention. In this paper the elasto–plastic action of cable net structures is investigated to determine overload behavior and factors of safety against collapse. It is found that whereas the geometric nonlinear behavior due to change of shape is reasonably easy to handle, the onset of non-linear material behavior causes some computational problems, at least when constant load steps are used.

Literature Review

The elastic analysis of cable net structures and shape finding procedures are well documented (ref (1)). A resume of the theory is given in this paper. Elasto – plastic behavior of cable nets has been studied (refs. 2, 3, 4, 5). In the solution techniques given in these references, the considerable discussion given to accelerating procedures suggests that some convergence problems were encountered, although not specifically mentioned. In ref. (2), Greenberg uses a convergence accelerator as follows. When a load increment is applied, a deflection r_0 is

---

[1]Professor, University of Queensland, Brisbane, Queensland, Australia

calculated using the tangent stiffness. The stiffness matrix is then assembled for the deflected position and the new deflection increment for the next iteration taken to be,

$$r = r_0 (\frac{k_{av}}{k_2})$$ (1)

where k_2 is the main diagonal term of the second stiffness matrix and $k_{av} = (k_1 + k_2)/2$. The procedure is reported to reduce the number of iterations. The tangent modulus is used for yielded members. In ref. (3), Jonatowski & Birnsteel use Newton – Raphson iteration on the direct stiffness method. Non-linear deformation behavior of both beams and cables is taken into account. No convergence problems are not mentioned although, because incremental loading is used, they can be eliminated by reducing the load step size. In ref. (4), Baron & Venkatesan deal only with elastic stiffness. In ref. (5), Saafan uses Newton – Raphson iteration with the direct stiffness method. Nonlinear stress strain behavior of the cables is taken into account and the tangent stiffness used. Convergence is accelerated by halving the incremental displacement for the first few iterations. The number of iterations treated in this way is a matter of judgment. The problem of instability of the iteration path using the tangent modulus is not mentioned except that the process above appears to have been used for that reason. The example of Baron & Venkatesan (4), which involves yielded but not fully plastic members, is solved by Saafan in 17 iterations. The method described herein solves it in 8 iterations to the same accuracy.

Theory

The force P_N in a given member N, is calculated from,

$$P_N = f (l - l_0)$$ (2)

where l is the distance between member nodes and l_0 is the unstrained length. For elastic behavior,

$$P_N = \frac{EA}{l_0}(l - l_0)$$ (3)

For a load increment $\{\Delta R_i\}$, the displacement increment is calculated,

$$\{\Delta r_i\} = [K_{Ti}]^{-1}\{\Delta R_i\}$$ (4)

The node geometry is now updated,

$$\{x_{i+1}\} = \{x_i\} + \{\Delta r_i\}$$ (5)

Then either equation 2 or 3 is used to calculate $P_{N(i+1)}$. The out of balance force on the nodes will be,

$$\{\Delta R_{i+1}\} = \{R\} - [a_{N(i+1)}]^T P_{N(i)}$$ (6)

The tangent stiffness $[K_T]$ used in eqn. (4), is generated by calculating the individual member stiffness matrices in global coordinates,

$$[k_T] = [k_E] + [k_G] \tag{7}$$

The elastic stiffness $[k_E]$ and the geometric stiffness $[k_G]$ matrices are calculated as follows for a single member. The member, nodes (i, j), has it's nodal coordinate vector,

$$\{x\}^T = \{x_i \; y_i \; z_i \; x_j \; y_j \; z_j\} \tag{8}$$

The length of the member is calculated,

$$l = \{(x_j - x_i)^2 + (y_j - y_i)^2 + (z_j - z_i)^2\}^{1/2} \tag{9}$$

and the member direction cosines,

$$\{c\}^T = \{c_x \; c_y \; c_z\} = \frac{1}{l}\{(x_j - x_i)(y_j - y_i)(z_j - z_i)\} \tag{10}$$

Similarly for the displacement vectors,

$$\{r\}^T = \{u_i \; v_i \; w_i \; u_j \; v_j \; w_j\} = \{r_i^T \; r_j^T\} \tag{11}$$

The updated nodal coordinates are given,

$$x_{i+1} = x_i + r_i \tag{12}$$

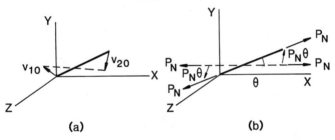

Figure 1 Rotation of member due to end displacements.

From Figure 1, the transformation from member nodal displacements to member distortion is given,

$$(l - l_0) = v_N = [-c^T \; c^T]\{r\} \tag{13}$$

The displacement at the end i, of a member may be expressed,

$$\{r_i\} = \{r_{i\|}\} + \{r_{i\perp}\} \tag{14}$$

The component parallel to the member has it's components,

$$\{r_{i\|}\} = \{c\}\{c\}^T\{r_i\} \tag{15}$$

Hence, substitution in equation (14) gives,

$$\{r_{i\perp}\} = [I_3 - \{c\}\{c\}^T]\{r_i\} \tag{16}$$

If δ, represents the vector components of the displacement of end j relative to end i,

$$\{\delta\} = [-[I_3 - cc^T]\ [I_3 - cc^T]]\begin{Bmatrix} r_i \\ r_j \end{Bmatrix} \tag{17}$$

Now the components $\{\theta\}$ of the rotation vector of the member are given,

$$\{\theta\} = \frac{1}{l}\{\delta\}$$

and the components of the vector force P_N due to rotation,

$$P_N\{\theta\} = \frac{P_N}{l}\{\delta\} \tag{18}$$

Applying the contragredient principle to eqns 3, 13, 17 and 18, gives the member stiffness matrices,

$$[k_E] = \frac{EA}{l_0}\begin{bmatrix} cc^T & -cc^T \\ -cc^T & cc^T \end{bmatrix} \tag{19}$$

$$[k_G] = \frac{P_N}{l}\begin{bmatrix} [I_3 - cc^T] & -[I_3 - cc^T] \\ -[I_3 - cc^T] & [I_3 - cc^T] \end{bmatrix} \tag{20}$$

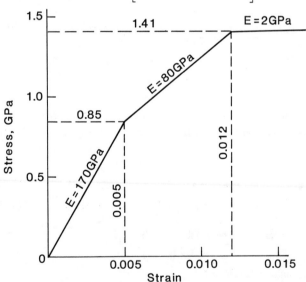

Figure 2   Cable Stress–Strain Curve Idealization.

Cable Properties

Cables are prestressed before use to about one half of their breaking force. Their stress–strain behavior is thus linear up to this point, and nonlinear for higher stress levels (refs. 16, 17). An ultimate strength or limit state analysis which includes the nonlinear behavior should result in design economies. The reasons being that,

(1) Cable tensions in a network usually will not increase in proportion to the load, because of the prestress at zero load.

(2) Plastic deformation in cables will increases the sag, which in turn tends to reduce cable tensions and finally,

(3) Cables undergoing plastic deformation under localized loads will, in most cases, shed some of their load to nearby cables, thus making use of the redundant nature of the cable net work.

The stress–strain curve for a cable may be modeled by the power law,

$$\varepsilon = (\frac{\sigma}{E}) + (\frac{\sigma}{B})^n \tag{21}$$

In equation (21) ε is the cable strain, σ the stress, E the initial Modulus of Elasticity, and B and n are constants determined experimentally. In the present work, this curve can be approximated by using from 2 to 6 straight lines, with the three line approximation being shown in Figure 2.

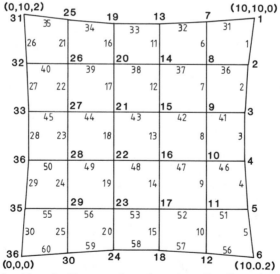

Figure 3   Plan of cable net, nodes and members. Coordinates in metres cable areas: edge − 0.5×10^{-3} m^2, mesh − 0.561×10^{-4} m^2

Table 1

Unstressed Member Lengths

Member No.	Length (metres)	Member No.	Length (metres)
1-5,26-30,31-35,56-60	2.04000	40	1.79377
6	1.82913	41	1.64787
10	1.82476	45	1.64547
11	1.69234	46	1.64430
15	1.69040	50	1.64904
20	1.69321	50	1.64904
21	1.82426	51	1.79305
25	1.82965	55	1.80013
36	1.79939	All others	2.00000

Iteration including plastic deformation

The inclusion of plastic deformation is an additional source of nonlinear behavior which leads to additional equilibrium iterations. Also, plastic members with nearly zero tangent stiffness, may cause the iteration to become unstable. It would appear that when a load increment is applied so that the stresses in some of the members exceed the yield stress, large plastic deformations are produced, causing relatively large global geometry changes. These changes, produce large out of balance nodal forces which, coupled with the reduced tangent stiffness of the plastic members, cause the solution for the next iteration to under shoot the equilibrium position. This process may diverge if the load step is too large. Stability of the iteration path can be achieved by using the initial elastic stiffness in all iterations. However, this can be time consuming and may require incremental loading with the recalculation of the stiffness at the beginning of each load step. In this study, a solution method is considered unsatisfactory if the number of iterations to convergence exceeds twenty. The problem is to devise a method by which the deflections of the net due to plastic deformation are allowed to build up more quickly than by elastic increments without causing the system to become unstable. The method presented, although given for a specific example, should nevertheless be fairly general, since the convergence characteristics appear to be stable over a wide range of values of the modulus selected. The basic idea is simply to use a modulus somewhere between the secant modulus (E_S) and the tangent modulus (E_T).

Example Analysis

A plan view of the cable net analyzed is shown in Figure 3 and the member lengths are given in Table 1.

The load cases considered are:

(1) Edge nodes 0.12 kN, interior nodes 0.16 kN, node 15, 90.0 kN.

(2) Load case 1 plus an additional 20 kN on node 15.

(3) Load case 1 plus an additional 40 kN on node 15.

The input starting position for the structure is a rough approximation of the elastic equilibrium position for load case 1. The same initial position was used for all runs. The convergence criterion used in all the analyses given in Table 2 is that no component of the out of balance force at any node is greater than 1.5 kN. In Table 2, the quantity in brackets for divergent (or slowly convergent), cases is the ratio of the maximum out of balance force component to 1.5 kN. Twelve different trials are given in Table 2.

Table 2

	Load Case		
Modulus Used	1	2	3
1. Secant Modulus E_S	19	19	15
2. $0.333 E_T + 0.667 E_S$	13	14	11
3. $0.5 E_T + 0.5 E_S$	11	10	8
4. $0.667 E_T + 0.333 E_S$	8	7	6
5. $0.75 E_T + 0.25 E_S$	7	6	5
6. $0.775 E_T + 0.225 E_S$	9	6	4
7. $0.785 E_T + 0.215 E_S$	>20(23.0)	>20(25.4)	>20
8. $0.8 E_T + 0.215 E_S$	>20(76.4)		
9. $0.9 E_T + 0.1 E_S$	diverged		
10. Tangent modulus E_T	diverged		
11. E_S for three iterations and 9/10 E_T + 1/10 E_S every fourth iteration	9	8	8
12. E at zero stress	>20(2.4)	>20(1.5)	>20(3.8)

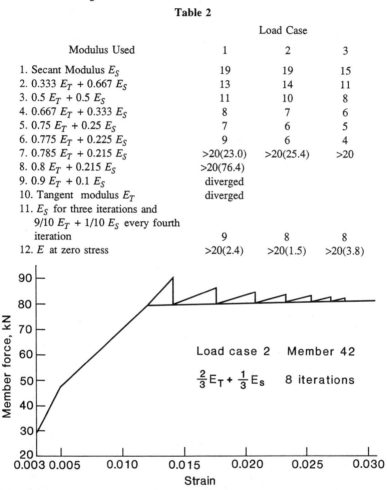

Figure 4   Force in member 42 during iteration cycle, Case 4 in Table 2.

From Table 2, Cases 10 and 12, it is seen that the tangent stiffness gave a divergent solution whereas E at zero stress takes more that 20 iteration cycles. On this basis, one can try to use an E which lies somewhere between these two extremes and which can be chosen in some predetermined way. The obvious choice is to use the secant modulus, E_S. The results are given in Case 1, in Table 2, and a clear improvement to 19 iterations is obtained. It is informative to plot the member force and member strain in member 42, which from Figure 2 is one of the members most highly stressed and therefore also undergoes plastic load shedding. One such graph is shown in Figure 4. When the same graph for E at zero stress is plotted it is seen that the steps are so small that by 20 cycles the process has failed to find the true strain for this member. When the secant modulus is used the process converges to a load slightly below the yield force in member 42. Based on these results it seems reasonable to try either of two strategies,

(1) Use the tangent modulus on some cycles and the secant modulus on the remainder.

(2) To use a modulus lying somewhere between the tangent modulus and the secant modulus.

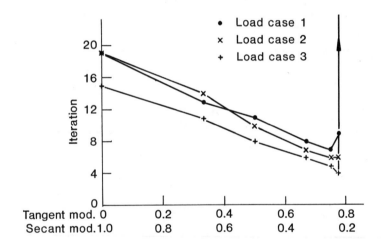

Figure 5   Cycles to converge for various ratios of E_T to E_S.

The results of the first suggestion are shown in case 11 of Table 2. This process converged satisfactorily when a modulus of $(0.9E_T + 0.1E_S)$ was used on every 4th iteration and E_S for the remainder. For the second suggestion, varying proportions of E_T to E_S were used. The various cases and the corresponding number of iteration cycles are given in cases 2 to 9 in Table 2. In all 12 different trials are given in Table 2. The force strain diagram for member 42 for the case 4, $(2/3E_T + 1/3E_S)$ is plotted in Figure 4. It is seen that as the iteration cycles take place, increasing amounts of plastic strain are picked up by member 42, indicating improved convergence of the process. The results of Table 2 are plotted in Figure 5 and these show that as E_T is approached, the cycles to convergence decrease until approximately $(0.78E_T + 0.22E_S)$ and then rapidly increase. In such a situation it is considered unnecessary to try for the optimum value and so the value of $(2/3E_T + 1/3E_S)$ was adopted.

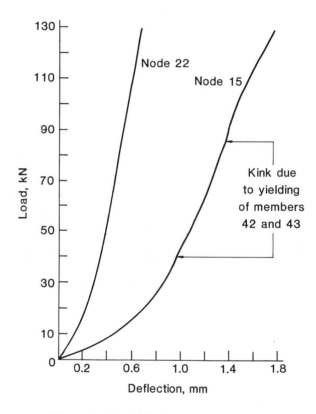

Figure 6   Load deflection curves–nodes 15 and 22.

The load deflection curves, for nodes 15 and 22 (see Figure 2), are plotted in Figure 6. It is seen that the deflection of the net greatly increases the stiffness. Even when members 42 and 43 become plastic, the net maintains considerable stiffness at the load point.

Finally in Figure 7 a perspective view of the net shows the local sagging at node 15 as the concentrated load is applied there.

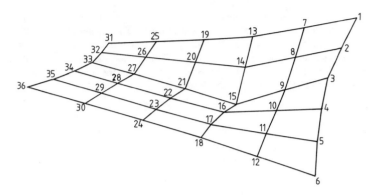

Figure 7   Perspective view of deflected net.

References

1 Argyris, J.H. and Scharpf, D.W. *"Large deflection analysis of prestressed networks"*, J. Of the Struct. Div., ASCE,Vol. 98, No. ST3, March 1972, pp. 633-54.

2 Greenberg, D.P. *"Inelastic analysis of suspension roof structures"*, J. of the Struct. Div., ASCE, Vol. 96, No. ST5, May, 1970, pp. 905-30.

3 Jonatowski, J.J and Birnstiel, C. *"Inelastic stiffened suspension space structures"*, J. of the Struct. Div., ASCE, Vol. 96, No. ST6, June 1970, pp. 1143-66.

4 Baron, F, and Venkatesan, M.S. *"Nonlinear analysis of cable and truss structures"*, J. of the Struct. Div., ASCE, Vol. 97, No. ST2, February 1971, pp. 679-710.

5 Saafan, S.A. *"Theoretical analysis of suspension roofs"* J. of the Struct. Div., ASCE, Vol. 96, No. ST2 February 197-0, pp. 303-405.

6 ASCE, Subcommittee on Cable-suspended Structures. *"Cable-suspended roof construction state-of-the art"*, J. of the Struct. Div. ASCE, Vol. 97, No. ST6, June 1971, pp. 1715-61.

Intermediate Support Cable Logging Systems

James W. Charland[1], Alan G. Hernried[2], and Marvin R. Pyles[3]

Abstract

Cable systems for the transportation of fresh cut logs often employ existing trees as intermediate supports for the main support cable. Failure of these support trees is life threatening and catastrophic to the logging operation. To improve safety and reliability, the maximum loads on the support trees must be predicted before logging operations commence. A model of a general cable logging system employing one set of intermediate support trees is developed. The cables are assumed to be massless. The pulleys are assumed to be massless and frictionless. The support trees are represented as linear springs. The model is described by simultaneous equations of static equilibrium and compatibility for the pulleys and cables, respectively. These equations are highly coupled and non-linear. An iterative approach is used to solve these equations, specifically a modification of Newton's algorithm which utilizes a finite difference approximation to the Jacobian. Determining the static equilibrium configuration of the cable logging system yields the cable tensions and the loads on the support trees directly. An example system is presented to illustrate the methodology.

Introduction

Cable suspension systems for the transportation of fresh cut logs are commonly employed when logging operations are conducted in rough or sensitive terrain. Such systems often use existing trees as intermediate support columns and tree stumps as cable anchors. The design of such systems is severely restricted by the availability of existing trees which may be used as intermediate supports or cable anchors. Of particular concern in the design of these systems is the loading on the trees used as intermediate supports. Failure of these trees during logging operations presents a

----

(1) Research Assistant, Dept. of Civil Engineering, Oregon State University, Corvallis, OR 97331

(2) Associate Professor, Dept. of Civil Engineering, Oregon State University, Corvallis, OR 97331

(3) Associate Professor, Dept. of Forest Engineering, Oregon State University, Corvallis, OR 97331

major hazard to the loggers, as well as a significant cost to the operation in down time and re-rigging of the cable logging system. This is particularly true when logging in second growth forests where the trees are smaller and have lower support capacity than those found in old-growth forests. An accurate method of analyzing prospective cable logging systems in the field, before they are constructed, would significantly reduce the threat of catastrophic failure of the intermediate support trees or the anchors. Previously proposed methods for the analysis and design of cable systems do not satisfy the particular analytical needs of cable logging systems.

Nishino (1989) presents a method for the design of complex cable networks for a given geometry of the cable system under load. The given geometry includes cable anchor locations as well as points where cables intersect. This geometry is fixed throughout the design process. Only the unstretched lengths and areas of the cables are variables in the design process. For a determinate cable system, the forces in each cable are obtained directly from static equilibrium equations. The areas and unstretched lengths of the individual cables are determined based on considerations of allowable stresses and elastic stretching. For indeterminate cable systems, an infinite number of design solutions satisfy the static equilibrium conditions, so equilibrium conditions alone are not sufficient. The set of all possible design solutions in the indeterminate case is all feasible designs. A feasible design is one where all cables are in tension. Where more than one feasible solution exists, optimization criteria may be specified based on considerations such as strength, weight, and stiffness of the entire cable structure. The optimum cable system, the design cable areas and unstretched lengths, is selected to satisfy the equilibrium conditions as well as the additional optimization conditions. Nishino's method, however, is concerned only with the cables, and does not incorporate non-cable structures. The behavior of the elastic support trees with the cable system is an important part of the overall cable logging system's response, and thus must be included in the analysis and design processes.

Peyrot (1978) presents a very general method for combining cable systems with other structural components. Elemental stiffness properties for each cable are determined by comparing changes in cable forces with small changes in each cable's geometry. The cable elements are then combined with other structural elements. The overall structure is solved using an iterative numerical solution technique. At each iteration, new stiffness properties for the cable elements are found. Such iterative techniques are necessitated by the substantial changes in cable behavior as the system geometry undergoes large changes.

The analysis of cable logging systems presents the complication of having a single cable run continuously over several supporting points, creating several separate cable spans. Compatibility conditions which apply to multiple cable spans simultaneously are not included in Peyrot's analysis. Also, the effective flexibility of the intermediate support trees may be treated in a very simple but accurate manner, obviating the need for the more general, and complicated, stiffness approaches used by Peyrot.

In this paper, a simple and straightforward method for the analysis of intermediate support cable logging systems is presented. An iterative approach is used to solve a set of non-linear simultaneous equations including force equilibrium and cable length compatibility. The method is implemented for the specific case of an intermediate support cable logging system. The approach is completely general, however, and would apply equally well to other cable logging system configurations.

Cable Systems

The cable systems treated here are composed of one or more cables and contain fixed (anchor) points and free (floating) points. The cables themselves are assumed to be massless, eliminating the effects of catenary sag, unstretchable, and to have no flexural stiffness. Throughout the analysis, the tension in each cable is assumed to be constant along the cable. The locations of the fixed points are known. Generally each cable is attached to fixed points at the two ends and a finite number of free points along the length of cable. The ends of a cable may also be connected to a fixed and free point, or two free points. Physically free points can consist of steel pulleys (assumed frictionless) attached by a short cable (strap) to the trees or a rigid device that supports one cable by another (jack or carriage). These devices will be explained in more detail later in the paper.

The behavior of cable systems considered here is governed by a set of non-linear simultaneous equations. At each free point, a vector equation describing static equilibrium can be written. In general, such an equation will contain as unknowns the magnitudes of the forces, the tensions in the cables and straps, as well as their directions. The direction of action of the forces is a function of the relative positions of the free and fixed points. Equilibrium equations alone are insufficient to determine the unknown forces and locations of the free points. These equations are supplemented by compatibility conditions; i.e. a geometric description of the length of each cable (which is known) in terms of the locations of the fixed and free points.

Figure 1a shows a simple cable system consisting of two cables. Cable 1 connects points A, B, and C and cable 2 connects points B and D. The system has two free points (B and D), two fixed points (A and C), and one constant loading force, F, whose direction and magnitude are known. In order to establish the final configuration of this system, eight unknown quantities must be determined. The unknowns are: the location of points B and D in space (6 unknowns) and the tensions of the two cables (2 unknowns). Eight equations are required to solve for the eight unknowns. These equations express the static equilibrium of points B and D, and the lengths of cables 1 and 2. The equations of static equilibrium are

$$T_1 \mathbf{u}_{AB} + T_1 \mathbf{u}_{BC} + T_2 \mathbf{u}_{BD} = \mathbf{0}; \ T_2 \mathbf{u}_{BD} + \mathbf{F} = \mathbf{0} \tag{1}$$

where T_1 and T_2 are the tensions cables 1 and 2, respectively; and $\mathbf{u}_{AB}$, $\mathbf{u}_{BC}$, and $\mathbf{u}_{BD}$ are unit vectors along AB, BC, and BD, respectively.

Equations (1) can be written in scalar form, yielding 6 equations, by equating components in the x, y, and z directions. The compatibility equations are

$$\left| \mathbf{r}_{AB} \right| + \left| \mathbf{r}_{BC} \right| = L_1; \left| \mathbf{r}_{BD} \right| = L_2 \tag{2}$$

where, $|\mathbf{r}_{AB}|$ is the magnitude of a position vector from A to B, and similarly for $|\mathbf{r}_{BC}|$, and $|\mathbf{r}_{BD}|$. The known lengths of cables 1 and 2 are L_1 and L_2, respectively.

In addition to this basic cable arrangement, the stiffness of the intermediate support trees may be included as a function of the deflection of a point in space relative to a reference point. The reference point is the location of the connection of the cable system to the undeformed tree. Cable systems which are combinations of elastic structural components with cables are termed combined cable systems.

Figure 1b shows a combined cable system similar to that shown in Figure 1a. This system has two cables and a linear spring. The spring is shown in its deformed configuration. The spring represents the bending stiffness of a tree. Without loss of generality, the undeformed length of such a spring is zero. Cable 1 connects points A, B, and C and cable 2 connects points B and D, as before. The spring connects points D and E. The system has two free points (B and D) and three fixed points (A, C, and E). With the addition of the spring force, there are nine unknown quantities to be found in determining the final configuration of this system. The unknowns are: the location of points B and D in space (6 unknowns) and the tensions of the two cables (2 unknowns), as well as the force in the spring (1 unknown). Nine equations are required in order to determine the nine unknowns. These equations express the static equilibrium of points B and D (6 equations), the lengths of cables 1 and 2 (2 equations), and the constitutive relation between the spring force and the relative motion of points D and E. The equilibrium equation of point B is identical to the first of Equations (1), while the equilibrium equation of point D is

$$T_2 \mathbf{u}_{BD} + F_s \mathbf{u}_{DE} = \mathbf{0}; \text{ and } F_s = k * \left| \mathbf{r}_{DE} \right| \qquad (3)$$

where F_s is the force in the spring, $\mathbf{u}_{DE}$ is the unit vector along DE, k is the linear spring constant, and $|\mathbf{r}_{DE}|$ is the deformation of the spring. Seven scalar equations result from Equations 1 and 3. As before, the two compatibility equations for the cables remain unchanged (Equations 2).

Solution Procedure

The simultaneous equations derived from force equilibrium and cable length compatibility are highly coupled and non-linear. The non-linearities result from the expressions for cable lengths and force components, which involve squares and roots of the unknown quantities. Because of these non-linearities, no closed form solution of the equations is possible. A numerical solution is therefore necessary.

The solution procedure is based on a variation of Newton's method for finding the roots of an n-dimensional system of equations. For one equation, $f(x) = 0$, Newton's method may be stated as:

$$x_{i+1} = x_i - f(x_i)/f'(x_i) \qquad (4)$$

where x_i is the $i\,th$ estimate for the root, and prime denotes derivation with respect to x. Here as with the multi-dimensional case, the quality of the initial guess for the root is critical. A good guess may converge, where a poor one may not. The root is found when $|f(x_i)| < \varepsilon$, where ε is some arbitrarily small positive number.

For multi-dimensional problems, $\mathbf{f(x)} = \mathbf{0}$, Newton's method is

$$\mathbf{x}_{i+1} = \mathbf{x}_i - \mathbf{A}^{-1} * \mathbf{B}_i \qquad (5)$$

where $\mathbf{x}_i$ is the $i\,th$ estimate for the root of $\mathbf{f}$, $\mathbf{B}_i = \mathbf{f(x_i)}$, and $\mathbf{A}$ is the Jacobian of $\mathbf{f}$, i.e.

$$\mathbf{A} = [A_{kl}] = \partial f_k / \partial x_l \qquad (6)$$

A variation of this technique of root finding has been implemented in the IMSL routine NEQNF, which is used here to solve the coupled, nonlinear system of algebraic equations. The program which solves the combined cable system is extremely computationally efficient, and a solution is generally found within a few iterations. A method for finding a good initial guess for the variables is incorporated into the program. The initial guess for the solution is based on selecting free point coordinates and cable tensions to satisfy cable length compatibility constraints and simplified equilibrium conditions, respectively. This initial guess algorithm will be discussed later in the paper.

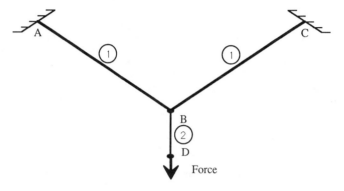

Figure 1a. Simple cable system with point load.

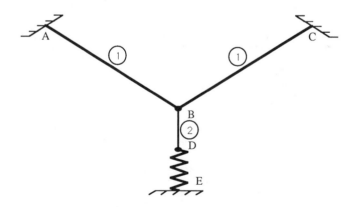

Figure 1b. Simple combined cable system.

Model of Intermediate Support Trees

The cable system attaches to the trees high above the ground, loading the trees axially and laterally. The trees may be modeled as cantilevered beam/columns. Comparisons on typical trees and loading configurations indicate that the axial deformations will be orders of magnitude less than the lateral deformations. It is therefore reasonable to ignore the axial deformations of the tree. This allows the tree to be modeled as a two-dimensional spring acting in the horizontal plane only and located at the strap attachment point. The equivalent spring stiffness is computed considering bending deformations in the horizontal plane. The effects of secondary bending are also ignored in assessing the behavior of the tree.

The lateral stiffness of the intermediate support trees is estimated using Euler beam theory. The values of Young's Modulus, effective length, and moment of inertia must be found. Appropriate values for Young's Modulus may be found in ASTM handbooks and in the literature. The effective length is found by measuring the height of the connection to the tree above ground level.

The moment of inertia of the support trees is computed assuming a circular cross section. The radius of the tree is not constant along the length. Assuming the radius varies linearly along the length leads directly to an expression for the moment of inertia $I(x)$ as a function of position along the tree. The lateral stiffness may be computed based on these assumptions.

Demonstration Problem

The methodology for the analysis of intermediate support cable logging systems discussed in the preceding sections is demonstrated on a simplified example. This particular example has been chosen to illustrate the methodology and to facilitate comparison to a closed form solution. Figure 2 shows the cable logging system in plan, while Figure 3 shows the elevation view. The set-up shown contains all the components of an actual cable logging system.

There are three principal cables in this type of logging system. These cables are denoted 1, 2, and 3 in Figures 2 and 3. The continuous cable running from point F to point J, connecting with points G and H is the skyline (Cable 2). The payload of the cable logging system is carried by the carriage (point G), which travels along the skyline. The continuous cable running between points A, B, C, D, and E is the intermediate support cable (Cable 1). The intermediate support cable supports the skyline by means of the intermediate support trees and a device called the jack. The jack is a rigid device which connects points C and H (Cable 6). Another long cable connects points G and I (Cable 3). This is the mainline, which pulls the carriage along the skyline. Short cables, called straps, connect points K and B, and points M and D (Cables 4 and 5). The two intermediate support trees are modeled as the springs connecting points K and L, and points M and N.

Points A and E are the left and right anchor points, respectively, of the intermediate support cable. Point L is a point in space which is coincident with point K when the left intermediate support tree is in its undeformed state. Point N is a point in space which is coincident with point M when the right intermediate support tree is in its undeformed state. Points F and J are the anchor points of the skyline. Point I is the end point of the mainline. The payload is pulled from this point. The coordinates of these points, given in Table 1, do not change in the three different load cases of the demonstration problem. The lateral stiffnesses of the support trees is 125.

Point B is the pulley connecting the intermediate support cable to the strap which connects to the left support tree. Point D is the connecting pulley on the right side.

Point C is the connection of the skyline support device, the jack, to the intermediate support cable. Point G is the location of the carriage, the device which rides along the skyline on pulleys and carries the payload. The external load on the cable logging system is applied at point G. Point H is the connection of the jack to the skyline. Points K and M are the deformed locations of the support tree connections.

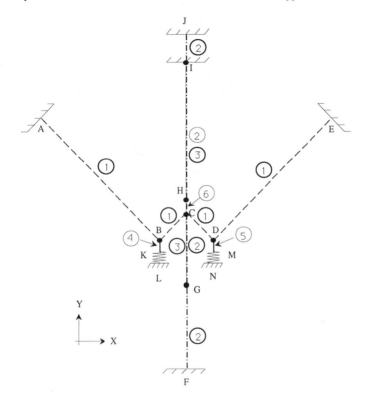

Figure 2. Plan view of demonstration cable logging system.

The dimensions of this example are unrealistic. In an actual cable logging system the straps and jack are generally much shorter than the other cables. Such small cables would be difficult to see on a scale drawing of an actual cable logging system. Similarly, the angles of the skyline between points F, G, and H and points G, H, and J are larger than they would be in an actual cable logging system. Another aspect of the proportions which is unusual is the height of the support trees versus the dimensions of the system in the direction of the skyline. In an actual cable logging system, the length of the skyline would be orders of magnitude greater than the height of the intermediate support trees.

The procedure for solving a cable logging system begins with determining the locations of the fixed points in the system (A, E, F, I, J, L, and N). The lengths of the individual cables must also be provided, along with the stiffnesses of the two support trees. As discussed in the previous section, determining the solution involves solving a system of equations for the unknowns. The unknowns are the locations of points B, C, D, G, H, K, and M; the forces in the six load bearing cables, and the forces in the two trees. The solution process must begin with a reasonable guess for each of these unknowns. The initial guess for the unknowns may be calculated automatically by the program, or it may be provided directly by the user.

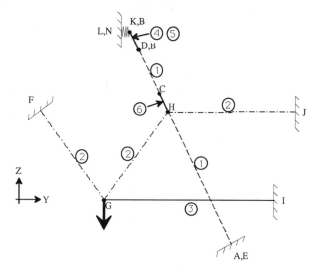

Figure 4. Elevation view of demonstration cable logging system.

The automatic initial guess algorithm determines starting values for the floating points by satisfying simplified compatibility conditions. Initial cable tensions are determined using this geometry and simplified equilibrium conditions. If the automatic guess does not lead to convergence, the user will have to apply further reasoning to make the initial guess more accurate. A more accurate guess should increase the chance for a successful solution to the system of equations.

One configuration of the cable logging system is analyzed. The external load on the cable logging system is 100 in the negative z direction. The geometry of the configuration corresponds to that shown in Figures 2 and 3. The cable lengths are shown in Table 1. The geometry of Figures 2 and 3 has been designed to facilitate hand analysis of the system, specifically the determination of the cable tensions.

The manual solution for the locations of the floating points and cable tensions are shown in Table 3. The computer solutions for the floating point locations and cable tensions are shown in Table 4. The manual and computer solutions are nearly identical, verifying the solution algorithm.

Conclusions & Further Research

A method is presented for determining the response of a system of cables exhibiting complex geometry combined with elastic structures. The method has been specifically developed for the intermediate support cable logging system and as such should be of direct use for designing or analyzing such cable logging systems. The method has been implemented in FORTRAN on a digital computer, and illustrated on an example combined cable system. The method is efficient and accurate, thus many alternate designs can be investigated before an actual cable logging system is constructed. This should prove to be an aid to loggers as they work to improve the safety and efficiency of their operations. Although the method has been developed and implemented exclusively for cable logging systems, it may prove to be a valuable tool in analyzing other types of cable structures as well.

Further research on this methodology is needed. In particular, the effect of the self-weight and catenary sag of the cables on system behavior should be investigated. Elastic extension of cables can also be included to further improve the accuracy of the method. A simplified treatment of the complex interaction of the jack and the skyline could also be included in the formulation.

Currently the method is directed more towards the analysis of a proposed cable logging system, and less toward the design of such a system. The specification of certain parameters of interest, for instance the height of the skyline between the intermediate support trees, and the subsequent determination of the required cable lengths, would make this program an even more valuable design tool. These modifications are relatively straight forward, and could be accomplished with little effort.

References

1. Nishino, F., R. Duggal, and S. Loganathan,"Design Analysis of Cable Networks," *Journal of Structural Engineering*, Vol. 115, No. 12. pp 3123-3139. (1988)

2. Peyrot, A.H. and A.M. Goulois, "Analyis of Cable Structures," *Computers and Structures* Vol 10. pp 805-813. (1979)

Acknowledgement

The support of the Center for Wood Utilization Research, grant number 92-34158-7137, is gratefully acknowledged.

Table 1. Configuration 1 fixed point locations and cable lengths.

Point	X	Y	Z	Cable	Length
E	2.70	3.00	0.00	A-B-C-D-E	13.23
N	0.50	0.40	4.80	F-G-H-J	8.00
A	-2.70	3.00	0.00	G-I	4.00
L	-0.50	0.40	4.80	K-B	0.45
F	0.00	-1.50	3.00	M-D	0.45
J	0.00	4.50	3.00	C-H	0.45
I	0.00	4.00	1.00		

Table 2. Computer generated initial guess.

Point	X	Y	Z	Cable	Tension
M	0.50	0.40	4.80	A-B-C-D-E	56.32
D	0.50	0.40	4.35	F-G-H-J	58.35
C	0.00	0.40	3.39	G-I	0.00
K	-0.50	0.40	4.80	K-B	56.32
B	-0.50	0.40	4.35	M-D	56.32
G	0.00	0.00	1.33	C-H	100.00
H	0.00	0.40	2.94	Spring L-K	56.32
				Spring N-M	56.32

Table 3. Hand solution for floating points and cable tensions.

Point	X	Y	Z	Cable	Force
M	0.50	0.80	4.80	A-B-C-D-E	30.62
D	0.50	0.80	4.40	F-G-H-J	62.50
C	0.00	1.30	3.40	G-I	0.00
K	-0.50	0.60	4.80	K-B	55.91
B	-0.50	0.80	4.40	M-D	55.91
G	0.00	0.00	1.00	C-H	55.91
H	0.00	1.50	3.00	Spring L-K	25.00
				Spring N-M	25.00
				Left axial	-50.00
				Right axial	-50.00

Table 4. Computer solution for floating points and cable tensions.

Point	X	Y	Z	Cable	Force
M	0.50	0.80	4.80	A-B-C-D-E	30.61
D	0.50	0.80	4.40	F-G-H-J	62.50
C	0.00	1.30	3.40	G-I	0.00
K	-0.50	0.60	4.80	K-B	55.90
B	-0.50	0.80	4.40	M-D	55.90
G	0.00	0.00	1.00	C-H	55.90
H	0.00	1.50	3.00	Spring L-K	25.00
				Spring N-M	25.00
				Left axial	-50.00
				Right axial	-50.00

Transient Response of Guyed Telecommunication Towers
Subjected to Cable Ice-Shedding

Ghyslaine McClure,[1] Member, ASCE, and Na Lin[2]

Abstract

This paper presents the results of a numerical study of the transient response of guyed telecommunication towers subjected to sudden ice shedding from the guy wires. Detailed nonlinear dynamic analyses were conducted for three guyed towers with heights of 24.4, 60.7 and 213.4 m, and with two, four and seven stay levels, respectively. Six ice-shedding scenarios were simulated for each structure, with a radial thickness of 40 mm on the cables and on the mast members, in still air conditions.

The transient response of the supports is examined for the first 4-5 s following ice shedding. Results indicate that the tilt rotation of the tallest support exceeds the assumed serviceability limit of 0.5°, whereas in all other cases the displacements and rotations are within the limits. In the most critical shedding scenarios, it was found that some of the internal forces in the mast were in the order of 50% of the estimated capacities in shear, bending and axial compression.

In general, the largest dynamic amplifications are observed for shedding scenarios that affect the top part of the mast where the antennae are attached. Further studies should focus on very tall towers (more than 100 m), and should address the combined effects of ice shedding and cable galloping. This is an even more complex problem that will require proper numerical modelling of damping.

[1]Asst. Prof., Dept. of Civ. Engrg. and Applied Mechanics, McGill University, Montréal, Québec, Canada, H3A 2K6.
[2]Structural Engineer, CANATOM, 2, Place Félix-Martin, Montréal, Québec, Canada, H2Z 1Z3.

Introduction

Although recent studies of cable ice-shedding have been conducted in reference to overhead power line applications (Jamaleddine et al. 1993), ice-shedding from the guy wires of telecommunication towers is a source of dynamic loads that has not been investigated yet. Considering that increasingly taller multilevel guyed towers are being built, and that these towers must meet stringent serviceability limits, ice-shedding effects on tower response deserve to be better understood. Could these transient effects jeopardize the tower's serviceability? Could they ultimately trigger tower instability and collapse when combined with adverse wind conditions? Or is their influence marginal? Since direct field measurements are not available and difficult to make, numerical modelling is essential to answer these questions.

Background

The effects of ice loads very often govern the design of telecommunication structures in countries with cold climates. It is well known that latticed structures and cables are susceptible to the accumulation of ice, and failures of guyed towers due to ice have occurred frequently under severe icing conditions, as summarized by Williamson (1973). Even though the static effect of the ice load is considered in design, the dynamic transient effects due to sudden ice shedding are not usually accounted for. Static effects of combined wind and nonsymmetric icing on guyed towers have been studied by Peil and Nölle (1991), in comparison to symmetric icing patterns. Results from over 6,000 parametric cases with different mast geometries, initial states, wind loads and wind directions have led to the conclusion that internal forces in the tower legs and cable tensions are governed by the fully symmetric icing case (full ice on all tower components). In addition, nonsymmetric icing on the guy cables requires consideration only when radial ice accretion exceeds 100 mm. Such results validate the current practice of many design standards to consider symmetric icing conditions only.

Davenport and his collaborators (Davenport 1986 and Novak et al. 1978) have investigated the dynamic response of guyed towers under combined wind and ice loadings for over 15 years. They have observed that although very large static effects can be caused only when ice accretion exceeds very large thicknesses (in the order of meters rather than millimeters), smaller thicknesses can destabilize the aerodynamic shape of the guy cables, and occasionally the mast, and large dynamic movements may result. Guy wire galloping may have dramatic effects on towers if dynamic interactions between cables and mast are resonant or amplified by the negative aerodynamic damping of the ice-laden cables. It seems reasonable to assume that large amplitude movements of the guy cables are likely to cause some ice shedding, which in turn will perturb the dynamic equilibrium of the structure.

Project Objective

    More insight is needed in the response of guyed towers to ice shedding before this phenomenon can be studied in combination with cable galloping. Therefore, the purpose of this project is to investigate these transient effects in still air conditions, in order to uncouple them from dynamic wind effects.

Modelling of Guyed Towers

    Detailed numerical simulations on three guyed towers were completed (Lin 1993) in order to evaluate the order of magnitude of the dynamic amplifications of the response due to sudden ice-shedding from the guywires. All simulations were carried out with the ADINA software (ADINA R&D 1987), which allows time-step integration of the geometrically nonlinear equations of motion.

    The towers modelled were three-legged welded lattice steel guyed masts, and were representative of those used in the Canadian telecommunication industry. Figure 1 shows the geometry of the tallest tower analysed, which is 213.4 m (700 ft) tall with seven stay levels. All models were sufficiently detailed to represent the dynamic cable tensions generated, and to allow cable-mast interactions to take place. The mast was modelled with beam-column elements with stiffness properties equivalent to those of the three-dimensional lattice structure: the effects of shear deformations and St.Venant's torsion were included, but warping torsion was not modelled. Validation of the equivalent mast properties was done through several static analyses and through an eigenvalue analysis on the complete mast model on rigid base and lateral supports.

    Accurate simulation of mast and cable interactions is achieved by using appropriate types and meshes of elements in the finite element model. Convergence analyses were carried out for each tower model, and it was found that eight to 38 beam-column elements were necessary to capture

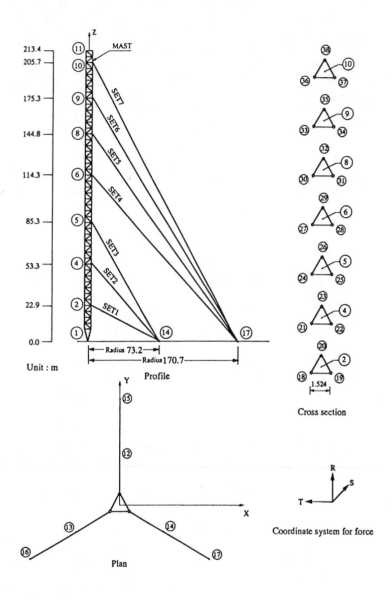

Figure 1.   Geometry of the 213.4-m tower analysed.

the five lowest frequencies and mode shapes of the masts.
Using three-node isoparametric elements, convergence
analyses on guy cable meshes were conducted by Guevara
(Guevara and McClure 1993) in a study related to seismic
analysis of guyed towers. They have indicated that 10 to
35 tension-only cable elements are required to model the
first five transverse modes of a guy cable. All cables
are assumed pretensioned to 10-12% of their ultimate
tensile strength.

     Six cable shedding scenarios were studied for each
tower, while the mast was assumed to remain fully iced in
all cases. The shedding scenarios for the tower of Figure
1 are illustrated in Figure 2; they are classified in two
series, and each series includes three scenarios. In
Series #2, two identical clusters of guy cables are shed
simultaneously, while in Series #1, only one cluster is
shed at a time.

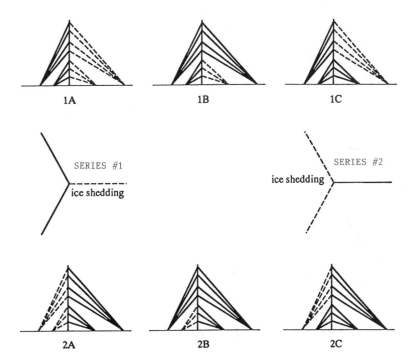

Figure 2. Ice-shedding scenarios for the 213.4-m tower.

It is recognized that one can imagine many more possible combinations of iced and bare cables, also varying the cable(s) to be shed, but these six particular scenarios were expected to yield extreme responses. It should also be noted that none of the scenarios studied generates torsion in the tower.

The ice is assumed uniformly distributed around the guy cables, with a radial thickness of 40 mm along the full cable length. The lumped mass technique was used so that the uniform ice load is concentrated at the nodes of the cable elements. Ice shedding is modelled by a sudden drop in external load on the shed cables.

Structural damping of a guyed tower in still air mainly consists of the internal damping of the guy cables and mast. The former, probably the most significant one, is due to frictional forces that are developed between the strands and individual wires of stranded cables, mostly when being subjected to transverse vibrations. This type of damping is both frequency dependent and nonlinear with respect to the amplitude of the motion, which makes it very difficult to model in a time domain analysis. The latter is relatively less, mainly due to frictional effects of the welded members in the mast. In view of the difficulties associated with realistic modelling of mast and cable damping for nonlinear analysis, damping effects were not included in the analyses conducted, except for one case where algorithmic damping was used (Newmark-ß operator with parameters $\delta = 0.55$ and $ß = 0.3$). The most severe dynamic effects under a sudden transient load such as ice shedding are expected to occur during the first few seconds (after a few fundamental cycles) following the event, in which it may be assumed that damping has not yet had a major influence on the tower response within such a short delay. Time histories were obtained for the first twelve seconds following ice shedding, but because physical damping is neglected in the numerical models, the analysis will be considered valid only during the first five seconds.

Results for the 213.4-m tower

The following elements of response were obtained: lateral displacements at the various stay levels, guy wire tension, horizontal shear force and bending moment in the mast at each stay level. Of particular interest are the following: the duration and magnitude of the peak lateral displacements that exceed the tower's serviceability criteria, the frequency content of the cable tension time histories, the dynamic amplification of cable tensions both in shed and unshed guywires, and the magnitude of the peaks and the frequency content of internal forces in the mast.   Detailed time histories and summaries of results can be found in Lin (1993).   Figure 3 reproduces four time histories obtained for the 213.4-m tower under shedding Scenario 2A.

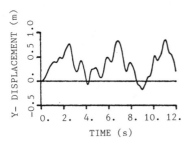

(a) Horizontal displacement
    at top of tower

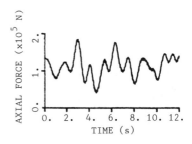

(b) Guywire tension in
    top cluster Set#7

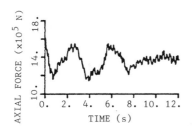

(c) Axial compression in mast
    force at Joint#2

(d) Horizontal shear
    force at Joint #8

Figure 3.   Time histories of the 213.4-m tower under ice-shedding Scenario 2A.

Lateral displacements at the top of the tower (antenna location) reach a peak value of the order of 850 mm (Figure 3a); the duration of the peak is about 2.3 s. The maximum tilt at the same location is 0.57°, and in this case, the dynamic effects of ice shedding are significant in terms of serviceability of the antenna. The maximum cable tension is 186 kN at the top guying level (Figure 3b) and corresponds to a dynamic amplification factor of 2.0.   This tension, however, is only in the order of 35% of the static ultimate tensile strength of the cable.

The dynamic amplification factors for the compression in the mast mostly range from 1.0 to 1.2, and a maximum of 1.4 is obtained at the top of the mast. Figure 3c is an example of time history at the lowest stay level. The maximum shear force (Figure 3d) and bending moment reach values that correspond to 23% and 50% of the estimated static capacity of the mast, respectively.

As expected, high frequency components are less important in the time histories of this tower than in the shorter ones. This is evidenced in the peak durations of the various response indicators, varying between 0.7 to 0.8 s in average, and corresponding to the four lowest natural frequencies of the mast. High frequencies exist mostly in the axial response of the mast, and have more influence at the upper stay levels.

Conclusions

Results of the numerical simulations indicate that the dynamic tensions generated in the guy cables are not critical since their peak values are within one third of their ultimate strengths. Dynamic forces in the mast appear relatively important, however, and cable vibrations tend to cause higher frequency components in the mast at the attachment points of ice-shed cables. The dynamic shear force (bending moment) in the shortest tower (medium tower) reaches 42% (52%) of the estimated capacity. For the tallest tower, the dynamic compressive force and bending moment in the mast also reach 50% of their respective estimated capacity. Investigation of the durations of the peak values of the response in the three towers indicates a direct correspondance with the natural frequencies of the lowest bending modes of the mast. Only the shortest tower shows important high frequency components, in the order of 10 to 30 Hz, which is attributed to the relatively large stiffness of this tower in both the vertical and transverse directions.

A more complete investigation of ice-shedding in guyed towers would require a better numerical model that could account for physical damping, especially if the combined effects of ice shedding and cable galloping are studied. So far, the largest dynamic amplifications have been observed for shedding scenarios that affect the top part of the mast, and further studies should concentrate on shedding scenarios involving mostly top clusters.

Acknowledgments

We wish to thank Mr. S. Weisman, P.Eng., of Weisman

Consultants Inc., Downsview, Ontario; Prof. M.S. Madugula of The University of Windsor, Ontario, and members of the CSA Technical Committee S37 on Antennas, Towers, and Antenna-Supporting Structures, for providing detailed data on the towers analysed. The project was supported by Grant No. OGP0121270 from the Natural Sciences and Engineering Research Council of Canada, for which we are grateful.

Appendix. References

ADINA R&D. (1987). "ADINA (Automatic Dynamic Incremental Nonlinear Analysis) Theory and modelling guide." Watertown. MA, Report ARD 87-8.

Davenport, A.G. (1986). "Interaction of ice and wind loading on guyed towers". Proceedings of the third international workshop on the atmospheric icing of structures, Vancouver, British Columbia, May 1986.

Guevara, E.I., and McClure, G. (1993). "Nonlinear seismic response of antenna-supporting structures." Computers & Structures, 47(4/5), 711-724.

Jamaleddine, A., McClure, G., Rousselet, J., and Beauchemin, R. (1993). "Simulation of ice-shedding on electrical transmission lines using ADINA." Computers & Structures, 47(4/5), 523-539.

Lin, Na. (1993). "Dynamic response of guyed antenna towers due to ice shedding." M.Eng. Project Report G93-15, Dept. of Civ. Engrg. and Applied Mechanics, McGill University, Montréal, Québec, Canada.

Novak, M., Davenport, A.G., and Tanaka, H. (1978). "Vibration of towers due to galloping of iced cables." ASCE J. Engrg. Mech., 104(EM2), 457-473.

Peil, U. and Nölle, H. (1991). "Nonsymmetric icing of guyed masts". Presented at a meeting of the IASS Working Group No.4 Masts and Towers, Stockholm, 10 p.

Williamson, R.A. (1973). "Stability study of guyed tower under ice loads." ASCE J. of Struct. Div., 99(ST12), 2391-2408.

Design and Construction of
The Cable-Stayed Wooden Space Truss Roof

Kei Sadakata[1]   Hong Wan[2]
Professor,   Graduate Student,
Toyohashi-University of Technology
Toyohashi 441, Japan

Abstract

The thinning must be enforced for healthy development of an artifical forest. The most effective use of thinning lumber is costruct of the space truss structure. This election systems can be composed a roof-plate structure of a large span buildings with small section lumbers.

The buildings of this report have been constructed in Aich-Prefecture Toei-Cho,Japan in 1991. Purpose of this facility is a forestory promotion,those are composed of three facility buildings. Scale of this buildings are differ but structural system has a same mechanism design, that is the cable stayed roof suspended from a independence big lumber column/pole.

Introduction

Classification of space truss plate by cofiguration method: Method of configuration the space truss structure can be divided in to two based on election system as described below. Refer to **Fig.1.**

A1: Construction of space truss units. A pyramid framwork is considered as one unit,and a truss plate is formed by continuing it in this method.

A2; A grid of plane truss beams is formed. Many plate truss beams are placed perpendicularly each other and a space grid plate is formed in this method.

Situation of experimental researches on space truss plate: The roof plate continuous pyramid skeleton unit. A pyramid frame work unit compreises eight equal length members connected together. At each connecting point (apex of framework),the truss members coming in three-dimentional directions are connected together. Therefore, design of the three-dimensional connectingm metal joint

at this nodal point is very important. The author developed a cylindrical ring joint(CRJ) in 1991,assembled a
space truss plate with 1/2 scale(270 cm X 270 cm with
plate depth of 636 cm) as shown in Fig.2 and performed
structural experiments.
     Experiments on a grid plate structure assembled by
intersecting the plane truss beams. The author produce a
cross-beam by intersecting the Virendeal-truss-beams in
cross shape. The connections were made by using DCS and
DCCS [4] connecting metal joints. Structural experiments
were performed,and the load bearing strength and the ratio
of load sharing were measured and used as design data for
buildings to be constructed (Fig. 3).
     A roof for bus station,Trial production and experi-
ment. The pyramid frames each having 60 cm for the length
of one side were continuously connected and assembled to
a space truss roof plate with the length of 540 cm x width
of 360 cm x depth of 43 cm. Truss members were thinning
lumber of Japanese CEDAR. This building is a bus stop for
passengers for Toyohashi Railway Co,Refer to Photo.1.

Building structure in suspension system

     System of suspension bridge and system of cable stay-
ed: When a distance between the points of support of a
truss type roof plate increases and the structure has a
large span, the bending stress in roof plate increases in
proportion to the square of span length. In this case,

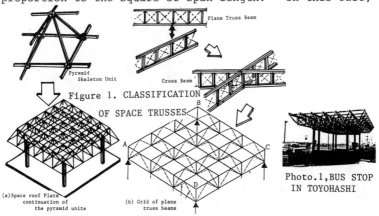

Figure 1. CLASSIFICATION OF SPACE TRUSSES

Pyramid Skeleton Unit
Plane Truss Beam
Cross Beam

(a)Space roof Plate continuation of the pyramid unite
(b) Grid of plane truss beams

Photo.1,BUS STOP IN TOYOHASHI

[1]Sadakata,Prof.Toyohashi University of Technology,Tempaku-
cho 2-1, Toyohashi 441, Aichi, JAPAN
[2]Hong wan,Graduate student,Architectural and Civil Engng.
Toyohashi University of Technology,Toyohashi 441,JAPAN

the increase in the depth of truss plate is limited by the
structure and material, so that it is better to suspend
the roof plate from the above/for example column top.That
is, the suspension structure system is introduced for essentially shortening the span.   This is a kind of hybrid
structure.

As suspension system hybrid structures in steel framing architecture,there two types that are well known; the
suspension bridge system(Burgo Paper Mill) by P.L.Nervi[1]
and cable stayed system (NEC Hall 7) by E.D.Mills[2]. In
the case of wooden structure,the cable stayed system is
better than the suspension type when considering the material and structural characteristics of lumber. Because
the roof plate is directly suspended from a column in the
cable stayed system,it is possible to increase the rigidity of structure frame.   This is important in the dis-
tricts where the earthquake resistant design must be always taken into account.

Introduction of cable stayed system to wooden
structure: For forming large space with a wooden structure is used by using wooden members with large cross sect-
ions(GLL,LVL).   If truss plates assembled with slender
lumber are used for a large span,then some problems will
be created such as increased deflection and creep displacement due to the use of many connecting joints. To
cope with this problem, normally a large span is divided
into smaller spans by supporting with several intermediate
points.   However, as long as a large space structure is
requested, columns cannot be used for supporting from underside.   Therefore, by converting the original thought,

Figure 2.EXPERIMENT:ROOF PLATE
CONTINUOUS PYRAMID SKELETON
UNITES

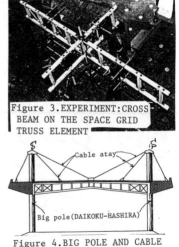

Figure 3.EXPERIMENT:CROSS-
BEAM ON THE SPACE GRID
TRUSS ELEMENT

Figure 4.BIG POLE AND CABLE
STAYED ROOF STRUCTURE

the author introduced a pulling structural system instead
of supporting (compression).

In the case of suspension system, a rope is lowered
from an upper point such as node S and tied to a particular
point on the roof plate for suspending it.   The main column is extended above the roof and said nodal point S is
located at the head of the column.   Thus, the length of
the main column is 9 to 10 m or more for a one-story building and the diameter must be 50 cm or more in the case of
log timber.   The author called it "Big-pole(DAIKOKU HA-
SHIRA) and Cable-stayed roof structure". Refer to Fig.4

For a group of buildings reported hear, Japanese Cedar
logs were used for the main columns with the length of 7 m
to 8 m and diameter of about 80 cm at the bottom end.

Unit and standardization of the structural system

"Suspension structural unit" and its Structural
system: Frame for suspending a space truss plate with
the length of side(L x L')by one big-pole was used as basic
unit(Fig.5).   If four basic units are combined together
as shown in the Fig.6, a self-standing structure can be
formed.   This is called the "Suspension Structural unit".

If the suspension structural units are continued in a
A Plane form in X and Y directions, it is possible to form
architectural buildings having a space with the scale corresponding to the respective uses.   According to this system, a large-scale structure can be formed by repeating
many times the basic units(a combination of one main column

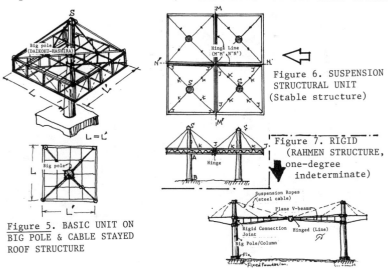

Figure 6. SUSPENSION
STRUCTURAL UNIT
(Stable structure)

Figure 7. RIGID
(RAHMEN STRUCTURE,
one-degree
indeterminate)

Figure 5. BASIC UNIT ON
BIG POLE & CABLE STAYED
ROOF STRUCTURE

and one roof plate).   If this basic unit is one part, the
mass production of large-scale wooden buildings can be
made possible by standardizing the columns and roof plates
as part elements.
        Conditions for realizing "Suspension Structural unit"
: The suspension structural unit(basic unit x 4)forms an
independent frame structure.   To resist with the horizontal loading (such as seismic force), basically a pin-hinge
joint is considered at the center of a beam as shown in
Fig.7, and it can be designed as a one-span rigid frame
structure with column bases fixed.   The vertical loads
(Dead Load and Snow Load)can be handled by suspending
the roof by the suspension ropes.   Therefore, the requirements for making this structural system valid are ,
        a) fixing column base, b) Rigid connection between
                               column and beams
        c) tensile force at suspention rope.
Also, the suspension rope is combined with said rigid-frame
structure and forms a hybrid structure (truss and rigid
frame structure/Rahmen structure).

Big pole and cable stayed roof structure(Design and construction of TOEI-Chō Forestry Center)

        Outline of the facilities of the forestry center:
Historical background. For soundly foresting an artificial
forest planted in the 1950's the thinning is indispensable.
        Since the age of trees is young, the diameter of effec-
tive cross section of thinning lumber is small(d=10 to 15cm

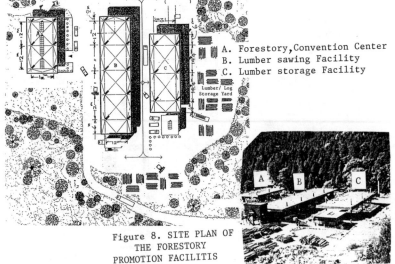

A. Forestory,Convention Center
B. Lumber sawing Facility
C. Lumber storage Facility

Lumber/ Log
Storage Yard

Figure 8. SITE PLAN OF
THE FORESTORY
PROMOTION FACILITIS

and also the effective length of members is limited to 3 to 4 m.   Therefore, the slender and thinning lumber is now used not much for structural members.

The TOEI Forestry Promotion Facilities were constructed as one of the measures for solving these problems and also for activating the forestry in Japan.   Because of this, its requirements were to build a wooden building by utilizing slender and thinning lumber produced locally for both the structural and finishing members.

Outline of the facilities.   These center facilities comprise,

(A) Forestry convention center(Assembly and Display)
(B) Lumber sawing facility
(C) Lumber storage facility

Layout and scale of each facility are shown in Fig.8.

Since it is possible to give flexibility to the floor plan design of buildings, these buildings in the group were all constricted by using the big-pole and cable roof system which is the continuation of "suspension structural unit".   Also, a large working space is required at (C)building/Lumber storage facility, so that the column interval was extended to 14.4 m.   In doing this, the the depth of beam was not increased and, instead, the column height of 3 m was adopted for the projected portion of column above roof level.   Moreover, in (C) , the column leg and column to beam connection were improved based on the experiences and experiment analysis data of(A) and (B) . Refer to Fig. 9.

Three buildings (A) TO (C)complise the assembly of suspension structural units of 1.5 unit,2.5 units and 1.5

Figure 9   LUMBER STORAGE FACILITY   (c)

modified unit respectively.   In this way, the concept
of the continuation of structural unit of above mentioned
was realized.

Structural design and the detail for the each portion

Configuration of roof truss plate (an irregular grid
plate) and suspension structure.   A structure in which
the main column passes through the center of roof plate
(L x L') is the " basic unit".   In the final design in
Fig. 10, L = 12 m, L' = 9 Or 12 m were used, it is"irregular basic unit".   Truss beam as the frame-work for the
grid plate is of Virendeel type, and the depth of beam is
100 cm at the end portion A and 90 cm at the connecting
portion at the center/point J .   Virendeel beem has no
diagonal brace , and the connecting joint between truss
chord member and strut member is simple, and the stress
conditions are simple.   However, it is necessary the rigidly connect the chord member to strut member, so that
the connecting system "DCS" was used joint which was previously reported[6].   For this beam, a cross section of
10 x 15 cm was used as chord member and 10 x10 cm was used
for strut member, kind of tree is Japanese cedar (locally
produced).   Grid interval (interval of V-beam) of roof
truss plate was set to L/4 = 3 m as standard.
A large bending moment act to the end portion of V-
beam during long-term/short-term loading.   Therefor,
steel rod braces were placed for reinforcement in ladder
frames at the end portion.   According to experiments[4],

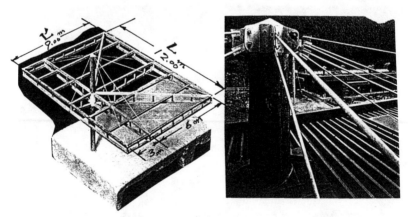

Figure 10   MODIFIED BASIC   Photo.2 DETAIL OF
            UNIT (Model)   THE BIG COLUMN HEAD

a moment resistance of almost twice larger can be expected
    Eight V- beams fixed to the main column in radial
form are respectively supported at two points: J(J'),it's
tips of one unit truss beam,and are intermediate points,
and 16 points in total are suspended by steel rope(in this
case a steel rod, 20 mm in diameter).   Other ends of
these ropes are all gathered at the column head.   After
lower wooden framing is assembled,tension is introduced
to the ropes while measuring the horizontal deflection on
roof surface.   Detail of connecting joints at the column
head is shown in Photo. 2.
    Detail of main big column and fixing on the column
leg and foundation.   An axial force N_T act to the column
head S during vertical loading and an axial force N_B act
to the connecting point A to beam during horizontal loading.   Therefore, the sum of N_T and N_B act to the column
leg as axial force.   Here, N_T is a component of the tensile force in the suspension rope, and N_B is equivalent
to shear force at beam end.
    For the main columns, it was possible to use the SUGI
/Japanese cedar logs 50 to 60 cm in diameter produced locally.   Since there no experimenting data for logs.
    One of the important points in structural design was
the method of construction for fixing the column leg to
concrete foundation.   The column leg fixing method is a
combination of two kinds of connection methods for handling two external foces: bending moment M_B and horizontal shear force Q_B acting to the column leg, by using respective mechanisms,Fig.11(a).   Firstly, traditional "attraction bolt technique" was used for M_B.   In this case,

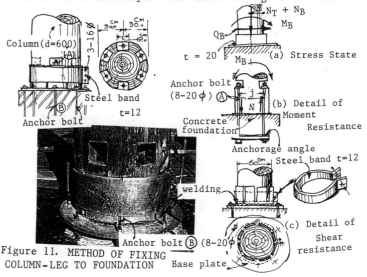

Figure 11. METHOD OF FIXING
COLUMN-LEG TO FOUNDATION

the tip of an anchor bolt A embedded in advance in a concrete foundation is anchored to the column leg by the
" tightening bolt".   Strength is determined by the shear
yield strength in the axial direction of lumber;in other
words, it is determined by the tighting length (ℓ) of
bolt. Refer to Fig.11(b).   Secondly, according to the
method to resist with Q_B, slits are made at equal intervals longitudinally in a short cylinder,and one end of
the cylinder is welded to a base plate (t= 20 cm), and leg
portion of the column is inserted into this cylinder.
Moreover, its outer periphery is tightened by a steel
band.   The base plate is embedded in foundation concrète.
by bolt B in advance, Fig.11(c).   This method prevents
the horizontal movement of column leg and improves the
initial rigidity for fixing column leg. Refer to Fig.11(d).

Structural design and construction of beam to column
connection joints.   As a method of connecting radially
the plane truss beams to a log/cylindrical column from 8
directions on a horizontal plane, the author developed the
following method: the log column is wound by steel ring
bands.   Finny plates in 8 directions are welded to this
ring.   And end portions of the upper and lower chord of
truss beam are attached to the finny plates and tightened
with bolts.

The point in design is to enhance the percentage of
rigid connection of beam to column joints.   Some of the
features of this connecting method is,(i) to divide the
steel ring into two pieces as shown in Fig.12(a),and the
column is sandwiched by the two half rings from both side
and tightened with bolts.   In addition, rag-bolts are

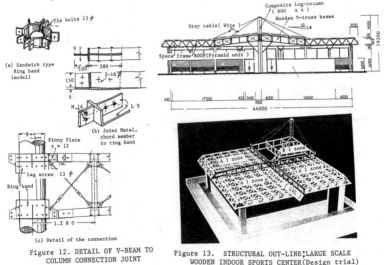

(a) Sandwich type
Ring band
(model)

(b) Joint Metal,
chord member
to ring band

(c) Detail of the connection

Figure 12. DETAIL OF V-BEAM TO
COLUMN CONNECTION JOINT

Figure 13.   STRUCTURAL OUT-LINE;LARGE SCALE
WOODEN INDOOR SPORTS CENTER(Design trial)

screwed in from the out side of the band for firmly adher-
ing the steel band to the column. (ii) Chord member of
V - beam is made of wood. In order to connect its end
portion to the finny plate of steel ring, detail of connecting parts as shown in Fig.12(b) are attached to the
end portion of the upper and lower chord members, and these
metal parts are connected to thefinny plates with bolts.
Though the chord member to finny plate joints are pin/hinge
connections, these upper and lower joints work in one united body and transmit to end moment of V -beam as "couple
of moment" of the axial force of the upper and lower chord
members, so that it is possible to form a moment resistant
connection having sufficient performance. Refer to Fig.12。

Design trial for large-scale wooden indoor sports center

        Outline of structural design: This building was de-
signed by in creasing the scale of "Basic Unit" of big -
pole and cable stayed roof structure(formed by big - pole
and one grid roof plate). And its periphery is surrounded by the structural type of space truss plate A1 stated
in befor section.
        In the structural design, the building has a square
44 x44 meter area. To cope with the vertical load, the
main column as cable stayed roof. This main column is a
sole supporting structure within the building area with
the exception of surrounding pillar. A space truss A1
having ⊔-shape with the width of 11 m at the periphery
respectively supports the inner edge by the outer frame in
A2 area and supports the outer edge by columns in a row.
With respect to the horizontal load such as seismic force,
the load is resisted by exterior wall and reinforced con-
creat buttresses. Since the rigidity on horizontal plane
of the space truss roof is high,sufficient resisting performance against earthquake. Refer to Fig. 13 .

1) P.L.Nervi, AESTHETICS AND TECHNOLOGY IN BUILDING,
   Harverd University Press, 1965
2) J.A.Throntou, The Design and Construction of Cable -
   Stayed Roofs, The Structural Engineer, p.275/284,Vol.
   62A, No.9. 1984
3) K. Sadakata, Study and Evaluation of Asismatic Performance of Neoconventional Wooden Frame Structures,
   Proc.of 1988 ITEC.P.73/81,1988,USA.
4) K.Sadakata, Experimental Study on The Practical use of
   Space Truss Structure Utilized Small Diameter Lumbers,
   Proc. of 1991 ITEC London, P.2.519/2.526, 1991

Concrete Shells Today

by Heinz Isler[1]

Abstract

Modern shells using experimental forms provide high
structural qualities. Their limitless repertory will
surely attract new attention of architects and clients.

With the invention of concrete it became possible to
reconsider the art of dome construction. The blocks of
stone were replaced by poured rock. Three new construction characteristics were created as a result:

1. Instead of having to set together a series of small
   building elements it became possible to pour large
   areas of monolithic sheets.

2. Reinforced concrete can withstand tensile forces.
   As a result other shell forms became possible besides the normal compression domes.

3. As a result of the genuine ability to withstand membrane forces (compression, tension, shear) it became
   possible to take on the various loadings in the same
   thin shell membrane. It was no longer necessary to
   rely on the bending strength of built up compression
   domes. Concrete shells could become "thin."

Many researchers such as Dischinger and Finsterwalder
prepared much theoretical basic information for the development of this new method of construction and thereby
opened up the field to the new building forms of that
time. The building methods were economic and came to be
used in many countries.

----
[1] Dipl. Ing. ETH/SIA, Prof. Dr. h.c., Ingenieur- und
    Studienbüro, Postfach, CH-3400 Burgdorf, Switzerland

1. Storage-Hall Walenstadt

2. Car Repair Nesslau

However, it is interesting to note that, in spite of
widespread use of shells, the repertory of form was extremely limited. Apart from cylinder sections, sphere
sections, a few cones and conoids or rotational forms very
few other shapes were used. A very fortunate extension to
these were provided by Felix Candela: he specialized in
hyperbolic paraboloids and developed his forms to perfection.

The principal characteristic of all these shells built
before 1960 was their geometric form. They all resulted
from our well known geometric-mathematical formulas.

However the strength of shell construction is to be
found in non-geometric forms. It is complicated and ardu-
ous to devise non-geometric forms analytically. There is
however another route which can lead to the creation of
new forms in an efficient way: the physical analogue or
experiment.

The "soap bubble" membrane created in a condition of
onesided higher pressure can lead to examples of forms
which are unique. This membrane possesses the feature
that in every direction and at every point the tensile
force is the same. A concrete shell using the same shape
has similar forces, and by inversion one can easily
change this from pure tension to pure compression.

The wonderful feature about this is that this basic
principle is valid for an unlimited number of shell
shapes, namely for any chosen variation or section into
which one can fit the "soap bubble" skin.

The examples shown in pictures 1 - 6 of concrete
shells created by the author represent some of the many
shells built since 1954 using this principle.

An even more interesting method of form finding is
the experiment with the "hanging membrane." This discovery by the author, quite by chance in 1955, has proven
to be extremely efficient. The "hanging membrane" principle can lead to statically very well balanced shells. The
point supports and the free edges can result in extremely
elegant and exciting shapes.

However in using such creative methods resulting in
new shapes it is necessary to take into account the danger of buckling. This tendency is not present in the experiments using hanging membranes as this membrane is
formed by tensile forces. The process of reversal which
also reverses the forces to compression tends to create
the buckling phenomenon. There are however steps which
can be taken to ensure buckling stability.

3. Factory Wangs

4. Production-Hall Madiswil

5. Shell Elements 20 x 22 m

6. Metal Treatement Egerkingen

Pictures 7 - 19 show examples of some of the projects constructed by the author using shells created on the basis of hanging membranes.

The two form finding methods discussed in this paper combined partly with prestressing show categories of concrete shells in which no tensile forces are present. This is the most important factor of all. As a result of this one can achieve, not only theoretically but practically, shells which do not crack and which are watertight, and which do not have any of the faults which are common in present-day concrete building. Buildings which are more than thirty years old are still problem free. Surface erosion of less than 2 mm points to a long maintenance-free life of these concrete shells.

These two characteristics, namely, the long-lasting maintenance-free life of the concrete shell together with the enormous choice of shapes available to the designer should ensure that modern concrete shell construction will be granted more attention in the future.

Pictures 20 - 30 show buildings with shapes found by other experimental methods (flow form, stretched membranes, combinations, etc.).

References

H. Isler: "New Shapes for Shells - Twenty Years after" Symposium 20 Anniversary of IASS 1979 in Madrid Vol. 71/72

H. Isler: "Third Decade of Structural Shells" Symposium 30 Anniversary of IASS 1989 in Madrid, Vol 1.

H. Isler: "Generating Shell Shapes by Physical Experiments". IASS Bulletin Vol. 34 (1993) n.1

H. Isler: "Longterm Behaviour of Shells" Seiken IASS Symposium Tokyo, Japan Oct. 1993

Ekkehard Ramm + Eberhard Schunk "Heinz Isler Shells" Karl Krämer Publications Stuttgart 1986 "Catalogue to a mobile exhibition".

All structural designs by the author.

7. Open Air Theatre Groetzingen

8. Inside View of Theatre

9. Musical Hall Stetten

10. Inside Musical Hall

11. Sport Center Norwich

12. Tennishall Crissier

13. Sportshalls Solothurn

14. Scaffolding

15. Gazoline Station Deitingen

16. Prestressing

17. Factory
    Geneva

18. Night View
of Winter Garden

19. North Facade

20. Airoplane Museum Dübendorf

21. Inside Museum

22. Shopping Center Bellinzona

23. Garden Center Ville Parisis

24. Garden Center Camorino

25. Inside Garden Center

26. Church
   Lomiswil

27. Monument
   Pully Lake
   of Geneva

28.
Garden Center
Ville du Bois
Paris

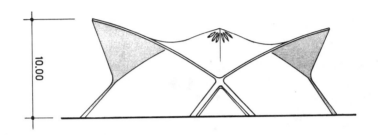

10.00

29. View of Villa

30. Villa in Geneva

NONLINEAR BEHAVIOR OF NORTHLIGHT REINFORCED CONCRETE SHELLS

P.Srinivasa Rao[1] and S.Duraiswamy[2]

ABSTRACT

The behavior of northlight shells is very sensitive to the boundary conditions at the top and bottom edges of the shell. The aim of this paper is to investigate the structural action of these shells at elastic, inelastic, cracking and ultimate load ranges with realistic boundary conditions at the top and bottom edges. The influence of construction method on the design and also the effect of concrete cracking, tension stiffening, nonlinear stress strain relations in compression, yielding of steel reinforcement and large deformations on the behavior of the shell are discussed. The full range behavior of a special 'S' type profile to suit the single shell situation and its superiority over the conventional profile are discussed. A brief description of the analytical model adopted in the Nonlinear finite element software used in this investigation is also presented.

1.0 INTRODUCTION

Northlight shells are widely used as roofing for industrial sheds. The cross-section of the shell is asymmetric and its center of gravity and shear center are not located on the same vertical line as shown in Fig.1. As a result, the shell is loaded by a twisting moment unless the adjacent shells are connected by the mullions of the glazing (window) frame before the application of the load (Ramaswamy 1986). Such unconnected shells act as "single shells" with very large tensile stress resultants and bending and twisting moments produced at the top edge of the shell. These require special care in the design of reinforcement. On the otherhand, if the mullions are provided before the application of load on the shell, then, a counteracting couple will be formed from the forces in the mullions thereby giving a better stability to the shell.

Whether the mullions are provided before or after the application of load on the shell depends on the construction method adopted(Ramaswamy 1986). Use of movable forms demands the mullions to be provided only after the deformations due to dead weight of shell have already taken place, forcing the shell to behave as a single shell at least for self weight. Hence, a clear understanding of the behavior of the

---

[1] Professor and [2] Research scholar, Structural Engineering laboratory, Indian Institute of Technology, Madras, 600036, India.

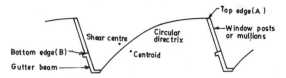

Figure. 1 Conventional northlight shell

shell under various situations is essential for correct design and detailing of the reinforcement.

Analytical solutions of this type of structures are difficult and time consuming because of the asymmetric shape and the interactions between top and bottom edges. Various design procedures (Mast 1962), considered the shell to have continuity over the mullions and performed elastic analysis by applying suitable boundary conditions to take care of this effect. P.W. Abeles et.al.(1963) discussed the effect of the mullions on the behavior of the shell, based on a model test. However, this work gives only a qualitative idea about the significance of this parameter. Elsawaf (1976) adopted finite element analysis with a special technique for analyzing an interior shell, by way of imposing constraint equations and performed elastic analysis of the shell. Srinivas (1989) and P.S. Rao et.al. (1988) have made a detailed investigation on the elastic behavior of Northlight shells. Michailescu et.al.(1960) proposed a special torsion free profile to suit the single shell situation. No analytical studies are available about the nonlinear - geometric and / or material - behavior of such shells.

The present paper gives the details of a recent analytical study (Duraiswamy 1993) carried out on the behavior of
  i) a 'Single shell' (without mullion support) with a circular directrix
 ii) a Single shell with a special 'S' shaped directrix and
iii) an 'intermediate shell ' (with mullion supports) with a circular directrix.

The entire range starting from the initial elastic stage going through the cracked stage upto the ultimate load stage has been studied using a Nonlinear finite element analysis program. Brief particulars of the nonlinear FEM are also presented.

2.0 ANALYTICAL MODELING

2.1 Finite element and material modeling

A general Nonlinear - material as well as geometric - Finite element analysis program has been developed for the analysis of reinforced and prestressed concrete shells with edge beams and validated by applying to a number of problems collected from literature for which experimental results are available (Duraiswamy 1993). For the purpose of consideration of material nonlinearity a Layered 9 noded Lagrangian shell element has been used for modeling the shell and a filamented linear beam element for modeling the eccentric edge stiffeners. Geometric nonlinearity was considered by a Total Lagrangian approach. The through thickness layering helped to consider the progressive cracking due to flexure and to model steel as a smeared out layer.

Each layer of concrete was assumed to be in a biaxial stress state. An Equivalent uniaxial strain approach proposed by Darwin et.al. (1977) has been adopted. The cracking of concrete was considered via a maximum tensile strength criterion. Tension stiffening, nonlinear stress - strain law in compression and aggregate interlocking have been duly accounted for. The steel was considered to be elasto - plastic.

An incremental iterative type of solution procedure has been adopted with a displacement control approach.

2.2 Analysis of intermediate shells

Rigorously accurate modeling of a multiple bay Northlight shell system in finite element method requires the discretisation of all shells and the mullions using shell, beam and truss elements (Srinivas 1989). But this approach is very costly and time consuming especially when performing nonlinear analysis. The stiffness of the mullions is very large in their own plane and hence it is realistic to assume that the displacements along the window plane at top edge of the shell (A) and bottom edge (B) are same (Fig.1). Using this condition, constraint equations have been developed and incorporated in the finite element solution following Abel(1979). This made the isolation of an interior shell possible and made the analysis simple.

3.0 RESULTS OF LINEAR ELASTIC ANALYSIS

According to normal design practice, the design of reinforcement for shells is generally based on stress resultants obtained in elastic analysis only. Hence for the purpose of studying the elastic behavior of these shells and to validate the finite element modeling of northlight shells including the procedure of imposing constraint equations, the following study has been made.

3.1 Linear elastic analysis of a single and intermediate shell of circular directrix

Using the program developed, a 30 m span northlight shell has been analyzed for single as well as intermediate shell conditions. The dimensions of the shell analyzed are shown in Fig.2 The deflections and stress resultants obtained using the program are compared with that of the results obtained with SAPIV package program for these two possible action of the shell. In the SAPIV analysis, three shells along with mullions have been analyzed together as a single unit with an assembly of shell elements and truss elements. Considering symmetry of the structure, only one half of the structure was taken for analysis in all cases. The loading considered was (1.0 D.L +1.0 L.L) with a L.L intensity of 1.10 kN/m^2 on shell area. The Young's modulus of concrete was assumed as 31000 N/mm^2.

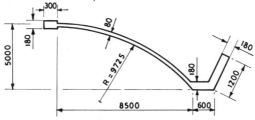

Fig.2 Details of 30m span shell

3.2 Discussion of results

The comparisons of the results for the single and intermediate shells are shown in Fig.3 to Fig.6. The results for the intermediate shell analyzed by applying constraint equations compare well with the results of SAPIV analysis taking all the units together.

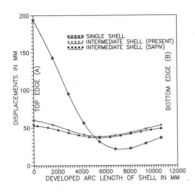

Fig.3 Vertical displacements

Fig.4 Longitudinal stress

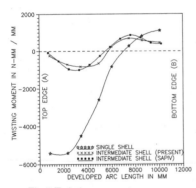

Fig.5 Twisting moment

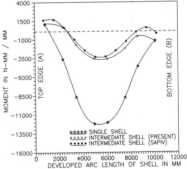

Fig.6 Transverse moment

The top edge of the single shell deflects vertically much more than the bottom edge. As a result the cross section of the shell gets rotated. Also, the longitudinal stress resultant, twisting moment and bending moment at the top edge are very large for a single shell compared to that for the intermediate shell. Significant differences in stress resultants are there at the other locations of the shell also. Fig.6 shows that the transverse moment acting on the single shell is several times more than that acting on the intermediate shell which is crucial in the design of transverse steel. As a whole, the single shell behaves unfavorably in comparison to an intermediate shell and it is evident that a shell designed, by mistake or otherwise, as an intermediate shell cannot satisfy the safety and serviceability requirements when acting as a single shell.

4. RESULTS OF NON – LINEAR ANALYSIS

4.1 General

A single and intermediate shell of conventional circular directrix and a shell of special 'S' curve directrix with single shell condition have been chosen such that all shells have same plan dimensions (Fig.7 and Fig.8) with a span of 20m and same intensity of insulation and live loads.

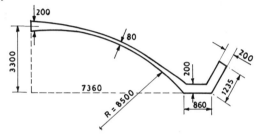

Fig.7 Details of 20m span shell

All the three shells were designed individually based on elastic analysis and they were analyzed upto collapse using the nonlinear analysis program described earlier. In the design of longitudinal steel, the allowable stresses in the steel bars were taken as proportional to their distance from the neutral axis. The load deformation behavior, cracking and steel yielding characteristics and also the failure mode and load factor obtained for each shell were studied.

The data for the shell design were as given below:
D.L = self weight + Insulation,   Insulation = 2.0 kN/m^2 on shell area
L.L = 0.75 kN/m^2 on shell area,   M 25 concrete and Fe 415 grade steel.

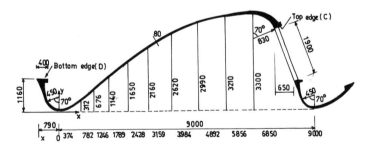

Fig.8 Details of 'S' curve shell

4.2 Circular directrix shells

In the analysis only one half of the shell was modeled, taking into account of symmetry, with 25 shell elements each having 6 concrete layers through the thickness. The valley beam was modeled with 20 nos. of eccentric beam elements with 100 filaments.The analysis was carried out by incrementing the vertical displacement at the topmost node (a dominant displacement component) till the load deformation path reached a horizontal plateau or a major displacement component reached a very large value.

The plots of load versus vertical deflection at the top of the shell for single and intermediate shell are shown in Fig.9 and Fig.10.

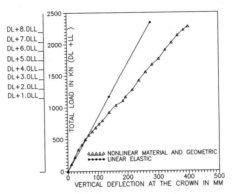

Fig.9 Single shell-Circular directrix

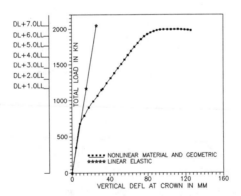

Fig.10 Intermediate shell-Circular directrix

Although both the shells sustained ultimate loads of same order, the single shell top edge underwent very large displacements. At the service load stage itself (1.0 D.L. +1.0 L.L.) the top edge of the single shell exhibited a span to deflection ratio of around 100 (20000 /195) which is far beyond the allowable limits. Hence it is very difficult to satisfy deflection serviceability in the case of single shells with circular directrix. Also since the top edge deflects much more in relation to the bottom edge, the cross section is subjected to large rotations (Fig.11).

The crack patterns obtained near ultimate load for both the shells are shown in Fig.12 and Fig.13. In **single shell** the initiation of cracking started, at a relatively

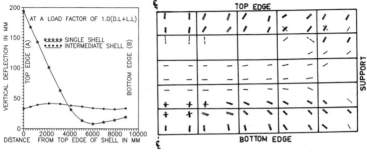

Fig.11 Vertical deflections   Fig.12 Crack pattern-Single shell

low load level of 0.35 (D.L. + L.L.) itself at the top edge of the shell due to longitudinal stress. Subsequently the cracked zone extended to a wider region. Cracking due to transverse bending moment started at a load level of 0.55 (D.L. +L.L.). At 0.675(D.L.+L.L.) shear cracks were initiated near the support. At about 0.75 (D.L. + L.L.) the bottom portion of the shell developed cracking. The yielding of the longitudinal steel reinforcement at top edge started at (1.0 D.L. + 6.0 L.L.) and extended all over the tension region, at top and bottom edge of shell, with a small further increment of load. The analysis was stopped at a load factor of (1.0 D.L. + 8.606 L.L.), since the deflection has already undergone a very large value and the longitudinal steel at midspan yielded completely warranting no further increase in load carrying capacity.

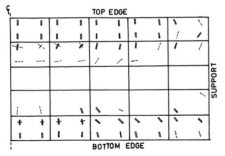

Notations :

— Cracked at top

--- Cracked at bottom

=Cracked through thickness

Fig.13 Crack Pattern-Intermediate shell

The deviation in load deformation paths, between linear elastic analysis and Non-linear analysis, that was observed even before cracking (Fig.9) seems to be due to the effect of geometric nonlinearity which caused a stiffening effect.

The **intermediate shell** deflects almost uniformly across the cross section avoiding large rotation of the shell (Fig.11). The deflections at any stage of loading are much less compared to that of a single shell and satisfaction of deflection criterion of serviceability is not a problem. Cracking started only at a fairly large load level of 0.6 (D.L. + L.L.) and is distributed at top as well as bottom edge of the shell and got stabilized rapidly. This implies a better distribution of stresses in all parts of the shell. The first yielding of the longitudinal steel started at top and bottom edges of the shell, at a load level of (1.0 D.L. + 5.5 L.L.). Upon further increase in load, more and more steel started yielding and the load deformation path reached a plateau at (1.0 D.L. + 7.0 L.L.) and the analysis was stopped.

In both single as well as intermediate shells, after the initiation of cracking , the nonlinear load deformation path deviates too far from that of the linear elastic load deformation path. At service load stage itself (1.0D.L+1.0L.L), the deflection obtained by nonlinear analysis is significantly larger than the elastic deflection.

4.3 'S' curve directrix shell

The cross section of the shell as shown in Fig.8 consists of a cycloid connected to its ends by two circular arcs. The speciality of this cross section is that the shear center and center of gravity fall almost on the same vertical line and hence the shell is free of torsional moment even when acting singly (Michailescu 1960).

The shell was discretised with 35 shell elements in one half and 6 layers through the thickness of each element. It was analyzed upto collapse assuming it to act as a single shell. The vertical deflection at top edge of the shell(C) is plotted against the load factor in Fig.14. The magnitude of deflections lie in between single and

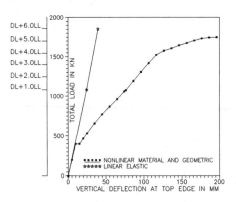

Fig.14 'S' Curve directrix shell

intermediate circular shell deflections but much smaller when compared to that of single circular shell. The rotation of the cross section also is very small (Fig.15) compared to single circular shell rotations. The load carrying capacity of this shell was almost like that of intermediate circular directrix shell.

The crack pattern developed near ultimate stage is shown in Fig.16.The first

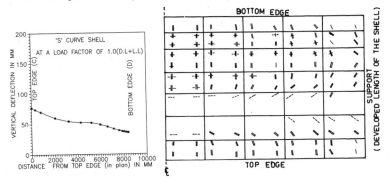

Fig.15 Vertical deflections   Fig.16 Crack pattern-'S' Curve shell

cracking load was 0.375 (D.L. + L.L.). The crack distribution and its progress were similar to that of an intermediate circular shell. Since the magnitude of transverse moment was small as in the case of an intermediate shell, the extent of longitudinal cracking was less. The first yielding of longitudinal steel started at a load level of (1.0 D.L. + 4.0 L.L) and yielding progressed upon further loading. After (1.0 D.L. + 5.5 L.L.), the load deformation path has reached a horizontal plateau and hence the analysis was stopped. Overall it can be stated that although the shell was loaded in a single shell situation it behaved almost similar to an intermediate shell of circular directrix.

5. CONCLUSIONS

1. The closeness of the results of the analysis of intermediate shell with constraint equations and with shells plus mullions taken together justifies the assumptions made in imposing the constraints.

2. The behavior of a single circular shell, without any mullion stiffness, in the elastic as well as non-linear range, is much inferior compared to the behavior of an intermediate shell with respect to deformation as well as stress state.

3. Consideration of the reduction in stiffness due to cracking of concrete is essential while computing the deflections and in checking for the satisfaction of deflection serviceability criterion for the Northlight shells. The deflection obtained through non-linear analysis even at working load stage was larger than the elastic analysis deflections by as much as 100% for an intermediate shell and by 70% for a single shell.

4. Single shell of circular directrix requires very heavy reinforcing in the longitudinal as well as transverse directions compared to intermediate shell to attain

same ultimate load carrying capacities.

5. Single shell of circular directrix though designed to carry same ultimate load as that of intermediate shell, deflects too much at top edge, especially after cracking, making it almost impossible to satisfy the serviceability criteria.

6. Extensive longitudinal cracks develop in the case of a single shell of circular directrix, due to a very large transverse bending moment, in comparison to intermediate shell and 'S' curve shell.

7. The behavior of 'S' type directrix shell even when acting as a single shell is comparable to that of an intermediate shell with circular directrix at all stages of loading. Both safety and serviceability requirements can easily be satisfied by adopting this type of shell. Hence when movable form works are to be employed, it is preferable to use this type of shell.

8. Even when northlight circular directrix shells are constructed with mullions, it seems desirable to make adequate provision in design for the possible action of the shells like single shell, because collapse of a portion of a shell or mullions make the shell to behave as single and may cause total collapse of the structure.

APPENDIX I. REFERENCES

1. Abel, J.F. and Shephard, M.S.(1979). "An algorithm for multipoint constraints in finite element analysis ",Int.J.Numer.Meth. in Engr.,14(3),464-467.

2. Abeles, P.W. and Lucas, J.(1963)." The Plastic behavior of a Prestressed concrete shell with and without antitwist adjustments (Based on a model Test) ", in proc. IASS symposium, Warsaw on Non-classical shell problems, pp.1084-1100.

3. Darwin, D. and Pecknold, A.(1977)."Nonlinear Biaxial stress-strain law for concrete ", J.Engrg.Mech.Div.,ASCE 103(2),229-241.

4. Duraiswamy, S.(1993)."Nonlinear behavior of concrete northlight shells", Ph.D Thesis (under preparation), Indian Institute of Technology, Madras, India.

5. Elsawaf, A.(1976). "An application of the semiloof shell and beam elements" in IASS World congress on space Enclosures, Building Research center, Concordia university, Montreal, July, pp.647-653.

6. Mast, P. E.(1962)."Design and construction of Northlight barrel shells", Journal of the American Concrete Institute, April, pp.481-523.

7. Michailescu, M. and Ungureanu, I.(1960)."A new shell form for prestressed sheds", IASS bulletin, no.3.

8. Ramaswamy, G.S.(1986)."Design and Construction of concrete shell roofs", CBS publishers and distributors, Delhi, India, pp.200-220.

9. Rao, P.S., Aravindan, P.K. and Srinivas, S.(1988). " Effect of Mullion stiffness on the behavior of Northlight cylindrical shells", in Proc. IASS symposium on Innovative applications of shells and spatial forms, Vol.1, Oxford and IBH publishing co. pvt. Ltd., pp.471-482.

10. Srinivas, N.S.(1989)."Effect of mullion stiffness and Prestress on the behavior of Northlight cylindrical shells ", M.S. Thesis, I.I.T, Madras, India.

Study on Reinforced Concrete Cylindrical Shells
with Edge Members under Impact Loading

Kazuhiko Mashita[1], Hidehiro Yoshitake[2],
Masaya Noguchi[2], and Masatoshi Hotta[2]

Abstract

The strength, cracking patterns and the other nonlinear dynamic behaviors of reinforced concrete roof-
type circular cylindrical shells influenced by edge
member stiffness and supporting conditions under impact
loading are investigated. Nonlinear dynamic finite
element procedures are applied to these study. The
impact loading is dynamic concentrated one, that
is applied to the center of shell surface. The results
of this study are compared to ones under static concentrated loading, which were obtained by Authors'
recent experiments. The interactions between the shell
and the edge members are investigated.

Introduction

This paper treats the nonlinear dynamic behaviors
of reinforced concrete roof-type cylindrical shells
influenced by edge members and supporting conditions
under impact loading. The effects of different stiffness of edge members, which consist of edge arches and
edge beams, are treated in this paper. Two different
types of supporting conditions are also considered. One
is a corner-type, which is supported at four corners,
and the other is a cantilever-type, which is sup-

----
[1]Professor of structural mechanics, Tokai University,
1117, Kitakaname, Hiratsuka, Kanagawa, 259-12, Japan
[2]Graduate student of engineering, Tokai University,
1117, Kitakaname, Hiratsuka, Kanagawa, 259-12, Japan

ported at two center points of edge beams. Each supporting point is roller-supported. The loading condition is impact loading at the center of shell surface,
which is dynamic concentrated loading. The results under
the dynamic concentrated loading are compared with
Authors' recent theoretical and experimental results
under static concentrated loading (Ref.3).

Dynamic analysis

Nonlinear transient dynamic finite element analyses
including concrete cracking and tension stiffening
effects are applied in this paper to investigate the
nonlinear behaviors of R/C cylindrical shells under
impact loading. In the discretization of shell models,
twenty-node hexahedral isoparametric finite elements are
applied to shells and edge members. In the idealization of R/C roof-type shells, the steel reinforcement is
incorporated in the concrete brick element by assuming
perfect bond. The reduced integration techniques, which
are effective to moderately thin shells, are adopted to
evaluate the stiffness of this element.

For evaluating the nonlinear behaviors of concrete
(Refs.1 and 2), a Drucker-Prager type yield function is
adopted and the compressive behaviors of concrete are
calculated using yield and strength limit surfaces. The
tensile behaviors of concrete are evaluated by considering the effects of tension cut-off and tension stiffening. A cracked shear modulus G_c is adopted to account
for shear stiffness in a smeared cracking model. The
reinforcing bars are considered as steel layers of
equivalent thickness and a bilinear idealization is
adopted to model the elasto-plastic stress-strain relationships.

6 small scale reinforced micro-concrete models of
R/C cylindrical shells are investigated in the this
study (Fig.1 and Tables 1, 2 and 3). The form and dimensions of shells are adopted taking into account of the
concrete shell models which were used in Authors' recent
experiments under static concentrated loading (Ref.3).
These models have the same 1120 mm (=L_1) x 1120 mm (=L_3)
square plan. The rise to chord-width ratio is 1/5. The
width of edge beams and edge arches is 40 mm. The
names of mathematical models under dynamic concentrated loading are D0R, DAR, DBR, D0C, DAC and DBC.
D0R, DAR and DBR are roller-supported at four corners,
while the others are roller-supported at two center
points of edge beams. The depth of edge members is 20 mm
or 40 mm. Each actual shell thickness is listed in

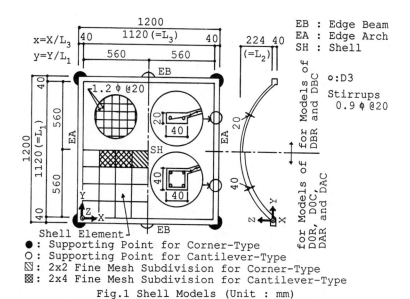

Fig.1 Shell Models (Unit : mm)

Table 1 Shell Models

Dynamic Analysis Model	Static Analysis Model	Static Test Specimen	Actual Thickness (mm)	Support Type	Material Type (See, Table 2)
DOR	SOR	TOR	10.3	Corner	MA
DAR	SAR	TAR	10.3	Corner	MB
DBR	SBR	TBR	9.3	Corner	MB
DOC	SOC	TOC	11.5	Cantilever	MA
DAC	SAC	TAC	8.9	Cantilever	MC
DBC	SBC	TBC	9.1	Cantilever	MC

Table 2 Material Constants of Concrete

Type	Young's Modulus ($\times 10^4$ MPa)	Poisson's Ratio	Ultimate Compressive Strength (MPa)	Ultimate Tensile Strength (MPa)
MA	2.02	0.193	55.7	2.9
MB	1.96	0.193	59.6	3.1
MC	2.23	0.175	53.1	3.4

Table 3 Material Constants of Steel Bar

Steel Bar Diameter (mm)	Young's Modulus ($\times 10^5$ MPa)	Yield Strength (MPa)	Ultimate Strength (MPa)
0.85	1.86	162	250
1.2	2.35	364	468
D3	1.98	334	487

Table 1, while the design shell thickness is 8 mm ($=T_0$).

The finite element idealizations are as follows. The cylindrical shell surface is discretized by two steps. At the first step, the surface is discretized by using 8x8 mesh of equal length in the horizontal plane, and at the second step the elements applied with the concentrated load are subdivided with the discretization of 2 x 2 mesh. Both in the corner-type shell and in the cantilever-type shell, only one quarter of it is considered in the analysis because of the double symmetries of it. The subdivisions of the edge members correspond with that of shell surface. The shell was reinforced with 1.2 mm diameter mild steel wires placed at 20 mm centers both ways. Edge members such as edge arches and edge beams were reinforced with four 3.0 mm diameter mild steel deformed bars placed in two layers. 0.85 mm diameter mild steel stirrups were placed at 20 mm centers.

The loading condition is dynamic concentrated loading in the vertical direction. Corner-type shells are loaded by one-point dynamic concentrated load acting on the center point (x= 0.5, y = 0.5) of whole shell surface. Each character x and y of loaded point (x,y) shows the coordinates, where L_1 is a chord width and L_3 is a shell length. Cantilever-type shells are loaded by two-point dynamic concentrated loads, which are symmetrically applied to two center points (x = 0.25, y = 0.5) and (x = 0.75, y = 0.5) of two half shells, where the symmetrical axis is a circular directrix connected with each center point of two edge beams. The intensity of dynamic loading calculated in this study is from 0.1 kN to 10 kN.

Static Analysis

Static nonlinear finite element analyses including concrete cracking and tension stiffening effects are

used to describe the nonlinear behaviors of R/C shells
under static concentrated loading. The static experimental study was conducted on reinforced micro-concrete
shell specimens(Ref.3). Those were loaded to failure
with static concentrated loading. The static load is
applied to the same position as the impact load.

Investigations

    Results of our study are shown in Tables 4, 5, 6, 7
and 8 and Figs. 2 and 3. In these tables, load intensi-
ties not only for the corner-type shells but for the
cantilever-type ones are shown for one concentrically
loaded point.

    Responses at the loading point   under impact load-
ing 0.1 kN are shown in Fig.2. In this figure, the
applied load is from 0.0 to 2.0 msec., where the maximum
value is 0.1 kN at 1.0 msec., that is a triangular load.
Figs.2(a) and (b) show vertical displacements   for
corner-type and for cantilever-type, respectively.

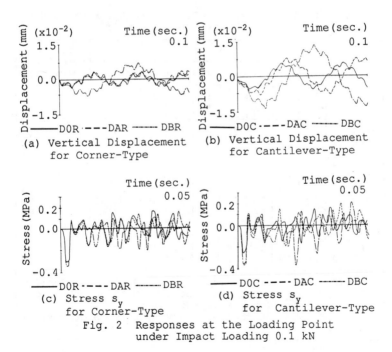

(a) Vertical Displacement
    for Corner-Type

(b) Vertical Displacement
    for Cantilever-Type

(c) Stress s_y
    for Corner-Type

(d) Stress s_y
    for Cantilever-Type

Fig. 2   Responses at the Loading Point
         under Impact Loading 0.1 kN

Figs.2(c) and (d) show stresses in the y-direction
for corner-type and for cantilever-type, respectively.

Ratios of load levels to the ultimate load of TOR (P_0
= 2.06 kN) under static loading are listed in Table 4.
In this table, ratios at the first crack load and at
the ultimate load for experimental and numerical models
are listed. In Tables 5, 6, 7 and 8 under dynamic loading, a Heaviside function is used as a time function
under dynamic loading. A time step length is set to be
0.1 msec. and dynamic responses are calculated from 0.0
msec. to 10 msec. The first crack under dynamic loading
for corner-type and for cantilever-type are listed in
Tables 5 and 6, respectively. Time until ultimate load
under dynamic loading from 0.1 to 10 msec. are listed in
Table 7. Ratio of the first cracking load under dynamic
loading to that for experimental specimen under static
one is listed in Table 8.

Fig.3 shows the cracking patterns under impact
loading, the intensity of which is 10 kN, that is an

Table 4   Ratio of Load Levels to the Ultimate Load of
TOR (P_0 = 2.06 kN)   under Static loading

Type	The First Crack Load Ratio (percent)										Ultimate Load Ratio (percent)
	Shell		Edge Arch				Edge Beam				
	Ts *1	Bs *1	Ta *2	Ba *2	Oa *2	Ia *2	Tb *3	Bb *3	Ob *3	Ib *3	
TOR	95	76	–	29	29	62	–	–	–	–	100
TAR	43	62	–	19	19	19	–	–	–	–	90
TBR	33	24	33	14	14	19	–	–	–	–	33
TOC	43	33	–	29	29	29	33	–	–	–	43
TAC	14	24	10	5	5	5	14	24	24	29	29
TBC	14	10	–	5	5	5	–	–	–	–	14
SOR	71	29	95	33	33	43	95	86	95	95	95
SAR	57	24	–	24	24	29	–	–	–	–	57
SBR	14	14	14	10	10	14	14	14	14	14	17
SOC	48	24	48	19	19	29	38	48	38	48	52
SAC	24	14	–	14	14	14	19	24	14	24	24
SBC	10	10	10	5	5	5	10	10	10	10	10

* 1 T(B)s   : On the Top (Bottom) Surface of Shell,
* 2 T(B,O,I)a : On the Top (Bottom, Outside or Inside)
                 Surface of Edge Arch,
* 3 T(B,O,I)b : On the Top (Bottom, Outside or Inside)
                 Surface of Edge Beam.

example for the highest load intensity in this paper. In this figure, the left-hand side and the right one show the top surface and the bottom one, respectively, while the upper side and the lower side show the edge member and the shell surface, respectively.

Under dynamic loading, different positions of crack occurrences are observed in comparison of the positions at the first crack under the lower load intensity with those under the higher ones. Although the positions of crack occurrences under the lower intensity of dynamic concentrated loading are rather similar to those under static loading, the cracks under the higher load intensity occur at the center of loaded position concentrically. Under impact loading 10 kN, the cracks occur on the whole shell surface, especially on the surface of DBR and DAC, which are a corner-type with small edge arch stiffness and a cantilever-type with

Table 5   The First Crack under Impact Loading
            for   Corner-Type(0.1 - 10 msec.)

Load	Shell Model								
	DOR			DAR			DBR		
(kN)	Shell	Beam	Arch	Shell	Beam	Arch	Shell	Beam	Arch
0.5	9.7 B	–	9.1 B	9.6 B	–	9.6 B	–	–	–
0.6	9.2 TB	9.3 TBI	8.7 B	9.2 B	–	9.0 B	3.8 B	–	–
0.7	3.3 B	9.0 TB	8.2 B	3.3 B	9.4 TBOI	8.3 B	3.3 B	–	9.5 B
0.8	3.2 B	8.6 TBOI	7.9 B	3.2 B	8.8 TBOI	7.9 B	1.7 B	5.2 TOI	4.1 B
0.9	0.6 B	4.1 TBO	4.1 TBI	1.9 B	4.4 TBOI	4.4 TBOI	0.6 B	4.0 T	3.8 B
1.0	0.6 B	3.5 TBOI	3.5 TBOI	0.6 B	3.4 T	3.4 TO	0.6 B	2.7 T	2.7 TBOI
5.0	0.2 TB	0.4 TBOI	0.4 TBOI	0.3 TB	0.5 TBOI	0.5 TBOI	0.2 TB	0.4 TO	0.4 TBOI
10.0	0.2 TB	0.3 T	0.3 T	0.2 TB	0.3 O	0.4 TBOI	0.2 TB	0.3 T	0.3 T

(Note) The upper number shows time   (unit:msecond), when the first crack occurs, and the lower character shows cracking surface, where T,B,O and I show Top, Bottom, Outside and Inside surface, respectively.

small edge beam one, respectively.

Conclusions

Within the scope of this study, the following conclusions are obtained. The nonlinear dynamic behaviors of reinforced concrete shells with different edge beam stiffness under impact loading are affected by the intensity of an applied load. Under the higher intensity of impact loading, the first cracks occur mainly on the shell surface, while under the lower intensity of impact loading the first crack occurs on the edge members as well as under static loading. The interactions between the shell and edge members under impact loading are greatly influenced by the supporting conditions.

Table 6 The First Crack under Impact Loading
for Cantilever-Type(0.1 - 10 msec.)

Load	Shell Model								
	DOC			DAC			DBC		
(kN)	Shell	Beam	Arch	Shell	Beam	Arch	Shell	Beam	Arch
0.3	-	-	-	-	-	-	-	-	10.0 B
0.4	-	-	9.6 B	10.0 B	-	-	9.9 TB	10.0 TBOI	9.1 B
0.5	8.6 B	9.8 TBOI	8.0 B	9.5 B	-	8.2 B	3.7 B	9.3 TBOI	4.0 B
0.6	8.2 B	8.9 TI	7.7 B	3.5 B	8.4 TBOI	5.9 B	3.5 B	4.3 TBOI	3.8 B
0.7	3.7 B	7.8 TBI	6.9 B	3.3 B	6.2 TO	5.7 B	2.2 B	4.1 TBOI	3.4 B
0.8	3.0 B	7.7 TI	5.6 B	0.6 B	4.0 TBOI	3.9 B	1.6 B	3.8 TBOI	3.2 B
0.9	2.9 B	3.9 T	3.9 TBOI	0.6 B	3.8 TBOI	3.8 TBOI	0.6 B	2.6 TO	2.6 TBOI
1.0	1.6 B	3.3 T	3.3 TBOI	0.5 B	2.0 T	2.0 TBOI	0.6 B	2.4 TBOI	2.4 TBOI
5.0	0.2 TB	0.4 TBOI	0.4 TBOI	0.2 TB	0.4 TBOI	0.3 T	0.2 TB	0.4 TBOI	0.3 TO
10.0	0.2 TB	0.3 T	0.3 TBI	0.2 TB	0.3 TOI	0.3 TBOI	0.2 TB	0.3 TO	0.3 TBOI

(Note) The upper numbers and the lower characters show time and position, which are the same as those shown in Table 5.

References

1. Hatano, T., "Dynamic behavior of concrete under impulsive tensile load," Central Research Institute of Electric Power Industry, C-6002, Tokyo, 1960.
2. Kupfer, H. B., Hilsdolf, H. K. and Rush, H., "Behavior of concrete under biaxial stresses," Proc. of ACI, Vol.66, No.8, Aug., 1969, pp. 656-666.
3. Mashita, K., "Ultimate strength analysis of R/C roof-type cylindrical shells with boundary members under concentrated loading," Proc. of the SEIKEN-IASS, Tokyo, Japan, Oct., 1993, pp. 331-338.

Table 7   Time until Ultimate Load under Impact Loading from 0.1 msec. to 10 msec.

Load	Time (msec.)					
(kN)	DOR	DAR	DBR	DOC	DAC	DBC
0.5	–	–	–	–	–	9.4
0.6	9.8	–	–	9.2	8.5	4.5
0.7	9.3	9.9	–	8.2	6.5	4.3
0.8	9.1	8.9	5.5	7.9	4.3	3.9
0.9	4.4	4.6	4.3	4.5	4.1	2.9
1.0	3.8	3.7	3.0	3.6	2.3	2.6
5.0	0.7	0.6	0.5	0.8	0.6	0.5
10.0	0.6	0.6	0.5	0.6	0.5	0.5

Table 8   Ratio of the First Cracking Load under Dynamic Loading to That for a Experimental Specimen under Static One.

Model	Ratio of the First Crack-ing Load	Time (ms)	Position			
			Dynamic Loading		Static loading	
			TBOI (*1)	SAB (*2)	TBOI (*1)	SAB (*2)
DOR	0.84	9.1	B	A	B:O	A:A
DAR	1.28	9.6	B:B	S:A	B:B:O	S:A:A
DBR	2.08	3.8	B	S	B:O	A:A
DOC	0.67	9.6	B	A	B:O:I	A:A:A
DAC	3.88	9.9	B	S	B:O:I	A:A:A
DBC	2.91	10.0	B	A	B:O:I	A:A:A

(*1) TBOI : T(Top Surface), B(Bottom Surface), O(Outside Surface), or I(Inside Surface)
(*2) SAB : S(Shell), A(Edge Arch) or B(Edge Beam)

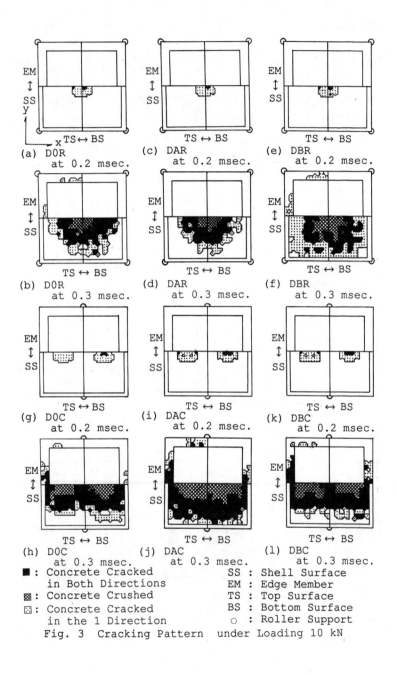

(a) DOR
at 0.2 msec.

(c) DAR
at 0.2 msec.

(e) DBR
at 0.2 msec.

(b) DOR
at 0.3 msec.

(d) DAR
at 0.3 msec.

(f) DBR
at 0.3 msec.

(g) DOC
at 0.2 msec.

(i) DAC
at 0.2 msec.

(k) DBC
at 0.2 msec.

(h) DOC
at 0.3 msec.

(j) DAC
at 0.3 msec.

(l) DBC
at 0.3 msec.

■ : Concrete Cracked
　　in Both Directions
▨ : Concrete Crushed
▥ : Concrete Cracked
　　in the 1 Direction

SS : Shell Surface
EM : Edge Member
TS : Top Surface
BS : Bottom Surface
○ : Roller Support

Fig. 3　Cracking Pattern under Loading 10 kN

NONLINEAR ANALYSIS OF A REINFORCED CONCRETE FOLDED PLATE STRUCTURE

Pattabhi Sitaram,[1] Stuart Swartz,[2] F. ASCE, and
Chidambaram Channakeshava[3]

ABSTRACT

This paper deals with the ultimate load analysis of a reinforced concrete (RC) folded plate shell subjected to dead load and monotonically increasing live loads, statically applied, using a nonlinear layered finite element technique. The nonlinear effects of concrete considered are cracking, tension stiffening, strain softening and aggregate interlock. A bilinear strain hardening model is used for steel reinforcement. Degenerate quadratic isoparametric 8-noded shell elements are employed using a layered discretization through the thickness. In order to obtain results which are independent of the choice of the finite element mesh, the tension softening behavior of the concrete is related to the fracture energy of concrete. A RC folded plate shell (inverted-U type) is analyzed and compared with available experimental results. The influence of fracture parameters on ultimate load and ductility are presented.

INTRODUCTION

In order to evaluate the current design procedures for RC structures, there is a growing demand for the development of suitable finite element (FE) models. One of the important goals of these models is to predict accurately, the load-deflection response up through the ultimate load. As

[1]Visiting Assistant Professor, Department of Civil Engineering, University of Missouri, Columbia, MO 65211
[2]Head, Department of Civil Engineering, Kansas State University, Manhattan, KS 66502
[3]Consulting Structural Engineer, #19, Kumara Krupa, Bangalore 560001, India

a result many material models have been applied to predict the response of RC structures (ASCE 1981; Scordelis 1990).

RC folded plates, Figure 1, should be designed to satisfy criteria in terms of serviceability and safety (strength). In most cases the design is based on linear elastic analysis assuming the materials to be uncracked, homogeneous, isotropic and linearly elastic.

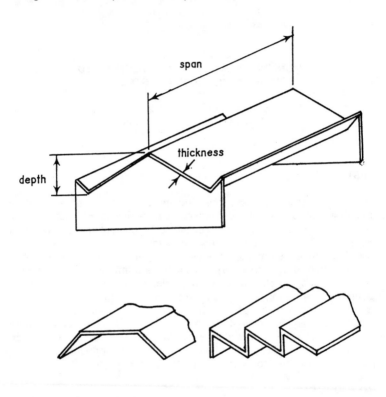

Figure 1. Typical Folded Plate Configurations

Ngo and Scordelis (1967) were the first to apply the FE method to RC structures by studying RC beams. The FE analysis based on layered approach has been used by many researchers (Hand et. al. 1972; Lin and Scordelis 1975; Barzegar 1988). However, none of these researchers used fracture energy concepts in modeling the post-peak tensile behavior of concrete. These concepts are very important since the results should be independent of the finite element mesh chosen, and it is known that using these concepts achieves this goal (Bazant 1981).

In recent years, increased interest has been manifested by civil engineers in the behavior of RC folded plate structures when stressed beyond the linear elastic range. Very recently Ahmad (1990) predicted the response of RC folded plate structures in the service level and inelastic ranges using a finite difference technique.

MATERIAL MODELING

Reinforced concrete is a composite material consisting of steel and concrete. The yield criterion employed by Figueiras and Owen (1984), Figure 2, is used. This criterion is in terms of two stress invariants I_1 and J_2 and involves two material parameters α and β

$$f(I_1, J_2) = \{\alpha I_1 + \beta(3J_2)\}^{1/2} = \sigma_o \tag{1}$$

where σ_o is equivalent effective stress taken as compressive strength from a uniaxial test. The material parameters α and β are obtained by fitting biaxial test results of Kupfer et. al. (1969) based on the assumption that the state of stress is not far from a biaxial one. These constants are $\alpha = 0.355\sigma_o$ and $\beta = 1.355$, and the yield function becomes

$$f(\sigma) = \{0.355\sigma_o(\sigma_x + \sigma_y) + 1.355[(\sigma_x^2 + \sigma_y^2 - \sigma_x\sigma_y) + 3(\tau_{xy}^2 + \tau_{xz}^2 + \tau_{yz}^2)]\}^{1/2} = \sigma_o \tag{2}$$

The associated flow rule is used to define the plastic strain increment, and the crushing criterion is given by replacing the stress invariants in the yield criterion by the strain invariants.

In tension, if the maximum principal stress in concrete reaches the maximum tensile strength f'_t of concrete, a crack is formed in a plane orthogonal to this stress, and concrete is no longer isotropic. The modulus of elasticity and Poisson's ratio are set to zero in a direction perpendicular to the crack plane and a reduced shear modulus is employed (Figueiras and Owen 1984). Because of bond effects, cracked concrete carries between cracks a certain amount of tensile force normal to the crack plane. This effect is known as 'tension stiffening'. Also concrete fracture is associated with the formation of microcracks involving a gradual decrease of tensile strength, and this phenomenon is known as 'tension softening'. This is shown in Figure 3 based on the model of (Figueiras and Owen 1987).

It is now widely accepted that to make the constitutive model objective with regard to the size of the finite elements used, the descending portion of the tensile stress-strain curve must be related to the fracture energy of concrete. The fracture energy is defined here as

$$G_f = H \, g_f = H\frac{1}{2}(\epsilon_m - \epsilon_t) \alpha f_t' \qquad\qquad (3)$$

where a and ϵ_m are tension softening parameters, ϵ_t the cracking strain, H the crack band width and g_f the shaded area shown in Figure 3. The crack band width is taken as the square root of the area associated with the integration point. Different values of a cause different shapes in the descending branch of the tensile stress-strain curve of concrete.

The reinforcing bars are considered as steel layers of equivalent thickness exhibiting a uniaxial response. The behavior is elasto-plastic with a bilinear idealization being adopted.

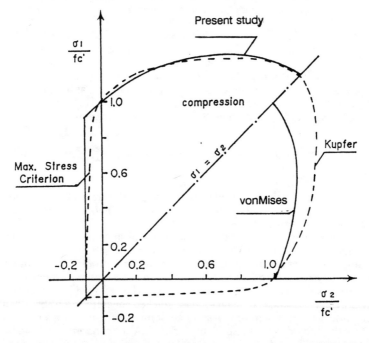

Figure 2. Stress-Strain Relationship for Concrete in Biaxial Stress Space (Figueiras and Owen 1984)

METHOD OF SOLUTION

An incremental-iterative technique is used for the nonlinear finite element analysis utilizing the tangent stiffness method. Both displacement and force convergence criteria are adopted.

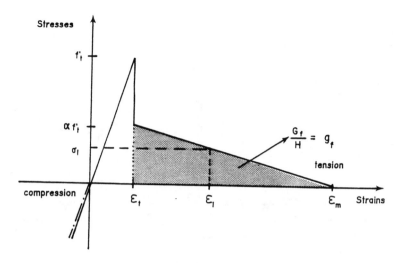

Figure 3. Behavior of Concrete in Tension (Figueiras and Owen 1984)

ANALYSIS OF A REINFORCED CONCRETE FOLDED PLATE STRUCTURE

The nonlinear FE method is applied to the inverted-U type RC folded plate structure shown in Figure 4, with transverse edges supported on end diaphragms and longitudinal edges free. The reinforcement pattern (longitudinal, transverse and inclined bars) and the experimental results, which were obtained by a simulated uniformly distributed load, are given in Pudhaphongsiriporon (1978). The material properties are specified in Table 1.

The analytical crack patterns for bottom surface, for one half of the structure, at load 105 psf and ultimate stage, and the corresponding cracks obtained by the experiment are shown in Figure 5. Cracking (FE) first occurred on the bottom surface at a load of 81 psf. These cracks were near the midspan section and oriented normal to the free edge. With further increase in load, these cracks propagated toward the fold, and also inclined cracks originated close to the quarter point of the free edge and near the end zones and propagated toward the diaphragm. The first set of analytical cracks occurred on the top surface on the top plate, at the midspan section, parallel and close to the fold (Sitaram 1993). As the load increased, these cracks propagated in an oblique manner toward the

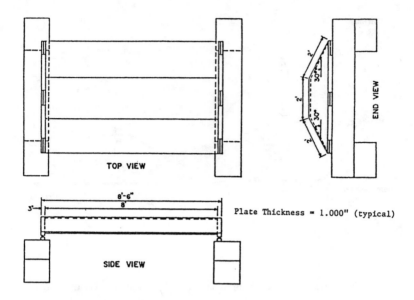

Figure 4. Principal Features of Pudhapongsiriporn's Models

Table 1. Material properties for the Pudhaphongsiriporn's model

CONCRETE	STEEL
Young's modulus, E_c = 3700 ksi	Young's modulus, E_s = 29200 ksi
Poisson's ratio, v = 0.17	Yield stress, f_y = 39.5 ksi
Comp. strength, f'_c = 4.95 ksi	Young's modulus, E'_s = 1000 ksi
Ult. comp. strain, ϵ_u = 0.0035	Ult. stress, f'_y = 53.1 ksi
Ten. softening parameter, a = 0.5	Threaded rod, area = 0.018 in^2
Fracture energy, G_f = 200 N/m	No. 2, dia. = 0.25, area = 0.049 in^2

end points of the structure. In all cases, the analytical crack patterns
agreed very well with those obtained experimentally.

Figure 6 includes the comparison between experiment and analytical
load-deflection responses at midspan section, free edge, for fracture
energy, G_f = 200 N/m and tension softening parameter, a = 0.5. The
analytical solution gave an ultimate load of 157 psf, and the experiment

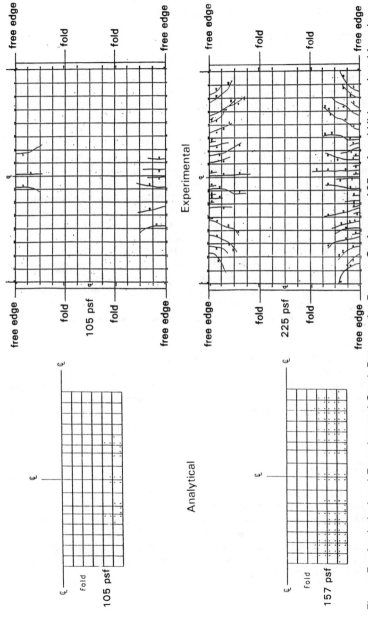

Figure 5. Analytical and Experimental Crack Patterns for Bottom Surface at 105 psf and Ultimate Load Levels

225 psf. Figure 6 contains the load-deflection response by the analytical solution up to one load level prior to the ultimate load.

PARAMETRIC STUDIES

The load-deflection response of reinforced concrete structures beyond the elastic limit is affected by several parameters. In the current study, the influences of fracture energy, G_f and the concrete tension softening curves on the behavior of reinforced concrete folded plate shell structures are investigated.

Influence of Fracture Energy

In the tensile modeling of concrete, to make the results mesh independent, the descending portion of the tensile stress-strain curve was related to the fracture energy (Fig. 3). The FE analyses were performed for fracture energies, G_f of 50, 100, and 200 N/m with a constant value of tension softening parameter, $a = 0.5$. Deflection results at midspan, free edge are shown in Figure 6. As G_f increases, the ultimate load capacity increases, although not in proportion to the change in fracture energy. As the fracture energy doubles from 50 to 100 N/m the ultimate load capacity increases by only 25%. When the fracture energy is again doubled [to 200 N/m] the ultimate load capacity increases by 26.9%. Also, increases in fracture energy make the load-deflection response stiffer, and it can be concluded that by using a high value of G_f the FE solution results would agree very well with the experimental results for the model.

Influence of Concrete Tensile Softening Curves

To study the influence of concrete tensile softening curves on the load-deflection response on the RC folded plate structure considered, three values of a (Fig. 3) equal to 0.5, 0.7 and 1.0 were used with the fracture energy kept constant at 200 N/m. Fig. 7 shows that as a decreases, the load-deflection response becomes softer, although the ultimate load capacity changes by only 2.5%.

CONCLUSIONS

The proposed FE method is capable of tracing the load-deflection response and propagation of cracks in a continuous manner up to ultimate load levels for monotonically increasing loads. The use of higher fracture energies increases the ultimate load capacity of RC folded plate structures, and also makes their responses stiffer. The shape of the descending portion of the tension softening curve of concrete considerably influences the ductility of the structure.

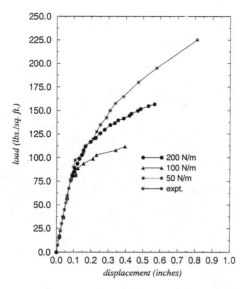

Figure 6. Influence of Fracture Energy, G_f, $a = 0.5$, Midspan Section, Free Edge

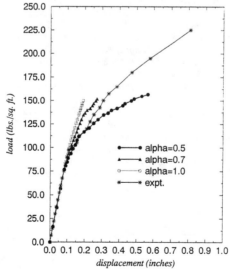

Figure 7. Influence of Tension Softening Parameter, a, $G_f = 200$ N/m.

CONVERSION FACTORS
1 in. = 25.4 mm
1 psf = 47.9 pa
1 ksi = 6890 kpa

REFERENCES

Ahmad, Y.N. (1990). "Analytical Study of a Long Span Reinforced Concrete Folded Plate." *Ph.D. Thesis*, Kansas State University.

ASCE Committee on Concrete and Masonry Structures. (1981). "Task Committee on Finite Element Analysis of Reinforced Concrete Structures: A State-of-the-Art Report on FE Analysis of Reinforced Concrete Structures," *ASCE Spec. Pub.*

Barzegar, F. (1988). "Layering of RC Membrane and Plate Elements in Nonlinear Analysis." *J. of Structural Engg.*, ASCE, 114, (11).

Bazant, Z.P. (1981). "Numerical Simulation of Progressive Fracture in Concrete Structures." *Proceedings of the International Conference on Computer-Aided Analysis and Design of Concrete Structures*, Part 1, Split, Yugoslavia, 1-17.

Bazant, Z. P., and Oh, B. H. (1983). "Crack Band Theory for Fracture of Concrete." *Materials and Structures (RILEM, Paris)*, 16, 731-742.

Figueiras, J. A., and Owen, D.R.J. (1984). "Nonlinear Analysis of Reinforced Concrete Shell Structures." *Proceedings of the International Conference on Computer-Aided Analysis and Design of Concrete Structures*, Part 1, Split, Yugoslavia, 509-532.

Hand, F. R., Pecknold, D. A., and Schnobrich, W. C. (1972). "A Layered Finite Element Nonlinear Analysis of Reinforced Concrete Plates and Shells." *Civil Engng. Studies*, SRS 389, University of Illinois, Urbana, IL.

Kupfer, H., Hilsdorf, K. H., and Rush, H. (1969). "Behavior of Concrete Under Biaxial Stresses, Proceedings." *ACI J*, 66, (8), 656-666.

Lin, C. S., and Scordelis, A. C. (1975). "Nonlinear Analysis of RC Shells of General Forms." *J. of Structural Engg.*, ASCE, 101, (3).

Ngo, D., and Scordelis, A.C. (1967)."Finite Element Analysis of Reinforced Concrete Beams." *ACI Journal,* 64 (3).

Pudhaphongsiriporn, P. (1978). "An Experimental Study of a Reinforced Concrete Folded Plate Structure." *Ph.D. Thesis*, Purdue University.

Scordelis, A.C. (1990). "Nonlinear Material, Goemetric and Time Dependent Analysis of Reinforced and Prestressed Concrete Shells." *Bulletin of the International Association for Shell and Spatial Structures*, 31-1, 2, (102,103),p.57-70.

Sitaram, P. (1993). "Ultimate Load Analysis of Reinforced Concrete Folded Plate Structures." *Ph.D. Thesis*, Kansas State University.

Spatial Structures Erected out of Thin
Shell Elements Fabricated on Plane
Followed by Bending

Vladimir V.Shugaev[1]

Abstract

    The paper offers the examples of spatial structures
of average spans (up to 24 m) and engineering structures
made of prefabricated curvilinear elements origionally
having a plane form.
    The plate bending is possible when the concrete is
mature and when it is coming fresh on a flexible   formwork immediately after manufacturing.
    Laboratory tests on models have proved high rigidi-
ty and strength of such structures.

Introduction

    A distinctive feature of development of reinforced
concrete spatial structures in Russia is that particular
attention was primarily paid to precast structures made
of prefabricated elements meant for wide-scale use   and
for unique large-span structures.
    Spatial structures are noted for the great diversi-
ty of surface configurations which, from the one   hand,
contributes to the development of expressive architectural forms and, from the other hand, makes   difficult
their sectioning into unified precast elements.   Under
these circumstances the configuration of structures and
their elements, provided the requirements to their bearing capacity are met, should be closely related to the

----
[1]Professor,Dr.Sc.,Research Institute for Concrete   and
Reinforced Concrete (NIIZhB) of Gosstroy R.F., 2nd In-
stitutskaja str. 6, 109428, Moscow, Russia

possibility of simplifying the production practice and
reducing the labour consumption during manufacturing
and erection of precast elements.
     One of the promising trends of development of thin-
-walled reinforced concrete spatial structures is their
manufacturing out of precast elements fabricated on plane followed by bending to give them a spatial form   of
the surface. Two main types of the said techniques   are
distinguished: bending of a plane matured concrete plate
and bending of a plane fresh reinforced concrete (ferro-
-cement, fibrous concrete) plate.
     Let us consider some examples of structures conta-
ining the elements manufactured with the use of the above two methods.

Spatial Structures of Elements
Manufactured by Bending of Plane
Matured Concrete Plates

     The Design Institute PI-1 (Saint Petersburg)   and
the NIIZhB (Moscow) have developed various alternatives
of shells made of flexible plates bent in erection stage.
These studies are based on the Italian experience   constructing water conveyance canals lined with thin reinforced concrete prestressed plates which are bent during
placing.
     In the alternative shown in Fig.1 the shell is   as-
sembled of 2.55 x 9 m plates 40 mm thick and   reinforced
concrete stiffening ribs of   -like section, 650 mm wide, with cylindrical outline and flanges to bear   the
plates on. Stiffening ribs are assembled into an   enlarged block with temporary tie rods. Such blocks are   placed on truss diaphragms. Plates are bent just when   they
are being placed on the stiffening ribs and bolted   to
them. Two adjacent plates in one block are welded in the
middle, butt straps being used in this case. Then   all
joints between plates are filled with concrete. Manufacturing of shell elements and experimental construction
have been effected in Saint Petersburg.
     Since the above mentioned plates are   manufactured
of ordinary reinforced concrete, only limited bending
from the plane was allowed, this bending depended on the
rated crack opening. Much greater bending from the plane
can be reached when flexible prestressed plates are used. The development of such plates and some structures
made of them in the NIIZhB was mainly oriented at   highly mechanized method of manufacturing plates on beds, the
prestressed reinforcement being placed with the help of
a wire winding machine.
     Among the most promising structures mention   should
be made of T-shaped thin-shell panels which have   under-

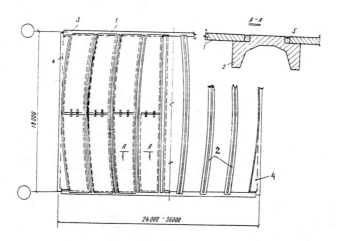

Fig.1 The shell consisting of plates bending in
      erection stage.

     1. matured concrete plate; 2. stiffening ribs;
     3. contour diafragm; 4. edge element-diafragm;
     5. cast-in-place concrete.

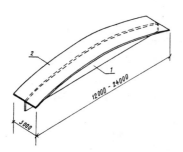

Fig.2 Composite thin-shell
T-shaped panel consisting
of flexible prestressed
plates.
1. rib of the panel;
2. bent plate as the panel
   flange.

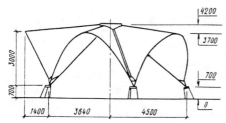

Fig.3 Precast multi-
wave ferrocement
shell for autopavi-
lion.

gone comprehensive experimental testing and have   been
used in practice / I /. The thin-shell panels are   assembled of plates one of which, being a vertical rib,
remains plane and the other one which serves as a flange gets bent in the process of assembly along the curve
following the outline of the upper rib face (Fig.2).
     The experience in manufacturing has proved that
due to prestressing it is easy to achieve noticeable
bending of plates without any cracks. After connection
of flanged plates and rib plates by welding with   the
use of embedded metals the shell plate starts   working
under load as a spatial system. The interconnection of
flanges of adjacent shell plates by welding with the
use of embedded metals and filling of joints with concrete also contribute to the spatial structure operation.
The tests of thin-shell panels have revealed their sufficient strength, crack resistance and rigidity.
     The above examples do not limit the achieved deve-
lopments and possibilities for designing of new structural forms whose main elements originate from plane thin
reinforced concrete plates bent in erection stage.
     The described trend is being developed in parallel
with another trend when plane plates are bent immediately after moulding.

Spatial Structures and Their Elements
Manufactured by Bending of Plane
Fresh Concrete Plates

     Three technological methods are distinguished   in
manufacturing curvilinear products by bending of plane
fresh reinforced concrete plates. These methods are in
many ways responsible for the shape of the spatial structure and its elements / 2 /:
     a) shell manufacturing by bending up an originally
plane plate with the help of an inflatable formwork,
where this plate was formed;
     b) manufacturing of curvilinear elements by wind-
ing of a fresh concrete plate on a rigid core;
     c) manufacturing of elements by free sagging or by
bending of plane fresh concrete plates on a flexible
formwork.
     In this paper we shall consider only the structu-
res manufactured with the used of "c" method since this
method has found wider use in construction practices.
     The process of bending fresh concrete plates is ba-
sed on the property of a fresh concrete layer placed on
a flexible formwork to undergo bending deformation under gravity without disturbing its integrity. Unlike
"a" method of shell manufacturing where cracks and concrete failure observed as a result of bending are   to

some extent can be eliminated by subsequent vibration,
in our case the lack of cracks is accounted for the
fact that over the whole concrete layer section in the
direction of bending the section reduction is of served. This reduction causes corresponding plastic flow
of concrete which grows from zero on the convex side of
the concrete layer to a certain maximum - on the concave side.

The studies made by the NIIZhB have shown that if
the concrete element is reinforced with wire cloth placed about the thickness of the element, in the process
of bending the element the wire cloth located closer to
the compressed concave surface gets displaced relative
to other layers. At high viscosity of concrete mix the
wire cloth cannot overcome its resistance and becomes
deformed.

Fabricating bent elements in case of free sag of
the flexible form with fresh concrete plate has receal-
ed that the deformation outward of the wire cloth and
concrete disintegration could be avoided should the wire cloth be placed at an angle of 30-45° to the bending
axis. Under this circumstance the wire cloth gets deformed in the process of bending due to the change in the
form of the wire cloth mesh.

In experimental specimens with indicated location
of the wire cloth no deformation outward of the wire
cloth and concrete disintegration were observed even at
the bending equaling to one half of the plate span value. In this case the plate was sagging along the curve
close to the funicular line. Similar results were obtained in fabricating bent concrete elements with disper-
sed reinforcement in the form of steel fiber. The results of studies have shown that this type of reinforcement is the best one for manufacturing thin-walled
structures by the bending method. Some new types of effective spatial structures of bent formed ferro-cement
and steel fibrous concrete elements have been elaborated on the basis of these studies. The designs of some
of them are given below.

The prefabricated ferro-cement shell (Fig.3) has
been designed for roofing pavilions though it can also
be used for commercial pavilions, small cafe, etc. The
shell is assembled of five curvilinear elements having
the form of conic shells. The curvilinear elements are
connected into a spatial structure by the upper monolithic ring and by welding embedded metals in the ribs.
The curvilinear elements of the shell 30 mm thick are
reinforced with double-layer wire cloth made of wire
1 mm in diameter, the mesh size being 8 x 8 mm,   and
with additional reinforcement in the form of separate
rods, wire meshs and cages.

Microconcrete of grade B30 is used for manufacturing the elements. The prefabricated curvilinear elements
are manufactured by bending of a plane fresh concrete
plate on a flexible formwork under gravity. In manufacturing the element the sides of the form are turned aro-
und the horizontal hinges of the form. As a result   of
this the flexible formwork becomes sagged and strained
in the process of bending. As soon as the concrete reaches the stripping strength the form sides are swung
out and the finished element is carried to the edge post
where it is turned into operation position.

The shell presented in Fig.4 is assembled out   of
numerous similar elements which do not need any edging
and are more gently sloping and alongated. The shell has
been designed for roofing the storage of loose materials.
The corrugated shell in the form of a truncated cone has
in plan the shape of a regular dodecagon, the base   diameter is 24 m and the height is 9.45 m. The shell is assembled of curvilinear elements 13.22 m long, from 1 to
6.2 m wide and the rise up to 0.8 m. They are placed into grooves of the upper and lower rings which are filled
with microconcrete. The curvilinear elements are interconnected by welding of embedded metals placed in   the
form of contour angles along the straight-line sides of
the curvilinear elements.
The curvilinear elements 30 mm thick are made   of
steel fibrous concrete with additional reinforcement by
welded steel mesh connected to the contour angles. The
shell elements are manufactured of microconcrete of grade B30.
It was proposed that the loose material should be
charged into the storage through the upper ring to which
a conveyer gallery comes, while the unloading should be
effected by materials-handling equipment through   the
opening in one of the shell corrugations.
The shell designs shown in Fig.3 and 4 have   been
tested till failure on large-scale reinforced   concrete
models. Curvilinear prefabricated elements of the   models were also made by the method of bending. Both structures have revealed high bearing capacity and have been
recommended for use.

An original design of a silo for grain storage
140 m in capacity has been worked out on the basis the
design given in Fig.4. It represents two similar truncated cones connected with each other by a large   base
(Fig.5). In the sections made by horizontal planes the
cone surface looks corrugated since it is assembled of
shell elements with convexity facing inside. Such design
favours static work of the silo walls sustaining the grain pressure and complete unloading of the silo without
any additional implements. The silo is assembled of 16

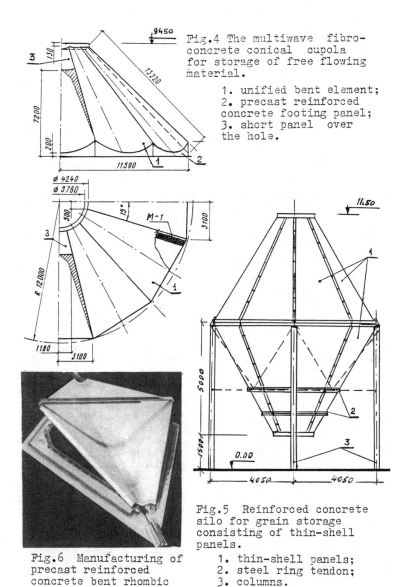

Fig.4 The multiwave fibro-concrete conical cupola for storage of free flowing material.

1. unified bent element;
2. precast reinforced concrete footing panel;
3. short panel over the hole.

Fig.5 Reinforced concrete silo for grain storage consisting of thin-shell panels.

1. thin-shell panels;
2. steel ring tendon;
3. columns.

Fig.6 Manufacturing of precast reinforced concrete bent rhombic element.

similar shell elements 35 mm thick reinforced with wire
cloth, the wire is 1 mm in diameter, the mesh size being 10 x 10 mm. The contour ribs 140 mm high have external angle reinforcement and a welded cage.

The lower truncated cone has steel circular tie-rods
meant to increase the rigidity and to perceive circular
forces. The silo is lifted a little over the ground surface and rests on four columns 300 x 300 mm in cross-section.

As proved by the analysis of the results of numerous studies aimed at dome sectioning into prefabricated
elements the minimum number of standard sizes was observed when the dome surface was dissected into triangular
elements. A set of two such elements forms a rhomb, their
combination also permits optimum dissection of the shell
surface into prefabricated elements.

Considerable advantage of the proposed technology
is in the fact that it allows using one form for manufacturing the elements with imparted various curvature in
cross direction. It offers the possibility of developing
the structures with various surface geometry using a limited number of forms (Fig.6).

Rhombic elements have been used in designing several dome-type buildings. One of them represents a steel-
-fibrous concrete shell with the circle diameter of 24 m,
the circle going through the axes of supports. The figure formed by the lines connecting the supports represents
a regular octagon.

The shell (Fig.7) is meant for roofing a market and
is designed for use in areas with seismicity magnitude
up to 9.

The distinctive feature of the above designs of
shells is that they are assembled practically without
scaffolding with the help of one central temporary erection pole to support assembly elements. The assembly
elements in the form of enlarged blocks are prefabricated out of four rhombic elements on the jig. The elements
are interconnected by welding embedded metals and casting in place concrete in spline joints. The whole shell
is assembled of curvilinear rhombic elements of three
types two of which are made in one formwork and differ
insignificantly in rise. The bearing element has increased thickness since it has to take up compressive and
bending forces.

One of the latest developments is a multipurpose
building whose design is given in Fig.8. The contractor
design have been elaborated for two types of buildings
with cells 12 x 30 and 18 x 30 m.

The structure is assembled from elements of four

Fig.7 Space roof struct-
ure for market consisting
of bent rhombic elements.

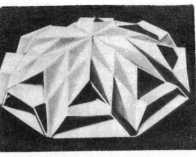

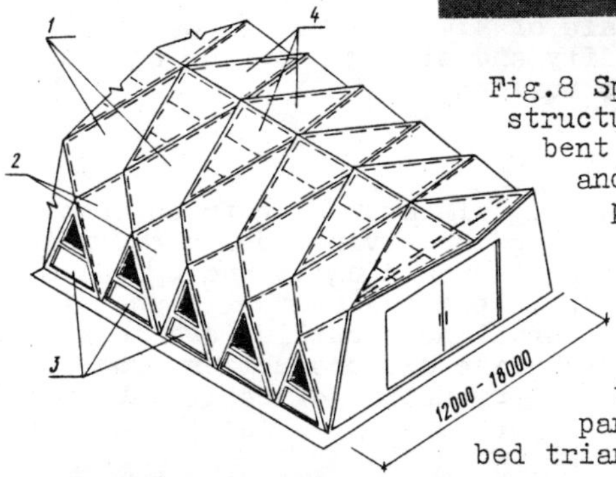

Fig.8 Space building
structure consisting of
bent shell elements
and flat ribbed
plates.
1. bent shell
roof elements;
2. bent shell
wall elements;
3. flat ribbed
triangular wall
panel; 4. flat rib-
bed triangular roof panel.

Fig.9 Test of the space building structure model.

types: two of them are plane triangular ribbed plates,
the other two are thin-walled bent elements ribbed along
the perimeter. The alternating plane and spatial elements
provide for unified structural solution of walls and roofing.
    The elements 20-25 mm thick have dispersed reinfor-
cement of two kinds: thin wire cloth or steel fiber. In
the plate ribs provision is made for combined reinforcement with application of welded cages having deformed
reinforcing steel bars. Microconcrete of grade B25   is
used for manufacturing prefabricated elements. The finite element method has been used in designing the building as a spatial system. The results of the design have
been confirmed by testing till failure on a reinforced
concrete model in scale of 1:4 (Fig.9). Thus, this test
has proved high rigidity and strength of the spatial
structure used in the building.

Conclusion

    The method of bending of matured or fresh concrete
plane flexible plates aimed at giving them a curvilinear form is one of the most promising technological methods which makes it possible to do away with use of
costly steel forms of curvilinear outline and to a great
extent simplify and mechanize all production processes
as well as to achieve the reduction of cost, labour consumption and to improve the quality of manufactured products.
    The experience in research, designing and construc-
tion has revealed that prefabricated unified bent elements can be successfully used in erection of structures
having expressive architectural forms, high economic efficiency and bearing capacity.   .

References

    1. Gambarov G.A., Ghitlevich M.B. Prefabricated roo-
fing panel out of thin prestressed plates, "Concrete and
Reinforced Concrete", 1981, No.7.
    2. Shugaev V.V. Spatial structures out of bent ele-
ments. Proceedings of NIIZhB "Studies of reinforced concrete thin-walled spatial structures", 1991.

Ultimate Strength of R/C Cooling Tower Shell

Takashi Hara [1], Shiro Kato [2] and Hiroshi Nakamura [3]

ABSTRACT

Reinforced concrete(R/C) cooling towers are the largest, thin shell structures. It is important to assess the ultimate strength of R/C cooling tower shells subjected to wind load. In the previous paper both the crack load and the ultimate strength of R/C cooling tower shell were analyzed focused on the effects of the reinforcing ratio considering geometric and material nonlinearities.

In this paper the influences of both the initial geometric imperfection and reinforcing ratio on the strength of the R/C cooling tower shells are analyzed. From the load displacement curves the initial crack strength and the ultimate strength are determined.

INTRODUCTION

Reinforced concrete (R/C) hyperbolic cooling towers are the largest, thin shell structures ever constructed. These towers stand more than 150m tall and have wall thickness of 0.20m-0.25m. Therefore these are thin shell structures. It is important to investigate these structures into the safety against the buckling under gravity load, wind load and earthquake load.

Most reinforced concrete cooling tower shells possess initial geometric imperfection arising from unavoidable constructional inaccuracy. These structures are also prone to cracking because these are the structures with a large surface area of concrete.

Following the collapse of cooling towers (Ferrybridge 1966, Ardeer 1974), numerous analytical studies on evaluating buckling behavior have been presented. Kemp(1976) investigated the collapse of the cooling tower at Ardeer and concluded that the stress distribution of the cooling towers and the reinforcement to be used were strongly influenced by the initial geometric imperfections. The results of

---

[1] Associate Professor, Dr., Tokuyama College of Technology. Kume-Takajo 3538 Tokuyama 745 Japan

[2] Professor, Dr., Toyohashi University of Technology, Hibarigaoka 1-1 Toyohashi 441 Japan

[3] Graduate Student M., Toyohashi University of Technology

these studies have led to the design recommendations such as IASS(1977,1979) and ACI-ASCE(1977,1984).

The finite element procedure enables us to analyze the nonlinear behavior of the reinforced concrete cooling tower shells. Mang(1983) focused on the the material and geometric nonlinearity of the reinforced concrete shells. He concluded that the ultimate strength of the cooling tower shells were not evaluated by the classical eigenvalue problems but must be defined by the nonlinear analysis considering the material nonlinearities such as the concrete cracks and the yielding of reinforcement. Milford(1984) studied the nonlinear behavior of the cooling tower shell focused on the stress redistribution after initial cracking and calculated the influence of the tension stiffening to the ultimate strength of the cooling tower shells. Meschk(1991) reported that cooling tower shells constructed in 1950s and 1960s show cracks caused by a temperature gradient and corrosion of steels due to concrete cracking. He calculated the ultimate strength considering aging effects. For these cooling towers which are failing in strength it is also important to present the reliable informations concerning the safety of the cracked reinforced concrete shell. Although the ultimate strength of the cooling towers have been studied by many researchers mentioned above many unsolved problems exist in this area.

In this paper the influences of both the initial geometric imperfection and the reinforcing ratio to the ultimate strength of the cooling tower shells are examined. In the numerical analysis the finite element procedure is adopted. To derive the finite element equations the geometric and material nonlinearities are taken into account based on the degenerated shell elements. The numerical model adopted herein is the Port Gibson tower which was already studied by many researchers (Mang 1983 and Milford 1984). To solve the nonlinear equations the displacement incremental procedure (Batoz 1979), which is useful for instability analysis, is adopted instead of the usual load incremental procedure.

NUMERICAL PROCEDURE

For the numerical analysis the finite element approach based on the isoparametric degenerated shell element is adopted. The geometrical nonlinearity is treated by using Green-Lagrange strain definition. Also so called 'layered approach' is employed here.

In the finite element analyses of reinforced concrete structures, to present the nonlinear behavior following assumptions are adopted (Hinton 1984, Hara 1993).

Concrete Behavior in Compression

Under a biaxial stress state, the yield function depends not only on the mean normal stress $I_1(\sigma_{oct})$ but also on the shear stress invariant $J_2(\tau_{oct})$. The yield condition of biaxial compressive concrete is expressed on the basis of Drucker Prager yield criterion.

$$f(I_1, J_2) = \sqrt{\beta(3J_2) + \alpha I_1} = \sigma_0 \qquad (1)$$

where parameters α and β depend on the material used. σ_0 is the equivalent stress as well. In this paper $\alpha = 0.355\sigma_0$ and $\beta = 1.355$ are adopted based on the experimental data by Kupfer(1969).

The normality of the plastic deformation rate vector to the yield surface is commonly assumed. For the hardening rule of the concrete the conventional Madrid Parabola is adopted. The crushing of the concrete is described as a strain control phenomenon and the condition is defined as like as a yield condition.

Concrete Behavior in Tension

It is assumed that the tensile crack of the concrete elements is occurred when the tensile principal stress exceeds the tensile ultimate strength. Before cracking the elements are considered to be isotropic elastic material. After cracking the assumption of smeared crack model is adopted to describe the element behaviors and in the analysis the orthotropic material behaviors are taken into account.

Fig.1 shows the stress strain relationship in tension. In this analysis so

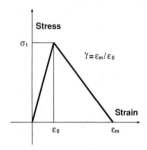

Figure 1: Stress strain relation of concrete in tension

called tension stiffening parameter is defined as follows.

$$\gamma = \frac{\epsilon_m}{\epsilon_0} \tag{2}$$

Material Properties of Reinforcing Bars

The reinforcing bars are considered as steel layers which have an equivalent thickness and the relation between stress and strain is assumed to be bilinear. Also the same relation is adopted for both compression and tension.

ANALYTICAL MODEL

Dimensions of the Model

To analyze the load displacement behavior and the ultimate strength of the reinforced concrete cooling towers subjected to gravity load and wind pressure Port Gibson cooling tower which was analyzed by many researchers (Mang 1983 and Milford 1984) is adopted herein.

Table 1: Dimensions of the shell

height	thickness	equivalent reinforcement		
z(m)	t(cm)	position ζ_t	hoop thickness ζ_{th}	meridional thickness ζ_{tm}
30.50	101.7	0.820	0.00666	0.00207
23.52	20.3	0.532	0.00351	0.00435
-7.26	20.3	0.532	0.00351	0.00971
-22.65	20.3	0.500	0.00351	0.01191
-38.04	20.3	0.529	0.00450	0.01099
-63.24	24.0	0.590	0.00380	0.00891
-88.44	28.7	0.678	0.00364	0.00736
-113.64	34.2	0.812	0.00514	0.00392
-120.00	76.2			

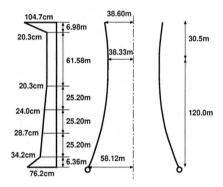

Figure 2: Dimensions of the shell

Fig.2 shows the geometric dimensions of the cooling tower shell. The configuration of the hyperbola is defined referring to Mang's analysis. Table 1 shows the shell thickness, the reinforcing position and the amount of the reinforcing at each shell height. The reinforcing position ζ_t is denoted as the ratio of the distance between reinforcing position and the shell mid surface to a half of shell thickness.

The amount of the reinforcing ζ_h and ζ_m is defined as the hoop reinforcing thickness and the meridional reinforcing thickness of inner or outer reinforcing layer, respectively. The amount of the reinforcement used in this model conforms not only to the IASS Recommendation (1977) but also to ASCE-ACI Committee 334(1977,1984). Table 2 shows the mechanical properties of the concrete and reinforcing bars used in this analysis.

Table 2: Material properties of the shell

CONCRETE		
Elastic Modulus	E_c	$2.827MN/cm^2$
Poisson's ratio	ν	0.175
Density	ρ	2.43
Compressive Strength	f_c	$3.447kN/cm^2$
Tensile Strength	f_t	$0.319kN/cm^2$

REINFORCEMENT		
Elastic Modulus	E_s	$20.59MN/cm^2$
Hardening Modulus	E_t	$1.03MN/cm^2$
Yield Strength	σ_y	$41.37kN/cm^2$
Ultimate Strength	σ_u	$62.05kN/cm^2$

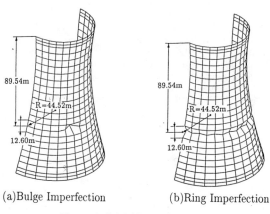

(a)Bulge Imperfection   (b)Ring Imperfection

Figure 3: Initial Imperfection

In the numerical analysis the shell is pin supported at the base and is free at the top. Also considering the symmetry of the loads and the geometric configuration the analyzed cooling tower is divided into two exact halves. Then a half of the shell is supported by the symmetric boundary condition which constrains the rotation around the meridional direction and has the free rotation around the hoop direction.

The shell is divided into 8 elements in the hoop direction and into 12 elements in the meridional direction. Therefore the shell has 96 elements. To define the element characteristics the layered approach is adopted. Each element has 8 concrete layers and 4 reinforcing steel layers.

The analyzed shell has two types of geometric initial imperfections shown in Fig.3. One is the bulge type of imperfection which has become the topic of the collapse of Ardeer tower (1974). The other is the ring type of imperfection which has discussed about the collapse of Fiddlers Ferry tower (1985). The position of the maximum amplitude of the imperfection is defined as the position where possesses the maximum defection of the elastic buckling patterns.

Wind Pressure
    The distribution patterns and the amplitude of the wind pressure are proposed by many researchers. In this paper to compare with the Mang's analyses(1983) the same wind pressure defined by him is adopted.

ULTIMATE STRENGTH OF R/C COOLING TOWER

    It is reported that the cooling towers constructed in 1950s and 1960s show cracks and corrosion of reinforcing bars caused by concrete cracking (Meschk 1991). It is important to analyze the influence of the reinforcing ratio as well as of the initial geometric imperfection to the ultimate strength of the cooling tower shell.
    To analyze the load deflection properties of the shell the response analyses are performed. In the analyses the reinforced concrete shells show the material nonlinearities such as concrete cracking and crushing. Therefore the response is calculated by use of the displacement incremental scheme (Batotz 1979). After gravity load is applied gradually wind pressure is loaded quasi statically.
    In the analysis load-displacement curves for shell structures with various reinforcing ratio as well as with initial geometric imperfection are examined.

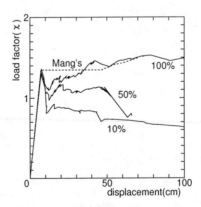

Figure 4: Load Displacement Relation($\gamma = 1$ Perfect Configuration)

Shell without Initial Imperfection

Fig.4 shows the load deflection curves of the shell The displacement denotes the horizontal component of the deformation at the throat on the windward meridian. In this model to evaluate the fundamental effects of the reinforcing ratio to the ultimate strength tension stiffening is ignored ($\gamma = 1$ see Eq.(2)). Also the model has perfect configuration. The parameter adopted herein is the reinforcing ratio. 100% denotes the designed reinforcement shown in Table.1.

The load deflection curves show the sudden drops when the cracks occur. In this figure the initial crack strength is few influenced by the reinforcing ratio. However the ultimate strength of the cooling tower after initial cracking is strongly influenced by the reinforcing ratio. The fewer the reinforcing ratio is the lower the ultimate strength is.

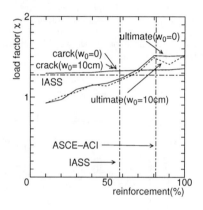

Figure 5: Crack load and ultimate strength ($\gamma = 1$ Bulge Imperfection)

Shell having Initial Imperfection

Fig.5 and 6 show the relation between the ultimate strength after cracking and the reinforcing ratio of the model. The abscissa and the ordinate shows the reinforcement ratio and the load factor respectively. In the figure $\chi=1$ means the designed wind load and the reinforcement ratio denotes the ratio of the amount of the steel used in the numerical analysis to the amount of the designed reinforcement.

In figures the vertical dotted dash lines denote the minimum reinforcing ratio which should be possessed in this model referring to the recommendations (IASS 1977, ASCE-ACI 1977,1984). The reinforcing ratio of the Port Gibson cooling tower in hoop direction is equal to the IASS recommendation while the reinforcing ratio in the meridional direction is about twice as much as recommended in IASS. In IASS(1977) and ACI-ASCE Committee 334 (1977, 1984) the shell reinforcement of this model in either direction should be not less than 0.25% and 0.35% of the concrete cross-sectional area, respectively.

In Fig.5 it is assumed that the model has the bulge type of initial geometric imperfection (see Fig.3(a)). The amplitude of imperfection is 10cm. IASS and ASCE-ACI Committee 334 recommends that the allowable imperfection amplitude should be less than 10cm and 5.4cm, respectively. In the figure the solid line denotes the results ignoring the initial geometric imperfection. The dotted line denotes the results considering the initial geometric imperfection. In the case of the initial crack load both results show the same tendencies. In the case of the ultimate load of the shell the results considering the bulge imperfection show smaller than those of the perfect shell. However these differences are within 4%.

The horizontal dotted dash line denotes the required strength under the wind pressure (IASS Recommendation 1979). The reinforced concrete shell must possess the load carrying capacity equal to 1.275 times the design wind load. In this model ASCE-ACI Committee suggests the sufficient load carrying capacity.

In Fig.6 it is assumed that the model has the ring type of initial geometric imperfection(see Fig.3(b)). The amplitude of imperfection is 10cm. The results show the same tendencies as shown in Fig.5. However there are some differences in initial crack load. The ultimate load of the shell having ring imperfection shows about 5% smaller than that of the perfect shell.

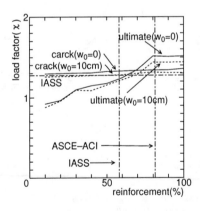

Figure 6: Crack load and ultimate strength ($\gamma = 1$ Ring Imperfection)

Strength of the Cracked Shell
        Meshk(1991) reported that cooling tower shells constructed in 1950s and 1960s show cracks caused by a temperature gradient and corrosion of steels due

to concrete cracking. In this case it is considered that the concrete has no tensile strength due to crack opening and the reduction of the reinforcing ratio is detected due to the corrosion.

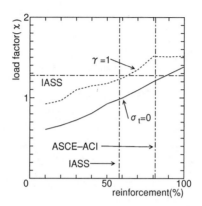

Figure 7: Crack load and ultimate strength ($\gamma = 1$)

Fig.7 shows the relation between the reinforcing ratio and the ultimate strength of the cooling tower shells. The solid line denotes the results having no tensile strength. The dotted line denotes the results obtained in Fig.5. The concrete having no tensile strength (see the solid line) shows the extreme results. However in the case of the cooling tower having the reinforcing ratio recommended (IASS 1977, ASCE-ACI 1977,1984) the strength under the design wind load is ensured even if the concrete loses the tensile strength.

CONCLUSIONS

In this paper the strength of R/C cooling tower shell is analyzed focused on the influences of the initial geometric imperfection and the reinforcing ratio to the initial crack load and the ultimate strength by use of the finite element method.
The conclusions obtained in this analyses are as follows.

1. The fewer the reinforcing ratio is the lower the ultimate strength is. Therefore the reinforcing ratio plays the important role of the strength of the R/C cooling tower shell.

2. The effect of the initial geometric imperfection on the strength of R/C cooling tower shell is not so large. The reduction of the ultimate strength is less

than 5% when the initial geometric imperfection are controlled within the allowable imperfection recommended.

3. The reinforcing ratio recommended in ASCE-ACI (1977,1984) is sufficient to get the ultimate strength recommended. In general IASS Recommendation (1977) shows fewer reinforcing ratio than ASCE-ACI.

REFERENCES

ACI-ASCE Committee 334 (1977). "Reinforced concrete cooling tower shells - practice and commentary." *ACI Journal* Vol.74, 22-31.

ACI-ASCE Committee 334 (1984). "Reinforced concrete cooling tower shells - practice and commentary." *ACI Journal* Vol.81, 623-631.

Batoz, J.L., and Dhatt, G. (1979). "Incremental displacement algorithms for non-linear problems." *International Journal for Numerical Methods in Engineering*, 1262-1267.

Hara T., Kato S., and Nakamura H. (1993). "Ultimate strength of R/C cooling tower shells under the influence of the reinforcing ratio." *Proc. of IASS-Symposium'* 343-352.

Hinton E. and Owen D.R.J. (1984). "Finite Element Software for Plates and Shells." *Pineridge Press Limited.*

Kemp, K.O. and Croll, J.G.A.(1976). "The rôle of geometric imperfections in the collapse of a cooling tower." *The Structural Engineer*, 54(1), 33-37.

Kupfer, H. and Hilsdorf, K.H. (1969). "Behavior of concrete under biaxial stresses. " *ACI Journal*, 66(8), 656-666.

Mang, H.A., Floegl, H., Trappel, F. and Walter, H. (1983). "Wind-loaded reinforced concrete cooling towers: buckling or ultimate load?" *Engineering Structures*, 5, 163-180.

Meschk M. Mang H.A. and Kosza P. (1991). "Finite element analysis of cracked cooling tower shell." *Journal of Structural Engineering*, 117(9), 2620-2635.

Milford, R.V. (1984). "Nonlinear behavior of reinforced concrete cooling towers." *Thesis of the Doctor of Philosophy, University of Illinois at Urbana Champaign The Graduate College.*

"Recommendations for the design of hyperbolic or other similarly shaped cooling towers." (1977) I.A.S.S. Working group No.3.

"Recommendations for reinforced concrete shells and folded plates." (1979),The Working Group on Recommendations of I.A.S.S.

"Report of the Committee of Inquiry into the Collapse of Cooling Towers at Ferrybridge, Monday 1 November 1965."(1966) London Central Electricity Generating Board.

"Report of the Committee of Inquiry into the Collapse of the Cooling Tower at Ardeer Nylon Works, Ayrshire on Thursday, 27th September 1973." (1974) London Imperial Chemical Industries Ltd., Petrochemical Division.

"Report on the Collapse of Cooling Tower B2 at Fiddlers Ferry Power Station on 13 January 1984." (1985) Central Electricity Generating Board.

REPAIR OF A CRACKED COOLING TOWER SHELL
BASED ON NUMERICAL SIMULATIONS

Thomas Huemer[1], Christian Kropik[1], Herbert A. Mang[2], Fellow, ASCE, and **Günther Meschke[1]**

Abstract

The present condition of a 30 years old cooling tower shell made of reinforced concrete is characterized by a relatively large number of long, thermally induced meridional cracks. For the purpose of designing the planned repair of the pre-damaged cooling tower by stiffening rings, a comprehensive numerical investigation has been performed by means of FEM. It is based on realistic material models for the concrete and the reinforcing steel. The degree and the spatial distribution of corrosion as well as the distribution of the cracks and the actual thickness of the shell are accounted for. Comparative ultimate load analyses have shown that a repair based on attaching two stiffening rings to the shell results in a sufficient degree of safety against collapse of the structure within the guaranteed residual lifetime of 25 years. The paper contains a description of the repair with special emphasis on the connection between the stiffening rings and the shell.

Introduction

This study is concerned with the design of the planned repair of a cracked cooling tower shell made of reinforced concrete. The shell was built 30 years ago in Ptolemaïs. Greece, as part of a 125-MW power station. Its dimensions are shown in Figure 2a. The present state of the concrete shell is characterized by an approximately uniform distribution of long, meridional cracks caused by thermally induced stresses. The latter were not considered adequately by the design specifications at the time the structure was built. According to a previous numerical study (Meschke, Mang & Kosza, 1991). the cooling tower shell will be sufficiently safe against structural failure even if no repair is carried out, provided corrosion does not result in a reduction of the diameters of the reinforcing steel bars larger than 10 %.

According to a survey carried out by the Public Power Corporation (PPC) of Greece. the actual thickness of the shell varies between 7.6 and 10.2 cm. The design thickness. however, is 10 cm. The corrosion-induced reduction of the diameters of the reinforcement bars in the meridional direction, i.e., parallel to the cracks, in the vicinity of the cracks, is approximately 16 %. The diameters of the circumferential bars are reduced by 17.5 - 30 %. The maximum crack width is 2.7 mm. Most of the cracks are membrane cracks.

Based on these findings and on numerical analyses by the Finite Element Method (FEM). a repair by means of attaching stiffening rings to the shell was planned. In the numerical investigation geometric and physical nonlinearity (cracking and plastification of concrete. yielding of the reinforcement) was considered. The state of damage was accounted for. This includes consideration of the spatial distribution of the cracks and

---

[1] Assistant. [2] Professor, Inst. f. Strength of Materials, Techn. Univ. Vienna. Austria

two different assumptions for the corrosion of the reinforcement, involving extrapolation from actual measurements to the year 2018 (end of the guaranteed residual lifetime of the cooling tower shell). The numerical investigation has shown that a repair based on attaching two stiffening rings to the shell results in a sufficient increase of the safety of the cooling tower against collapse within the guaranteed residual lifetime. The design of the stiffening rings and their connection with the shell by means of special bolts are described.

Constitutive Models for Concrete and Steel

The numerical representation of cracked concrete is based on the "fixed-crack" concept in the context of the "smeared-crack" approach. Cracks will begin to open normal to the direction of the maximum principal stress, if this stress reaches the tensile strength f_{tu}. Secondary cracks are restricted to the direction perpendicular to the primary crack. After crack initiation tensile stresses are gradually released according to a post-peak stress-strain relationship which is assumed to be linear (Figure 1a).

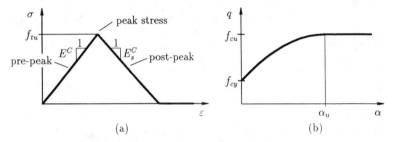

Figure 1: σ-ε-Diagram for Uniaxial Tension and Hardening Characteristic for Uniaxial Compression: (a) σ-ε-Diagram for the Pre- and the Postcracking Regime; (b) Uniaxial Compressive Hardening Characteristic

The ratio of the slope of the ascending part of the $\sigma - \varepsilon$ diagram (Figure 1a) to the absolute value of the slope of the descending part, $E^C/|E_s^C|$, is taken as $\kappa = 2600$: $2000 = 1.3 : 1$ for the shell concrete. This is a sufficiently conservative assumption, taken into account that the maximum value of the dissipated energy per unit area of the fracture surface in the FE-mesh of the shell, $D_{f,max} = 0.0073$ kNm/m^2, is considerably smaller than the respective experimentally obtained average value for the fracture energy of concrete, $G_f \sim 0.1$ kNm/m^2 (Reinhardt, Cornelissen & Hordijk 1986). $D_{f,max}$ is computed from

$$D_{f,max} = \frac{(1 + \frac{E^C}{E_s^C})\, l_{c,max} f_{tu}^2}{2\, E^C},\qquad(1)$$

(Dahlblom & Ottosen 1990), where $l_{c,max} = 5.1$ m is the maximum length of the cracked region in the direction normal to the crack planes (Figure 2). The more brittle concrete behaves under tensile stresses, the smaller the value of the dissipated strain energy will be.

		Shell	Rings
	E^C	$2,600$ kN/cm^2	$3,700$ kN/cm^2
	ν^C	0.20	0.20
Concrete	f_{tu}	0.180 kN/cm^2	0.291 kN/cm^2
	f_{cu}	1.70 kN/cm^2	2.70 kN/cm^2
	$|E_s^C|$	$2,000$ kN/cm^2	$2,000$ kN/cm^2
	E^C	$20,600$ kN/cm^2	$20,600$ kN/cm^2
Steel	ν^C	0.30	0.30
	σ_y^{ST}	40.0 kN/cm^2	55.0 kN/cm^2

Table 1: Shell, Stiffening Rings: Material Properties

The residual shear transfer across the cracks, resulting from the roughness of the crack face and from dowel action of the reinforcing bars, is considered by means of the shear modulus G_C, given as (Cedolin and Dei Poli 1977)

$$G_C = 0.25 \frac{E^C}{2(1+\nu^C)}(1 - \frac{\varepsilon_C}{\bar{\varepsilon}_C}), \qquad \bar{\varepsilon}_C = 0.004 , \tag{2}$$

where ε_C denotes the crack strain normal to the crack and $\bar{\varepsilon}_C$ represents a limiting value of ε_C. For $\varepsilon_C \geq \bar{\varepsilon}_C$, aggregate interlock is neglected.

The material parameters for concrete and for the reinforcement employed for the shell and the rings, respectively, are summarized in Table 1.

Closing and re-opening of cracks is considered by means of a secant unloading and reloading branch in the $\sigma - \varepsilon$ diagram for uniaxial tension of concrete (Meschke, Mang & Kosza 1991).

The ductile behavior of concrete under compression is accounted for by an elasto-plastic strain-hardening Drucker-Prager material model. The respective yield criterion is defined as

$$f_{DP}(\boldsymbol{\sigma}, q) = \sqrt{J_2} - \kappa I_1 + \frac{q}{\beta} \leq 0 , \tag{3}$$

where q represents the hardening parameter and κ and β are material parameters calibrated to the uniaxial (biaxial) compressive strength of concrete f_{cu} (f_{cb}). It is assumed that $f_{cb} = 1.16 f_{cu}$. Hardening is taken into account by a quadratic hardening law according to Figure 1b, where f_{cy} is the yield limit and α_u is the ultimate plastic compressive strain.

A linearly elastic – ideally plastic constitutive law is assumed for the reinforcing steel. The values of the yield stress σ_Y^{ST} of the reinforcement of the shell and of the stiffening rings are listed in Table 1.

In the numerical analyses the meridional and the circumferential reinforcement bars are smeared to mechanically equivalent, thin layers of steel which only have an axial stiffness in the respective direction.

Finite Element Model

All computations were performed with the FE Program MARC on a Hewlett-Packard HP 9000/755 workstation. The numerical investigation is based on a load-controlled, incremental-iterative procedure. An updated-Lagrangian approach was employed for consideration of geometric nonlinearity. Figure 2 contains two FE meshes used at

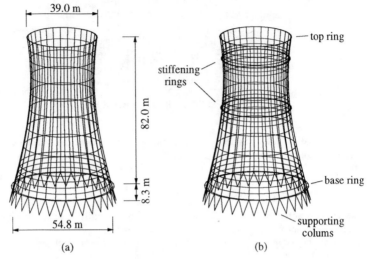

Figure 2: FE Meshes: (a) Coarse; (b) Fine

different stages of the numerical investigation. The spatial distribution of the initial cracks is unsymmetric. Hence, specification of symmetry conditions would have been inadmissible.

The coarse mesh (Figure 2a) is used for a comparative numerical investigation for determination of the optimum location and the number of the stiffening rings. It consists of 450 bilinear thick-shell elements with the displacements and rotations as nodal degrees of freedom. Each element is subdivided into 13 layers.

The fine mesh (Figure 2b), consisting of 990 elements, is employed for the numerical prediction of the structural safety of the repaired cooling tower at the end of its guaranteed residual lifetime. This discretization is characterized by a local refinement of the original mesh in the vicinity of the two concrete rings. The stiffening rings are discretized by 150 elements each. The flexibility of the supporting columns is considered by beam elements (Figure 2) with extensional, flexural and torsional stiffness.

The distribution of the meridional cracks, as obtained from the survey of the shell, is accounted for in the FE model by a reduction of f_{tu} at the integration points in the vicinity of these cracks. The opening of the cracks is triggered by the application of a thermal load history, representing a winter-summer cycle, prior to applying the wind load such that the standard wind load (VGB Guideline 1990) is multiplied by a dimensionless factor λ which is increased incrementally (Meschke, Mang & Kosza 1991). A temperature difference of $\Delta T_W = 45°$ C. representing winter conditions. is applied first. Then, the temperature of the outer surface is increased from $-22.5°$ C to $+2.5°$ C. resulting in a temperature difference of $\Delta T_S = 20°$ C. representing summer conditions. In conjunction with the local variation of f_{tu}, the temperature load induces

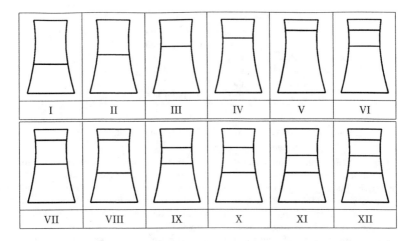

Table 2: Investigated Locations of Stiffening Rings

a nonuniform crack pattern at the outside surface of the cooling tower which approximately corresponds to the crack distribution obtained from the survey. Because of the inclusion of summer conditions in the thermal loading, the same crack pattern is induced at the inner surface. Because of the nonuniform distribution of the cracks, the value of λ corresponding to the collapse of the shell, i.e., the safety coefficient, depends on the direction of the wind load. In a comparative numerical study the worst direction of this load was evaluated. This direction was used for all numerical analyses.

Optimum Number and Location of Rings

In an extensive numerical study the influence of the number and the location of the stiffening rings on the structural safety of the repaired cooling tower was investigated. The analyses are based on the coarse FE-mesh (Fig.2a), using the material parameters listed in Table 1. The stiffening rings (dimensions of a preliminary design: width=80 cm, thickness=28 cm) were discretized by only one element in the radial direction. The "worst case assumption" for the corrosion of the reinforcement at the specified end of the lifetime of the structure is characterized by a 55 % reduction of the diameters of the reinforcement bars in the vicinity of the existing cracks (Case B). This value was obtained from extrapolation of the largest value of the diameter reduction at present to the respective value in the year 2018.

The optimum position of one single ring was determined first. The chosen levels of the ring above the base of the concrete shell are (in meters): 32.8, 43.0, 52.0, 61.0 and 70.0 m (I – V in Table 2). Table 3 contains the respective values of the safety coefficients λ_c, λ_y^S and λ_u obtained for Cases BI ... BV (The notation BI means that Case B is combined with ring location I). λ_c refers to the level of the crack plateau in the load-displacement curve (Figure 3), λ_y^S is the coefficient of safety against the onset

Case	Loc. of Ring	λ_c	λ_y^S	λ_u
B	unstiffened	0.905	0.800	0.910
BI	32.8 m	0.915	0.800	0.965
BII	43.0 m	1.005	0.800	1.010
BIII	52.0 m	1.040	0.800	1.215
BIV	61.0 m	1.075	0.900	1.155
BV	70.0 m	0.995	1.005	1.045

Table 3: Values of λ_c, λ_y^S and λ_u for One Stiffening Ring (Coarse Mesh)

of yielding of the shell reinforcement, and λ_u denotes the coefficient of safety against structural collapse. As far as λ_u is concerned, Case BIII has turned out to be the most effective option. For this case, an increase of the value of λ_u for the unstiffened shell by 33.5 % was obtained. With decreasing level of the stiffening ring the value of λ_u decreases significantly. For Case BI the increase of the value of λ_u in consequence of attaching the stiffening ring to the shell is only 6.0 %. According to Table 3, the value of λ_y^S is largest for a stiffening ring located at 70 m above the base of the shell (Case BV). For this case, however, the residual safety margin between λ_y^S and λ_u is very small.

Next, combinations of two stiffening rings (VI - XI in Table 2) were investigated. The respective values of λ_c, λ_y^S and λ_u are listed in Table 4. Remarkably, the most

Case	Loc. of Rings	λ_c	λ_y^S	λ_u
BVI	70.0/52.0 m	1.155	1.310	1.350
BVII	70.0/43.0 m	1.155	1.215	1.575
BVIII	70.0/32.0 m	1.075	1.205	1.340
BIX	61.0/43.0 m	1.175	0.800	1.430
BX	61.0/32.0 m	1.105	0.900	1.265
BXI	52.0/32.0 m	1.045	0.800	1.205

Table 4: Values of λ_c, λ_y^S and λ_u for Two Stiffening Rings (Coarse Mesh)

effective layout (Load Case BVII) does not include the optimum location obtained for one stiffening ring (Case BIII). For two stiffening rings located at levels of 70.0 m and 43.0 m, respectively, an increase of the value of λ_u for the unstiffened shell by 73.1 % was obtained. It is noteworthy that the optimum levels of the stiffening rings with respect to λ_u are not identical to the optimum levels with regards to λ_c and λ_y^S, respectively.

Three stiffening rings (Case BXII) cause an increase of λ_u by approximately 107 %.

Based on the described comparative investigation, the use of two ring stiffeners (layout according to Case BVII) was found to be the optimum choice for the repair of the cooling tower. It will result in the increase of the maximum sustainable gradient wind load from $\bar{v}_G = 135.0$ km/h to $\bar{v}_G = 167.4$ km/h.

Although the value of the coefficient of safety against collapse, $\lambda_u = 1.575$, is smaller than the value required by VGB (1990), $\lambda_{u,VGB} = 1.75$, it must be borne in mind that the former value is based on a "worst case assumption" for the corrosion of the reinforcement in the year 2018. The actual coefficient of safety against structural collapse within the guaranteed residual lifetime will be significantly larger. From ultimate load analysis based on a corrosion-induced diameter reduction of 37 % in the year 2018 (Case A) (Mang, Huemer, Kropik and Meschke 1993), the value of the safety coefficient λ_u is obtained as 1.79. Furthermore, it is emphasized that all assumptions concerning the material parameters for concrete (f_{tu}, E_s^C) and steel (σ_Y^{ST}) are conservative. It was also assumed that no action will be taken to protect the reinforcement from further corrosion.

Structural Safety of the Repaired Cooling Tower Shell

The numerical investigation of the structural safety of the repaired cooling tower shell was performed on the basis of the refined mesh (Figure 2b). The shape of the cross-section of a ring and the location of the reinforcement bars are illustrated in Figure 4.

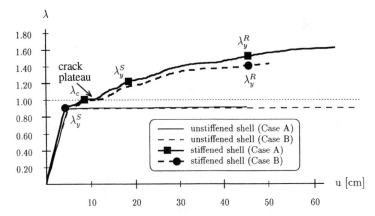

Figure 3: Load–Displacement Diagrams for the Unstiffened and the Stiffened (Repaired) Cooling Tower for Two Different Assumptions for the Corrosion in the Year 2018 (Fine Mesh)

The horizontal displacement u, plotted as a function of λ (Figure 3), was evaluated for a point located on the windward meridian at a level of 61.0 m above the base of the shell. λ_y^R and λ_y^S refer to the onset of yielding of the reinforcement in the stiffening rings and in the shell, respectively. It follows from Table 5 that the increase of the value of λ_u (coefficient of safety against structural collapse) in consequence of the repair of the cooling tower will be 77.6 % in the year 2018 for the assumed evolution of the corrosion of the reinforcement according to Case A. For Case B the respective increase of the value of λ_u will be 58.6 %.

The influence of an earthquake and of a nonuniform temperature distribution on the load-carrying capacity of the repaired cooling tower was also investigated (Mang.

Case	λ_c	λ_y^S	λ_y^R	λ_u
A (unstiffened shell)	0.900	0.915	—	0.915
B (unstiffened shell)	0.900	0.800	—	0.905
A (stiffened shell)	1.010	1.225	1.535	1.625
B (stiffened shell)	0.995	0.905	1.410	1.435

Table 5: Values of λ_c, λ_y^S, λ_y^R and λ_u for the Unstiffened and the Stiffened (Repaired) Cooling Tower for Two Different Assumptions for the Corrosion in the Year 2018 (Fine Mesh)

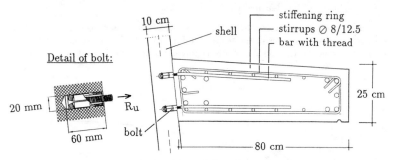

Figure 4: Cross-Section Through Ring Located at h=43.0 m

Huemer, Kropik & Meschke 1993). The obtained coefficients of safety against collapse of the shell are larger than the respective values in Table 5.

Design of the Stiffening Rings

The relatively difficult access to the levels of the two stiffening rings was an important parameter for the design of the rings. From examination of several alternatives it was concluded that stiffening rings made of reinforced cast-in-situ concrete, bolted to the shell, would be the most economic of repair of the cooling tower shell. Figure 4 shows a cross-section of the stiffening ring at a level of 43.0 m above the base of the shell. The ring is attached to the shell by bars with threads, screwed to the bolts which are drilled into the shell. The chosen bolts (HILTI MSC-IR M12*60) are so called "automatic undercutting bolts". They are made of no-corroding steel.

The bolts are suitable for cracked concrete as well as for dynamic loads. Moreover, they cause only small spread pressures. This is a requirement for bolts drilled into a relatively thin, cracked concrete shell. The ultimate axial force R_u of these bolts is equal to 18.0 kN. The ultimate circumferential shear force C_u is equal to 12.0 kN.

Figure 5 contains distributions of the radial force r and the circumferential force c per unit length, transmitted from the cooling tower shell to the stiffening rings located at 43.0 m (ring 4300) and 70.0 m (ring 7000), respectively, above the base of the shell.

For the transfer of the distributed forces r and c from the shell to the ring by bolts, the truss analogy is used (Figure 6). According to Figure 6, for n bolts per meter,

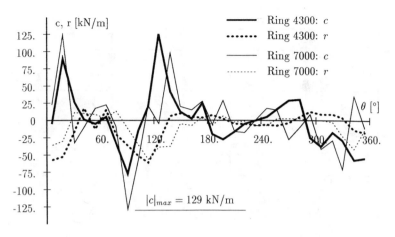

Figure 5: Distributed Radial and Shear Forces Transferred from the Shell to the Stiffening Rings

the axial force in the bars screwed to the bolts, S, and in the concrete struts, D, respectively, is obtained as

$$S = S_R + S_C = R + C \tan \alpha , \quad D = D_R + D_C = \frac{1}{2}\frac{R}{\sin \alpha} + \frac{C}{\cos \alpha} . \tag{4}$$

where α is the inclination of the struts and $R = r/n$, $C = c/n$. Assuming $\alpha = 45°$,

$$S = R + C , \quad D = \sqrt{2}\left(\frac{1}{2}R + C\right) . \tag{5}$$

According to Figure 6, the axial force in the bars screwed to the bolts is obtained as $S = S_R + S_C$. It is assumed that the circumferential component of D_C is transferred by means of friction from the shell to the stiffening ring. This assumption requires that the contact surface between the shell and the stiffening ring is cleaned and sandblasted in order to enhance friction in this surface. It follows that the bolts only need to be designed for the axial force S.

For each ring, 4 bolts per meter located near the top of each ring and 4 bolts per meter located near the bottom of each ring are chosen (Figure 4). The axial force associated with c_{max} is a compressive force (see Figure 5). Hence, $S_{max} < |C_{max}|$. It follows from

$$|C_{max}| = \frac{|c|_{max}}{n} = \frac{129.0}{8} = 16.13 \, \text{kN} < R_u = 18.0 \, \text{kN} \tag{6}$$

that the chosen dimensioning of the bolts is admissible.

Summary and Conclusions

FE ultimate load analyses based on realistic constitutive modelling of reinforced concrete were used as a tool for the design of the repair of a cracked cooling tower shell

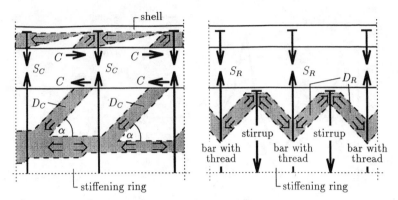

Figure 6: Truss Analogy for Determination of Internal Forces in the Stiffening Rings (Top View)

by means of stiffening rings. The weakening of the shell, resulting from a large number of long, meridional cracks, was taken into account. Two different assumptions for the corrosion of the reinforcement were made. They are based on extrapolations from present measurements to the end of the guaranteed residual lifetime of the shell in the year 2018. With regards to the "worst case assumption" for corrosion it was found that, out of a number of different arrangements of one or two stiffening rings made of reinforced cast-in-situ concrete, two stiffening rings, located at levels of 43.0 m and 70.0 m above the base of the shell, result in the largest increase of the coefficient of safety against structural collapse, namely, in 58.6 %. The connection between the rings and the shell was accomplished by undercutting bolts. Because of the relatively difficult access to the levels of the stiffening rings, the proposed repair of the shell is considered to be an economically effective means to increase the coefficient of safety of the cooling tower shell against structural collapse. The economic efficiency is based on the small number of stiffening rings and on the mode of attaching the stiffening rings to the shell.

References

Cedolin, L. and Dei Poli, S. (1977). "Finite element studies of shear-critical R/C beams." *J. Eng. Mech. Div.*, ASCE, 103(3), 395-410.

Dahlblom, O. and Ottosen, N. (1990). "Smeared crack analysis using generalized fictitious crack model." *J. Eng. Mech.*, ASCE, 116(1), 55-76.

Mang, H.A., Huemer, Th., Kropik, Chr. and Meschke, G. (1993). "Design of the repair of the cooling tower III Ptolemaïs SES.", Report, Institute for Strength of Materials, Technical University of Vienna, Vienna, Austria.

Meschke, G., Mang. H.A. and Kosza P. (1991). "Finite element analyses of cracked cooling tower shell." *J. Struc. Eng.*, ASCE, 117(9), 2620-2639.

Reinhardt, H.W., Cornelissen, A.W., Hordijk, D.A. (1986). "Tensile tests and failure analysis of concrete", *J. Struct. Eng.*, ASCE, 112(11), 2462-2477.

VGB-Guideline (1990)."Structural design of cooling towers." Part 2, Essen, Germany.

Design for Durability for Large Reinforced Concrete Shells

Wilfried B. Krätzig[1], Konstantin Meskouris[2], Karsten Gruber[3]

Abstract

Classical design concepts restrict themselves to sufficient structural safety, using linear analysis methods. Durability questions are mainly treated independently. Modern computer simulations based on realistic material descriptions admit unified approaches of structural safety and durability, as this paper demonstrates by example of large reinforced shells.

1. Motivation

In engineering design practice reinforced and prestressed concrete structures are generally analyzed as elastic models. The advantage of a linear elastic analysis is the validity of the principle of superposition, due to which stresses or stress-resultants of single load-cases can be evaluated separately and then superposed to load combinations of extremes [ACI 1989, EC2 1992].

A severe disadvantage of linear elastic design concepts is the lack of information after inelastic actions have taken place. Neither stress redistributions nor damage evolutions due to concrete cracking, yielding of reinforcement, creep or local fracture can be considered in this contexts. In consequence, assessments of reliability and durability, determinations of life-cycles and maintainance strategies cannot be obtained. Such investigations require inelastic, i.e. materially nonlinear analysis concepts, often combined with consideration of large deformations.

The present paper describes the material modelling and nonlinear analysis for reinforced concrete shells. It further elaborates on the application of these techniques to the improvement of their reliability and durability in order to lay a sound basis for the application of secondary protection means.

[1]o. Prof. Dr.-Ing., [2]Prof. Dr.-Ing., [3]Dipl.-Ing., all: Institute for Statics and Dynamics, Ruhr-University Bochum, D-44780 Bochum

897

2. Nonlinear Shell Analysis

2.1 Material Points and Discretized Shell Models

Nonlinear time-independent responses $P(V)$ are simulated by application of global incremental-iterative solution techniques to the discretized structure, whereby nonlinear material properties are defined in single material points. The global solution techniques are based on the tangential stiffness equation:

$$\mathbf{K_T} \cdot \overset{+}{\mathbf{V}} = \mathbf{P} - \mathbf{F}_i \;\; \rightarrow \;\; \overset{+}{\mathbf{V}} = \mathbf{K_T^{-1}} (\mathbf{P} - \mathbf{F}_i) \tag{2.1}$$

in which $\mathbf{K_T}$ denotes the tangential stiffness matrix, $\mathbf{P}$ the total applied load, $\mathbf{F}_i$ the internal nodal force vector and $\overset{+}{\mathbf{V}}$ the vector of increments of the nodal degrees of freedom. From this structural level the connection of global degrees of freedom to element ones stored during the assemblage process leads to the finite element level, and from there the inverse discretization process to the respective structural model (shells, plates, rods, etc.). Having reached this level, the kinematic field assumption (linear normal strain, etc.) of the selected model transfers to specific material points.

The different nonlinear material properties of the reinforced concrete shell, such as elasto-plastic actions with compression softening, tension cracking, yielding of reinforcement and nonlinear bond, are defined in material points in the form of an initial value problem:

$$d\tau^{ij} = C^{ijkm} d\gamma_{km} \tag{2.2}$$

For simplicity, neighborhoods Δh of finite thickness between selected material points – see Fig.1 – are defined as material layers, across which all physical properties vary linearly. By numerical integrations over the shell thickness h according to the definitions of stress-resultants

$$n^{\alpha\beta} = \int_{-h/2}^{h/2} \mu\, \mu_\rho^\beta\, \tau^{\alpha\rho}\, d\Theta^3 \;, \qquad m^{\alpha\beta} = \int_{-h/2}^{h/2} \mu\, \mu_\rho^\beta\, \tau^{\alpha\rho}\, \Theta^3\, d\Theta^3 \tag{2.3}$$

incremental constitutive laws for the shell model can be set up:

$$dn^{(\alpha\beta)} = \overset{1}{E}{}^{\alpha\beta\lambda\mu}\, d\alpha_{\lambda\mu} + \overset{2}{E}{}^{\alpha\beta\lambda\mu}\, d\beta_{\lambda\mu}, \; dm^{(\alpha\beta)} = \overset{2}{E}{}^{\alpha\beta\lambda\mu}\, d\alpha_{\lambda\beta} + \overset{3}{E}{}^{\alpha\beta\lambda\mu}\, d\beta_{\lambda\mu}, \tag{2.4}$$

in which the following abbreviations have been introduced:

$$\overset{n}{E}{}^{\alpha\beta\lambda\mu} = \int_{-h/2}^{h/2} C^{\alpha\beta\lambda\mu}\, (\Theta^3)^{n-1}\, d\Theta^3 \;, \quad n = 1, 2, 3 \tag{2.5}$$

and $\alpha_{\lambda\mu}$ as the 1st, $\beta_{\lambda\mu}$ as the 2nd strain tensor [Basar 1985].

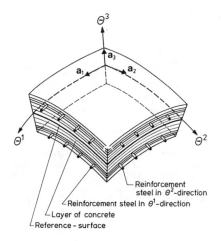

Θ^3

a_3

a_1 a_2

Θ^1 Θ^2

└ Reinforcement
   steel in Θ^2-direction
└ Reinforcement steel in Θ^1-direction
└ Layer of concrete
└ Reference - surface

Fig. 1: Sandwich idealization of reinforced concrete
shell

From here we return to the structural level as follows: The strain increments $d\alpha_{\lambda\mu}$, $d\beta_{\lambda\mu}$ in (2.4) are connected with the incremental displacement field of geometrically nonlinear shell theories [Basar 1989] for large deformations and rotations. The discretization process is based on the concept of high-precision NACS shell elements [Harte 1986] as displacement field approximations. Finally the tangential stiffness equation (2.1) is formulated with

$$K_T = K_{ep} + K_{\sigma\,ep} + K_{u\,ep} \ , \qquad F_i = F_{i\,ep} + F_{\sigma\,ep} \ , \qquad (2.6)$$

elasto-plastic (ep) contributions in all its constituents.

2.2 Local and Global Solution Techniques

Sub-incrementation techniques represent the most simple of the local solution procedures which solve the constitutive initial value problem (2.2). The natural drift-off of numerical integration schemes applied from the real solution curve can be controlled by automatic sub-incrementation techniques [Zahlten 1990]. In addition, for specific material properties there exist more advanced algorithms for improved accuracy, e.g. return-mapping algorithms for elasto-plastic materials with yield conditions [Ortiz 1986].

For the incremental-iterative solution of the tangential stiffness equation (2.1) two classes of global solution algorithms are available: the Newton methods [Matthies 1979] and the arc-length methods [Riks 1979, Decker 1980]. The following numerical simulations are carried out by a symmetrized Riks-Wempner alternative as global algorithm and the sub-incrementation technique as local one.

3. Material Modelling of Reinforced Concrete

3.1 Nonlinear Phenomena

Reinforced concrete is a highly nonlinearly reacting composite. Its response is characterized by the following nonlinear phenomena:

- nonlinear stress-strain relation in the compression domain;
- tension cracking at a relatively low stress level;
- elasto-plastic behavior of the reinforcement;
- nonlinear bond between reinforcement and concrete.

Each of these phenomena can be responsible for stress redistributions in connection with essentially irreversible material processes adding to the damage evolution in the structure. In numerical simulations all four phenomena can be modelled at different levels of accuracy as indicated in Fig. 2.

3.2 Overview over Models for Concrete in Compression

In existing computer codes for ultimate state design the stress-strain behavior of concrete in the compression domain is often modelled as uniaxial, even for applications in plate and shell structures. The most simple model is the nonlinear elastic, parabolic-linear curve [EC2 1992] in the first line of Fig. 2, with or without softening phase for monotonic or cyclic loading.

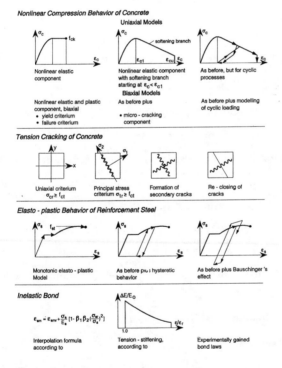

Fig. 2: Nonlinear model components of reinforced concrete in different levels of accuracy

All these idealizations are incapable of describing point-oriented stress redistributions, for which at least 2-dimensional elasto-plastic models have to be applied. Such models have to include biaxial yield conditions and biaxial failure envelopes [Chen 1975] as indicated in Fig. 2. On an advanced level the softening branch can be modelled by a microcracking component [Bazant 1979], and for cyclic loading endochronic models [Darwin 1976] have been successfully employed.

3.3 Tension Cracking

Concrete fails under tension in a brittle-like manner leading to stiffness reductions in the material points affected. For the analysis of complete structures, cracking is currently modelled in a "smeared" manner, neglecting localization phenomena.

As Fig. 2 indicates, a simple uniaxial criterion may be applied, allowing for cracking after stresses exceed the tension strength of concrete. More advanced models employ principal stress criteria, by use of which cracks can be described with inclined orientation to a given frame of reference. According to Fig. 3 the stiffness coefficients of the constitutive tensor or- thogonal to an assumed crack are deleted; the shear stiffnesses can be com- pletely or partly retained (aggregate interlock, $\mu < 1$). In this manner also secondary crack patterns can be described, orthogonal to the previous ones. In the most sophisticated model [Zahlten 1990], closing of cracks can be incorporated by means of a kinematic criterion as well as reopening in the original or a new direction (crack-memory, crack-rotation). Such a model has been employed in the examples of this paper.

Criterion of cracks :

$$\begin{aligned} \tau_{I,II} &< f_t : \quad \text{uncracked} \\ &\geq f_t : \quad \text{cracked} \end{aligned}$$

with $\tau_{I,II}$: principal stresses
       f_t : tensile strenght of concrete

Material tensor for concrete cracked in one direction:

$$\begin{bmatrix} 0 & 0 & 0 & 0 \\ 0 & \mu \cdot C^{1212} & \mu \cdot C^{1221} & C^{1222} \\ 0 & \mu \cdot C^{2112} & \mu \cdot C^{2121} & C^{2122} \\ 0 & C^{2212} & C^{2221} & C^{2222} \end{bmatrix}, \quad 0 < \mu < 1$$

Material tensor for concrete cracked in two directions:

$$\begin{bmatrix} 0 & 0 & 0 & 0 \\ 0 & \mu \cdot C^{1212} & \mu \cdot C^{1221} & 0 \\ 0 & \mu \cdot C^{2112} & \mu \cdot C^{2121} & 0 \\ 0 & 0 & 0 & 0 \end{bmatrix}, \quad 0 < \mu < 1$$

Fig. 3:   The material tensor C
          in cracked points

3.4 Steel Reinforcement and Bond

All modells for reinforcing bars should be able to describe the elasto-plastic behavior of steel, if ultimate strength problems are considered. The simplest possible model is a bilinear one for monotonically increasing loads. More realistic models simulate the real stress-strain behavior as experimentally recorded. For cyclic loading they must be able to consider Bauschinger's effect. Fig. 4 demonstrates the model used in the following simulations. The fully rigid bond of the classical

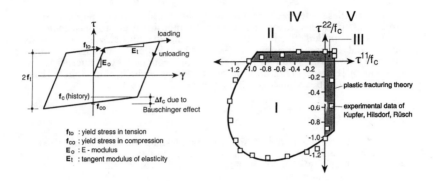

Fig.4: Elasto-plastic material law   Fig. 5: Biaxial failure envelope for concrete
        for steel

concrete design concepts seems inacceptable for the intended level of accuracy. According to Fig.2, a first acceptable approximation is the interpolation formula given in [EC2 1992], while the most widely used model is the "tension stiffening" concept after [Gilbert 1978].

3.5 Description of the Concrete Model Applied

The mechanical behavior of concrete in material points depends strongly on the local state of stress. Fig. 5 differentiates five domains related to the biaxial failure envelope in the principal stress plane. In domains I, II and III, the concrete is macroscopically uncracked. It exhibits cracks in one direction in domain IV, whereas it is cracked in two directions in the domain V.

The nonlinear behavior of concrete in the domains I, II and III will be modelled according to the elasto-plastic fracturing theory, applied to concrete by [Bazant 1979]. As pointed out in Fig. 6, this model subdivides each stress increment due to a certain incremental strain into three components:
- an elastic stress increase (el),
- the plastic stress decrease (pl) and
- the stress decrease due to microcracking(fr):

$$d\tau^{ij} = d\tau_{el}^{ij} + d\tau_{pl}^{ij} + d\tau_{fr}^{ij} \tag{3.1}$$

The material behavior of concrete is described excellently by this model. For example, Fig. 5 demonstrates its high accuracy by comparison of the failure envelope of this model with experimental results from [Kupfer 1969].

The elastic component is a classical linear hyperelastic one, namely:

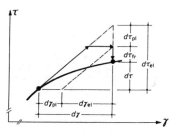

Incremental stress state:

$$d\tau^{ij} = d\tau_{el}^{ij} - d\tau_{pl}^{ij} - d\tau_{fr}^{ij}$$

Incremental stress state formulated by material tensor:

$$d\tau^{ij} = (C_{el}^{ijlm} - C_{pl}^{ijlm} - C_{fr}^{ijlm}) \cdot d\gamma_{lm}$$

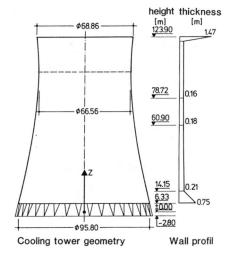

Cooling tower geometry   Wall profile

Fig. 6: Elasto-plastic fracturing theory   Fig. 7: Geometry of a cooling tower shell

$$d\tau_{el}^{ij} = [G(a_{il}\, a_{jk} + a_{ik}\, a_{jl}) + (K - \frac{2}{3}G)\, a^{ij}\, a^{lk}]\, d\gamma_{lk} = C_{el}^{ijkl}\, d\gamma_{kl} \qquad (3.2)$$

with G as elastic shear modulus, K as elastic bulk modulus and a^{ij} as contravariant metric tensor of a material point. The plastic properties of concrete employed in this model are governed by the classical Drucker-Prager yield criterion. Well-known steps of derivation lead to the plastic stress decrease:

$$d\tau_{pl}^{ij} = \frac{(\frac{G}{\tau}S^{ij} + \beta Ka^{ij})(\frac{G}{\tau}S^{lk} + \beta'Ka^{lk})}{h + G + \beta\beta'K}\, d\gamma_{lk} = C_{pl}^{ijkl}\, d\gamma_{lk} \qquad (3.3)$$

in which h abbreviates the plastic hardening modulus, β the plastic dilatancy factor and β' the coefficient of internal friction. Finally, the microcracking component can be understood as the complete duality to the plastic component in the strain space of material points. Similar steps as for the previous component lead to the incremental stress decrease due to microcracking:

$$d\tau_{fr}^{ij} = \Phi\,(\frac{e^{ij}}{2\gamma} + \frac{\alpha}{3}\, a^{ij})(\frac{e^{lk}}{2\gamma} + \frac{\alpha'}{3}\, a^{lk})\, d\gamma_{lk} = C_{fr}^{ijkl}\, d\gamma_{lk} \qquad (3.4)$$

with Φ as the fracturing modulus, α as the fracturing dilatancy factor and α' as the coefficient of fracturing friction.

4. Applications of Nonlinear Shell Analyses

The appication of the above presented nonlinear analysis for reinforced concrete shells will now be demonstrated by example of the shells
of large natural draught cooling tower. Fig. 7 shows the geometry and wall
thickness distribution of this shell. The durability and quality of reinforced
concrete shells are influenced by enviromental impacts to which the structures are exposed or by internal reasons within the material itself.

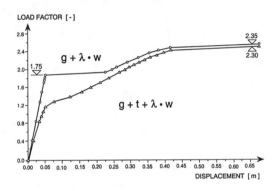

Fig. 8: Laod-displacement paths for the perfect cooling tower shell

Fig. 8 depicts the load-displacement paths of the normal displacement
of the above presented shell, designed according to the actual main rules
[BTR 1991]. The actions of deadweight and temperature remain constant,
the wind loads are proportionally increased introducing a load-factor λ until
failure. The load-displacement path of the load-combination g + w can be
supdivided into two domains. In the first linear part, the shell is nearly
crackfree, owing to the tensile strength of the concrete. At the end of this
domain first horizontal cracks occur in luff of the structure. After that, in
the second part, cracking spreads over the upper parts of the shell, leading
to large displacements. The failure of the cooling tower at the end of these
curves is characterized by the yielding of single reinforcement bars.

Additional thermal loading in the combination g + t + w softens the
shell early on but has no considerable influence on the level of failure.
Fig. 9 demonstrates the deformation of the cooling tower just before failure.
Large bulges from the upper rim down to the waist of the shell are typical
for this high level of wind-loads.

Fig. 11 depicts the influence of steel corrosion on the load-displacement behavior of the cooling tower structure. In Fig. 10 the distribution
of "smeared cracks" over the outer shell surface for the load g + 2.2 w is
demonstrated. An increase of the minimum wall-thickness or the amount
of reinforcment seems to be the only convenient method, to guaranty a

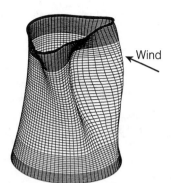

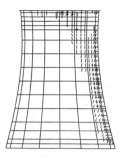

Fig. 10: "Smeared cracks" of the outer face for g + 2.20 w

Fig. 9: Deformation of the cooling tower shell for g + 2.30 w

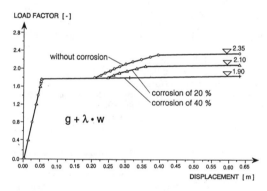

Fig. 11:
Load-displacement paths of the cooling tower, damaged by corrosion of steel

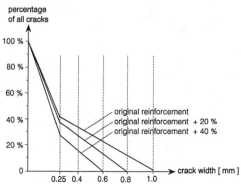

Fig. 12:
Diminuation of crack width of the outer face for g + 1.80 w by an increased reinforcement

sufficient durability for the whole life-time of the shell. Fig. 12 illustrates the positive influence of an increased amount of reinforcement on the diminution of the crack widths of the above presented cooling tower shell.

References

ACI Committee 318, 1989: Building Code Requirements for Reinforced Concrete (ACI 318-89) and Commentary (ACI 318R-89), American Concrete Institute, Detroit.

Basar, Y., Krätzig, W.B., 1985: Mechanik der Flächentragwerke. Verlag Friedr. Vieweg & Sohn, Braunschweig/Wiesbaden.

Basar, Y., Krätzig, W.B., 1989: A Consistent Shell Theory for Finite Deformations. Acta Mechanica 76, 73-87.

Bazant, P.Z., Kim, S.-S., 1979: Plastic Fracturing Theory for Concrete. ASCE, Journ. Eng. Mech. Div. 105, 407-428.

BTR, 1991: Structural Design of Cooling Towers (BTR-Bautechnik bei Kühltürmen). VGB Technical Association of Large Power Plant Operators.

Chen, A.C.T., Chen, W.F.,1975: Constitutive Relations for Concrete. ASCE, Journ. Eng. Mech. Div. 101, 465-481.

Darwin, D., Pecknold, D.A., 1976: Analysis of RC Shear Panels under Cyclic Loading. ASCE, Journ. Struct. Div. 102, 355-359.

Decker, D.W., Keller, H.B., 1980: Solution Branching - A Construction Technique. In: H.P. Holmes (ed.): New Approaches to Nonlinear Problems in Dynamics, 53-69. SIAM Philadelphia.

EUROCODE EC2, 1992: Common Unified Rules for Concrete Structures. Final draft, British Cement Association, Slough.

Gilbert, R.I., Warner, R.F.,1978: Tension Stiffening in Reinforced Concrete Slabs. ASCE, Journ. Struct. Div. 104, 1885-1900.

Harte, R., Eckstein, U.,1986: Derivation of Geometrically Nonlinear Finite Shell Elements via Tensor Notation. Int. Journ. Num. Meth. Engg. 23, 367-384.

Krätzig, W.B., Zhuang, Y., 1992: Collapse simulation of reinforced concrete natural draught cooling towers. Eng. Struct. 14, 291-299.

Krätzig, W.B., Gruber, K., Zahlten, W., 1992: Numerical Collapse Simulations of Large Cooling Towers Checking their Safety and Reliability. Report No. 92-3, Inst. for Struct. Eng., Ruhr-University Bochum.

Kupfer, H.B., Hilsdorf, H.K., Rüsch, H., 1969: Behavior of concrete under biaxial stress. Journ. Amer. Concrete Inst. 66, 656-666.

Matthies, H., Strang, G., 1979: The Solution of Finite Element Equations. Int. Journ. Num. Meth. Engg. 14, 1613-1626.

Ortiz, M., Simo, J.C., 1986: An Analysis of a New Class of Integration Algorithms for Elastoplastic Constitutive Relations. Int. Journ. Num. Meth. Engg. 23, 353-366.

Riks, E., 1979: An Incremental Approach to the Solution of Snapping and Buckling Problems. Int. Journ. Solids Struct. 15, 529-551.

Zahlten, W., 1990: A Contribution to the Physically and Geometrically Nonlinear Computer Analysis of General Reinforced Concrete Shells. Techn. Report No 90-2, Inst. for Struct. Engg., Ruhr-University Bochum.

Punching Shear of Reinforced Concrete Shell Structures

Gajanan M. Sabnis[1] and Amjad A. Shadid[2]

Abstract

In reinforced concrete shells punching shear failure may take place, if heavy concentrated loads are applied. Such failure is undesirable, since it does not provide adequate warning and ductility. There is a no design aid available for the punching shear. This paper presents the results of a multiple regression analysis of the available data and effect of the various parameters on punching shear failure loads. The analysis resulted in a simple yet reliable equation to give adequate results for more accurate design of the curved members. It indicates that the present ACI code provisions for punching shear strength are considerably conservative.

Introduction

The reinforced concrete shell structures are used for several types of structures such as bridge piers, offshore structures, towers, etc. These structures may be subjected to the concentrated loads possibly resulting in a punching shear failure. The concentrated loads may be due to the collision of ships, breaking waves, floating ice and dropped objects. Due to the magnitude of such concentrated load, construction of such walls (or slabs) needs huge amounts of material using the present conservative methods in the codes. Since the punching shear mechanism in shells is not well understood, the one in a plane slab is used. However, the curvature of shell may extend the failure surface beyond the edge of the load at the compressive surface of the slab. This distance may be up to twice the slab depth and the angle of inclination of the cracks varying from 20°-45° [7].

The failure characterized by the punching shear is limited by a failure surface running through the slab from the column or loaded area as

1, 2 Professor and Research Assistant, Respectively, Department of Civil Engineering, Howard University, Washington, D.C.

indicated in Figure 1.a. The main reinforcement is not very effective in
preventing this type of failure, which consequently may reduce the
ultimate load below the flexural capacity of the slab. In the case of
punching shear failure of shells, the load introduces compressive
membrane forces and the actual wall thickness is larger due to curvature.
The stress due to concentrated load is symmetrical as in a flat slab. The
inclined cracks, however, will not be developed simultaneously in all
direction, due to the curvature. These cracks will result in failure in the
direction where membrane stresses are the smallest. (Figure 1.b)

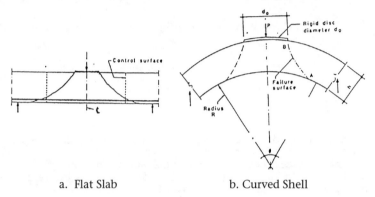

a. Flat Slab   b. Curved Shell

Figure 1   Punching Shear Problem in Flat Slab and Shell

The punching shear provisions in design codes are not directly
applicable to the curved concrete elements, because they are based on flat
slabs and therefore, result conservatively in larger thickness. A limited
study is available which deals with the punching shear strength for
curved members. This indicates a need to find a simple and convenient
method of design. Sabnis and Shadid [8] studied test results from McLean
[4] and compared the punching shear provision in the ACI code and an
equation developed by Solanki and Sabnis [5]. This paper presents the
continuation of the previous work and conclusions from studies of more
extensive work [7].

In this paper a logical approach based on a statistical analysis of the
available results is presented, which should result in a simpler yet realistic
method for the design of curved shells.

Factors Affecting Punching Shear Strength

Number of factors affect punching shear strength of both flat and curved
shells. These are reviewed in detail in Ref. [7] and are briefly mentioned
here. They include: concrete strength; amount of flexural reinforcement;
slab or shell thickness; size of the loaded area; shape of the loaded area;
interaction of shear and flexure; and radius and angle of shell (curvature).

Prediction of Punching Shear Strength

Many equations have been proposed for predicting punching shear resistance of concrete slabs using some of the above factors, which result in a large variation in the predicted strength. These are cited here and their details are discussed in Ref. [7]. Various models for punching shear were developed by Kinnunen et al., Nielsen, Anis, Ashtari, Braestrup and Morley using theory of plasticity. Loseth, Slatto and Syvertsen, Bruggeling and Ostlander, Caldwell and Billington and Braestrup, Nielsen, Jensen and Bach. Birdy, Bhula, Smith and Wicks, Regan and Hamadi and McLean studied the problem for Arctic concrete platforms using scaled models and developed prediction equations. Recently, Solanki and Sabnis showed that the curved shell differs from slab, in which the load introduces compressive membrane forces and also the increase in actual wall thickness due to curvature. The following equation showed a much better prediction, in which new parameters related to membrane stresses and angle of curved shell were introduced.

$$P_u = \pi \, \mu \, h(d_0 + h) \sqrt{f_c^*} \tag{1}$$

where :

μ = $\sqrt{(1 + f_m/f_c)}$

f_m = effective membrane stresses = $f_c' \sin\theta \cos\theta$

f_c = $10\sqrt{f_c'}$ psi $[0.85\sqrt{f_c'}$ MPa$]$

f_c^* = $4.22\sqrt{f_c'}$ psi $[0.35\sqrt{f_c'}$ MPa$]$

h = thickness of shell

d = diameter of the loaded area

θ = angle of curved shell element

Present Code Provisions

The design for punching shear for flat slabs is specified in various codes of practice or specifications. The provisions of three major codes of practice for concrete design are summarized here and as before discussed in detail in Ref. [7]. ACI 318-89 code [1], the European CEB-FIP code [2], and the British code CP110 [3]. It should be noted that there are no provisions for punching shear in curved concrete slabs.

ACI, CEB-FIP and CP110 indicate that strength by the CEB-FIP code is the most conservative. Furthermore, strength using CP110 code is more conservative than the ACI code, but the trend is reversed for larger reinforcement ratios and thicker slabs. It should also be noted that there are no provisions in any of the codes except the CEB-FIB code to account for the effect of in-plane compressive forces. The ACI code does not show any effect on punching shear strength of the amount of flexural reinforcement or the thickness of slab.

Method of Analysis

The analysis was carried out using 68 available test results [7]. As discussed earlier, the variables in this study can be: R, t, A, ρ, f'_c, u and θ. Multiple regression analysis was conducted for R, t, A, ρ and f'_c only, due to the insufficient data for the other two variables. In a general form, V can be expressed as a function of R, t, A, ρ and f'_c in the form of:

$$V = k. R^{b1}.t^{b2}.A^{b3}.\rho^{b4}.f'^{b5}_c \qquad (2)$$

Once the mathematical form is selected as described above the next step is to transform the data into a form suitable for this analysis. The model which will be used in this analysis is:

$$\text{Log } V = B_0 + B_1.\text{Log}R + B_2.\text{Log}(R/t) + B_3.\text{Log}A + B_4.\rho + B_5.\text{Log } f'_c \qquad (3)$$

Range of variables used in this analysis is shown below in Table 1.

Table 1 Range of Different Parameters used
in the Statistical Analysis

Variable	Unit	Minimum	Maximum
R	in.	10	216
R/t	-	4	36
Loaded Area	in^2	4	254.47
% Steel	-	0	9.48
f'_c	psi.	4,249	9,805
V_{test}	Kips	14.61	627.5

Discussion of Results

By applying the multiple linear regression analysis to all the data, the best model determined in this thesis has the following parameters :

$$B_0 = -0.564 \qquad B_3 = 0.357$$
$$B_1 = 1.093 \qquad B_4 = 0.014$$
$$B_2 = -0.159 \qquad B_5 = 0.331$$

Eq. 3 can be rewritten as:

$$V = 0.273 \, (R^{1.093}) \, ((R/t)^{-0.159}) \, (A^{0.357}) \, (\rho^{0.014}) \, (f'^{0.331}_c) \qquad (4)$$

Simplifying:

$$V = 0.273 \ (R^{0.089}) \ (t^{-0.159}) \ (A^{0.356}) \ (\rho^{0.013}) \ (f_c^{'0.33}) \qquad (5)$$

In general this model gave the best results when compared with the original prediction by different authors. The mean for V_{test}/V_{pr} was 1.015 compared with 1.297 for V_{test}/V_{auth}, the coefficient of variation (COV) was 4.32% compared with 72.2% and the standard deviation was 0.208 compared with 0.849. This equation gave a better fit with the test results compared with the original prediction by the authors. Effect of different variables was carried out for all the models obtained compared with the original prediction of the authors. This detailed work [5, 7] suggests that: a - the effect of the percentage of shear reinforcement and the effect of the shell thickness had *small* effect on the result; and b - the *most* effective factors were the loaded area, the radius of the shell, and the compressive strength of concrete. This model (V_{prop}) used to analyze the data and to obtain an equation which will be simpler and usable. The equation is:

$$V_{prop} = 0.106 \ \sqrt{R} \ \{R/t\} \ . \ ^{3}\sqrt{\{A.f_c'\}} \ . \ 10^{0.01\rho} \qquad (6)$$

When this equation was applied to the data, the predictions obtained are shown in Table 2 in which the mean for V_{test}/V_{prop} was 1.05, COV was 9%, and the standard deviation of 0.29. The results are slightly different than those obtained using equation 6, but still can be considered as a good prediction of punching shear strength.

In order to simplify equation 6, the loaded area was replaced by $A = \{a^2/4\} . \pi$ (for a circular loaded area) and $A = a^2$ (for a square loaded area), where: $a = R.\sin\beta$, and β = angle of the diameter of a circular loaded area or angle of the side length of a square loaded area. By applying these modifications to the same data, following equations were obtained:

a - for a circular loaded area:

$$V_{final_1} = 0.054 \ . \ R \ . \ t \ . \ \sin\beta \ . \ \sqrt{f_c'} \ . \ 10^{0.01\rho} \qquad (7)$$

and b - For a square loaded area:

$$V_{final_2} = 0.059 \ . \ R \ . \ t \ . \ \sin\beta \ . \ \sqrt{f_c'} \ . \ 10^{0.01\rho} \qquad (8)$$

Eqs. 7 and 8 provide an easy form to find the punching shear strength of shells, but it is less accurate than equation 6. As shown in Table 2 For V_{test}/V_{final} the mean was 1.14 , COV of 16%, and the standard deviation was 0.4. Eqns. (7, 8) still gave better results than those by the original authors.

Using the same procedure for shells without shear reinforcement, the following equation was obtained for V_{final} in kips:

$$V_{final} = 0.014 . R . t . \sin\beta . \sqrt{f'_c} \qquad (9)$$

Table 2 summarizes the test results and the different prediction equations to predict punching shear of shells for comparison as follows: V_{test} (test results); V_{auth} (original authors' prediction); V_{prop} (initial proposed equation); V_{final} (simpler form of V_{prop}); and $V_{c,ACI}$ (prediction by ACI code). These predictions indicate that equation 3 gives the best statistical result, but simplified form 9, is acceptable can be modified easily in the future as more test results become available. It also indicates that the ACI code prediction is very conservative. Figs. 2.a and b show the two final versions of these equations. It can be concluded further that the proposed Eqs. 7, 8 and 9 provide useful tool in predicting punching shear strength of reinforced concrete shells.

Conclusions

Tests of punching shear strength of curved slabs are relatively few in number, and the variables affecting shear strength have not been studied systematically. Therefore, it is not possible at this time to develop new design procedures which agree fully with the observed mechanisms of failure; this paper offers a statistical solution based on test data and gives a good correlation. Following conclusions are drawn from this study.

1. The shear strength increases as the shell curvature increases. The effect of curvature was more (about ten times) pronounced, when shear reinforcement was not included in the analysis.

2. Increase in the R/t ratio decreased the punching shear resistance of the shell.

3. Increase in the size of the loaded area increased the punching shear strength of the shell; however, its effect combined with that of the radius of the shell was more pronounced.

4. Increase in the amount of shear reinforcement increased the punching shear strength.

5. The larger the compressive strength of concrete, the greater was the punching shear strength. This effect was enhanced when R/t ratio was not included in the analysis.

6. Eqs. 7, 8 and 9 can be used for punching shear in reinforced concrete shells, much better than those available in the present ACI code.

References:

1. ACI Committee 318, "Building Code Requirements for Reinforced Concrete (ACI 318-89)," and Commentary--ACI 318R-89 American Concrete Institute, Detroit, Mich., 1989.

2. CEB-FIP, "Model Code for Concrete Structures," Comite Euro-International du Beton et Federation International de la Preconstrainte, London, 1978.

3. CP110: Part 1, "British Unified Code of Practice for the Structural Use of Concrete," British Standards Institute, London, 1972.

4. McLean, D.I., "A Study of Punching Shear in Arctic offshore Structures," Ph.D. Thesis, Cornell Univ., 1987.

5. Solanki, H. and Sabnis, G.M., "Punching Shear Strength of Curved Slabs," Indian Concrete Journal, July 1987, pp. 191-193.

6. Shadid, Amjad A., and Sabnis, G.M, Discussion of "The Paper Punching Shear of Behavior of Lightweight Concrete Slabs and Shells," by McLean, D., et. al., ACI Structural Journal, May-June 1991.

7. Shadid, A. A., "Punching Shear of Reinforced Concrete Shell Structures" M.E. Thesis, Department of Civil Engineering, Howard University, Washington, D.C., August 1992.

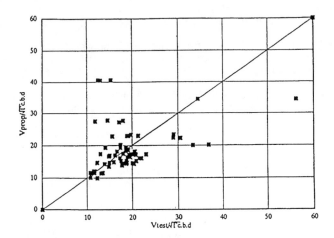

Figure 2.a Comparison between the Test results and Predicted
Values by the Proposed Model, V_{prop}

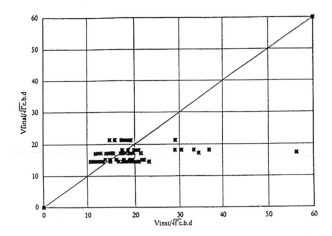

Figure 2.b Comparison between the Test results and Predicted
Values by the Proposed Model, V_{final}

Table 2 Comparison between V_{Test} and Predictions by Various Models, V_{auth}, V_{Prop}, V_{final} and V_{ACI}

specimen	V_{p1} (kip)	$\dfrac{V_{p1}}{\sqrt{f'_c}\,b\,d}$	$\dfrac{V_{test}}{V_{p1}}$	V_{prop} (kip)	$\dfrac{V_{prop}}{\sqrt{f'_c}\,b\,d}$	$\dfrac{V_{test}}{V_{prop}}$	V_c(ACI) (kip)	$\dfrac{V_c}{\sqrt{f'_c}\,b\,d}$	$\dfrac{V_{test}}{V_c(ACI)}$	V_{final} (kip)	$\dfrac{V_{final}}{\sqrt{f'_c}\,b\,d}$	$\dfrac{V_{test}}{V_{final}}$
1.	414.50	14.10	1.51	677.00	23.03	0.93	176.37	6.00	3.56	507.63	17.27	1.24
2.	404.50	13.95	1.12	663.40	22.88	0.69	174.00	6.00	2.61	495.20	17.08	0.92
3.	405.46	14.08	1.35	663.71	23.04	0.82	172.81	6.00	3.16	494.29	17.16	1.11
4.	407.32	14.05	1.37	666.77	22.99	0.84	174.00	6.00	3.20	497.71	17.16	1.12
5.	405.14	14.27	1.38	661.61	23.30	0.85	170.19	6.00	3.29	490.42	17.27	1.14
6.	481.03	17.31	0.89	1129.33	40.65	0.38	166.70	6.00	2.56	479.79	17.27	0.89
7.	469.77	17.30	0.72	1105.32	40.71	0.31	162.93	6.00	2.08	466.02	17.16	0.73
8.	472.18	17.25	0.76	1111.05	40.60	0.32	164.19	6.00	2.18	469.65	17.16	0.76
9.	32.25	36.28	0.95	30.72	34.55	1.00	5.33	6.00	5.73	15.15	17.31	1.99
10.	32.48	36.15	1.56	30.93	34.43	1.64	5.39	6.00	9.38	15.35	17.31	3.25
11.	37.01	15.67	1.21	34.85	14.76	1.28	14.17	6.00	3.16	35.02	14.83	1.28
12.	47.41	15.63	1.14	42.23	13.93	1.18	18.19	6.00	2.97	46.02	15.18	1.17
13.	32.89	18.08	1.08	30.01	16.49	1.15	10.92	6.00	3.25	27.61	15.18	1.29
14.	32.89	18.08	1.05	30.01	16.49	0.98	10.92	6.00	3.17	27.61	15.18	1.25
15.	44.88	16.06	0.87	39.96	14.30	0.96	16.77	6.00	2.34	42.42	15.18	0.92
16.	31.16	18.58	0.81	28.42	16.95	0.89	10.06	6.00	2.52	25.45	15.18	1.00
17.	31.16	18.58	0.88	28.42	16.95	0.96	10.06	6.00	2.72	25.45	15.18	1.08
18.	49.95	15.22	0.98	44.50	13.56	1.10	19.68	6.00	2.49	49.79	15.18	0.98
19.	34.64	17.60	1.23	31.62	16.07	1.34	11.81	6.00	3.60	29.87	15.18	1.42
20.	34.64	17.60	1.25	31.62	16.07	1.37	11.81	6.00	3.67	29.87	15.18	1.45
21.	28.69	17.29	1.21	25.30	15.25	1.37	9.95	6.00	3.48	24.27	14.63	1.43
22.	28.69	17.29	1.09	25.30	15.25	1.23	9.95	6.00	3.13	24.27	14.63	1.28
23.	28.69	17.29	1.02	25.30	15.25	1.16	9.95	6.00	2.94	24.27	14.63	1.20
24.	47.05	14.18	0.81	40.16	12.10	0.95	19.91	6.00	1.92	48.53	14.63	0.79
25.	47.05	14.18	0.83	40.16	12.10	0.97	19.91	6.00	1.95	48.53	14.63	0.80
26.	19.90	20.02	0.84	17.98	18.09	0.92	5.96	6.00	2.79	14.54	14.63	1.14
27.	19.90	20.02	1.03	17.98	18.09	1.14	5.96	6.00	3.43	14.54	14.63	1.41
28.	31.43	16.50	1.22	27.74	14.57	1.38	11.43	6.00	3.34	27.86	14.63	1.37
29.	31.43	16.50	1.06	27.74	14.57	1.24	11.43	6.00	3.01	27.86	14.63	1.23
30.	31.43	16.50	1.00	27.74	14.57	1.22	11.43	6.00	2.97	27.86	14.63	1.22
31.	51.55	13.53	0.86	44.04	11.56	1.17	22.86	6.00	2.26	55.72	14.63	0.93
32.	51.55	13.53	1.01	44.04	11.56	1.01	22.86	6.00	1.95	55.72	14.63	0.80
33.	21.80	19.11	1.01	19.71	17.28	1.12	6.84	6.00	3.22	16.69	14.63	1.32
34.	21.80	19.11	1.21	19.71	17.28	1.33	6.84	6.00	3.84	16.69	14.63	1.58

Table 2 Comparison between V_{Test} and Predictions by Various Models, V_{auth}, V_{Prop}, V_{final} and V_{ACI} (continued)

specimen	V_{p1} kip	$\dfrac{V_{p1}}{\sqrt{fc}.b.d}$	$\dfrac{V_{test}}{V_{p1}}$	V_{prop} kip	$\dfrac{V_{prop}}{\sqrt{fc}.b.d}$	$\dfrac{V_{test}}{V_{prop}}$	$V_o(ACI)$ kip	$\dfrac{V_o}{\sqrt{fc}.b.d}$	$\dfrac{V_{test}}{V_c(ACI)}$	V_{final} kip	$\dfrac{V_{final}}{\sqrt{fc}.b.d}$	$\dfrac{V_{test}}{V_{final}}$
35.	67.97	12.13	0.89	56.96	10.16	1.07	33.62	6.00	1.81	81.97	14.63	0.74
36.	31.03	16.61	0.91	27.38	14.66	1.03	11.21	6.00	2.51	27.32	14.63	1.03
37.	21.52	19.24	0.68	19.46	17.39	0.75	6.71	6.00	2.18	16.37	14.63	0.89
38.	50.89	19.24	0.94	19.46	17.39	1.04	6.71	6.00	3.01	16.37	14.63	1.24
39.	31.03	13.62	0.79	43.17	11.64	0.93	22.42	6.00	1.80	54.65	14.63	0.74
40.	69.95	16.61	0.74	27.38	14.66	0.84	11.21	6.00	2.05	27.32	14.63	0.84
41.	31.93	11.95	1.03	58.63	10.02	1.23	35.11	6.00	2.05	85.60	14.63	0.84
42.	52.37	16.37	1.25	28.19	14.45	1.41	11.70	6.00	3.40	28.53	14.63	1.39
43.	31.93	13.42	0.82	44.74	11.47	0.96	21.41	6.00	1.83	57.07	14.63	0.75
44.	52.37	16.37	1.15	28.19	14.45	1.30	11.70	6.00	3.13	28.53	14.63	1.28
45.	61.91	13.42	0.98	44.74	11.47	1.15	23.41	6.00	2.20	57.07	14.63	0.90
46.	67.12	22.13	0.85	50.99	18.22	1.03	16.79	6.00	3.12	59.43	21.24	0.88
47.	58.93	21.23	0.85	55.31	17.50	1.04	18.97	6.00	3.02	67.14	21.24	0.85
48.	64.24	22.13	0.86	48.97	18.56	1.03	15.83	6.00	3.20	56.05	21.24	0.90
49.	35.93	19.50	0.89	53.09	16.11	1.08	19.77	6.00	2.90	69.99	21.24	0.82
50.	58.93	27.24	1.07	48.97	23.39	1.24	7.92	6.00	4.84	28.02	21.24	1.37
51.	74.93	17.88	0.90	61.29	14.86	1.08	19.77	6.00	2.67	69.99	21.24	0.75
52.	52.79	20.31	0.74	44.08	16.61	0.90	22.14	6.00	2.49	78.38	18.20	0.70
53.	52.76	20.79	0.96	44.05	17.36	1.14	15.23	6.00	3.17	46.20	18.20	1.09
54.	53.80	20.80	1.00	44.92	17.37	1.20	15.22	6.00	3.47	46.15	18.20	1.14
55.	54.25	20.59	0.98	45.30	17.20	1.17	15.67	6.00	3.36	47.53	18.20	1.11
56.	38.21	20.51	1.01	32.62	17.12	1.21	15.87	6.00	3.47	48.14	18.20	1.14
57.	38.79	23.64	1.56	33.11	20.18	1.83	9.70	6.00	6.15	29.41	18.20	2.03
58.	42.62	23.46	1.42	35.53	20.03	1.66	9.92	6.00	5.55	30.08	18.20	1.83
59.	43.22	23.20	0.81	36.03	19.34	0.97	11.02	6.00	3.12	33.43	18.20	1.03
60.	31.17	23.03	0.76	36.57	19.20	0.91	11.26	6.00	4.14	34.14	18.20	0.96
61.	31.46	26.23	1.17	26.57	22.36	1.29	7.13	6.00	4.82	21.62	18.20	1.59
62.	182.84	17.43	0.68	26.82	22.26	1.37	7.23	6.00	5.07	21.93	18.20	1.67
63.	190.95	17.24	1.00	289.32	27.58	0.43	62.95	6.00	2.87	179.15	17.08	0.69
64.	187.38	17.61	1.03	301.61	27.23	0.63	66.45	6.00	3.01	190.16	17.17	1.00
65.	182.48	17.65	0.83	295.29	27.75	0.65	63.84	6.00	2.45	183.71	17.17	1.05
66.	164.85	15.39	0.92	288.12	27.86	0.53	62.04	6.00	2.36	177.55	17.17	0.86
67.	155.67	16.03	1.09	207.46	19.36	0.73	64.29	6.00	2.92	182.95	17.08	0.83
68.				195.41	20.12	0.87	58.28	6.00		166.78	17.17	1.02
mean			1.02			1.05			3.13			1.14
var			0.04			0.09			1.39			0.16
std			0.21			0.29			1.18			0.40

Stability Analysis of Stiffened Plates and Shells

U. Jäppelt, H. Rothert [1]

Abstract

A stability analysis of stiffened plates and shells by using degenerated finite elements is presented. The studies are based on a thin shell theory, large deflections, but on an elastic material law and small strains. Weight optimization by applying stiffeners results in sensitive stability problems because eigenvalues may cluster. Such problems can only be treated numerically by using reliable and robust solvers and path-following algorithms.

1. Introduction

The load-bearing capacity of shells is strongly increased by adding stiffeners. Despite the many investigations to determine the most economical pattern for these stiffeners only few accurate results are available. The limit load capacity of stiffened structures depends on many parameters, such as shell geometry, imperfections and cross-section, eccentricity, slenderness and location of the stiffener.

When designing such stiffened shell structures one has to realize the existence of different failure modes, such as local and global buckling or overall buckling. In this connection local buckling means the failure of the stiffeners or the sheet between the stiffeners alone, while global buckling means the failure of the shell together with one or more stiffeners. It has been shown in experiments (e. g. Eßlinger [8]) that narrow panelled shells fail due to overall buckling, while the wide panelled shells fail due to local buckling. To obtain the optimal structure one has to find that shell thickness and that stiffener spacing which leads to simultaneous global and local buckling. In the same way different cross–sections of the stiffener (e.g. T– or L– cross–sections) can be designed. Bushnell proves

---

[1] Univ.-Prof. Dr.-Ing. H. Rothert, Dipl.-Ing. U. Jäppelt, Institut für Statik, Universität Hannover, Appelstraße 9a, 30167 Hannover, Germany

in [4] that it is necessary to incorporate the interaction between the shell and the stiffener. He demonstrates that the consideration of interaction leads to the prediction of lower buckling loads than those obtained from simple formulas in which each type of buckling is considered separately.

A second problem concerning stiffened shells is the accurate modelling of the physical properties in order to guarantee realistic numerical results. For shells with closely spaced stiffeners the influence of the stiffener may be incorporated by an orthotropic material law (Rothert/Dehmel [11]; Walker/Sridharan [19]). However, for a given shell with "smeared" stringers there is a minimum number of stringers which are needed to ensure that overall buckling is the failure mode, and not local buckling ([15]). For sparsely stiffened shells the "smeared method" is not valid and it becomes necessary to discretize shells and stringers separately.

In the present paper two models of stiffeners are compared. Most of the previous applications of the finite element method used shell or plate elements to describe the stiffener discretely. The results presented in the following are obtained for degenerated shell elements of the "assumed stress" family that are coupled with degenerated rod elements of the Mindlin type (e.g.: Bernhardi [3]; Diack [5]; Ferguson/Clark [7]; Weimar [20]; Wetzel [23]). The mixed formulation together with the inclusion of rod eccentricities, moderate rotations, and the warping of the stiffeners proves to be a realistic modelling, especially of the situation in the interface of the shell and the stiffener.

The numerical results presented will also include a comparison of the critical loads of a cylinder where the stiffeners are either located at the inside or the outside of the cylinder. Different ratios of the stiffener slenderness $\frac{d_s}{t_s}$ are studied and the influence of the imperfections will be shown. To prove the validity of the model used by the authors the convergence of the solution with increasing numbers of finite elements has been studied.

2. Numerical Solution Procedure

2.1 The Finite Elements Employed

The basic assumptions made in the analysis are as follows: large deflection theory of thin shells, moderate rotations, elastic material law and small strains. The displacement field for the doubly degenerated isoparametric eccentrically located rod element (Fig. 1a) is given by

$$u_i = \sum_k h_k u_i^k + \sum_k h_k \left(\frac{t}{2}a_k + e_t\right) \Delta n_{ti}^k + \sum_k h_k \left(\frac{s}{2}b_k + e_s\right) \Delta n_{si}^k , \qquad (2.1)$$

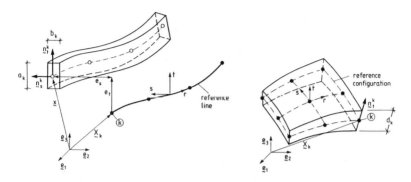

Figure 1: a) Degenerated rod element, b) degenerated shell element

and for the degenerated isoparametric shell element (Fig. 1b) by

$$u_i = \sum_k h_k u_i^k + \Theta \sum_k h_k \frac{d_k}{2} \Delta n_{ti}^k . \tag{2.2}$$

By applying the assumption that the shape of the cross-section of the rod element will not change, it is possible to describe the rotations of n_{ti}^k and n_{si}^k with three rotational degrees of freedom. Now each node can be described by the three global components of the displacement u_i, v_i and w_i at the reference line nodes in addition to three local rotations α_i, β_i and γ_i. Each node of the shell element is described in the well-known formulation of Bathe [2] with three components of the displacement and only two components of the rotation. By coupling shell and eccentrically located rod elements the stiffness of the third rotational degree of freedom γ_k remains free and uncoupled. The procedure for coupling has been published in detail in [12].

The computation of the tangent stiffness matrix $\mathbf{K}_T$ as well as the load vectors follows the standard finite element procedure. The tangent stiffness matrix is given by

$$\mathbf{K}_T = \mathbf{K}_l + \mathbf{K}_g , \tag{2.3}$$

with the linear stiffness matrix

$$\mathbf{K}_l = \int_V \mathbf{B}_L^T \, \mathbf{C} \, \mathbf{B}_L \, dV , \tag{2.4a}$$

and the geometrical stiffness matrix

$$\mathbf{K}_g = \int_V \mathbf{B}_{NL}^T \, \mathbf{S} \, \mathbf{B}_{NL} \, dV . \tag{2.4b}$$

The internal force reaction is given by

$$f = \int_V \mathbf{B}_L^T \, \hat{\mathbf{S}} \, dV \, . \tag{2.4c}$$

2.2 Stability Analysis

The necessary condition for elastic stability is that the second variation of the potential energy is zero. This leads to the criterion:

$$\det (\mathbf{K}_T) = 0. \tag{2.5}$$

The calculation of $\det (\mathbf{K}_T) = 0$ is equivalent to the solution of the eigenvalue problem

$$[\mathbf{K}_T - \omega_j \mathbf{I}] \, \phi_j = 0 \, , \tag{2.6}$$

where ω_j is the jth eigenvalue, and ϕ_j is the corresponding eigenvector. The classical approach in the analysis of elastic stability involves the linearization of the prebuckling state. This leads to an eigenvalue problem for the load parameter. When proportional loading is applied the stability criterion can be written as

$$\mathbf{P} = \lambda \, \hat{\mathbf{P}} \, , \tag{2.7}$$

with the reference vector of the load $\hat{\mathbf{P}}$ and with the scalarvalued load-intensity parameter λ. This leads to the nonlinear eigenvalue problem

$$\left[\mathbf{K}_e + \lambda_j(\mathbf{K}_g^L + \mathbf{K}_u^L) + \lambda_j^2(\mathbf{K}_g^{NL} + \mathbf{K}_u^{NL})\right] \phi_j = 0 \, . \tag{2.8}$$

Several steps of approximation are possible. To obtain the linear stability equations it is assumed that the higher order nonlinear part of the stiffness matrix $\mathbf{K}^{NL}$ is neglected. In the total Lagrangian formulation one obtains

$$\left[{}_0^t\mathbf{K}_e + {}_0^t\mathbf{K}_u + \lambda_j \, {}_0^t\mathbf{K}_g^L\right] {}^t\phi_j = 0. \tag{2.9}.$$

For the numerical results presented in this paper we used a better approximation given by

$$\left({}_0^t\mathbf{K} + \lambda_j \, {}_0\Delta\mathbf{K}\right) \phi_j = 0 \, , \tag{2.10}$$

with the increment of the stiffness matrix ${}_0\Delta\mathbf{K} = {}_0^{t+\Delta t}\mathbf{K} - {}_0^t\mathbf{K}$. In this formulation the updated and the total Lagrangian formulation are equal (Ramm [9]).

2.3 Nonlinear Finite Element Analysis

To analyse the complicated load - deflection behaviour of stiffened plates and shells there are some tools available which enable the application of the path following to localize critical points and the path after bifurcation (Rothert/Gebbeken [13]).

For the incremental-iterative solution of the nonlinear equations the arc-length method proposed by Wessels in [22] from the authors' institute will be used.

The current stiffness parameter (CSP) and the determinant method (DET) are used to find critical values, such as snap-through points and bifurcation points. The current stiffness parameter is the ratio of the normalized energy in the first solution step and the normalized energy of the current state:

$$\text{CSP} = \frac{\Delta_1 \mathbf{v}^T \mathbf{P}}{\Delta_i \mathbf{v}^T \mathbf{P}}. \tag{2.12}$$

At a snap-through point the current stiffness parameter **and** the determinant have the value zero. At a bifurcation point only the value of the determinant becomes zero.

EXAMPLE 1: Hinged arc: bifurcation/snap-through problem
This problem is documented in [9] and [18], with which we can compare our results. The load - deformation curve is shown in Fig. 3, the values of the determinant and the current stiffness parameter are added. The curves of the current stiffness parameter and the determinant show the localization of the bifurcation point and the snap-through point.

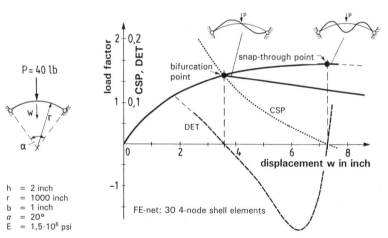

Figure 2: Hinged arc a) system and loading
b) load vs. displacement curve, CSP, DET

3. Convergence Study

Two examples are presented to illustrate the accuracy and economic advantage
of the coupling of the shell and eccentrically located rod elements.

EXAMPLE 2: Stiffened panel: displacement analysis
The deflections of the eccentrically stiffened panel subjected to an end load
are considered as an illustration of the coupling. The geometry and material
properties of the panel are given in Figure 3a. With a concentrated end force
of 25.2 N, the deflection calculated with the beam theory is 0.907 mm and the
normal stress at the lower bound is $-225.88 \frac{N}{mm^2}$. The convergence for various
mesh refinements is shown in Figure 3b and c.

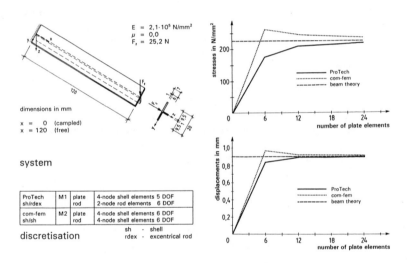

Figure 3: Eccentrically stiffened panel, a) system,
b) normal stresses for x = 0, c) displacements for x = 120

EXAMPLE 3: Stiffened panel: buckling analysis
The panel shown in Figure 4 has been analysed by Bernhardi in [3] using different elements for the stiffener. Our own results agree well with those of
Bernhardi. However, it has to be noted that the problem of plate buckling with
thin stiffeners cannot be handled by the rod element presented here.

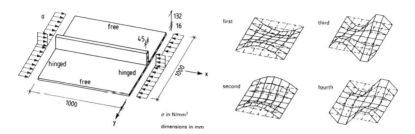

Figure 4: Eccentrically stiffened panel, a) system, b) buckling modes

4. Isotropic and Orthotropic Cylinder

In order to reduce the CPU time it is more efficient to perform the calculation of
a cylindrical shell with subsystems. With respect to the critical buckling pattern
this subsystem was chosen in accordance with the expected wave numbers. The
imperfections are also assumed in accordance with the critical buckling pattern,
which has to be calculated by an eigenvalue analysis, as was described in chapter
2. Another possibility for determining the subsystem of isotropic or orthotropic
cylinders without a numerical eigenvalue analysis is the use of a nomogram,
which has been described in detail in [17]. For thin simply supported circular
shells under axial compression we obtained the nomogram for the stability
analysis, while the critical loadpoint and the slenderness ratio $\frac{l}{r}$ is drawn in a
double logarithmic scale versus the geometric curve parameter

$$ K = \sqrt{\frac{l}{r} \sqrt{\frac{t}{r}}} \, . \tag{4.1}$$

The nomogram provides the lowest critical load P_{cr} and the critical wave number n_{cr} for buckling modes with $m = 1$ as it will be shown in example 4. It is
worth mentioning, that different cylindrical shells with identical parameter K
show the same buckling behaviour, i.e. the same number of critical circumferential waves.

EXAMPLE 4: Cylinder: buckling analysis
For a simply supported stiffened cylinder with the smeared thickness $t_m =$
1.04 mm as given in Figure 5, we obtain a critical axial buckling stress of
$\sigma_{cr} = 377.4$ (381) $\frac{N}{mm^2}$ and a critical wave number of $n_{cr} = 9$ (9) by using
the nomogram. This coincides with the values calculated by the ECCS [1],
given in brackets. Assuming a critical wave number of 10 for the FE analysis

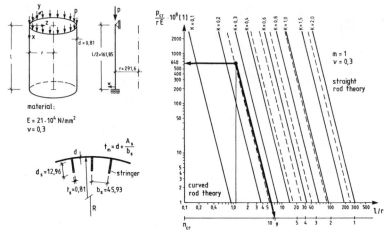

Figure 5: Cylinder, a) System, b) Nomogram for $\nu = 0.3$, m = 1

the cylinder can be represented by an 18 degree subsystem. This preliminary investigation greatly reduces the CPU time for parametric studies.

5. Parametric Studies

Different parametric studies have been carried out for the cylindrical shell presented in chapter 4. The influence of the location of the stiffeners has been examined.

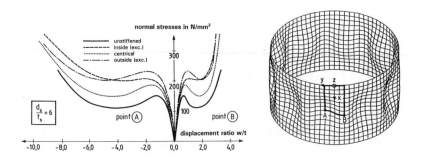

Figure 6: a) Variation of the location of the stiffeners
b) buckling pattern for $\sigma_{cr} = 195 \frac{N}{mm^2}$

Previous studies (e.g. [8], [16]) showed that outside stiffeners are more efficient than inside ones. The parametric studies at the cylindrical shell UC9 [19] have been carried out for four cases: the shell being unstiffened, having central stringers and with inside or outside stiffeners. The results for the stiffener slenderness $\frac{d_s}{t_s} = 6$ are given in Figure 6. It can be noted that the positive effect of the outside stiffeners will decrease and the load nearly reaches the post-buckling load of the shell stiffened on the inside when the displacement ratio $\frac{w}{t}$ has a certain value. The enhancement of the buckling load can be seen clearly. By adding 10% to the material one obtains a 29% higher buckling load when using centrical stiffeners, and an increase of 74% if eccentrically located stiffeners are used. The influence of the imperfection and the ratio of the stiffener slenderness are presented in Figure 7.

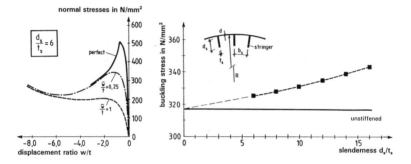

Figure 7: a) Variation of the imperfection,
b) variation of the slenderness of the stiffeners

Acknowledgement:
The substantial support of the German National Science Foundation (Deutsche Forschungsgemeinschaft) is gratefully acknowledged. The authors would like to thank Mr. Urrutia–Galicia for carrying out the nomogram.

References

[1] ECCS – Techn. Working Group 8.4: *Buckling of Steel Shells.* Europ. Rec., 4th Ed. (1988)

[2] Bathe, K.-J.: *Finite Element Procedures in Engineering Analysis.* Englewood Cliffs, Prentice-Hall (1984)

[3]   Bernhardi, O.: *Eine geometrisch nichtlineare Finite-Element-Formulierung für die Idealisierung exzentrischer Aussteifungen.* Inst. für Schiffbau, Konstruktion und Statik, RWTH Aachen (1985)

[4]   Bushnell, D.: *Buckling of shells – pitfall for designers.* AIAA Journal 19 (1981) 1183–1226

[5]   Diack, A.: *Beitrag zur Stabilität diskret längsversteifter Kreiszylinderschalen unter Axialdruck.* Bericht Nr. 3, Institut für Baustatik, Universität Stuttgart, (1983)

[6]   Eßlinger, M.: *Ein Verfahren zur theoretischen Untersuchung des Beul- und Nachbeulverhaltens dünnwandiger Kreiszylinder mit eingespannten Rändern.* DLR - Forschungsbericht, 68-70 (1968)

[7]   Ferguson, G. H.; Clark, R. D.: *A variable thickness, curved beam and shell stiffening element with shear deformations.* Int. J. Num. Meth. Eng. 14 (1979) 581–592

[8]   Neut van der, A.: *The general instability of stiffened cylindrical shells under axial compression.* Nationaal Luchtvaart-Laboratorium, Rep. (1946)

[9]   Ramm, E.: *Geometrisch nichtlineare Elastostatik und Finite Elemente.* Institut für Baustatik Bericht Nr. 76–2, Universität Stuttgart 1976

[10] Rothert, H.; Dickel, T.; Renner, D.: *Snap-Through Buckling of Reticulated Space Trusses.* ASCE, J. of Structural Division, Vol. 107, No. St1 (1981) 129-143

[11] Rothert, H.; Dehmel, W.: *Nonlinear Analysis of Isotropic, Orthotropic and Laminated Plates and Shells.* Comp. Meth. in Applied Mech. and Eng., 64 (1987) 429-446

[12] Rothert, H.; Gebbeken, N.; Jäppelt, U.: *Novel Aspects in Finite Element Modelling of Stiffened Shells.* IASS (1993)

[13] Rothert, H.; Gebbeken, N.: *On Numerical Results of Reticulated Shell Buckling.* Int. J. of Space Structures, Vol. 7, No. 4 (1992) 299-319

[14] Rothert, H.; Wetzel, M.: *Zur Berechnung versteifter Platten- und Schalenstrukturen.* SFB 151-Berichte Nr. 23, Ruhr-Universität Bochum, Hrsg.: Meskouris/Niemann, (1992) B11-B19

[15] Singer, J.; Haftka, R.T.: *Buckling of discretely stringer–stiffened cylindrical shells and elastically restrained panels.* AIAA Journal 13 (1975) 849–850

[16] Thielemann, W.; Eßlinger, M.: *Über den Einfluß der Exzentrizität von Längssteifen auf die axiale Beullast dünnwandiger Kreiszylinderschalen.* Der Stahlbau 11 (1965) 332-333

[17] Urrutia–Galicia, J. L.; Rothert, H.; Jäppelt, U.: *Stability analysis of axially compressed circular cylindrical shells.* (to be published)

[18] Wagner, W.: *Zur Behandlung von Stabilitätsproblemen der Elastostatik mit der Methode der Finiten Elemente.* Forschungs– und Seminarberichte aus dem Institut für Baumechanik und Numerische Mechanik Bericht Nr. F 91/1: Universität Hannover 1991

[19] Walker, A.C.; Sridharan, S.: *Analysis of the behaviour of axially compressed stringer-stiffened cylindrical shells.* Proc. Instn. Civil Engrs., Part 2, 69 (1980) 447-472

[20] Weimar, K.: *Ein nichtlineares Balkenelement mit Anwendung als Längssteifen axial belasteter Zylinder.* Bericht Nr. 10, Institut für Baustatik, Universität Stuttgart, (1989)

[21] Wessels, M.: *Das statische und dynamische Durchschlagproblem der imperfekten flachen Kugelschale bei elastischer rotationssymmetrischer Verformung.* Mitt. -Nr. 23, Inst. für Statik, Univ. Hannover (1977)

[22] Wetzel, M.: *Nichtlineare Berechnung versteifter elastischer Platten und Schalen mittels der Finite-Elemente-Methode.* Mitt. -Nr. 40-91, Institut für Statik, Universität Hannover (1991)

EFFICIENT NONLINEAR ANALYSIS OF COLUMN-SUPPORTED CYLINDERICAL SHELLS

P.L. Gould[1], R.V. Ravichandran[2] and S. Sridharan[3]

Abstract

This paper presents a finite element methodology for the study of geometrically nonlinear behavior of column-supported cylinder. Due to the discrete nature of the column supports, the shell is locally deviant from its axisymmetric geometry. As an alternative to the conventional two-dimensional finite element discretization of the shell domain, a local-global analysis technique is used which takes advantage of the axisymmetric characteristics of the shell, thus striving to achieve computational simplicity in terms of modeling and analysis efforts. An example of a column-supported cylinder is considered to demonstrate the use of local-global method. The use of this method for the study of the effect of localized imperfection on the load carrying capacity is also demonstrated.

Introduction

Large elevated silos can generally be classified as thin walled shells of revolution. They are cylindrical in shape with a conical hopper bottom,and are supported on discrete number of columns at the transition zone between the cylinder and the cone.

The granular materials, such as grains, which are generally stored in such silos exert internal pressure and a downward meridional traction due to friction along the inside surface of the shell. This, in addition to the discrete nature of the column supports, results in very high meridional compressive stresses in the vicinity of the column-shell junction. Also, the circumferential distribution of the stresses is far from uniform in this vicinity. Generally, in shells of revolution, buckling is the predominant mode of failure when nonuniform axial stress distribution is involved. Accordingly, buckling and nonlinear collapse behavior often govern the design of

---

[1] Harold D. Jolley Professor and Chairman, [2] Former Doctoral Student and
[3] Professor, Dept. of Civil Engineering, Washington University, St. Louis, MO 63130.

these shells. The critical load can be expected to be much lower than for a continuously supported shell. In addition, geometrical imperfections are known to play an even more vital role in reducing the load carrying capacity.

In view of the above, the nonlinear analysis of cylindrical thin shells of revolution supported on discrete column supports assumes significant importance.

Previous Work

Earlier studies on column supported silos concentrated on linear analysis and the determination of prebuckling stress distributions (Gould et al. 1976; Ory and Reimerdes 1987; Rotter 1982). Some classical solutions can be found in the literature addressing the problem of shell buckling under circumferentially varying loads (Bijlaard and Gallagher 1959; Jones 1966; Libai and Durban 1976, 1977). All of these works were based on linear bifurcation analysis of perfect shells. Since shells in general behave in a nonlinear manner under local loads, and are highly imperfection sensitive, the conclusions reached in these studies are of limited practical use.

In more recent studies, all of the influencing factors were duly accounted for in a straight forward manner by using the finite element method. Teng and Rotter (1990) presented such a study in which the geometric nonlinear analysis of column-supported cylinders were considered. Parametric effects of factors such as the number of column supports and the presence of imperfections on the ultimate load carrying capacity were observed. The effects of combined material and geometric nonlinearities on similar column-supported cylinders were studied by Guggenberger (1991). Here, the stiffened shells were considered as well, and some knock-down factors useful for design were proposed.

The above mentioned two studies used commercial finite element analysis packages LUSAS (1989) and ABAQUS (1989) respectively. Eventhough the geometry of the problems considered was predominantly axisymmetric, the programs required the use of complete two-dimensional discretization. This is because of the fact that the shell has lost its axisymmetric characteristic, albeit locally, due to the discontinuous nature of the column supports. This requires significant computational resources in terms of storage and CPU time, particularly for a nonlinear analysis.

Fortunately, there is a special purpose computer program to efficiently treat shells of revolution with local deviations such as the column-supported cylinder (Han and Gould 1986). Developed at the Department of Civil Engineering, Washington University, St. Louis, this program has been undergoing continuous upgradation, and is based on the local-global analysis scheme. The capability of this package has been well documented in literature for linear elastic (Han and Gould 1984, 1986) and localized plasticity (Lin and Gould 1987) problems of shells of revolution. The local-global methodology consists of partitioning the shell into an axisymmetric zone and a local zone. The local zone which is nonaxisymmetric due to some geometric deviation or nonaxisymmetric loads is modeled using the isoparametric general shell elements (two-dimensional discretization). The remaining axisymmetric zone is modeled using the rotational shell elements (one-dimensional discretization). This hybrid discretization method, when coupled together using either transitional elements or a transformation scheme (Ravichandran et al. 1992), offers a robust single-pass analysis procedure for shells of revolution with local deviations. Recently, this methodology was further utilized to address the problem of shells of revolution with localized collapse mechanisms (Ravichandran et al. 1992). In this, all of the problem nonlinearities were assumed to emanate from the local zone which also contained some nonaxisymmetric imperfections. The axisymmetric zone remained linear elastic resulting in significant savings in the computational effort.

The geometric characteristic and the collapse modes of the column supported silos make them ideal candidates for the use of the local-global model for the nonlinear analysis. The objective of this paper is to demonstrate the application of the local-global for this problem. A cylindrical shell supported on eight columns, and is subjected to a uniform downward meridional traction is chosen as the example. The effect of the imperfection on the load carrying capacity is also studied.

Analytical Model

Figure 1 shows a shell of revolution divided into substructures I and II. Substructure I is the axisymmetric zone wherein the geometry of the structure permits the use of rotational shell elements. Thus, in this zone, the discretization is one dimensional. The nonaxisymmetric mode of deformation in this zone is captured through the use of appropriate Fourier harmonics in the semi-analytical formulation. In this study, the degenerated isoparametric axisymmetric shell elements (Ahmad et al. 1968) are used to model the axisymmetric zone.

Substructure II is the local zone where there is deviation from axisymmetry of the shell. This happens, in the case of column-supported cylinders, due to the discontinuous nature of the column supports. Two- dimensional degenerated isoparametric shell elements are used to model this zone. Since this zone also contains much of the problem nonlinearity due to the local effects, accurate modeling is necessary. The Heterosis shell element (Hughes and Liu 1981) is employed in this study. Experience indicates that this is a satisfactory element to predict the collapse load accurately in shell problems (Ravichandran and Sridharan 1989).

In order to achieve compatibility between the local and axisymmetric zones, a transformation procedure is adopted. First, a transformation matrix is constructed expressing the compatibility between the discrete displacement components of the nodes of the local zone elements that lie on the common nodal circle and the Fourier harmonic amplitudes of the displacement components of the common nodal circle. This matrix, then, is used to transform the condensed stiffness matrix and the load vector corresponding to the entire local zone. The transformed stiffness matrix and load vector, now in terms of the Fourier coefficients of the displacements of the common nodal circle, can be directly assembled with those of the rest of the shell and solved. More details of this procedure along with the nonlinear solution algorithm for localized collapse problems are documented in literature (Ravichandran et al. 1992).

Recently, this local-global strategy was further enhanced by considering the geometric nonlinearity effects in the axisymmetric zone as well. Despite the coupling between the harmonics in the nonlinear regime, the advantages of the rotational shell elements such as the one-dimensional discretization are still retained by adopting the pseudo load method and a special conjugate gradient type nonlinear solution algorithm (Wunderlich et al. 1985). Performance of this enhanced version of the local-global strategy has been demonstrated through numerical examples (Ravichandran 1993).

Nonlinear Analysis of Column-supported Cylinder

Figure 2 shows the idealized column-supported cylinder considered in this paper, which is subjected to an uniform meridional traction on the inside surface. This load closely approximates the frictional force exerted by the granular solid stored in silos. This example has been extensively studied by Teng and Rotter (1990) and provides a basis for evaluating the performance of the present local-global model. Figure 3 shows the layout of the local-global finite element mesh.

For an axisymmetrically loaded cylinder supported on n number of columns, there exist 2n rotational symmetry axes around the circumference. Therefor only 1/2n of the cylinder needs to be modelled in the local zone. In this study, n is assumed to be equal to 8. Sixty four 9-noded Heterosis elements are used for modeling the local zone. Due to the nonuniform circumferential distribution of stresses near the column supports, the mesh is more refined in this vicinity than the region away from it. Eight quadratic axisymmetric shell elements are used to model the axisymmetric zone, and the effects of geometric nonlinearity are included. The following boundary conditions are adopted: at the top and bottom edges of the cylinder all displacement components except the vertical components are constrained; symmetry boundary conditions are imposed along the nodes on the vertical edges of the local zone. The columns are assumed to be rigid supports.

Behavior of Perfect Cylinder

Figure 4 shows the relationship between the meridional traction and the inward radial displacement of the point at z/t=60 above the column center. The results obtained correspond to the case whwrein the nonaxisymmetric deformation of the axisymmetric zone is characterized by two Fourier harmonics namely 0 and 8. An out-of-balance force norm of 1% is used for terminating the Newton-Raphson iterations within each load increment. The comparison between the present results and those obtained by Teng and Rotter (1990) shows and excellent agreement. At higher load levels, the local-global model predicts a slightly stiffer response, due to which the computed critical load is higher. This difference is expected to vanish on mesh refinement.

Behavior of Imperfect Shell

Next, for the same shell considered above, an axisymmetric meridional imperfection is imposed, as shown in Figure 5, which is of the form (Teng and Rotter 1990)

$$\delta = \delta_o \, e^{-\frac{\pi z}{\lambda}} \left[\sin\left(\frac{\pi z}{\lambda}\right) + \cos\left(\frac{\pi z}{\lambda}\right) \right] \tag{1}$$

where

$$Z = Z - Z_o \tag{2}$$

This imperfection resembles a typical weld depression in circumferentially welded steel cylinders. The maximum amplitude δ_o is the amplitude located at a height $z/t=60$ from the top of the column. Again, the radial inward displacement is plotted against the applied meridional traction. Figure 6 shows the results obtained by the present local-global model in comparison to those obtained by Teng and Rotter (1990). The agreement can be seen to be excellent.

Conclusions
        This paper explored the possibility of utilizing a local-global nonlinear analysis technique for studying the geometrically nonlinear collapse behavior of column-supported cylinders. The objective is to minimize the modeling and computational effort by the judicious use of two-dimensional and one-dimensional discretization simultaneously. From the numerical study, it can be concluded that this technique is an effective analysis tool for the study of nonlinear collapse of column-supported cylinders. Besides its simplicity, the method offered solutions with commendable accuracy in comparison to complete two-dimensional discretization.

Appendix: References

ABAQUS (1989), Theory Manual and User's Manual, Version 4.8, Hibbit, Karlsson & Sorensen, Inc., Providence, Rhode Island.

Ahmad, S., Irons, B.M. and Zienkiewicz, O.C. (1968), "Curved thick shell membrane elements with particular reference to axisymmetric problems." in *Proc. Second Conf. Matrix Methods in Structural Mechanics*, Wright-Patterson A.F. Base, Ohio.

Bijlaard, D.L. and Gallagher, R.H. (1959), "Elastic instability of a cylindrical shell under arbitrary circumferential variation of axial stresses." *J. of Aerospace Sciences*, 27, 854-858.

FEA (1989), LUSAS User's Manual and LUSAS Theory Manual, Version 9, Finite Element Analysis, Ltd., Surrey, U.K.

Gould, P.L., Sen, S.K., Wong, R.S.C. and Lowrey, D. (1976), "Column supported cylindrical conical tanks." *J. Struct. Divn., ASCE*, 102(ST2), 429-447.

Guggenberger, W. (1991), "Buckling of cylindrical shells under local axial loads." in *Buckling of Shell Structures, on Land, in the Sea and in the Air*, (Ed.) J.F. Jullien, Elsevier Applied Science, N.Y.

Han, K.J. and Gould, P.L. (1984), "Shells of revolutions with local deviations." *Int. J. Numer. Methods Engrg.*, 20, 305-313.

Han, K.J. and Gould, P.L. (1986), "Local-global analysis of shells of revolution." in *Finite Element Methods for Plates and Shells*, Hughes, T.J.R. et al. (eds.), Pineridge press, Swansea, U.K.

Hughes, T.J.R. and Liu, W.K. (1981), "Nonlinear finite element analysis of shells: part I, three dimensional shells." *Comp. Methods Appl. Mech. Engrg.*, 26, 331-362.

Jones, D.J. (1966), "On the linear buckling of circular cylindrical shells under asymmetric axial compressive stress distribution." *J. of Royal Aeronautical Society*, 1095-1097.

Libai, A. and Durban, D. (1976), "A method for approximate stability analysis and its application to circular cylindrical shells under circumferentially varying loads." *J. of Appl. Mech.*, ASME, 40, 971-976.

Libai, A. and Durban, D. (1977), "Buckling of cylindrical shells subjected to nonuniform axial loads." *J. Appl. Mech.*, ASME, 44, 714-720.

Lin, J.S. and Gould, P.L. (1987), "Shells of revolution with localized plasticity." *Comp. Methods Appl. Mech. Engrg.*, 65, 127-145.

Ory, H. and Reimerdes, H.G. (1987), "Stresses in and stability of thin walled shells under non-ideal load distribution." *Proc. Int. Colloquium on Stability of Plate and Shell Structures,* Gent, Belgium, April 6-8, 555-561.

Ravichandran, R.V. and Sridharan, S. (1989), "A comparative study of the performance of isoparametric shell elements in geometrically nonlinear analysis." *Research Report NO. 82*, Department of Civil Engineering, Structural Division, Washington University, St. Louis.

Ravichandran, R.V., Gould, P.L. and Sridharan, S. (1992), "Localized collapse of shells of revolution using a local-global strategy." *Int. J. Numer. Methods Engrg.*, 35, 1153-1170.

Ravichandran, R.V. (1993), "Nonlinear Local-Global Analysis of Shells of Revolution.", *D.Sc Dissertation*, Dept. of Civil Engineering, Washington University, St. Louis.

Rotter, J.M. (1982), "Analysis of ringbeams in column-supported bins". *Proc. Eighth Australasian Conference on the Mechanics of Structured and Materials,* University of Newcastle, August.

Teng, J.G. and Rotter, J.M. (1990), "A study of buckling in column-supported cylinders." in *Contact Loading and Local Effects in Thin-walled Plated and Shell Structures*, Proc. IUTAM Symposium, Prague, 52-61.

Wunderlich, W., Cramer, H. and Obrecht, H. (1985), "Application of ring elements in the nonlinear analysis of shells of revolution under nonaxisymmetric loading." *Comp. Methods Appl. Mech. Engrg.*, 51, 259-275.

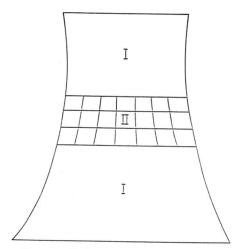

Figure 1. Shell of revolution with substructures

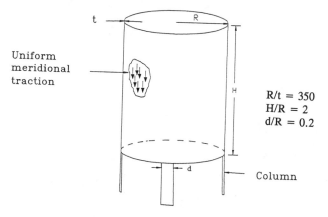

$$R/t = 350$$
$$H/R = 2$$
$$d/R = 0.2$$

Figure 2. Column-supported cylinder

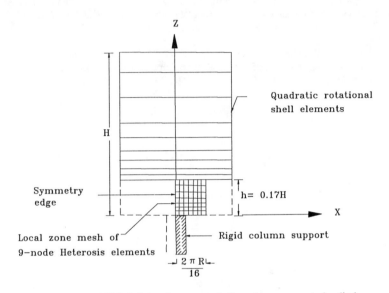

Figure 3. Local-global finite element mesh for column-supported cylinder

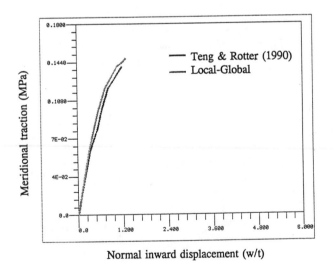

Figure 4. Perfect Cylinder on 8 columns

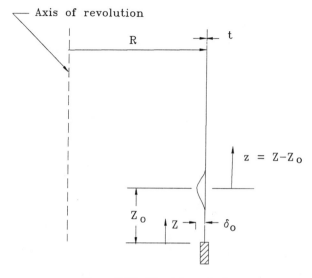

Figure 5. Meridional imperfection

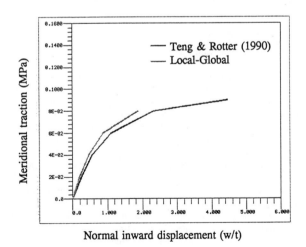

Normal inward displacement (w/t)

Figure 6. Imperfect cylinder on 8 columns

Elasto-Plastic Buckling Design for Shallow Cylindrical Shells under External Pressure

Seishi Yamada [1] and James G. A. Croll [2]

Abstract

Pressures providing elastic-plastic buckling conditions for thin shallow cylindrical shell panels are presented. Classical simple support boundary conditions have been adopted for the shells. First surface yield analysis using the elastic, fully non-linear, Ritz method and the von Mises hypothesis of plastic yield, has indicated broadly two types of failure: a mid-shell failure which is very sensitive to imperfections, and; a corner failure which can be closely approximated from a linear bending stress analysis. For mid-shell failure an imperfect, reduced stiffness, model is shown to provide a conservative estimation to the elasto-plastic lower bound pressures.

1. Introduction

A critical design condition for cylindrical barrel vault roof structures is that of providing adequate resistance to buckling under the combined actions of self-weight, snow loading, and, in seismic regions, the possibly amplified gravity forces arising from a dynamically induced response. Inter-stiffener shell elements within orthogonally stiffened cylindrical shells will for pressure loading involve a similar design condition. However, there is still no clear consensus on how such shallow cylindrical shells should be designed to prevent buckling.

It is widely recognized that under pressure loading the nonlinear snap buckling exhibited by even geometrically perfect shallow cylindrical shell panels is considerably more complex than the behavior of a complete cylinder. A reinterpretation of cylindrical panel buckling has shown, however, that they too can be modeled in terms of a classically bifurcating system. It was demonstrated in Yamada and Croll (1989) that classical bifurcation pressures are approached when geometric imperfections are chosen so that they effectively nullify the inherent loading imperfections. In this context the loading imperfection can be consider to be the bending distortions, to the otherwise uniform radial contraction, offered by the longitudinal supporting edges. Lower bounds to the imperfection sensitive elastic buckling are then provided by a 'reduced stiffness' extension to the classical

[1] Associate Professor, Toyohashi University of Technology, Toyohashi 441, JAPAN
[2] Professor, University College London, London WC1E 6BT, U.K.

bifurcation pressures. What Yamada and Croll (1989) did not consider is how to include in design the effects on buckling of material plasticity.

The present paper provides a simple, theoretically based, design methodology for elasto-plastic buckling of shallow pressure loaded cylindrical shell panels. It builds upon the conclusions from Yamada and Croll (1989), that the reduced stiffness method provides lower bounds to the imperfection sensitive elastic buckling loads. An imperfect form of the reduced stiffness modeling will be used to develop lower bound to the onset of plasticity.

2. The Shell Model

The present model is identical to that used in Yamada and Croll (1989). It has a longitudinal length l, wall thickness t, radius r and central angle ϕ. Uniform external pressure q is taken to be positive inward as shown in Fig.1. Displacement components in the longitudinal, x, circumferential, y, and normal (positive inward), z, directions, are denoted by u, v and w, respectively. Elastic nonlinear responses have been found using a Ritz analysis of the Donnell approximation for geometrically imperfect shallow cylindrical shell panels. Classical simple supported edge conditions (the so-called SS3 conditions) relate closely to many full scale applications, have been adopted as boundary conditions. Displacement functions (u, v, w) are taken as linear combinations of a large number harmonic functions, each of which separately satisfies the boundary conditions. The resulting nonlinear algebraic equations are solved using a step by step, incremental Newton-Raphson iteration method. For each of the nonlinear paths repeated in what follows, care has been taken to achieve convergence of both the functional approximation, through the use of the requisite numbers of approximating harmonics, and also the iteration procedure, by ensuring convergence of the Newton-Raphson solution scheme.

3. Elastic Nonlinear Analytical Results

3-1. Elastic Buckling Lower Bound

A number of recent studies (Yamada 1991; Goncalvas and Croll 1992; Yamada and Croll 1993; Yamada et al. 1993; Kashani and Croll 1993) have shown that for initial imperfection sensitive shell structures, a reduced stiffness analysis gives simple, theoretical lower bound estimates of experimental elastic buckling loads. In the reduced stiffness method the theoretical lower bound membrane stiffness (or energy) components, which contribute to the loss of load-carrying capacity in the

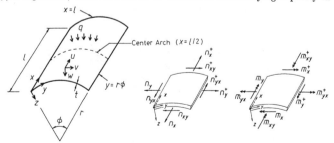

Fig.1 Geometry and internal stress and moment resultants

postbuckling response, are eliminated (Yamada and Croll 1993). It is a fairly straight forward matter to identify in the modes characterizing classical buckling the energy components which are at risk through the erosive effects of mode coupling catalyzed by the presence of initial imperfections.

In Fig.2 typical elastic nonlinear paths show the loss of load-carrying capacity after buckling and the initial geometric imperfection sensitivity for a shell having geometry defined by the Batdorf parameter and aspect ratio as $Z\equiv \sqrt{(1-\nu^2)}l^2/(rt)=$ 200 and $a\equiv l/(r\phi)=1.0$, respectively. For these results, as well as all subsequent results, the Poisson's ratio ν was taken equal to 0.3. The imperfection modes are simply defined in the present case as follows,

$$w^0 = tW^0_{1,2}\sin(\pi x/l)\sin[2\pi y/(r\phi)] + tW^0_{1,3}\sin(\pi x/l)\sin[3\pi y/(r\phi)] \qquad (1)$$

Figure 2(a) shows a symmetric deflection analysis for which the anti-symmetric imperfection $W^0_{1,2}=0$; the results are similar to these by Yamada and Croll (1989) in which the symmetric imperfection mode adopted was of the same form as the exact, linear bending deflection mode. We obtained the same total 60 degrees of symmetric-displacement freedom as in Yamada and Croll (1989). It is apparent

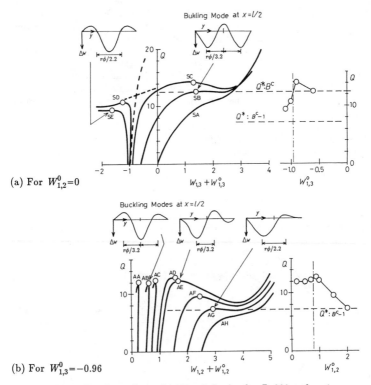

(a) For $W^0_{1,2}=0$

(b) For $W^0_{1,3}=-0.96$

Fig.2 Elastic nonlinear buckling behavior for $Z=200$ and $a=1$

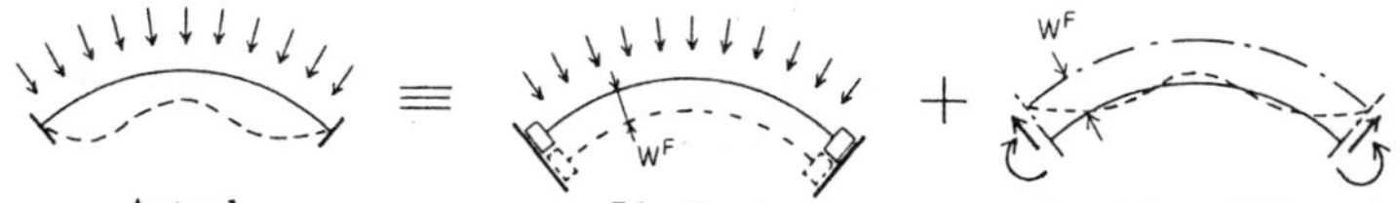

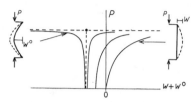

Actual   Idealized   Load Imp. Effect

Fig.3 Idealized model with uniform fundamental state

Fig.4 Elastic behavior of eccentrically axial compressed columns

from Fig.2(a) that, around $W_{1,3}^0=-0.96$ there is a transitional behavior associated with an underlying bifurcation. At this level of geometric imperfection the loading distortions caused by the longitudinal edge restraint to the otherwise uniform radial deformation would be effectively cancelled. In this case the effective loading imperfection would be $W_{1,3}^l=0.96$. As the total equivalent imperfection $W^e \equiv W^0 + W^l$ is decreased it is clear that there is a very severe loss of load capacity for small changes in W^e.

Holding the symmetric geometric imperfection constant at $W_{1,3}^0=-0.96$, Fig.2(b) illustrates what then happens when a component of anti-symmetric imperfection is added; for the cases reported in Fig.2(b) a further 13 anti-symmetric degrees of freedom have been added. Up to $W_{1,2}^0=0.8$ the shell continues to be biased into a symmetric buckling mode. At $W_{1,2}^0=0.8$ there is again a sudden jump in behavior associated with an underlying bifurcation into anti-symmetric buckling mode. Anti-symmetric imperfections $W_{1,2}^0>0.8$ show once again a heightened level imperfection sensitivity.

Application of the reduced stiffness method to the pressure loaded cylindrical panel requires the concept of an idealized model as shown in Fig.3. The actual pre-buckling deformation of the geometrically perfect cylindrical panel may be considered to comprise two parts: an idealized uniform radial deformation w^F due to the pressure loading; and, a loading imperfection resulting from the distortions at the longitudinal edge restraint. It is the existence of this loading imperfection that required a symmetric geometric imperfection of $W_{1,3}^0=-0.96$ to produce an effectively zero total equivalent imperfection $W_{1,3}^e$ in Fig.2(a). A similar behavior would be displayed by an axially loaded column depicted in Fig.4. In this analogous case a negative geometric imperfection would be required to counterbalance the loading imperfection associated with the eccentric load. When so balanced a bifurcation behavior would occur.

The idealized membrane fundamental state is identical to the case of a pressure loaded complete cylinder and is given by $n_y^F=-qr$, $n_x^F=n_{xy}^F=0$. Then the reduced stiffness critical pressure associated with circumferential half-wave numbers n are given in Yamada and Croll (1989) as

$$Q^* = (B+B^{-1})^2 \qquad (2)$$

where: the nondimensional pressure parameter $Q \equiv qrl^2/(\pi^2 D)$; flexural stiffness $D \equiv Et^3/[12(1-\nu^2)]$; $E=$ Young's modulus; and, $B \equiv an$ is a normalized circumferential half-wave number.

The reduced stiffness analytical results are illustrated by two dotted level lines in Fig.2; the one is for the idealized classical critical mode which are assumed as continuous values, in this case $B^c = 3.2$; the other is for $B^c - 1 = 2.2$. Figure 2(a) shows that the present nonlinear analytical lower bound, for total imperfection $W^0_{1,3} + W^e_{1,3} > 0$, is approximately equal to $Q^*(B^c)$, and also that the dominant circumferential half-wave number in the buckling mode is B^c. For total imperfection $W^e_{1,3} \equiv W^0_{1,3} + W^i_{1,3} < 0$ the dominant symmetric buckling mode is lengthened to $B^c - 1$. Snap buckling loads an paths 'SD' and 'SE' can be observed to approach Q^* for mode $B^c - 1$. In Fig.2(b) the lower bound is again in very good agreement with the lower reduced stiffness critical load for the longer circumferential wave-length, $Q^*(B^c - 1)$. In this case this is due to mode coupling caused by the larger amplitude geometric imperfection, $W^0_{max} = -3.29$.

Shown in Fig.5 are summaries of the two categoried lower bounds and associated buckling mode. The broken line demonstrates the classical upper bound for the idealized model, while the solid lines give the associated reduced stiffness critical loads for modes B^c and $B^c - 1$. Open dots represent the lowest observed nonlinear snap buckling loads for incremental buckling into modes B^c, while closed dots represent the lowest loads when buckling is into the sub-dominant mode $B^c - 1$. In each case these modes would correspond with the lowest symmetric and anti-symmetric critical buckling loads. With the exception of new results for $Z = 40, 200$ the nonlinear numerical studies for Fig.5 have been fully reported in Yamada and Croll (1989). It is apparent that the reduced stiffness prediction provides fairly convincing lower bounds to the buckling of pressure loaded shallow cylindrical shells. They would provide safe but not overly conservative estimates for the elastic buckling collapse.

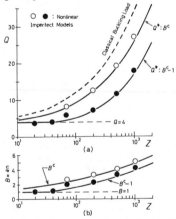

Fig.5 Comparison of elastic lower bound and its buckling mode between the reduced stiffness analysis and the nonlinear numerical experiments

3-2. First Surface Yielding Load by Elastic Nonlinear Analysis

In the present study, the sets of elastic nonlinear equations have been solved using a step-by-step process in which either load or displacement are used as a control parameter. This incremental analytical procedure make it convenient to compute at any stage along the equilibrium path the distributions of stress and moment resultant components. It is then possible to derive the well-known effective stress n_{eff} at any equilibrium state. A sufficiently small mesh interval ($l/200$ in x-direction and $r\phi/200$ in y-direction) of the sampling points for the stress computation, provides a very close approximation to the maximum of the effective stress. The von Mises yielding condition for 2-dimensional body may be written as

$$N_{eff} = \sqrt{(N_x \pm M_x)^2 - (N_x \pm M_x)(N_y \pm M_y) + (N_y \pm M_y)^2 + 3(N_{xy} \pm M_{xy})^2} = 1 \quad (3)$$

where $N_{eff} \equiv n_{eff}/(\sigma_Y t)$ is non-dimensional effective stress, σ_Y uniaxial material yielding stress, $(N_x, N_y, N_{xy}) \equiv (n_x, n_y, n_{xy})/(\sigma_Y t)$ non-dimensional stress resultants, and $(M_x, M_y, M_{xy}) \equiv 6(m_x, m_y, m_{xy})/(\sigma_Y t^2)$ non-dimensional moment resultants. In Eq.3, positive and negative signs are, respectively, for the inside and outside surface of the shell.

The various dots in Fig.6 show the maximum effective stress behavior of elastic nonlinear stress analysis using $E/\sigma_Y = 875$ for the cases of imperfections corresponding with paths 'SA', 'SC', 'AG' and 'AH' in Fig.2. In this figure the geometric parameters Z and a are fixed as 200 and 1.0, respectively; then the other geometric parameters are obtained using the ratio of radius to thickness, r/t, as $l/r = \phi = 14.48/\sqrt{r/t}$. The open dots indicate that the maximum point is at the corner on the inside surface, while the close dots are for the point on the outside surface at mid-shell. For relatively thick shells, low r/t, corner failure proceeds that at mid-shell; this failure is largely independent of the initial geometric imperfection. As the shells become thinner, r/t increases, mid-shell failure represents the limiting condition. At mid-shell the maximum bending stresses are dependent upon the imperfection level. It is for this reason that first surface yield shows high sensitivity to imperfection level. With the normalized form used in Fig.6 the non-dimensional yield condition $n_y^F = -\sigma_Y t$ would occur at the squash load

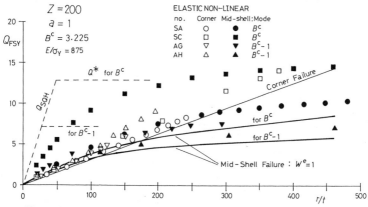

Fig.6 Effects of geometric parameters on the first surface yield pressure

$$Q_{SQH} = (12/\pi^2)\sqrt{1-\nu^2}(\sigma_Y/E)(r/t)Z \qquad (4)$$

It is clear that the elasto-plastic knockdowns from either the squash load Q_{SQH} or the lower bound elastic buckling Q^* are considerable.

4. Elasto-Plastic Lower Bound for Mid-Shell Failure

4-1. First Surface Yielding Load, Q_{FSY}

One of the most difficult design aspect in shell buckling has been the prediction of an appropriate allowance for additional knockdowns arising from the interaction between elastic and plastic nonlinearities. The simplicity of the reduced stiffness method makes it a particularly convenient basis for the predict of plastic collapse design estimates (Croll 1981,1982). On the basis of a reduced stiffness buckling model, any imperfection introduction into the shell will, at a prescribed pressure level, provide upper bounds of the incremental deformation compared with that predicted for the exact shell behavior. This means that incremental stress components found using the reduced stiffness model will, at this prescribed pressure level, be upper bounds of those occurring in the exact behavior. Consequently, the load for first yield using the reduced stiffness method will be a lower bound of the exact first yield occurrence.

For a shallow cylindrical shell a similar approach is possible when the total equivalent imperfections, involving the loading induced imperfection and the geometric imperfection is used to calculate mid-shell surface yielding. The non-dimensional idealized fundamental stress state is

$$N_y = -Q/Q_{SQH}, \qquad N_x = N_{xy} = 0 \qquad (5)$$

In the reduced stiffness model, the incremental deflection is assumed to be related to total equivalent imperfection $W^e \equiv w^e/t$ through

$$w = tW^e \left[Q/(Q^*-Q)\right] \sin(\pi x/l) \sin[n\pi y/(r\phi)] \qquad (6)$$

Neglecting the periodic components of membrane stress, which would be lost in the elastic postbuckling erosion of stiffness, the maximum surface stress occurs when the fundamental membrane stress combines with the maximum flexural stresses arising from

$$M_x = (1+\nu B^2)F, \quad M_y = (\nu+B^2)F, \quad M_{xy}=0, \quad \text{for } F=6\,W^e Q/[(Q^*-Q)Q_{SQH}] \qquad (7)$$

Substituting Eqs.5 and 7 into Eq.3, results in a nonlinear algebraic equation for Q, whose numerical solution gives Q_{FSY}. The results are the solid curves included in Fig.6.

4-2. First Full Plastic Collapse Load, Q_{FFP}

While the first yield criterion described above would provide lower bounds to the limiting elastic behavior, it may not have a direct relationship to the loads at which collapse is eventually precipitated. The central arch of shell could evidently continue to support pressure until the yielding on the crests and troughs spreads throughout the thickness. Once such a full plasticity state is attained there is little conceivable way that the shell could continue to support the pressure without greatly increasing the deformations. With such increased deformations inevitably involving further increases in moment, the average circumferential stress and therefore pressure would need to be reduced if the stresses are to remain on the full plasticity surface for this cross-section. The resulting drop-off in pressure resistance for this central arch implies that first full plasticity would be likely to provide a close indication of incipient collapse.

Full plasticity would from Fig.7 be attained when

$$M_x = 3t^*(1-t^*)(\sigma_{xi}-\sigma_{xe})/\sigma_Y , \quad N_x = [t^*\sigma_{xi}+(1-t^*)\sigma_{xe}]/\sigma_Y \quad (8a)$$

$$M_y = 3t^*(1-t^*)(\sigma_{yi}-\sigma_{ye})/\sigma_Y , \quad N_y = [t^*\sigma_{yi}+(1-t^*)\sigma_{ye}]/\sigma_Y \quad (8b)$$

where $t^* \equiv t_n/t$ is the elevation of the non-dimensional neutral surface from the inside face.

Making use of Eq.8 and the von Mises criteria at both the inside and the outside surfaces

$$\sigma_{xi}^2 - \sigma_{xi}\sigma_{yi} + \sigma_{yi}^2 = \sigma_Y^2 , \qquad \sigma_{xe}^2 - \sigma_{xe}\sigma_{ye} + \sigma_{ye}^2 = \sigma_Y^2 \quad (9)$$

may be shown to result in the non-dimensional forms,

$$\alpha_0 + 3\alpha_1 t^* + 9\alpha_2(t^*)^2 = 0 , \quad \alpha_0 - 3\alpha_1(1-t^*) + 9\alpha_2(1-t^*)^2 = 0 \quad (10)$$

where

$$\alpha_0 = M_x^2 - M_x M_y + M_y^2 , \qquad \alpha_1 = 2N_x M_x + 2N_y M_y - N_x M_y - N_y M_x$$

$$\alpha_2 = N_x^2 - N_x N_y + N_y^2 - 1$$

Equation 10 gives the relations,

$$t^* = 0.5 - \alpha_1/(6\alpha_2) \quad (11)$$

In this case, to use Eqs.5 and 7 results in a nonlinear algebraic equation for Q, whose numerical solution gives Q_{FFP}.

The first full plasticity conditions obtained for the case of $W^e = 1$ are shown by solid full lines in Fig.8. In this figure the geometric parameters a and ϕ are fixed as 1.0 and $\pi/3$, respectively: then the other geometric parameters are $l/r = a\phi = \pi/3$ and $Z = 1.046(r/t)$. As Z reduces there is a relatively greater reserve in strength after first yield.

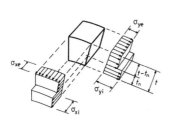

Fig.7 Stress block on circumference attaining fully plastic state

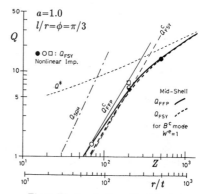

Fig.8 Elasto-plastic critical loads

5. Corner Failure Criterion

5-1. First Surface Corner Yielding Load, Q_{FSY}^c

In the case of panel structures, it would be necessary to consider a corner failure mechanism. Also for shallow cylindrical shells having SS3 boundary conditions, it would be possible for first corner yielding to occur before that at mid-shell. These corner failure pressures would be affected by the load-induced imperfections, but only moderately by the geometric imperfections. Consequently, it is possible to provide estimates of the corner failure pressures through the linear bending stress analysis defined as the modified model in Yamada and Croll (1989). When the nonlinear term are omitted, the linear equations give

$$u = Qt(l/r) \sum_m \sum_n U_{mn} \cos(m\pi x/l) \sin[n\pi y/(r\phi)] \qquad (12a)$$

$$v = Qt(l/r) \sum_m \sum_n V_{mn} \sin(m\pi x/l) \cos[n\pi y/(r\phi)] \qquad (12b)$$

$$w = Qt \sum_m \sum_n W_{mn} \sin(m\pi x/l) \sin[n\pi y/(r\phi)] \qquad (12c)$$

$$U_{mn} = W_{mn}(B^2 - \nu m^2)m/(\pi S) \ ; \qquad V_{mn} = -W_{mn}[B^2 + (2+\nu)m^2]B/(\pi S) \ ;$$

$$W_{mn} = 16Z/[\sqrt{1-\nu^2}\pi^4 mnT] \ ; \qquad S = (B^2 + m^2)^2 \ ; \qquad T = S + 12Z^2 m^4/(\pi^4 S)$$

The wave numbers corresponding with the function representations of Eq.12 are m=1,3,5,.....,31 and n=1,3,5,.....,31.

At the corner $(x=y=0)$, only shear stress components exist for the present boundary condition. In this case Eq.3 predicts

$$Q_{FSY}^c = Q_{SQH} \ / \ [\sqrt{3}(M_{xy}^c + N_{xy}^c)] \qquad (13)$$

$$M_{xy}^c = 6(1-\nu)\sum_m \sum_n mBW_{mn} \ ; \quad N_{xy}^c = 12\sqrt{1-\nu^2}Z \sum_m \sum_n m^3 BW_{mn}/(\pi^2 S) \qquad (14)$$

Superimposed upon Figs.6 and 8 are the first surface corner yielding loads Q_{FSY} associated with the present linear stress analysis. They show the good agreement with nonlinear numerical experiments indicated by open dots.

5-2. First Full Plastic Collapse Load at the Corner, Q_{FFP}^c

A first full plastic collapse load at the corner can be derived in a similar way to that at mid-shell. From Eq.14 this would be given as

$$Q_{FFP}^c = \sqrt{3}t^* Q_{SQH} \ / \ (M_{xy}^c + 3t^* N_{xy}^c) \qquad (15)$$

where t^*=0.5$(1-\xi+\sqrt{1+\xi^2})$ and $\xi = M_{xy}^c/(3N_{xy}^c)$. Superimposed upon Fig.8 is the first full plastic collapse load at the corner, Q_{FFP}^c. It shows that the common logarithm of Q_{FFP}^c is approximately in proportional to that of the geometric parameters Z and r/t.

6. Discussion

Shallow cylindrical shell buckling is heavily dependent upon the high levels of loading imperfection, resulting from the effect of edge constraints defined in Fig.3. This consideration should be reflected in the elasto-plastic design estimate for buckling. Tolerance levels for imperfection would accordingly need to be larger than these for a complete cylinder.

In addition, the elasto-plastic design for the shallow cylindrical shell panel has the other check points not only on the above corner failure mechanism but also on a re-distribution of nonlinear components in an elastic large deformation path. It is clear from Fig.2 that even quite large changes in geometric imperfection have surprisingly little effect on buckling loads; indeed for both symmetric and anti-symmetric imperfections the elastic buckling loads actually increase for increases in geometric imperfection. This should not lead to the interpretation that these shells are insensitive to geometric imperfections. Rather, it implies that the inherent loading imperfection for this particular shell geometry is so large that it swamps any additional effects from geometric imperfections. However, even for this geometry there are the additional imperfection sensitivities associated with mode shift in the postbuckling. For both the symmetric and anti-symmetric geometric imperfections there are critical thresholds at which buckling experiences a sudden additional sensitivity to small changes in geometric imperfection. For the example of Fig.2 this occurred when $W_{1,3}^0 = -0.96$ or $W_{1,2}^0 = 0.8$. But for other geometries (for instance, for $Z=500$ in Fig.6 of Yamada and Croll 1989), this sensitivity to geometric imperfections is exhibited when imperfections are very much smaller.

7. Conclusions

A fully-nonlinear Ritz analysis of imperfect shallow cylindrical shell panels under external pressure has been developed for the purpose of confirming the elasto-plastic reduced stiffness imperfect model. For a mid-shell failure, the present numerical experiments on the first surface yield according to a von Mises criterion, have for varying levels of geometric imperfection shown the reduced stiffness elasto-plastic criterion to provide lower bounds to material failure. The elasto-plastic design of the shallow cylindrical shell panels needs to also check the plate-panel behavior characterized by the edge corner failure. This form of failure is likely to be critical for short, thick cylindrical shell panels. This approach could be useful in the plastic collapse design of many other shell buckling problems.

Appendix. References

1. Croll, J.G.A. (1981). "Lower bound elasto-plastic buckling of cylinders." *Proc. Instn Civil Engrs*, Part 2, Vol.71, 235−261.

2. Croll, J.G.A. (1982). "Elasto-plastic buckling of pressure and axial loaded cylinders." *Proc. Instn Civil Engrs*, Part 2, Vol.73, 633−652.

3. Goncalvas, P.B., and Croll, J.G.A. (1992). "Axisymmetric buckling of pressure-loaded spherical caps." *Journal of Engineering Mechanics*, A.S.C.E., Vol.118, 970−985.

4. Kashani, M., and Croll, J.G.A. (1993). "Lower bounds for the overall buckling of spherical spacedomes." *Journal of Engineering Mechanics*, A.S.C.E. (to be published).

5. Yamada, S., and Croll, J.G.A. (1989). "Buckling behavior of pressure loaded cylindrical panels." *Journal of Engineering Mechanics*, A.S.C.E., Vol.115, 327−344.

6. Yamada, S. (1991). "Relationship between non-linear numerical experiments and a linear lower bound analysis using finite element method on the overall buckling of reticular partial cylindrical space frames." *Computer Applications in Civil and Building Engineering, Proc. 4th I.C.C.C.B.E.*, Kozo System Inc., Tokyo, 259−266.

7. Yamada, S. and Croll, J.G.A. (1993). "Buckling and postbuckling characteristics of pressure loaded cylinders." *Journal of Applied Mechanics*, A.S.M.E., Vol.60, 290−299.

8. Yamada, S., Uchiyama, K., and Croll, J.G.A. (1993). "Theoretical and experimental correlations of the buckling of partial cylindrical shells." *Proc. SEIKEN−IASS Symp.*, Tokyo, 151−158.

Topology Optimization of Plate and Shell Structures

K. Maute[1] and E. Ramm[2]

Abstract

Topology optimization is understood as a general principle for the genesis of structures. Instead of special micro structure models and the method of homogenization a direct approach is chosen in the present study to get the local stiffness as a function of density. The density distribution of an isotropic material is used as design variable to describe the body of a structure by a simple 1/0 decision (material or no material) in a relaxed formulation. The proposed method of topology optimization of continuum structures is based on a linear FE analysis. It is embedded into a conventional mathematical programming scheme. Therefore, its generality allows to handle a wider range of optimization problems. Unlike using optimality criteria methods optimization problems with different objectives and constraints can be solved. Furthermore, it is possible to combine material based topology optimization with other kinds of structural optimization, e.g. shape optimization. Selected examples for plate and shell structures demonstrate the advantages and disadvantages of this approach.

1. Introduction

The potential of optimization in the design of structures is more and more perceived and utilized. On the one hand, this is due to strongly improved and user orientated optimization software. On the other hand, structural optimization helps to manage increasing demands of challenging design problems. The purpose of structural optimization is to improve an initial design by variation of its geometrical and material properties with regard to a set of prescribed objectives and constraints.

There are three classes of optimization problems depending on the geometrical and material properties which are allowed to vary during the optimization process. In sizing problems 1D geometrical parameters, like cross sectional dimensions or plate thicknesses, and material parameters, like fibre orientation in composites, are optimized. In 2D and 3D shape optimization the position of joints in discrete structures or the boundaries of continuum structures are varied to reach an "optimum". Finally, in topology optimization problems the connectivity between structural members of discrete structures or between domains of continuum structures can be varied (fig.1). For discrete structures, like trusses, the variation of connectivity means to generate or to

---

[1] Dipl.–Ing., Research Assistant, Institut für Baustatik, University of Stuttgart, Germany
[2] Professor, Dr.–Ing., Institut für Baustatik, University of Stuttgart, Germany

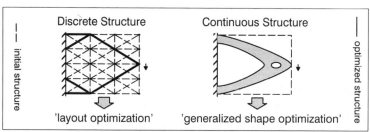

Fig.1 : *Topology optimization of discrete and continuum structures*

eliminate structural members between existing joints, but also to define new joints or to remove existing joints. Analogous, for continuum structures the variation of connectivity means to separate or to join together structural domains and to build up or to reduce structural domains. In addition to pure boundary variation, topology optimization provides the possibility to generate new holes and to eliminate existing gaps. This is the reason why topology optimization of continuum structures is called generalized shape optimization and topology optimization of discrete structures layout optimization (Rozvany, Zhou, Sigmund 1992).

There are two different possibilities to describe the body and therefore also the topology of a homogenous structure (fig.2). Describing a structure geometrically it is defined by internal and external boundaries ("geometrical description"). From a material point of view the topology is defined by a simple 0/1 decision for each point in the design space, whether there is material (1) or not (0) ("material description"). It is obvious, that different optimization problems result depending on which possibility is chosen .

Although topology optimization has gained substantial popularity in the last five years, more than 100 years ago people considered already the problem of optimal design. Based on Maxwells work (1860) Michell published in 1904 an initial paper about layout theory. Hemp (1973) used Michell's work to evolve a method designing truss structures. Later, Prager and Rozvany (1977) worked on layout problems for quasi continuous structures consisting of truss–like members. Lately, new interest in topology optimization was awaked by Bendsøe and Kikuchi (1988) introducing a new method in topology optimization. This so called "microhole approach", which is explained in section 2 of this paper, is based on the material description of structures and can be used for a special kind of optimization problems. For linear elastic plates some researchers studied optimal design problems using either thickness or special stiffening parameters as design variables (Cheng and Olhoff 1981, Rozvany et al. 1987). For plate and shell structures Suzuki and Kikuchi (1991) used the "microhole approach" solving maximal stiffness problems by optimality criteria methods. New examples for maximal stiffness plates and shells were presented by Tenek and Hagiwara in 1993.

Up to now, since all known methods for topology optimization use optimality criteria methods, they are limited in their present formulation to a few special optimization problems. However, embedding a material based topology optimization approach into a mathematical programming scheme design problems with a variety of objectives and constraints can be solved. Consequently in this study, a simple approach for topology optimization is used as an initial step and implemented in a general mathematical program-ming setting. In section two, the material based topology optimization approach is explained considering the well known methods using micro structure models. In section three, an alternative approach is introduced. Further, in section four it is outlined how topology optimization can be carried out by mathematical programming and which new possibilities are provided by this concept. Finally, in section 5 a few examples for plate and shell structures show advantages and disadvantages of this method.

2. Material based topology optimization

For structures of one homogeneous isotropic material the material description leads to a discrete valued parameter function χ defining the structure in the design space (Bendsøe 1989).

$$\chi_{(x)} = \left\{ \begin{array}{ll} 0 & \rightarrow \quad \text{no material} \\ 1 & \rightarrow \quad \text{material} \end{array} \right. \qquad x \in \Omega \qquad (1)$$

For numerical optimization this indicator function is discretized and parametrized by piecewise constant shape functions. This yields an integer optimization problem using indicator parameters on a macroscopic level as design variables. However, Kohn and Strang (1986) stated that this formulation of a general shape optimization problem is not well posed. This means, that on the one hand, optimization results are strongly dependent on the chosen discretization, and on the other hand, the integer formulation compromises many artificial local minimum. Moreover, solving integer optimization problems is very costly considering the number of design variables that are necessary to describe at least approximately the basic framework of a structure.

The above mentioned problems can be solved by transferring the integer problem into a continuous one (relaxation). The discrete valued parameter function becomes a continuous distribution of a new parameter. This new parameter can be clearly interpreted as a normalized density. The relationship between the limits of density (0/1) and material behavior is obvious. On one hand, the properties of a structural point with density equal to 1 are identical with those of the basic homogenous material. On the other hand, density values equal to 0 simply means no material and therefore no stiffness.

$$\chi_{(x)} = \left\{ \begin{array}{l} 0 \\ 1 \end{array} \right. \rightarrow \quad \text{Relaxation} \quad \rightarrow \quad \varrho_{(x)} \; : \; 0 \le \varrho_{(x)} \le 1 \quad x \in \Omega \qquad (2)$$

Now, there are two possibilities to interpret the intermediate density values. Following Rozvany, Zhou and Birker (1992) density values between 0 and 1 can simply be seen as an intermediate stage of the optimization process. In principle the relationship between material properties and density in the intermediate range is arbitrary. Since intermediate values are actually not

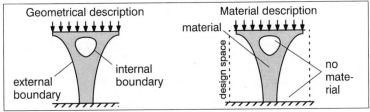

Fig.2 : *Geometrical and material description of structures*

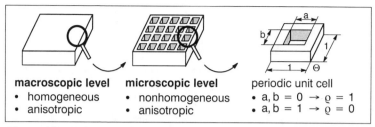

macroscopic level	microscopic level	periodic unit cell
• homogeneous	• nonhomogeneous	• a, b = 0 → ϱ = 1
• anisotropic	• anisotropic	• a, b = 1 → ϱ = 0

Fig.3 : Microhole approach by Bendsøe and Kikuchi 1988

permissible at the optimum, the optimization problem has to be posed such that the final optimized design contains only 0/1 density values.

The second possibility is to interpret the intermediate density values physically by micro structure models. In this case two levels can be distinguished. On a macroscopic level, properties, like density or Young's modulus, define the effective material behavior. These macroscopic properties can be derived from a micro structure on a microscopic level. For topology optimization micro structures are thought to be anisotropic inhomogeneous composites made of only one basic homogeneous material. The structure of these micro models and therefore the macroscopic material properties are defined by a few geometrical parameters, like length a, depth b and orientation Θ of the unit cell for the "microhole approach" (fig.3). Consequently, intermediate density values correspond to porous material. Based on this micro models macroscopic properties are computed either analytically or numerically by the method of homogenization (Sanchez−Palencia 1980) .

Computing the material stiffness coefficients C_{ijkl} as functions of density for the "microhole approach" by the homogenization method approximately an exponential dependency (3) turns out.

$$C_{ijkl\,(\varrho)} = C^o_{ijkl}\,\varrho^\mu \qquad \mu = 3....9 \qquad C^o_{ijkl} : \text{homogeneous material} \qquad (3)$$

Maximizing structural stiffness for a given mass this functional results in a penalization of intermediate density values, i.e. the design space is clearly subdivided into voids and solid structural elements. Only a few parts consist of porous material. In a following postprocessing step the framework of the optimal structure can be determined very easily from the density distribution in design space (fig.4). Exactly this feature of micro structure models is desired in topology optimization because optimizing topology does not mean optimizing material composition of structures but generating the optimal layout of structures of one or more given materials.

3. An artificial approach

Up to now the best results in topology optimization were obtained using micro structure models. However, these models are based on complex, in general implicit relationships between microscopic and macroscopic parameters. Furthermore, since the number of microscopic design variables defining macroscopic material properties is extremely high, only optimality criteria (OC) methods are used. The advantage of these methods is to solve big optimization problems with hundreds of optimization variables very efficiently. But only simple optimization problems with mostly one constraint and easily computable gradient information can be handled using OC methods. This is the reason

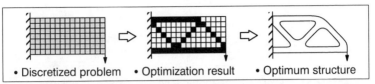

• Discretized problem   • Optimization result   • Optimum structure

Fig.4 : Material based topology optimization

why until now, primarily maximum stiffness problems for given mass have been studied. Mathematical programming (MP) is commonly used if more complex optimization problems are to be solved. These methods are applicable for all kinds of optimization problems with several constraints. Since the computational effort needed to solve optimization problems by MP is much higher compared with OC methods, a simplified approach with less design variables is required to embed topology optimization into a mathematical programming scheme. For such an approach equation (3) is simplified. Using only macroscopic properties an artificial relation between Young's modulus and (design variable) density is defined:[1]

$$E_{(\varrho)} = E^{0}\, \varrho^{\mu} \qquad \mu = 3....9 \qquad (4)$$

Without the complexity and the big number of design variables, which micro structure models contain, this isotropic approach has the same demanded feature, i.e. it leads almost to a 0/1 material distribution in the entire design space for maximum stiffness optimization problems. Using this artificial approach the number of design variables can be reduced to a third compared with the "microhole approach".

Moreover, using micro models special micro structures have to be developed and homogenized for each mechanical problem, i.e. plate, slab, shell or brick problem. In comparison, the simplified approach can be directly applied to all kinds of mechanical problems. The well known constitutive laws for the corresponding mechanical problems can be used, because Young's modulus of an isotropic material is the only variable. In particular, since it is quite costly to develop special constitutive laws from micro models for plates and shells, this feature is very advantageous.

Of course, the physical meaning of the intermediate density value is lost within this artificial approach. But, if only 0/1 density values are permissible in topology optimization, a physical interpretation of the intermediate stages is absolutely not necessary. This approach is motivated by the results of micro models but interprets density values between 0 and 1 simply as an intermediate stage during the optimization process.

4. Mathematical programming scheme in topology optimization

A wide range of optimization problems can be defined as follows :

minimize the objective $f_{(s)}$ (5)
subject to

equality constraints $g_{j\,(s)} = 0 \,,\; j = 1,..,m_e$
inequality constraints $g_{j\,(s)} \leq 0 \,,\; j = m_e + 1,..,m$
bounds for optimization variables $\mathbf{s}_L \leq \mathbf{s} \leq \mathbf{s}_U$

---

[1] see also Bendsøe (1989) introducing this formula as direct approach

$$\frac{dt}{ds} = \frac{\partial t}{\partial s} + \frac{\partial t}{\partial u}\frac{du}{ds} \qquad f, g \in t \tag{6}$$

Numerical DSA	Semi–analytical DSA	Analytical DSA
$\dfrac{dt}{ds} \approx \dfrac{\Delta t}{\Delta s}$	$\dfrac{du}{ds} = K^{-1}\left(\dfrac{\Delta R}{\Delta s} - \dfrac{\Delta K}{\Delta s}\right)$	$\dfrac{du}{ds} = K^{-1}\left(\dfrac{dR}{ds} - \dfrac{dK}{ds}\right)$

• **u**.. state variable vector   • **K**.. system stiffness matrix   • **R**.. load vector

Fig.5 : Discrete sensitivity analysis (DSA)

Diverse objectives and constraints are possible in this formulation. Common objectives are minimum weight or maximum stiffness. But also tuning to a single frequency or maximizing critical load factors are used as objectives. In addition, multi–objective optimization problems can be handled by MP. Inequality constraints in stresses and displacements are introduced to get reliable structures. Sometimes it is necessary to assign equality constraints, e.g. mass has to be prescribed when structural stiffness is maximized (Bletzinger, Ramm 1993).

Moreover, MP provides not only the flexibility to handle diverse objectives and constraints, but it allows also combinations of different types of optimization variables in one optimization problem. For topology optimization, the way to go is to combine design variables from a geometrical and at the same time from a material description. In other words, a variable density distribution and a variable shape together can be used to find the optimal layout of a structure. The advantages of this combination are demonstrated in the third example of section 5.

Solving an optimization problem by first order methods gradient information on objectives and constraints with respect to optimization variables is required. Discrete sensitivity analysis (DSA) can be carried out either numerically, semi–analytically or analytically (fig.5). The most efficient method to compute gradient information is the analytical version. Although analytical DSA requires a high programming effort, this method has been preferred for topology optimization, because the number of optimization variables is still high.

Using function values and gradient information optimization problems can be efficiently solved by non–linear programming methods. Within MP Lagrange methods are the most sophisticated numerical optimization techniques (Ramm, Bletzinger, Reitinger 1993). These SQP (sequential quadratic programming) algorithms are based on an iterative solution of the Kuhn–Tucker conditions using an extended Newton–Raphson procedure (Schittkowski 1981).

Embedding material based topology optimization into this mathematical programming scheme only two modifications have to be introduced. First, based on the artificial approach explained before, density is simply used as optimization variable in this general formulation. Second, the exponential relationship between Young's modulus and density is implemented as a special constitutive law in the FE code. Therefore, topology optimization is handled as a sizing problem using an artificial material.

5. Examples

The following academic examples are chosen to illustrate the concept of the proposed method. The used technique to solve material based topology

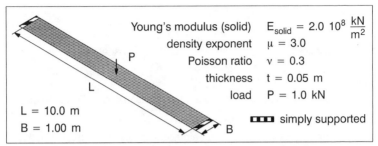

Young's modulus (solid) $E_{solid} = 2.0 \cdot 10^8 \frac{kN}{m^2}$

density exponent $\mu = 3.0$

Poisson ratio $\nu = 0.3$

thickness $t = 0.05$ m

load $P = 1.0$ kN

$L = 10.0$ m
$B = 1.00$ m

▢▢▢ simply supported

Fig.6 : Simply supported slab

optimization problems is briefly explained. First, a mesh is generated for the given design space. This mesh is used as discretization of the optimization problem and as FE mesh. The density of each finite element is used as optimization variable. The density distribution of one finite element is assumed to be constant. Except of example 5.3c, only the density distribution in the design space varies, but the mesh is invariant during the optimization process. In all examples, 2x2 reduced integrated, isoparametric, 8–node shell elements are used for the linear analysis. The sensitivity analysis is exclusively carried out analytically. The optimization problem is solved by an iterative SQP–method.

5.1 Topology optimization for a slab

The first example is used to demonstrate the effect of different constraints on the optimization result. The design space is a simply supported slab (fig.6). The center of the slab is loaded by a vertical load P. Due to the symmetry of the structure and the load only one quarter of the slab has to be analyzed. The FE mesh, which is identical to the optimization discretization, consists of 122 elements. The objective of the topology optimization problem is minimum weight.

First, the maximum vertical displacement in the loaded node is restricted to $v_{max}=0.025$m. This optimization problem is equivalent to a maximum stiffness problem, because the restriction of the displacement of the loaded node means that the reduction of the structural stiffness is limited. At the beginning of the optimization process, the design space consists of material with maximum density. The final result is shown in fig.7a. The mass in the

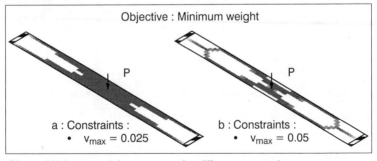

Objective : Minimum weight

P

P

a : Constraints :
• $v_{max} = 0.025$

b : Constraints :
• $v_{max} = 0.05$

Fig.7 : Minimum weight structures for different constraints

optimized design space is reduced by 39%. An almost 0/1 density distribution is attained. To demonstrate the effect of decreasing stiffness on the optimal topology the maximum displacement of the loaded node is now limited to $v_{max}=0.05m$. The result of the modified problem is shown in fig.7b. The mass in the optimized design space is reduced by 65%. It is noticed that for low volume structures the porous domains in design space increase.

5.2 Layout of stiffeners for a stadium roof

Topology optimization cannot only be used to find the optimum framework of a structure itself, but it also allows to determine the optimum layout of reinforcements by extra structural elements like stiffeners. In this example, the shape of a stadium roof is prescribed. For the loading case "snow", i.e. uniform vertical load, the layout of the reinforcement for maximum stiffness has to be found. The structural situation is shown in fig.8a. The design space for the reinforcement is identical to the given membrane roof. Due to symmetry only one half of the shell structure has to be analyzed. The roof and therefore the design space is discretized by 150 shell elements. The stiffness of one FE node results from the corresponding roof and the related reinforcement stiffness.

The objective of the optimization problem is maximum stiffness. The reinforcement mass in the design space is prescribed, i.e. at the beginning of the optimization process the design space consists of equally distributed material of 30% of the maximum density. In the optimum, the stiffness of the roof design can be improved by 97% compared to the initially homogenous reinforcement design. The dead load of the reinforcement structure is not considered. The optimal layout of the reinforced structure is shown in fig.8b. Since there are still porous domains in the optimum design space, a clear layout of the stiffeners cannot be determined.

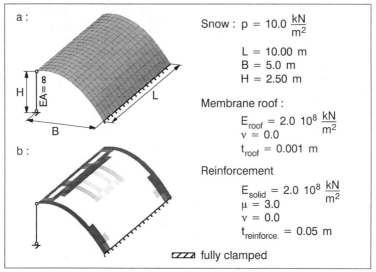

a :

Snow : $p = 10.0 \frac{kN}{m^2}$

$L = 10.00$ m
$B = 5.0$ m
$H = 2.50$ m

Membrane roof :

$E_{roof} = 2.0 \ 10^8 \frac{kN}{m^2}$
$\nu = 0.0$
$t_{roof} = 0.001$ m

b :

Reinforcement

$E_{solid} = 2.0 \ 10^8 \frac{kN}{m^2}$
$\mu = 3.0$
$\nu = 0.0$
$t_{reinforce.} = 0.05$ m

▨▨ fully clamped

Fig.8 : Reinforcement of a stadium roof

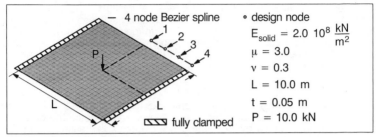

Fig.9 : Quadratic slab

5.3 Combination of shape and material based topology optimization

In the last example, a combination of shape and material based topology optimization is provided. The structural situation of the design space is shown in fig.9. The quadratic slab is clamped on two edges and loaded by a single load in the center. Due to symmetry only one quarter of the slab is analyzed. The design space is discretized by 144 shell elements. The objective is minimum weight. The vertical displacement of the loaded node is restricted to v_{max}=0.05m. At the beginning of the optimization process the design space consists of material of maximum density.

First, the optimization problem is solved by pure boundary variation technique, i.e. shape optimization. The shape parametrization is defined by one 4−node Bezier spline (fig.9). Due to the continuity of the symmetric structure design nodes 1 and 2 are linked. In the optimum of the shape optimization (fig.10a) the weight is reduced by 82.2 %. The remarkable feature of shape optimization is the smoothness of structural shape. The disadvantage of classical shape optimization is the lack of the possibility to generate new holes. The result of material based topology optimization alone (fig.10b) shows that by generating new holes the objective can be reduced by 83.4%. But this result is lacking of smoothness. A non−optimal postprocessing step is necessary. Combining both techniques it is possible to get optimum structures with new holes and, at least, smooth external boundaries. For this, a variable shape and a variable density distribution is introduced. The results of the combined optimization is shown in fig.10c. The reduction of mass is 84%. As a further step also internal boundaries can be smoothed.

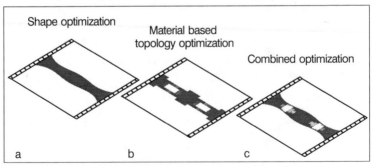

Fig.10 : Shape versus material based topology optimization

6. Conclusions

Topology optimization was embedded into a mathematical programming scheme allowing to consider a broad range of objectives and practice relevant constraints. For this reason a material based isotropic approach was presented using directly the macroscopic density as design variable. Compared with well known micro structure models this artificial approach needs less optimization variables and is based on a simplified relation between density and stiffness. The presented plate and shell examples show that this approach leads to reasonable results. However, due to the lack of clearness and smoothness these results can often only be used as a conceptional design idea instead of a clearly defined structural layout.

This is the reason why it is intended to introduce an additional penalty function to penalize intermediate density values combined with an adaptive optimization mesh generation. A further challenging task will be to improve efficient smoothing methods for internal as well as external boundaries. As an initial step classical shape optimization can be applied to smooth external shapes.

Acknowledgements

This work is part of the research project SFB 230 "Natural Structures − Light Weight Structures in Architecture and Nature" supported by the German Research Foundation (DFG) at the University of Stuttgart. The support is gratefully acknowledged.

References

Bendsøe, M. (1989). "Optimal shape design as a material distribution problem." *Structural Optimization*, 1, 193−202.

Bendsøe, M., and Kikuchi, N. (1988). "Generating optimal topologies in structural design using a homogenization method." *Comp. Meth. Appl. Mech. Eng.*, 71, 197−224.

Bletzinger, K.−U., and Ramm, E. (1993). "Formfinding of shells by structural optimization." *Engineering with Computers*, 9(1), 25−35.

Cheng, K.−T., and Olhoff, N. (1981). "An investigation concerning optimal design of solid elastic plates." *Int. J. Solid Structures*, 17, 305−323.

Hemp, W.S. (1973). *Optimum structures*. Clarendon, Oxford, U.K.

Kohn, R.V., and Strang, G. (1986). "Optimal design and relaxation of variational problems." *Comm. Pure Appl. Math.*, 39, 1−25 (Part I), 139−182 (Part II), 353−377 (Part III).

Maxwell, C. (1869). *Scientific papers II*. Univ. Press, Cambridge, U.K.

Michell, A.G.M. (1904). "The limits of economy of material in frame structures." *Philos. Mag.*, 8, 589−597.

Prager, W., and Rozvany, G.I.N. (1977). "Optimal layout of grillages." *Int. J. Struct. Mech.*, 5(1), 1−18.

Ramm, E., and Bletzinger, K.−U., Reitinger, R. (1993). "Shape optimization of shell structures." *IASS Bulletin of the Int. Ass. for Shells and Spatial Structures*, 34(2),103−122.

Rozvany, G.I.N., Ong, T.G., Szeto, W.T., Sandler, R., Olhoff, N., and Bendsøe, M. (1987). "Least−weight design of perforated elastic plates." *Int. J. Solids Structures*, 23(4), 521−536.

Rozvany, G.I.N., Zhou, M., and Birker, T. (1992). "Generalized shape optimization without homogenization." *Structural Optimization*, 4, 250−252.

Rozvany, G.I.N., Zhou, M., and Sigmund, O. (1992). "Topology optimization in structural design." *Research report in Civil Engineering*, 59, University of Essen, Germany.

Sanchez−Palencia, E. (1980). *Non−homogeneous media and vibration theory*. Springer, Berlin, Germany.

Suzuki, K., and Kikuchi, N. (1991). "Generalized layout optimization of three−dimensional shell structures." *Geometric aspects of industrial design*, V. Komkov, ed., SIAM, Philadelphia.

Tenek, H.L., and Hagiwara, I. (1993). "Optimal plate and shell topologies using thickness distribution or homogenization. To be printed in Comp. Meth. Appl. Mech. Eng.

Schittkowski, K. (1981). "The nonlinear programming method of Wilson, Han and Powell with an augmented Lagrangian type line search function." *Numerische Mathematik*, 38, 83−144.

The Structuring of a New Architectural Landscape

Karen Van Lengen[1]

Abstract

Structure plays an increasingly important role in the language of architecture design. Examples from the 2nd half of the 20th century, including transportation systems and buildings, sport halls, and other large public buildings, show how the increasing global demand for large scale construction will promote new architectural expressions dependent on the design of structure for their psychological and aesthetic meaning.

----

[1] Karen Van Lengen, Architects, New York, NY, USA

TENSILE STRUCTURES HIGHLIGHT NEW DENVER AIRPORT
Horst Berger, F.ASCE   Edward M. DePaola M.ASCE

Abstract

The tensile membrane structures of the Landside Terminal of the new Denver International Airport are the most recent and most significant application of tensile structure design. The advanced structural technology used in the design of the long-span tensile roof over the Great Hall of the main terminal helps create an extraordinary building. This paper describes the design, engineering and construction of the fabric structures.

History and Development of the Technology

The tensile roof structure for the Great Hall is the latest in a series of major fabric structures erected in the last two decades. These include the world's largest roof structure, the Haj Terminal of the Jeddah International Airport, the roof of the Riyadh Stadium, large enough to fit the Astrodome inside its support pylons, and numerous sports and performing arts facilities.

Suspension bridges, from the Brooklyn to the Verrazano Narrows are spectacular examples of the development of a technology which demonstrates the power of tensile structure principles for super-large spans.

Horst Berger is a principal design consultant to **Severud Associates,** New York, NY and is the designer of the roof and enclosure structures at the Denver International Airport.

Edward M. DePaola is a principal at **Severud Associates,** New York, NY and is the principal-in-charge of the enclosure structures.

The first dramatic use of this principle for an architectural application occurred with the construction of the Raleigh Arena (Severud 1953). The design was for a pure tensile cable net surface developed from two sets of intersecting, parabolic cables curved in opposite directions and restrained by two intersecting arches. All future developments of tensile architecture derive from this origin.

The Munich Olympic Stadium (Otto 1972) was based on a more tent-like form and introduced cable net technology. Fabric was originally introduced into permanent building construction for a number of stadium-size roofs (Geiger/Berger 1972-1982) using pneumatic structures. Fabric tensile structures followed this development. A recent structure is the roof of the San Diego Convention Center (Berger 1989) which has a clear span of 92 M (300 ft) with supports on two sides only, leaving two ends completely open over the full width of the building.

The roof of the Riyadh Stadium (Berger 1985) covers 427,600 square meters (4.6 million square feet), with a record span between its supporting columns of 247 M (810 ft). Any one of the ten modules of the Haj Terminal (Berger 1981) of the Jeddah Airport covers an area of 42,750 square meters each (460,000 sq ft). All of these structures basically cover outdoor spaces, whereas the Denver roof covers an enclosed space.

The tensile structures of the Denver Airport enclose an important public building which is in use 24 hours a day every day of the year. The roof structure, consisting of a series of tent-like modules supported by two rows of masts extending for a length of over 305 M (1000 ft) covers a space almost six times that of New York's Grand Central Station.

Architecture and Engineering

Working with architects C.W. Fentress J.H. Bradburn and Associates, Denver CO, an overall roof shape was developed that would meet several challenges. The fabric roof design, developed as a change to an earlier conventional roof scheme, resulted in the reduction of the cost of the facility to fit the budget, the improved energy efficiency of the building, and the cutting of the construction time sufficiently to make the opening date of late 1993 possible.

Since the construction schedule required that the east and west parking garages were to be erected simultaneously with the terminal, erectability with restricted access was an important criteria. The lightweight characteristics of a fabric structure satisfied this criteria admirably.

One of the prime considerations was to create a shape that would become recognized world-wide and remain a powerful symbol that would be mindful of the Rocky Mountains and identify the City of Denver as an international world hub (See Figure 1).

Due to the translucency of the roof membrane and the huge expanse of the cable supported glass walls which surround it, the terminal space is flooded with daylight. In the night, the light transmitted outward can be seen from miles around. It is this active use of daylight which is also the principal feature of its energy efficiency.

Figure 1 - Overall View

Within the full perimeter glass walls, the roof consists of two layers of fabric approximately 600 mm (24 inches) apart. The inner liner has the purpose of providing thermal insulation and acoustic absorption. This double layer roof design has a translucency of approximately 7%. Combined with the light transmission from the glass walls the result is light levels far in excess of those possible with artificial lighting for most daytime conditions while saving the related energy costs. The low heat absorption and related high reflection of the surface membrane reduce heat build-up due to sunlight. Night radiation due to the translucency combined with heat storage of the large space and its components further reduce the load on the air conditioning system. Only the heating load in winter nights is higher than for a conventional building. The net result is a substantial savings in energy consumption and operating costs while producing a more comfortable and attractive interior environment.

Landside Terminal Roof

The Landside Terminal consists of three adjacent terminal spaces forming a continuous space of 274 M (900 ft) in length and 67 M (220 ft) in width. Seventeen tent-like modules, spaced 18.3 M (60 ft) apart, are supported by two rows of masts with a spacing of 45.7 M (150 ft) between them. The typical masts rise 31.7 M (104 ft) above the main floor of the terminals. Two groups of four masts each, located at the "borders" between the three terminals, rise slightly higher than the typical masts to a height of 38.1 M (125 ft). This design creates the desired image of the mountains while helping to create an extraordinary interior space (Figure 2).

Figure 2 - Interior View

In many ways the configuration of the roof structure applies the simplest and most direct tensile structure forms. Tensile structures consist mainly of members which can carry loads in tension only. Their directional nature requires therefore that specific members are designated and appropriately shaped and stressed to carry specific types of load. In this roof structure the ridge cables carry downward loads, such as the weight of the structure and the snow load. The valley cables resist the wind load which is predominantly upward or outward suction. The fabric membrane is stretched between these ridge and valley cables and bordered by edge catenary cables, forming a double curved geodesic surface which is in equilibrium at a predetermined prestress level.

The ridge cables are draped over the masts and anchored to the rigid roof structures which cover the spaces adjacent to the great hall. The valley cables arch clear above the width of the full interior space between the east and west clerestory walls (Figure 3). Their anchorage consists of the combination of a vertical tie-downs six feet outside the clerestory walls and sloped anchor cables which are attached to the same anchor points as the ridge cables. The tie-down cables are adjustable in length and are detailed to allow the stressing of the tensile structure.

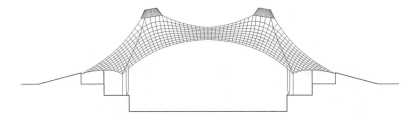

Figure 3 - Cross-sectional view

Anchorage on the north and south ends of the building where the roof overhangs by one full module each, require special anchor structures consisting of sloping struts and vertical tie-down cables. Internal prestress and the downward loads from the weight of the structure, from wind, and from snow, are carried by the vertical masts which rest directly on main columns of the building structure below. Large oval steel rings near the top of the mast form the upper edge of the fabric and support mast top units which contain service elements such as exhaust fans, electrical service and FAA lighting.

A system of secondary cables located within the surface of the fabric membrane serves a double function. It consists of a number of cables located orthogonal to the ridge and valley cables and of diagonal cables connecting the rings at the mast tops with the "octopus" connectors located at the top end of the vertical tie-downs. These cables are located and shaped to avoid high stresses in the fabric at certain critical load cases. They also provide the structure with redundancy by being able to handle the force carried by the stresses of the adjacent fabric in case of a fabric tear or during a fabric replacement.

This then forms the total structural system which is continuous over the full length of the landside terminal roof cover. Though the roof membrane has a total surface area of 35,000 square meters (377,000 sq.ft) no expansion joints are required because of the flexible nature of the folded fabric structure system which easily absorbs temperature deformations with little or no impact on the stress levels.

A second layer of fabric, the inner liner, is constructed with its own set of ridge and valley cables approximately 600 mm (24 inches) below the outer membrane. As described previously, its main functions are for insulation value and acoustic absorption. Around the periphery it is connected to the upper members of the window wall framing system.

Loads and Structural Behavior

Because of the light weight of the structural system and its considerable flexibility, the magnitude, distribution and effect on the structure of the superimposed loads such as wind and snow required careful examination. Since lightweight tensile structures absorb loads by change of shape as much as by change in stress, the analysis had to be carried out using non-linear analysis methods which took full account of the deformed state of the system under load. Figure 4 shows a network of a computer model used for shaping and analyzing the structure, produced with proprietary computer programs. The network as illustrated describes half the length of the roof.

The dead weight of the structure is less than 0.1 kN/sq.m (2 lbs/sq.ft.), (approximately 1/15th the weight of a steel framed roof system and 1/40th the weight of a roof framed with concrete structure). Consequently the superimposed loads, particularly snow loads, can reach magnitudes of 20 times the dead load.

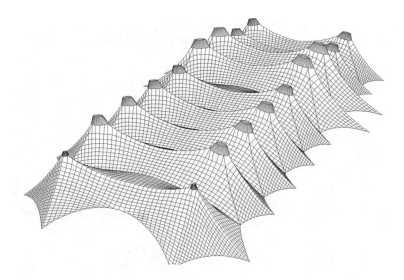

Figure 4 - Computer generated mesh

Both wind and snow load conditions were studied by special environmental load consultants RWDI, Ontario Canada using a combination of statistical evaluation of weather data over long time periods, model testing, and computer analyses. Wind load distributions were determined on a rigid model in a boundary layer wind tunnel. A number of representative loadcases were developed including the effects of gusts.

Snow loads were of particular concern. The magnitude of snow loading the roof would be expected to experience was established by computer simulation using recorded weather data over a 25 year period. The effect of the roof's geometry on snow drifting was studied by model testing (Figure 5). The analysis of the roof structure was based on a number of critical snow load configurations from this loading study. Nowhere was the roof designed for loads less than the most stringent requirements of the Denver Building Code.

Components of the Fabric Structure Roof

The most important component of this roof structure is the fabric membrane. Both the outer membrane and the inner liner are made of teflon-coated fiberglass, the material most commonly used over the last twenty years for permanent building applications.

Figure 5 - Snow simulation model.

In addition to having sufficient structural capacity these materials bring to the project the secondary properties which are critical for the design of this building. Teflon-coated fiberglass is non-combustible and satisfies the Building Code's requirements for fire resistance. It is durable: the oldest existing tensile structure using teflon coated fiberglass is now twenty years old, showing only a slight decrease in its strength characteristics. The seamless membrane panels spliced with reliable mechanical connections at the main cable lines produce a waterproof roof requiring substantially less maintenance than conventional roofs. The inert nature of the teflon surface rejects dirt and needs only infrequent cleaning.

The cable system which supports and reinforces the fabric membrane consists of high strength galvanized bridge strand encased with a continuous PVC coating for extended corrosion protection. Connections between fabric and cable are made by continuous aluminum clamping systems as shown in Figure 6. Masts, top rings, and connection details are welded structural steel construction.

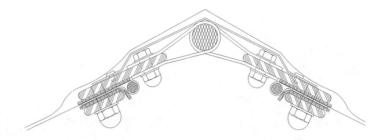

Figure 6 - Typical clamping detail

The Construction Process

The construction of the roof began with the erection of the masts which were held in position by temporary erection cables. The mast top units which consisted of the main fabric support rings and the fiberglass masttop enclosures were erected next. The installation of the fabric/cable membrane system followed, starting from the north end of the terminal. Sections of the fabric/cable membrane were assembled on the ground, lifted into position, and connected to the masts, anchors, and the previously installed sections.

Typically, these types of structures are prestressed by pulling or pushing up on the high points. For the Denver roof, this would have required the design of adjustable masts with a run of almost 1.2 M (4 feet). The answer to the problem was an innovative solution using the opposite of the normal procedure: pulling down on the low points. Prestressing could then be performed at an easily accessible location (rather than at the top of a 38.1 M (125 ft) mast, with a lower force and with less travel than with conventional prestressing procedures.

Stressing of the membrane structure proceeded in stages following a distance behind the installation process. To accommodate the stressing process, the main connectors located in the line of the valley cables just outside the clerestory walls were detailed to allow the installation of jacks which were used to shorten the vertical tie-downs and thereby put the structure under prestress. These connectors, used to resolve the forces from the catenarys, the valley cables and the mast stay cables became an architectural feature as well (See Figures 7 and 8).

Figure 7 -
Weldment Connector
prior to installation.

Figure 8 -
Weldment Connector
in place.

Glass Wall Support Structures

The entire perimeter of the main terminal is enclosed with glass walls shaped to fit the configuration of the roof cover which extends beyond. The east and west walls are formed by clerestory windows of a triangular shape supported by tubular steel frames and vertical cable trusses similar to stayed sailboat masts.

At the south end of the terminal, the glass wall extends from the main floor to the underside of the fabric roof with a maximum height of 19.8 M (65 ft) and a horizontal span of 67.0 M (220 ft). Since the fabric roof floats over the top, the wall was designed as a free standing, two-way, prestressed cable truss system. The vertical members are the principal load carrying components. A horizontal cable truss located 11.3 M (37 ft) up from the bottom of the wall forms a symmetrical double bow string truss and acts as lateral load distributor and unifier for the vertical elements. The inner liner is attached at its uppermost element.

Vertical members, spaced 4.6 M (15 ft) apart, have a common shape with varying lateral dimensions dictated by the shape of the horizontal bowstring cables and corresponding to the variation in wall height. Two main cables in the form of an A-frame fix the main node point at the elevation of the horizontal bowstring truss. Horizontal struts extend inward and outward from this main node point, defining the shape of the bowstring cables and producing the intermediate support points for two outermost cables running vertically on the inside and outside. Forming inward bows towards the vertical spine member to which they are connected by a series of horizontal ties, these cables complete the wall bracing system (See Figure 9).

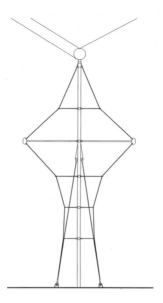

Figure 9 - South Wall cross-sectional elevation

Closure Panels

The roof itself undergoes large deformations both under wind and snow load conditions as is typical for a tensile membrane structure which absorbs loads by shape change as much as by stress increases. In order to close the gap between the rigid walls and the flexible membrane roof, circular, pneumatic closure panels where designed which are flexible enough to accommodate the movements without exerting unacceptable forces on the fabric or the wall but strong enough to resist the wind loads.

Curbside Canopies

Sidewalks adjacent to the terminal are also covered by continuous fabric structure canopies cantilevering over the departures level roadway for protection against rain and snowfall. Echoing the variation of the geometry of the main roof, the modules over the main entrances are slightly taller than the typical bays.

The masts have a regular spacing of 9.2 M (30ft) except for the main entrances where their spacing widens to 18.4 M (60 ft). A set of horizontal and vertical struts on top of the masts holds the fabric membrane up and out. Tie-down points in midbay between the piers hold the structure down. The tie-down points also provide the drainage for rain and melting snow. This condition together with the requirement to avoid large snow slides being directed onto the roadway determined the shape of the structure. Clearances over and around components of the terminal building put further limitations on the geometry.

Figure 10 - Curbside Canopy

Conclusion

The innovative use of a new technology provided a means of enclosing space efficiently, economically and elegantly. The simplicity of the structure will be obvious even to the layman, and the clear structural lines will be part of an enjoyable architectural experience created by a truly integrated effort by architects and engineers.

The roof of the new Denver International Airport, designed with the latest advances in design techniques and constructed with the materials of an advanced technology, will become a powerful symbol that identifies the City of Denver as an important international landmark.

DENVER INTERNATIONAL AIRPORT TENSILE ROOF CASE STUDY
THE FABRICATION AND CONSTRUCTION PROCESS

Martin L. Brown

Abstract

Case study of the Fabric Contractor's scope of work, including the planning, fabrication and construction phases of the tensioned fabric roof system at the new Denver International Airport. Includes the physical modelling and computer modelling required, the fabrication, the installation of the system and some of the associated problems.

Introduction

The tensile membrane fabric roof structure enclosing the Great Hall area of the New Denver International Airport is truly a milestone project for the tensile structure industry. It unites structural engineering with architecture to produce a magnificent and expansive interior space.

The fabric roof measures approximately 300 by 1000 feet in plan. It is supported by 34 masts of approximately 100 feet in length, has a surface area of about 380,000 square feet, and uses literally miles of structural steel cable. The dramatic peaks and valleys give it a unique shape emulating the Rocky Mountains that are synonymous with Denver and provide a striking backdrop to the new airport's western view.

This paper will provide a case study of the fabrication and construction phases of the project. The topics discussed in detail are the initial planning, the computer modeling, the fabrication, and the installation of the tensile roof system.

----

1   Project Manager, Birdair, Inc. 65 Lawrence Bell Drive, Amherst, NY 14221, United States of America

Initial Planning

One of the challenges that must be overcome to successfully
construct a fabric roof of this magnitude, is determining a safe
method to accomplish the installation, in particular, the fabric
panels and rigging. Each bay of the structure is comprised of over
20,000 square feet of fabric. The risk of wind damage during
fabric lifting is extremely high, if not performed properly. The
roof is also vulnerable during the time period when the fabric is
partially installed. During this period, the fabric has only
partial pre-stress and therefore less inherent stability. It will
be subjected to loading conditions that are completely different
from the design conditions of the completed structure. To overcome
these hazards, extensive planning and analysis requiring both
physical and computer modeling techniques are used.

The first step of the installation planning was to construct a
working physical model. The physical model is used to
qualitatively study the installation and formulate a preliminary
plan. In the case of the Great Hall roof, we constructed a 1/8"
scale model of half of the structure. The model represented all
the major structural components of the fabric roof system and the
primary surroundings that would be present during construction.

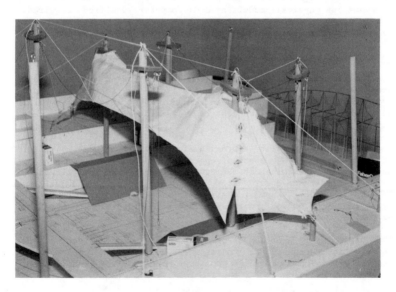

Figure 1: Physical Model. A working physical
model is used to develop and test the
installation procedure.

Working with scale replicas of the fabric assemblies, we tested out different methods and sequences of fabric packaging, handling, rigging, and hoisting. We worked with the physical model until we had schemes that we believed were physically possible to achieve and could be accomplished safely in the field. Later, the same physical model was sent to the field where it was used on site to help refine procedures and instruct the installation crews.

Computer Modeling

After the qualitative work was completed with the physical model, and a general plan had been established, computer models were built to perform the quantitative structural analysis. Large deflection finite element method analysis software is used for this work. The computer models are required for both the construction planning and the fabrication detailing. Three general types of models are required; overall system models, installation models, and fabric pattern models.

The overall system models are used to represent as much of the entire system as possible, in order to get an understanding of the overall behavior and structural interaction of the system as a whole. In the Great Hall fabric roof, the behavior and equilibrium of the various components are all inter-related. The system model is used to determine the geometrical configuration and pre-stress forces that will work in equilibrium together to produce the desired architectural and structural performance.

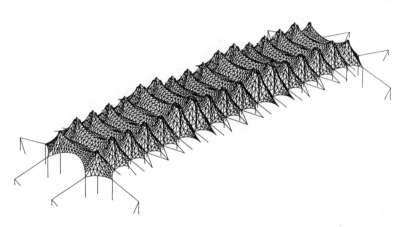

Figure 2: System Model. A "coarse mesh" system
computer model is used to quantify
overall behavior.

To make the installation models, a portion of the system model appropriate to represent a particular stage of construction is used. The installation rigging and temporary guying systems that will be present are added to the model. The pre-stress forces and geometry are modified to better represent the real conditions. The installation models are used to design the installation rigging and check the permanent roof components during the different construction phases. It is not unusual to uncover problems that were not possible to determine during the design phase when the final installation sequence was not known.

The pattern models are used to produce a very precise representation of the final geometry of the membrane and cables. These models will be used to produce the fabric cutting patterns and final cable fabrication lengths. A different pattern model was built for each bay of the Denver Airport. The pre-stress forces and boundary geometry established through work with the system models are used in the input data to these models. A much "finer mesh" is used to better represent the actual geometry. The software used to generate the pre-stressed equilibrium shape of the membrane also pulls the node lines (later to become seam lines) onto geodesic curves (ie. shortest path curves) along the membrane surface. This insures optimal seam locations both from a fabrication and aesthetic perspective.

Fabrication

The patterns are produced on the computer by laying sections of the model down into 2-D. The pattern data is then transferred to the fabrication shop electronically where a wide-area plotter plots the templates full-scale on paper. A typical template is 12-feet wide

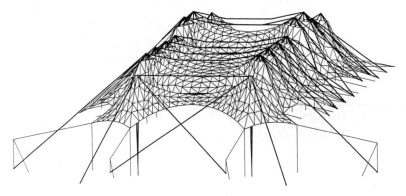

Figure 3: Installation Model.   An installation
computer model is used to analyze and
design the temporary rigging and
partially installed roof.

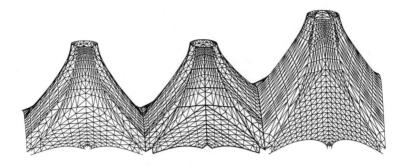

Figure 4:   Pattern Model. "Fine mesh" computer models are used
             to generate the precise shape for patterning the
             fabric. The mesh lines (later to become seams) are
             established along geodesic curves on the surface.   An
             example of these is shown with heavy lines in the
             center bay.

to match the fabric roll width, and up to 100-feet long.   Fabric
panels are cut from the templates and heat-welded together in the
shop to form "assemblies". Each bay of the Great Hall roof
consisted of 4 fabric assemblies.   Each fabric assembly was
individually rolled or folded and then packaged for shipment to the
site.

Installation

At the time installation of the roof systems began, the concrete
structure of the Great Hall building was complete up to the 5th
level.   Level 5 is the floor level for the primary Great Hall
space.   It was used during installation as a staging area and work
surface for both men and equipment.   Designed for live loads as
much as 250 PSF, it was able to support up to 40-ton cranes,
provided load-distributing mats were used.

The masts were delivered to site in one piece.   Top weldments,
rigging, and miscellaneous hardware were attached while the masts
were on the ground.   The masts were then erected using conventional
boom cranes located outside of the building.   In the completed
structure, the masts are stabilized by the fabric roof system and
associated cables that are located within the shape of the membrane
surface.   They have no external guy cables and therefore must be
allowed to pivot on spherical bearings at their bases.   Temporary
guy cables were required to stabilize the system during
installation.   As there was no place to position guy cables that
would not interfere with fabric installation later, temporary mast
top extensions were bolted to the masts to provide a place to
attach the guying system.

The guying system and partially-erected fabric subjected the masts
to loading conditions and bending moments in the upper sections
that the masts would not be able to carry. The problem was
analyzed and solved using the installation computer models
discussed earlier. The solution used was to add temporary stay
cables to work in conjunction with the truss rings (similar to the
stay cables on a boat mast) and remove the bending movement in the
masts.

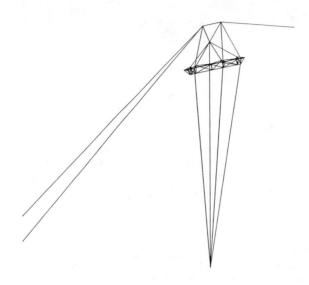

Figure 5: Stay Cable Rigging. A computer model
is used to design the temporary stay
cable rigging and mast extensions.

The truss rings were delivered to the site in two pieces, set
around the mast bases (at the Level 6 elevation), and then welded
together. The mast top units, skylights, and mechanical equipment
were then assembled on the rings. Hoisting of all the rings (two
at a time) was accomplished with a large drum hoist secured in one
location on Level 5. The drum hoist cables traveled through a
series of sheave blocks and fairleads up to Level 6, over to the
appropriate mast, up the mast, and into a block and tackle system
to produce the required mechanical advantage. Using this system,
the rings together with their mast top units were sequentially
hoisted in pairs.

Figure 6: Mast and Truss Ring Installation.   The
masts are guyed with external temporary
cables.   The truss rings are   assembled
around   the   mast   bottoms   and   then
winched up into position.

Figure 7: Fabric Hoisting.   The fabric was lifted
with a drum hoist secured on the   level
5 slab (seen in   the foreground).   Two
hydraulic cranes were used to assist.

As the ring assembly and hoisting proceeded, the outer fabric
installation began.   The fabric assemblies were unrolled on the
Level 5 slab, and installed one bay at a time.   The perimeter
clamping hardware and cables (ridge, valley, etc.) were attached to
the fabric while down on the slab.   The fabric was positioned such
that the two halves of a bay rested together, one on the top of the
other prior to hoisting.   The primary hoisting was performed using
the same winch that was used to hoist the truss rings.   The hoist
cables were attached to each end of the bay's ridge cable which
were lifted towards the rings.   Two hydraulic cranes (positioned on
the Level 5 slab) were also used to assist.   As the ridge cable was
lifted, the fabric bay went with it.   Once the ridge cable was
pinned, the fabric bay was spread open and attached at the valleys
to the neighboring bays.

Figure 8: Partially Installed Roof

Following completion of the outer fabric the clerestory framing was installed. This work was erected from the inside of the building using hydraulic cranes situated on the Level 5 slab. An air-inflated expansion joint was installed to close the space between the outer fabric and the rigid clerestory framing.

Figure 9: Elevation of the Installed Roof

The liner was installed after the clerestory glazing was complete, and an interior space was protected from the weather. It was erected sequentially in much the same manner as the outer fabric. However, being much lighter and protected against the wind, small electric winches were used instead of the large drum hoist. A temporary dust barrier was installed with the liner to minimize dust accumulation on the fabric that would be produced by the finishing trades to follow.

The fabrication and construction of the roof system took the efforts of more than 300 people, over a time period of approximately three years. The fabric roof will become a landmark to the City and County of Denver, recognized worldwide for its unique architecture and the magnificent space it creates.

Figure 10:   Aerial View of the Roof

History and Development of Fabric Structures

R.E. Shaeffer[1]

Traditional tent Forms

The tent is man's oldest dwelling except for the cave. Evidence of mammoth bones and tusks used as supports for animal hides has been found at sites verified to be more than 40,000 years old in the Ukraine region. Rings of stones used as ballast for the edges of conical tepee forms have been found at ancient sites in northern Asia and North America (1).

The tent has been the dwelling in one form or another for most nomadic peoples from the Ice Age to the present. Vegetation permitting, the most common supports for tents were tree branches or the trunks of saplings. The heavier of these were sometimes left behind because of transportation problems. The skin or *velum* of early tents utilized animal hides or, less frequently, birch bark pieces or latticed leaf fronds. Gradually, possibly starting as early as 10,000 years ago, these were replaced by felt or woven materials, such as wool or canvas (2). Contemporary materials include aluminum, fiberglass, and steel for the supporting elements and highly sophisticated synthetic fabrics for the velum.

Until quite recently most tents consisted of three basic forms: the conical or tepee shape, the widespread *kibitka* or *yurt* which has cylindrical walls and a conical or domical roof [Fig. 1], and the "black" tent which has the velum tensioned into saddle shapes [Fig. 2]. The black tent gets its name from the goats' hair used to

[1]   Kibitka.   [2]   "Black" tent   (after Drew).

[1]Professor of Architecture, Florida A&M University,
Tallahassee, Florida, 32307

weave the velum. (The gable-roofed ridge type tent saw little use in ancient times but became a popular and durable military form beginning in the 18th Century. It could be considered as an adaptation of the kibitka form to a rectangular plan.)

Of the three basic forms the conical tepee form is the oldest and saw widespread use across northern Europe, northern Asia, and North America. The conical kibitka shape dates to 2000 B.C. or earlier and has been the world's most popular dwelling form (3). The same shape executed in vines and straw is found throughout Africa and South America. This tent form developed in a wide band from the eastern Mediterranean region to Mongolia. Its shape has been the one most copied or adapted for later tents. For example, the parasol roof shape found in the tents of the military and the royalty of most European and Asian nations in the 14th to the 18th centuries came directly from the kibitka (4).

The "black" tent is probably about as old as the kibitka form and like it, is still much used today. The loosely woven cloth permits the passage of air and its dark color provides a high degree of shade, appropriate for its use in hot arid climates. It developed in Asia from Iran to Afghanistan and later spread to northern Africa.

One can easily contrast the black tent of the warmer arid regions and the tepee shape of the northern climates. The steeply sloped sides of the latter form do not easily collect snow and provide a natural chimney for the necessary fire within. The low profile and shallow slopes of the black tent make it resistant to the desert winds.

Of the three basic shapes the black tent is the only one in which the form is not completely determined by its supporting framework. In the first two, the velum serves only as a barrier to the elements and is not an integral part of the structural system. In the black tent, however, the amount of tension (or prestress) in the velum establishes its scalloped form and provides stability for the supporting elements. In this manner and because of its basic anticlastic (opposite in sign) principal curvatures, it is highly related, from a structural standpoint, to contemporary tensioned fabric architecture (5).

Another structurally interesting tent form is the "envalet" popular in the Catalunya region of Spain for several decades near the turn of this Century. These had a clear span of about 30 meters and were erected annually for village festivals. Tall wood poles were placed around the perimeter of the roof and ropes were suspended across the rectangular plan so that the fabric could be suspended from above (6) [Fig. 3].

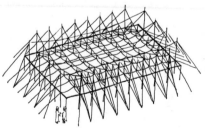

[3]   "Envalet" tent (after de Llorens).

One of the largest tents ever constructed was the one used in 1925 for the thirty-ninth National Congress of India led by Mahatma Gandhi. It provided shade in a single space for more than 20,000 delegates and visitors. Poles were spaced about 30 meters on center to support the handwoven cloth (7).

The largest wall tents were the traveling circus "big tops" popular in the U.S. from the early 19th Century. Harnessed elephants were often used to pull the supporting poles into place as the tents were set up and taken down many times in the course of a single season. In the 1950s these reached their maximum size covering more than one hectare. Shortly thereafter, circuses abandoned the tents as more cities were able to provide a rigid-roofed civic center or coliseum.

Air Structures

The air-supported roof provides an economical way to achieve long spans. Such structures were first proposed by William Lanchester of England in 1917 for use as field hospitals. He received a patent but never constructed one. In 1946 Walter Bird pioneered the radome; the first one was constructed of neoprene-coated fiberglass with a diameter of 15 meters [Fig. 4]. By the 1960s, his Birdair Company was building them with spans of more than 60 meters using a laminated Dacron fabric with a Hypalon coating (8).

In 1958 Walter Bird constructed the McBac Arts Center Theatre in Boston. Designed by architect Carl Koch and engineered by Weidlinger Associates, it was intended to be erected each summer. The roof consisted of an air-inflated lense-shaped "pillow" supported by a steel compression ring. (The Birdair Company grew to construct almost all of the large fabric structures in the U.S. in the last 30 years.)

The 1970 World's Fair site in Osaka, Japan provided the impetus for rapid developments in fabric structures. The poor soil conditions and the threat of seismic shaking both suggested the use of lightweight structures. From a structural standpoint the most significant building at the Fair was the U.S. Pavilion designed by the architectural firm of Davis and Brody and engineer David Geiger of the Geiger-Berger firm [Fig. 5]. The low-profile cable-restrained air-supported structure was made of vinyl-coated fiberglass spanning to an oval-shaped concrete compression ring. It provided 139 x 78 meters of column-free exhibit space. By using a super-ellipse for the ring and a diagonal cable pattern, Geiger was able to greatly reduce the bending forces in the ring. This simple innovative structure was actually the result of major budget cutbacks which had sacked two previous designs by the architects.

[4]   Walter Bird and the
       first radome, 1946.

[5]   U.S. Pavilion, Osaka World's Fair, 1970.

At about this time, Harold Gores of the Educational Facilities Laboratory (EFL), an arm of the Ford Foundation, was looking for ways to provide temporary college athletic facilities to accommodate the arriving babyboomers. The search was on for a fabric for use in air-supported roofs that was very strong but resistant to both fire and ultraviolet deterioration. A team of John Effenberger (DuPont), Malcolm Crowder (Owens-Corning), John Cook (Chemical Fabrics Corp.) and David Geiger proposed using fiberglass coated with polytetrafluroethylene (PTFE), better known as teflon. (This material had been developed by NASA for space suits.) Hal Gores convinced the presidents of two private colleges to gamble on the new structural system. The Steve Lacy Field House at Milligan College in Tennessee was constructed in 1972-75. It was a cable-restrained insulated roof with a diameter of 65 meters. The Thomas H. Leavey Activities Center at Santa Clara College in California was completed in 1973 and consisted of two oval-shaped structures, the larger being 91 x 59 meters in plan. From the outset the Milligan roof encountered difficulties with back-up generators and high utility costs and was replaced by a rigid steel frame in 1986. The smaller of the Santa Clara bubbles was recently dismantled but the larger one is still in use.

In 1975, the Silverdome at Pontiac, Michigan was completed measuring 220 x 159 meters providing a clear span exceeding those of the Astrodome and the Superdome. Smaller college facility domes were constructed in the next few years: UNIDome at the University of Northern Iowa (1976), Dakota Dome at the University of South Dakota (1979), O'Connell Center at the University of Florida, and the Sundome at the University of South Florida (both 1980). Following Pontiac, five more of the large domes were built: Carrier Dome at Syracuse University (1980), Metrodome in Minneapolis (1982), B.C. Place in Vancouver (1983), Hoosier Dome in Indianapolis (1984), and Tokyo Dome (1988). Almost all of these were engineered by David Geiger and built by Birdair.

Many of these air-supported facilities have proven to be difficult to maintain under bad weather conditions. In some cases the hot-air snowmelt systems were inadequate and in other cases the air pressure control systems were not sophisticated enough. Most of them have suffered from accidental deflation, some more than once, with the attendant plethora of law suits. (It should be pointed out that the lawsuits involved material damage and not personal injuries.) Perhaps the most spectacular failure occurred when the fabric of the Silverdome was almost completely destroyed by high winds and very heavy wet snow in 1985. With the advent of the cable dome (discussed later) it is unlikely that more of these large air-supported roofs will be built. As the existing fabric reaches the end of its life (25-30 years?), most of the roofs will be replaced by some type of a cable-and-strut structure or other lightweight system. It is worth noting that while these giant air structures are headaches for the owners and operators, the fans love them! They have become symbols of pride for most of the cities in which they have been constructed.

Cable Nets

The forerunners of contemporary tensioned fabric structures were cable net structures. Perhaps the most influential is the first one, the J.S. Dorton Arena in Raleigh, North Carolina designed by architect Matthew Nowicki and engineer Fred

Severud in 1951. This is the structure which Frei Otto says had a significant impact upon him when he visited Severud's office in New York City as a student. Other early cable roofs include Eero Saarinen's Yale University Hockey Rink, again engineered by Severud (1957), the French Pavilion at the Brussels World's Fair, designed by Rene Sarger (1958), and the Sydney Myer Music Bowl in Australia, designed by architect Robin Boyd and engineer Bill Irwin (1958). A highly sophisticated cable net and trussed arch system was used in 1982 by engineer Jorg Schlaich for the Munich ice skating rink in Germany.

Tensioned Fabric Structures

The modern tensioned fabric era began with a small bandstand designed and built by Frei Otto for the Federal Garden Exhibition in Cassel, Germany in 1955. He built several more complicated canopies for various exhibitions including the entrance pavilion and a dance pavilion at the Cologne Federal Garden Exhibition in 1958. Because he lacked a fabric of sufficient strength, these canopies were limited in span to around 25 meters or less. Perhaps his best known works are two large cable nets. With architect Rudolph Gotbrod, he designed the elegant German Pavilion for EXPO '67 in Montreal, Canada and with architect Behnisch and Partners, the huge Olympic Stadium for the Munich 1972 Olympics. He was also a consultant with Gotbrod on the King Abdul Aziz University Sports Hall, engineered by the British firm, Buro Happold in 1979. There is no doubt that Frei Otto had considerable influence as a pioneer in the development of tensile structures.

With the advent of teflon-coated fiberglass, architect John Shaver of Salina, Kansas was able to convince the president of LaVerne College in California to construct the new Student Activities Center and Drama Lab using a mast-supported tensioned fabric roof [Fig. 6]. Unfortunately, the translucency of the fabric could not be used to advantage because, at that time, the local building code required an overly conservative burn-through test and an opaque insulating liner had to be added to the underside of the roof in order to pass this test. Nevertheless the "Supertents", as the complex is called, is a favorite with the students. This landmark building reached its twentieth birthday last year and tests indicate the fabric is in excellent shape.

[6]   LaVerne College Student Activities Center, 1973.

From 1968 to 1983 Horst Berger and David Geiger were partners. Geiger worked mostly with air-supported structures and Berger with tensioned fabric membranes. In 1976 Horst Berger, working with the architectural firm of H2L2, designed two significant fabric structures for the Bicentennial celebration in

Philadelphia. The Folklife Pavilion spanned 21 meters using fourteen 17-meter tall vertical masts in two parallel rows. The Independence Mall Pavilion was the larger one covering over 4000 square meters using eight tilted masts in two rows for a clear span of approximately 35 meters. It was one of the largest tensioned fabric spans in the world at the time it was constructed. Both of these structures used vinyl-coated polyester fabric. They were the first of many successful Berger designs using a ridge-and-valley geometry.

The largest fabric roof to date is the Haj Terminal Building at Jeddah, Saudi Arabia used to provide shade for the hundreds of thousands of pilgrims who make the journey to Mecca each year [Fig. 7]. It was designed by the architect-engineer firm of Skidmore-Owings-Merrill with Horst Berger as a consultant. It consists of 210 identical cone-shaped canopies square in plan, each measuring 45 meters on a side. It covers approximately 47 hectares and was completed in 1981.

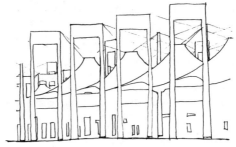

[7]   Haj Terminal, Jeddah, Saudi Arabia, 1981.

Berger acted as the structural engineer or as a design consultant for several noteworthy pavilions including: Queeney Park skating and tennis canopy in St. Louis, Missouri (1978), New Florida Festival structure at Sea World, in Orlando, Florida (1980), Tennessee State Amphitheater at the World's Fair in Knoxville (1982), and the Crown Center Ice Rink and Performing Arts Center in Kansas City, Missouri (1983). All but the Sea World building had exterior masts using trussed tubes to provide column-free space on the interior. With architect L. Gene Zellmer, Berger designed the Bullocks Fashion Island department store in San Mateo, California in 1981. The innovative frame had eight masts and a cantilevered edge. In that same year, Zellmer designed the single-masted Good Shephard Lutheran Church in Fresno, California. (It is not clear to the author and others as to why the translucent nature of fabric has not found its way into the designs of many churches and other religious buildings.)

In 1983, the Geiger-Berger firm collaborated with the Chandler/Kennedy Group to design the unique Lindsay Park Aquatic Center in Calgary, Canada. It has a 128-meter trussed arch which acts as a central spine to support fabric and cables. The recreational facility is well insulated (R16), but is still highly translucent.

In 1985, Berger worked with architects Ian Fraser and John Roberts to create the huge King Fahd Stadium in Riyadh, Saudi Arabia [Fig. 8]. It has a diameter of 288 meters and consists of 24 identical cantilevered modules. The center is open in keeping with international regulations regarding natural turf. Pictures make the stadium appear deceptively small; even with its central opening it covers one-third more area than the Pontiac Silverdome. The two-meter-diameter masts are almost 60 meters tall. It is arguably the best of Berger's many tensioned fabric designs.

[8]   Riyadh Stadium, Saudi Arabia, 1985.

For EXPO'86 in Vancouver, the Zeidler/Roberts Partnership Architects collaborated with Horst Berger to design Canada Place, an exhibition building on the city's harbor waterfront. The roof structure covers an area 135 meters by 55 meters and consists of five modules in a ridge-and-valley pattern with two rows of five masts. The structure is unique in that the five modules are skewed significantly with respect to the building's long dimension. This was done in order to reduce the span which the fabric had to accommodate from ridge to ridge. The building houses a convention center, a cruise ship terminal, a hotel, and an IMAX theater and has become a landmark for the city.

In 1989, Horst Berger designed an elegant canopy for the roof deck of Arthur Erikson's new San Diego Convention Center. Spanning almost 100 meters, it provides shade and rain protection for special exhibits, concerts and banquets. It consists of five ridge-and-valley modules each having a pair of flying struts, i.e., vertical masts which do not deliver their loads to the base level, but are suspended in the air by cables. Running down the middle of the span is a unique "fly" system that covers openings in the main roof and accentuates the sail-like nature of the structure.

In 1990, Horst Berger designed the Cynthia Woods Mitchell Performing Arts Center in Woodlands, Texas outside of Houston [Fig. 9]. The pavilion roof covers 3000 seats and the stage and provides the focal background for a possible 7000 more people on the amphitheater lawn. It is a classic undulating minimal surface form using exterior trussed masts to hold up the three peaks. The use of tie backs to the ground is avoided by using interior compression struts (also trussed) to pick up the horizontal components of the membrane edge forces.

[9]   Woodlands, Texas Amphitheater, 1990.

Also in 1990, the Chene Park Amphitheater was constructed on the Detroit riverfront [Fig. 10]. The fabric roof, designed by architect Kent Hubbell and engineer Bob Darvas, provides shelter for 6000 seats and uses three masts, each topped by a fan-shaped trussed arch. The roof itself approximates a quarter circle in plan with a large space truss truncating the narrow end above the stage.

[10]   Chene Park, Detroit Amphitheater, 1990.

For the 1990 Olympics the Italian National Olympic Committee completely rebuilt their stadium to give it a capacity of 85,000 seats. The elliptical roof in the form of a ring covers all but the playing field, left uncovered as per Olympic rules. The fabric-covered steel frame cantilevers 45 meters toward the center from the outer ring, which is 314 by 220 meters in plan. The huge (yet light in appearance) roof is supported on 80 rows of reinforced concrete columns which are located under the outer ring to provide clear sight lines for all seats. The architectural firm was Italproggeti and the engineer was M. Majowieki (9).

Also in 1990, the architect Yatsui and the engineer M. Kawagachi collaborated to design the unusual roof of the Para Disso Amusement Park in Osaka, Japan [Fig. 11]. The arched roof is suspended from two rows of seven inboard masts which pierce the roof. Each mast picks up the steel roof arches at six locations in a manner reminiscent of the Spanish envalet tents. Cable-restrained fabric spans approximately 20 meters from arch to arch in anticlastic form. The building contains several swimming pools, a giant water slide and a large indoor beach (10).

In 1991, Todd Dalland of FTL Associates in New York City designed the Carlos Moses Music Pavilion. This portable orchestra shell covers more than 300 square meters and arrives at the site on seven trucks, five of which provide support for the erected structure. The hydraulically operated external masts enable the structure to be erected without cranes in under six hours and dismantled in the same

[11]   Para Disso Amusement
        Park, Osaka, 1990.

[12]   Pier Six Concert
        Pavilion, Baltimore, 1992.

amount of time. In its first summer the structure served 30 performances in 16 park locations throughout the city.

In 1992, the new Pier Six Concert Pavilion in Baltimore Inner Harbor was completed [Fig. 12]. This superbly detailed structure, also designed by Todd Dalland, provides seating for 3400 concertgoers. At the stage end the fabric attaches to a curved concrete beam and makes a unique transition to the metal roof of a masonry building. The curvilinear structure provides a welcome contrast to mostly geometric angular forms of the other new buildings of the harbor.

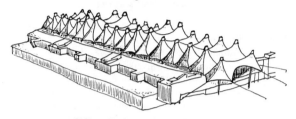

[13]   Denver Airport, 1993.

At the end of 1993, the Great Hall of the Denver Airport was completed [Fig. 13]. The fabric roof covers approximately 14 hectares including the enclosed landside terminal, which is 300 by 70 meters in plan. The architectural firm of Fentress and Bradburn selected Horst Berger with Severud Associates in New York to create the roof structure. The roof membrane consists of two layers of PTFE-coated fiberglass located 600 mm apart. The inner layer provides thermal insulation and acoustic absorbency.

The vertical enclosure consists of a glass curtain wall cantilevered upward from the main floor by a system of cables and struts, in some cases as much as 18 meters. The closure system between the glass walls (having relatively limited deformation capability) and the fabric roof (needing to sustain large deformations under wind and snow loading) utilizes a continuous inflated tube, more than a meter in diameter. Many see the Denver Airport as a test case for large tensioned fabric structures. Located in an area of significant snowfall and other adverse weather conditions, its success could mean the development of many large fabric enclosure schemes.

Cable Domes

The latest technology for long-span roofs is found in the fabric-covered cable dome, a structural system based upon Buckminster Fuller's tensegrity dome developed in the 1950s. The basic scheme is circular in plan using radial trusses made of cables except for vertical compression struts. Circular hoops provide the bottom chord forces [Fig. 14].

The Fuller dome developed instability under unsymmetrical loading conditions and this problem prevented any practical use until it was overcome by design changes introduced by David Geiger and David Chin (of Geiger's office). The first successful cable domes were constructed in Seoul, Korea for the 1986 Asian Games and later

[14]   Cable dome schematic.

used for the 1988 Olympics. The Gymnastics Arena was 120 meters in diameter and the Fencing Arena, 90 meters.

The first cable dome in the United States was the Geiger-designed Redbird Arena on the campus of Illinois State University. It is elliptical, 90 by 77 meters in plan, and heavily insulated between the outer structural fabric and inner fabric liner. It was completed in 1989 and is unusual in design in that it has only one tension hoop between the inner tension ring and the perimeter compression ring. This visually emphasizes the peaks created by the vertical struts and gives the roof a more crown-like appearance.

In 1990, the Suncoast Dome was completed in St. Petersburg, Florida. The stadium was designed by HOK, Architects and the roof was engineered by the Geiger firm. It is the first cable dome designed for baseball and the 210-meter-diameter roof is tilted 6 degrees to provide more seating behind home plate and the infield area. The structure uses four hoops to support the fiberglass fabric and its acoustic liner. Unfortunately, the city of St. Petersburg, at this writing, has not yet been able to attract a major league team.

The largest cable dome to date is the Georgia Dome in Atlanta completed in 1992 [Fig. 15]. Designed for football it has an oval plan, 235 by 186 meters, with a 56-meter-long truss running down the middle. The design of the roof was accomplished by the engineer Matthys Levy of Weidlinger Associates. It has three elliptical hoops between the truss and the compression ring. It is different from previous cable domes in that the ridge cables of each radial "truss" do not lie in the same plane but form triangles with the vertical strut system. This results in a diamond pattern of hyperbolic paraboloids for the fabric roof panels.

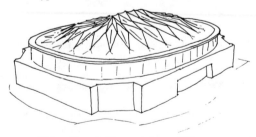

[15]   Georgia Dome, Atlanta, 1992.

Convertible Roof

Perhaps the most unusual fabric roof is the convertible one atop the Montreal Stadium [Fig. 16]. The stadium design was originally done by the French architect, Roger Taillibert, and the structure was to be completed in time for the 1976 Olympics. A giant inclined tower 168 meters tall forms a "skyhook" from which 26 cables suspend the fabric roof to cover a 200 by 120 meter opening in bad weather and retract the roof at other times to permit an exposed playing field. The building encountered much difficulty during construction with faulty prestressing being discovered in the base of the tower and deterioration in some of the base details. Numerous cost overruns caused a delay of several years. The building was finally completed through the efforts of the SNC-Lavalin engineering firm. The upper portion of the tower frame was changed from concrete to steel and the compression ring around the fabric-covered opening was completely redesigned. The building was completed in 1987 using a polyurethane and PVC-coated Kevlar fabric (11).

[16]   Montreal Stadium, 1987.

CITATIONS

(1)   E.M. Hatton, The Tent Book, Houghton Mifflin
        Co., Boston, 1979, pg 6.
(2)   E.M. Hatton, pg 4.
(3)   Philip Drew, Tensile Architecture, Westview Press,
        Boulder, CO, 1979, pg 31.
(4)   E.M. Hatton, pg 27.
(5)   J.J. de Llorens and A. Soldevila, "the 'Envalet', a Big Dancing Tent for Local
        Holidays in Catalunya (Spain)," Proceedings of the First International
        Conference on Lightweight Structures in Architecture, Sydney, 1986, pg 42.
(6)   Philip Drew, Tensile Architecture, Westview Press,
        Boulder, CO, 1979, pg 31.
(7)   E.M. Hatton, pg 38.
(8)   Walter Bird, "Air Structures - Early Development and Outlook," Proceedings
        of the First International Conference on Lightweight Structures in Architecture,
        Sydney, 1986, pg 554.
(9)   Stadio Olympics, comitato Olimpico Nazionale Italiano, Rome, 1990.
(10)   Membrane Structure, Taiyo Kogyo Corporation, Osaka, 1991, pg 124.
(11)   Luc Lainey, "Montreal Olympic Stadium," Proceedings of the International
        Association for Shell and Spatial Structures - Canadian Association of Civil
        Engineers International Congress, Toronto, 1992, pg 521.

The Body Language of Tensile Structures

Todd Dalland, AIA
Principal, FTL/Happold, New York

Introduction

It is said that about 60% of the information communicated in face-to-face conversations between people is non-verbal.

(This is quite a testament to the importance of the visual arts in relation to the rationalist humanities which can be expressed with words and numbers alone.)

Verbal communication is based on language, which is rational.

Non-verbal communication is based on sensory information having to do with sight, hearing, smell or touch, and is intuitive.

What is the connection between the visible surface of an object and the invisible meaning it contains?

The visual portion of information which is communicable in a non-verbal manner shall, for the purposes of this paper, be called Body Language.

- Reading is 0% Body Language
- Radio and telephone are 0% Body Language
- Dance is 100% Body Language
- Graphics (without text) is 100% Body Language (two-dimensional)
- Fashion (clothing design) is 100% Body Language
- "Style" in the visual arts is 100% Body Language

A building can be considered a"body" and its visual appearance can be considered as "non-verbal, intuitable information".

Since for the most part buildings do not use spoken or written words to communicate their messages (functional, structural, cultural or artistic) to the people who use or see them, it follows that Body Language is the primary mode employed by buildings (or rather their designers; engineers and architects) to communicate.

Over time a great deal has been learned by designers and building users about the Body Language of traditional buildings which are made of familiar materials such as stone, wood, concrete and steel, which embody familiar structural forces such as compression and bending, and are visually represented with familiar forms such as Platonic and Cartesian ones.

The ability to establish complex relationships between forms, spaces, colors, materials, details, etc. is often required to communicate to the senses the minimum quantity of relevant information necessary for satisfactory architectural/engineering communication. How does this happen in tensile buildings?

Tensile structures are a relatively new building type which use an unfamiliar construction material: flexible composite fabric, in an unfamiliar structural condition: pure tension, in an unfamiliar geometry: negative doubly-curved surfaces.

As opposed to spaces where Cartesian forms are constructed using materials in compression and bending, the intrinsic message of these relatively new and uncommon forms and structural principle and their communication potential has not yet been impressively exposed and absorbed by designers and society.

It follows that the Body Language of tensile structures is currently largely unfamiliar to building designers and users, is actually still in a relatively immature state, and represents an area of design which can benefit from additional study.

It would be about codifying conscious and subconscious psychological reactions associated with the experiencing and perception of architectural spaces where the operative sensual information-givers are doubly-curved forms constructed of materials in tension.

The purpose of this paper is to attempt to assist in the maturisation of intuitive communications skills in the business of tensile structures design. As well as text based skills, visually dependant ones are required.

(What are the secondary informing elements, after the primary elements of skin and supporting structure configuration, in the medium of tensile buildings? Translucency of fabric, configuration of ridge and valley cables, configuration of fabric fields bounded by cables, orientation of seam lines, curve transformations of seam lines, degree of inclination of masts and cables, degree of curvature of fabric membrane surfaces...)

Ethical norms for physical ugliness and beauty are based on configuration of form. Subtle variations of form can change information from ethical to unethical. From purity to pollution.

Note: Not to be considered as relevant to this paper are the aspects of buildings which employ written text, literal graphics or literal forms to communicate, such as:

- Words written on buildings, such as: "Neither snow, nor rain, nor heat, nor gloom of night stays these couriers from the swift completion of their appointed rounds," chiseled across the cornice of the U.S. Post Office Building on 8th Ave. and 34th St., in New York City.

- "Moving" numbers, letters, words and sentences on buildings such as the regularly updated information about current events and "the news" at the Times Building in Times Square in New York City where the building's designer has placed a band of illuminated and programable light bulbs around the entire building.

- Large painted or photographic murals such as those commissioned for the lobbies of the buildings at Rockefeller Center in New York City which depict scenes of American life to illustrate the building's theme of "New Frontiers and the March of Civilisation" .

- The building shaped like a duck for a Long Island fast food drive-in, or the building shaped like a Sphinx (half man and half lion) at Giza, Egypt, in devotion to a God of the same description.

Body Language is real it is not a pseudo-science.

Good Body Language improves quality of life and financial return.

About Body Language and its relationship to Building

Body language is making meaning apparent without the literalness of language.

Mime, the performance art as practiced by Marcel Marceau, is instructive about Body Language. Only facial expressions and body positions are used to communicate. The human body, along with clothing and make-up, is the medium. Unlike most buildings, though, in mime the medium (the human body) is free to move, change shape, change position in space, and change location in space.

As Mr. Marceau states: "The art of mime is an art of metamorphosis. It's not stronger than words. You cannot say in mime what you can say better in words. You have to make a choice. Mime is an art beyond words. It is the art of the essential. And you cannot lie. You have to show the truth. The public has to understand immediately what mind you have, what situation you are in. That's why it's so hard to describe mime. You have to see it."

The relationship of idea to form is largely inarticulable (it can't be described by words alone, nor quantified mathematically with numbers alone.)

Body Language is rationally ungraspable information.

How does one communicate the relationship of soul to body? Of imagination to reason? Text to image?

A building is similar to a human body in that both are "communicating bodies" and both comprise three-dimensional compositions of form and light (color). They have different attributes, though, as the important dimensions of time and motion are added.

In mime, the viewer is stationary and the communicating body moves through space, whereas in building, the communicating body is stationary and the viewer moves through space.

In mime, each time the performer moves, new information is revealed as facial expressions, body positions and props change. Elaborate stories can be "told" with considerable depth and subtlety, with the addition of time or motion

In building, new information is revealed each time the viewer moves in or around the building, also enabling rich and complex stories to be recorded.

Compositions of form and light = information (ideas, feelings and emotions).

Changing compositions of form and light = information streams.

Forms communicate feelings.
In building, form is inseparable from structure.
Form equals structure.

Structure = visual information (artistic, and cultural)

Structural forces cause emotional reactions.

As one begins to walk around or inside a building one begins to experience the stream of information put there by the designers.

Note: Some aspects of buildings do represent actual changes of form in time such as retractable roofs, moveable walls, on-off glass walls, moving (re-locatable) buildings, temporary buidings, and tensile roofs which deflect in response to live loads. Also, as the sun moves in time across the sky, its changing direction, angle and intensity can change the form/light composition of a building.

The complete and successful transmission of a designer's information, as expressed in the Body Language of a building, to a viewer or user (which has the potential to be a rich and satisfying experience for the viewer or user) would include and be the result of the following:

- viable messages to be communicated
- Body Language communication skills on the part of the designers
- implementation of the design information into the construction of the building
- Body Language cognitive skills on the part of the viewer

There are several basic types of information which can be communicated this way. The Porta delle Suppliche, Florence, Uffizzi, 1574, by architect Bernardo Buontalenti is used here as a building example to demonstrate these types of information. In quotes are what Buontalenti's responses could have been as designer of the building:

1. Functional information:

    Pediment, two columns, lintel and door = "I wanted to direct circulation to this place and make it clear that an important entrance is located here."

2. Cultural information:

    Pediment, two columns, lintel and door = "The client and I are intentionally associating ourselves with the achievements of Classic (Roman and Greek) architecture and civilization by selecting this historical style of construction for use in our building."

3. Structural information:

    Pediment, two columns, lintel and door = "In the type of heavy masonry construction (stone and concrete materials with some limited iron connector and tension elements) typically used for important civic works in Classical building, exterior walls are load bearing walls and all large penetrations or openings such as doors and windows involve the use of lintel beams in bending, and/or columns and arches in compression."

4. Information inherent to the building medium selected by the designer (i.e. the medium is the message):

    - Character and feeling of masonry materials: hard, heavy, massive, rigid, opaque, sound proof, cold and damp (or hot and dusty), permanent, impenetrable, unmoveable, powerful.
    - Visual direction of primary motion lines of structural force: downward due to gravity, or upward due to resistance of gravity.
    - Vertical columns carrying compressive dead loads straight to the ground (while resisting buckling)
    - Horizontal lintels transferring vertical loads in right angle paths (while resisting bending) to vertical members.

5. Artistic information:

> Pediment, two columns, lintel and door = "I have
> demonstrated an understanding of Classic proportion and
> symmetry in an original composition which conveys the
> feelings of grandeur, elegance, order and virtuosity. Further,
> by taking the scrolled pediment, breaking it in half, and then
> switching the position of the two halves I have provoked
> additional ideas, thoughts and feelings (messages)
> including:"

> - Things should not always be taken for granted!
> - Things placed out of order can convey a disturbing
>   feeling...
> - Things placed out of order can draw attention to
>   themselves inviting closer scrutiny.
> - Things placed out of order intentionally can actually
>   convey not only a knowledge of the correct order, but
>   suggest something more. Is "correct" actually correct?
>   Correct enough? Not correct at all?
> - The pediment is actually non-structural since it can be
>   removed, re-configured and re-attached without the
>   building falling down.
> - Experiments with the random or arbitrary re-
>   arrangement of things can yield pleasant surprises;
>   the reversed scroll forms create a very unexpected and
>   interesting pair of "wings".

What subconscious tectonic associations with tensile structures do we possess?

Artistic information can be a conscious or subconscious comment about
a building's functional, cultural, structural or medium inherent
information, or about its contextual site-related information, or it can be
a philosophical or emotional idea about any other aspect of the human
experience.

Creating art releases the emotional tensions of being rational and regenerates the spirit to enjoy beauty in life.

Artistic information can provide relief from the demands of daily life,
and it can create a social forum for change.

A building which will stand on a site for a long time will be communicating the same message and information to the same viewers and users over and over again for a long time. In such a case the information designed into the body of the building should be long-lived information, not hastily considered information or throw-away information.

Note: Sometimes a building's users or viewers make changes to the building which can be considered as reverse communications back to the building's designer.

About Body Language and Tensile Structures

As Buckminster Fuller might say, "Tension is one of the basic building forces of the Universe, along with compression, bending and torsion."

As Peter McCleary might say, "Inventing and mastering the tectonics of tensile forms and materials, in ways as scholarly and sensual as those achieved by designers over the last several millenia for bending and compression forms and materials, is the goal."

As Bill Katavoulos might say, "Buildings are moving from mass to membrane. Building skins are simply environmental filters. Composite fabrics can be engineered to filter any combination of things a designer could want. Air, water, moisture, liquids, particles, chemicals, odors, heat, light, radiation, electricity, sound, etc. "

Mature and functional tensile structure building messages and means to communicate them are not yet part of the automatic subconscious awareness and ability of our culture.

Is this why some early tensile structure designs provide poor quality information or indecipherable information to the public? And even unintentional dis-information in some cases?

What are the unfamiliar materials properties and loading characteristics of tensile structure forms and details? Unlike almost any other construction technology, they move significantly (visually and dynamically) under wind and snow loading. Unlike almost any other construction technology, the primary materials are flexible. Tensile buildings weigh a small fraction of what other buildings weigh, and many of their materials are translucent.

How can this medium and message of lightness and motion be expressed and poetically enhanced by the designer, as opposed to being ignored, negated or worse?

Prestress forces and lateral loads play a much greater role in tensile structures than they do in conventional structures where gravity loads often dominate. What alternate real and perceptual relationship to the ground and the sky does the tensile building have as a result? What are the unique psychological expressions of non-gravity construction and how are they made manifest in material, form and detail?

Where are the spatial layerings and transparencies in tensile buildings, the arcades and porticos and courtyards? Where is the cogent expression of detailing? Which is the cornice, the eave, and the triglyph in a tensile building?
Where is the facade? Which side is the front of the building? How does a tensile building say "entrance"? What is a column in a tensile building and what is the meaning of the expression of its capital and base?

Once a vocabulary is established in a subjective and abstract language generally recognizable by many, perhaps more designers will be able to communicate more complex tensile information in a more articulate and interesting manner)

When tensile buildings do become familiar to us, will we have learned something new or remembered something old?

This line of questioning is not wholly facetious. If it is inappropriate to ask classical questions of a new architecture then the point is that even new architectural questions must be discovered before answers possessing comparable levels of richness and delight will be forthcoming.

Analysis of the Body Lang1/6/94uage of several built examples of Tensile Buildings:

Where are the expressions of insightful metaphors concerning the human experience in the tensile buildings that have been built to date?

The second part of this paper comprises an intuitive visual analysis of a variety of successful and less successful tensile buildings using photographic slides of projects by different designers. The following aspects of each project will be discussed in the presentation:

- Membrane/structure configuration
- Details
- Interface between tensile building and connecting "hard" buildings, and/or contextual relationship to site.

THE NATURE OF STRUCTURAL MORPHOLOGY
and some interdisciplinary examples

Ture Wester[1]

Abstract

Among members of IASS, the most common definition of Structural Morphology is the study of interaction between geometrical form and structural behavior. The most significant structural morphologists of our century, dealing with Structural Morphology in this sense, all seem to have been inspired by and relate their work to the morphology of structures in nature (e.g. Felix Candela, Antoni Gaudi, Heinz Isler, Pier Luigi Nervi, Frei Otto, Le Ricolais, Eduardo Torroja etc.). On this basis the present paper will try to approach some central qualities of Structural Morphology through examples evolved in nature and developed by the human intellect, and through this to underline the importance of the work in the future.

Introduction

Since the start of the IASS Working Group No.15 in Copenhagen in September 1991, there has been a vivid discussion, inside and outside the group, on the meaning of *Structural Morphology*. In the following I will try to give an answer to this question, not as a square definition, but as an approach to the topic.

Morphology simply means the *Study of Form*. Form is usually interpreted as type, geometrical shape or surface character. The word *Structural* is much more ambiguous. Any kind of organized physical or psychical matter has a structure, and hence *Structural Morphology* could be applied to anything containing order. It seems appropriate now to try to place some limitations on the idea as a topic in the IASS environment.

---

[1]Associate Professor, Royal Danish Academy of Fine Arts, School of Architecture, Kongens Nytorv 1, DK-1050 Copenhagen K, Denmark.

The first and most important part of the answer is to realize the connection. If *Structural Morphology* is the given name, then *IASS* or *Spatial Structures* is the family name. In other words *Structural Morphology* means the **Study of Structural Form (Structure = load-carrying elements)**. This definition seems to be valid to point out the center of gravity for the idea, but of course not good for its limitations. We could easily find research work which could be classified directly to this, but we would miss important and relevant aspects if we restricted ourselves to the above interpretation.

A definition or guideline which seems to cover the needs for our group is that *Structural Morphology* is the study of *Form* related to the *Load-Carrying Elements* of a structure seen in perspective of one or more of the following subjects: Forces, materials, resources, energy, production methods, functions, aesthetics, cost, perception, harmony, ecology, etc.

From this it is obvious that the topic is often an interdisciplinary one, and may besides Architecture and Engineering, involve disciplines such as Biology, Physics, Philosophy, Anthropology, Archaeology, Sculpturing, Crystallography, Geometry, History etc.

Even without the interdisciplinary aspect it seems that all IASS Working Groups deal more or less with Structural Morphology, even if it is not their declared topic. This fact puts Working Group No.15 in a very central position, and gives us the possibility to emphasize a common interest for all WGs.

Examples

As described above, the form related to the forces in the structure is the central topic for *Structural Morphology*. The extensive work done in the field of *Form Language*, very much inspired by the new possibilities by using computers, might be extended to a *Form and Force Language* (Wester,1992). This extremely interesting field seems unlimited and open to many different interpretations. It seems that many group members would be able to contribute considerably to this topic, which might be one of the major tasks for the group.

In order to make the presented ideas more substantial, a few *Structural Morphology* relevant interdisciplinary subjects will be discussed in the following. The examples are more or less accidental and just sketches. But they indicate that the idea of *Structural Morphology* is broad and diverse, and goes beyond the usual studies of IASS - but still it is important to stay with the definitions as stated above.

Two Indonesian ethnic building types

Two ethnic groups, both living in Indonesia, the Samosir Bataks and the Tana Toraja people are claimed to originate from the same group, previously located at the present border area between Thailand, China, Laos and Vietnam. This group is said to have migrated to Indonesia in the 14th-15th century, the Bataks to Northern Sumatra and the Tana Toraja to Sulawesi (Celebes).

Fig.1 Samosir Batak village. The roofs are made of palm fibers and of corrugated iron sheets.

Fig.2 Batak house with tiled roof.

Both groups have the same characteristic sagging - or saddleback - roof (fig.1,2,3). It is obvious that this shape can easily be utilized in the structure. The reason for the shape, according to the local myths and the scanty literature, is never explained as beeing connected to the structural efficiency, but e.g. as remains from the ships by which they migrated or as symbols of horns from the sacred and very important buffalo. From the outside it is not easy to judge the structural behavior of the roof, but the inside shows that the Batak roof consists in each end of four interconnected timber poles to fix the two end points for the beam which is so slender that it sags and forms a chain curve. The horizontal forces from the chain are transferred by the pyramidal pole structures. The roof covering is then supported by the chain and the facades. The shape facilitates the original palm tree fiber roofing, the newer clay roof tiles and the contemporary corrugated iron sheet roof plates. The roof appears as a very slender, highly efficient and elaborate structure. The ordinary Batak house does not seem to have as high a status as it does in Tana Toraja. The roof of the Tana Toraja house, on the contrary, is a much more heavy and material consuming structure, where the overhang is moulded and constructed as a fairly complicated system of cantilevered beams (fig.4).

Fig.3 Typical Tana Toraja houses in the central part of South Sulawesi.

Fig.4 Tana Toraja house under construction. The saddle shape is achieved
by cantilevered beams.

Fig.5 Old Tana Toraja house with palm
fiber roof and extra support at the
cantilever.

Some very old houses reveal the structural complexity as the tip of the roof tends to hang - and needs an extra vertical support (fig.5).

Now, the interesting point is to verify these observations, to work out a proper description of the structural morphology of these houses, to find out why, how and when the two similar looking roof types developed differently and to describe the structural system for the original and common roof type. In order to get solid answers to these questions there needs to be interdisciplinary studies between engineers/architects and technical ethnographers and archaeologists. The material efficient Batak type has been of interest to comteporary architects and engineers, e.g. Frei Otto (see Happold et al, 1987). The ecological and technical angle of the materials used for the structures and joints is also of great interest. It is worth thinking about that these structures have been developed slowly through centuries. Probably to continuously better adaptation to the environment and to continuously more sophisticated technical performance.

It should be mentioned that other Batak clans in the same Sumatran area use houses, some long-houses (eg. the Karo Batak), with a straight roof. Other ethnic minority groups from other parts of Indonesia are also claimed to be of the same origin.

The dome-shaped mud houses in the Near East deserts

The described type (fig.6) enables dwelling in arid and semi-arid areas where no timber is available. What is needed is just mud, created as a mixture of the desert dust and water, often brought to the surface through kilometers of underground channels, called *Karez* systems. The mud is used for sun-dried bricks and mortar for joining the bricks. By using a special building technique, formwork is not needed at all (fig.8). This so-called *pitch-brick* dome type is more than 6,000 years old and needs highly skilled craftsmen. The method is rightly called the most ingenious innovation in the history of building (Van Beek,1988). As can be seen in fig.6&7, the shape of the domes are closer to be paraboloids than spheres, which increases the structural efficiency as the material can only transfer compression.

Fig.6 Typical desert village near Tashkurghan in the northern part of Afghanistan.

This means that it may be built lighter and hence load the surrounding walls with smaller horizontal forces (Wester,1993). Types from other areas of this part of the world are more spherical in shape. Seen from the viewpoint of structural efficiency, ecology, natural resources, energy (except human labour!), harmony, adaptation to the landscape - i.e. structural morphology in many of its aspects, it is an extraordinary elegant and clever solution, probably far better than what we do today in our civilization. This ancient technique, which has survived up until today, could, with some modernization, probably be adapted to other cultures (e.g. the Western culture) in an advantageous way.

Fig.7 Domes covering a public bath in North Afghanistan.

Fig.8 The ancient technique of dome construction without any formwork. A desert village near Herat in West Afghanistan.

Configurations in nature with low structural redundancy

It is significant that many very fascinating structures found in our surrounding nature are extremely close to beeing statically determinate. It might be because engineers and architects are attracted by these often beautifully well-organized special structural configurations. At least it seems that it is these kinds of biological structures which are very often referred to when pointing at the

perfection of the structures of Mother Nature. A viewpoint which is often supported by Darwin's theory of evolution through the "Survival of the Fittest", a
refinement of structural performance through millions of years. A theory which is
close to the above mentioned evolution of the Batak roof structure. However, we
must not forget that there are structures in nature which seem to be inefficient,
unadapted and unreasonable. These so-called "Lucky Monsters" which have survived without evolutionary changes through millions of years, need other explanations for their presence. The colossal spectrum of configurations of nature often
makes one think that whatever theory you want to prove by using examples from
nature, can be proved. It seems that whatever quality you want to connect to nature
seems possible.

The orb cobweb

This fascinating configuration (fig.9,10) which has very little redundancy
when regarded as a pre-tensioned plane pure bar-and-node structure.

Fig.9 A spider's orb web overloaded
with morning dew.

Fig.10 The regularity of the web of
this tropical spider is astonishing.

Regarded as a plane lattice structure, all normal inner nodes are 4-valent,
i.e. kinematically neutral (Wester,1992). The central node is n-valent, and all n
peripheral nodes are 3-valent, which together again gives 4 on average, i.e.
kinematically neutral. The central node has, so to say, exactly as much extra
stability as the peripheral nodes lack in stability, i.e. they balance each other. To
secure such a structure you need four strings including the one to tension them all.
Therefore, if the orb-web has at least four threads attached to the surroundings
there are just enough threads to make it kinematically stable - a very beautiful
configuration, indeed. If the orb-web is subjected to a point load perpendicular to
the plane of the web, the web will deform in such a way that it creates the best
possible shape for carrying the load. The elasticity of the threads can absorb much
energy if it is a dynamic load. This structural performance reflects exactly the
advantage for the spider to rapidly crawl around on a vertical stable structural net,
and at the same time to catch its prey with high kinematic energy in the structurally most perfect way. This is pure honey for a structural morphologist.

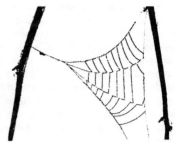

Fig.11 The overloaded plane orb web becomes a spatial structure.

Fig.12 The triangular web is geometrically a slice of the orb web, but structurally very different.

If the web is overloaded by e.g. the morning dew, some of the threads would be in compression, but as this is not possible, the web escapes by moving into the third dimension (fig.9,11) and forms a model of the inverse of a perfectly shaped grid shell as developed by e.g. Frei Otto and his collaborators or free-edged concrete shells by Heinz Isler or sculptural masonry structures by Gaudi.

It is equally interesting that the spider's triangular shaped web in fig.12 is kinematically unstable in all directions, which enables the spider to entangle its prey in the web when moving it as a childs jump rope.

Other types of spider cobwebs have been inspirations and design models for the saddle shaped pretensioned tent structures.

The spongy tissue inside our bones

Where the 4-valent nodes are kinematically neutral for plane 2-D structures, the 6-valent node is neutral for the 3-D structure (Wester,1992).

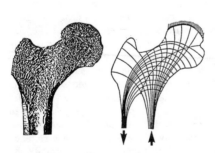

Fig.13 The structure of the inside of the upper part of our thigh bone. Natural and calculated.

Fig.14 The internal structure of our heel bone fits perfectly with the arch of our feet.

A look into the enlarged ends of our tubular bones (fig.13,14) shows a 6-valent orthogonal 3-D structural grid. Regarded as a bar-and-node system, it will hence form a kinematically neutral configuration. It is obvious that the periphery of such a system will lack a number bars for keeping it stable, if not extra bars are placed on e.g. the periphery itself. This is exactly what the so-called *compacta*, the compact bone surface, can perform. The total configuration regarded as a 3-D lattice structure forms therefore a slightly redundant structure. The geometry of this spongy tissue follows closely the trajectories of the forces. The tissue has therefore the optimal position for high structural efficiency. This was discovered through the well known interdisciplinary structural morphological studies by Meyer and Culmann in Zürich 1866 (Thompson,1988).

The almost "just stable" situation enables possibly internal changes in the structural geometry without the occurrence of unwanted secondary stresses. Such changes are observed in the bones of astronauts and other individuals with abnormal loading of these structural elements. These theories might be simulated and verified by computer models, and inspire the architectural design of highly efficient bar-and-node structures.

The sea urchin test

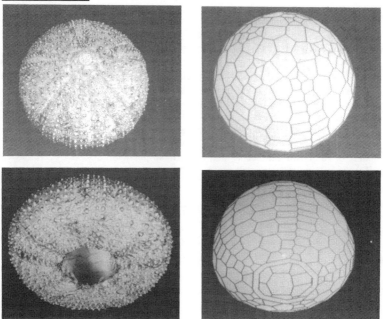

Fig.15 Sea urchin test compared to its computerized counterpart. Note the distinct 3-valent patern at the inside of the test.

Another "just stable" configuration is the shell (or test) of the sea urchin. If analyzed as a pure plate structure, where all stabilizing forces are shear forces between the plates which are rigid in their own plane, the test turns out to be a statically determinate pure plate structure. The "normal" plate is seen to be hexagonal on average, or - what is geometrically eqivalent - the vertices between the plates are 3-valent (fig.15), which means kinematical neutrality just as a 6-valent node does for the 3-D lattice structure, eg. the previously mentioned bone structure. At the SFB-230 conference in Stuttgart 1991 three papers dealt with this fascinating structure (Philippi *et al*,1991. Telford,1991. Wester,1991). The mechanism of growth, the maintaining of stability during growth, the strength of the shape, the distribution of stresses etc. in connection with the evolutionary history was discussed between biologists, architects and engineers. Basic questions of e.g. the growth and the stability during growth have not yet been solved.

In order to verify these observations, it is necessary to involve biologists. It is my own experience, in the case of the sea urchin, that the interdisciplinary exchange of ideas is not only for the benefit of both parts but extremely inspiring, and often results in a better understanding of the structural morphology of our surrounding nature.

Conclusion

It is hoped that the definitions stated in the first part of this paper will help to guide our group into the important and fruitful field of research of *Structural Morphology*. The idea is also to encourage interdisciplinary studies, where there is so much to be done with our special knowledge and scientific background in order to bring deeper insight in many different fields, but not least to get inspiration and new ideas for the design of the future architecture of structures in all its aspects. In order to accomplish this we need all our personal resources, the use of all the latest scientific achievements in our own and adjacent fields and all the most recent computerized techniques. But without common sense, simple physical models of cardboard and chains, and furthermore possibilities to test our thoughts and theories in practise, we will end up with useless pure academics.

References

HAPPOLD,E., DICKSON,M., SUTHERLAND,I. (1987).
A School for Woodland Industry - a study in the engineering use of green roundwood timber. *Proceedings of the International Conference on Non-Conventional Structures.* B.H.V. Topping (ed). Vol.1, pp.107-114. Civil-Comp Press, London.

PHILIPPI,U., NACHTIGALL,W. (1991).
Constructional Morphology of Sea Urchin Tests. *Proceedings of the II. International Symposium of the SFB-230.* Part I, pp.183-191. Universität Stuttgart and Tübingen. Stuttgart.

TELFORD,M., (1991).
Echinoids: Domes, Pneus, or something else. *Proceedings of the II. International Symposium of the SFB-230.* Part I, pp.103-109. Universität Stuttgart and Tübingen. Stuttgart.

THOMPSON,D'ARCY. (1988).
On Growth and Form. Abridged ed. Edited by Bonner, J.T. Cambridge University Press.

WESTER,T., (1991).
Structural Dualism and Sea Urchins. *Proceedings of the II. International Symposium of the SFB-230.* Part III, pp.177-182. Universität Stuttgart and Tübingen. Stuttgart.

-       (1992). An Approach to a Form and Force Language based on Structural Dualism. *Proceedings of the First International Seminar on Structural Morphology.* Motro and Wester (eds). pp.13-24. Université Montpellier II.

-       (1993). Efficient Faceted Surface Structures. *Proceedings of the Fourth International Conference on Space Structures.* Parke and Howard (eds). University of Surrey. Thomas Telford Services, London.

Architecture and Lightweight Structures:
a Methodical Design Approach

Vinzenz Sedlak*

Abstract

Lightweight structures such as membrane structures, shells, space-grid, lightweight trussed and suspended structures have become an integral part of the international building scene. While a wide range of constructed examples exists, the general quality of architectural integration with the building complex and environment is, with few exceptions, arguable. To improve architectural design quality , accessability to the vast repertoire of lightweight structures must be improved and a methodology be developed that architects and designers can employ during the conceptual stage of the design process. Current research at LSRU focusses on the hypothesis that there is a logical grammar which rules the relationships between design application, building form, structural type and building envelope and that can be used as a predictive design tool. Based on a morphology of structures developed by the author [Sedlak, 1987], a typological method has been established and its validity has been checked by systematic examination of a database containing 200 Australian projects constructed over the last 100 years. Statistical evaluation of this database shows the interrelationship that exists between the three main aspects of lightweight structures design: shape, structure and architectural application.

Introduction

The development of increasingly sophisticated and lightweight building materials has lead to a growing use of lightweight structures in architecture and engineering in Australia and Overseas [Sedlak, 1987]. Although often identified with sports stadia, exhibition halls or sound shells, lightweight structures are finding increasing usage in more conventional and commercial applications such as domestic dwellings and shopping centres. Contrary to established beliefs, the assumption that lightweight structures are for unique building types is, therefore, unfounded.

* Associate Professor and Director, Lightweight Structures Research Unit
School of Architecture, The University of New South Wales
PO Box 1 Kensington NSW 2033 Australia

The tendency in contemporary usage of lightweight structures is to use the materials as "high tech" alternatives to conventional construction. However, this approach denies the cost, time and structural savings inherent in using lightweight structures as integrated building concepts in their right. The failure to understand the economies inherent in the use of lightweight structures is closely linked to the fact that a soundly-based applied design theory relating application, building envelope, structure and building form iis presently lacking. The development of this theory has been based on the following research methodology involving four basic tools:

1. a morphology of structures [Sedlak, 1987].
2. several separate typologies related to significant parameters of this morphology (building use=application, building shape, structure type, material and shape, building envelope=cladding type and material) [Loh, 1989; Sedlak Debello and Loh, 1991; Sedlak and Loh, 1992; Sedlak, 1993].
3. a project-related database of executed lightweight structures [Sedlak et al, 1987-93].
4. a methodology for systematic statistical evaluation of the database of executed projects [Loh, 1989; Sedlak and Loh; 1992].

The chosen procedure involves analysis, the identification of principle design parameters, with a sufficiently large and therefore representative sample of lightweight structure buildings in Australia and Overseas. By applying the above typologies in a consistent, readily understood and reproducible form (mnemonic codes) a particular lightweight structure building can be first analysed and thus identified. The subsequent statistical analysis involves frequencies, variance of distribution and interdependence between different variables in order to establish patterns of uses of particular design parameters (shape, structure, envelope) related to building application.

The Typologies

A working basis was established for the systematic identification of lightweight structures by establishing the following separate typologies:

Typology of Building Use:
Based on the ICONDA format (International Construction Database) buildings were classified according to 30 main- and 240 sub-categories, for statistical evaluation purposes they were rationalised into 13 categories:

Column 2

Applications		
	assembly	industrial
	commercial outlet	multipurpose
	dwelling private	outdoor stage
	environmental protection	transport
	exhibition/museum	others
	feature	dwelling public
	indoor sports	

One of the six attributes below specify the part of the project that is the lightweight structure, defined in terms of the degree of protection provided from climatic and other environmental influences:-

Column 3
Application Attributes
building   - a close-fit envelope providing environmental protection;
enclosure   - a loose-fit envelope providing only filtered protection;
canopy   - an open sided loose-fit envelope;
roof   - a horizontal cover of a building;
floor   - the interior base plane of a building;
others

Typology of Building Envelope Shape:
Building volumes can be created either within singular shapes (unit-shapes: for instance: prism, hemisphere) or within additions of unit-shapes (aggregate-shapes). Eight principle unit-shapes were established (for instance: cones-pyramids, cylinders(vaults)-prisms, spheres(domes)-polyhedra). According to position (horizontal-vertical), curvature of the surface (concave-convex, domical-saddle) and other parameters a range of typical attributes to the principle shapes were established. An example for the typology attributes of the unit-shape "Cone" is given in Figure 1.
Aggregate-shapes were identified according to typical arrangements of the unit-shapes that create the aggregate:

Column 4		**Column 5**	
Building Volume Shapes	cone	Shape Attributes 1	unit
	cylinder		aggregate
	dome		
	prism		
	pyramid		
	sphere		
	vault		

For the purpose of statistical analysis, we simplified and re-grouped the established building volume shapes into seven groups as shown above.
Polyhedra were eliminated and instead became a shape attribute to the primary shape, such as "dome polyhedral" or "vault polyhedral".
Two categories of Shape Attributes 1 were chosen in order to identify shape more specifically: Column 5 describes whether the building volume shape is a:
        unit   - the one single primary shape; or an
        aggregate - the assembly of various units to form a complex shape.

Column 6			
Shape Attributes 2	saddle	folded	
	synclastic	concave	
	pitched (raked)	curved	

describe the surfaces that border building volume shapes.

Typology of Structure Type and Structure Shape:
A range of 17 principle structure types was established also their known variations and shapes: beam, arch, frame, truss/truss-frame, cable structure,

Primary shape	Attrib 1	Attrib 2	Attrib 3	Attrib 4	Attrib 5
(Circular)	scallop	ridged	oblique	saddle	inverted
Triangular	umbrella		raked	synclastic	inclined
Square			truncated		
Rectangular					
Pentagonal					
Hexagonal					
Octagonal					
Polygonal					

Figure 1   Cone Shape Attributes (Typology of Building Volume Shapes)

Attrib 3 Structure Shape	Attrib 4 Arrangement	Attrib 5 Structure	Attrib 6 Support Position	Attrib 7 Support 1	Attrib 8 Support 2	Attrib 9 Boundary
calotte	2-way	non-prestressed	external	arch	arch	cable
cone	3-way	prestressed	internal	arch-truss	arch-truss	rigid
cone elliptical saddle	diagonal 2-way		perimeter	cable	beam	
cone rectangle saddle	parallel			cable-truss	cable	
cone saddle	parallel 2-way			cable/mast	cable/mast	
cone saddle ridge	peripheral			frame	cable/tie	
cone square	radial			mast	column	
cone square saddle	sequential			ridge-beam	frame	
cone trunc saddle	sequential vertical			space-frame	mast	
hypar				tie	tie	
planar(roof) hypar				tree-mast	truss	
planar(roof) saddle				truss-grid		
prism: vault saddle						
prism: mansard saddle						
prism pitched saddle						
prism rectangle saddle						
pyramid square						
pyramid square saddle						
saddle						
vault semicirc saddle						

Figure 2   Tent Structures Attributes (Typology of Structure Types)

beam-grid, truss-grid, space-grid, skin/frame, shell, shell-grid, folded surface structure, cable net, tent, air-structure, space-frame, spatial-net. An example for the typology attributes of "Tents" is given in Figure 2.
For the purpose of statistical analysis (by chi-square test) we rationalised the above-mentioned principle types into seven categories:

Column 7		Column 8	
Structure Types	air structure & tent	Structure Attributes	arch
	arch & frame		stayed
	beam & truss		suspended
	cable		frame
	folded & shell		grid & rib
	space grid		net
	skin/frame		truss
			rafter:joist, ridge, purlin

Column 8 gives primary attributes to structure types: for instance, cable was made an attribute to net, stayed, suspended or truss; membrane structures were made attributes to arch, frame, grid, net or truss.
Applying these typologies to the classification of existing projects of Australian lightweight structures lead to useful overviews of the range of possibilities and variations of application, shape and structure. Figure 3 shows a sample page of Cone-shaped Tent structure Buildings of different architectural applications.

Typology of Structural Material:
a systematic listing of materials was established [Sikora, 1992].

Typology of Cladding Type and Cladding Material:
The building envelope is divided into roof and wall cladding and these were identified according to their main parameters: filter function, build-up, support function, position relative to the structure, continuity [Sikora, 1992].
As the research project progresses additional typologies will be established such as for Joints and Connections, Environmental Control Measures, etc.

The Database

Approximately 400 lightweight structures projects (200 Australian and 200 International examples) have been fully documented [Sedlak et al, 1987-93] constituting the :   LSRU Database of Lightweight Structures*.
A typical layout sheet for the project "Sydney Mobile Stage"is shown (Fig. 4).

Data fields fall into the following categories:
1.   General project information in clear text:
      project title, application, location (state/town/country), cost;
      Authors: architect, engineers, contractors and client/owner;
      Description of the main structural/constructional features of the project.

AUSTRALIAN LSAE DATABASE
AUST CURRENT DB ALL.082

15 October 1993

Project Name SYDNEY MOBILE STAGE THE DOMAIN Sydney
Application Outdoor stage canopy (temporary)
Year 1983   C 2.1 20.05 C 2.2 07.06   C 2.3   **LSAE No** U 03
State:PCode New South Wales   **Cost A$** 400,000
Town:Sub Sydney: City   **Country** Australia

Sel	SEL	Sketch
Length:Dia m	36	
Length:Dia ft	118	
Width m	26	
Width ft	85	
Height m	14	
Height ft	46	
Span m	42	
Span ft	138	
Plan Area sqm	830	
Plan Area sqft	8934	
Roof Area sqm	1340	
Roof Area sqft	14424	
Volume cum	11371	
Volume cuft	399520	

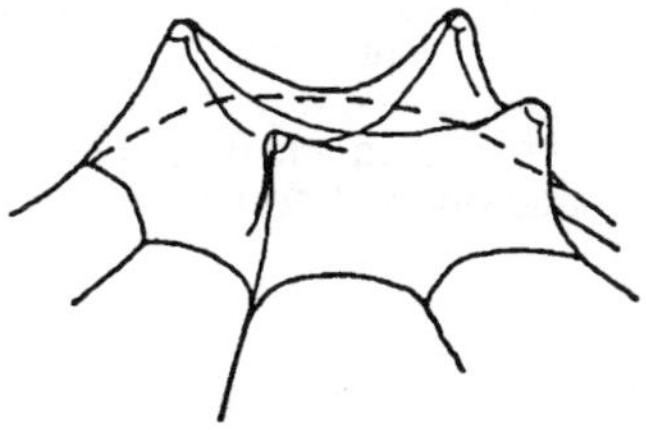

Building Shape Prism saddle:7s
Structure Type Tent* int.cable/ mast/ perim.cable/ tie
Structure Shape *Cone saddle: ridged4/ 2-way
Structure Material Plastic: CF (PVC•PES)* ST/ CHS: ST/ ST/ ST
Cladding Type Memb Struc (Roof) Ext: S U
Cladding Material Plastic:CF (PVC•PES)

Architect Surface & Spatial Structures:V Sedlak Pty Ltd
Engineer 1 George Clark & Assoc
Engineer 2 Peter Kneen Pty Ltd
Contractor 1 B W Bilsborough & Sons Pty Ltd
Contractor 2 Barrier Cons Ind
Client:Owner NSW Dept of Cultural Affairs

Ref 1 Clark G, Sedlak V 1984 Mobile Stage for open-air performaces, Sydney, NSW. Paper presented at 1984 Convention 'Coated fabrics for Membrane Structures', Melbourne. MSAA- 30th & 31st May
Ref 2 Davis B, Sedlak V 1984 The Membrane Structures Field in Australia - its Development and its Future. International Symposium on Architectural Fabric Structures, Orlando, Florida, USA, Nov 1984.
Ill Ref 1 LSRU photo lib: 7/86(9, 35)
Ill Ref 2 LSRU photo lib: 9A/86(22)

Description
The membrane canopy provides protection for performers from sun, wind and rain during outdoor events such as concerts, theatre, opera and ballet. The canopy is a large sand coloured prestressed tension membrane structure commissioned by the N.S.W. Government. Structurally designed for high wind loads while being readily demountable, it's components are easily stacked, transported and erected. Erection time is three days, dismantling time 1.5 days including site preparation, transport and storage. Annual erection and dismantling costs are A$20,000.

The canopy is an anticlastically shaped membrane surface supported by an internal diagonal cable system composed of two intersecting main cables that are supported by four internal masts and a central "flying strut" at the top intersection. Mast staying-cables at the four corners connect the mast tops to perimeter ground anchorages. The membrane is bordered by catenary edge cables and is supported from each mast top with cable-bordered rosettes and at the corners cables stress the surface to nine anchorages. Resulting dimensions are 35.9m x 26.1m, 13.7m internal height at the centre, 19.5m front mast height and 16.9m rear mast height. The material used is a PVC coated polyester fabric 2/2 weave construction 900g/sq m. flame-retardant to Australian Standard 1530 and stabilised for high ultra violet exposure.

Figure 4 LSRU Project Database Typical Datasheet: "The Mobile Stage, Sydney"

2. Classification of the project in terms of typologies described previously: a mnemonic coding system was developed that allows rapid classification in a concise, readily understood format. (for statistical evaluation numerical codes were assigned).
3. Dimensional information: basic dimensions, span, areas and volumes
4. Pictorial representation: sketch
5. References: literature and illustration

The database allows instant access to a wide range of project information and as such is a very useful design aid in its own right. It can be accessed from any data field and multiple searches can be conducted to obtain specific information or development trends. A variety of layouts can be produced to display these data in almost any required format. Database information has been continuously extended and we plan to include specific information such as construction detailing, contract planning, environmental data etc.

Statistical Evaluation

Firstly a preliminary statistical check-analysis was conducted, frequencies were established in each of the search categories, then significant correlations between building use, shape and structure type were established using non-parametric statistical analysis (the chi-square test)
The outcome is described in the following:

Data Establishment
The selection of projects was based on two main criteria:-
-their perceived quality in terms of innovation, application and execution at the time of construction.
-they being significant representations of the range of structure types in Australia, developed from the 1760s until to date.
Influenced by the above criteria the sample group has the specific frequency distribution of structure types as illustrated in Figure 5 .
We explain this particular distribution as follows:-
-a major emphasis was given to the lightweight structure part of the application, be it the entire building, the roof, the wall or the floor
-significant historical developments of structure types were considered ranging from linear structures such as   beams, posts, trusses, arches and frames to surface structures such as folded structures, shells and membrane structures
-any examples which indicated a structural response to general cultural, socio-economic preferences in Australia were included such as a high demand for a certain building type, application and/or a particular spatial quality which could be provided quite readily by a particular structure type (for instance, tent structures).
-in addition personal preferences were accommodated

In summary, we accepted the following shortcomings in our data selection:

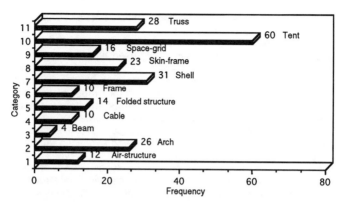

Figure 5.   Frequency of Structure Types: Australian Projects

1. there could be a difference between a sample group of Australian projects and an international one
2. the data sample contains a large proportion of tents and is therefore influenced by the characteristics and general application of this particular structure type

The subsequent stage of this research will investigate a larger sample including international projects and utilise a wider range of structure types for comparison.

The Statistical Database
The aim of the research is to identify interrelationships between building volume shape, structure type and application. These main parameters form the three primary categories of variables. Their attributes form a further set of four categories of variables. The primary and secondary variables are listed in the previous section on typologies:- there are a total of 8 categories which were identified as Column 1 to Column 8:
Column 1 contains the LSAE code for the purpose of identification only.
Columns 2 to 8 describe primary and secondary variables of application (building use), building volume shape and structure type.

Initial chi-square test runs discovered that the 200 sample observations from the LSAE Database were small when compared to the large number of preliminary categories of variables. Moreover, there were a number of cells in each matrix which had zero or one/two observations only. They were not significant enough to stand as individual categories, yet they affected the values significantly and thus the level of significance of data. These preliminary categories were consolidated, combined and re-classified into related groups to form larger, more generalised categories.

Two examples of simple chi-square relationships between primary categories of variables are now presented:
Correlation of Application to Structure Type:
Membrane structure - is the most widely used type forming 31.44% of all structure types. 42.1% of all membrane structures are environmental protection facilities, 26.7% are outdoor stages, with 20.0% each indoor sports and multipurpose facilities. On the other hand, most of the applications categories show high proportions of membrane structures: outdoor stages (80.0%), feature structures (75.0%), environmental protection facilities (54.5%), indoor sports and multipurpose facilities (37.5% each), commercial outlets (34.6%), assembly facilities (32.0%) and exhibition facilities (24.0%). Note that tents form the largest proportion of all structure types in the sample.
Folded & Shell - make up 19.6% of structure types. 33.4% of the examples were found in the categories of Others (mainly experimental structures), 30.3% are assembly facilities, 25.0% industrial facilities and 21..2% private dwellings. High proportions of the following applications were found: others (57.1%), assembly facilities (40.0%), feature structures and multipurpose facilities (25.0% each), industrial facilities (21.4%), private dwellings, exhibition facilities and outdoor stages (20.0% each).
Arch & Frame - are 15.7% of structure types used. Mainly transport facilities, private dwellings and exhibition facilities are represented. High proportions of this structure type correspond to the historical development of arched structures in transport and exhibition facilities. The vault which is formed by parallel arrangement of arches is the most widely used building volume shape for the following buildings: private dwellings (28.6%), commercial outlets (26.9%) and exhibition facilities (24.0%).
Beam & Truss - form 14.0% of structure types. Most are industrial, transport, environmental protection facilities and private dwellings: 42.9% of all industrial facilities and 33.4% of all transport facilities .
Skin/Frame - forms 10.0% of structure types. 31.6% of all transport facilities, 20.0% of all private dwellings and 20.0% of all exhibition facilities are skin/frame structures.
Space grid - forms only 7.0% of all structure types and are most widely used in environmental protection, industrial facilities and others.
Cable - constitutes a mere 2.2% of structure types.
Their frequency is low and only occurred in 6 of 13 categories of application.

Correlation of Building Volume Shape and Structure Type:
Structure types with their inherent shape characteristics cause the variations within each shape categories for buildings. This makes, for instance, one hemispherical dome different from the other, when one is an air supported membrane and the other is a folded surface structure. These characteristics also cause the variations of spatial qualities within a shape category; for instance a glazed shell dome encloses a space that varies from a space enclosed within a masonry dome. The spatial qualities, especially the amount of light entering the interior space, create two entirely different experiences.

Examples illustrating the alternative structural types for each primary shape category of Australian lightweight structures are as follows:-

Cones	Skin	Prisms	Pyramids	Vaults
Arch	Space frame	Air structure	Air structure, frame	Air structure
Cable net		Beam grid	Air structure, truss	Arch
Folded grid	Domes	Beam & post	Cable net	Arch frame
Folded shell	Air structure	Cable net	Folded shell	Arch truss
Grid	Arch	Cable stayed	Frame grid	Folded grid
Shell	Arch frame	Cable truss	Grid	Folded shell
Skin	Arch truss	Folded shell	Shell	Grid
Space frame	Folded grid	Folded plate	Skin	Skin
Tent	Folded shell	Frame	Space grid	Space frame
	Grid	Grid	Tent	Tent arch support
Cylinders	Shell	Shell		
Air structure	Skin	Skin	Spheres	
Arch	Space frame	Space frame	Air structure	
Beam grid	Tent, arch frame	Space truss	Folded shell	
Cable net		Tent	Grid	
Cable truss	Polyhedra		Skin	
Folded shell	Folded shell		Shell	
Frame grid	Frame			
Grid	Grid			
Shell	Skin			

By giving the range of available structure types for certain building shapes and vice versa as related to existing structures this finding is useful information for the conceptual design of lightweight structures. It has been applied by the author for the initial generation of design variants in architecture courses.

An Example of Statistical Evaluation:

From the correlation analysis conducted between Columns 2 to 8 we chose the following example for illustration of the type of statistical information that can be extracted from the database. The example also served as evidence for the principal validity of the adopted research methodology. It should be noted that the analysis is preliminary in character in sofar as the statistical method has yet to be optimised by application of other non-parametric tests which may be better suited depending on the information required from case to case.

First we check Column 3: Application Attributes for the highest occurrence:
Buildings - 35.1%   Enclosures - 20.4%   Canopies - 13.0%
Roofs - 31.2%   Floors - 0.4%   Others - 1.7%

Using "buildings" we check chi -squares of Columns 2/3:
Correlation of Application to Application Attributes for application:
Assembly - 7.8%   Commercial outlet - 9.1%   Dwelling private - 27.3%
Environmental protection - nil   Exhibition/museum - 14.3%
Feature - nil   Indoor - 9.1%   Industrial - 9.1%
Multipurpose - 5.2%Outdoor stage - nil   Transport - 10.4%
Others - 1.8%

FREI OTTO EXHIBITION

Exhibition pavilion enclosure
            C 2.1 20.01   C 2.2 7.05   C 2.3

Cone saddle*5/2-way:os
Tent-net*int.mast/tie/perim.rope
Cone saddle*:parall 2-way

TODD STREET MALL CANOPY

Feature & environmental protection canopy
            C 2.1 14.06   C 2.2 20.03   C 2.3

Cone saddle:os
Tent*int.mast/perim.cable/masts
*Saddle:8s saddle:6s/parall

Figure 3   Examples of Cone-Shaped Tent Structures

CAPTAIN PHILIP PREFABRICATED HOUSE

Dwelling building (prefabricated)
            C 2.1 24.01   C 2.2 7.09   C 2.3

Prism rect gabled
Frame:rafter-joist*perim.wall(stud)
*:pitched(roof)/parall

THE LOREN IRON HOUSE

Dwelling building
            C 2.1 24.01   C 2.2 7.09   C 2.3

Prism rect gabled:concave
Frame:rafter-ridge*perim.column/wall
*:pitched:curved(roof)/parall

MCINTYRE HOUSE KEW

Dwelling building
            C 2.1 24.01   C 2.2   C 2.3

Prism rect gabled
Truss-frame:cantilever*perim.A-frame supp.
*:pitched/parall

CLARKE HOUSE ASCOT

Dwelling building
            C 2.1 24.01   C 2.2   C 2.3

Prism rect:horiz
Truss-frame:cantilever*perim.column
*:Modified-Howe/periph

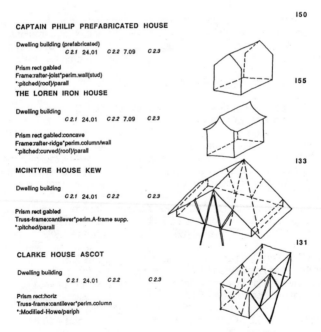

Figure 6   Examples of Lightweight Structure Private Dwellings:
              Prismatically Shaped Frame Structure Buildings.

Thus lightweight structure "buildings" are mainly:-
Dwelling private - 27.3% (ie. 21 of 77 examples)
Exhibition/museum - 14.3% (ie. 11 of 77 examples)
Transport - 10.4% (ie. 8 of 77 examples)

Relating now "dwelling private" to the most commonly found structure type we
find by Correlation of Application to Structure Type: Columns 2/7 :
Arch/frame - 28.6%   Folded/shell - 20.0%   Skin/frame - 20.0%

Examining "dwelling private" for the most frequent building volume shape we
find by Correlation of Application to Building Volume Shape: Columns 2/4:
Prism - 70.6%   Dome - 17.7%

Conclusion from the above:
The most common application within the database is "building" which accounts
for 35.0% of the sample. Within "building", the most common application type
is "dwelling private" which accounts for 27.3%. Private dwellings using
lightweight structures tend to be mainly arch and frame structures (28.6%).
The analysis clearly establishes that 70.6% of private dwellings in Australia
using lightweight structures favour prismatic shapes. The chi-square analysis
proved to be significant beyond the 0.0 level of significance.
This finding suggests that there is a statistically significant tendency for private
dwellings to be predominantly prismatic. This conclusion is in agreement with
established knowledge and supports the validity of the employed method. Four
historically different examples from the database which demonstrate this
finding are:- Captain Philips Prefabricated House (1788), the Loren Iron
House, Melbourne(1854), the McIntyre House, Kew (1957) and the Clarke
House, Ascot Brisbane (1986) (see Figure 6.)

As a second example we are now interested in the "Building Volume Shape"
first in relation to "application" and then in relation to "shape attribute":
Building volume shapes that occur in sufficient numbers are cones, domes,
prisms, vaults and to a lesser extent cylinders.The following frequencies result:
Prism - 53.4%   Vault - 20.4%   Dome - 15.1%   Cone - 8.4%

Checking "prism" across the application categories by
Correlation of Application to Building Volume Shape: Columns 3/4
we find (from the most popular to the least popular):
Dwelling private - 70.6% (24 of 34 examples)
Industrial - 69..2% (9 of 13 examples)
Environmental protection - 64.2% (27 of 42 examples)
Assembly - 56.5% (13 of 23 examples)
Commercial outlet - 52.0% (13 of 25 examples)
Outdoor stage - 50.0% (3 of 6 examples)
Transport - 50.0% (9 of 18 examples)
Multipurpose - 50.0% (4 of 8 examples)
Dwelling public - 50.0% (3 of 6 examples)

Exhibition/museum - 36.0% (9 of 25 examples)
Indoor sports - 33.4% (3 of 9 examples)
Feature - 25.0% (2 of 8 examples)
Others - 12.5% (1 of 8 examples)

From Column 4/6: Correlation of Building Volume Shape to Shape Attribute 2 we find that "prisms" are predominantly "pitched" and "saddle", together approximately 82% of all examples of prismatic structures in the sample:
Pitched - 48.7%   Saddle - 32.5%   Folded - 10.0%
Concave - 6.2%   Synclastic - 2.5%

Conclusion from the above:
Prisms are the most commonly used building volume shapes in Australian lightweight structures. Approximately 70% of private dwellings and also 70% of industrial facilities are prismatic. Prisms are predominantly pitched and/or saddled and together they account for approximately 82% of all prismatic shapes within the sample. A historical example of the saddled prismatic dwelling is the Loren Iron House, Melbourne built in 1854 (see Figure 6.).

Examining the relationship between shape attribute and application:
Columns 2/6: Correlation of Application to Shape Attribute 2 we find that onlly saddle, pitched, folded and concave shapes are sufficiently represented:
Saddle - 46.0% (51 of 111 examples)   Pitched - 35.1% (39 of 111 examples)
Folded - 15.3% (17 of 111 examples)   Concave - 3.6% (4 of 111 examples)

Saddle is the most commonly used surface shape.
Related to "application" , saddle shapes are used as follows:-
100% of outdoor stages   100% of feature structures
70% of exhibitions/museums   69% of environmental protection structures
Note that there were only 5 examples found in each of the first two categories and that most of the examples were tents.

Conclusion from the above evaluation:
Lightweight structures in terms of surface shape (shape attribute 2) are predominantly saddle types with outdoor stages, feature structures, exhibition/ museum facilities and environmental protection facilities being the most common examples. 82% of these saddle structures are membranes.
Two examples are The Mobile Stage, Sydney, 1984 (Fig.4) and Todd Street Mall Canopy, Alice Springs, 1987 (Fig.3).

Conclusion

From an ongoing research study into a methodology for conceptual architectural design of lightweight structures we presented intermediate results. Based on a morphology of structures and on a database of executed lightweight structures as powerful tools for analysis and synthesis, typologies were developed that identify building use, building volume shape and structure type.

A first statistical evaluation of a selected range of Australian lightweight structures established proof of validity of the original hypothesis namely that relationships exist between these three principle parameters of architectural design. Upon closer examination of a substantially larger and diverse sample of data and with optimised statistical evaluation we expect these relationships to lead to a set of predictive patterns that, together with catalogues of design choice established by the separate typologies, can then be used as definitive design aids.

The methodical approach adopted will also lend itself readily to subsequent development of a knowledge based system of design aids in the future.

Acknowledgments

The research project has been supported by a research grant from the Australian Research Council (ARC, 1992-93) and by a Special Projects Grant from the Faculty of Architecture, University of New South Wales (1991).

We thank Associate Professor Peter Bycroft for his guidance in the statistical evaluation of this project. Also many undergraduate students of the School of Architecture contributed with their work to this project at different stages, we thank them all.

References

Loh. S.K. (1989):
"A Vocabulary of Shape and Structure Type in Australian Applications of Lightweight Structures"
BArch Diss. LSRU/School of Architecture, The University of New South Wales, Dec.1989
Sedlak, V. (1987):
"The Morphological Approach to the Teaching of Structures" Proc. Int.Conf.on Lightweight Structures
in Architecture, Sydney 1986,Unisearch Ltd., The University of New South Wales, Vol.3, pp.1154-1187
Sedlak, V. (1987):
"Lightweight Structures in Australia" Proc. Int.Conf.on Lightweight Structures in Architecture,
Sydney 1986, Unisearch Ltd., The University of New South Wales, Vol.3, pp.1188-1210
Sedlak, V. (1987):
"Recent Lightweight Structures in Australia" Proc.Int.Conf.on Non-Conventional Structures, London,
December 1987, Civil Comp Press, London, Vol.1, pp.251-262
Sedlak, V et al.(1987-93):
"The Database of Lightweight Structures in Architecture and Engineering " Macintosh-based database
of executed Australian and International projects from 1850 to 1993, LSRU/The University of New South Wales
Sedlak, V. (1990): "
Architecture and Textile Structures-a Question of Design Choice" Proc. Int.Symposium "Textile Composites
in Building Construction" FITAT Lyon July 1990, Editions Plurals Paris, Vol 2, pp 3-20
Sedlak, V., Debelo, S. & Loh, S.K.(1991):
"A Typology of Lightweight Structures"
Draft Report, Special Project Grant, LSRU/The University of New South Wales, 248 pgs
Sedlak, V. and Loh, S.K.(1992):
"A Study into a Design Methodology for Lightweight Structures: a Typology of Shape and Structure"
Proc.2nd Int.Symposium on Natural Structures, SFB230 Universities of Stuttgart and Tübingen,
Stuttgart Oct.1991, Vol 3, pp 117-129
Sikora, W.J.(1992): "A Typology of Cladding Types and Materials", Report, LSRU/School of Architecture/
The University of New South Wales, Aug.1992, 38 pgs.

Tensegrity Structures: Filling The Gap Between Art And Science

Michele Melaragno[1], IASS Member

Abstract

Underlining the role of structuralism in contemporary architecture this paper addresses the application of structural morphology in the context of an educational experience. Focussing on the imaginative creations of B.Fuller, the *tensegrity* has been chosen as a conceptual prototype from which a morphology class could start a process of manipulation of ideas aiming to gap the scientific and the creative components of design.The abstract values of a geometrical composition are filtered through a process which eventually ends with physical structures in the context of an architectural program. Specifically the tensegrity considered as a modular space structure is explored first for the design of a peculiar urban composition that gives form to a vertical park that responds to a detailed program. Secondly, after the introduction of a case study The Georgia Dome , the class is exploring the tensegrity dome per se, understanding its vocabulary and applying it to specific solutions.

Introduction

Filling the gap between the art and science of building, the

---

[1]Professor of Architecture and Building Sciences, College of Architecture, University of North Carolina at Charlotte, Charlotte, North Carolina, 28223

structuralists in their stronghold occupy a position of extreme relevance in this century of technological and scientific dominance.   Gaudi, Torroja and Candela from Spain, Nervi and Morandi from Italy, Le Ricolais and Duchatau from France, Maillart and Isler from Switzerland, Otto from Germany, and Fuller, Saarinen, Tedesko and Geiger from the United States, are some of the leading structuralists who have energized a century of explorations through an architecture of substance both in form and meaning.   Through their work we know that a new spirituality has emerged through the generation of new ideas that will continue to expand in an innovative identifiable field that has found its place among design theorists.   Much more so in fact than the plurality of various design schools of thought which invariably fluidify in constant mutation, structuralism stands out on a solid foundation of rationalism that depends on immutable principles insensitive to ephemeral fads. Structuralism in fact, responds to the evolution of science and technology and to the fundamental changes of lifestyles and the consequential demands that follow them with a rigorous logic, so that one would feel to be on solid ground when using structuralism per se as a departing point for formulating the premises for a design theory.

On this basis from the perspective of the platform that structuralists have embraced, the present writing follows a design sequence that from geometrical conceptions finds a resolution in structures and eventually arrives to the physical entity of buildings.   Scaling down from broad generalizations to specifics, this paper illustrates the tensegrity compositions of Buckminster Fuller as the starting point for manipulating geometrical forms in a structural morphology class.   It then pursues practical applications into space frameworks which eventually can be applied to useful facilities: a vertical urban park as well as tensegrity domes.

A Morphology Class

Outside of the design studio, not for choice but for a practicality of facts, a structural class steps out from the rigid lecture format to expand into the world of design inspired by a conceptual reality of immutable principles that geometry and mechanics have disclosed to the inquisitive mind. Transcending conventional programs stemming from pure utilitarian demands, the three-dimensional archetype of Buckminster Fuller emerges from its platonic status in the world of ideas and lends itself into the materialization of structural artifacts. A vertical urban park is taken as the inspiring prototypes for the application of tensegrities to contemporary forms of steel frameworks that display their constructivistic character. However preliminary to the individual applications, the 3-D morphological concept of Fuller s creativeness is first explored.

In the search for a geometrical shape capable to generate a 3-D assembly with a minimum number of compression members, a unique filiform assembly emerged just like a platonic idea that wanted to be born and manifest itself. The majority of this array of slender longitudinal members can in fact be fabricated with tensile wires, while only few require to be built in a more consistent manner so that they will be able to carry compression forces when the rest are pulled tight to stabilize the assembly. In other words, this skeletal composition emerges as a continuous flow of tension members and a few sparse compression members which stand in isolation from one another and interconnected only by tensile wires. The compression members with their stiffer appearance seem in reality to be floating in space without touching one another. Such a peculiar composition responded quite well to the name given it by its inventor or discoverer: tensegrity .

The structure, consisting of six compression members only and of twenty-four tension members, responds to geometrical law that relates together the number of members m and the number of joints j into a mathematical relationship that could be written as: $m=3j-6$

More specifically this expression indicates that m is the exact number of members necessary for the structure to be stable. Should the members in fact be less than m the structure would be unstable. By the same token if the number of members in excess of m the structure would have redundant members which are not necessary for its stability. Thus, for the specific case of tensegrity one could see that substituting 12 in place of j, the expression gives $m=30$, verifying that a tensegrity responds indeed to such a law.

The original tensegrity consisting of 6 compression members groups them into three couples. Each couple is characterized by the length of the members and spacing between the two. Thus, when a regular tensegrity is constructed the three couples are usually equal to one another, obtaining therefore 3-D symmetry. Yet, when the structure is observed critically for exploring other different possibilities, one could immediately see that each of the couples can vary in length and spacing creating a variety of individual moduli. Further more the symmetry could also be broken by moving the point of confluence along any of the three axes. Then having attained a variety of moduli, these can be obviously arranged in various combinations and grouping, progressing vertically or horizontally along the x,y or z axes.

With such an introduction the morphology class can explore these avenues by experimenting in the construction of individual moduli following a sequence of numerous experimentation that integrate the moduli into more complex organizations. It is only after a substantial amount of time spent in these exercises that the application of tensegrity moduli to real structures can begin. Thus, at this point various assignments are progressively explored considering a vertical urban park.

A VERTICAL PARK

A tensegrity structure with its inherent openness finds an ideal application for the framework of an outdoor facility including mostly non-enclosed floor slabs supported at different levels to create a multi-story outdoor feature to be used as a park in the dense habitat of an urban center. Greenspaces that have traditionally been used as recreational parks by landscaping the ground may give way to a new concept of expanding vertically rather than horizontally employing a man-made structure when land is not available. This in fact may be the case in many urban agglomerates where the cost of land is prohibitive for low density uses as a park would automatically imply.

Following the verticality of typical urban structures, the interpretation of a greenspace expanding also vertically find a logical justification even if at first glance the concept of an urban park is strongly attached to the soil on the ground. Naturally however floral and arboreal vegetation have long standing traditions of having flourished in various types of containers to create all sorts of pensile roof gardens totally unconnected with the terrain below, so that the vertical urban park supported by an artificial structure should not appear too bizarre, although certainly may strike as unusual.

The tensegrity archetype developed in this context harmonizes with unequivocal clarity the rest of the structures of the conventional city in which typical articulated frame skeletons of steel or concrete dominate the spectrum of new constructions. Capitalising on the character of multi-story frames which can be fully open to the sunshine and the outdoor air, a vertical multi-story park framed as a tensegrity would be particularly effective because of this complete openness. These structures in fact respond with the greatest efficiency to an open air assembly of usable spaces. Minimizing in fact the number of compression members to an absolute unsurpassed minimum, the majority of the essential structural members are thin tensile components almost invisible due to their higher slenderness as in the case of steel ropes. The tensegrity moduli hereby studied fit with incomparable ease the basic requirements that these structures imply.

The essential component of the program for an urban park involves quite a number of basic parameters which are hereby briefly discussed.

- The whole framework is assumed to be developed out of the assembly of two or more tensegrity moduli that expand vertically and horizontally in one or more than one direction. Each modulus includes three basic axes of orientation x, y, and z, assuming z as the vertical direction. In this organization longitudinal axial members in compression can have equal lengths or different ones as the designer chooses to adopt. The horizontal axis at different levels are obviously supporting the various floors while the vertical ones   constitute the supports emerging from the ground bearing compressive loads which in turn need to be stabilized by the tension members for guaranteeing the necessary   equilibrium for the entire structure.
- The various levels are interconnected by staircases to

provide adequate circulation either by isolated structures or by stairs that develop inside the hollow vertical members or around the outside of them. More than one to satisfy fire ordinances, the stairs act as inclined links that bridge the various decks sustaining a flow of pedestrian traffic.

While the tension members tie the structure together   in one structural tensile continuum, the compression members carrying the load can be stressed into bending also since this is within the nature of the loads. The compression members may benefit substantially from expanding the cross-sectional configurations acting as beams.

The anticipated general population of users for these facilities is subdivided into three groups for practical design purposes. Each group will reflect a different level of activity that is practically associated with age. In the distribution of spaces among such three groups considerations must of course be given to the efforts of climbing stairs if a vertical mechanical transportation system is not provided.

In structuring the function of a park the horticultural aspects for the displaying and maintaining various types of plants will involve special requirements including planting containers, trellises, watering devices, wind protection, differentiation between intense sunlight exposure and shading devices, etc. Particular attention is also dedicated to the proper uses of shading apparatus such as canvas awnings or canvas sails to be stretched in various orientation to attain the desired effects very much similar to the rigging of sailing vessels where the trimming of the canvas is adjusted as necessary. To this effect the major structural tensile member that crosses diagonally the space from joint to joint can be used as the main supports for these secondary infrastructures of shading membranes. These aspects in which colorful sails could easily be envisioned is particularly underlined for its specific contribution to the overall character of the facility.

The compression members of the original tensegrity expressed by a pair of vertical members and two pairs of horizontal ones have to be logically interpreted in accordance to their structural role. The two vertical members in the module could in fact include a massive volumetric configuration due to the heavy loads in compression that they carry and could also consequentially be built as major towers. The other two horizontal pairs carrying compression and bending at the same time may for instance be designed as slender trusses by capitalizing on the efficiency of this system and on the openness that trusses also display. Such trusses could either be totally beneath the decks or could be above them so that the decks could hang from the lower chord allowing the trusses to act as railing systems around the perimeter of the decks.

The structural decks which develop geometrically along longitudinal paths due to the overall geometry of the structure would logically suggest one way systems spanning across the supporting trusses, allowing the possibility of using any desired material including permanent concrete slabs as well as replaceable wooden decks.

The types of amenities and ancillary facilities to be envisioned with this miniature vertically extended park is practically open to a large number of options that would automatically emerge in the creative process. However, the picturesque theme of structuring an outdoor recreational facility would definitely bring to mind a kaleidoscopic reflection of images by recollecting from experiences found worldwide. For instance, the colorful ambience of the urban parks of Europe and Asia can bring to mind an animated crowds with ice cream cones and flying balloons, melodic tunes, flying pigeons, in which everybody is an actor and a spectator at the same time, running or promenading or basking in the sun.

THE TENSEGRITY DOME

This last assignment is based on the tensegrity dome which is an application of tensegrity already well developed through the many structures built already. To this effect, the latest of such structures is synthetically introduced in the class as the springboard from which the student can explore the concept for his/her practical application to enclose a space with the same structural vocabulary. The Georgia Dome completed in 1992 in Atlanta, Georgia is located on a site where the 1996 Olympics will take place. This dome with its elliptical configuration is covered by a roof with an area of approximately 400,000 square feet extending 630 feet (193 meters) along the minor axis and 748 feet (228 m) along the major axis. Descriptions of this structure in the technical literature refers to it as the first Hypar-Tensegrity dome since its surface includes sections of a hyperbolic paraboloid. Designing architects for the project were Weidlinger Associates, New York, while the construction, engineering, fabrication and direction per se was carried out by Bird Air. In its basic schematic simplification a tensile dome of this kind will have a compression ring at the base in opposition to the typical tension ring that conventional domes have. Then peculiar to these domes are concentric hoops in tension at different levels. In this particular case this dome employs three of such hoops. Above the last hoop crowning the dome is a 184 ft. truss along the longer axis of the ellipse. Such a dimension is measured between the first and the last of the vertical compression members within the truss. This truss includes nine vertical compression members of equal length measuring 35 feet each between top and bottom chord, some of which extends partially above the upper chord to support an additional upper cable that completes the exterior shape of the dome. The lower chord of this structure is carried by the last tension ring and stands up vertically because of the vertical compression members of the truss itself, giving the feeling that the truss is floating in space as the rest of the structure does. This truss because of its special loading condition is stressed in such a manner so that the bottom chord as well as the upper chord and the diagonal members are all in tension while only the vertical members are in compression. Consequently top and bottom chords and diagonals are made out of tension cables creating indeed a unique truss pre se, while only the compression members consists of steel pipes. Each hoop carries 26 vertical compression members (half of the 52 columns at the ground) extending vertically from one hoop to the next. Such compression members are then stabilized by diagonal cables and are so located so that like in any tensegrity structure they do not touch each other and give the visual illusion to be floating in space. These vertical posts (for a total of 78) vary in length including 61 feet for members over the first hoop, 80 feet for those over the second hoop and 49 feet for those over the third hoop. These members with a maximum diameter of 24 inches are sealed hollow tubes for preventing any internal erosion. The cables per se varying between ten different sizes from 1.25 to 4.9 inches (10 cm) in diameter consisted of two types: parallel strands and wire ropes. Parallel strands more ridged and more difficult to handle in the field, yet less expensive were used for tension hoops as well as for the top and bottom chord and diagonals of the truss, while the more flexible wire ropes were used for the rest. These cables were delivered to the job in continuous lengths that reached up to 1200 ft. wound around spools during their transport to the site. Precut in the factory to the exact length with their end-fittings attached, the cables were ready to be installed reducing the cost in comparison to what was usually done before in other jobs when the cutting took place in the field. The total length of the cable is in the order of magnitude of approximately 5 miles. The joints connecting the vertical post and the cables weighted up to 2 tons each and carried the patented name weldments . The joints housed the cables in square cut grooves in their upper part and clamp them with a cover that is tightly bolted to the main body of the joints. Between the cables and the plate an aluminum bar acting as a filler is used. Notice that the joints on the upper part of the post differ substantially from those at the lower end of the post because while the former are at the surface level, the latter are connected to the tension hoops. The compression ring at the base of the dome consist of a 26 feet wide 8 feet deep concrete structure supported by 52 columns. The connectors because of their sizeable weight (up to 2 tons) were indeed a challenge. In previous tensegrity domes they were built out of cast steel but in this particular case they were fabricated with welded steel parts. The membrane roof itself stretched over the network of steel cables consists of 114 diamond shaped teflon coated fiberglass panels shaped along the profile of the hyperbolic paraboloid surface and clamped to the cables themselves. The membrane by the trade name of SHEERFILL is partially transparent allowing 15% of daylight to pass through creating a glowing surface over the space below. The individual membrane panels which differ from one another reached dimensions up to 80 x 180 ft. and were rolled around steel pipes for easy transportation. When arrived to the job site, each roll was lifted by cranes passing through the network of cables. Then the individual membrane panel was unrolled and fastened to the cables with aluminum hardware. Along the seams between one panel and the next, 18 in. wide strips of membrane was heat bound to guarantee a water proof seam.

The erection procedure for the roof carried out by Bird Air corporation consisted in assembling most of the cable network on the ground including the central truss and then lifting it up to the level of the compression ring. The lifting was done using individual jacks located on top of each column supporting the concrete ring. The jacks pulled the cables during an operation that lasted approximately one week lifting the whole assembly in eight intermediate steps. At this point a second operation that lasted several months included the erection of the three tension hoops and the supporting vertical posts. To minimize the load to be lifted, the 184 ft. central truss was supported by two cranes rather than hanging from the cables. Note worthy is to recognize the level of complexity that these structures require during their erection and the consequential risk that may be involved. In this project for instance the failure of one of the connectors on the upper hoop caused the collapse of one of the posts that was part of a platform supporting some workers, killing one who fell down to ground level below and injuring two others on October 17, 1991.

The overall geometrical configuration of the dome include a variety of criteria of major interest. For instance, the individual diamond shaped panels of the roof membranes were curved into a convex and concave paraboloid configurations, and a hyperbolic paraboloid was formed by lifting two of the corners of the diamond. The overall curvature of the dome was attained by the size and the vertical spacing of the concentric hoops. Starting from the compression ring to the first hoop above it the slope includes an angle of 45 degrees and then it gradually reduces from hoop to hoop. In this manner since the load increases from the top down it was logical to increase the slope as one proceeds downwards because in so doing the tensile forces will increase at a lower rate.

In presenting the G Dome for it s structural significance, some reservations are made concerning full acceptance of it s architectural vocabulary so that in all fairness while the structural concept is fully endorsed in this work with major enthusiasm the overall delineation of the dome and it s relationship with the rest of the structure is questioned. The cusps projecting on the exterior of the surface are reminiscent of an old melancholic idea of the circus tent totally antithetical to the dynamics of a new iconography portraying tensile stresses in a more vibrant and exciting morphological expression.

THE ASSIGNMENT

The present assignment requires the structuring of a tensegrity dome that will use the same basic concepts in a personal interpretation of the theme. To this effect it is noticed that the tensegrity dome departs from the conventional dome structures in many points. These include: a. the base ring is always in compression in this case rather than being in tension as for conventional domes; b. the dome consists of polygonal hoops at various levels supported vertically by compression members; c. the compression members as in all tensegrities are not in direct contact with one another and constitute a small minority in contrast with the rest of the members in tension; d. the vertical compression members are stabilized by diagonal members in tension. In addition to such characteristics it is pointed out that the compression members may be structured in different ways to become other than single solid members but could be structured differently such as composite members of lighter construction. The connectors at both ends of the compression members constitute also a basic element of design in need of special attention since these constitute a critical point involving patented systems. Further explorations expected in the assignment could involve domes springing from the grounds or placed over a sub-structure beneath. Also, special attention is called upon finding other solutions that diminish that tent-like connotation that this structure projects with it s protruding cusps and sagging membranes as also typically found in Frei

Otto s structures in Germany.   Tensile structures in fact do not necessarily require the use of such vocabulary, but could definitely express themselves with different morphologies.

CONCLUSION

Tensegrities that have inspired the largest roof structures in the world: respectively the Sun Coast Dome in St. Petersburg, Florida (1990) and the Georgia Dome in Atlanta, Ga (1992) which has been discussed in this work, are still in their infancy of development and retain all the fascination that come from the simplicity and articulation of their geometry not yet fully explored.   The two themes that have been analyzed in these structural morphology exercises have been presented in this work to emphasize the role that structural morphology itself plays in architectural education and in the forum of ideas which is open to all those who are interested in architectural research.

The tensegrity, still in it s puberty stage, not yet developed to it s maturity, continues to retain it s mystery similarly to an organism that develops to reach it s potential.   It s major features include it s inherent simplicity of the geometry and the efficiency contained in a structure that has no redundancies in terms of number of members.   From structural analysis to design, and from design back to structural analysis in a periodical switching between the two approaches, the art and science of design unrolls in a symbiotic methodology, gaining a proliferation of ideas causing enthusiasm.   Exploring tensegrities through such two applications is just like opening a crack in a window that faces quite a broad and extensive view.   Many other topics are ready to be developed such as the articulation of tensegrity towers for religious buildings, for civic displays, for campus chimes, for electric power distribution, etc.   Also a variety of applications for building frames offers a prolific spectrum for practical applications.   Then of course the tensegrity dome points to quite a diversity of potential morphologies which can follow an overall configuration different from spherical surfaces, exploring for instance cylindrical, conical, hyperbolic-paraboloidal surface, etc., projecting quite a complex spectrum of possibilities.

Space Frames and Polyhedra

J. François Gabriel[1]

Abstract

Polyhedra are easy to build and they can be useful as architectural matrices. Relatively small polyhedra can be built as faceted structures; very large ones can be built with space frames.

There is a family of polyhedra whose geometry is congruent with space frames. These polyhedra are of particular interest to architects because they can form clusters (or infinite structures), which lend themselves to the design of large and complex buildings.

Structurally, space frames are the ideal means for the actualization of these polyhedra. Space frames can also be the means to organize the interior spaces within, i.e., the architectural spaces.

The 12-C Network

Certain polyhedra are congruent with space frames.

The space frame we have in mind is an aggregate of octahedral and tetrahedral frames where twelve members meet on every node. Thus its name, the 12-connected network.

Three polyhedra can be obtained from portions of the 12-connected network. They are the cuboctahedron (which we call CO), the truncated tetrahedron (TT), and the truncated octahedron (TO).

[1]Professor, Syracuse University School of Architecture, 103 Slocum Hall, Syracuse, New York 13244-1250, USA

Figure 1. A six-story structure made of an aggregate of TOs, TTs and COs. In this model, the middle is occupied by one single TO, to which four TTs and six COs are attached. The COs share the six square faces of the TO. Four of the eight TOs' hexagonal faces are shared with TTs. The other faces (one of which is clearly visible in the middle of the model) could receive four more TTs but their orientation would be reversed in relationship to the ground.
The nodes of the polyhedra line up on horizontal planes at even intervals, creating a natural place for floors.

Depending on its orientation in space, the 12-C network carries either two-directional or three-directional chords. We are interested here in the latter. In that case the TO can be described as a four-layer space frame—or a three-story structure if we are to make our polyhedra habitable. Both the CO and the TT will be two-story structures. These three polyhedra can be assembled to form an infinite structure (Fig. 1).

The Honeycomb Pattern

The habitability of space frames has been demonstrated in a number of essays (Gabriel, 1991). Simply stated, the division of space made by diagonals between two chords amounts to a honeycomb pattern. While a honeycomb does satisfy the basic requirements for shelter, it does not provide the variety of spaces, large and small, that even a simple architectural program requires. Some ways to introduce spatial

Figure 2. The same configuration as in Fig. 1 but formed by the 12-C network. The TO that is at the core of the model is a three-story space entirely free of structural members.

Stairs have been inserted in the model. They can unfold freely in the TO and do not interfere with diagonals elsewhere. They require a modification of the chords consistent with the honeycomb pattern (Nooshin, 1991).

variety must be found. Several solutions to this problem have already been proposed (Gabriel, 1985). It is tempting to see in the congruency of our three polyhedra with the 12-C network another means to break away from the straitjacket of the honeycomb. We suggest here the elimination of the space frame within the TO as a means to introduce "breathing space" in the matrix (Fig. 2).

The 12-C network is structurally redundant, especially in multi-layer configurations. Two of our polyhedra are formed by tetrahedra and octahedra connected face to face: they are rigid. The CO, however, is not. It is formed by

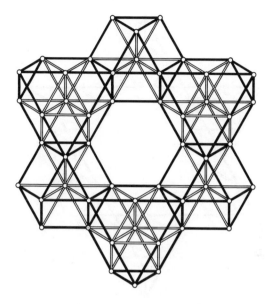

Figure 3. Top view of the story directly above the ground floor in the configuration shown in Fig. 1. The lower third of the TO is surrounded by the upper level of three COs and the lower level of three TTs. In this drawing, as well as in the others, the edges of the polyhedra are shown in solid black. The 12-C network runs continuously through the polyhedra that surround the TO but not through the TO itself.

whole tetrahedra but only halves of octahedra. Although more research needs to be done in this direction, we believe that the rotational risk inherent in the CO would be eliminated by the stabilizing effect of the adjacent TTs (Figs. 3, 4, 5, 6).

Although we elect here to eliminate the 12-C from the TOs, it must be noted that we could, instead, eliminate the 12-C network from all COs, or even from both the COs and the TTs. This would leave only TOs as structural elements and habitable matrices. In that case, a TO would be attached to 12 others, through 12 of its 24 peripheral nodes. Breathing space would be found in the now empty (that is, free of the 12-C network) COs and TTs.

Breathing Space

What we call breathing space must be clarified. As we pointed out, a three-

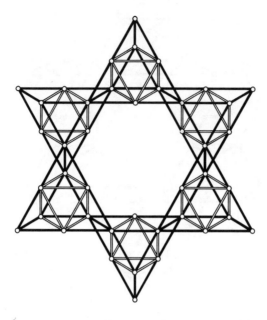

Figure 4. Top view of the story directly above that shown in Fig. 3. The middle third of the TO is surrounded by the upper level of three TTs. Three more TTs are added here (they were not included in Figs. 1 and 2). These are upside down and share a triangular face with the COs underneath.

way space frame is habitable in the basic form of a honeycomb, but it is a rather cramped form. Once the 12-C network is eliminated from a polyhedron, a two- or three-story space is created. Two possibilities are then available to the architect. The empty polyhedron will become a larger room, or it will be used as an exterior space. The decision depends on many factors: the functional requirements of the building, the width and depth of the structure, i.e., its relative density, the climate, the need for vistas and, we hope, aesthetic considerations.

It must be further noted that, free or not of the 12-C network, a polyhedron could still be used as an outdoor space. Indeed, the designers (ideally an engineer-architect team) have the option of eliminating the 12-C network altogether from these polyhedra that will be used as rooms. To compensate, the structural space frame will be found on the exterior of the buildings. Of course, a building program that should only require double- or triple-height spaces is unusual. It is more likely that large and small rooms will be wanted and that the 12-C network will be useful both inside and outside the buildings, but not everywhere (Fig. 7).

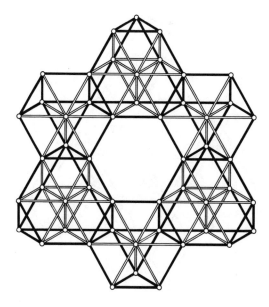

Figure 5. Top view of the story directly above that shown in Fig. 4. The upper third of the TO is surrounded by the upper level of the upside down TTs and the lower level of three COs.

Architectural Considerations

This brings us to the notion of the 12-C network as a means to organize interior space. The 12-C network and an infinite structure of TOs, TTs and COs are congruent, but when they are used architecturally, a geometric conflict appears. The honeycomb pattern does resolve one conflict, that between diagonals and the upright position of human beings. But it creates another.

In the orientation we have selected, none of the polyhedra's faces are vertical. Whether square, triangular or hexagonal in shape, the faces are all leaning in or leaning out. When the honeycomb pattern comes into contact with the face of a polyhedron, a choice must be made. If that face is shared between two polyhedra, the decision is easy to make, for the honeycomb pattern runs uninterruptedly from one polyhedron to the next and the face need not exist as a physical element. If the face with which the honeycomb pattern comes into contact is one that separates the interior of the building from the exterior, then the problem is not as simple. Some of the questions one must ask are: How practical is a leaning wall? What use is it? Does it have any significance? Can it be redressed? <u>Should</u> it be redressed?

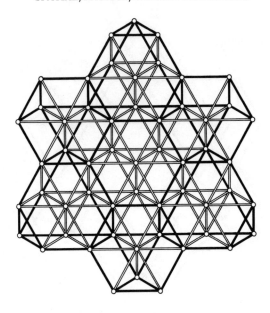

Figure 6. This is the same configuration as in Fig. 5 but here the 12-C network has expanded across the TO.

Designers intrigued by polyhedra have been troubled by these questions for years. The problem is interesting precisely because it can be solved in many ways but there is no perfect solution. That is always the case with architectural problems (Fig. 8).

Conclusion

Space frames and polyhedra used jointly present interesting design problems. They open new horizons in architecture and urban design and they can solve old problems in new, better ways. Above all, they provide an architectural language which enriches our understanding of the relationship between architectural form and space.

References

Gabriel, J. F. "Space Frames: The Space Within—A Guided Tour," International Journal of Space Structures, Vol. 1, No. 1, 1985, pp. 3-12.

Figure 7. A design for a community house using polyhedra in conjunction with the 12-C network. The outside view (above) shows that the geometric integrity of the polyhedra is respected. The cross-section (top) illustrates the use of the 12-C network to span a large room as well as the flow of uninterrupted space underneath.

Figure 8. A study model exploring the architectural potential of the honeycomb pattern (with vertical dividers) used in conjunction with polyhedra (with non-vertical faces).

References (cont.)

Gabriel, J. F. "The Architecture of Space Frames," International Journal of Space Structures, Vol. 6, No. 4, 1991, pp. 287-295.

Nooshin, H. Studies in Space Structures. Brentwood, Essex: Multi-Science Publishing Co., Ltd., 1991, pp. 69-86.

Acknowledgements

The designer of the project illustrated in Fig. 7 is Marcello DiGeronimo. The study model illustrated in Fig. 8 was made by Stephen L'Heureux. All drawings and photographs are by the author.

The author wishes to express his gratitude to the dean of the School of Architecture at Syracuse University, Bruce Abbey, for his financial support.

A Quasicrystal for Denmark's Coast

Tony Robbin

Abstract

The first true quasicrystal structure will have been built in the fall of 1993 at COAST (The Center for Art, Science, and Technology at the Danish Technical University). With 600 aluminum nodes and 1000 22 inch aluminum rods as well as colored, mirrored and half-mirrored acrylic plates, it will be exquisitely and magically responsive to both changes in light and the viewer's movement.

The Visual Fascination of Quasicrystals

Quasicrystals are non-repeating patterns and thus an apparent paradox: we usually think of patterns as composed of regularly repeating elements. In quasicrystals there is regularity: all the nodes, edge lengths, and two dimensional shapes are identical, all the joints are in rows, and only two similar three dimensional shapes make up the entire structure. Yet the pattern is non-repeating and consequently counter-intuitive and intriguing. Equally fascinating is the fact that quasicrystals have five-fold, three-fold, and two-fold symmetry at the same time. Experimental physicists know that have a sample of quasicrystal material when they shoot x-rays through it and obtain a pentagonal scatter pattern, and then turn the sample to obtain a scatter pattern of isometric triangles and hexagons, and turn it again to obtain patterns with right angles. Knowing that photons and x-rays are the same thing, I realized that such structures would cast shadow patterns that transmogorify as the sun passes overhead.

These visual effects can best be seen in the project study of a quasicrystal dome of 700 vertices (fig 1). From underneath, the dome appears to be a five-pointed star pattern. Amazingly, the dome appears to made up of triangles and hexagons when one

----

Tony Robbin, Independent Artist
423 Broome Street, New York, NY 10013, USA

is under the dome looking slightly to the left, and it appears to be squares when one is under the dome looking slightly to the right. But when one goes to where the triangles are, only pentagonal stars can be seen. As the sun passes overhead, from 10 AM, to Noon, to 2 PM the patterns on the floor transform to these computer plots. Every quasicrystal dome, spaceframe, or vault has this two-fold, three-fold, and five-fold symmetry.

Although a short tradition of studying similar such icosahedrally symmetric structures for use in architecture exists with S. Baer, K. Miyazaki, and H. Lalvani, it is only when new mathematical tools are borrowed from solid state physics that it is practical to design true, large scale quasicrystal structures, or to discover their unique properties. Baer's dodecahedral nodes are necessary but not sufficient to construct quasicrystals: an algorithm is required to insure that the nodes are correctly connected. True, if one already knows what a quasicrystal looks like, one can use dodecahedral nodes to replicate it and improvise upon it in a way that might be faithful to the mathematics, but Baer did not discover quasicrystals from his nodes and rods nor make structures with the visual and mathematical properties of true quasicrystals. On the basis of Baer's example, Miyazaki assembled the golden zonohedra from dodecahedral nodes and rods, and speculated about the further assembly of these sub-units into larger groups. Again, the main opportunity of building structures that change their apparent shape as the viewer passes through the structure was missed, probably because Miyazaki had no large unit models or computer models of these structures at this time (the early 1980s). Lalvani has done impressive work in generating large scale quasicrystals, even inventing new mathematical tools on his own to do this. Like Miyazaki and myself, he understands the relationship between quasicrystals structures and the geometry of four spatial dimensions. Yet Lalvani's concentration on studying the families of quasicrystals has not given him the opportunity to build structures or consider their visual properties in detail.

An additional fascination of quasicrystals is their so called inflation/deflation property. Inflation here means that the cells of quasicrystals (either the two dimensional rhombuses that are the unit cells in the two dimensional case or the two rhombohedra that are the cells in three dimensions) can be sub-divided with self-similar units so that a similar but not identical pattern is created. This endless self-similar foliation is not unlike self-similarity in chaos theory or fractal theory, except that quasicrystal tessellations fill two dimensional and three dimensional space exactly. In the three dimensional case, which is the one that concerns architects and engineers, one ratio between the edge lengths of the original and deflated cell is Tau cubed to one, where Tau is the golden ratio (1.61803...). This three dimensional inflation/deflation was discovered by the Japanese physicists T. Ogawa, (Ogawa 1985). In an attempt to be faithful to the mathematics, I have included smaller unit cells, with edge lengths is this ratio, as non structural elements in the design. Having cells in two scales and having nodes in two scales, suggests that what we have in front of us at COAST is a manifold - a space with mathematical properties, rather than a singular object.

The Structural Challenge of Quasicrystals

Although I have a theory (Robbin 1992) that with the use of a stretched skin, quasicrystals could be especially efficient structures, the COAST structure is stabilized by the use webs that add bending stiffness to the nodes and by using plates in the structure. Such devices are necessary because skeletal quasicrystal structures are inherently deficient in rigidity. The solution to stiffening quasicrystal structures that occurs to most engineers when they first experience them is to triangulate the rhombic faces, and to pass a compression member between opposite corners of the rhombohedral cells that are inevitably engendered. Such a solution, advocated even by Baer, ruins the quasicrystal nature of the structure and defeats the visual opportunities that we have been discussing, and so it was rejected out of hand by both Erik Reitzel, the engineer, and the myself.

At COAST, the stiffening webs between rods are not co-planar with the rods, rather they are normal to the rods; these stiffening webs become partial outer shells of the dodecahedral nodes, further emphasizing both the icosahedral symmetry of the structure and, at the same time, making reference to inflation/deflation symmetries of the structures. These stiffening webs are all the same size and shape, and consequently mass produced. Where they are needed, the pentagon-shaped stainless steel webs are slipped on the rods, by means of a hole in their center, before the rods are attached to the nodes. The plates are then welded together edge to edge forming a corner resistant to bending.

Additionally rigidity is added to the structure by transparent, colored plates. For example, four-cell rhombic dodecaherdra occur frequently as sub-units in the structure, and although six planes are needed to firm up all 15 nodes in this sub-unit, only two planes adds remarkable stiffness to the sub-unit in three perpendicular directions. Planes could also be made of metal tracery, or highly polished stainless, or mirrored glass; these strengthening devices further enrich the possibilities of design.

Three transparent material were studied for plates: glass, polycarbonate such as Lexan, and acrylic such as Plexiglass or Acrylite. Glass presents a number of difficulties and opportunities as a material for the planes. Compared to plastics, glass is expensive; it cannot be trimmed on site; and it is more dangerous should it break. On the other hand, glass is more stable and more strong: it has a static modulus of elasticity 50 times greater than polycarbonate and a thermal expansion coefficient only 1/10 as large. The color of plastics may not be constant in exterior uses. With the cooperation of a manufacturer of laminated glass, the artist might have access to the plastic laminate on which to silk screen deflated patterns of the quasicrystal. Finally, there exists glass that changes its color depending on the angle of illumination, the angle of viewing, and the angle of transmission of light. Such panels would have the fluid and protean quality of the structure itself. For these reasons, glass was selected as the material of choice for exterior structures. Although Lexan has high impact strength, it has a modulus of elasticity of 340,000 psi compared to a modulus of elasticity of 450,000 psi for Plexiglass. Acrylics also have a harder surface than Lexan making them more resistant to scratching. Since the primary stress on the plates is tension, acrylic is the preferred material for interior work.

The Quasicrystal at COAST

In the original plan for the COAST structure, the site was to be a North facing wall which will always be in shadow (fig 2). Behind this wall, COAST's main permanent facility will be built, and in front of the wall is a main pedestrian and automotive street. With engineer Erik Reitzel, I decided that the quasicrystal structure should do four things: 1) It should reflect sunlight onto the stone wall - a moving display on the wall as the sun moves; 2) the structure should reveal its 5 3 2 symmetries as the viewer passes by on the road (fig 2); 3) light should be reflected down into the structure so that the structure glows against the dark wall - suggesting the energy and creativity going on behind the wall in the COAST facility; and 4) although we do not emphasize it, the structure should be safe to climb upon, as it may provide an irresistible, though illegal, shortcut into the building.

Even at noon in mid-summer the sun does not pass overhead in Lyngby, Denmark, and so a structure oriented primarily for a floor shadow effect would not be effective. On the other hand, most visitors to the University are forced to pass by the structure. How wonderful to have a structure, without moving parts, that would dramatically change its appearance in the 20 seconds or so that it would take to pass by in a bus. Fortunately there is an orientation for quasicrystals that allows for this visual effect, and also allows some planes to be horizontal to the earth (for what become stair treads and perches up the structure) and also allow for some planes to be perpendicular to the earth, so that sunlight can be reflected onto the wall by mirrored and half-mirrored panels. As the sun moves through the day, the patterns on the wall would slide and transform, and the different symmetries of the structure would be emphasized.

Because of unexpected time constraints due to the closing of the Danish Culture Fund, the original project could not be completed, and work began instead on an interior quasicrystal structure to hang in the three story high atrium of the university's administration building (fig 3). Standard parts were made by a machinist, and assembled on site in about two weeks. An interior structure hung from the roof and supported by the walls of the architecture is less of a test of quasicrystal geometry for use in architecture than an exterior project would have been. Nevertheless, all the visual effects mentioned above are present: the viewer moves under, through, around, and on top of the structure seeing the three symmetries and the a-periodic nature of the structure; the sun casts different patterns from the same structure as its light passes through the glass roof at different angles during the day; and artificial light traces these same patterns on dark days. Moreover, freed from some of the structural constraints, more emphasis can be given to the inflation/deflation properties of quasicrystal.

In conclusion, the easy and fruitful collaboration between artist (myself) and engineer (Erik Reitzel) must be mentioned. There has been an interplay between engineering and aesthetics throughout the project, with some of the best engineering ideas coming from the artist and some of the best artistic ideas coming from the engineer.

References

1.BAER S. Structural System. U. S. Patent Office, n. 3,722,153, Mar 27, 1973.

2.de BRUIJN N. Algebraic theory of Penrose's non-periodic tilings of the plane. Ned. Akad. Weten Proc. Ser. A . 1981.

3.LALVANI H. Building Structures Based on Polygonal Members and Icosahedral ymmetry. U. S. Patent Office, n. 4,723,382, Feb. 9, 1988.
see also:
LALVANI H. Non Periodic Space Structures. Space Structures, Vol. 2, #2, 1986-7, p.93-108.
LALVANI H. Non Periodic Space Filling with Golden Polyhedra. Proceedings vol 1, The First International Conference on Light Weight Structures, Sidney Australia, August 1986, LSA 86 p. 202-211.

4.MIYAZAKI K. and TAKADA I. Uniform Ant-hills in the World of Isozonohedra. Structural Topology 4, 1980.

5.OGAWA T. On the Structure of a Quasicrystal. Journal of the Physical Society of Japan, Vol. 54, No. 9, September 1985.

6.PETERSON, I. Shadows and Symmetries. Science News,vol 140 nos. 25 & 26, December 21, 1991. p.408-410.

7.REITZEL E. Kvasikrystaller ved DTH, Forproject 254-01, October 1992, Reitzel F.R.I, Kobenhaven.

8.ROBBIN T. Fourfield:Computers, Art, & the 4th Dimension, Bulfinch/Little Brown & Co. New York 1992.
see also:
ROBBIN T. Quasicrystal Architecture, Proceedings of the IASS Symposium, Arkitekskolen Forlag: Copenhagen, 1991.
ROBBIN T. A Quasicrystal for Denmark's Coast, Proceeding of the Fourth Iternational Conference on Space Structures, University of Surrey, Guilford, 1993.

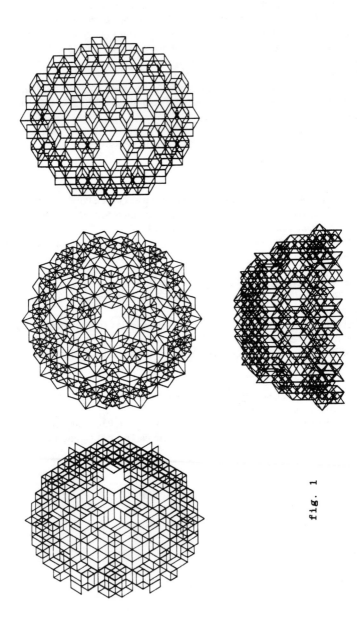

fig. 1

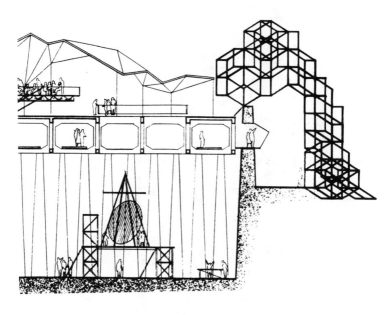

fig. 2

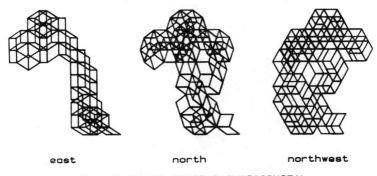

east   north   northwest

THREE VIEWS OF COAST'S QUASICRYSTAL

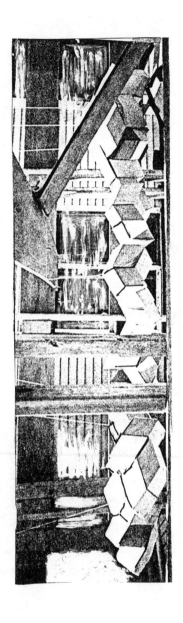

Fig 3

Hyper-Geodesic Structures : Excerpts from a Visual Catalog

Haresh Lalvani[1]

Abstract
A class of hyper-geodesic surface structures is presented. The surfaces of known polygons, polyhedra, plane and space tessellations, higher dimensional polytopes can be subdivided into parallelograms. The subdivisions are portions of non-periodic tilings having edges parallel to an n-star, a star of n vectors, and are projections from n-dimensional space. The tilings are used to subdivide regular polygons or their fundamental regions, and the subdivided polygons are then used as faces of all regular polytopes. A repertory of plane-faced and curved space structures derived this way is shown in a visual catalog. The derived structures are alternatives to the well known geodesic surface structures used in architecture.

Introduction
This paper follows author's earlier work [Refs.1-4] and presents a special class of hyper-geodesic surfaces as examples of n-dimensional space surface structures for potential application in architecture. Higher dimensional structures, when projected to 2- and 3-dimensional space, provide candidate single-layered, multi-layered and multi-directional hyper-space frames, hyper-geodesic spheres, cylindrical hyper-vaults, hyper-hyperbolic structures, and so on. The structural and architectural properties of these hyper-structures remains to be explored, but new geometries always open up new possibilties for architecture. The classification of these structures is based on the

---

[1] Professor, School of Achitecture, Pratt Institute, Brooklyn, New York 11205

classification of regular polytopes and their various subdivisions. The classification of polytopes follows the "2-dimensional periodic table" described for regular periodic surfaces - flat plane, spherical and hyperbolic plane [Ref.5], recently extended by the author into a "higher-dimensional periodic table" to include higher-dimensional polytopes in Euclidean and hyperbolic space [Ref.6].

Hyper-Geodesic Surfaces

Hyper-geodesic surfaces are the higher dimensional analog of the well-known 2- and 3-dimensional geodesic surfaces commonly used in architecture. The best known example of the latter is Fuller's geodesic sphere. Parallel work in this extension to higher dimensions has come from Miyazaki, Mosseri and Sadoc, and Lalvani [Ref.7]. The class of hyper-geodesic surfaces presented here is composed of rhombii or parallelograms [Refs.8 and 9] derived from n-vector stars, or n-stars, which are the generators of n-dimensional structures [Ref.1]. The parallelograms are produced by taking all pairs of vectors from the set of n vectors which make up the n-star, such that the sides of the parallelograms are equal and parallel to the selected vector pair. The face angles of parallelograms equal the angles between the vectors. Though the n-stars may have unequal vectors at unequal angles, the simplest orderly structures are those which are derived from n-stars based on regular polygons and obtained by joining the center of a regular polygon to its vertices. The rhombii obtained this way are neatly summarized in a Table of Rhombii given in Ref.10.

These rhombii are used to generate non-periodic tilings by various techniques. Included amongst these tilings is the well-known Penrose tiling (n=5 case) composed of two golden rhombii, and its many relatives derived from other values of n (see, for example, [Ref.10]). The tilings are 2-dimensional projections from n-dimensional space, and the vertices of the tilings can be defined by n-dimensional Cartesian co-ordinates. Portions of these tilings are used to subdivide regular polygons which are then used as faces of polytopes. Such subdivisions, in their plane and curved states, are examples of a new class of hyper-geodesic surfaces.

Examples of subdivided polygons are shown first. These are followed by examples of subdivided curved polygons. The planar and curved polygons are used as faces of the five regular polyhedra and their corresponding spherical subdivisions. Extensions to higher polytopes is shown next, followed by examples of miscellaneous other structures like hyper-geodesic cylinders, torii and curved space labyrinths.

Examples of Subdivided Polygons

Figures1-5 show examples of 3-, 5-, 6-, 7- and 14-sided regular polygons subdivided into rhombii. Their fundamental region, the right-angled triangular fundamental region shown in dotted line, is defined by the mid-point of the polygon, the mid-point of its edge and the vertex. In a p-sided polygon, it is (1/2p)th region of the polygon. Here the region is subdivided into rhombii. This subdivided region is then repeated by symmetry operations to derive the complete polygon to obtained a subdivided polygon. The triangle (Fig.1) and the hexagon (Fig.2) are composed of three different rhombii from n=6, the pentagon (Fig.3) is the Penrose case composed of two golden rhombii from n=5, and the heptagon (Fig.4) and the 14-sided polygon (Fig.5) are composed of three rhombii from n=7. Fig.6 shows the possibility of subdividing each rhombus in the fundamental region periodically into smaller rhombii, Fig.7 shows a triangulated version, and Fig.8 shows the use of parallelograms with unequal sides.

Figs.9-17 show various subdivided polygons having different fundamental regions. Fig.9 is a decagon composed of the two golden rhombii (n=5); it fundamental region is 1/10th of the polygon. The subdivision is a variant of the Penrose tiling. Figs.10-12 show a sequence of three squares composed of two rhombii (n=4); the subdivision is obtained by a gnomonic aperiodic growth sequence. Fig.13 is an octagon obtained by truncating a square, the next higher stage from the previous sequence. Fig.14 is an asymmetric subdivision of a triangle (n=6) obtained by re-arranging the rhombii of Fig.1. In Fig.15, six such triangles form an asymmetrically subdivided hexagon by re-orienting the triangles differently. Fig.16 shows an asymmetrically divided pentagon from the two golden rhombii (n=5) by re-arranging the rhombii within a symmetrically subdivided pentagon. Fig.17 is obtained by re-arranging the rhombii in Fig.5.

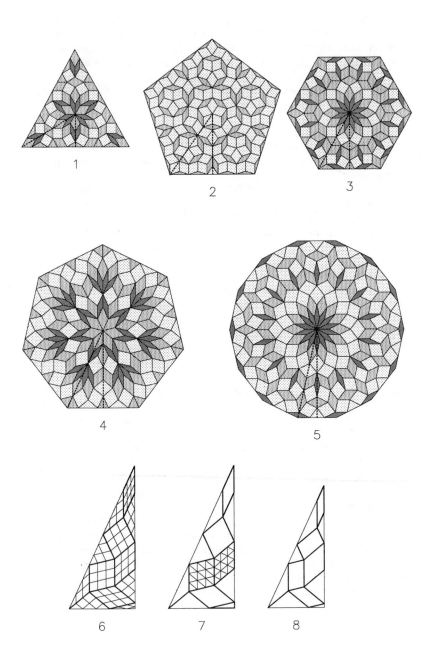

1

2

3

4

5

6

7

8

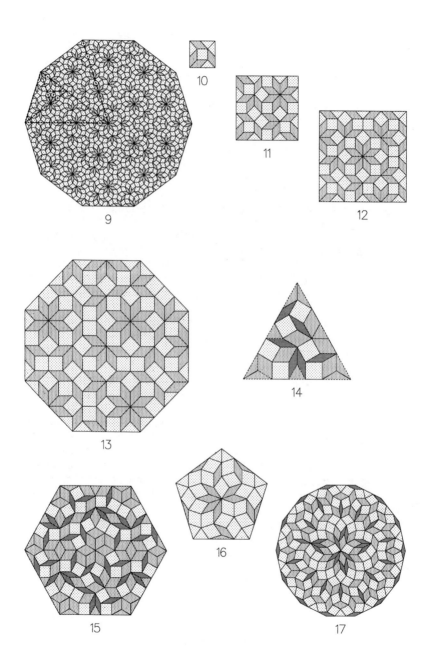

9

10

11

12

13

14

15

16

17

Examples of Subdivided Curved Polygons

The subdivisions of plane polygons can be used to subdivide curved polygonal surfaces, or alternatively by projecting plane subdivisions on to curved surfaces. Figs. 18-25 show curved versions of plane polygons shown earlier. Fig.18 is a saddle-shaped variant of the square in Fig.12. Fig.19 is a double-barrell vault also based on Fig.12. Fig.20 is a spherical pentagon corresponding to Fig.16. Fig.21 is a curved hexagon with a columnar support point derived from Fig.3 and triangulated. Fig.22 is a curved hexagon with a pinnacle top derived from Fig.16. Figs.23-25 are spherical and ellipsoidal shells based on Figs.4, 5 and 17, respectively.

Examples of Subdivided Polyhedra and Hyper-Geodesic Spheres

These subdivided plane and curved polygons can be used as faces of polyhedra, plane and space tessellations, higher dimensional polytopes, and other space structures with plane or curved surfaces. The remaining figures show miscellaneous examples of subdivided polyhedra, new geodesic surfaces derived from these polyhedra, and cells of some higher-dimensional polytopes.

Figs.26-37 show examples of various subdivided polyhedra and their spherical counterparts. These offer attractive alternatives to the Fuller's geodesic domes which are based on periodic subdivisions of the triangular face of a polyhedron, commonly into smaller triangles.   Figs.26-28 show the tetrahedron, octahedron and the icosahedron, respectively,   where each triangular face is subdivided symmetrically and identically for the three cases.   Figs.29 and 30 are the spherical analogs of the latter two.   The cube of Fig.31 and its spherical case in Fig.32 are composed of the square of Fig.12 re-oriented differently on adjacent faces. the pair of Figs.33 and 34 are similarly obtained by using the subdivided pentagon of Fig.16 and re-orienting each pentagon to produce randomly subdivided dodecahedron. Fig.35 is a spherical subdivision of dodecahedron such that each pentagonal face is a portion of the Penrose tiling with true 5-fold symmetry. Fig.36 is a triangulated version of Fig.35 viewed along its 3-fold axis of symmetry. Fig.37 is the fundamental region of the sphere in Fig.30, where each rhombus is raised into a rhombic pyramid suggesting a technique for double-layered versions (when the

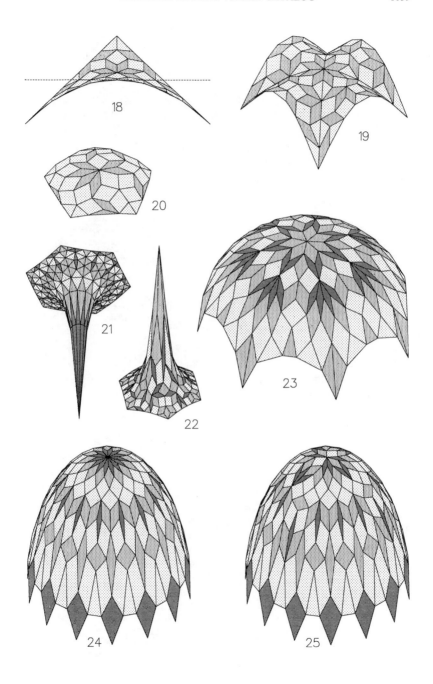

tops of the pyramids are connected appropriately) of all plane-faced and curved structures shown here.   Clearly, the structures could also be made into multi-layered structures in various ways, the simplest being the conversion of rhombii into rhombic prisms or frustums of rhombic pyramids.

The recursive property of non-periodic tilings can be used to subdivide the fundamental regions, and the polygons, into larger and larger portions of non-periodic tilings and thus produce structural surfaces which are "smoother" or structures which span large distances.   Interesting cases are the "series" of geodesic spheres from the Penrose tiling, one of which was shown in Fig.35, or of the cubes based on the sequence of squares of Figs10-12.

Examples of Subdivided Higher-Dimensional Polytopes
All subdivided regular polyhedra described so far, and others derived in a similar way,   can be used as cells of higher-dimensional polytopes. The tetrahedron of Fig.28 could be the cell of the n-dimensional 'simplex' as shown with the 5-cell in Fig.38.   The octahedron of Fig.29 could be the cell of the n-dimensional 'cross-polytope', the cube of Fig.33 could be the cell of n-dimensional cubes or of n-cubic 'honeycombs' as shown with one cell in Fig.39.   Similarly, the icosahedron of Fig.30 could be the cell of the 4-dimensional 600-cell.   One such cell is shown in Fig.40.   Fig.41 shows one dodecahedral cell of the 4-dimensional 120-cell, and Fig.42 shows one subdivided dodecahedral cell of the hyperbolic polytope in which four or more such cells meet at an edge in higher space.   Other hyperbolic cases using hyperbolic versions of the subdivided polyhedra shown are interesting possibilities.   Similar subdivisions of hyperbolic polygons also provides interesting tilings in Poincare's disc model of the hyperbolic plane.

Other Structures
A variety of other hyper-geodesic surfaces can be derived in a similar way.   The 4-sided subdivided polygons (e.g. Fig.12) can be rolled into a cylinder, bent into a Mobius strip, or curved into a torus.   The 8-sided subdivided polygon can make a double torus (a technique suggested by Coxeter).   Fig.43 shows an interesting cylinder subdivided into a variant of the Penrose tiling.   The two

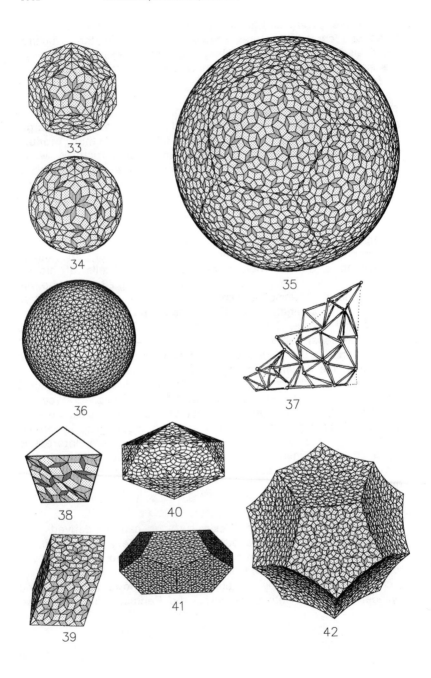

33

34

35

36

37

38

40

39

41

42

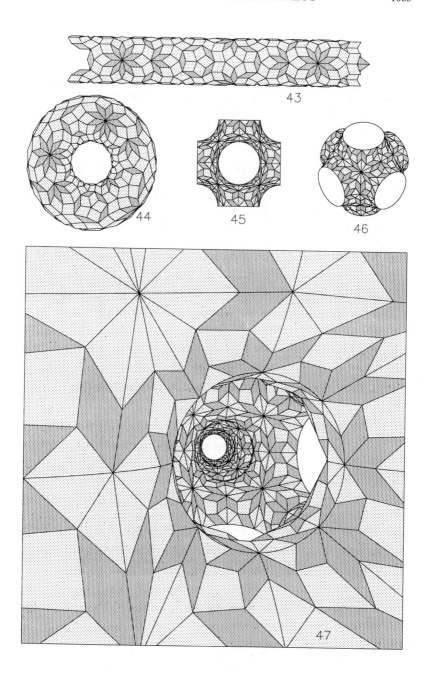

43

44

45

46

47

ends of this torus are joined to make a torus in Fig.44. Non-periodic subdivisions of other surfaces include the new classes of subdivided Schwartz surfaces and other curved space labyrinths. One example of the subdivided Schwartz surface is shown in Figs.45-47. The individual cells are shown in Figs.45 and 46 and are composed of eight saddle hexagons, each subdivided into rhombii. The subdivisions could be with or without symmetry. Fig.47 is an interior view of this surface obtained by repeating the individual cells in a periodic array.

Multi-layered and multi-directional versions of all these subdivisions are other exciting possibilities. These follow from the spatial subdivision of the entire 3-dimensional fundamental region of polyhedra, and the higher-dimensional fundamental region of higher polytopes.

Author's Note : This project is supported in part by a grant from the Graham Foundation for Advanced Studies in Fine Arts (1992-93). The concepts in this paper were first presented at the *Art and Mathematics* Conference at SUNY, Albany, in June 1992. All images in this paper have been executed by Neil Katz of Skidmore, Owings and Merrill, New York.

References :
1.　Lalvani, H. Non-periodic Space Structures, *Space Structures* Vol.2 No.2
　　　1986-87.
2.　Lalvani, H. Non-periodic space-fillings with Golden Polyhedra, In:
　　　Proceedings, First International Conference on Lightweight Structures in Architecture, Sydney, Australia, 1986.
3.　Lalvani, H. Morphological Aspects of Space Structures, In : *Studies in Space Structures*, ed. H. Nooshin, Multi-Science Publ. U.K. 1991.
4.　Lalvani, H. Towards n-Dimensional Architecture, In: *Proceedings : First Intl. Conf. on Structural Morphology*, Montpellier, France, 1992.
5.　Lalvani, H. Continuous Transformations of Subdivided Periodic Surfaces, *Space Structures*, Vol. No. 5, Nos. 3 and 4, 1990.
6.　Lalvani, H., Unpublished, first presented at the *Fourth International Conference on Space Structures*, University of Surrey,Guildford, UK, September 1993.
7.　See articles by Lalvani, Miyazaki, and Mosseri and Sadoc, *Space Structures*, Vol. No. 5, Nos. 3 and 4, 1990.
8.　Lalvani,H., First presented in the lecture *Higher Dimensional Structures* at 'Art and Mathematics' Conference, SUNY, Albany, June 1992.
9.　Lalvani, H., Non-Periodic Subdivision of Polygonal Surfaces, U.S.Patent Application, June 1993.
10. Lalvani,H., Continuous Transformations of non-periodic Tilings and Space-fillings, In :*Five-fold Symmetry*, ed. I. Hargittai,World Scientific, 1992.

Morphology of Reciprocal Frame Three-Dimensional
Grillage Structures

John C. Chilton[1], Ban Seng Choo[2] and Jia Yu[3]

Abstract

This paper is an introduction to the morphology of structures using the Reciprocal Frame (also known as Mandala Dach in Germany) which is a patented three-dimensional beam grillage structural system currently used primarily in roof construction. The principle of the Reciprocal Frame (RF) system is described and some of the varied geometries that are attainable using the principle are explored. Finally, some alternative methods of cladding the basic configuration are proposed and some architectural possibilities are discussed.

Introduction

Beam grillages have been used to span medium to long distances using individual elements shorter than the full span for many centuries. For instance there are several examples of medieval floors supported on four beams, arranged as shown in Figure 1, and this is quite a common configuration for framing stairwells. However, these grillages are generally restricted to a flat plane, unlike the Reciprocal Frame described in this paper, which generates a three-dimensional structure.

---

[1] Lecturer, School of Architecture, University of Nottingham, University Park, Nottingham, NG7 2RD, United Kingdom.

[2] Lecturer, Department of Civil Engineering, University of Nottingham, University Park, Nottingham, NG7 2RD, United Kingdom.

[3] Research Student, School of Architecture, University of Nottingham, University Park, Nottingham, NG7 2RD, United Kingdom.

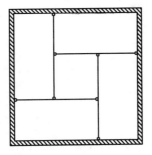

Figure 1. Traditional horizontal grillage of four beams.

Full scale RF structures of up to 11 metres in diameter have been built in the U.K. since 1988. A similar configuration of timber beams, 7 metres in diameter, has been used recently for the roof of the Artopolis at Kumamoto, in southern Japan, described in Japan Architect (1993). At Lausanne in Switzerland a salt storage building of 26 metre span was constructed with 11 tapered, glued laminated beams, using an analogous planar grillage as reported by Natterer (1991). Earlier this century, examples were also constructed by the Spanish architect José Maria Jujol, who worked with Gaudí, at Casa Negre, San Juan Despí, Barcelona in 1915 and Casa Bofarull, Pallaresos, Tarragona 1913-1918. Both works are portrayed in Flores (1982). The potential use of these structures in large span roofs has also been described previously by Chilton (1992).

The Reciprocal Frame

The name Reciprocal Frame, given to the system by the U.K. patentee, derives from the way in which each beam in the grillage both supports and in turn is supported by the other beams in the structure (reciprocally). No compression or tension ring is necessary, as the beams resist vertical loads by bending action. Its geometry, where a closed circuit of beams is formed, is described in detail below. As it is used mainly in roof construction, the configuration has considerable visual impact and can appear very dynamic when viewed from floor level. This is because the primary beam structure seems to be rotating about an axis, in empty space, at the centre of the roof. Additionally, there is no apparent means of support for the inner ends of the beams and this generates a visual tension in the structural form. Experience has shown that this appeals to architects who are keen to explore the spatial qualities of the roof form. In Germany, some architects have named the structure 'Mandala Dach', which translates as Mandala Roof, due to the resemblance of the configuration in plan to mandalas used as an aid to meditation in some Eastern religions.

Basic Reciprocal Frame geometry

Each beam in the basic Reciprocal Frame grillage is placed tangentially around a central closed curve so that it rests upon the preceding beam and this procedure is continued until the ring is complete, with the first beam resting on the last. An enclosed polygon, which need not be regular, is formed, with a set of radiating beams equal in number to the sides in this polygon. The simplest example of a stable planar beam structure constructed from elements shorter than the span is a system of two lapped beams (Figure 2(a)) but a minimum of three beams is required for a stable three-dimensional form (Figure 2(b)). The only restriction on the maximum number of beams is that of physically assembling them at the central polygon; in the limit a continuum results. If the central polygon is regular, the plan view of the beams is similar in appearance to the lines forming the iris of a camera shutter (Figure 2(c)).

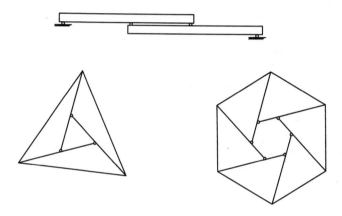

Figure 2. Structures with beams shorter than the span (a) elevation on two lapped beams, (b) plan on minimum, three beam RF three-dimensional grillage and (c) plan showing similarity to iris of a camera lens.

The grillage beams are also supported near the ends remote from the central polygon and these supports are located on an outer polygon, which again need not be regular. As the beams rest upon each other, there is a rise from the outer to the inner polygon and a three-dimensional structure is formed. Although the RF has been described as a grillage of beams resting on each other, the principle can be applied equally to configurations of lattice trusses and also to systems where adjacent beams or trusses hang from each other (i.e. where the central polygon is below the plane of the outer supports). In some ways, the structure appears to be similar to some of the tensegrities of Richard Buckminster Fuller but here the "compression" bars of the tensegrity are in contact with each other, and act in bending, and there are no tension elements.

The simplest structures using the Reciprocal Frame are usually polygonal or circular in form but even in these basic cases there are several variables to be considered, namely:-

- the number of beams - which does not necessarily have to equal the number of sides of a polygonal building
- the length of beams
- the size of the outer polygon or circle
- the size of the central polygon or circle
- the rise of the roof from the outer supports to the central polygon
- the direction of rotation (clockwise or anti-clockwise).

Some of these variables are mutually dependent. For instance, if beams of similar length and rise are specified with a set dimension vertically between the beam centres where they intersect, the size of the inner and outer polygons will depend on the number of beams comprising the complete circuit.

Alternative geometries

Additional exciting architectural possibilities can be obtained if some of the conditions of regularity are relaxed. For example, alternative plan forms (either regular or irregular) may be considered for the perimeter supports and central polygon, the angles in plan between adjacent beams may be varied around the circuit, the point of contact between adjacent beams may be varied (and, consequently, the slope, length and rise of each beam) etc. Also, the height of supporting columns and walls can be varied at will. Some examples of alternative RF roof beam configurations, for a variety of building plan forms are shown in Figure 3.

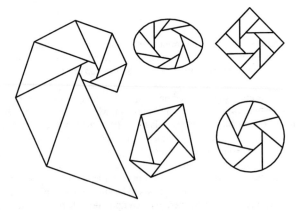

Figure 3. Examples of alternative RF roof beam configurations, for a variety of building plan forms.

To counter the potential rotation of the central polygon under load, and the possibility of progressive collapse of the basic structure, two RFs (one clockwise and one anti-clockwise) may be used together, connected at the inner polygon. Under load each of these two RFs will rotate in opposite directions, thus, if they are of similar stiffness the rotational effects cancel each other out completely.

So far in this paper, it has been assumed that the upper end of each beam terminates when it reaches the central polygon, however, there is no reason why this should be the case. Potentially rewarding architectural forms may be derived by allowing the beams to pass beyond the first intersection. For example, a "crown of thorns" effect can be achieved if the extended beams project above the roof envelope, especially if the extensions are of different length. For regular polygonal plans, if the beams continue so that their highest points are directly above the perimeter of the supporting polygon, a series of such structures may be stacked to form a rather unstable and spring-like tower.

Alternative roof covering systems

There are several alternative ways of covering RF beam grillages some of which exploit the morphology of the structure both internally and externally whilst others conceal the structure from the outside to maximise the visual impact on entering the building.

To date, the structures constructed in the U.K. have had a facetted roof form of inclined planes, with a vertical step at each beam. However, the roof of the Artopolis in Kumamoto, Japan has a fairly conventional conical covering with a system of purlins laid across the grillage beams and rafters set radially. In the facetted alternative, flat roof panels are fixed to the top of one beam and to the side of the adjacent beam on which the first beam rests. To achieve this the edge resting on the beam must be packed with a tapered fillet and the opposite edge has to be chamfered. At Kumamoto, the purlins do not span directly between adjacent beams but the straight beams are connected at points slightly offset from the primary beam centrelines to form a polygon (see Figure 5(a)).

If equivalent positions on adjacent beams are connected by a series of straight lines a slightly warped, hyperbolic paraboloid surface is generated as can be seen in Figure 4. The formation of a double-curved surface suggests that a tensile membrane weatherproof envelope, between the beams, may be appropriate.

As yet the authors know of no examples using lightweight membranes for the weatherproof envelope, however, there are several ways that these could be used, in addition to that described above. For example, a central mast suspended from the intersections of the inner polygon could support a conical membrane tensioned by perimeter cables fixed between the beam supports. Alternatively, the RF structure could be exposed on the outside, with the apex of a conical membrane suspended below the inner polygonal ring, again tensioned by edge cables between the outer beam ends. Individual panels of the RF can also be covered with tensile membranes of a variety of forms. For instance, a small conical cap over the central polygon and separate triangular membranes between adjacent beams. These can easily be formed to a double curved surface by struts supported by rods or cables from the corners of the triangular panel.

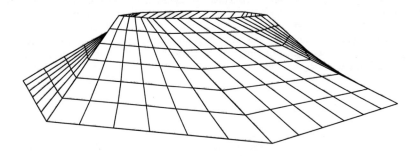

Figure 4. Hyperbolic paraboloid surface formed between RF beams.

Covering the central polygon

An important consideration is how to form the structure over the central polygon. A simple cap with radial rafters may be provided, and this was used in the Artopolis at Kumamoto as well as in the roofs constructed so far in the UK (see enlarged detail of the Kumamoto roof shown in Figure 5(b)). However, an interesting alternative is to construct a smaller version of the same beam grillage with rotation in either the same or opposite direction (Figures 6(a) and (b)). Of course, with this configuration ever smaller RFs can be stacked on top of each other until the size of the central void is such that it is impractical to construct another RF to fill it. Alternatively, a small dome may be provided or, as suggested above, a tensile membrane.

Load distribution in the Reciprocal Frame

As the RF is usually only supported at the perimeter, in a regular circular or polygonal form carrying a uniformly distributed vertical load the beam reactions are all equal to the total roof load divided by the number of beams. However, when a point load is applied to an individual beam it will be partially carried by all of the beams in the grillage and the individual beam reactions will depend on the position of the load and geometry of the grillage. For example, in a five beamed RF, where, in plan, the beam intersections occur at a distance 2m from the supports along beams 3m long, half (5kN) of a 10kN load applied at the mid-length of one beam will be transmitted to the adjacent beam. In turn two thirds of this load (3.333kN) will be carried by the next beam, 2.222kN by the next etc., until some part of the initial load is returned to the first beam to be further distributed around the circuit. This load distribution property is beneficial in situations where asymmetrical loads are encountered, such as wind loading or partial snow loading.

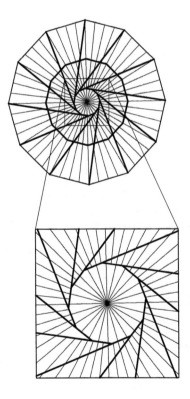

Figure 5. (a) Plan of the Artopolis roof at Kumamoto and (b) enlarged detail of the centre of the roof.

Figure 6. Plan views of systems of stacked RF structures with (a) all anti-clockwise RFs and (b) alternating clockwise and anti-clockwise RFs.

Post-tensioned RF structures

One of the disadvantages of this structural form is that large deformations that occur, due to changes in geometry. Just as loads are transmitted from beam to beam around the grillage, the deflection of any one beam changes the elevation of the end of the beam resting on it and its inclination. Therefore, the elevation and inclination of all beams in the structure will be affected by the deflection of all beams, including itself.

It is, however, possible to post-tension combinations of RF structures to limit movements under variable loads. The tower structure, described earlier, which can be formed by stacking RFs of a particular type, may be stabilised by tensioning down the perimeter intersection points of beams. Any load applied to the top of the tower will then only reduce the forces in the tensioning elements and any resulting vertical movements will be minimal.

Similarly, if two equivalent RF structures, one orientated normally and one inverted, are connected together at the outer support polygon and pre-tensioned by pulling the inner polygons closer together, any additional load applied to the combined structure (up to the pre-stress load) will cause only limited movement.

Retractable RF structures

A form of retractable roof may be constructed from the RF with beams or trusses fixed in position, but pinned about their vertical axis, at the perimeter supports. As each beam within the grillage rests on the adjacent beam, provision of a connection that permits the upper beam to slide across the lower beam and both beams to rotate relative to each other, will allow the roof to open like the iris of a camera lens. This principle is easy to envisage for the basic RF grillage but there are considerable problems to be overcome to realise this in practice and these are currently under investigation at the University of Nottingham.

Foldable RF structures

Further interesting possibilities exist if suitable full pin joints are provided about the vertical axis at the intersection points of the beams. In this case, unless appropriate restraints are provided at the perimeter supports, RF structures with more than three beams are mechanisms in the horizontal plane, thus they have the potential to become foldable or moveable structures. For example, by moving alternate beams under each other, a basic regular RF structure with 10 beams can form either a regular hexagon at the centre or a 5 pointed star (as shown in Figures 7(a) and (b)) or any one of an infinite number of intermediate configurations.

It is possible to conceive of a roof form that can be transformed at the whim of the building owner to achieve a new dynamic architectural form but, as for the retractable structure described above, there are several practical problems to be overcome in the selection and detailing of suitable cladding systems.

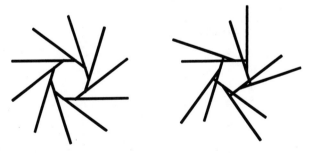

Figure 7. Ten beam foldable RF structure in alternative forms (a) with central decagon and (b) with central five pointed star.

Conclusions

From the many alternative RF structures described briefly above, it can be seen that there is considerable potential for exploration of this particular structural form. It may be that, despite their obvious visual appeal, such structures are rarely used because of practical difficulties in determining their geometry and construction details. To overcome this first problem, a series of AutoLISP procedures have been developed to draw the basic beam grillages and further procedures are currently being produced to generate more complex configurations. Once the basic three-dimensional beam grillage has been drawn, it is then possible to investigate alternative roof forms for the areas between the beams.

The authors hope that the concise introduction to the morphology of Reciprocal Frame beam grillages given in this paper will stimulate practising architects and engineers to further investigate the potential of this intriguing structural form.

Appendix - References

1.Chilton J. C. and Choo B. S. (1992). "Reciprocal Frame Long Span Structures" Innovative Large Span Structures ed. Srivastava, Sherbourne and Roorda, The Canadian Society for Civil Engineering, Montreal, Canada, pp. 100-109.

2.Flores C. (1982). Gaudi, Jujol y el Modernismo Catalan, Aguilar, p. 269 and p. 303

3.Japan Architect 1993-1 Annual p. 67.

4.Natterer J., Herzog T. and Volz M. (1991). Holzbau Atlas Zwei, Institut für Internationale Architektur, Munich, p. 179.

I.P.L. SPACE TRUSSES: STRUCTURAL PERFORMANCE AND ANALYSIS

(COMPUTER AIDED DESIGN OF THE SPATIAL STRUCTURES ON THE PARANET -
MASSIVELY PARALLEL DISTRIBUTED COMPUTER)

M. Burt[1] , D. Shriftailig[2] , V. Babaev[3] , D. Beilin[4] , A. Bogdanov[5]

ABSTRACT

The utilization of the fast working massively parallel computers in solving intense numerical problems, in the field of structural engineering left disappointed many of practical engineers. The solving large systems of equations appears to be effective, while pre-processing a tedious task taking time which is not comparable with fast working solvers. This paper presents a method tools for effective pre-processing. By developing this method the effectiveness of the solvers can be improved as well creats a positive momentum in the CAD, CAE engineering applications. The method described herein, although it was developed on the background of the Framed Structures can be applied to a wide range of continuum problems in various branches of the modern numerical computing.

INTRODUCTION

The PARANET computer utilizes network of parallel processor with a distinctive feature that this network can be reconfigurated in real-time. In other terms the Topology of the network formed by processor links can be altered whenever it is necessary. The alternation take place either under software or it is being triggered by a flow of data. The consequences of this dynamic behavior that the great flexibility can be achieved in "mirroring" a complexities of a problem under design onto a "current" configuration. This also enable a dynamic allocation of the network resources among the processes. This results in a network which may consist of the processor clusters forming a unit(workstation), a clusters of units, thus forming so called a distributed computer. The resources can be allocated, dynamically per process, or group of processes.

1- Prof., D.Sc.; 2- D.Sc., Lecturer; 3-5 - Ph.D., Researchers, Faculty of Architecture and Town Planning, Technion, Haifa 3200, Israel

There are two approaches in "paralleling" a problem. One purely algorithmic approach know from everyday experience, when facing a problem of performing a complex process of production a conveyer was invented. The idea is a simple one of gaining processing speed by breaking the entire production process into a finite number of simple operations performed in parallel by workers on a production line. Each worker performs one simple operation in parallel with his fellows workers, a collection of operation forms a complex process. The conveyer in the language of computer scientists is termed the Pipeline. The Pipeline suffers from the same problems, and it has the same advantages as any automated manufacture line in the car industry, or any other industry. Usually when the line is started not all workers are pre-occupied, since their turn to work is still ahead, so called the "Fill-time". Again when the line is emptied, while the operation is being terminated there is a period when some workers are out of work since no more operations are required during the time while line is being emptied. Redundancy at the Beginning and at the End could be significant if a large number of small operations (workers) is assumed, or "large in size elements" are being moved through a pipeline. The time required to Fill and to Empty the Pipeline is no the only disadvantage, there are some algorithms which do not lend themselves to the easy pipelining and in some instances it is simply impossible to decompose the process into a set of effective operations. [3]

This is the point where the other method of parallelling such a problem,difficult for pipelining, steps in. This method is being called "The Geometrical Partition" or "Break-up". The idea of the "Partition", "Break-up" was widely exploated by Mathematicians, Structural Engineers who developed mathematical models describing continuum as a set formed by a partition of a solution domain into a Finite set of elements, and super elements formed by several simple elements.

CONFIGURATION GENERATORS

The "Partition" and "Break-up" are tools which have very sensible appeal to structural engineers as the basic concepts of their professional "tool set". But the "Geometrical subdivision" has also very deep sense for the Parallel Programming, and for developing software applications based on Parallel Programming. The luck of success the Parallel Computer experience is related to the fact that many existing applications leave the problem of paralleling the problem to the user which may find such applications very difficult to use in practice. **The methods we introduce and describe hereby are capable to revolutionize the CAD and CAE of large, complex structures,** either having large number of structural elements, or requiring extremely fine discretization scheme leading to a large number of discrete elements, consequently large size. These methods are closely related to Pre-Processing(Input), Processing, Post-processing(Output), therefore they may effect not only the design of the spatial structures, but in general any form of engineering design.

The example we exersice in our paper is based on a discrete framed structure, however it can be easily extrapolated on a continuum problems. The first stage of the design is generation of the structural Topology-Geometry. A global geometrical boundary condition are conveniently described graphically by any set of planes, curved surfaces.   A regular 3D mesh is generated using E.P.R. - Elementary Periodic Region, [1].   By asserting node-joint in E.P.R and applying the multiplication rules a 3D mesh is being generated. It is being cut, distorted(bended, stretched), to meet the specified geometrical boundary conditions. The topology-connectivity of a regular structure, generated by applying symmetry group(multiplication), follows directly from the orbitals of the symmetry group typical for a given E.P.R. When cut by a boundary planes it is allway cut along the joints leaving edges integral. The distortions introduce a variation of length of the bars. Those variations are computed and recorded, in fact the distortion process can be restricted by constraints defined by the user, and limiting number of possible variation of length to a some desirable number. In any case even if there are hundred, thousands of lengths, the table is still reasonably small. The other cases of changes introduced into the original topology may be inspired by a necessity to modify the structure in certain regions following, structural requirements. For instance a Densification of structural members-bars. This kind of distortion results not only in introducing members of a different length, but also changing the joints, by adding additional bars which are incident with the given joint and neighboring joints. Thus another reference table containing joint "valencies" different from a regular joints, (valency is the number of bars incident with a joint), is being generated. This reference table is very small, and it cannot exceed 24 elements, in practice it always will be much smaller not exceeding 12 elements. While generating the 3D mesh, and distorting it to a specified boundary conditions only the coordinates of nodes are kept, and the updated references tables (members/lengths), (joints/valency).   Hence, topological information, adjacency, structure, is not being generated. If we have a structure having 1,000,000 nodes-joints, and the maximal valency is 12, a vector 12,000,000 elements of lengths 48 Megabytes is required to describe the adjacency.

The topology can be "re-generated" from unordered set of coordinates, each time it is required. The generation of Topology is being reduced to a "Region of Interest", a "Substructure" or other portions of a structural geometry.

The effective algorithm was developed - Algorithm ADJ1. ADJ1 algorithm is the subject of patent application and is not enclosed in herein description. The ADJ1 is numerically intense algorithm, however, when executed on a parallel system having 32 processors, (this is the configuration the authors of this paper used in their R&D program), a structure having more than 1,500,000 members can be processed in less than one second.

As it has been indicated above the dimensions of the window, the region of interest, are chosen to have engineering sense of a "Substructure", "Super Element". Clearly, that the dimensional parameters, or the principle of "moving" this window, we may also use the term "cutting into" instead of "moving, will define the algebraic objects describing the structural features of a window - a "Substructure", "Super-element".

But the most important impact is imposed by the way in which the global structure is cut into pieces by windows, because this pattern will describe the global features of the structure.

The process of "windowing", cutting the in structure into recongnizible pieces of the substructures, super-elements is performed by node driving the network, or cluster processors, driving a cluster of processors, a network formed by clusters. Each processor has the Norm-table, the Joint-table, and the DOCS-table (the typical direction cosines). The tables are used to generate stiffness matrix and other algebraic objects describing mechanical features of a structure under analysis, design. It is evident, that when such a structure is a pin joined space truss, the underlying algebraic objects contain parameters such as member; crossection area, modulus of elasticity, and norm(length), these parameters can be taken out from the stiffness matrix as a coefficients. It is also very clear, that the remaining portion of the elements in the stiffness matrix are determined by structural topology (elements which are non-zero, i.e., are adjacent), and structural geometry (direction cosines). Let us imagine a structure in which all members are of the same length, indeed we have many geometries in the structural lexicon which has this feature, all of them are made of the same material, and initially all of them have similar crossection. Thus, the stiffness and many other mechanical features are mainly influenced by structural Topology - Geometry.

This is the basis on which structures having distinct Topology-Geometry can be compared, i.e., their structural effectiveness. We are indicating this property as an important tool of the modern structural design, although it is not an imperative of our discussion.

The topology essential for construction of the algebraic objects is being generated in the each subsequent "window"-"super-element", while the geometry is specified by the coordinates. Since the global indexing system does not exist, as well as the local indexing system when the algorithm ADJ1 is applied the location of the joints within the window become very important. The window is a three dimensional prism, (with faces which are not necessarily planar), Each time a joint is being processed by ADJ1 the test is being performed to verify the location of the joint within the window. The "internal" joints are indexed first the joints laying on the faces,(boundaries of the window), are indexed last. In fact the actual indexing does not take place, and the algebraic structures are generated on fly and condensed. More complex procedure is related to indexing of the neighboring joints, joints common to the distinct processors. This is very important since because of the "topological neighborhood" processors processing same joint must exchange data (boundary results) among themselves.

If a priory, user cuts the structure into windows, super-elements, and then windows are "centrally" distributed among workers-processors, than everything seems to be logically simple. If, however, this is done by a computer, the intelligent processing is required. This is still far away from so called Artificial Intelligence, because there is the "inherited topological intelligence" in the structure itself. Even it is cut in pieces, the topology, connectivity among the pieces can be processed in a similar way as the connectivity of a single members is being processed. In other words the neighboring pieces have something in common, i.e. boundary faces, boundary joints.

The establish the Processor active links, through which the exchange of data will be carried out while solving the global structure, a supplementary algorithm ADJ2 is applied. This algorithm carries out the re-constitution of the global topology - connectivity between the windows, super-elements. This algorithm also performs the "follow-up" procedure of the storage and retrieval throughout the classical "forward" and "backward" substitution.

The processors forming either network, or clusters configurations, inject the boundary information into "large cycles" (Hamiltonian cycles, a cycle which passes through each vertex-processor only once). Each processor "pick-up" the information which is relevant to a decision making concerning his neighbors, and records the network indices of neighboring processors.

The typical windows, super-elements are shown in Fig. 1. The structure APS, (Aeroelectrical Power Station), tower; 1070 m. high, and having the 200 m radius (Fig. 2, shown a fragment of I.P.L. space truss), was broken-up into 88 different super-elements, processed by a network of parallel processors split into two clusters of 16 processors each. The resulting algebraic structures were effectively processed and resulting displacements determined for a global structure made up from a typical super-elements.

REFERENCES

1. Burt M., 1982, " The Wandering Vertex Method ", Structural Topology, No.6, pp 5-12.

2. Burt M., 1989, " Low density I.P.L. space frames for wide-span and high rise structures " , Proceedings IASS Symposium, Osaka, vol. 3, pp. 1-10.

3. INMOS Limited " OCCAM 2 Reference Manual ", Prentice Hall International, 1988, p. 133.

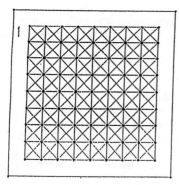
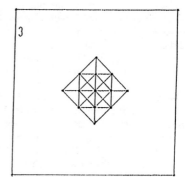
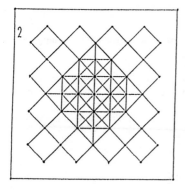
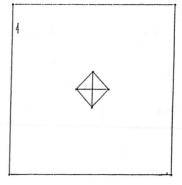

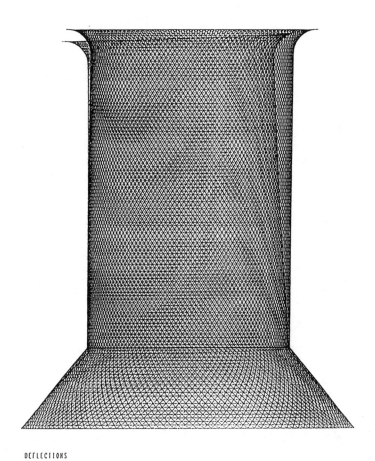

DEFLECTIONS

Fig.2

STUDY OF INTERACTION BETWEEN PNEUMATIC SPHERICAL AND CYLINDRICAL FORM SHELLS SUBJECTED TO WIND PRESSURE

D.BEILIN[1], V.POLYAKOV[2]

Abstarct

In the design of pneumatic membrane structures, analysis of stress and deformations of structures subjected to wind is among the most important requirements. For the purpose of studying the wind pressure distribution and shell shape alteration under nondisturbance air flow, wind tunnel tests were performed. The large-scale air supported shell models with truncated spherical and air inflated model with semi-cylindrical model forms were tested. The tests were carried out in the super critical streamline area (Re = 10^6 - 10^7) corresponding to the auto model process.

In the course of the experiment the distribution of wind pressure through shell surface, aerodynamic performances, changes of inside surplus pressure and displacements of the shell surface points were studied.

The paths and magnitudes of the principal tension stresses, "critical" inner pressure giving rise to unidirectional areas of stress state and the boundaries of this areas were established. Based on results of experimental investigations practical method of the soft shell stress-strain estimation are proposed.

---

[1] Ph.D., Eng. Senior Researcher, TECHNION - Israel Institute of Technology, Haifa, Israel

[2] Mg.Sc, Eng. Senior Researcher, Research Institute of Assembly and Special Works in Construction, Moscow, Russia

INTRODUCTION

The interaction of the soft shells with stationary air flow is a central problems of stress-strain state analysis since wind load is a dominating factor of external force influence. On this interaction the shape of shell is changed with the result that it has effect on the flow parameters and consequently on the distribution of the pressure over the shell surface. This phenomenon produces the "feedback" of sort between streamlined thin-walled structure and flow.

This new shape may be absolutely different from originally with result that aerodynamic of soft shell is subjected to essential changes. That is the reason that a comprehensive information about aerodynamic pressures and forces may be brought only by wind tunnel testing of the flexible models.

Experimental investigations [1-4] reveal the general streamline pattern of spherical and cylindrical air-supported shells as a result of which diagrams of wind pressure distribution over the shell surface and the aerodynamic forces and moments factors were obtained. However many aspects of true behavior of the soft shells in air flow, specific for realizable structures remain up to now unsolved for lack experimental data or their incorrectness. Among these are primarily the investigations of shell shape alteration and estimation of influence the number of factors depending from technological process during the structure manufacturing on shell shape.

The precise calculation which takes into consideration the all disturbance factors associated with interaction the soft shells and air flow, including geometrical and physical nonlinearity is a complicate problem of aeroelasticity. Then this very important to carry out the estimation method for determination of soft shell stress-strain state which should be taken in to account the stress in uniaxial zones and size of these latter.

In this article the investigation results performed with aim to study the stress-strain state of the spherical and cylindrical forms soft shell subjected to an air flow paralleled to plane of fastening are brought.

RESULTS OF EXPERIMENTAL INVESTIGATIONS

The objects of tests were a row of pneumatic air-supported shells with initial shape as a truncated sphere with diameter D = 4.2 m and high (H) to D ratio 0.3, 0.5, 0.7 and 0.82 and also semi- cylindrical air inflated two-layered shell with spherical ends D =4 m and length L =10 m. The shells were manufactured from the one layered rubberize by spreading capron fabric with thickness h = 0.6 mm (E_1 = 0.78.10^5 N/m , E_2 = 0.72.10^5 N/m).

The tests of shells were being pursued in wind tunnel with open working part and elliptical nozzle with area 264 m^2,

providing the maximal speed of a flow 65 m/s.Shield with shell
fixed thereon was placed on the upper platform of aerodynamic
balance in horizontal position. Overall model dimensions were
not beyond the section of the flow core (core charging was
3.7 %) The alteration of model slip angle in the range 0-180°
was fulfilled by rotation of the balance.

The tests of the air-supported shells were carried out in
the wide range of character parameter ϑ changes: ϑ = (P/q) =
0.1÷ 34 where P- inner surplus pressure, q-velocity head of
flow). The inner pressure was changed from 50 to 3000 Pa in
250÷500 Pa intervals. Flow velocity was altered in range 12÷
45 m/s. By this means the wind tunnel investigations were
being conducted in super critical Reinolds numbers: Re =
(2.5 ÷ 12).10^{6}, corresponding to auto model process of
stream-line flow such that border lay of flow in the immediate
vicinity of a separation point is become turbulent over
reasonably a wide range of ϑ.

In the design of investigation program the fulfilling of a
partial similarity by numbers Re,N and Eu is provided .

The influence of velocity head on the distribution of wind
pressure factors, aerodynamic characteristics, changes of the
inner pressure and shell shape , and also forces in fastening
model elements are studied during the progress of experiment.

Determination of aerodynamic factors (c_x,c_y,c_z,m_x,m_y, and
m_z) was performed by the use of 6-th components aerodynamic
balance that mounts tested structure.

Deflections of the surface shell points were measured in
geodesic coordinates by the stereophotogrammetry method with
synchronous receipt of the pressure distribution diagram.

Experimental procedure was following: in the shell (air -
supported) a surplus pressure was produced and wind tunnel
was put in prescribed velocity regime. Further, the slip angle
varied discretely in specified range, after which wind
pressure, weight coefficients and deformation of the
surface were measured for each individual position of the shell
and for each level of pressurization. Thereafter slip angle was
changed and all cycle of experiment was repeated.

In the process of flow around soft shells by streamlined
undisturbed air flow paralleled to base, the local pressure
jumps were noted in separation and aerobraking points .As this
take place the separation points along main meridian are
displaced in flow direction at the angle to 30°; with increase
in shell rigidity they are come nearer to zenith.

At the same time in latitudinal direction the separation
occurs not perpendicularly to flow inherent to rigid shells,
but at angle up to 60° to flow. As this take place on the
"shade" (lee) side in response to wake the suction is seen over
the whole shell surface. Notice that suction value is 1.5÷2
times more than for rigid body. The latter is explained by
variations of separation points coordinates. A decrease in
structure rigidity produced by the changes of velocity head to
surplus pressure ratio leads to the decrease of pressure
amplitude in the suction zone and to rise of active pressure

zone. Plots of experimental values of maximal wind pressure factors $\bar{p}_{max}$ for the truncated spherical shells is illustrated on the Fig. 1.

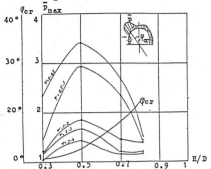

Fig.1 Relationship between wind pressure factors $\bar{p}$ and azimuthal angle ϕ_{cr} on the one hand and ratio H/D on the other at different parameters of stiffness Ψ.

It can be seen on this figure also that as H/D is increased "critical" azimuthal angle ϕ_{cr} corresponding $\bar{p}_{max}$ grows.

The obtained wind pressure factor $\bar{p}$ diagrams along air-supported shell surface in the range covered of Ψ changes show that total flow braking ($\bar{p}$ =1) zone causing the initiation of uniaxial stress state area arises at $\Psi \leq 2$ .For spherical shell with (H/D) = 0.82 this zone is about 5% of middle section at Ψ= 1 ÷ 2 and reach 13% at $\Psi \leq 0.5$ (Fig.2). Thus the formation of of uniaxial zone in air-supported shell modeled shapes which use as inhabited shelters should be expected in all range of wind velocities and inner pressure.

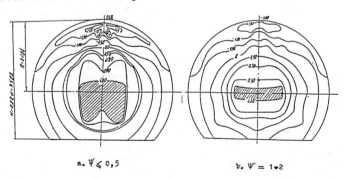

a. $\Psi \leqslant 0,5$ b. $\Psi = 1÷2$

Fig.2 Diagrams of wind pressure factors for truncated spherical shell (H/D) = 0.82

Wind-tunnel testing of semi-cylindrical air-inflated two layered model with spherical ends (L= 10 m, H= 2.0 m, P= 20 kPa , $\mathcal{V}$= 16÷222) made possible to obtain the diagrams of wind pressures under the action of air flow in crosswise and at angle. The diagrams of pressure distribution along middle shell cross section at angle of attack α = 0°, 30°, 60° and 90° are shown on the Fig 3.

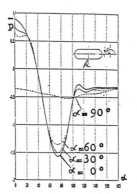

Fig.3 Distribution of wind pressure factor along middle section of the air inflated semi-cylindrical shell

As might be expected , the values of coefficient $\bar{p}$ are become critical at α ≅ 0° as a result of great model stiffness. It is worth noting that for air- supported shells with the same geometry the increased values of $\bar{p}$ in suction zone are fixed at angle of attack α =30° [4].

The interrelation between outer and inner pressure on the shall surface on the one hand and displacements (forming) in air flow on the other is one of the important problem of experimental study. Fig. 4 illustrates curves of radial displacement (w) of main meridian located in stereoscope vision range versus angle ϕ along meridian (ϕ is measured from vertical axes of shell). This displacement are most important from the viewpoint setting of safe deformation zones in design of pneumatic shells, used as dome of engineering construction and inhabited shelters from atmospheric influences. Maximal values of w/R (R- radius of cutting form of shell) for $\mathcal{V}$ ≤ 2 and (H/D) ≥5 exist at ϕ = $\pi/2$; as $\mathcal{V}$ increases the displacement curves are notable smoothed along the meridian.

Of special interest is the formation of uniaxial zones on the deformed shell surface. On the assumption that formation of this zones is connected with the conversion to zero the surface curvature, one can see that beginning of the uniaxial zones formation in equatorial streamlined area of model corresponds to $\mathcal{V}$ in range 1÷2. On further decreasing of $\mathcal{V}$ the formation of folded zone ("wind spoon") is observed. It covers a high part of shell surface at surplus pressure 50÷500 Pa.

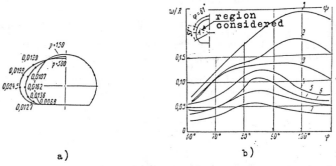

a)   b)

Fig.4 Radial displacements of truncated spherical shell:
           a- from inner pressure; b- in air flow
1- $\phi \leq 0.5$; 2- $0.5 < \phi \leq 1$; 3- $1 < \phi \leq 2$; 4- $2 < \phi \leq 3$; 5- $3 < \phi \leq 5$; 6- $5 < \phi \leq 10$;
7- $10 < \phi \leq 34$

Thus, during flow about the shell by air stream the change
of shell form takes place depending on the characteristic
parameter $\mathcal{V}$. In deformed shell the three main areas can be
picked out (Fig.5):
      - the area where shell form is kept close to spherical
($5 \leq \mathcal{V} \leq 30$) in this case (w/R) ≤ 0.05;
      - the area of form distortion conforming to displacements
up to 10 % of shell radius ($2 \leq \mathcal{V} \leq 5$) and
      - the area of formation and development of uniaxial zones
($\mathcal{V} \leq 2$).

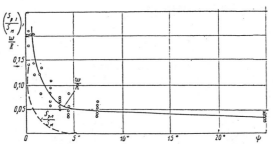

      Fig.5   Relationship between (w/R) ratio and $\mathcal{V}$ for
spherical shells

Therefore for choosing of safe deformation zones of shell, can
be used plot on Fig.5. On this figure also is shown the
relationship between area of full braking of flow $S_{(p=1)}$
and area of maximal shell cross section S_M ratio and parameter
$\mathcal{V}$.This can be seen graphically that the relationship closely
resemble the analogous dependence (w/R)f($\mathcal{V}$).

The relationship between dynamic pressure of flow q and maximal displacement for semi-cylindrical shell is shown on Fig. 6. In the framework of designated wind velocities (v) the radial displacements of measured points have tendency to monotonous rise with increasing velocity. The greatest displacements are observed on the windward shell side in point "1". Displacements on the lee side where suction was occurred, were moderate. Calculation of displacements of this shell executed with consideration for it orthotropy by torus type FE agreed satisfactory with experimental results .

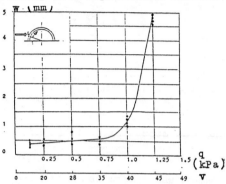

Fig.6 The interrelation between dynamic pressure of flow q and radial displacements of point "1" for air inflated semi-cylindrical shell.

ESTIMATION OF STRESS-STRAIN STATE OF SPHERICAL AIR-SUPPORTED SHELLS

1. Setting up a problem

Let us enter into the space Cartesian coordinates x,y,z , which are related with center of sphere and wind flow direction and on shall - angular coordinates ϕ, β (Fig.7).

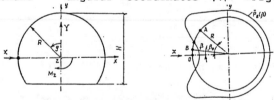

Fig.7 Design scheme and diagram of wind pressure factors during the flow about truncated spherical shells

It is reasonable safe to suggest that at flow about the soft shell deplanations of horizontal cross sections are absent.

(This hypothesis was confirmed by experimentally in the all range $\mathcal{V} \le 10$). During the flow about sphere the observed separation point A is arranged in particular on the shell "big circle" ($\phi = \pi/2$) at angle β_o to flow direction (Fig.7).

Let us consider dimensionless coefficients of vertical and horizontal components of aerodynamic force $c_{,x}$ c and longitudinal moment m_z and also pressure factors $\bar{p}(\beta)^y$ on the shell "big circle" from the formulae:

$$c_x = \frac{X}{\pi R^2 q_-} \quad ; \quad c_Y = \frac{Y}{\pi R^2 q_-} \quad ; \quad m_z = \frac{M_z}{\pi R^3 q_-} \quad ;$$

$$\bar{p}_B(\beta) = \frac{p_{i_B}(\beta) - p_-}{q} \tag{1}$$

where: R - radius of "big circle"; X - drag force ; Y - lift force; M_z-longitudinal moment; p_{i_B} = pressure of vortex less flow in point B of horizontal plane; p_- and q = static pressure and velocity head of potential flow at infinity (Table 1, Fig.7).

Table 1

Experimental values of aerodynamic characteristics averaged over ranges of shell stiffness factor $\mathcal{V}$

H/D	$\mathcal{V}$	c_x	c_y	m_z	β_o rad.
1	2	3	4	5	6
0.82	0.5	0.2940	0.1744	0.0506	1.5708
	0.5÷1	0.3048	0.2030	0.0509	1.1168
	1 ÷ 2	0.2993	0.2082	0.0552	1.1138
	2 ÷ 3	0.3017	0.2221	0.0604	1.5708
	3 ÷ 4	0.3007	0.2082	0.0638	1.5708
	4 ÷ 5	0.2892	0.2206	0.0599	1.5708
	5 ÷ 10	0.2921	0.2191	0.0664	1.1168
0.7	0.5	0.3040	0.2412	0.0418	-
	0.5÷1	0.3149	0.2557	0.0536	1.1778
	1 ÷ 2	0.3107	0.2609	0.0673	1.1778
	2 ÷ 3	0.3030	0.2533	0.0712	1.1778
	3 ÷ 4	0.3043	0.2507	0.0716	1.5708
	4 ÷ 5	0.3131	0.2731	0.0745	1.1778
	5 ÷ 10	0.3108	0.2685	0.0896	1.5708
0.5	0.5	0.1192	0.0996	0.0263	1.0738
	0.5÷1	0.1167	0.0924	0.0180	1.0738
	1 ÷ 2	0.1207	0.1005	0.0154	1.0738
	2 ÷ 3	0.1126	0.0854	0.0107	1.0738
	3 ÷ 4	0.1059	0.0837	0.0094	1.0738
	4 ÷ 5	0.0999	0.1009	0.0142	1.0738
	5 ÷ 10	0.0974	0.0445	0.0057	1.0738

Table 1
(continue)

1	2	3	4	5	6
	0.5	0.0191	0.0756	0.0231	1.0738
	0.5÷1	0.0202	0.0777	0.0202	1.4658
	1 ÷ 2	0.0180	0.0774	0.0182	1.0738
0.3	2 ÷ 3	0.0200	0.0829	0.0178	1.0738
	3 ÷ 4	0.0193	0.0820	0.0172	0.6808
	4 ÷ 5	0.0211	0.0909	0.0176	1.4658
	5 ÷ 10	0.0181	0.0811	0.0149	1.0738

2.   Principal stresses and critical values of V

Drag force X and lift force Y cause on the horizontal cross circle of shell ($\phi = \pi/2$) normal σ_r, circular σ_θ and tangential τ stresses with a knowledge of which principal tensions $T_{1,2}$ in point B can be found. Relationship between inner pressure and velocity head in "critical" state can be obtained from the assumption that uniaxial zone $T_2(\beta) = 0$ is formed on the length OB.

Displacements of the shell surface points disposed along diametral section perpendiculared to flow which were derived by stereophotogrammetry method was compared with diagrams of pressure coefficients. This comparison show the proximity of distribution of this values on the angle β. Based on this [5] will be consider that

$$\sigma_r(\beta) = a(1 - \frac{H\cos\beta_o}{D}) + b\cos\beta \qquad (2)$$

where a,b - undetermined coefficients.

By analogy to [5], taking into consideration that separation point is arranged on "big circle" at angle β to flow direction, we obtain:

$$a = \frac{PR}{2h} + \frac{Y}{2\pi R_o h} \qquad (3)$$

where R_o - radius of shall basement, h- thickness of shell.

The value of b (Eq.2) may be determined from condition the equality of longitudinal moment M_z to stresses moment σ_r relative to axis Z.

$$M = \int_{-\pi}^{\pi} \sigma_r(\beta) R^2 h \cos\beta \, d\beta \qquad (4)$$

Substituting Eq.1 in Eq.3,4 and after integration obtain:

$$b = \frac{R \, m_z q_\sim}{h} \; ; \quad h\sigma_r(\beta) = \frac{R(d_{11} P + d_{12} q_\sim)}{2} \qquad (5)$$

where:
$$d_{11} = 1 - \frac{H\cos\beta_o}{D} \quad ; \quad d_{12}(\beta) = \frac{R\ c_y d_{11}}{R_o + 2\ m_z}$$

For determination of circular stresses σ_θ we use Laplace formula:

$$\frac{h}{R}\ (\sigma_r + \sigma_\theta) = p_{iB} - p_B \tag{6}$$

where: $p_B = P + p_\sim$ - surplus pressure in point B at velocity head q_B.

From Bernoulli equation it might be assumed that
$$p_\sim + q_\sim = p_B + q_B \ ,$$
whence taking into consideration $q_B = q_\sim (1-\bar{p}_B)$ follows that
$$p_{iB} - p_B = P - \bar{p}_B q_\sim \tag{7}$$

Substituting Eq.7 in Eq.6 after transformation obtain:
$$h\sigma_\theta = \frac{R(d_{21} P + d_{22} q_\sim)}{2} \tag{8}$$

where:
$$d_{21} = 1 + \frac{H\cos\beta_o}{D}; \quad d_{22}(\beta) = -2\bar{p}_B(\beta) - d_{12}(\beta)$$

Let us assume that tangential stresses are distributed along the circle $y = 0$ according to the relation:
$$\tau(\beta) = \begin{cases} f\ \sin\ y\beta & \text{at } 0 \le \beta \le \frac{\pi}{y} \ ; \\ 0 & \text{at } \frac{\pi}{y} \le \beta \le \pi \ . \end{cases}$$

The choosing of approximating function in this form is connected with condition of continuity of $\tau(\beta)$ at $\beta = \pm\pi$. The coefficient y pick out so that function $\tau(\beta)$ was maximum in separation point $\beta = \beta_o$. One can readily see that

$$y = \frac{\pi}{2\beta_o} \begin{cases} = 1 \text{ at } (\pi/2) = \beta \\ \ne 1 \text{ at } (\pi/2) = \beta_o \end{cases} \tag{9}$$

The constant f is defined from condition:
$$X = \int_{-\pi}^{\pi} \tau hR\sin\beta\ d\beta$$

After integration taking into consideration Eqs.1-9 we find that
$$f = \frac{c_x\ PRq_\sim}{2gh}$$

where:

$$g = \int_0^{2\beta_o} \sin\beta\ \sin y\beta\ d\beta = \begin{cases} \dfrac{\pi}{2} & \text{at } \beta_o = \dfrac{\pi}{2} \\[2mm] \dfrac{\sin 2(y-1)\beta_o}{2(y-1)} - \dfrac{\sin 2(y+1)\beta_o}{2(y+1)} & \text{at } \beta_o \ne \dfrac{\pi}{2} \end{cases} \tag{10}$$

Finally:

$$\tau h = \frac{R\, d_{32}\, q_{\sim}}{2} \;\; ; \quad d_{32}(\beta) = \frac{c_x\, \pi\, \sin\gamma\beta}{g} \;\; ; \quad 0 \le \beta \le \frac{2\pi}{\gamma} \; .$$

Principal stresses can be computed from the formula:

$$\hat{T}_{1,2} = \frac{h}{2}\left\{\, [\sigma_r + \sigma_\theta \pm (\sigma_r - \sigma_\theta)^2 + 4\tau^2\,]^{0.5}\right\} \tag{11}$$

Substituting Eqs. 5,8 and 10 in Eq.11 and equaling $\hat{T}_2$ to zero after routine transformation obtain the quadratic equation relative to characteristic of stiffness $\mathscr{V}$:

$$d_{11}d_{21}\mathscr{V}^2 + (d_{11}d_{22} + d_{12}d_{21})\mathscr{V} + (d_{12}d_{22} - d_{32}^2) = 0 \tag{12}$$

For determination of critical numbers of $\mathscr{V}^*$ a aerodinamical parameters from Table 1 in different diapasons of $\mathscr{V}_i \le \mathscr{V} \le \mathscr{V}_{i+1}$ must be substitute in coefficients of Eq.12 and find positive roots $\mathscr{V}^*$.(Negative root will correspond to the formation of uniaxial zone at suction). Critical values $\mathscr{V}^*(\beta)$ for each angle β are defined as $\max \mathscr{V}_i(\beta)$ where "max" is taken through the whole i so that $\mathscr{V}_i \le \mathscr{V}_i(\beta) \le \mathscr{V}_{i+1}$. In Table 2 are listed the values $\max \mathscr{V}^*(\beta)$ for each shell.

A curves $\mathscr{V}^*(\beta)$ (Fig.8) admit to define a borders of uniaxial stress state along β-coordinate in all values of P and $q_{\sim}$.

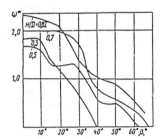

Fig.8 Relationship between "critical" stiffness parameter $\mathscr{V}^*$ and latitudinal angle β

Experimental data display that tensions reach maximal values on the bigger horizontal cross section of shell in separation point. The principal tension in separation point can be obtained from Eq.11 with regard to Eqs.5,8 and 10. After transformation we have:

$$\hat{T}_1 = R\left\{\frac{P}{2} - 0.5\, \bar{P}_B(\beta_0)\, q_{\sim} + [(m_1 P + m_2 q_{\sim})^2 + (m_3 q_{\sim})^2\,]^{0.5}\right\} ,$$

where: $m_1 = \dfrac{H \cos\beta_0}{2D}$; $m_2 = \dfrac{\bar{P}_B(\beta_0) - d_{12}(\beta_0)}{2}$; $m_3 = \dfrac{c_x\, \pi}{2g}$.

It can easily be shown that:

$$\hat{T}_{1\,max} \le R\,(n_1 P + n_2 q_{\sim}) \tag{13}$$

where: $n_1 = 0.5 + m_1$; $n_2 = |m_2| + m_3 - \bar{P}_B (\beta_0)$.

The values of n_1 and n_2 are listed in Table 2. By comparison, the IASS Recommendations [6] give for all H/D :$n_1 = 0.5$; $n_2 = 1.5$. displacement and corresp o nding ly decreasing of active interaction zone, obtained by us values $n_1 > 0.5$.

On the experimental data base the relation between change of inner pressure in shell δP and head velocity q was clarified. Analysis show that in interesting zone

$$\frac{(\mathcal{V} \leq \mathcal{V}^*)\delta P}{P} \leq 5$$

and thus change of inner pressure δP has not effect on shell behavior.

Table 2

The coefficients n_1 , n_2 (Eq.13) and critical stiffness $\mathcal{V}^*$

$\dfrac{H}{D}$	$\mathcal{V}^*$	n_1	n_2
0.82	2,37	0.68	1.29
0.7	2.00	0.63	1.36
0.5	1.70	0.61	0.84
0.3	1.84	0.57	0.97

The method of estimation of soft shell stress-strain state outlined above was used for design real air-supported shelters with diameter 27 m.

References

1. Beger G., Macher E.- Results of Wind Tunnel Tests on Some Pneumatic Structures. *Proc. of the I Intern. Colloq. Pneumatic Structures*, Stuttgart, 1967.
2. Neemann H - Wind Tunnel Experiments on Aeroelastic Models of Air-Supported Structures: Results and Conclusion - *Proc. of IASS Int. Symp. on Pneumatic Structures*, 1972
3. Beilin D., Polyakov V *at el.*- The Application of Stereophotogrammetric Method to Investigation of Stress-Strain State of Soft Shells in the Air Flow - *Scientific Notes of TSAGI*, v.13, n°8, 1982, Moscow
4. Srivastava N.K., Turkkan N, Dickey R - Wind Tunnel Study of a Flexible Membrane Structure. *Proc. 3 Intern. Conference on Space Structures*, Guildfor, UK,1984
5. Alekseev S. - About Analysis of the Soft Spherical Shell in Liquid Flow- *Mechanic of Solid*, n°3, 1967, Moscow
6. IASS Recommendation for Air-Supported Structures, *Working Group n°7*, *IASS*, 1985, Madrid
7. Beilin D., Levitin M., Polyakov V - About Interaction of Spherical Soft Shells with Air Flow Paralleled to the Base - *Structural Mechanic and Design of Constructions*, n°1, 1991, Moscow.

SUBJECT INDEX
Page number refers to first page of paper

AUTHOR INDEX
Page number refers to first page of paper

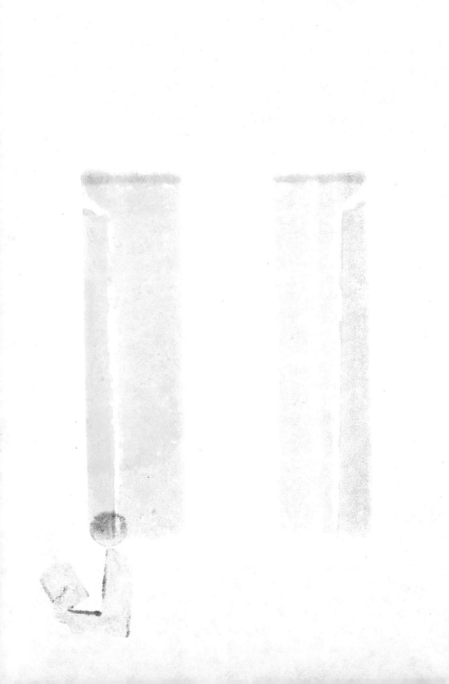

DEMCO